Lecture Notes in Computer Science　　　10116

Commenced Publication in 1973
Founding and Former Series Editors:
Gerhard Goos, Juris Hartmanis, and Jan van Leeuwen

Editorial Board

More information about this series at http://www.springer.com/series/7412

Chu-Song Chen · Jiwen Lu
Kai-Kuang Ma (Eds.)

Computer Vision – ACCV 2016 Workshops

ACCV 2016 International Workshops
Taipei, Taiwan, November 20–24, 2016
Revised Selected Papers, Part I

Springer

Editors
Chu-Song Chen
Institute of Information Science
Academia Sinica
Taipei
Taiwan

Jiwen Lu
Tsinghua University
Beijing
China

Kai-Kuang Ma
School of Electrical and Electronic
    Engineering
Nanyang Technological University
Singapore
Singapore

ISSN 0302-9743   ISSN 1611-3349   (electronic)
Lecture Notes in Computer Science
ISBN 978-3-319-54406-9   ISBN 978-3-319-54407-6   (eBook)
DOI 10.1007/978-3-319-54407-6

Library of Congress Control Number: 2017932642

LNCS Sublibrary: SL6 – Image Processing, Computer Vision, Pattern Recognition, and Graphics

Printed on acid-free paper

This Springer imprint is published by Springer Nature
The registered company is Springer International Publishing AG
The registered company address is: Gewerbestrasse 11, 6330 Cham, Switzerland

Preface

It is our great pleasure to present the workshop proceedings of three LNCS volumes, which contain the papers carefully reviewed and selected from the 17 workshops that were held in conjunction with the 13th Asian Conference on Computer Vision (ACCV), during November 20–24, 2016, in Taipei, Taiwan. There are 134 papers selected from 223 papers submitted to all the 17 workshops as listed below.

1. New Trends in Image Restoration and Enhancement (NTIRE): 14 papers
2. Workshop on Assistive Vision: 6 papers
3. ACCV 2016 Workshop on Hyperspectral Image and Signal Processing: 6 papers
4. Computer Vision Technologies for Smart Vehicle: 7 papers
5. Spontaneous Facial Behavior Analysis: 8 papers
6. 3D Modelling and Applications: 16 papers
7. 4th ACCV Workshop on e-Heritage: 4 papers
8. Multiview Lip-Reading Challenges: 5 papers
9. Workshop on Facial Informatics (WFI): 11 papers
10. Discrete Geometry and Mathematical Morphology for Computer Vision: 4 papers
11. Workshop on Mathematical and Computational Methods in Biomedical Imaging and Image Analysis: 15 papers
12. International Workshop on Driver Drowsiness Detection from Video: 6 papers
13. Workshop on Meeting HCI with CV: 6 papers
14. Workshop on Human Identification for Surveillance (HIS) Methods and Applications: 8 papers
15. Benchmark and Evaluation of Surveillance Task (BEST): 9 papers
16. The Third Workshop on Computer Vision for Affective Computing (CV4AC): 3 papers
17. Workshop on Interpretation and Visualization of Deep Neural Nets: 6 papers

The workshop topics are related to computer vision and its applications, interdisciplinary themes with other application areas, as well as challenges or competitions. Every workshop handles its own paper submission system, and each paper is reviewed by two to three reviewers. We thank all the workshop organizers for their great efforts in holding these successful workshops. We also thank the help of the publication chairs in making this publication possible.

November 2016

Chu-Song Chen
Jiwen Lu
Kai-Kuang Ma

Organization

W01: 3D Modelling and Applications

Chia-Yen Chen	National University of Kaohsiung, Taiwan
Min-Chun Hu	National Cheng Kung University, Taiwan
Li-Wei Kang	National Yunlin University of Science and Technology, Taiwan
Chih-Yang Lin	Asia University, Taiwan
Tang-Kai Yin	National University of Kaohsiung, Taiwan
Guo-Shiang Lin	Da-Yeh University, Taiwan
Chia-Hung Yeh	National Sun Yat-Sen University, Taiwan

W02: 4th ACCV Workshop on e-Heritage

Katsushi Ikeuchi	Microsoft Research Asia, China
El Mustapha Mouaddib	Université de Picardie Jules Verne, France
Takeshi Masuda	AIST, Japan
Takeshi Oishi	The University of Tokyo, Japan

W03: ACCV 2016 Workshop on Hyperspectral Image and Signal Processing

Keng-Hao Liu	National Sun Yat-sen University, Taiwan
Wei-Min Liu	National Chung Cheng University, Taiwan

W04: Benchmark and Evaluation of Surveillance Task (BEST)

Xiaokang Yang	Shanghai Jiao Tong University, China
Chong-Yang Zhang	Shanghai Jiao Tong University, China
Bingbing Ni	Shanghai Jiao Tong University, China
Lin Mei	The Third Research Institute of the Ministry of Public Security, China

W05: Computer Vision Technologies for Smart Vehicle

Li-Chen Fu	National Taiwan University, Taiwan
Pei-Yung Hsiao	National University of Kaohsiung, Taiwan
Shih-Shinh Huang	National Kaohsiung First University of Science and Technology, Taiwan

W06: Discrete Geometry and Mathematical Morphology for Computer Vision

Jean Cousty   Université Paris-Est, ESIEE Paris, France
Yukiko Kenmochi   Université Paris-Est, CNRS, France
Akihiro Sugimoto   National Institute of Informatics, Japan

W07: International Workshop on Driver Drowsiness Detection from Video

Chen-Kuo Chiang   National Chung Cheng University, Taiwan
Shang-Hong Lai   National Tsing Hua University, Taiwan
Michel Sarkis   Qualcomm Technologies Inc., USA

W08: Large-Scale 3D Human Activity Analysis Challenge in Depth Videos

Gang Wang   Nanyang Technological University, Singapore
Amir Shahroudy   Nanyang Technological University, Singapore
Jun Liu   Nanyang Technological University, Singapore

W09: Multiview Lip-Reading Challenges

Ziheng Zhou   University of Oulu, Finland
Guoying Zhao   University of Oulu, Finland
Takeshi Saitoh   Kyushu Institute of Technology, Japan
Richard Bowden   University of Surrey, UK

W10: New Trends in Image Restoration and Enhancement (NTIRE)

Radu Timofte   ETH Zurich, Switzerland
Luc Van Gool   ETH Zurich, Switzerland
Ming-Hsuan Yang   University of California at Merced, USA

W11: Spontaneous Facial Behavior Analysis

Xiaopeng Hong   University of Oulu, Finland
Guoying Zhao   University of Oulu, Finland
Stefanos Zafeiriou   Imperial College London, UK
Matti Pietikäinen   University of Oulu, Finland
Maja Pantic   Imperial College London, UK

W12: The Third Workshop on Computer Vision for Affective Computing (CV4AC)

Abhinav Dhall   Abhinav Dhall, University of Waterloo, Canada
Roland Goecke   University of Canberra/Australian National University,
                         Australia
O.V. Ramana Murthy   Amrita University, India
Jesse Hoey   University of Waterloo, Canada
Nicu Sebe   University of Trento, Italy

W13: Workshop on Assistive Vision

Chetan Arora   Indraprastha Institute of Information Technology,
                         Delhi, India
Vineeth N. Balasubmanian Indian Institute of Technology, Hyderabad, India
C.V. Jawahar   International Institute of Information Technology,
                         Hyderabad, India
Vinay P. Namboodiri   Indian Institute of Technology, Kanpur, India
Ramanathan Subramanian   International Institute of Information Technology,
                         Hyderabad, India

W14: Workshop on Facial Informatics (WFI)

Gee-Sern (Jison) Hsu   National Taiwan University of Science
                         and Technology, Taiwan
Moi Hoon Yap   Manchester Metropolitan University, UK
Xiaogang Wang   Chinese University of Hong Kong, Hong Kong,
                         SAR China
Su-Jing Wang   Chinese Academy of Science, China
John See   Multimedia University, Malaysia

W15: Workshop on Meeting HCI with CV

Liwei Chan   National Chiao Tung University, Taiwan and Keio
                         Media Design, Japan
Yi-Ping Hung   National Taiwan University, Taiwan

W16: Workshop on Human Identification for Surveillance (HIS): Methods and Applications

Wei-Shi Zheng   Sun Yat-sen University, China
Ruiping Wang   Institute of Computing Technology, Chinese Academy
                         of Sciences, China

Weihong Deng   Beijing University of Posts and Telecommunications,
                                      China
Shenghua Gao   ShanghaiTech University, China

W17: Workshop on Interpretation and Visualization of Deep Neural Nets

Alexander Binder   Singapore University of Technology and Design,
                                      Singapore
Wojciech Samek   Fraunhofer Heinrich Hertz Institute, Germany

W18: Workshop on Mathematical and Computational Methods in Biomedical Imaging and Image Analysis

Atsushi Imiya   Chiba University, Japan
Xiaoyi Jiang   Universität Münster, Germany
Hidetaka Hontani   Nagoya Institute of Technology, Japan

Contents – Part I

Workshop on Assistive Vision

Hyperspectral Image and Signal Processing

Spontaneous Facial Behavior Analysis

3D Modelling and Applications

Contents – Part II

4th ACCV Workshop on e-Heritage

Multi-view Lip-Reading Challenges

Workshop on Facial Informatics (WFI)

Discrete Geometry and Mathematical Morphology for Computer Vision

Workshop on Mathematical and Computational Methods in Biomedical Imaging and Image Analysis

Contents – Part III

Workshop on Meeting HCI with CV

Workshop on Human Identification for Surveillance (HIS) Methods and Applications

Benchmark and Evaluation of Surveillance Task (BEST)

**The Third Workshop on Computer Vision for Affective
Computing (CV4AC)**

Workshop on Interpretation and Visualization of Deep Neural Nets

New Trends in Image Restoration and Enhancement (NTIRE)

Blind Image Deblurring Using Elastic-Net Based Rank Prior

Hongyan Wang, Jinshan Pan, Zhixun Su$^{(\boxtimes)}$, and Songxin Liang

School of Mathematical Sciences, Dalian University of Technology, Dalian, China
zxsu@dlut.edu.cn

Abstract. In this paper, we propose a new image prior for blind image deblurring. The proposed prior exploits similar patches of an image and it is based on an elastic-net regularization of singular values. We quantitatively verify that it favors clear images over blurred images. This property is able to facilitate the kernel estimation in the conventional maximum a posterior framework. Based on this prior, we develop an efficient optimization method to solve the proposed model. The proposed method does not require any complex filtering strategies to select salient edges which are critical to the state-of-the-art deblurring algorithms. Quantitative and qualitative experimental evaluations demonstrate that the proposed algorithm performs favorably against the state-of-the-art deblurring methods.

1 Introduction

Blind image deblurring has been witnessed significant advances in the vision and graphics community with the last decade as it involves many challenges in problem formulation and optimization. The goal of blind image deblurring is to recover a clear image and a blur kernel from a blurred input. The blur process is usually modeled by convolution when the blur is uniform:

$$B = I \otimes k + n, \tag{1}$$

where B denotes the observed blurred image, I denotes the latent clear image, k denotes the blur kernel (a.k.a., point spread function (PSF)), n denotes the additive noise, and $\otimes$ denotes convolution operator. This problem is highly ill-posed because only the blurred image B is known and we can find many different pairs of I and k to satisfy the blur model (1).

To make this problem tractable, most of deblurring methods usually make assumptions on blur kernels and latent images. The assumptions usually play critical roles in deblurring methods. The sparsity of image gradients [1–3] is one kind of image prior in image deblurring. Levin et al. [4] analyze that the blind deblurring methods based on the sparsity of image gradient prior (e.g., hyper-Laplacian prior [2]) tend to favor blurred images over clear images, especially for those algorithms formulated within the maximum a posterior (MAP) framework.

Electronic supplementary material The online version of this chapter (doi:10.1007/978-3-319-54407-6_1) contains supplementary material, which is available to authorized users.

C.-S. Chen et al. (Eds.): ACCV 2016 Workshops, Part I, LNCS 10116, pp. 3–17, 2017.
DOI: 10.1007/978-3-319-54407-6_1

To overcome this limitation, some new image priors that favor clear images over blurred images have been proposed, such as normalized sparsity prior [5], internal patch recurrence [6], dark channel prior [7]. Another kinds of deblurring methods [8,9] use additional heuristic edge selection to estimate blur kernel to overcome aforementioned limitations as sharp edges usually work well in MAP based methods according to analysis of [4]. However, these methods usually fail when sharp edges are not available.

Different from existing methods, we propose a novel image prior by exploiting similar patches of an image. We develop an elastic-net regularization of singular values computed from the similar patches of an image to guide the kernel estimation.

The contributions of this work are as follows:

– We propose a novel image prior for blind image deblurring. The prior is based on an elastic-net regularization of singular values computed from the similar patches of an image.
– We develop an efficient numerical optimization method to solve the proposed model, which is able to converge well in practice.
– We analyze that the proposed prior favors clear image over blurred images, which is able to facilitate kernel estimation.

2   Related Work

In this section we discuss the most relevant algorithms and put this work in the proper context. As blind image deblurring is an ill-posed problem, it requires additional information to constrain the solution space. Fergus et al. [1] use a mixture of Gaussians to learn image gradient prior via variational Bayesian inference. Shan et al. [3] develop a certain parametric model to approximate the heavy-tailed natural image prior, where the deblurring processing is performed in a MAP framework. Comprehensive analysis by [4] shows that variational Bayesian based deblurring methods (e.g., [1]) are able to remove trivial solutions in comparison to other approaches with naive MAP formulations.

To overcome the limitations of naive MAP formulations, numerous deblurring methods based on MAP formulations have been developed with different likelihood functions and image priors [3,5–7,10,11].

As pointed by Levin et al. [4], sharp edges usually work well in MAP framework, some edge selection based methods have been proposed [8,9]. These methods introduce an additional step in the conventional MAP framework and achieve better results as evidenced by [9]. However, these edge selection methods are based on the assumption that strong edges exist in the latent images and involve some heuristic image filters, e.g., bilateral filter [8] and shock filter [8,9,12]. These methods are likely to fail when the sharp edges are not available.

Instead of selecting sharp edges from blurred images, recent exemplar-based methods [13–15] exploit information contained in both a blurred input and example images from an exemplar dataset. However, querying a large external dataset is computationally expensive.

Nonlocal self-similarity prior has been widely used in image denoising problem [16–19]. Dong et al. [17] establish the relationship between nonlocal self-similarity prior and low-rank prior in image denoising. As low-rank prior is usually modeled by a standard nuclear norm [17,20] which may shrink all the singular values fairly, Gu et al. [18] propose a weighted nuclear norm in image denoising. Ren et al. [21] introduce an enhanced low-rank prior combined by intensities and gradient maps and employ a weighted nuclear norm minimization method for optimization.

Since the standard nuclear norm has a weak convexity, which makes the algorithm sensitive to the observed data, Kim et al. [22] enforce a strong convex penalty of F-norm regularized term of singular values and propose an elastic-net regularization to enhance the stabilities of the algorithm.

In this work, we propose a new image prior for blind image deblurring. The proposed prior is built over weighted nuclear norm, giving a less shrinkage for large singular values to help preserving major structures, together with a convex F-norm to achieve a stable and robust solution. Most importantly, the proposed prior favors clear images over blurred images, which is able to facilitate blur kernel estimation.

3 Proposed Prior

In this section, we first give an overview of the deblurring model and then present the proposed image prior for image deblurring. We also give some explanations about the prior.

Our deblurring method is based on the conventional MAP framework, which is usually modeled as

$$\min_{I,k} \|I \otimes k - B\|_2^2 + \alpha \varphi(I) + \gamma \psi(k), \tag{2}$$

where the first term is the data term, $\varphi(I)$ and $\psi(k)$ are regularized terms formulated from image prior and kernel prior, respectively; and α and γ are weights.

For the kernel regularization $\psi(k)$, we follow the existing deblurring methods [8,15] and adopt $\psi(k) = \|k\|_2^2$. As the regularization $\varphi(I)$ plays a critical role in image deblurring, we propose an effective prior based on similar patches of an image and an elastic-net regularization of singular values to define $\varphi(I)$ for blur kernel estimation.

3.1 Nonlocal Self-similarity Based Regularization

Natural images usually contain recurrent patterns. This important property has been widely employed in many low-level vision tasks, e.g., image restoration [16–19,23]. One classic strategy to model this property is nonlocal self-similarity which is widely used in image restoration. Dong et al. [17] show the relationship between nonlocal similarity and low-rank priors, and propose an image denoising method based on low rank priors.

To deal with the fairly shrinking limitations of standard nuclear norm (e.g., [17]), Gu et al. [18] propose a weighted nuclear norm model for denoising which is defined as

$$\min_{X} \frac{1}{\eta} \|X\|_{\omega,*} + \|Y - X\|_F^2, \tag{3}$$

where Y denotes a matrix stacked by some similar patches of a noisy image, X denotes the low rank matrix to be restored. $\|X\|_{\omega,*}$ denotes the weighted nuclear norm of matrix X, which is defined as the sum of all the singular values with different positive weights ω_j:

$$\|X\|_{\omega,*} = \sum_j |\omega_j \sigma_j(X)|, \tag{4}$$

where $\sigma_j(X)$ denotes the j-th singular value of X. By assigning different weights on singular values, model (3) performs effectively in image denoising.

To enhance the stability of the algorithm, Kim et al. [22] propose an elastic-net regularization of singular values

$$\phi(X) = \lambda_1 \|X\|_* + \frac{\lambda_2}{2} \|X\|_F^2. \tag{5}$$

The elastic-net regularization of singular values have been successfully applied on several real-word problems including non-rigid motion estimation, photometric stereo, and background modeling problems. In this paper, we find that this prior favors clear images over blurred images. This intriguing observation makes us propose a new blind image deblurring method.

3.2 Proposed Prior

We propose a new image prior which is also based on nonlocal self-similarity. Specifically, the prior is defined as

$$\varphi(I) = \lambda_1 \sum_i \|I_i\|_{\omega,*} + \frac{\lambda_2}{2} \sum_i \|I_i\|_F^2, \tag{6}$$

where I_i denotes the matrix stacked by patches which are similar to i-th patch in image I, λ_1 and λ_2 are parameters.

By the Singular Value Decomposition (SVD) of I_i, (6) has an equivalently form which can be written as

$$\varphi(I) = \lambda_1 \sum_i \sum_j |\omega_j \sigma_j(I_i)| + \frac{\lambda_2}{2} \sum_i \sum_j |\sigma_j(I_i)|^2. \tag{7}$$

where $\sigma_j(I_i)$ denotes the j-th singular value of I_i.

The proposed prior is important for kernel estimation for the following reasons. The weighted nuclear norm term in (6) is not only a low-rank constraint, but also can make an unfair thresholding effect over singular values for preserving salient structures. The second term is the strong convex F-norm penalty term. Together with this term, both a thresholding effect from ω and a shrinkage from λ_2 will be imposed on singular values. This may take proper correction on suboptimal solution during alternating minimization, furthermore, a possible poor solution can be avoided.

With the proposed prior, we can solve (2) by iterating alternatively between intermediate latent image recovering and kernel estimation.

4   Optimization

In order to obtain the solutions of I and k from the blurred image B based on the image prior proposed in Sect. 3, we alternatively solve

$$\min_I \|I \otimes k - B\|_2^2 + \alpha\varphi(I), \tag{8}$$

and

$$\min_k \|I \otimes k - B\|_2^2 + \gamma\psi(k). \tag{9}$$

In the following, we present the optimization details about these two problems.

4.1   Intermediate Latent Image Estimation

As $\varphi(I)$ in (8) involves a hybrid norm, it is difficult to obtain the solutions directly. To solve I from (8), we use half-quadratic splitting technique proposed by [24,25]. By introducing an auxiliary variable J, the minimization problem (8) can be rewritten as

$$\min_{I,J} \|I \otimes k - B\|_2^2 + \alpha(\lambda_1 \sum_i \|J_i\|_{\omega,*} + \frac{\lambda_2}{2} \sum_i \|J_i\|_F^2) + \beta\|I - J\|_2^2. \tag{10}$$

where β is a positive weight penalizing the similarity of I and J. If β is larger enough, the solutions of (10) will approximate to those of (8).

For (10), we alternatively solve the following two tractable subproblems:

$$\min_J \alpha\lambda_1 \sum_i \|J_i\|_{\omega,*} + \frac{\alpha\lambda_2}{2} \sum_i \|J_i\|_F^2 + \beta\|I - J\|_2^2, \tag{11}$$

and

$$\min_I \|I \otimes k - B\|_2^2 + \beta\|I - J\|_2^2. \tag{12}$$

For a fixed I, the minimization problem (11) can be equivalently expressed as

$$\min_J \sum_i \frac{2\alpha\lambda_1}{\alpha\lambda_2 + 2\beta}\|J_i\|_{\omega,*} + \|J_i - \frac{2\beta}{\alpha\lambda_2 + 2\beta}I_i\|_F^2. \tag{13}$$

Algorithm 1. Solving Intermediate Latent Image I

Input: Kernel k.
  Initialize I from the previous iteration.
  for $l = 1 \to L$ **do**
    solve J by minimizing (13).
    solve I by minimizing (12).
  end for

We note that (13) is a weighted nuclear norm minimization problem [18]. Thus, we can obtain the solution of (13) by

$$J_i = U_i S_\omega^i(\Sigma_i)V_i^T, \tag{14}$$

where U_i, Σ_i and V_i are derived from the SVD of matrix $\frac{2\beta}{\alpha\lambda_2+2\beta}I_i$, $S_\omega(\Sigma_i)$ denotes the result of weighted soft-thresholding operating over the diagonal singular value matrix Σ_i, whose j-th diagonal element is

$$(S_\omega(\Sigma_i))_j = (\Sigma_i)_j - \omega_j. \tag{15}$$

The weighted vector ω is chosen inversely proportional to $\sigma(J_i)$, and its j-th element is defined by

$$\omega_j = c\sqrt{m}\big/(\sigma_j(J_i) + \epsilon), \tag{16}$$

where c is a positive constant, ϵ is a small number to avoid a zero denominator, and $\sigma_j(J_i)$ is the k-th singular value of J_i. Since J_i is unavailable in (16), we initialize it according to [18] by

$$\sigma_j^{(0)}(J_i) = \sqrt{\max\left(\sigma_j^2(I_i) - 2m\alpha\lambda_1/(\alpha\lambda_2 + 2\beta), 0\right)}, \tag{17}$$

where $\sigma_j(I_i)$ is the j-th singular value of I_i. Then a low rank matrix J_i corresponding to i-th patch in J can be obtained.

  Once J and k are both known, model (12) is a least squares problem. Thus, we can get its closed-form solution through fast Fourier transforms (FFTs) according to [3,9]

$$I = \mathcal{F}^{-1}\left(\frac{\overline{\mathcal{F}(k)}\mathcal{F}(B) + \beta\mathcal{F}(J)}{\overline{\mathcal{F}(k)}\mathcal{F}(k) + \beta}\right), \tag{18}$$

where $\mathcal{F}(\cdot)$ and $\mathcal{F}^{-1}(\cdot)$ denote the FFT and the inverse FFT, and $\overline{\mathcal{F}(\cdot)}$ denotes the complex conjugate of $\mathcal{F}(\cdot)$.

  The main process for solving I is summarized in Algorithm 1.

4.2 Kernel Estimation

Given a fixed I, we can estimate blur kernel by (9). We note that kernel estimation methods based on gradients have been shown to be more accurate according to [8]. Thus, we estimate the blur kernel in gradient domain by

$$\min_k \|\nabla I \otimes k - \nabla B\|_2^2 + \gamma\|k\|_2^2. \tag{19}$$

Algorithm 2. Kernel Estimation Algorithm

Input: Blurred image B and kernel size k_1, k_2.
  Initialize I with B.
  for $t = 1 \rightarrow T$ **do**
    solve for k using (20).
    solve for I using Algorithm 1.
  end for

As (19) is a least squares problem, we can get its closed-form solution by

$$k = \mathcal{F}^{-1}\left(\frac{\overline{\mathcal{F}(\partial_x I)}\mathcal{F}(\partial_x B) + \overline{\mathcal{F}(\partial_y I)}\mathcal{F}(\partial_y B)}{\mathcal{F}(\partial_x I)^2 + \mathcal{F}(\partial_y I)^2 + \gamma} \right), \tag{20}$$

where ∂_x and ∂_y denote the horizontal and vertical differential operators, respectively.

After obtaining blur kernel k, we set its negative elements to be 0, and normalize k so that k satisfies that the sum of its elements is 1.

Similar to most of state-of-the-art deblurring methods, we adopt a coarse-to-fine manner using an image pyramid to help kernel estimation [1,3,5,11], as shown in Fig. 1. The main steps for the proposed kernel estimation algorithm are shown in Algorithm 2.

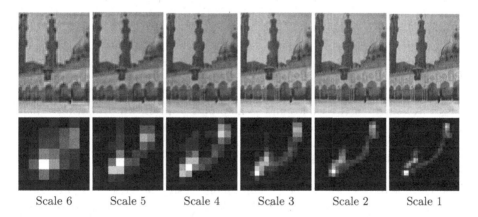

| Scale 6 | Scale 5 | Scale 4 | Scale 3 | Scale 2 | Scale 1 |

Fig. 1. Deblurring process in a coarse-to-fine manner: intermediate estimated images I and kernels k at each scale level.

4.3   Final Deblurring

Once the blur kernel k is determined, the latent image can be estimated by a number of non-blind deconvolution methods. In this paper, we employ a hyper-Laplacian prior $L_{0.8}$ by Levin et al. [2] to recover the latent image, which can be formulated as

$$\min_{I} \|I \otimes k - B\|_2^2 + \mu \|\nabla I\|^{0.8}. \tag{21}$$

We use the iterative reweighted least square (IRLS) method to solve (21).

5   Analysis and Discussion

In this section, we provide more insight and analysis on how the proposed algorithm performs on image deblurring. In addition, we also discuss the limitations of the proposed method.

5.1   Effectiveness of the Proposed Prior

According to the analysis of [4], most priors used in image deblurring favor blurred images over blurred ones. If these priors are used in the conventional MAP framework, the trivial solutions are likely to be obtained. We note that the work by [5] proposes a normalized sparsity prior which favors clear images over blurred ones. This prior is based on the image gradients which usually model the disparity between pairs of pixels. As patch-based prior has been shown to be effective, Michaeli and Irani [6] propose a new image prior based on internal patch recurrence.

Different from aforementioned work, we consider the similar patches of an image and develop an elastic-net regularization of singular values computed from similar patches. We find that the proposed prior favors clear images over blurred ones. This property ensures that the proposed method performs well under the conventional MAP framework.

To verify this property, we synthesize 54 blurred image from 9 natural images with 6 kernels, and compute the value of the proposed prior for each image. We show the comparison of prior values between clear images and blurred images in a logarithmic scale in Fig. 2. The comparison results show that the blurred images have high energy values under the proposed prior, which indicates that the proposed prior favors clear images over blurred images.

As mentioned in Sect. 3.2, the weighted nuclear norm term in (6) is helpful to preserve salient structures, and the F-norm term enhances the stabilities of the algorithm. Compared with the intuitional low-rank constraint term in the prior, the role of the F-norm term seems ambiguously. In fact, the F-norm term is very critical in the prior. We find that the F-norm term helps to favor clear images over blurred images. To verify we also compute the energy values of the F-norm term for clear images and blurred images respectively. The average energy value of blurred images is nearly 10 times larger than the corresponding value of clear images. To better understand the role of the F-norm term, we show a visual example in Fig. 3. The images on the rightmost column are final deblurred images. Intermediate kernels estimated by the proposed prior are more clean and accurate and the deblurred image restored by the proposed prior is visually much clear.

To further evaluate the effectiveness of the proposed prior, we compare the proposed prior with other priors (e.g., Total Variation [26], normalized sparsity

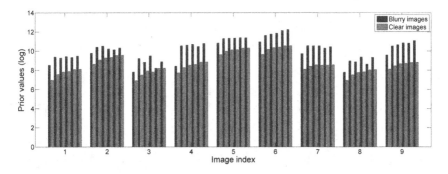

Fig. 2. The statistical results of the proposed prior on 54 pair of test images. The proposed prior favors clear images over blurred images.

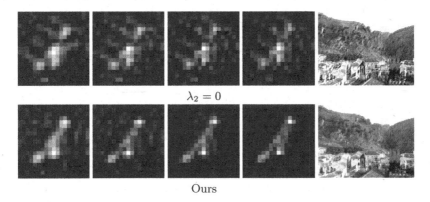

Fig. 3. Effectiveness of the F-norm term.

prior [5]) commonly used in image deblurring. We choose 21 blurred images from the dataset introduced by [13] for test and use PSNR to evaluate the quality of each restored images. The results shown in Fig. 4 demonstrate the effectiveness of the proposed prior. Moreover, we find that the deblurring method without using the F-norm term in (6) does not perform well, which further indicates the effectiveness of the proposed prior.

5.2 Convergence Property

As the proposed objective function is highly non-convex, a natural question is whether our optimization method converges (to a good local minimum). We quantitatively evaluate the proposed method on the benchmark dataset by [4]. We compute the average kernel similarity metric proposed by [27] of 32 blurred images from 4 images with 8 kernels. This metric measures the similarity between estimated kernels and the ground truths kernel via computing maximum response of normalized cross-correlation with some possible shift. Figure 5 shows

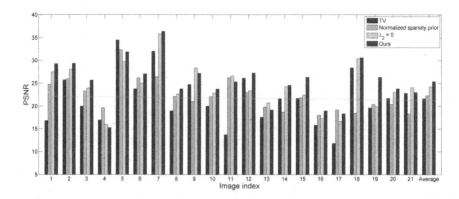

Fig. 4. Comparisons with the commonly used priors in blind image deblurring. The proposed prior is able to generate high-quality deblurring results. The average PSNR values of all the images are shown on the rightmost column.

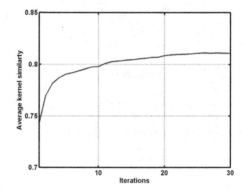

Fig. 5. Convergence property of the proposed algorithm.

that the proposed method converges after less than 20 iterations, in terms of the average kernel similarity values [27].

5.3 Limitations

Although the proposed method is able to help blur kernel estimation, it is not without limitations. If an image contains rich textures and these rich textures are located in most regions or nearly whole image, nonlocal self-similarity does not hold for this case. This is mainly because that there are not enough good similar patches to form a low-rank matrix due to the variety of rich textures. As a result, the proposed method is likely to fail. Figure 6 shows a failure example.

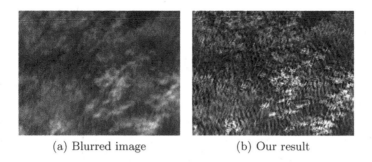

(a) Blurred image   (b) Our result

Fig. 6. One failure example. The blurred image contains rich details which do not hold for the proposed prior due to the variety of rich textures.

6   Experimental Results

In this section, we examine the proposed method on both synthetic and real-word blurred images and compare it to state-of-the-art natural image deblurring methods. The proposed algorithm is implemented in MATLAB on a computer with an Intel Xeon E5630 CPU and 12 GB RAM. The kernel estimation process takes about 12 min for a 255×255 image with a 29×29 kernel without code optimization.

Parameters Setting. In all experiments, we empirically set $\mu = 0.002$, $\alpha = 1$, $\gamma = 2$, $\lambda_2 = 0.05$, $\beta = 0.025$, and $T = 5$. The parameter λ_1 is set according to blur kernel size and weighted vector parameter c are set according to [18].

6.1   Quantitative Evaluation on the Dataset of [4]

We evaluate the proposed algorithm on the benchmark dataset by Levin et al. [4]. This dataset contains 32 blurred images of size 255×255 from 4 ground truth images with 8 blur kernels.

The error ratio metric proposed by [4] is used to evaluate the restored results, which is defined as

$$\zeta = \frac{\|I_e - I_g\|_2^2}{\|I_k - I_g\|_2^2},\tag{22}$$

where I_e is the recovered image by the estimated blur kernel, I_k is the recovered image by the ground truth blur kernel, and I_g is the ground truth clear image.

For fair comparison, we use the provided code from other methods to generate the blur kernels. The final deblurring results are all generated by the non-blind deconvolution method [2]. We compare our method with state-of-the-art methods including Krishnan et al. [5], Levin et al. [11], Cho and Lee [8], Xu and Jia [9]. Figure 7 shows the cumulative error ratio where higher curves indicate more accurate results. The proposed algorithm generates better results than state-of-the-art methods on the dataset [4].

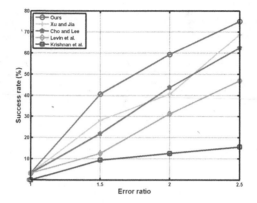

Fig. 7. Quantitative evaluation on the dataset by Levin et al. [4].

We show one example from the test dataset in Fig. 8 for discussion. The normalized sparsity-prior-based method [5] generates deblurred images with significant blur residual. The result generated by Levin et al. [11] contains fewer details. Although the result by Xu and Jia [9] contains much clearer textures,

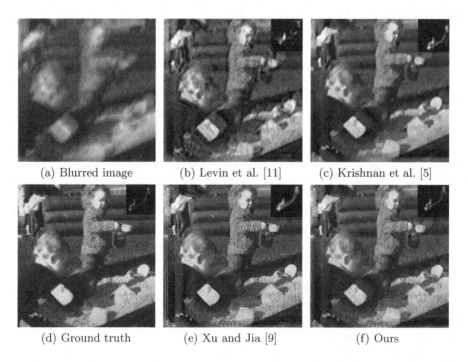

(a) Blurred image   (b) Levin et al. [11]   (c) Krishnan et al. [5]

(d) Ground truth   (e) Xu and Jia [9]   (f) Ours

Fig. 8. Visual comparisons using one image from the dataset by Levin et al. [4]. The proposed method generates much clearer images.

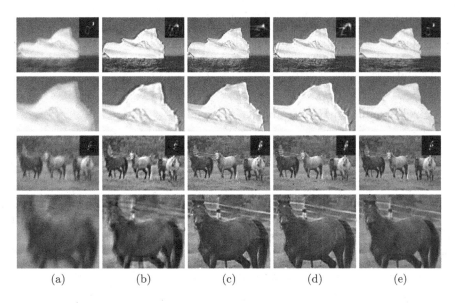

Fig. 9. Two examples from the dataset by Sun et al. [13]. (a) Blurred images. (b) The deblurred results by Krishnan et al. [5]. (c) The deblurred results by Shan et al. [3]. (d) The deblurred results by Xu and Jia [9]. (e) Ours.

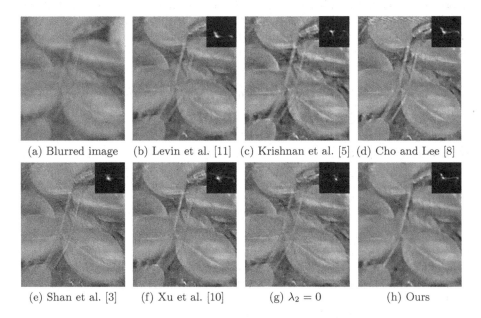

Fig. 10. Comparisons on a real natural image. The recovered image by the proposed algorithm is visually more pleasing.

there are significant ringing artifacts around image boundaries. Our method generates much more clearer image with fine details.

We further test our method using other examples as shown in Fig. 9. Compared to other deblurring methods, our method generate the images with fine details and fewer ringing artifacts.

6.2   Deblurring on Real Images

We further evaluate the proposed method on real captured images and compare it with state-of-the-art deblurring methods [3,5,8,10,11]. Figure 10 shows a real blurred example. The recovered images by state-of-the-art deblurring methods [3, 5,8,10,11] still contain significant blur effect as shown in Fig. 10(b)–(f), while the proposed method generate a much clearer image. We note that the method without using F-norm term does not perform well as shown in Fig. 10(g), which further indicates the effectiveness of the proposed prior. More results can be found in supplemental material.

7   Conclusion

In this paper, we propose a new image prior for image deblurring. The proposed prior exploits similar patches of an image and is based on an elastic-net regularization of singular values. We show that the proposed prior favors clear images over blurred image, which helps kernel estimation within conventional MAP framework. We also develop an efficient algorithm to solve the proposed model. Both quantitative and qualitative experiments show that the proposed method performs favorably against the state-of-the-art deblurring methods.

Acknowledgements. This work has been partially supported by National Natural Science Foundation of China (No. 61572099, 51379033, and 51522902) and National Science and Technology Major Project (No. ZX20140419 and 2014ZX04001011).

References

1. Fergus, R., Singh, B., Hertzmann, A., Roweis, S.T., Freeman, W.T.: Removing camera shake from a single photograph. ACM Trans. Graph. **25**, 787–794 (2006)
2. Levin, A., Fergus, R., Durand, F., Freeman, W.T.: Image and depth from a conventional camera with a coded aperture. ACM Trans. Graph. **26**, 70 (2007)
3. Shan, Q., Jia, J., Agarwala, A.: High-quality motion deblurring from a single image. ACM Trans. Graph. **27**, 73 (2008)
4. Levin, A., Weiss, Y., Durand, F., Freeman, W.T.: Understanding and evaluating blind deconvolution algorithms. In: CVPR, pp. 1964–1971 (2009)
5. Krishnan, D., Tay, T., Fergus, R.: Blind deconvolution using a normalized sparsity measure. In: CVPR, pp. 233–240 (2011)
6. Michaeli, T., Irani, M.: Blind deblurring using internal patch recurrence. In: Fleet, D., Pajdla, T., Schiele, B., Tuytelaars, T. (eds.) ECCV 2014. LNCS, vol. 8691, pp. 783–798. Springer, Cham (2014). doi:10.1007/978-3-319-10578-9_51

7. Pan, J., Sun, D., Pfister, H., Yang, M.H.: Blind image deblurring using dark channel prior. In: CVPR (2016)
8. Cho, S., Lee, S.: Fast motion deblurring. ACM Trans. Graph. **28**, 145 (2009)
9. Xu, L., Jia, J.: Two-phase kernel estimation for robust motion deblurring. In: Daniilidis, K., Maragos, P., Paragios, N. (eds.) ECCV 2010. LNCS, vol. 6311, pp. 157–170. Springer, Heidelberg (2010). doi:10.1007/978-3-642-15549-9_12
10. Xu, L., Zheng, S., Jia, J.: Unnatural l0 sparse representation for natural image deblurring. In: CVPR, pp. 1107–1114 (2013)
11. Levin, A., Weiss, Y., Durand, F., Freeman, W.T.: Efficient marginal likelihood optimization in blind deconvolution. In: CVPR, pp. 2657–2664 (2011)
12. Money, J.H., Kang, S.H.: Total variation minimizing blind deconvolution with shock filter reference. Image Vis. Comput. **26**, 302–314 (2008)
13. Sun, L., Cho, S., Wang, J., Hays, J.: Edge-based blur kernel estimation using patch priors. In: ICCP, pp. 1–8 (2013)
14. Hacohen, Y., Shechtman, E., Lischinski, D.: Deblurring by example using dense correspondence. In: ICCV, pp. 2384–2391 (2013)
15. Pan, J., Hu, Z., Su, Z., Yang, M.H.: Deblurring text images via l0-regularized intensity and gradient prior. In: CVPR, pp. 2901–2908 (2014)
16. Buades, A., Coll, B., Morel, J.M.: A non-local algorithm for image denoising. In: CVPR, pp. 60–65 (2005)
17. Dong, W., Shi, G., Li, X.: Nonlocal image restoration with bilateral variance estimation: a low-rank approach. IEEE Trans. Image Process. **22**, 700–711 (2013)
18. Gu, S., Zhang, L., Zuo, W., Feng, X.: Weighted nuclear norm minimization with application to image denoising. In: CVPR, pp. 2862–2869 (2014)
19. Xu, J., Zhang, L., Zuo, W., Zhang, D., Feng, X.: Patch group based nonlocal self-similarity prior learning for image denoising. In: ICCV, pp. 244–252 (2015)
20. Cai, J.F., Candès, E.J., Shen, Z.: A singular value thresholding algorithm for matrix completion. SIAM J. Optim. **20**, 1956–1982 (2010)
21. Ren, W., Cao, X., Pan, J., Guo, X., Zuo, W., Yang, M.: Image deblurring via enhanced low-rank prior. IEEE Trans. Image Process. **25**, 3426–3437 (2016)
22. Kim, E., Lee, M., Oh, S.: Elastic-net regularization of singular values for robust subspace learning. In: CVPR, pp. 915–923 (2015)
23. Wang, S., Zhang, L., Liang, Y.: Nonlocal spectral prior model for low-level vision. In: Lee, K.M., Matsushita, Y., Rehg, J.M., Hu, Z. (eds.) ACCV 2012. LNCS, vol. 7726, pp. 231–244. Springer, Heidelberg (2013). doi:10.1007/978-3-642-37431-9_18
24. Geman, D., Reynolds, G.: Constrained restoration and the recovery of discontinuities. IEEE Trans. Pattern Anal. Mach. Intell. **14**, 367–383 (1992)
25. Geman, D., Yang, C.: Nonlinear image recovery with half-quadratic regularization. IEEE Trans. Image Process. **4**, 932–946 (1995)
26. Perrone, D., Favaro, P.: Total variation blind deconvolution: the devil is in the details. In: CVPR, pp. 2909–2916 (2014)
27. Hu, Z., Yang, M.-H.: Good regions to deblur. In: Fitzgibbon, A., Lazebnik, S., Perona, P., Sato, Y., Schmid, C. (eds.) ECCV 2012. LNCS, vol. 7576, pp. 59–72. Springer, Heidelberg (2012). doi:10.1007/978-3-642-33715-4_5

Single Image Dehazing Using Fixed Points and Nearest-Neighbor Regularization

Shengdong Zhang and Jian Yao[✉]

Computer Vision and Remote Sensing (CVRS) Lab,
School of Remote Sensing and Information Engineering, Wuhan University,
Wuhan, Hubei, People's Republic of China
jian.yao@whu.edu.cn
http://cvrs.whu.edu.cn/

Abstract. Natural images captured in bad weather conditions often suffer from poor visibility. Dehazing, the process of removing haze from a single input image or multiple images, is a crucial task in image and video processing, which is quite challenging because the number of freedoms is lager than the number of observations. In this paper, we propose a novel method to reduce the block artifacts and halos for single image dehazing, which replaces the widely used soft matting and contextual regularization. We first find some fixed points in a maximum filter and then apply a Nearest-Neighbor (NN) regularization to recover a smooth transmission map. Compared with the state-of-the-art single image dehazing methods, the experimental results on some typical and challenged images demonstrate that our method can produce a high-quality dehazed image and recover the fine detail information and vivid color from the image haze regions.

1 Introduction

Natural images captured in outdoor scenes are often suffered by bad weather conditions such as air particles, water droplets, fog, haze, smoke, rain or snow. In such cases, atmospheric phenomena such as haze and fog caused by atmospheric absorption and scattering will greatly reduce the visibility and quality of the images captured in the scenes. The light received by camera sensors from the object in the scene is attenuated along the light propagating direction. The arriving light is fused with the air-light [1] - the ambient light reflected into the line of sight by air particles. Vividness, visibility and contrast of a natural image captured in outdoor under these bad weather conditions are dramatically degraded, which make it hard to recognize the objects existed in this image farther away from the camera center. As shown in Fig. 1(a), the degraded image loses color fidelity and contrast and the distant objects in the image are hard to be distinguished. As stated in [2], the degradation is spatially variant because the scattering depends on the distance between the camera and the objects existed in the scene.

Restoring images captured in outdoor under the bad weather conditions has caused great interesting in the past decade. Haze removal is quite needed in

© Springer International Publishing AG 2017
C.-S. Chen et al. (Eds.): ACCV 2016 Workshops, Part I, LNCS 10116, pp. 18–33, 2017.
DOI: 10.1007/978-3-319-54407-6_2

(a)   (b)   (c)

Fig. 1. An illustration example for our proposed single image dehazing approach: (a) The input hazy image; (b) Our recovered transmission map; (c) The dehazed image. (Color figure online)

computer vision applications and commercial/computational photography. First, removing haze can improve the visibility of the image and correct the color distortion caused by the air-light. In general, the haze-free image is more visual friendly. Second, most computer vision algorithms, such as feature detection, filtering and image analysis, usually assume that the input image is haze-free. The performance of computer vision algorithms and advanced image editing ones will be greatly degraded by the low contrast and visibility images. Last, bad weather conditions will cause traffic accidents due to the poor visibility. If the haze removing techniques can be efficiently utilized for the car navigation, the image visibility can be greatly improved and the traffic accidents are possibly reduced to some extent. However, defogging or haze removal is a very challenging because the amount of scattering depends on the unknown distances of the camera and the scene points and the air-light is also unknown. Obviously, this is a quite ill-posed problem especially when the input is only a single hazy image. Therefore, those early proposed methods for haze removal rely on multiple images of the same scene and the additional depth information [3–5]. For example, in [3], the depth discontinuities in the scene and scene structure were computed by using the changes in intensities of scene points under different weather conditions. In [4,6], the polarization based methods were proposed to remove image haze through two images taken with different degrees of polarization. Kopf et al. [5] developed a depth-based method to dehaze an image by using the scene depth information, which are directly accessible in the geo-referenced digital terrain or city models. Although all these methods can produce good results, but the constrain on input limits the applications of these methods.

Recently, removing haze from a single image attracted much attention and made significant progresses due to its wide applications. A lot of single image haze removal methods have been proposed based on the assumption or stronger priors. Comparing haze-free images and haze ones, Tan [7] found that the haze-free images have high contrast than the haze ones. Based on this observation, he proposed an interesting single image haze removal method by maximizing the local contrast of the restored image. This method can yield a visually compelling results. However, the restored image often exists color shift, block artifacts and significant halos because this method might not be physically valid. Fattal [8] proposed to remove haze from a color image under the assumption that the surface texture and the transmission are locally uncorrelated, which is used to solve the air-light-albedo ambiguity. This method is physical sound and can yield compelling results. However, it is time-consuming and cannot deal with the grayscale images. In addition, it could fail to deal with dense-hazy images. Fattal [9] proposed another method for single-image dehazing, which relies on pixels of small patches in natural images, which often show a one-dimensional distribution in the RGB color space. In this method, a local model was proposed to explain the color-lines in the context of hazy scenes and recovering the scene transmission by considering the lines' offset from the origin. A new Markov random field was described for yielding complete and regularized transmission.

He et al. [2] proposed a novel dark channel prior based dehazing method, which is a basis of the dark-object subtraction methods [10]. The dark channel prior means that in most local patches there exists often at least one color (RGB) channel containing some pixels with very low intensities. With this prior, He et al. estimated the thickness of haze locally from the dark-channel pixels found within a local patch. This method is very simple, physically sound and can produce impressive results even in images with heavy haze. However, it cannot deal with the sky images quite well because the dark channel pixels are possibly unavailable in those bright image regions. In addition, it is time-consuming. Recently, a lot of methods have been proposed to overcome the weakness of the Dark Channel Prior (DCP). For example, He et al. [11], Tarel and Hautiere [12] and Tarel et al. [13] used "median of median filter" and guided filtering to replace soft matting. Carr and Hartley [14] combined the scene geometry and the dark channel prior to estimate the transmission, and used the alpha-expansion optimization technique to recover a smooth transmission map. Gibson and Nguyen [15] proposed a new dark channel prior for removing haze from the image. Unlike the traditional dark-channel prior that assumes a zero minimal value, the new prior searches for the darkest pixel in each ellipsoid. Recently, Zhu et al. [16] proposed a new, simple and powerful prior-color attenuation prior based on statistics of natural images. A linear model for the scene depth of the fog image with the color attenuation prior was created to solve the model parameters via supervised learning. With the resulting depth map, the air-light can be easily recovered. Li et al. [17] proposed to decompose the simplified dark channel of the hazy image into a detail layer and a base layer based on edge-preserving decomposition. The base layer was used to estimate the transmission map for restoring the haze-free images.

To effectively remove haze from a single image, some methods based on the principle of image fusion have been proposed recently. Ancuti et al. [18, 19] show the effectiveness of the fusion-based method for removing haze from a single input image. The fusion-based method proposed in [19] first pre-processed two original hazy image inputs by applying a contrast enhancing and a white balance procedure. Then it blended the resulting inputs by computing three measurements (weight maps). Furthermore, it adopted a multiscale strategy to reduce the artifacts. This method can produce good results similar to and even better than complex state-of-the-art techniques. In contrast, Wang et al. [20] used a multiscale depth fusion method to recover a haze-free image. A new technique was proposed to estimate the depth by considering the influence of noise. They developed an inhomogeneous Laplacian-Markov random field with edge-preserving and smooth constrains for multiscale depth fusion to estimate a depth map and produce an improving contrast and vivid color result.

In this paper, we propose to replace the widely used soft matting [2] and contextual regularization [21] in estimating the transmission map for haze removal by first finding some fixed points in a maximum filter and then applying a Nearest-Neighbor (NN) regularization. Figure 1 shows an illustrative example for our proposed dehazing method. Experimental results on some typical and challenged images illustrate that our method can produce a high-quality dehazed and vivid color image.

2 Background

2.1 Haze Imaging Model

In computer graphics and computer vision, the haze imaging model presented in [1], which describes the formation of a hazy image, is widely used as following:

$$\mathbf{I}(\mathbf{x}) = \mathbf{J}(\mathbf{x})t(\mathbf{x}) + A(1 - t(\mathbf{x})), \tag{1}$$

where $\mathbf{x}$ is a pixel location, $\mathbf{I}$ is the observed haze image, $\mathbf{J}$ is the haze free image, $\mathbf{I}(\mathbf{x})$ and $\mathbf{J}(\mathbf{x})$ stand for the intensities of the point $\mathbf{x}$ in $\mathbf{I}$ and $\mathbf{J}$, respectively, A is the air light (or atmospheric light) of the haze image, and t is the transmission coefficient, which describes the probability of the light reflecting from object is not scattered and absorbed by air particles. Recovering a haze free image $\mathbf{J}$ from the observed haze image $\mathbf{I}$ is equal to solve A and t from $\mathbf{I}$.

As we can see, the longer light traveling before reaching a camera, the more scattered and attenuated. Thus, we can express the transmission coefficient t as follows:

$$t(\mathbf{x}) = e^{-\beta d(\mathbf{x})}, \tag{2}$$

where $d(\mathbf{x})$ represents the distance between the camera and the scene point corresponding to $\mathbf{x}$, and β represents the attenuation coefficient of the atmosphere (often set to 1). So from Eq. (1), we find that the contribution of the air-light term A is more important when the observed scene is far away from the camera.

2.2   Nonlinear Filtering

Nonlinear filtering is often used to get a good bound transmission for each pixel. There are two kinds of filters: pixel-wise and patch-based. Nishino et al. [22], and Caraffa and Tarel [23] applied a pixel-wise nonlinear operation to recover a finer estimation of the transmission map, and then imposed a Markov regularization using the contextual constrain. The fine estimation prevents from the block artifacts and halos of the patch-based filter, but more constraints are needed to recover a high accurate transmission map. The white objects and the farther objects make it hard to estimate the transmission map accurately. He et al. [2] applies a patch-based filter to solve the white objects problem. While the patch-based filter can solve the white object problem, both block artifacts and halos incurred by the patch-based filter need to be further solved.

2.3   Fixed-Point Theorem

The fixed-point theorem [24] has been proposed for a long time. In mathematics, a fixed-point theorem is a result saying that a function F will have at least one fixed point (a point x for which $F(x) = x$), under some conditions on F that can be stated in general terms. In this paper, we define the point x as a fixed one if it satisfies the following equation:

$$F(x) = x. \tag{3}$$

2.4   Non-local Principle

Non-local principle has attracted a lot of attention for its broad applications, such as denoise [25] and matting [26,27]. As stated in [25], the nonlocal principle is that for a given distorted image, the pixel can be restored by taking a weighted sum of the pixels that have similar appearance, where the weights are given by a kernel. In our proposed single image dehazing approach, we apply the similar principle, but we make some modifications so as to better suit the regularization problem. These modifications will be described in details in Sect. 3.4.

3   Our Approach

3.1   Atmospheric Light Estimation

There exist a lot of methods to estimate the atmospheric light, i.e., the parameter A in Eq. (1). He et al. [2] proposed a method to estimate A based on the dark channel prior. Kim et al. [28] also proposed an approach based on the quadtree based subdivision, which may be more suitable for our application because the method proposed by He et al. [2] is based on the dark channel prior, which is not required in our proposed method. However, we still adopt He et al.'s method because of its efficiency. We first compute the dark channel prior of a hazy image. Second we pick up the top 0.1% of the dark channels. Third we find out the pixel with highest intensity as the atmospheric light. According to Eq. (2), when the objects are far away from the camera, the depth is very large and the transmission of the objects is close to 0. According to the haze imaging model in Eq. (1), when the transmission is equal to 0, the observed object intensity in the hazy image is equal to the atmospheric light, i.e.,

$$\mathbf{I}(\mathbf{x}) = \mathbf{J}(\mathbf{x})t(\mathbf{x}) + A(1 - t(\mathbf{x})) = A. \qquad (4)$$

This method also works well when the image doesn't have pixels at infinite distance because t in most haze-opaque regions is very small and the influence of sunlight is so weak that can be ignored.

3.2   Lower Bound of Transmission

In order to get an initial transmission value of some pixel in an image, we rewrite Eq. (1) as following:

$$\mathbf{J}(\mathbf{x}) = \frac{\mathbf{I}(\mathbf{x}) - A(1 - t(\mathbf{x}))}{t(\mathbf{x})}. \qquad (5)$$

In the RGB color space, the intensities of the image pixel fall in range $[0, 255]$. In order to facilitate the calculation, we normlize the intensities into range $[0, 1]$ correspondingly. In this way, we get the following equation:

$$0 \le \frac{\mathbf{I}(\mathbf{x}) - A(1 - t(\mathbf{x}))}{t(\mathbf{x})} \le 1, \qquad (6)$$

which can result in the following two equations:

$$t_1(\mathbf{x}) \ge \frac{\mathbf{I}^c(\mathbf{x}) - A^c}{1 - A^c} \quad \text{and} \quad t_2(\mathbf{x}) \ge 1 - \frac{\mathbf{I}^c(\mathbf{x})}{A^c}, \qquad (7)$$

where the superscript 'c' represents one of the RGB channels. The minimum transmission in the above two inequations are written as:

$$\hat{t}_1(\mathbf{x}) = \frac{\mathbf{I}^c(\mathbf{x}) - A^c}{1 - A^c} \quad \text{and} \quad \hat{t}_2(\mathbf{x}) = 1 - \frac{\mathbf{I}^c(\mathbf{x})}{A^c}. \qquad (8)$$

Thus, the lower bound of the transmission can be defined as:

$$t_b(\mathbf{x}) = \max\left(\hat{t}_1(\mathbf{x}), \hat{t}_2(\mathbf{x})\right), \qquad (9)$$

which is a special case of the lower bound of the transmission used in [21] with the parameters $C_0 = 0$ and $C_1 = 255$ as:

$$t_b(\mathbf{x}) = \min\left(\max\left(\frac{A^c - \mathbf{I}^c(\mathbf{x})}{A^c - C_0}, \frac{A^c - \mathbf{I}^c(\mathbf{x})}{A^c - C_1}\right), 1\right). \qquad (10)$$

3.3   Fixed Points of Filtering

Nonlinear filtering is a popularly used operation to obtain the transmission map from the chromaticity of a hazy image. In our approach, we use the maximum-minimum filter as used in [21] to get a transmission as:

$$\tilde{t}(\mathbf{x}) = \min_{\mathbf{y} \in \Omega(\mathbf{x})} \max_{\mathbf{z} \in \Omega(\mathbf{y})} t(\mathbf{z}), \qquad (11)$$

where $\Omega(\mathbf{x})$ and $\Omega(\mathbf{y})$ represent local patches centered at the points $\mathbf{x}$ and $\mathbf{y}$, respectively.

Based on the dark channel prior proposed by He et al. [2], we can estimate the transmission value of each pixel. Then, a rough transmission map can be recovered via the maximum-minimum filtering. To recover a more accurate transmission map, we first need to find the stable points whose transmission values are more accurate. The fixed-point theorem can be used to find such these stable points. For a given pixel $\mathbf{x}$ in a hazy image, if we apply the maximum-minimum filter on $\mathbf{x}$ with its lower bound transmission value $t_b(\mathbf{x})$, $\mathbf{x}$ is a fixed point if the following condition is satisfied:

$$F(t_b(\mathbf{x})) = t_b(\mathbf{x}), \qquad (12)$$

where the function F stands for the maximum-minimum filter. Figure 2 shows an example of fixed points found with two different filter patch sizes, from which we can observe that the use of a smaller patch size will generate more fixed points.

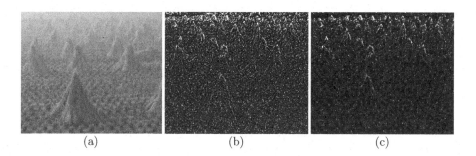

(a)   (b)   (c)

Fig. 2. An example of fixed points extracted from the input hazy image (a) with two different filter patch sizes, 7×7 (b) and 17×17 (c).

3.4   Nearest-Neighbor Regularization

Figure 3 shows two hazy images with their corresponding transmission maps, from which we can observe that the transmission map is very smooth except abrupt depth jumps. As mentioned before, the dehazing problem is severely under-constrained. Therefore, we need to introduce some assumptions on the natural transmission map. As stated before, the non-local principle considers a

Fig. 3. Hazy images in the left and their corresponding transmission maps in the right.

pixel as a liner combination of similar pixels in neighborhood. In the dehazing context, we consider an assumption that similar pixels have same transmission value. As stated in [9,21], our assumption is quite reasonable and widely used in most dehazing methods.

To efficiently measure the similarity between two image pixels, we need to consider the color similarity and spatial variation and smoothness. Given an image point $\mathbf{x}$, we define its feature vector for similarity measurement as follows:

$$f(\mathbf{x}) = (R, G, B, \lambda x, \lambda y)^\top, \qquad (13)$$

where R, G and B represent the intensities of $\mathbf{x}$ in three RGB channels, respectively, x and y are the spatial coordinates of $\mathbf{x}$, and λ is balancing factor.

Although we use the same feature as in [27], but dehazing is different from matting, so for we need to modify the non-local principle widely used for matting.

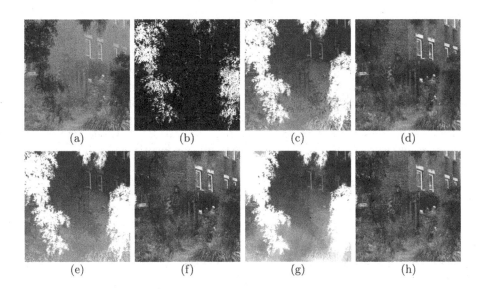

Fig. 4. The dehazing results with different values of λ: (a) the input hazy image; (b) fixed points; (c)–(d) the estimated transmission map and the dehazed image with $\lambda = 1$; (e)–(f) the estimated transmission map and the dehazed image with $\lambda = 2$; (g)–(h) the estimated transmission map and the dehazed image with $\lambda = 10$.

(a)   (b)   (c)

Fig. 5. The dehazed images with the use of the RGB (b) and HSV (c) color spaces from the input hazy image (a). (Color figure online)

In this paper, we treat the transmission values of fixed points as accuracy and reasonable, and we propose to recover the transmission values of the remaining points by finding their nearest neighbours from the set of fixed points based on the constructed k-d tree with the 5-dimensional feature vectors defined in Eq. (13). Given the atmospheric light A and the recovered transmission, a haze free image can be solved using the following equation:

$$\mathbf{J(x)} = \frac{\mathbf{I(x))} - A}{t(\mathbf{x})} + A. \qquad (14)$$

Figure 4 shows the dehazing results with different values of the balancing factor λ, from which we can observe that the large value of λ will result in a more smooth transmission map, which is quite consistent with our assumption.

The HSV color space can be alternate of the RGB color space for similarity measurement. Figure 5 shows the dehazed images with the RGB and HSV color spaces, respectively, from which we observe that some hazy image regions cannot be clearly removed with the use of the HSV color space. By contrast, the RGB color space will produce a better dehazed image. So, the RGB color space is used in our method.

3.5   Our Dehazing Framework

The framework of our proposed single image dehazing method is summarized in Algorithm 1.

Algorithm 1. Our proposed single image dehazing framework.

Input: The hazy image.
Output: The haze free image.
1: Compute the atmospheric light A using Meng et al.'method;
2: Compute the lower bound of the transmission;
3: Find the fixed points with the maximum-minimum filter;
4: Recover the transmission map by the nearest neighbour (NN) regularization;
5: Recover the haze free image using the atmospheric light A and the recovered transmission map.

4   Experimental Results

In order to evaluate the effectiveness of the proposed dehazing method, we tested it on a lot of hazy images and compared it with some state-of-the-art methods [2,5,7–9,12,19]. In all examples presented in this section, we assume the atmospheric light estimation is given by using the air-light vector A calculated by [21] or selecting the densest hazy pixel. All the results of other methods come from the Fattal's homepage[1].

4.1   Tests on Typical Examples

Figure 6 shows the dehazed result on an example image with inhomogeneous fog. Figure 7 shows the dehazed results on three example images with homogeneous fog. For these example images we used the patch size 17×17 and $\lambda = 2$. As we can see from Figs. 6 and 7, we can observe that our dehazed images present full details and vivid color information recovered from the fog or hazy images. As stated in [21], the fog in the image of "Tiananmen" cannot be regarded as homogeneous. This case often happens due to the large area of sky in the image. For this hazy image, we use the atmospheric light A selected by manual, the values were set as $[0.7961, 0.7529, 0.6588]$ for the RGB channels, respectively. We also draw the same conclusion with [21] that the transmission map represents the density of the hazes or fogs in the captured image.

4.2   Visual Comparison

We also compared our method with several state-of-the-art methods. Figure 8 illustrates an example with the comparisons between our method and the other two methods: He et al.'s [2] and Fattal's [9]. For this example, we used the parameters with $A = [0.53, 0.53, 0.53]$, the patch size 40×40, and $\lambda = 10$. The A value was used in the Fattal's work [9]. From Fig. 8, we can see that our method can restore the finer details and vivid color from the hazy regions.

Figure 9 shows the dehazed result using our method on a heavy hazy image by comparison with the other two recent methods. From Fig. 9, we can observe

---

[1] Available at http://www.cs.huji.ac.il/~raananf/projects/dehaze_cl/.

Fig. 6. The dehazed result on the "Tiananmen" image using our method: (Left) the input hazy image; (Middle) the recovered transmission map; (Right) the dehazed image. (Color figure online)

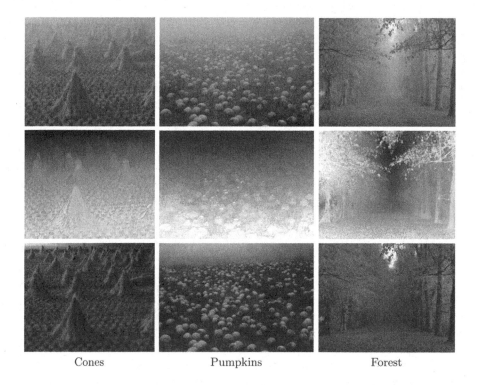

Cones   Pumpkins   Forest

Fig. 7. The dehazing results on three typical images: (Top) the input hazy images; (Middle) the recovered transmission maps; (Bottom) the dehazed images. (Color figure online)

that our result on this heavy hazy image is comparable in terms of visual quality to those results in [19,29]. For this example, we used the parameters with $A = [0.63, 0.62, 0.62]$, the patch size 40×40, and $\lambda = 10$. As we can see from Fig. 9, the Ancuti et al.'s method [19] and Choi et al.'s method [29] cannot remove the haze completely. In contrast, our method can restore the detail information and vivid color well.

(a) the hazy image   (b) He et al.'s   (c) the Fattal's   (d) ours

Fig. 8. An illustrative comparison with the He et al.'s and the Fattal's methods. (Color figure online)

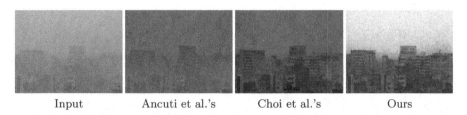

Input   Ancuti et al.'s   Choi et al.'s   Ours

Fig. 9. An illustrative comparison with the Ancuti et al.'s and the Choi et al.'s methods.

Figures 10 and 11 show the comparison of our method with the other six state-of-the-art methods on "ny12" and "ny17" images, respectively, which are widely used for testing the performance of the dehazing method. Tan's method [7] can greatly enhance the image visibility and recover the image details. However, the colors in the restored images are often over enhanced. Since this method is not a physical sound and the transmission may thus be underestimated. Tarel et al.'s method [12] is a filtering based method. Its greatest advantage is real time while its dehazed images are not quite visual pleased. The Fattal's method [8] estimates the transmission based on the sufficient color information. When the haze is very heavy, the transmission may be wrongly estimated. He et al.'s method is a statistics based method, it can produce a very good result. But it may be fail in regions with many depth jumps. Ancuti et al.'s method is a fusion-based method and can produce a visually pleasing result. For the "ny17" image, in order to reduce the noise of the result image we restrict the transmission value between 0.2 and 1, and our result can recover a high quality and rich details from the hazy image. In order to equally compare the result, we used the value of A reported in [9].

4.3   Quantitative Comparison

Based on the above results on the "ny12" and "ny17" images, we conducted a quantitative comparison using the blind assessment of [30]. Hautière et al. [30] computed the ratio between the gradients of the input image and the restored one. This method is based on the concept of the visibility level, which is widely used in lighting areas. Table 1 shows the quantitative comparisons on the "ny12"

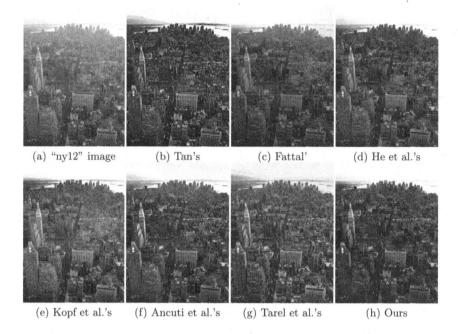

(a) "ny12" image   (b) Tan's   (c) Fattal'   (d) He et al.'s

(e) Kopf et al.'s   (f) Ancuti et al.'s   (g) Tarel et al.'s   (h) Ours

Fig. 10. Comparisons with the state-of-the-art methods on the "ny12" image.

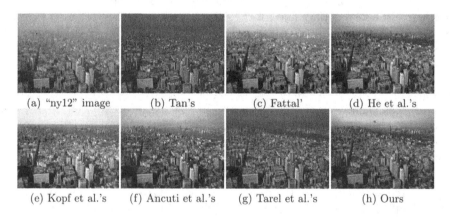

(a) "ny12" image   (b) Tan's   (c) Fattal'   (d) He et al.'s

(e) Kopf et al.'s   (f) Ancuti et al.'s   (g) Tarel et al.'s   (h) Ours

Fig. 11. Comparisons with the state-of-the-art methods on the "ny17" image.

and "ny17" images, in which the indicator "e" represents the newly visible edges ratio after restoration, the indicator "$\sum$" represents the percentage of pixels which are completely white or dark after restoration, and "$\bar{r}$" represents the mean ratio of the gradients at visible edges.

From Table 1, we can observe that all tested methods (including our method) produce a small value of the indicator $\bar{r}$, which represents the ratio of the pixels

Table 1. Quantitative comparison on the "ny12" and "ny17" images based on three indicators, e, $\sum$ and $\bar{r}$ used in [30].

Image	Tan's			Fattal's			Kopf et al.'s			He et al.'s		
	e	$\sum$	$\bar{r}$	e	$\sum$	$\bar{r}$	e	$\sum$	$\bar{r}$	e	$\sum$	$\bar{r}$
ny12	−0.14	0.02	2.34	−0.06	0.09	1.32	0.05	0.00	1.42	0.06	0.00	1.42
	Tarrel et al.'s			Ancuti et al.'s			Choi et al.'s			Ours		
	0.07	0.0	1.88	0.02	0.00	1.49	0.09	0.00	1.56	0.06	0.00	1.36
Image	Tan's			Fattal's			Kopf et al.'s			He et al.'s		
ny17	−0.06	0.01	2.22	−0.12	0.02	1.56	0.01	0.01	1.62	0.01	0.00	1.65
	Tarrel et al.'s			Ancuti et al.'s			Choi et al.'s			Ours		
	−0.01	0.0	1.87	0.12	0.00	1.54	0.03	0.00	1.49	0.14	0.00	1.82

being completely white or dark. On the one hand, the indicator e shows that most methods remove some of the visible edges, Only our method, He et al.'s [2] and Ancuti et al.'s [19] are positive values of the indicator e for these two test images. On the other hand, for the indicator $\sum$, the measurement [30] yields small values of the indicator $\sum$ for our results, which means our method doesn't have the problem of over-saturation. The values of our results are close to $\bar{r} = 1$, which show that the local contrast was restored moderately, He et al.'s [2], Kopf et al.'s [5] and Fattal's [8] methods also have this feature. In contrast, Tarel et al.'s [12] and Tan's [7] methods increase the local contrast too strongly and as a result these methods have a high values of the indicator $\bar{r}$. In general the method with low values of the indicator $\bar{r}$ show less artifacts and spurious. To the best of our knowledge, the blind assessment of [30] is the only method designed to produce a quantitative explanation for defogging operation. The indicators can give some explanations to the level of restoration and degradation, which can be used to measure the performance of the dehazing method.

5 Conclusion

Haze removal is an important task in image and video editing and results in a great challenge for computer vision. In this paper, we proposed a novel method to reduce the block artifacts and halos for single image dehazing. First, we proposed a simple method to estimate the transmission for some pixels based on the fixed-point theorem. Then we introduced a new regularization method based on the assumption that the transmission map varies smoothly. Experimental results on some typical and challenged images illustrate that our method can produce good results comparative to and even better than the more complex state-of-the-art methods.

Acknowledgment. This work was partially supported by the National Natural Science Foundation of China (Project No. 41571436), the National Natural Science Foundation of China under Grant 91438203, the Hubei Province Science and Technology

Support Program, China (Project No. 2015BAA027), the Jiangsu Province Science and Technology Support Program, China (Project No. BE2014866), and the South Wisdom Valley Innovative Research Team Program.

References

1. Harald, K.: Theorie der horizontalen Sichtweite: Kontrast und Sichtweite, vol. 12. Keim & Nemnich, Munich (1924)
2. He, K., Sun, J., Tang, X.: Single image haze removal using dark channel prior. IEEE Trans. Pattern Anal. Mach. Intell. **33**, 2341–2353 (2011)
3. Narasimhan, S.G., Nayar, S.K.: Contrast restoration of weather degraded images. IEEE Trans. Pattern Anal. Mach. Intell. **25**, 713–724 (2003)
4. Shwartz, S., Namer, E., Schechner, Y.Y.: Blind haze separation. In: IEEE Computer Society Conference on Computer Vision and Pattern Recognition (CVPR), vol. 2, pp. 1984–1991 (2006)
5. Kopf, J., Neubert, B., Chen, B., Cohen, M., Cohen-Or, D., Deussen, O., Uyttendaele, M., Lischinski, D.: Deep photo: model-based photograph enhancement and viewing. ACM Trans. Graph. (TOG) **27**, 32–39 (2008)
6. Schechner, Y.Y., Narasimhan, S.G., Nayar, S.K.: Instant dehazing of images using polarization. In: IEEE Computer Society Conference on Computer Vision and Pattern Recognition (CVPR), vol. 1, pp. 325–332 (2001)
7. Tan, R.T.: Visibility in bad weather from a single image. In: IEEE Computer Society Conference on Computer Vision and Pattern Recognition (CVPR), pp. 1–8 (2008)
8. Fattal, R.: Single image dehazing. ACM Trans. Graph. (TOG) **27**, 1–9 (2008)
9. Fattal, R.: Dehazing using color-lines. ACM Trans. Graph. (TOG) **34**, 13 (2014)
10. Chavez, P.S.: An improved dark-object subtraction technique for atmospheric scattering correction of multispectral data. Remote Sens. Environ. **24**, 459–479 (1988)
11. He, K., Sun, J., Tang, X.: Guided image filtering. IEEE Trans. Pattern Anal. Mach. Intell. **35**, 1397–1409 (2013)
12. Tarel, J.P., Hautiere, N.: Fast visibility restoration from a single color or gray level image. In: IEEE International Conference on Computer Vision (ICCV), pp. 2201–2208 (2009)
13. Tarel, J.P., Hautière, N., Caraffa, L., Cord, A., Halmaoui, H., Gruyer, D.: Vision enhancement in homogeneous and heterogeneous fog. IEEE Intell. Transp. Syst. Mag. **4**, 6–20 (2012)
14. Carr, P., Hartley, R.: Improved single image dehazing using geometry. In: Digital Image Computing: Techniques and Applications (DICTA), pp. 103–110 (2009)
15. Gibson, K.B., Nguyen, T.Q.: An analysis of single image defogging methods using a color ellipsoid framework. EURASIP J. Image Video Process. **2013**, 1–14 (2013)
16. Zhu, Q., Mai, J., Shao, L.: A fast single image haze removal algorithm using color attenuation prior. IEEE Trans. Image Process. **24**, 3522–3533 (2015)
17. Li, Z., Zheng, J.: Edge-preserving decomposition-based single image haze removal. IEEE Trans. Image Process. **24**, 5432–5441 (2015)
18. Ancuti, C.O., Ancuti, C., Hermans, C., Bekaert, P.: A fast semi-inverse approach to detect and remove the haze from a single image. In: Kimmel, R., Klette, R., Sugimoto, A. (eds.) ACCV 2010. LNCS, vol. 6493, pp. 501–514. Springer, Heidelberg (2011). doi:10.1007/978-3-642-19309-5_39
19. Ancuti, C.O., Ancuti, C.: Single image dehazing by multi-scale fusion. IEEE Trans. Image Process. **22**, 3271–3282 (2013)

20. Wang, Y., Fan, C.: Single image defogging by multiscale depth fusion. IEEE Trans. Image Process. **23**, 4826–4837 (2014)
21. Meng, G., Wang, Y., Duan, J., Xiang, S., Pan, C.: Efficient image dehazing with boundary constraint and contextual regularization. In: IEEE International Conference on Computer Vision (ICCV), pp. 617–624 (2013)
22. Nishino, K., Kratz, L., Lombardi, S.: Bayesian defogging. Int. J. Comput. Vis. **98**, 263–278 (2012)
23. Caraffa, L., Tarel, J.P.: Markov random field model for single image defogging. In: IEEE Intelligent Vehicles Symposium (IV), pp. 994–999 (2013)
24. Kakutani, S.: A generalization of Brouwer's fixed point theorem. Duke University Press, Durham (1941)
25. Buades, A., Coll, B., Morel, J.M.: A non-local algorithm for image denoising. In: IEEE Computer Society Conference on Computer Vision and Pattern Recognition (CVPR), vol. 2, pp. 60–65 (2005)
26. Lee, P., Wu, Y.: Nonlocal matting. In: IEEE Conference on Computer Vision and Pattern Recognition (CVPR), pp. 2193–2200 (2011)
27. Chen, Q., Li, D., Tang, C.K.: KNN matting. IEEE Trans. Pattern Anal. Mach. Intell. **35**, 2175–2188 (2013)
28. Kim, J.H., Jang, W.D., Sim, J.Y., Kim, C.S.: Optimized contrast enhancement for real-time image and video dehazing. J. Vis. Commun. Image Represent. **24**, 410–425 (2013)
29. Choi, L.K., You, J., Bovik, A.C.: Referenceless prediction of perceptual fog density and perceptual image defogging. IEEE Trans. Image Process. **24**, 3888–3901 (2015)
30. Hautière, N., Tarel, J.P., Aubert, D., Dumont, E.: Blind contrast enhancement assessment by gradient ratioing at visible edges. Image Anal. Stereology **27**, 87–95 (2008)

Robust Noisy Image Super-Resolution Using ℓ_1-norm Regularization and Non-local Constraint

Bo Yue[1], Shuang Wang[1(✉)], Xuefeng Liang[2(✉)], and Licheng Jiao[1]

[1] Key Laboratory of Intelligent Perception and Image Understanding of Ministry
of Education, International Research Center for Intelligent Perception
and Computation, Xidian University, Xi'an, China
shwang@mail.xidian.edu.cn
[2] IST, Graduate School of Informatics, Kyoto University, Kyoto, Japan
xliang@i.kyoto-u.ac.jp

Abstract. Conventional coupled dictionary learning approaches are designed for noiseless image super-resolution (SR), but quite sensitive to noisy images. We find the cause is the commonly used ℓ_2-norm coefficients transition term. In this paper, we propose a robust ℓ_1-norm solution by introducing two sub-terms: *LR coefficient sparsity constraint term* and *HR coefficient conversion term*, which are able to prevent the noise transmission from noisy input to output. By incorporating our simple yet effective non-linear model inspired by auto-encoder, the proposed ℓ_1-norm dictionary learning achieves a more accurate coefficients conversion. Moreover, we bring the non-local similarity constraint from pixel domain to the sparse coefficients optimization. The improved sparse representation further enhances SR inference on both noisy and noiseless images. Using standard metrics, we show that results are significantly clearer than state-of-the-arts on noisy images and sharper on denoised images.

1 Introduction

The goal of single image super-resolution (SISR) is to reconstruct a high resolution (HR) image from a low resolution (LR) input. This problem is inherently ill-posed, thus very challenging in computer vision. To solve it, different forms of prior knowledge have been explored. In particular, learning-based (example-based) strategy [1–12] has received a great attention in the past decade. The dominant approach utilizes the sparse representation to model the relation between the sparse coefficients of LR and HR patches over an over-complete dictionary pair $(\mathbf{D}_l, \mathbf{D}_h)$. It first extracts overlapped patches from the LR image, which are then encoded as the higher dimensional sparse vectors with respect to a LR dictionary $\mathbf{D}_l$. Next, the LR sparse coefficients are projected to HR ones via a learned mapping function. Finally, the HR sparse coefficients are passed into

Electronic supplementary material The online version of this chapter (doi:10.1007/978-3-319-54407-6_3) contains supplementary material, which is available to authorized users.

© Springer International Publishing AG 2017
C.-S. Chen et al. (Eds.): ACCV 2016 Workshops, Part I, LNCS 10116, pp. 34–49, 2017.
DOI: 10.1007/978-3-319-54407-6_3

a HR dictionary $\mathbf{D}_h$ for reconstructing HR patches. To learn the coefficients conversion, various assumptions have been imposed on the underlying mapping model from the initial sparse representation invariance [1–3,8], through to the linear mapping [4,5], and finally to the statistical dependence [6,9]. This task is carried out through the ℓ_2-norm coefficients transition term $\|\mathbf{W}_l\alpha_l - \mathbf{W}_h\alpha_h\|_F^2$ in the shared objective function (1) by optimizing the projection matrices ($\mathbf{W}_l$, $\mathbf{W}_h$) under varied assumptions. With the ℓ_2-norm minimizing the mapping error, these models achieve quite convincing performances for the noiseless image SR. Unfortunately, LR images often contain certain noise in the real-world applications. We find all above approaches are sensitive to noise, even the noise is weak. Figure 1 demonstrates the noise sensitivities of one representative of sparse representation approach (Yang et al. [1]), two other recent approaches (Wang et al. [4] and A$^+$ method [8]) and the proposed method.

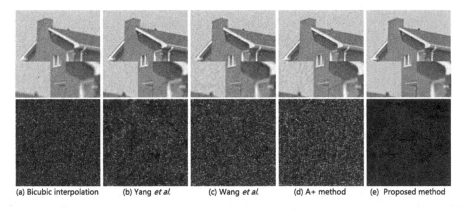

(a) Bicubic interpolation   (b) Yang *et al.*   (c) Wang *et al.*   (d) A+ method   (e) Proposed method

Fig. 1. SR results on a noisy LR image. The input is with Gaussian noise with standard deviation of 5. **Top:** SR outputs. **Middle:** the magnified local patches of SR output. **Bottom:** the noise residual maps visualizing the differences between the SR outputs of the noiseless LR inputs and their noisy ones, where the residual noise (white dots) is magnified by 10 times. (a) LR input interpolated by Bicubic operator; (b) Result by Yang et al. [1]; (c) Result by Wang et al. [4]; (d) Result by A$^+$ approach [9]; (e) Result by proposed method.

In this paper, we explore the mechanism of noise sensitivity of conventional approaches by studying the convex objective function (1) in Sect. 3.1. One can see that the coefficients conversion is carried out by the ℓ_2-norm due to its simplicity as manifested by a closed-form and unique solution. This, however, is by no means a declaration the ℓ_2-norm is the truly best choice for noisy image SR. Our analysis reveals the ℓ_2-norm coefficients transition term causes the sparsity of LR sparse coefficients poorly maintained, and further encodes noise from the LR sparse coefficients into the HR sparse coefficients. A 2D geometric interpretation is given in Fig. 2(a) by solely considering the solution to LR sparse coefficients. Owing to the convexity of ℓ_2-norm, the solution/intersection mostly takes place off the axis with all non-zero coordinates, which is not sparse.

Our *main contribution* is the following upgrade of coefficients transition term. To well maintain the sparsity of LR coefficients, the intersection is desired to take place on the axis. We find the ℓ_1-norm is a better choice, please refer to Fig. 2(b). Nevertheless, simply replacing the ℓ_2-norm by a ℓ_1-norm to the entire transition term brings two problems. Firstly, it leads to an inaccurate calculation of HR coefficients, because ℓ_1-norm restrains the non-zero sum other than the mean squared error in ℓ_2-norm. To address this problem, we introduce two auxiliary variables to split the transition term into two sub-terms: LR coefficient sparsity constraint term with ℓ_1-norm and HR coefficient conversion term with ℓ_2-norm in Sect. 3.2. Secondly, the presence of two ℓ_1-norm terms (the transition term and the sparsity-inducting term) cannot apply the standard optimization techniques to sparse-coding. To alleviate this problem, we propose a smooth proximal gradient method to approximate the transition term, and then are able to employ the efficient solvers, e.g. Fast Iterative Shrinkage Thresholding algorithm [13] in Sect. 3.4 (**Update** α_l, α_h).

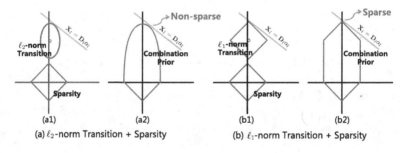

(a1)   (a2)   (b1)   (b2)

(a) ℓ_2-norm Transition + Sparsity   (b) ℓ_1-norm Transition + Sparsity

Fig. 2. The combination of transition term and sparsity-inducing term acting on the state in sparse-coding. (a) ℓ_2-norm regularization: (a1) The ellipse represents the ℓ_2-norm coefficients transition term, the diamond denotes the sparsity-inducing term. (a2) The joint solution takes place off the axis, which is not sparse. (b) ℓ_1-norm regularization: (b1) The upper diamond represents the ℓ_1-norm coefficients transition term. (b2) The joint solution takes place on the axis, which is much sparser. (Color figure online)

Our *second contribution* is a non-linear coefficients conversion learning. After applying ℓ_1-norm on LR coefficient and ℓ_2-norm on HR coefficient, the coefficients conversion learning becomes more complicated. Either the sparse representation invariance [1–3,8] or the linear mapping [4,5] is incompetent to carry out an accurate relation modeling task, which is crucial to the dictionary learning SR approaches. Instead, we propose a non-linear method inspired by the auto-encoder that possesses an encoding process and a decoding process. It is simple yet effectively embedded into our optimization procedure in Sect. 3.2.

Our *third contribution* is a non-local constraint over HR sparse coefficients. Our approach regularizes the HR coefficients conversion using ℓ_2-norm by assuming the HR image has had no noise already during optimization. Unfortunately, it does not always hold in the iterative updates. Thus, we introduce a non-local constraint into the dictionary learning that correlates the similarity among non-local patches to regularize the HR sparse coefficients updates in Sect. 3.3. This constraint is able to distinguish the texture from noise effectively and results in a better noise suppression.

We show improved PSNR and SSIM over the competing sparse-coding approaches for a wide range of noise levels on three benchmark datasets. In particular, given noise level up to 20, ours degrades only 2.53 dB and 0.1140 comparing with the least decreases (5.96 dB and 0.3345) of others. In configuration of denoising + SR, our method also outperforms others on denoised inputs by improving PSNR 0.68dB, SSIM 0.0187 at noise level 20. To show the generalization, our method is compared on the noiseless data against state-of-the-arts as well. The recovered images are superior to all others including SRCNN [14].

2 Related Work

Among the sparse-coding SR approaches, Yang's approaches [1,2] are the most widely used, which assume that the sparse representation is invariant over the LR and HR dictionary pair. However, this simple assumption restricts its ability of the SR recovery. After that, there have been several attempts to go beyond the invariance assumption to improve the flexibility for accurate relationship learning. Wang et al. [4] proposed a semi-coupled dictionary training model by considering a linear mapping between the LR-HR coefficients. Similar to Wang's work, a beta process model [6] was proposed to learn the dictionaries whose sparse representations have the same sparsity but different values. It reported a more consistent and accurate mapping result. Alternatively, Huang et al. [5] presented a joint model which incorporated a common feature space learning into the coupled dictionary scheme to better describe the relationship. Recently, A^+ approach [8] exploits the same invariance assumption as Yang's works and combines the dictionaries with neighbor embedding methods to obtain improvements in both quality and speed. Dai et al. [15] jointly learned a collection of invariant mapping functions to alleviate the inability of a universal regressor for modeling the complex relationship.

Notwithstanding the demonstrated success in the noiseless image SR, none of the conventional sparse-coding approaches is robust to noisy LR images. To our best knowledge, very few works address on this issue. A straightforward thought could do image denoising first, and super-resolve the denoised image afterward. Our test shows that many high frequency details are inevitably lost by denoising process, which is absolutely important to SR process. Singh et al. [16] tried to integrate the merits of image denoising and image SR. They first super-resolve the noisy LR image and the denoised LR image. Then, a convex combination of the denoised HR image and noisy HR image is exploited to obtain the final HR image. This approach achieves a better performance but has limited improvement.

The conventional sparse-coding approaches assume that each image patch could be sparsely represented in the transform domain and is independent from others. This is a typical idea of using the local prior. Recently, another trend, non-local regularization [17], is applied to alleviating the ill-condition of image SR. It breaks the independent constraint and recovers pixels from all similar patches in the image. To bring sparse-coding and non-local regularization together, approaches [4,18] super-resolve the LR patches independently, followed by a non-local processing.

Unlike all previous approaches, we provide a ℓ_1-norm solution to suppress the noise, meanwhile, put a non-local constraint into our dictionary learning framework. As was expected, our method out-competes previous approaches on noisy data by a big margin.

3   Robust Coupled Dictionary Learning

This section describes our solution for noisy image SR. We start by analyzing the cause of noise sensitivity in conventional approaches that use ℓ_2-norm to regularize the relation between the sparse coefficients (α_l, α_h) in LR and HR images. The ℓ_2-norm is found to be the reason of encoding the noise from α_l to α_h. Next, we provide our ℓ_1-norm solution by introducing two auxiliary variables to make the objective function solvable. To model the relation between α_l and α_h more precise under the new regularization, we design a non-linear mapping approach. Afterward, we apply the non-local constraint into our dictionary learning framework to further improve the performance. Finally, we detail the optimization procedure followed by the learning and the inference algorithms.

3.1   The Cause of Noise Sensitivity

It has been known that noiseless images can be approximated as a linear combination of a few elementary atoms, which is named sparse representation/coding. This find inspired and laid the foundation of the coupled dictionary learning methods for SR. Their general objective function is formulated as follows:

$$\underset{\mathbf{D}_l, \mathbf{D}_h, \mathbf{W}_l, \mathbf{W}_h, \alpha_l, \alpha_h}{\operatorname{argmin}} \frac{1}{2}(\|\mathbf{X}_l - \mathbf{D}_l \alpha_l\|_F^2 + \|\mathbf{X}_h - \mathbf{D}_h \alpha_h\|_F^2) \\ + \lambda\|\mathbf{W}_l \alpha_l - \mathbf{W}_h \alpha_h\|_F^2 + \mu(\|\alpha_l\|_1 + \|\alpha_h\|_1), \tag{1}$$

where α_l and α_h are the sparse coefficients over the LR-HR dictionary pair $(\mathbf{D}_l, \mathbf{D}_h)$, respectively, $\|\alpha_l\|_1$ and $\|\alpha_h\|_1$ are the sparsity-inducing terms. Since $\mathbf{X}_l$ and $\mathbf{X}_h$ are in two different spaces, one of the tasks in function (1) is to establish the relation to link sparse coefficients in two spaces. The *coefficients transition term* $\|\mathbf{W}_l \alpha_l - \mathbf{W}_h \alpha_h\|_F^2$ is designed to learn the relation, where $\mathbf{W}_l$ and $\mathbf{W}_h$ are the projection matrices who project α_l and α_h into a common feature space. λ and μ are the regularization parameters to balance the terms. Several algorithms have been proposed to solve the objective function (1), such as the joint dictionary training algorithm [1,3,7], the coordinate descent algorithm [4,5] and the Bayesian algorithm [6].

In practice, most LR images contain certain noise. Due to its randomness, noise cannot be approximated as a sparse combination using the same set of atoms for image coding. Thus, the sparsity-inducting term $\|\alpha_l\|_1$ in the function (1) is expected to suppress the noisy. Extensive experiments using the conventional approaches, however, do not show the desired efficacy.

By analyzing the convex solution of function (1), we consider that the ℓ_2-norm on the coefficients transition term makes the sparsity of α_l poorly maintained and causes the failed noise suppression. The 2D geometric interpretation is shown in Fig. 2. Let's only consider the LR space because the HR images are assumed to be free of noise. The ellipse (ℓ_p ball) in Fig. 2(a1) represents $\|\mathbf{W}_l\alpha_l - \mathbf{W}_h\alpha_h\|_F^2$, the diamond denotes $\|\alpha_l\|_1$, and the green line is $\mathbf{X}_l = \mathbf{D}_l\alpha_l$. When we jointly solve them, the combination of ellipse and diamond becomes the convex shape in Fig. 2(a2). One can see the intersection (the problem solution α_l) of the red curve and the green line takes place off the axis which is not sparse, with all non-zero coordinates. The noise in the LR images, therefore, is transmitted/encoded into α_h through the coefficient transition term.

To suppress the noise, the intersection of the green line and the ℓ_p ball is expected to take place on the axis. This situation happens only if $0 < p \le 1$. While $p < 1$, the solution is no longer convex and difficult to be solved. Thus, ℓ_1-norm becomes a better choice which is convex and the tendency to sparsity we are referring to. As we move the coefficient transition term from ℓ_2-norm towards ℓ_1-norm, Fig. 2(b) shows the intersection takes place on the axis, where most coordinates are zeros. Thus, it leads to a sparser solution. More specifically, we are able to prevent the noise from α_l to be encoded into α_h estimation using a ℓ_1-norm coefficient transition term.

3.2 ℓ_1-norm and Coefficients Conversion Learning

However, simply applying a ℓ_1-norm to the coefficient transition term brings two problems. In this section, we focus on the first problem of an inaccurate α_h calculation, and provide our solution in below. For the second one of optimizing two ℓ_1-norm terms in the objective function (1), we propose a smooth proximal gradient algorithm to solve it in the Sect. 3.4 (**Update** α_l, α_h).

Applying a ℓ_1-norm to the entire coefficient transition term leads to an inaccurate HR coefficients calculation, $\alpha_h = \mathbf{W}_h^{-1}\mathbf{W}_l\alpha_l$, because $\|\mathbf{W}_l\alpha_l - \mathbf{W}_h\alpha_h\|_1$ restrains the non-zero sum rather than the mean square error in ℓ_2-norm. This is one of the reasons that ℓ_2-norm is widespread in various problems. To solve it, we introduce two auxiliary variables $\mathbf{P}_h$ and $\mathbf{P}_l$ to split the transition term into two sub-terms: *LR coefficient sparsity constraint term* with ℓ_1-norm for ensuring the sparsity, and *HR coefficient conversion term* with ℓ_2-norm for the accuracy guarantee. The new definition is given as follows:

$$\|\alpha_l - \mathbf{T}_l^T\mathbf{P}_h\|_1 + \|\alpha_h - \mathbf{T}_h^T\mathbf{P}_l\|_F^2 \quad s.t. \quad \mathbf{P}_h = \mathbf{T}_h\alpha_h, \mathbf{P}_l = \mathbf{T}_l\alpha_l, \quad (2)$$

where $\mathbf{T}_h$ and $\mathbf{T}_l$ are the transition matrices of α_h and α_l respectively. With the new transition term ready, we note that the joint dictionary learning approaches in [1–3,8] are special cases when $\mathbf{W}_h = \mathbf{W}_l = I$, where I is a unit matrix. The approaches in [4,6] are similar but by having $\mathbf{W}_h = I$ only. They assume that the relations should be linear. In this problem, the LR images come with noise but our expected HR images have no noise. Although we condition α_l by the ℓ_1-norm, α_l and α_h locate in different sparse representation spaces which unlikely have a linear mapping relation. In this paper, our idea is to model the relations non-linearly using an encoding process and a decoding process similar to auto-encoder. Specifically, we encode the input (α_l/α_h) non-linearly, and decode the latent representation to the output (α_h/α_l) linearly. Due to this non-linearity and ℓ_1-norm coefficient transition term, our algorithm is able to converge at a sparser α_l. Thus, the function (2) becomes:

$$\|\alpha_l - \mathbf{T}_l^T \mathbf{P}_h\|_1 + \|\alpha_h - \mathbf{T}_h^T \mathbf{P}_l\|_F^2 \qquad s.t. \quad \mathbf{P}_h = soft(\mathbf{T}_h \alpha_h), \mathbf{P}_l = soft(\mathbf{T}_l \alpha_l), \quad (3)$$

where the non-linear mapping is done by a soft-thresholding operator $soft(\cdot)$. In this situation, the auxiliary variables $\mathbf{P}_h$ and $\mathbf{P}_l$ are regarded as the latent representations in the auto-encoder network, which is obtained by projecting the input to the feature space. In fact, our coefficients conversion learning via auto-encoder is a generalization of the conventional linear mapping model.

3.3 Non-local Constraint on HR Sparse Coefficient

When introducing the function (3), we assume the HR image $\mathbf{X}_h$ has been free of noise and then regularize α_h by ℓ_2-norm. Nevertheless, the iterative optimization does not guarantee this, particularly, at the first few iterations. The encoded noise in α_l still affects α_h. To make our solution converge at a better result, the sparsity of α_h is also desired to be regularized in both the dictionary learning and the image SR inference.

It has been found that the local image structures tend to repeat within a large region. These similar patches at different locations in the image are regarded as multiple observations of the target patch. Such non-local similarity provides additional information for estimating the target patch and inspired the non-local means (NLM) methods [17,19] that had been applied to image denoising. Thus, a non-local constraint, which correlates the similarities among patches, can be used to regularize the sparse decomposition and then produce a superior result.

Since α_l has been constrained by ℓ_1-norm to ensure sparsity, we therefore incorporate the non-local constraint (the correlation of patch similarities) only into regularizing the HR sparse coefficients update during the optimization, which is formulated as:

$$\underset{\alpha_h}{\operatorname{argmin}} \sum_{i=1}^{n} \|\alpha_h^i - \sum_{j=1}^{n} \mathbf{N}_{ij} \alpha_h^j\|_F^2 = tr(\alpha_h \mathbf{L} \alpha_h^T), \quad (4)$$

where $\mathbf{N}$ is the weight matrix, $\mathbf{N}_{ij}$ denotes the similarity between patches $\mathbf{x}_i$ and $\mathbf{x}_j$, and $\mathbf{L} = (\mathbf{I}_{n \times n} - \mathbf{N})^T (\mathbf{I}_{n \times n} - \mathbf{N})$. We construct $\mathbf{N}$ through connecting every patch to its k most similar image patches and compute the weights of connected patches by Gaussian kernel function [17]. In addition, we do this on the HR output only because the HR image has more details, the weight matrix constructed from it is more accurate. Finally, the non-local constraint $tr(\alpha_h \mathbf{L} \alpha_h^T)$ is employed to regularize our dictionary learning and the image SR inference to further enhance the SR performance.

3.4 Optimization and Inference

By incorporating all the above new constraints on our coupled dictionary learning, the proposed method needs to solve the following optimization problem:

$$\underset{\mathbf{D}_l, \mathbf{D}_h, \mathbf{T}_l, \mathbf{T}_h}{\operatorname{argmin}} \frac{1}{2}(\|\mathbf{X}_l - \mathbf{D}_l \alpha_l\|_F^2 + \|\mathbf{X}_h - \mathbf{D}_h \alpha_h\|_F^2) + \mu(\|\alpha_l\|_1 + \|\alpha_h\|_1)$$
$$+ \lambda(\|\alpha_l - \mathbf{T}_l^T \mathbf{P}_h\|_1 + \|\alpha_h - \mathbf{T}_h^T \mathbf{P}_l\|_F^2) + \gamma tr(\alpha_h \mathbf{L} \alpha_h^T) \qquad (5)$$
$$s.t. \ \ \mathbf{P}_h = soft(\mathbf{T}_h \alpha_h), \mathbf{P}_l = soft(\mathbf{T}_l \alpha_l).$$

While the objective function (5) is not jointly convex to $\mathbf{D}_l$, $\mathbf{D}_h$, $\mathbf{T}_l$, $\mathbf{T}_h$, α_l and α_h, it is convex with respect to each of them if the remaining variables are fixed. Given the training data $\mathbf{X}_l$, $\mathbf{X}_h$ and $\mathbf{L}$, we apply a coordinate descent algorithm (as shown in Algorithm 1[1] to optimizing the dictionaries $\{\mathbf{D}_l, \mathbf{D}_h\}$, transition matrices $\{\mathbf{T}_l, \mathbf{T}_h\}$ and coefficients $\{\alpha_l, \alpha_h\}$, respectively. We now discuss how to update these variables in each iteration.

Update $\mathbf{D}_l, \mathbf{D}_h$: We first apply the algorithm of joint dictionary learning [1] to the initialization of $\mathbf{D}_l$ and $\mathbf{D}_h$ for the optimization process. When updating the two dictionaries during each iteration, we consider the sparse coefficients $\{\alpha_l, \alpha_h\}$ and transition matrices $\{\mathbf{T}_l, \mathbf{T}_h\}$ as constants. As a result, the problem of (5) can be simplified into the following forms:

$$\underset{\mathbf{D}_l^{k+1}}{\operatorname{argmin}} \|\mathbf{X}_l - \mathbf{D}_l^{k+1} \alpha_l^k\|_F^2 + \eta \|\mathbf{D}_l^{k+1} - \mathbf{D}_l^k\|_F^2,$$
$$\underset{\mathbf{D}_h^{k+1}}{\operatorname{argmin}} \|\mathbf{X}_h - \mathbf{D}_h^{k+1} \alpha_h^k\|_F^2 + \eta \|\mathbf{D}_h^{k+1} - \mathbf{D}_h^k\|_F^2. \qquad (6)$$

After updated by an iterative scheme (e.g. the Conjugate Gradient method), the dictionaries $\mathbf{D}_l$ and $\mathbf{D}_h$ are column normalized to avoid any trivial solutions.

Update α_l, α_h: Similar to dictionary updates, the transition matrices $\{\mathbf{T}_l, \mathbf{T}_h\}$ and dictionaries $\{\mathbf{D}_l, \mathbf{D}_h\}$ are fixed when we calculate the solutions of sparse coefficients $\{\alpha_l, \alpha_h\}$. Thus, the objective functions are written as follows:

$$\underset{\alpha_l}{\operatorname{argmin}} \frac{1}{2}\|\mathbf{X}_l - \mathbf{D}_l \alpha_l\|_F^2 + \lambda\|\alpha_l - \mathbf{T}_l^T \mathbf{P}_h\|_1 + \mu\|\alpha_l\|_1,$$
$$\underset{\alpha_h}{\operatorname{argmin}} \frac{1}{2}\|\mathbf{X}_h - \mathbf{D}_h \alpha_h\|_F^2 + \lambda\|\alpha_h - \mathbf{T}_h^T \mathbf{P}_l\|_F^2 + \mu\|\alpha_h\|_1 + \gamma tr(\alpha_h \mathbf{L} \alpha_h^T). \qquad (7)$$

---

[1] The matrix $\mathbf{L}$ (non-local constraint) is applied to regularize α_h as the function (4).

Besides the standard sparse-coding formulation, here is an additional coefficient sparsity constraint term $\|\alpha_l - \mathbf{T}_l^T \mathbf{P}_h\|_1$ with respect to α_l. So the commonly used minimization algorithms are no longer applicable. We propose a smooth proximal gradient method motivated by [20,21]. Specifically speaking, the non-smooth ℓ_1-norm coefficient sparsity constraint term is transformed into its dual domain, where it can be approximated by a smooth ℓ_∞ term using Nestrov's algorithm. Since the output is convex and differentiable with respect to α_l with a sparsity constraint, we are able to solve it efficiently using a proximal method, e.g. Fast Iterative Shrinkage Thresholding Algorithm (FISTA) [13].

Thus, the approximate gradient $\|\alpha_l - \mathbf{T}_l^T \mathbf{P}_h\|_1$ over α_l is:

$$\nabla_{\alpha_l} \|\alpha_l - \mathbf{T}_l^T \mathbf{P}_h\|_1 \approx S(\frac{\alpha_l - \mathbf{T}_l^T \mathbf{P}_h}{\rho}), \tag{8}$$

where ρ is a smoothness parameter in Nestrov's method, $S(\mathbf{x})$ is a function projecting $\mathbf{x}$ onto an ℓ_∞-ball,

$$S(\mathbf{x}) = \begin{cases} \mathbf{x}, & -1 \leq \mathbf{x} \leq 1 \\ 1, & \mathbf{x} > 1 \\ -1, & \mathbf{x} < -1 \end{cases}. \tag{9}$$

Finally, the gradient with respect to α_l is written as:

$$\nabla_{\alpha_l} h_1(\alpha_l) \approx \mathbf{D}_l^T (\mathbf{D}_l \alpha_l - \mathbf{X}_l) + \lambda S(\frac{\alpha_l - \mathbf{T}_l^T \mathbf{P}_h}{\rho}), \tag{10}$$

where $h_1(\alpha_l) = \frac{1}{2}\|\mathbf{X}_l - \mathbf{D}_l \alpha_l\|_F^2 + \lambda \|\alpha_l - \mathbf{T}_l^T \mathbf{P}_h\|_1$.

For the optimization of α_h, we also use the proximal methods like FISTA. Its gradient is given as follows:

$$\nabla_{\alpha_h} h_2(\alpha_h) \approx \mathbf{D}_h^T (\mathbf{D}_h \alpha_h - \mathbf{X}_h) + 2\lambda(\alpha_h - \mathbf{T}_h^T \mathbf{P}_l) + \gamma \alpha_h (\mathbf{L} + \mathbf{L}^T), \tag{11}$$

where $h_2(\alpha_h) = \frac{1}{2}\|\mathbf{X}_h - \mathbf{D}_h \alpha_h\|_F^2 + \lambda \|\alpha_h - \mathbf{T}_h^T \mathbf{P}_l\|_F^2 + \gamma tr(\alpha_h \mathbf{L} \alpha_h^T)$.

Update $\mathbf{T}_l, \mathbf{T}_h$: When updating the transition matrices, only the terms associating with $\{\mathbf{T}_l, \mathbf{T}_h\}$ are considered in the optimization. With fixed $\{\mathbf{D}_l, \mathbf{D}_h\}$ and $\{\alpha_l, \alpha_h\}$, we solve the following problem for update,

$$\begin{aligned} \underset{\mathbf{T}_l^{k+1}}{\arg\min} \|\alpha_l - \mathbf{T}_l^{k+1}{}^T \mathbf{P}_h\|_1 + \eta\|\mathbf{T}_l^{k+1} - \mathbf{T}_l^k\|_F^2, \\ \underset{\mathbf{T}_h^{k+1}}{\arg\min} \|\alpha_h - \mathbf{T}_h^{k+1}{}^T \mathbf{P}_l\|_F^2 + \eta\|\mathbf{T}_h^{k+1} - \mathbf{T}_h^k\|_F^2. \end{aligned} \tag{12}$$

This can be done by the Conjugate Gradient method as well.

Inference: Once the optimization is complete, the derived model is employed for image SR reconstruction. We first partition the LR image into overlapped patches, and initialize the sparse coefficients α_l of $\mathbf{X}_l$ via solving:

$$\underset{\alpha_l^0}{\arg\min} \frac{1}{2}\|\mathbf{X}_l - \mathbf{D}_l \alpha_l^0\|_F^2 + \mu\|\alpha_l^0\|_1. \tag{13}$$

Algorithm 1. Dictionaries and transition matrices learning

Input: Training data matrices $\mathbf{X}_l \in \mathbb{R}^{p_1 \times q}$ and $\mathbf{X}_h \in \mathbb{R}^{p_2 \times q}$, non-local matrix $\mathbf{L} \in \mathbb{R}^{q \times q}$.

1. Initialize $\mathbf{D}_l^0$, $\mathbf{D}_h^0$, α_l^0 and α_h^0 by [1], transition matrices $\mathbf{T}_l^0$ and $\mathbf{T}_h^0$ as $\mathbf{I}$.
2. Let $\mathbf{P}_l^0 \leftarrow soft(\mathbf{T}_l^0 \alpha_l^0)$ and $\mathbf{P}_h^0 \leftarrow soft(\mathbf{T}_h^0 \alpha_h^0)$, $k = 0$.

while not converged **do**

    3. Update $\mathbf{D}_l^{k+1}$ and $\mathbf{D}_h^{k+1}$ by (6) with α_l^k and α_h^k derived from the previous iteration.

    4. Update α_l^{k+1} and α_h^{k+1} by (7) with $\mathbf{D}_l^{k+1}$, $\mathbf{D}_h^{k+1}$, $\mathbf{T}_l^k$ and $\mathbf{T}_h^k$.

    5. Update $\mathbf{T}_l^{k+1}$ and $\mathbf{T}_h^{k+1}$ by (12) with $\mathbf{D}_l^{k+1}$, $\mathbf{D}_h^{k+1}$, α_l^{k+1} and α_h^{k+1}.

    6. $\mathbf{P}_l^{k+1} \leftarrow soft(\mathbf{T}_l^{k+1} \alpha_l^{k+1})$ and $\mathbf{P}_h^{k+1} \leftarrow soft(\mathbf{T}_h^{k+1} \alpha_h^{k+1})$.

    7. $k \leftarrow k + 1$.

end while

Output: $\mathbf{D}_l$, $\mathbf{D}_h$, $\mathbf{T}_l$ and $\mathbf{T}_h$.

Algorithm 2. Image super-resolution inference

Input: Dictionaries $\mathbf{D}_l$ and $\mathbf{D}_h$, transition matrices $\mathbf{T}_l$ and $\mathbf{T}_h$, a low-resolution image $\mathbf{Y}$.

1. Extract LR patches $\mathbf{X}_l \in \mathbb{R}^{p_1 \times M}$ from LR image $\mathbf{Y}$ with 1 pixel overlap in each direction.
2. Initiate α_l^0 by (13) and α_h^0 by (14).
3. Let $\mathbf{P}_l^0 \leftarrow soft(\mathbf{T}_l^0 \alpha_l^0)$ and $\mathbf{P}_h^0 \leftarrow soft(\mathbf{T}_h^0 \alpha_h^0)$.
4. Initialize: $\mathbf{X}_h^0 \leftarrow \mathbf{D}_h \alpha_h^0$, $\mathbf{L}^0$ computed from $\mathbf{X}_h^0$, $k = 0$.

while not converged **do**

    5. Update α_l^{k+1} and α_h^{k+1} by (7) with $\mathbf{X}_h^k$, $\mathbf{L}^k$, $\mathbf{P}_l^k$ and $\mathbf{P}_h^k$.

    6. Update $\mathbf{P}_l^{k+1} \leftarrow soft(\mathbf{T}_l^k \alpha_l^{k+1})$, $\mathbf{P}_h^{k+1} \leftarrow soft(\mathbf{T}_h^k \alpha_h^{k+1})$, $\mathbf{X}_h^{k+1} \leftarrow \mathbf{D}_h \alpha_h^{k+1}$ and $\mathbf{L}^{k+1}$ computed from $\mathbf{X}_h^{k+1}$.

    7. $k \leftarrow k + 1$.

end while

8. Put the patches $\mathbf{X}_h$ into a high-resolution image $\mathbf{X}$.

Output: HR image $\mathbf{X}_h$.

Once α_l is calculated, we initialize α_h in the derived feature space:

$$\alpha_h^0 = \mathbf{T}_h^T soft(\mathbf{T}_l \alpha_l^0). \tag{14}$$

After iteratively updating α_h through the objective function (5), we have $\mathbf{X}_h = \mathbf{D}_h \alpha_h$ as the final SR output. The procedure is given in Algorithm 2 (See footnote 1).

4   Experiments

To validate the robustness of our method to noise and the effectiveness on detail recovery, we first conduct the assessment on two settings: 1. a direct comparison with state-of-the-arts on a variety of synthesized noisy images, 2. an indirect comparison with others using denoised images, where denoising algorithm is

BM3D [22]. These competing methods are Yang's approach based on the sparse representation invariance [1], Wang's approach based on a semi-coupled dictionary [4], He's approach [6] based on the Beta process, Dai's approach (JOR) [15] based on jointly optimized regressors and A^+ approach [8] based on an anchored neighborhood regression. To show the generalization of our method, we also do a comparison on the noiseless images with the SRCNN [14,23] that uses the deep convolutional networks to achieve a state-of-the-art performance. To have a better insight into our proposed method, we deliberately create two versions: **proposed1**, a partial version of only optimization on the ℓ_1-norm coefficients transition term to suppress noise; **proposed2**, a full version including both the ℓ_1-norm and the non-local constraint for improving SR quality. All these approaches share the same training datasets with an exception to SRCNN who uses the ImageNet which is much greater than ours. The quantitatively validations are calculated in terms of Peak Signal-to-Noise Ratio (PSNR) and Structural Similarity Index (SSIM) from their HR outputs.

4.1 Experimental Datasets and Parameters Setting

To have noisy LR images, we blur and down-sample HR images using Matlab function "imresize()" with a scaling factor 3, and then add Gaussian white noise with varied standard deviations. For the training dataset, 500,000 LR-HR image patch pairs are randomly extracted from BSD200 that is the complement of BSD100 in BSD300 [24], where the LR patches are interpolated up to the same size as the HR training data. The HR patch size is fixed as 9×9 with one pixel overlap between adjacent patches. For the test datasets, we employ three commonly used datasets (Set5, Set14, BSD100) from [7]. For the parameters in objective function (5), $\{\lambda, \mu, \gamma\}$ is set to $\{0.5, 0.1, 0.1\}$. The dictionary size is defined as 1024 by a line search strategy. Moreover, our approach finds $k = 10$ most similar patches to calculate the weight matrix $\mathbf{N}$ for the non-local constraint.

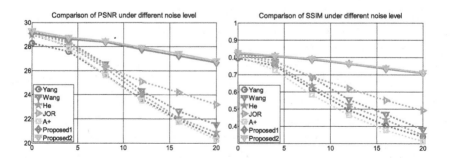

Fig. 3. PSNR and SSIM of seven methods when noise level varies from 0 up to 20.

4.2 Experiment on Noisy LR Inputs

We compare the proposed method against five state-of-the-art approaches [1,4,6,8,15] on noisy and denoised images from Set5, Set14 and BSD100, and plot PSNR, SSIM results in Figs. 3 and 4 respectively when varying the standard deviation of Gaussian noise from 0 up to 20 with step size 4. From Fig. 3 one can see other five approaches degrade much quicker when noise increases. In contrast, propose1 and propose2 perform well across all noise level, especially at higher ones. Although, Fig. 4 illustrates others have a better performance on denoised data, our proposed method still archives an obvious improvement than them without additional denoising. Figure 5 shows the visual results of a

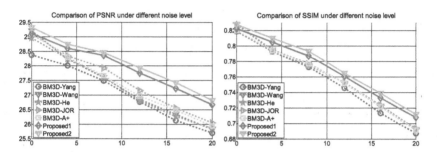

Fig. 4. PSNR and SSIM of seven methods, where the inputs of other five have been denoised, and noise level varies from 0 up to 20.

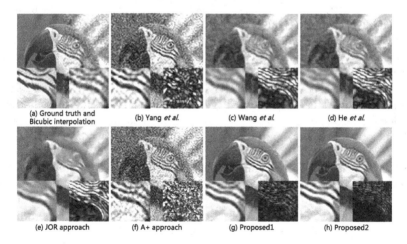

Fig. 5. Results of the LR image *parrots* with a noise level of 20. **Top:** Visual SR results. **Bottom:** *left.* Magnified local SR patches. *right.* Noise residual maps visualizing the difference between SR patches of the noiseless LR inputs and their noisy ones, where the residual signal (white dots) is magnified by 10 times. The noise signal is magnified by 10 times. (a) Ground-truth (left) and its interpolated LR version (right); (b) Yang's approach [1]; (c) Wang's approach [4]; (d) He's approach [6]; (e) JOR approach [15]; (f) A^+ approach [8]; (g) Proposed1; (h) Proposed2.

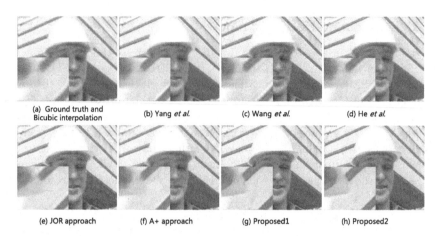

Fig. 6. Results of the LR image *foreman* with a noise level of 12. Please refer to Fig. 5 for the description of subfigures.

noisy input with noise level 20 and Fig. 6 of a denoised input when original noise level is 12. In Fig. 5, one can see that the five sparse-coding approaches generate plausible details and sharper edges along *parrots* beak, but are poor at suppressing the noise, particularly on the smooth regions. JOR generates relatively pleasing result by incorporating multiple mapping functions, and outperforms the similar approach [8]. Our proposed1 (Fig. 5g) produces visually comparable result that has little remaining noise. Moreover, the proposed2 with the non-local constraint (Fig. 5h) further improves SR performance by finely synthesizing the high-frequency details and eliminating artifacts. Figure 5 shows all other competing approaches produce unacceptable results, where the noise is magnified rather than suppressed. The reason is that the noise in LR input is first upscaled by interpolation, then encoded into HR result through the ℓ_2-norm coefficients transition term. Wang's approach [4] generates a relatively finer result due to a non-local post-processing. However, it does the sparse-coding and non-local constraint regularization separately, which restricts the noisy SR performance. Dai's approach (JOR) [15] produces an over-smoothed HR result because its class-label selection becomes ambiguous when the noise is severe. In contrast our methods (Fig. 5g and h) successfully suppress the noise and produce visually pleasing SR results. Moreover, We do a comparison on denoised data. The synthesized noisy images are first denoised by BM3D algorithm, and then inputted into five competing methods. Thanks to denoising, they all produce rather clear result with little noise in Fig. 6 but tend to smooth high frequency signals as shown along the brim of hat. By contrast, our method produces much sharper edges and more faithful details because of the inherent ℓ_1-norm who preserves sparsity of input signal. More visual comparison can be found in supplementary document, Figs. D1–D4.

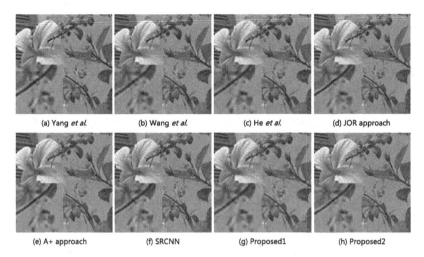

(a) Yang *et al.*   (b) Wang *et al.*   (c) He *et al.*   (d) JOR approach

(e) A+ approach   (f) SRCNN   (g) Proposed1   (h) Proposed2

Fig. 7. The SR results of the noiseless LR image *flower*. The lower left is a magnified local patch. (a) Yang's approach [1]; (b) Wang's approach [4]; (c) He's approach [6]; (d) JOR approach [15]; (e) A$^+$ approach [8]; (f) SRCNN [14]; (g) Proposed1; (h) Proposed2.

4.3   Experiment on Noiseless LR Inputs

To verify the effectiveness of our proposed method to noiseless images, we also compare it against the aforementioned five approaches plus the SRCNN [14], and summarize the average PSNR and SSIM scores on three test datasets in Table 1. One can see our methods are still able to outperform all other sparse-coding approaches across all datasets, where the proposed1 is slightly inferior to the SRCNN. However, by applying the non-local similarity constraint, our proposed2 outperforms SRCNN over 0.2 dB in PSNR and 0.0020 in SSIM. Figure 7 shows a visual comparison. One can see that Yang's approach [1] produces blurred textural details and zigzag edges because of their simple assumption of the invariant sparse representation. He's approach [6] is very competitive in terms of visual quality compared to Yang's by relaxing the representation relationship assumption to a beta process prior for the coupled dictionary learning. Wang's approach [4] is a large improvement by introducing a more complicated assumption that there is a linear mapping between α_l and α_h. Although it produces a better HR image with many fine details, some unpleasant artifacts still can be found along major edges. JOR approach [15] improves the SR performance significantly compared to Yang's approach [1] by involving multiple regressors. Using a similar framework (sparse representation invariance assumption) of Yang's, A$^+$ approach [8] combines the learned sparse dictionaries and neighbor embedding to improve the SR performance. However, noticeable zigzags are created along dominant edges, and some fine image structures are missing. The SRCNN generates much sharper edges and suppresses noticeable artifacts in the SR result. Our proposed1 produces a slightly worse reconstruction quality with a few artifacts along the edges shown in Fig. 7g. But, our proposed2 recovers a HR image with

Table 1. Average PSNR and SSIM on test data set Set5, Set14, BSD100.

Benchmark		Yang [1]	Wang [4]	He [6]	JOR [15]	A$^+$ [8]	SRCNN [14]	Proposed1 (ours)	Proposed2 (ours)
Set5	PSNR	31.42	32.46	32.38	32.55	32.59	32.75	32.65	**32.94**
	SSIM	0.8682	0.9045	0.9037	0.9064	0.9088	0.9090	0.9083	**0.9105**
Set14	PSNR	28.31	29.26	28.97	29.09	29.13	29.30	29.15	**29.44**
	SSIM	0.7954	0.8204	0.8175	0.8194	0.8188	0.8215	0.8178	**0.8223**
BSD100	PSNR	27.64	28.32	28.15	28.17	28.13	28.36	28.24	**28.49**
	SSIM	0.7774	0.7961	0.7953	0.7951	0.7948	0.7966	0.7954	**0.7975**

richer details shown in Fig. 7h, superior to all other SR approaches including SRCNN. It suggests that the non-local similarity constraint effectively improves the stability of sparse decomposition, which benefits the SR performance.

5   Conclusions

We propose a robust dictionary learning for noisy image SR, in which our ℓ_1-norm solution on coefficients transition term prevents the noise to be transmitted from noisy LR input to HR output. By incorporating the non-local constraint on HR sparse coefficient into our dictionary learning framework, the improved sparse representation further enhances SR inference. Results show that our proposed method outperforms the-state-of-arts on both noisy, denoised and noiseless data.

Although ℓ_1-norm effectively suppresses the noise, it requires sufficient iterations during SR inference. In our next work, we will try to use the dictionaries and transition matrices learned within much fewer iterations as the initialization of a deep neural network framework. It may reduce the training cost of the network and speed up the SR inference simultaneously.

Acknowledgments. This work is supported by the National Basic Research Program (973 Program) of China (No. 2013CB329402), the Fund for Foreign Scholars in University Research and Teaching Programs (the 111 Project) (No. B07048), the Program for Cheung Kong Scholars and Innovative Research Team in University (No. IRT 15R53), and JSPS Grants-in-Aid for Scientific Research C (No. 15K00236) for funding.

References

1. Yang, J., Wright, J., Huang, T.S., Ma, Y.: Image super-resolution via sparse representation. IEEE. Trans. Image Process. **19**, 2861–2873 (2010)
2. Yang, J., Wang, Z., Lin, Z., Cohen, S., Huang, T.: Coupled dictionary training for image super-resolution. IEEE. Trans. Image Process. **21**, 3467–3478 (2012)
3. Zeyde, R., Elad, M., Protter, M.: On single image scale-up using sparse-representations. In: Boissonnat, J.-D., Chenin, P., Cohen, A., Gout, C., Lyche, T., Mazure, M.-L., Schumaker, L. (eds.) Curves and Surfaces 2010. LNCS, vol. 6920, pp. 711–730. Springer, Heidelberg (2012). doi:10.1007/978-3-642-27413-8_47
4. Wang, S., Zhang, L., Liang, Y., Pan, Q.: Semi-coupled dictionary learning with applications to image super-resolution and photo-sketch synthesis. In: CVPR, pp. 2216–2223. IEEE (2012)

5. Huang, D.A., Wang, Y.C.F.: Coupled dictionary and feature space learning with applications to cross-domain image synthesis and recognition. In: ICCV, pp. 2496–2503. IEEE (2013)
6. He, L., Qi, H., Zaretzki, R.: Beta process joint dictionary learning for coupled feature spaces with application to single image super-resolution. In: CVPR, pp. 345–352. IEEE (2013)
7. Timofte, R., De, V., Van Gool, L.: Anchored neighborhood regression for fast example-based super-resolution. In: ICCV, pp. 1920–1927. IEEE (2013)
8. Timofte, R., De Smet, V., Van Gool, L.: A+: adjusted anchored neighborhood regression for fast super-resolution. In: Cremers, D., Reid, I., Saito, H., Yang, M.-H. (eds.) ACCV 2014. LNCS, vol. 9006, pp. 111–126. Springer, Heidelberg (2015). doi:10.1007/978-3-319-16817-3_8
9. Peleg, T., Elad, M.: A statistical prediction model based on sparse representations for single image super-resolution. IEEE. Trans. Image Process. **23**, 2569–2582 (2014)
10. Schulter, S., Leistner, C., Bischof, H.: Fast and accurate image upscaling with super-resolution forests. In: CVPR, pp. 3791–3799. IEEE (2015)
11. Timofte, R., Rothe, R., Van Gool, L.: Seven ways to improve example-based single image super resolution. arXiv preprint arxiv:1511.02228 (2015)
12. Kim, J., Lee, J.K., Lee, K.M.: Accurate image super-resolution using very deep convolutional networks (2015). arXiv preprint arxiv:1511.04587
13. Beck, A., Teboulle, M.: A fast iterative shrinkage-thresholding algorithm for linear inverse problems. SIAM J. Imaging Sci. **2**, 183–202 (2009)
14. Dong, C., Loy, C.C., He, K., Tang, X.: Image super-resolution using deep convolutional networks. IEEE. Trans. Pattern Anal. Mach. Intell. **38**, 295–307 (2015)
15. Dai, D., Timofte, R., Van Gool, L.: Jointly optimized regressors for image super-resolution. In: Computer Graphics Forum, Wiley Online Library, pp. 95–104 (2015)
16. Singh, A., Porikli, F., Ahuja, N.: Super-resolving noisy images. In: CVPR, pp. 2846–2853. IEEE (2014)
17. Buades, A., Coll, B., Morel, J.M.: A non-local algorithm for image denoising. In: CVPR, pp. 60–65. IEEE (2005)
18. Zhang, K., Tao, D., Gao, X., Li, X., Xiong, Z.: Learning multiple linear mappings for efficient single image super-resolution. IEEE. Trans. Image Process. **24**, 846–861 (2015)
19. Protter, M., Elad, M., Takeda, H., Milanfar, P.: Generalizing the nonlocal-means to super-resolution reconstruction. IEEE. Trans. Image Process. **18**, 36–51 (2009)
20. Chen, X., Lin, Q., Kim, S., Carbonell, J.G., Xing, E.P., et al.: Smoothing proximal gradient method for general structured sparse regression. Ann. Appl. Stat. **6**, 719–752 (2012)
21. Chalasani, R., Principe, J.C.: Deep predictive coding networks. arXiv preprint arxiv:1301.3541 (2013)
22. Dabov, K., Foi, A., Katkovnik, V., Egiazarian, K.: Image denoising by sparse 3-d transform-domain collaborative filtering. IEEE Trans. Image Process. **16**, 2080–2095 (2007)
23. Dong, C., Loy, C.C., He, K., Tang, X.: Learning a deep convolutional network for image super-resolution. In: Fleet, D., Pajdla, T., Schiele, B., Tuytelaars, T. (eds.) ECCV 2014. LNCS, vol. 8692, pp. 184–199. Springer, Heidelberg (2014). doi:10.1007/978-3-319-10593-2_13
24. Martin, D., Fowlkes, C., Tal, D., Malik, J.: A database of human segmented natural images and its application to evaluating segmentation algorithms and measuring ecological statistics. In: ICCV, pp. 416–423. IEEE (2001)

CNN-GRNN for Image Sharpness Assessment

Shaode Yu[1,2], Fan Jiang[1], Leida Li[3(✉)], and Yaoqin Xie[1(✉)]

[1] Shenzhen Institutes of Advanced Technology,
Chinese Academy of Sciences, Shenzhen, China
yq.xie@siat.ac.cn
[2] Shenzhen College of Advanced Technology,
University of Chinese Academy of Sciences, Shenzhen, China
[3] School of Information and Electronic Engineering,
Chinese University of Mining and Technology, Xuzhou, China
lileida@cumt.edu.cn

Abstract. Image sharpness is key to readability and scene understanding. Because of the inaccessible reference information, blind image sharpness assessment (BISA) is useful and challenging. In this paper, a shallow convolutional neural network (CNN) is proposed for intrinsic representation of image sharpness and general regression neural network (GRNN) is utilized for precise score prediction. The hybrid CNN-GRNN model tends to build functional relationship between retrieved features and subjective human scores by supervised learning. Superior to traditional algorithms based on handcrafted features and machine learning, CNN-GRNN fuses feature extraction and score prediction into an optimization procedure. Experiments on Gaussian blurring images in LIVE, CSIQ, TID2008 and TID2013 show that CNN-GRNN outperforms the state-of-the-art algorithms and gets closer to human subjective judgment.

1 Introduction

With the massive dissemination of smart phones, digital images become a major source of information acquisition, knowledge sharing and entertainment. Since an image is prone to various kinds of distortions, massive attentions have been paid to the assessment of perceptual image quality [3,17,19]. Subjective image quality assessment (IQA) is the most straightforward approach. However, it is laborious and biased. Relatively, objective evaluation of visual image quality enables impartial judgment and has made remarkable progress in the past decade [19,31,37]. Objective IQA involves full-reference (FR), reduced-reference (RR) and no-reference based methods. To FR metrics, the test image is compared to its reference image for a relative score and both images refer to the same content; to RR metrics, what can be used is a set of features from the reference; but to the no-reference metrics, no prior knowledge is available at all. Although FR and RR metrics have achieved high-level performance [20–22,29,38–40], in most possible situations, the reference image or information is not easy to access. Therefore, no-reference or blind IQA is more useful in real-world applications.

© Springer International Publishing AG 2017
C.-S. Chen et al. (Eds.): ACCV 2016 Workshops, Part I, LNCS 10116, pp. 50–61, 2017.
DOI: 10.1007/978-3-319-54407-6_4

Among various kinds of distortions, sharpness is a key factor regarding image readability and scene interpretation. It is common to find a blurred image because of camera out-of-focus, relative target motion and lossy image compression. Sharpness is inversely related to blur which is typically characterized by the spread of edges and accordingly the attenuation of high frequency components. Karam et al. introduced Just Noticeable Blur (JNB) and pooled local contrast and edge width in each blocks into a probability summation model [5] and later advanced it with the Cumulative Probability of Blur Detection (CPBD) [23]. Vu et al. [36] combined two features, high-frequency content with the slope of local magnitude spectrum and local contrast with total variation, and proposed the Spectral and Spatial Sharpness (S3) index. Vu et al. [35] developed a Fast Image SHarpness (FISH) metric that weighted the log-energies of coefficients in the discrete wavelet domain. Hassen et al. [6] used the strength of Local Phase Coherence (LPC) to assess image blur because blur disrupts the LPC structure. Sang et al. [30] took advantage of the shape of Singular Value Curve (SVC) to measure the extent of blur, since blur can cause attenuation of singular value in images. Bahrami and Kot [1] utilized Maximum Local Variation (MLV) of each pixel with its eight-connected neighbors and computed the standard deviation of the ranking-weighted MLVs as the sharpness score. Li et al. [14] employed discrete moments to characterize image blur and implemented it with discrete moments for Blind Image BLur Estimation (BIBLE), and they also designed a blind SPArse Representation based Image SHarpness assessment (SPARISH) utilizing dictionary learning of edge patterns [16].

Convolutional neural networks (CNNs) have revolutionized image representation [2,12] and achieved marked success in Imagenet large scale visual recognition challenge [10,28,33]. To quantify image quality, CNNs have also been studied. Kang et al. [9] designed a shallow CNN that integrated feature learning and score regression into an optimization process for image quality estimation. Hou and Gao [7] recast BIQA as a classification task and developed a saliency-guided deep framework for feature learning. Li et al. [15] used Shearlet transform for spatial features and utilized deep networks for score prediction. Hou et al. [8] proposed a deep learning model pre-trained by deep belief net and discriminatively fine-tuned for blind image quality estimation. Lv et al. [18] explored the local normalized multi-scale difference of Gaussian response as features and used deep neural network as the pooling strategy. To our knowledge, no deep learning architectures have been delicately designed for image sharpness estimation.

A hybrid model, CNN-GRNN, is proposed in this paper. It utilized a single-feature-layer CNN for intrinsic representation of image sharpness and used the general regression neural network (GRNN) [32] for precise score prediction. This model tends to build functional relationship between retrieved features and subjective human scores by supervised learning. It is known that what CNNs have learned in the first feature layer is mainly about edge structures, and interestingly, image sharpness is found to be characterized by the spread of edges. Thus, utilizing a single-feature-layer CNN for image sharpness representation is feasible. Moreover, small sample sets make deep learning networks hard to converge and also increase the risk of over-fitting. Consequently, a shallow CNN is reasonable for image sharpness assessment. It should be noted that [9] is also a shallow CNN. While quite different from [9], our CNN is further simplified. Moreover, we use GRNN to replace the multi-layer perceptron (MLP) in the original CNN architecture for score regression. On one side, GRNN works well on small training samples [32]. On the other side, it has demonstrated superior performance on image quality estimation with limited features [13]. In addition, a novel training strategy is utilized. This strategy can lessen time consumption and work in a random search alike way.

2 CNN-GRNN

The procedure of CNN-GRNN for image sharpness estimation can be described as following. Given a gray scale image, local contrast normalization is performed and a number of no-overlapping patches are sampled. Based on these patches, a simplified CNN is trained by supervised learning to build functional relationship between the image patch and its subjective score. After several epochs of training, extracted CNN features are sent to GRNN for score regression as well as the performance validation. If the validated accuracy achieves the pre-defined value, CNN-GRNN is deemed to be ready for test.

2.1 Local Contrast Normalization

Local contrast normalization is used to unearth image structures as in [9]. Here, assuming $I(i,j)$ is the pixel intensity at (i,j), its normalized value $\widetilde{I}(i,j)$ is computed in Eq. (1) where C is a positive constant ($C = 10$) to avoid dividing by zero, while $(2P + 1)$ and $(2Q + 1)$ are the window size.

$$\begin{cases} \widetilde{I}(i,j) = \dfrac{I(i,j) - \mu(i,j)}{\sigma(i,j) + C}, \\[2mm] \mu(i,j) = \dfrac{1}{(2P+1) \times (2Q+1)} \displaystyle\sum_{p=-P}^{p=P} \sum_{q=-Q}^{q=Q} I(i+p, j+q), \\[2mm] \sigma^2(i,j) = \displaystyle\sum_{p=-P}^{p=P} \sum_{q=-Q}^{q=Q} (I(i+p, j+q) - \mu(i,j))^2, \end{cases} \quad (1)$$

A small size of the normalization window improves the algorithm performance [9,20] and we set $P = Q = 3$. One image "parrots" after the preprocessing is shown as the input in Fig. 1. It is found that local contrast normalization highlights image structures and textures.

2.2 A Shallow CNN

An extremely simple CNN is illustrated in Fig. 1. It consists of one feature layer before fully connecting to MLP. The feature map layer involves image convolution and down-sampling for feature retrieval. The CNN is shallow and works

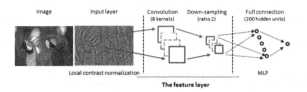

Fig. 1. A shallow CNN architecture with essential blocks. It embeds 8 convolutional kernels and after pooling, it fully connects to MLP with 200 hidden units.

independently when training. The convolutional network can be viewed as a feature extractor and a score predictor and both parts are trainable [24]. The extractor retrieves discriminating features from image patches with convolutional filtering and pooling. After several epoches of back-propagation, the extracted features in the hidden units are passed to GRNN for validation. Note that in the training phase, MLP acts as the score predictor; while in the validation phase, GRNN is the predictor.

2.3   GRNN

GRNN is powerful in numerical regression [32]. It takes a single pass through a set of m learning instances $\{(X_i; Y_i)\}_{i=1}^m$ and needs no iterative training. The first layer of GRNN (see Fig. 2) is fully connected to the pattern layer and the number of units in the input layer depends on the total number of the observation parameters. In pattern layer, each neuron presents a training pattern, and its output measures the distance between the input and the stored patterns. Each unit in the pattern layer is linked to two neurons in the summation layer, one as the numerator and the other as denominator in Eq. (2). In the end, the prediction from the summation layer is passed to the output layer.

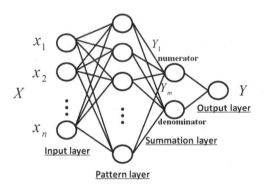

Fig. 2. A semantic description of GRNN. It is composed of 4 layers and performs score prediction after the mini-CNN achieves the number of iteration in training.

For an input feature of $X = (x_1, x_2, ..., x_n)^T$, the output of GRNN is shown in Eq. (2), where n is the feature dimension ($n = 200$), x_i is the ith observation parameter, Y_i weights the ith neuron in the pattern layer and the numerator neuron, and σ is the spread parameter. The distance $(X - X_i)^T (X - X_i)$ measures how well each training sample X_i can represent the point of predication X.

$$\begin{cases} Y = \dfrac{\sum_{i=1}^{n} Y_i e^{-t_i}}{\sum_{i=1}^{n} e^{-t_i}}, \\ t_i = \dfrac{(X - X_i)^T (X - X_i)}{2\sigma^2}. \end{cases} \tag{2}$$

The σ controls the smoothness of data distribution modeling. A large σ leads to a multivariate Gaussian distribution, while a small one allows non-Gaussian distribution. If samples are sufficient, σ can be estimated. However, in this study, the number of training samples is limited and the estimated σ may be biased. Thus, we followed [13] and adjusted the σ experimentally.

2.4 Learning Strategy

The learning strategy is target-oriented (see Fig. 3). In a database, the images are randomly grouped according to a proportion of 3:1:1 for training, validation and test, respectively. As shown in Fig. 3, firstly, in the training samples, a number of patches in normalized images are sent to the shallow CNN. With several epoches of back-propagation, the extracted·CNN features are sent to GRNN for validation. If the validation achieves the pre-defined performance (θ), CNN-GRNN is deemed as well-trained and it is ready for real-world applications. Theoretically, the higher precision the validation is, the better the performance is. While in practice, θ should be tuned according to the state-of-the-art performance, or else it becomes unattainable or hard to reach.

3 Experiments

3.1 Software and Platform

Codes are implemented with Matlab and running on a Win7 system (8 Intel(R) Cores(TM) of 3.7 GHz and 8 GB DDR RAM). The source CNN code is downloaded from https://github.com/rasmusbergpalm/DeepLearnToolbox [25] and GRNN is with the embedded function *newgrnn*.

3.2 Gaussian Blur Images

Gaussian blurring images are collected from LIVE [31], CSIQ [11], TID2008 [26] and TID2013 [27]. LIVE and CSIQ respectively contain 29 and 30 images with different content and a total of 145 and 150 distorted images (each with 5 blurring levels). Differential mean opinion scores (DMOS) is applied for image scoring. Both TID2008 and TID2013 have 25 reference images and use mean opinion scores (MOS) for each distorted images, except that each image in TID2008 is blurred with 4 levels and in TID2013 with 5 levels.

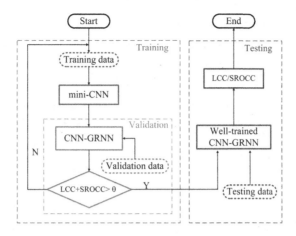

Fig. 3. A target-oriented learning strategy for CNN-GRNN. It will stop training when the validation of CNN-GRNN achieves the overall performance better than θ. That means CNN-GRNN is well-trained and ready for test.

3.3   Experiment Design

Images are randomly grouped for training, validation and testing, herein, 20/4/5 in LIVE, 20/5/5 in CSIQ and 16/4/5 in TID2008 and TID2013, respectively. Likewise, ten rounds of evaluation are carried out and the overall performance is concerned. After testing the well-trained CNN-GRNN (feasibility or intra-database experiment), we further estimate its generalization capacity with other image databases (extensibility or inter-database experiment).

In this study, nine state-of-the-art algorithms for no-reference image sharpness assessment are compared to CNN-GRNN. These metrics include JNB [5], CPBD [23], S3 [36], FISH [35], LPC [6], SVC [30], MLV [1], BIBLE [14] and SPARISH [16]. No change is made of these source codes.

3.4   Parameter Settings

After the model is designed, one difficulty we face is parameter setting. Because no knowledge is available, on the one hand, some parameters are determined once for all, including the size of image patches ([16 16]), the size of the convolutional filtering kernels ([7 7]), the number of the kernels (8), the ratio of the down-sampling operation (2) and the number of iteration (120). On the other hand, three key parameters (the number of patches, the spread parameter σ and the threshold for validation θ) are cautiously tuned. The selection of the patch number and the spread parameter is based on the experimental evaluation. After a series of experiments on LIVE, the number of patch per image is set as 200, the spread parameter $\sigma = 0.01$ and $\theta = 1.8$.

3.5 Performance Criteria

A nonlinear regression is first applied to map the predicted scores to subjective human ratings using the four-parameter logistic function as in Eq. (3) where s and $Q(s)$ are the input score and the mapped score, and x_i $(i = 1, 2, 3, 4)$ are determined during the curve fitting.

$$Q(s) = \frac{x_1 - x_2}{1 + exp(-\frac{s - x_3}{x_4})} + x_2. \tag{3}$$

Two criteria are recommended for IQA performance evaluation by the video quality experts groups (VQEG, http://www.vqeg.org). One is Pearson linear correlation coefficient (LCC) which evaluates the prediction accuracy, and the other is Spearman rank-order correlation coefficient (SROCC) which measures the prediction monotonicity. Both LCC and SROCC values range in $[0, 1]$ and a higher value indicates a better performance. Moreover, average LCC and SROCC on all data sets are calculated as the overall performance. Furthermore, the human subjective scores are linearly normalized before the model training.

4 Results

4.1 Feasibility Experiment

LCC is shown in Table 1 and values in bold indicate the best predication accuracy. It is observed that CNN-GRNN consistently outperforms other BISA metrics on TID2008 and TID2013, followed by BIBLE and SPARISH, and it rivals to BIBLE on LIVE and CSIQ with slightly inferior performance. The overall performance reveals that CNN-GRNN is the best with an average of 0.9433 over all four data sets, followed by BIBLE (0.9251) and SPARISH (0.9218).

SROCC is listed in Table 2 and values in bold reveals the best predication monotonicity. It is also found that CNN-GRNN consistently outperforms all other BISA metrics both on TID2008 and TID2013, rivals BIBLE [14] and SPARISH [16] on LIVE and but is not so good on CSIQ. While among all metrics, CNN-GRNN respectively ranks top-3, top-1 and top-1 on LIVE, TID2008 and TID2013. Again, the overall performance reveals CNN-GRNN the top-ranking metric for blind image sharpness estimation, followed by BIBLE and SPARISH.

Among the involved BISA metrics, CNN-GRNN achieves the best on overall performance by averaging scores over four databases. In general, CNN-GRNN rivals to BIBLE and SPARISH, while it completely outperforms the other metrics. Moreover, experimental results on TID2008 and TID2013 indicate that CNN-GRNN triumphs over all other BISA metrics. In addition, both Tables 1 and 2 indicate that GRNN improves the prediction performance over MLP with the same features. As we know, MLP is a feed-forward artificial neural network (ANN) model and its parameters are optimized iteratively by back-propagation. GRNN is also an ANN model, but it needs no iterative training. In particular, GRNN shows strength on blind IQA with limited features. Based on the results in this study, we might conclude that GRNN seems better than MLP on numerical regression.

Table 1. Evaluation of predication accuracy (LCC) on Gaussian blurring images. The highest LCC value on each database is in bold for observation.

Methods	LIVE	CSIQ	TID2008	TID2013	Overall
JNB [5]	0.8161	0.8061	0.6931	0.7115	0.7567
CPBD [23]	0.8955	0.8822	0.8236	0.8620	0.8658
S3 [36]	0.9434	0.9107	0.8542	0.8816	0.8975
FISH [35]	0.9043	0.9231	0.8079	0.8327	0.8670
LPC [6]	0.9181	0.9158	0.8573	0.8917	0.8957
SVC [30]	0.9416	0.9319	0.8556	0.8762	0.9013
MLV [1]	0.9429	0.9247	0.8583	0.8818	0.9019
BIBLE [14]	**0.9622**	**0.9403**	0.8929	0.9051	0.9251
SPARISH [16]	0.9595	0.9380	0.8891	0.9004	0.9218
CNN	0.8965	0.9265	0.9064	0.8927	0.9055
CNN-GRNN	0.9570	0.9353	**0.9439**	**0.9370**	**0.9433**

Table 2. Evaluation of prediction monotonicity (SROCC) on Gaussian blur images. The best prediction performance is highlighted in bold.

Methods	LIVE	CSIQ	TID2008	TID2013	Overall
JNB [5]	0.7872	0.7624	0.6667	0.6902	0.7266
CPBD [23]	0.9182	0.8853	0.8414	0.8518	0.8742
S3 [36]	0.9436	0.9059	0.8480	0.8609	0.8896
FISH [35]	0.8808	0.8941	0.7828	0.8024	0.8400
LPC [6]	0.9389	0.9071	0.8561	0.8888	0.8977
SVC [30]	0.9343	0.9055	0.8362	0.8589	0.8837
MLV [1]	0.9312	**0.9247**	0.8548	0.8787	0.8974
BIBLE [14]	**0.9607**	0.9132	0.8915	0.8988	0.9160
SPARISH [16]	0.9593	0.9141	0.8869	0.8927	0.9133
CNN	0.8685	0.8739	0.8865	0.8705	0.8748
CNN-GRNN	0.9438	0.8990	**0.9263**	**0.9031**	**0.9181**

4.2 Extensibility Experiment

The performance of inter-database prediction is shown in Fig. 4. With previously well-trained model from one database (60% images for training, 20% for validation and 20% for testing) in the feasibility experiment, we further tested its generalization capacity on other three databases.

In Fig. 4, different color bars correspond to different databases and the height indicates the predication performance. The proposed CNN-GRNN model trained on LIVE shows best extensibility performance on CSIQ, followed by TID2013

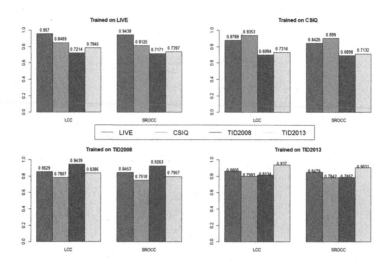

Fig. 4. Score prediction in extensibility experiment. The generalization capacity of well-trained models in the feasibility experiment is validated on other three databases.

and TID2008. And the model trained on CSIQ shows the best extensibility performance on LIVE, followed by TID2013 and TID2008. While on the databases with MOS scores, the hybrid model trained either on TID2008 or TID2013 shows the best extensibility performance on LIVE. Score mapping makes this kind of generalization ability in inter-database experiments understandable. Besides, it should be noticed that LIVE seems the database to be easiest predicted using well-trained CNN-GRNN on other databases.

4.3   Limitations

Several limitations should be mentioned here that will help practitioners and pioneers move forward. The first limitation comes from the algorithm implementation. Since the hybrid CNN-GRNN model is built with MATLAB, hyperparameter optimization and large-scale comparison become quite difficult. Consequently, we can't afford an exhaustive searching of optimal parameter settings. In practice, we found that each epoch costs about two minutes and about four to five whole procedure can result in a well-trained CNN-GRNN. With more epoches, this model can be full of potential to perform better. Second, under the sub-optimal parameter settings, we compared nine state-of-the-art metrics based on simulated Gaussian blurring images in four databases and no real-life blurring images are tested. With the accessibility of BID2011 [4] and CID2013 [34], it will be interesting to explore the proposed algorithm for more general and more "real life" applications. Moreover, no study was carried out between CNN-GRNN and the CNN architecture in [9] to show how much performance is gained on CSIQ, TID2008 and TID2013. On Gaussian blurring images in LIVE, the CNN in [9] reaches 0.953 and 0.962 on LCC and SROCC which are close to our results of 0.957 and 0.944, respectively. That means both networks achieve comparable performance. On the other hand, it indicates the proposed learning strategy can reduce time consumption, while how much the load reduced is not quantified, because of different code implementation, parameter settings and software platform.

4.4 Future Works

Further improvements can be made from several aspects. Above all, color information can be combined into feature extraction, because human visual system is extremely sensitive to colors. Moreover, deeper CNNs might enhance the performance, while the computational complexity is increased in hyper-parameter optimization. Furthermore, incorporating existing handcrafted features might enhance algorithm performance. Last but not the least, adding advanced technologies into the network can be a good choice.

5 Conclusion

A shallow convolutional neural network is proposed for objective blind image sharpness assessment in combination with the general regression neural network. The hybrid system outperforms state-of-the-art algorithms on four data sets. It also shows acceptable generalization capacity. Further improvements can be made from optimal parameter settings and addition information incorporation.

Acknowledgement. The authors would like to thank reviewers for their valuable suggestion that has helped to improve the paper quality. This work is supported from National Natural Science Foundation of China (Grant No. 81501463 and 61379143), Guangdong Innovative Research Team Program (Grant No. 2011S013), National 863 Programs of China (Grant No. 2015AA043203), the Shenzhen Fundamental Research Program (Grant Nos. JCYJ20140417113430726, JCYJ20140417113430665 and JCYJ20150401145529039), the Qing Lan Project of Jiangsu Province and the China Postdoctoral Science Foundation (Grant No. 2016M590827).

References

1. Bahrami, K., Kot, A.C.: A fast approach for no-reference image sharpness assessment based on maximum local variation. IEEE Sig. Process. Lett. **21**(6), 751–755 (2014)
2. Bengio, Y., Courville, A., Vincent, P.: Representation learning: a review and new perspectives. IEEE Trans. Pattern Anal. Mach. Intell. **35**(8), 1798–1828 (2013)
3. Chandler, D.M.: Seven challenges in image quality assessment: past, present, and future research. ISRN Sig. Process. **2013**, 1–53 (2013)
4. Ciancio, A., Costa, A.D., da Silva, E., Said, A., Samadani, R., Obrador, P.: No-reference blur assessment of digital pictures based on multifeature classifiers. IEEE Trans. Image Process. **21**(3), 934–945 (2012)

5. Ferzli, R., Karam, L.J.: A no-reference objective image sharpness metric based on the notion of just noticeable blur (JNB). IEEE Trans. Image Process. **18**(4), 717–728 (2009)
6. Hassen, R., Wang, Z., Salama, M.M.: Image sharpness assessment based on local phase coherence. IEEE Trans. Image Process. **22**(7), 2798–2810 (2013)
7. Hou, W., Gao, X.: Saliency-guided deep framework for image quality assessment. IEEE Multimedia **22**(2), 46–55 (2015)
8. Hou, W., Gao, X., Tao, D., Li, X.: Blind image quality assessment via deep learning. IEEE Trans. Neural Netw. Learn. Syst. **26**(6), 46–55 (2015)
9. Kang, L., Ye, P., Li, Y., Doermann, D.: Convolutional neural networks for no-reference image quality assessment. IEEE Conference on Computer Vision and Pattern Recognition, pp. 1733–1740 (2014)
10. Krizhevsky, A., Sutskever, I., Hinton, G.E.: Imagenet classification with deep convolutional neural networks. In: Advances in Neural Information Processing Systems, pp. 1097–1105 (2012)
11. Larson, E.C., Chandler, D.M.: Most apparent distortion: full-reference image quality assessment and the role of strategy. J. Electron. Imaging **19**(1), 11006 (2010)
12. LeCun, Y., Bengio, Y., Hinton, G.: Deep learning. Nature **521**, 436–444 (2015)
13. Li, C., Bovik, A.C., Wu, X.: Blind image quality assessment using a general regression neural network. IEEE Trans. Neural Netw. **22**(5), 793–799 (2011)
14. Li, L., Lin, W., Wang, X., Yang, G., Bahrami, K., Kot, A.C.: No-reference image blur assessment based on discrete orthogonal moments. IEEE Trans. Cybern. **46**(1), 39–50 (2016)
15. Li, Y., Po, L., Xu, X., Feng, L., Yuan, F., Cheung, C.H., Cheung, K.W.: No-reference image quality assessment with shearlet transform and deep neural networks. Neurocomputing **154**, 94–109 (2015)
16. Li, L., Wu, D., Wu, J., Li, H., Lin, W., Kot, A.C.: Image sharpness assessment by sparse representation. IEEE Trans. Multimedia **18**(6), 1085–1097 (2016)
17. Lin, W., Kuo, C.: Perceptual visual quality metrics: a survey. J. Vis. Commun. Image Represent. **22**(4), 297–312 (2011)
18. Lv, Q., Jiang, G., Yu, M., Xu, H., Shao, F., Liu, S.: Difference of Gaussian statistical features based blind image quality assessment: a deep learning approach. In: IEEE Conference on Image Processing, pp. 2344–2348 (2015)
19. Manap, R.A., Shao, L.: Non-distortion-specific no-reference image quality assessment: a survey. Inf. Sci. **301**, 141–160 (2015)
20. Mittal, A., Moorthy, A.K., Bovik, A.C.: No-reference image quality assessment in the spatial domain. IEEE Trans. Image Process. **21**(12), 4695–4708 (2012)
21. Mittal, A., Soundararajan, R., Bovik, A.C.: Making a "completely" blind image quality analyzer. IEEE Sig. Process. Lett. **20**(3), 209–212 (2013)
22. Moorthy, A.K., Bovik, A.C.: A two-step framework for constructing blind image quality indices. IEEE Sig. Process. Lett. **17**(5), 513–516 (2010)
23. Narvekar, N.D., Karam, L.J.: A no-reference image blur metric based on the cumulative probability of blur detection (CPBD). IEEE Trans. Image Process. **20**(9), 2678–2683 (2009)
24. Niu, X., Suen, C.: A novel hybrid CNN-SVM classifier for recognizing handwritten digits. Pattern Recogn. **45**(4), 1318–1325 (2012)
25. Palm, R.B.: Prediction as a candidate for learning deep hierarchical models of data. Technical University of Denmar (2012)
26. Ponomarenko, N., Lukin, V., Zelensky, A., Egiazarian, K., Astola, J., Carli, M., Battisti, F.: TID2008 - a database for evaluation of full-reference visual quality assessment metrics. Adv. Mod. Radioelectron. **10**(4), 30–45 (2009)

27. Ponomarenko, N., Jin, L., Ieremeiev, O., Lukin, V., Egiazarian, K., Astola, J., Vozel, B., Chehdi, K., Carli, M., Battisti, F., Kuo, C.C.J.: Image database TID2013: peculiarities, results and perspectives. Sig. Process. Image Commun. **20**, 57–77 (2015)

28. Russakovsky, O., Deng, J., Su, H., Jonathan, K., Satheesh, S., Ma, S., Huang, Z., Karpathy, A., Khosla, A., Bernstein, M., Berg, A., Li, F.: Imagenet large scale visual recognition challenge. Int. J. Comput. Vis. **115**(3), 211–252 (2015)

29. Saad, M.A., Bovik, A.C., Christophe, C.: Blind image quality assessment: a natural scene statistics approach in the DCT domain. IEEE Trans. Image Process. **21**(8), 3339–3352 (2012)

30. Sang, Q., Qi, H., Wu, X., Bovic, A.C.: No-reference image blur index based on singular value curve. J. Vis. Commun. Image Represent. **25**(7), 1625–1630 (2014)

31. Sheikh, H.R., Sabir, M.F., Bovik, A.C.: A statistical evaluation of recent full reference image quality assessment algorithms. IEEE Trans. Image Process. **15**(11), 3440–3451 (2006)

32. Specht, D.F.: A general regression neural network. IEEE Trans. Neural Netw. **2**(6), 568–576 (1991)

33. Szegedy, C., Liu, W., Jia, Y., Sermanet, P., Reed, S., Anguelov, D., Erhan, D., Vanhoucke, V., Rabinovich, A.: Going deeper with convolutions. In: IEEE Conference on Computer Vision and Pattern Recognition, pp. 1–9 (2015)

34. Virtanen, T., Nuutinen, M., Vaahteranoksa, M., Oittinen, P.: CID2013: a database for evaluating no-reference image quality assessment algorithms. IEEE Trans. Image Process. **24**(1), 390–402 (2015)

35. Vu, P.V., Chandler, D.M.: A fast wavelet-based algorithm for global and local image sharpness estimation. IEEE Sig. Process. Lett. **19**(7), 423–426 (2012)

36. Vu, C.T., Phan, T.D., Chandler, D.M.: S3: a spectral and spatial measure of local perceived sharpness in natural images. IEEE Trans. Image Process. **21**(3), 934–945 (2012)

37. Wang, Z., Bovik, A.C.: Reduced- and no-reference image quality assessment. IEEE Sig. Process. Mag. **28**(6), 29–40 (2011)

38. Wang, Z., Bovik, A.C., Sheikh, H.R., Simoncelli, E.P.: Image quality assessment: from error visibility to structural similarity. IEEE Trans. Image Process. **13**(4), 600–612 (2004)

39. Yu, S., Zhang, W., Wu, S., Li, X., Xie, Y.: Applications of edge preservation ratio in image processing. In: IEEE International Conference on Signal Processing, pp. 698–702 (2014)

40. Zhang, L., Zhang, L., Mou, X., Zhang, D.: FSIM: a feature similarity index for image quality assessment. IEEE Trans. Image Process. **20**(8), 2378–2386 (2011)

Model and Dictionary Guided Face Inpainting in the Wild

Reuben A. Farrugia[1(✉)] and Christine Guillemot[2]

[1] University of Malta, Msida, Malta
reuben.farrugia@um.edu.mt
[2] INRIA Rennes-Bretagne-Atlantique, Rennes, France

Abstract. This work presents a method that can be used to inpaint occluded facial regions with unconstrained pose and orientation. This approach first warps the facial region onto a reference model to synthesize a frontal view. A modified Robust Principal Component Analysis (RPCA) approach is then used to suppress warping errors. It then uses a novel local patch-based face inpainting algorithm which hallucinates missing pixels using a dictionary of face images which are pre-aligned to the same reference model. The hallucinated region is then warped back onto the original image to restore missing pixels.

Experimental results on synthetic occlusions demonstrate that the proposed face inpainting method has the best performance achieving PSNR gains of up to 0.74 dB over the second-best method. Moreover, experiments on the COFW dataset and a number of real-world images show that the proposed method successfully restores occluded facial regions in the wild even for CCTV quality images.

1 Introduction

Human faces captured in unconstrained acquisition conditions are often partially occluded by other objects, long hair, beard and other accessories such as sunglasses and hats. Research in cognitive sciences has shown that the recognition ability of humans degrades when considering non-frontal [1] and partially occluded faces [2,3]. Moreover, a recent study conducted by Klare *et al.* [4] revealed that state-of-the-art automated face recognition software still achieve low accuracies under these difficult conditions.

Image inpainting can be used to restore the occluded areas within an image and can be divided into three main categories: (i) Diffusion-based, (ii) Exemplar-based and (iii) Global based methods. *Diffusion-based inpainting* [5] methods adopt smoothness priors via partial differential equations (PDEs) to diffuse local structures from the non-occluded to the occluded parts of an image. These methods are naturally well suited for completing small regions but inappropriate to conceal large areas. *Exemplar-based inpainting* methods [6] are local greedy methods that approximate the missing pixels within a patch using a combination of neighbouring patches with similar texture. These methods exploit the

© Springer International Publishing AG 2017
C.-S. Chen et al. (Eds.): ACCV 2016 Workshops, Part I, LNCS 10116, pp. 62–78, 2017.
DOI: 10.1007/978-3-319-54407-6_5

self similarity within an image and manage to better recover the missing texture. The third family of methods, often referred to as *Global inpainting* methods [7] introduce a global prior (*e.g.* low-rank, least squares or sparse) on the whole image. These methods are known to work well when the missing areas are not too large. A complete review on image inpainting can be found in [8].

Face inpainting is a special case of inpainting where the image to be restored is known to be a face. Most face inpainting approaches found in the literature [9–15] adopt a dictionary of frontal faces as global prior to restore the occluded face images. Min and Dugelay [16] demonstrated that global methods are not suitable to conceal sparse occlusion and propose to use Fields of Experts [17] to inpaint sparse facial occlusions. However, the limitation of this approach (and all generic inpainting methods) is that it does not exploit the facial structure and thus fails to restore large occlusions. Moreover, none of the above mentioned methods are suited to restore non-frontal face images, which limits their application in practice. More recently, Burgos-Artizzu *et al.* [18] presented a face inpainting method which can be used to conceal non-frontal images. This method used a trained sub-dictionary per cluster to cater for different pose and expression variations. However, each sub-dictionary is of very large size and computationally intensive to train. Moreover, their training methodology does not guarantee that the pose of the face to be inpainted is well represented by one of the clusters.

The contribution of this paper is two-fold. We first present a *face-frontalization method*[1] that synthesizes the frontal pose of a face acquired in the wild. Unlike the method in [19], this work adopts a simple 2D warping method to get a first approximation of the frontal view and then applies a Robust Principal Component Analysis (RPCA) based method to suppress warping distortions. Unlike existing RPCA-based face denoising [20,21] and alignment [22] methods, our method does not need multiple representations of the same test subject. In fact, our method considers that the training images (which do not include the test subject) to be samples on a manifold and tries to push the distorted frontal face on this manifold. We also present a novel *local patch-based face inpainting algorithm* which exploits the facial structure when computing the inpainting process. The proposed system is different from the work in [18] since it does not need to be trained a-priori and its application is not limited to the poses considered during the training-phase. It differs from existing face inpainting methods [9–16] since it is a local method which ensures that the reconstructed face images is locally smoother and in general more natural. It diverges from local patch-based image inpainting methods such as [6,23–26] since the missing pixels are hallucinated using a dictionary of aligned faces. This work is somewhat related to face super-resolution [27,28] which is not suitable for inpainting.

---

[1] Frontalization is a terminology recenty introduced in [19] to refer to the process of synthesizing a frontal view of a person whose original pose is unconstrained.

2   Related Work

2.1   Face Inpainting

Hwang and Lee [14] presented the first work on face inpainting, where they used morphable models to restore partially occluded faces. In their work, the authors approximate the deformation required for the non-occluded region, and then use the same deformations to synthesize the occluded region. Mo *et al.* [9] replaced the morphable model with a dictionary of frontal face images and derived the combination weights using non-negative least squares (NNLS) on the non-occluded part, which are then used to reconstruct the occluded face region. Several Principal Component Analysis (PCA) [10,13,15] based schemes have been proposed to extend the work of Mo *et al.*, where they try to restore a face image using a linear combination of eigenfaces.

The authors in [12] adopt a Bayesian formulation unifying the occlusion detection and recovery stages. The occlusion is detected using Graph Cuts while the inpainting is computed using Maximum a Posteriori (MAP) estimation. Another iterative strategy was presented by Zhou *et al.* [11] where Markov Random Fields (MRF) models were used to detect the support of the occluded region while sparse coding was used to approximate the reconstructed face. Min and Dugelay [16] employed RPCA to detect the sparse occlusions which are then inpainted using FoE.

A Pose and expression invariant face inpainting algorithm was recently presented in [18]. This method clusters more than 16,000 face images based on their landmark's positions obtained using Robust Cascaded Pose Regression (RCPR) [29] and a dictionary is learned for each cluster using an online Sparse Coding method described in [30]. The landmarks and occlusion mask of the test face image are first extracted using RCPR. The landmarks are used to determine the closest cluster, and the corresponding dictionary, and the reconstructed face is computed using sparse coding. The training is performed off-line, but is very computationally intensive. Moreover, the pose variations considered in this method are limited by the poses represented by the clusters.

2.2   Overview of Exemplar Based Inpainting

Exemplar based Inpainting methods exploit the self-similar property of natural images to propagate texture patterns from non-occluded pixels to fill holes. The seminal work of Criminisi [6] divides the input image in a known part Φ and an unknown part Ω. As shown in Fig. 1, the pixels in Ω that are neighbours of Φ form the front line, denoted by $\delta\Omega$. The target patch Ψ_p centred at p on $\delta\Omega$ is selected using a patch order selection scheme. The patch Ψ_p has two non-overlapping parts: the known region $\Psi_p^k = \Psi_p \cap \Phi$ and the unknown region $\Psi_p^u = \Psi_p \cap \Omega$. The missing pixels in Ψ_p are approximated using template matching.

There are several extensions to the Criminisi inpainting algorithm. Neighbour Embedding methods try to estimate the unknown pixels using a linear combination of several best matching patches, instead of simply copying the pixels of

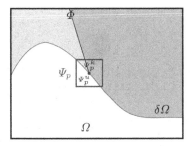

Fig. 1. Notation diagram of Criminisi's algorithm [6].

the best matching patch. The authors in [23] use non-local means to give higher weights to patches which are more similar to the known samples of the input patch. Sparse approximation was used instead in [24,25] where they try to derive a least squares approximation of the known part under a constraint of sparsity on the weights. Several Neighbour Embedding methods were considered in [26].

3 Proposed Face Inpainting in the Wild System

We introduce here an approach which allows pose and orientation independent inpainting of facial images. The schematic diagram of the proposed method is illustrated in Fig. 2, where in this example we are going to inpaint the cigar from the actor's mouth (marked in green). This method assumes that the landmark points (marked by red crosses) are available (either manually inserted or else detected using a state-of-the-art landmark detection algorithm such as [29,31]). The face region $\mathbf{F}$ is then segmented from the background region $\mathbf{B}$ based on the peripheral landmark points marked with a dotted blue line. Only the occlusion on the face segment $\mathbf{F}$ is concealed by the proposed system, since this contains the most important and discriminative information[2]. We will show in Sect. 4.5 that the occluded region in $\mathbf{B}$ can be restored using off the shelf inpainting algorithms since these areas are generally smooth and easier to conceal by exploiting the self-similar texture present in the neighbourhood of Ψ_p.

The standard Delaunay triangulation method is computed from the set of P landmark locations. Each landmark triplet from the face image $\mathbf{F}$ forms a triangle, as shown in solid blue lines. A reference face shape model $\mathbf{S}$ represents the shape of a frontal 2D-face whose P landmark points are normalized such that the distance between the eye centres (inter-eye distance) is of 40 pixels[3]. The frontal pose of the test subject $\hat{\mathbf{F}}$ is synthesized using the proposed

---

[2] The accuracy of the segmentation process is dependent on the number of landmarks used. In this example, one can use more landmark points to segment the lower part of the face region (and possibly the entire occluded region) in segment $\mathbf{F}$.

[3] An inter-eye distance of 40 pixels is sufficient for identification. Nevertheless, this method is not affected by this resolution and higher (or lower) resolutions can be configured.

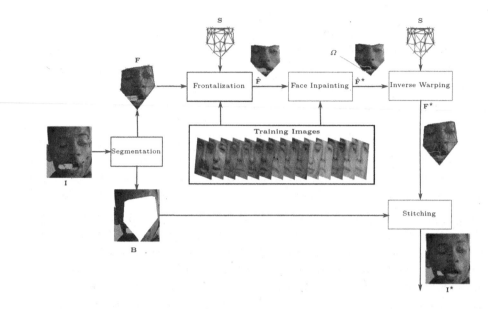

Fig. 2. Schematic diagram of the proposed face inpainting in the wild method (Color figure online).

frontalization method, which employs a list of training face images aligned such that the inter-eye distance is set to 40 pixels (see Sect. 3.1). The alignment of the training images can be done off-line since it is independent from the pose of the facial image being processed. The *Face inpainting process* then employs the same list of training face images to fill the occluded region Ω (see Sect. 3.2) which hallucinates the missing pixels to restore $\hat{\mathbf{F}}^\star$.

Only the recovered region Ω, and therefore the corresponding triangles marked in red, is used to warp the inpainted region from the frontal view $\hat{\mathbf{F}}^\star$ onto the original image $\mathbf{F}$ to fill its missing pixels. The resulting inpainted face image $\mathbf{F}^\star$ is made of the known part of $\mathbf{F}$ and the warped inpainted region Ω from the frontal view $\hat{\mathbf{F}}^\star$. The inpainted face segment $\hat{\mathbf{F}}^\star$ and background images are then simply stitched together to form the restored image $\mathbf{I}^\star$.

3.1   Face Frontalization

As mentioned above, the face image $\mathbf{F}$ is partitioned into triangular regions using the Delaunay triangulation method. Similarly, the face shape model $\mathbf{S}$ is partitioned into triangles. We use the 21 facial landmark points defined in [32] since it caters for both affine (rotation, translation and scaling) and more complicated 3-dimensional deformations from the frontal view. The first part of the face frontalization method proposed here is to transfer the pixels that reside within each triangle in $\mathbf{F}$ onto the corresponding triangle in $\mathbf{S}$ using piecewise affine transformation, to get the warped face $\mathbf{F}_w$. This warping process does not manage to recover all the pixels within $\mathbf{F}_w$ because of scale and occlusions caused by the facial pose, and it therefore contains a number of missing pixels. Moreover, $\mathbf{F}_w$ contains warping errors which can be significant when the face orientation deviates considerably from the frontal pose. The missing pixels can be recovered using the face inpainting algorithm described in Sect. 3.2, while the suppression of warping distortions will be considered in the remaining part of this subsection.

We define $\mathbf{s}_w$ to represent the support of known pixels in $\mathbf{F}_w$ and denote a matrix $\mathbf{M} \in \mathbb{R}^{|\mathbf{s}_w| \times N}$ where each column-vector corresponds to a vectorized representation of the N training images and the rows that are not contained in $\mathbf{s}_w$ are removed. We also denote a vector $\mathbf{f}_w = \hat{\mathbf{f}}_w + \mathbf{e}_w$ to represent the vectorized form of $\mathbf{F}_w$, where $\mathbf{e}_w$ stands for the warping error and $\hat{\mathbf{f}}_w$ is the noise-free version of the frontal face which is unknown. The matrix $\mathbf{M}$ and vector $\mathbf{f}_w$ are concatenated to form an augmented matrix $\mathbf{M}_a$, where without loss of generality $\mathbf{f}_w$ is placed as the first column-vector in $\mathbf{M}_a$.

Several studies have shown that face images reside on a low-dimensional manifold [33,34], and this property was extensively exploited in face analysis literature [35,36]. This observation suggests that if we vectorize a number of aligned and undistorted face images and stack them into a matrix $\mathbf{M}$, then $\mathbf{M}$ can be approximated by a low-rank matrix. However, given that $\mathbf{f}_w$ is distorted by warping errors, this vector will be linearly independent from all the other column vectors of $\mathbf{M}$, which will inherently increase the rank of $\mathbf{M}_a$. We therefore decompose the augmented matrix $\mathbf{M}_a$ using the Robust PCA (RPCA) [20]

$$\text{minimize } ||\mathbf{A}||_* + \lambda ||\mathbf{E}||_1 \text{ subject to } \mathbf{A} + \mathbf{E} = \mathbf{M}_a \qquad (1)$$

where $||\mathbf{A}||_*$ denotes the nuclear norm of $\mathbf{A}$ that corresponds to the sum of its singular values, $||E||_1$ denotes the l_1-norm of $\mathbf{E}$ and λ is a regularization parameter. The authors in [20] recommend to set the regularization parameter

$$\lambda = \frac{C}{\sqrt{\max{(m^2, N)}}} \qquad (2)$$

where C is a positive constant and the resolution of the frontal image is $m \times m$ ($m = 67$ in this work). This problem is convex and can be solved in polynomial time. In this work we solve this problem using the inexact Augmented Lagrange Multiplier method [37].

The rationale behind this formulation is that the column-vectors in $\mathbf{M}$ are points that reside on a low-dimensional manifold. The vector $\hat{\mathbf{f}}_w$ is an undistorted face image, which is unknown, resides on the same manifold. However, the observed warped face image $\mathbf{f}_w$ which contains warping errors does not reside on the manifold. By forcing the matrix $\mathbf{A}$ to be low-rank we are pushing the warped face image $\mathbf{f}_w$ on the low-dimensional manifold while the distortion caused by warping will go in the error matrix $\mathbf{E}$. The frontal denoised face image is then simply the first column-vector in $\mathbf{A}$, where the frontal face image $\hat{\mathbf{F}}$ is derived by reshaping it to an $m \times m$ image.

The use of RPCA was considered in the past for denoising [20,21] and alignment [22]. However, opposed to our proposed frontalization method, these methods are applied on a set of face images of the same subjects, i.e. RPCA is applied directly on a matrix which consists of different representations of the same subject. On the other hand, our approach considers the training images to represent points on a low-dimensional manifold, where the subject being denoised is not present in the list of training images.

3.2   Patch Based Inpainting of Frontalized Face Images

Figure 3 depicts the outline of the proposed face inpainting method. At each iteration, the patch Ψ_p with the highest priority is chosen for inpainting, where p represents the position of the patch. The priorities of each patch residing on the frontal line are assigned based on the product of three terms

$$P(p) = C(p)D(p)E(p) \tag{3}$$

where the confidence term $C(p)$ is a measure of reliable information surrounding the pixel at position p, the data term $D(p)$ reflects the presence of some structure perpendicular to the front-line, and the edge term $E(p)$ that caters for structure which is non-orthogonal to the front line. More information and detail on how these terms are computed can be found in [26].

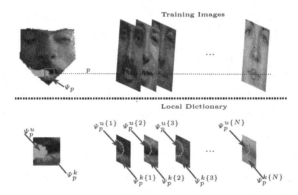

Fig. 3. Schematic diagram of the proposed local face inpainting algorithm.

Given a training set of N facial images, we extract a patch at position p from each training image, where we denote $\Psi_p^{\{i\}} i \in [1, N]$ to represent the collocated patch extracted from the i-th image. To simplify the notation, we divide the training patches $\Psi_p^{\{i\}}$ into a known part $\Psi_p^{k\{i\}} = \Psi_p^{\{i\}} \cap \Psi_p^k$ and an unknown part $\Psi_p^{u\{i\}} = \Psi_p^{\{i\}} \cap \Psi_p^u$. In other words, $\Psi_p^{k\{i\}}$ ($\Psi_p^{u\{i\}}$) corresponds to the collocated pixels from the i-th training image which are known (unknown) in Ψ_p.

The proposed method employs two dictionaries: $\mathbf{D}_p^k$ which is used to derive the linear combination weights and $\mathbf{D}_p^u$ which is used to approximate the missing pixels. The dictionary $\mathbf{D}_p^k$ ($\mathbf{D}_p^u$) is constructed by stacking the N vectorized representation of $\Psi_p^{k\{i\}}$ ($\Psi_p^{u\{i\}}$). The next step is to search for K neighbours that best represent the known part Ψ_p^k within $\mathbf{D}_p^k$. We define the support of the K-nearest neighbours by $\mathbf{s}_p$ such that $\mathbf{D}_p^k(\mathbf{s}_p)$ correspond to the K column-vectors which are closest in Euclidean space to the vectorized representation of Ψ_p^k. We then derive the combination weights $\mathbf{w}^\star$ that minimize the following constrained optimization problem

$$\mathbf{w}^\star = arg\,min_{\mathbf{w}} \|\Psi_p^k - \mathbf{D}_p^k(\mathbf{s})\mathbf{w}\|_2^2 \text{ subject to } \sum_{i=1}^{K} w_i = 1 \qquad (4)$$

This formulation corresponds to the locally linear embedding (LLE) introduced in [33] for data dimensionality reduction, which has an analytical solution. Note that the sum to unity constraint is important since this ensures that the reconstructed pixels preserve the original bit depth. Once the combination weight vector $\mathbf{w}^\star$ is found, the same weights are applied in the linear combination of the "unknown" pixels[4] in the dictionary $\mathbf{D}_p^u(\mathbf{s}_p)$, where the support $\mathbf{s}_p$ is used to choose the K column vectors from $\mathbf{D}_p^u$. The unknown part Ψ_p^u is therefore approximated using

$$\Psi_p^{u\star} = \mathbf{D}_p^u(\mathbf{s}_p)\mathbf{w}^\star \qquad (5)$$

This process is iterated until all occluded pixels are inpainted, and the final solution is denoted by $\hat{\mathbf{F}}^\star$. The accuracy of this method is dependent on the accuracy of the frontalization method employed. In fact, if the frontalized image contains warping deformations, these will propagate within the missing region causing sever distortions in the resulting inpainted face.

We emphasize here that the LLE inpainting method in [26] approximates missing pixels using a combination of spatial neighboring patches with similar texture. However, it does not exploit the contextual information of the facial component being inpainted, and thus fails to reconstruct natural face images. The proposed method (see Fig. 3) finds collocated patches present within an aligned dictionary to find the neighboring patches suitable for reconstruction i.e. the right eye patches are used to inpaint the right eye of a test subject.

4   Experimental Results

The training images used in the following experiments are derived from the Color Feret dataset [38]. All these images are registered based on their eyes and mouth centre coordinates such that the inter-eye distance is of 40 pixels and the resolution is $m \times m$ pixels, where in all experiments $m = 67$. For all the experiments, subjects wearing glasses were removed[5], resulting in a total of $N = 2081$ facial images to be used for training.

---

[4] These pixels are known in the dictionary $\mathbf{D}_p^u$ since the training images do not have occlusions. However, these pixels are collocated with the unknown pixels within the patch being inpainted Ψ_p.

[5] Subjects wearing glasses were removed since we want to use the training images to synthesize people without facial occlusions.

The frontalization method was tested using the LFW [39] dataset, while the inpainting process was evaluated using both AR [40] and the COFW [29] datasets. The *AR dataset* was used to evaluate the performance of the proposed face inpainting on synthetic occlusions where the ground-truth is available. For these experiments, one image per subject is considered, where subjects wearing glasses were removed resulting in a total of 84 test images. These images were aligned to the training images based on the eye and mouth center coordinates. The *COFW dataset* is a more challenging dataset which contains images with various poses and occlusions. However, this dataset cannot be used for objective evaluation since the ground truth (non-occluded face) is not available. This dataset is mainly used to qualitatively assess the performance of various face inpainting algorithms. The proposed system was also evaluated on a number of real-world images (see Sect. 4.5) presenting more challenging scenarios.

Unless otherwise stated, we assume that the occluded region (called mask in the sequel) is manually marked, since the automated detection of occlusion is not in scope of this work. The automated detection of the occlusion mask introduces inaccuracies (see Fig. 8 for example) which will degrade the quality of the restored image. In order to be able to fairly assess the performance of the proposed approach with respect to existing methods, it was decided to use the same known occlusion mask. In addition, the landmark points are assumed to be manually defined by the user. The code is made available on-line[6].

4.1 Frontalization Analysis

Figure 4 compares the faces warped using piecewise affine warping (2nd column) and the final frontalized face where warping artefacts are suppressed using our modified RPCA method (3rd column). Pixels which are left empty in the frontal view are treated as holes and inpainted using the method described in Sect. 3.2. In this experiment we consider a number of images from the LFW dataset [39], and set $C = 2.3$ since it was found to give satisfactory performance. One can also notice that images that are warped using piecewise affine transformation without denoising contain significant warping artefacts that negatively affect the inpainting process. In contrast, it can be seen that the proposed RPCA denoising method manages to remove a substantial portion of the noise induced by warping. It can be noticed from the last row of images that our modified RPCA method treats occlusions as noise deviating the frontal face from the low-dimensional manifold and partially suppresses these occlusions automatically without having the occlusion mask.

4.2 Face Inpainting Parameter Selection

In this subsection we investigate the effect of two parameters, namely the number of neighbours K and the patch size, which affect the performance of the

---

[6] Face Inpainting Demo: https://goo.gl/ws3NG4.

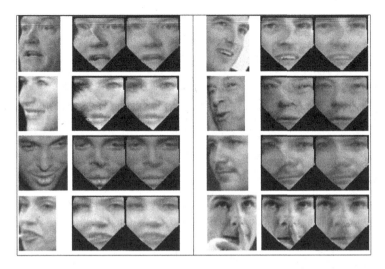

Fig. 4. The first column from each cell shows the cropped images from the LFW dataset. The second column represents the piecewise affine warped face image and the fourth column were restored using our modified RPCA method.

Eyes Balaclava Random Bars Chains Mouth

Fig. 5. Occlusion masks used for this experiment.

proposed face inpainting algorithm. For this experiment we considered six different masks: (i) Both eyes are occluded (Eyes), (ii) the face is covered by a balaclava (Balaclava), (iii) 85% of the pixels are randomly occluded (Random), (iv) the face is occluded by vertical bars (Bars), (v) the face region is occluded by a chains link fence (Chains) and (vi) the mouth region is covered (Mouth). Figure 5 show the masks used for this experiment where the 84 test images from the AR dataset (described above) were used. The test images are already aligned with the training images and therefore the forward and inverse warping processes were disabled. Figure 6 depicts how the quality, measured in terms of PSNR, is affected by the number of neighbours K and the patch size. It can be seen from Fig. 6 (a) (which was obtained by fixing the patch size to 15) that the PSNR increases steadily up till around $K = 500$ beyond which no substantial gain in performance is observed. On the other hand, the results in Fig. 6 (b) (which was obtained by fixing $K = 500$) show that the performance is quite independent of the patch size. These results seem to generalize for all six masks considered. To get a compromise between quality and complexity we set $K = 500$ and a patch-size of 15 for the following experiments.

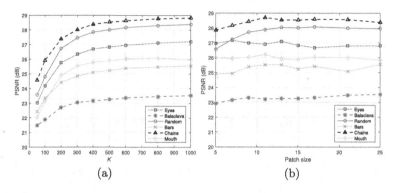

Fig. 6. Evaluating the performance of the proposed face inpainting method in Sect. 3.2 as a function of (a) number of candidates K and (b) patch size.

4.3   Synthetic Occlusion

In this experiment we use the same set of 84 images from the AR dataset and masks discussed in Sect. 4.2. Once again, given that the test images are aligned to the training images, the forward and inverse warping were disabled. We compare our proposed face inpainting method (described in Sect. 3.2) with four other baseline methods [9–11,16]. The parameters used by these baseline methods were optimized using a methodology similar to the one described in Sect. 4.2. Qualitative tests were conducted on frontal faces because all baseline methods [9–11,16] were designed for frontal faces, and testing on non-frontal faces will penalize the baseline methods not our proposed framework.

Table 1. Quality analysis using the PSNR quality metric.

Method	Eyes	Balaclava	Random	Bars	Chains	Mouth
Mo *et al.* [9]	25.19	22.30	22.98	23.51	23.56	25.81
Wang & Tao [10]	25.92	17.28	25.59	21.10	27.02	20.07
Zhou *et al.* [11]	26.37	22.91	25.48	24.67	26.86	24.25
Min & Dugelay [16]	20.22	14.10	27.38	18.62	27.79	14.13
Ours	**26.84**	**23.26**	**27.99**	**25.24**	**28.52**	**25.88**

The results in Table 1 show that the method of Zhou *et al.* [11] provides good performance, especially when the occlusion mask is not sparse. On the other hand, the method of Min & Dugelay [16] performs very well only when the error is sparse. Nevertheless, the proposed method achieves the best performance for all types of occlusions, achieving averaged PSNR gains of up to 0.74 dB relative to the second-best method.

4.4   Real Occlusion

To asses the ability of the proposed method in the wild we used the testing images from the COFW dataset. Here, and in the following subsection, we adopt the full system described in Sect. 3. Since the baseline methods against which we are comparing are not able to perform face inpainting in the wild, we have extended the reference methods to handle different poses by integrating them within our framework *i.e.* the test image is frontalized using our method described in Sect. 3.1, the frontalized image is inpainted using either our approach (Sect. 3.2) or one of the baseline methods, and the reconstructed region is warped back to restore the missing pixels. Figure 7 shows some results with the COFW dataset, where it can be seen that the method of Min & Dugelay does not perform well on all these images. This is attributed to the fact that their method (like all exemplar-based inpainting methods including [26]) does not employ the facial structure when inpainting and therefore its application is limited to conceal very small occlusions. It can be seen that the images reconstructed using the methods in [10,11] are generally noisy while those recovered by Mo *et al.* [9] provide discontinuities at the boundaries between the known and inpainted part of the face image. On the other hand, the images recovered using the proposed method are more natural and of higher quality. It must be noticed that the proposed method conceals regions which intersect with the face region $\mathbf{F}$ but not the background region $\mathbf{B}$. Nevertheless, it will be shown in Sect. 4.5 that the missing pixels in $\mathbf{B}$ can be recovered using off the shelf inpainting methods.

Given that the code of Burgos-Artizzu *et al.* [18] was not released we could not compare it with our method directly. Nevertheless, we use the same images and occlusion mask adopted by the authors in [18][7]. Figure 8 compares our proposed scheme with that of Burgos-Artizzu *et al.* The face images reconstructed using their approach show visual discontinuities (see left image) and ringing artefacts (see right image). Opposed to this, our proposed method reconstructs smoother and visually more pleasing facial images.

4.5   Real-World Examples

In this experiment we picked up a number of images from the Internet where the faces had a wider range of pose variations, quality (include CCTV quality images) and image resolutions (up to 1 Mega pixel) compared to those in the COFW dataset. The occlusion on the face region $\mathbf{F}$ were inpainted using the proposed method while the occlusion on the non-facial part $\mathbf{B}$ was concealed using the LLE based inpainting algorithm proposed in [26].

Figure 9 shows that the proposed method manages to reconstruct the occluded part of the face quite effectively, even when the pose is almost profile. This contrasts with the method in [18] where their approach is restricted to poses considered during training. Apart being successful for image editing (see

---

[7] The images provided by Burgos-Artizzu *et al.* [18] were in grayscale, and therefore only results on grayscale images are presented here.

Original   Ours   Mo *et al.*   Wang & Tao   Zhou *et al.*   Min & Dugelay

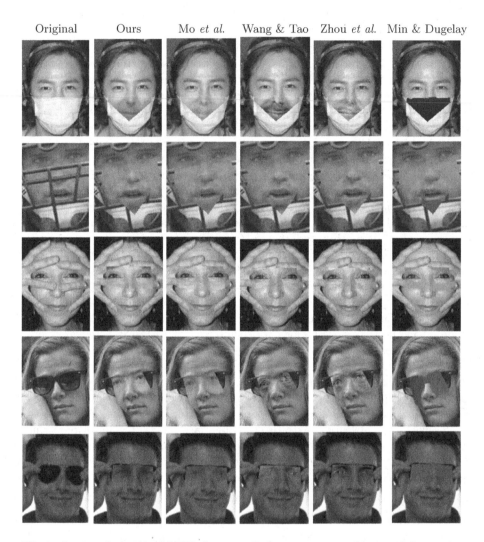

Fig. 7. Images from the COFW dataset which were recovered using different face inpainting algorithms best viewed by zooming onto the images.

(a)   (b)

Fig. 8. Original image (left), the results obtained using the Burgos-Artizzu *et al.* [18] (center) and the proposed face inpainting method (right).

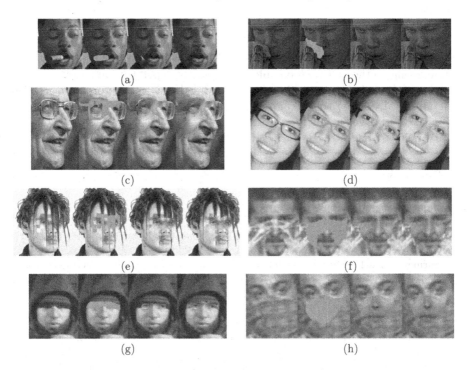

(a)

(b)

(c)

(d)

(e)

(f)

(g)

(h)

Fig. 9. Original image (*left*), the original image with the region to be inpainted marked in pure green (*center-left*), the inpainted image to remove the occluded region intersecting the frontal face (*center-right*) and the final inpainted image where the background was inpainted using the LLE method in [26] (*right*) (Color figure online).

Fig. 9 (a) – (d)), the proposed method was effective in reconstructing face images captured by CCTV (see Fig. 9 (f) – (h)) which are quite noisy and reveals the face which is appropriately camouflaged to reduce identification (see Fig. 9 (e)). The face inpainting process took between 5 – 11s on a standard machine (Intel Core i7 with 8 GB of memory), using MATLAB.

5   Comments and Conclusion

In this paper we propose a novel face inpainting algorithm that can be used to restore occluded facial regions in the wild. This method introduces two innovative components: (i) it presents a frontalization method that synthesizes a frontal view of the face in an unconstrained pose and (ii) it describes a local patch-based face inpainting algorithm. Experimental results show the importance of using our modified RPCA method to suppress warping artefacts introduced by the piecewise affine warping step, especially when the pose is almost profile. RPCA in our method differs from its classical use for denoising since it can denoise subjects not included in the dictionary. Experimental results also show that the proposed face inpainting in the wild system manages to inpaint different occlusions and reconstructs images of higher quality. The main advantages of the proposed method are (i) it can inpaint faces with a wider range of pose variations than existing methods, (ii) its complexity is mainly dependent on the number of neighbours K and the size of the holes to be filled, (iii) the restored images are generally locally smoother and of higher quality. This method was also evaluated on real-world images at different orientations, resolutions and quality, including CCTV quality images taken during riots in UK.

References

1. Hancock, P.J., Bruce, V., Burton, A.: Recognition of unfamiliar faces. Trends Cogn. Sci. **4**, 330–337 (2000)
2. Terry, R.L.: How wearing eyeglasses affects facial recognition. Curr. Psychol. **12**, 151–162 (1993)
3. Yarmey, A.D.: Eyewitness recall and photo identification: a field experiment. Psychol. Crime Law **10**, 53–68 (2004)
4. Klare, B.F., Klein, B., Taborsky, E., Blanton, A., Cheney, J., Allen, K., Grother, P., Mah, A., Burge, M., Jain, A.K.: Pushing the frontiers of unconstrained face detection and recognition: IARPA Janus benchmark A. In: IEEE Conference on Computer Vision and Pattern Recognition (CVPR), pp. 1931–1939 (2015)
5. Bertalmio, M., Sapiro, G., Caselles, V., Ballester, C.: Image inpainting. In: Annual Conference on Computer Graphics and Interactive Techniques, pp. 417–424 (2000)
6. Criminisi, A., Perez, P., Toyama, K.: Region filling and object removal by exemplar-based image inpainting. IEEE Trans. Image Process. **13**, 1200–1212 (2004)
7. Liu, J., Musialski, P., Wonka, P., Ye, J.: Tensor completion for estimating missing values in visual data. IEEE Trans. Pattern Anal. Mach. Intell. **35**, 208–220 (2013)
8. Guillemot, C., Meur, O.L.: Image inpainting : overview and recent advances. IEEE Sig. Process. Mag. **31**, 127–144 (2014)
9. Mo, Z., Lewis, J., Neumann, U.: Face inpainting with local linear representations. In: British Machine Vision Conference, pp. 37.1–37.10 (2004)
10. Wang, Z.M., Tao, J.H.: Reconstruction of partially occluded face by fast recursive PCA. In: International Conference on Computational Intelligence and Security Workshops, pp. 304–307 (2007)
11. Zhou, Z., Wagner, A., Mobahi, H., Wright, J., Ma, Y.: Face recognition with contiguous occlusion using markov random fields. In: IEEE International Conference on Computer Vision, pp. 1050–1057 (2009)
12. Lin, D., Tang, X.: Quality-driven face occlusion detection and recovery. In: IEEE Conference on Computer Vision and Pattern Recognition, pp. 1–7 (2007)
13. Hosoi, T., Nagashima, S., Kobayashi, K., Ito, K., Aoki, T.: Restoring occluded regions using FW-PCA for face recognition. In: IEEE Conference on Computer Vision and Pattern Recognition Workshops, pp. 23–30 (2012)
14. Hwang, B.W., Lee, S.W.: Reconstruction of partially damaged face images based on a morphable face model. IEEE Trans. Pattern Anal. Mach. Intell. **25**, 365–372 (2003)
15. Liwicki, S., Tzimiropoulos, G., Zafeiriou, S., Pantic, M.: Euler principal component analysis. Int. J. Comput. Vis. **101**, 498–518 (2012)
16. Min, R., Dugelay, J.L.: Inpainting of sparse occlusion in face recognition. In: IEEE International Conference on Image Processing, pp. 1425–1428 (2012)

17. Roth, S., Black, M.J.: Fields of experts. Int. J. Comput. Vis. **82**, 205–229 (2009)
18. Burgos-Artizzu, X.P., Zepeda, J., Clerc, F.L., Perez, P.: Pose and expression-coherent face recovery in the wild. In: IEEE International Conference on Computer Vision Workshop, pp. 877–885 (2015)
19. Hassner, T., Harel, S., Paz, E., Enbar, R.: Effective face frontalization in unconstrained images. In: IEEE Conference on Computer Vision and Pattern Recognition, pp. 4295–4304 (2015)
20. Candès, E.J., Li, X., Ma, Y., Wright, J.: Robust principal component analysis? J. ACM **58**, 11:1–11:37 (2011)
21. Yuan, Z., Xie, X., Ma, X., Lam, K.M.: Color facial image denoising based on RPCA and noisy pixel detection. In: IEEE International Conference on Acoustics, Speech and Signal Processing, pp. 2449–2453 (2013)
22. Peng, Y., Ganesh, A., Wright, J., Xu, W., Ma, Y.: RASL: robust alignment by sparse and low-rank decomposition for linearly correlated images. IEEE Trans. Pattern Anal. Mach. Intell. **34**, 2233–2246 (2012)
23. Wong, A., Orchard, J.: A nonlocal-means approach to exemplar-based inpainting. In: IEEE International Conference on Image Processing, pp. 2600–2603 (2008)
24. Xu, Z., Sun, J.: Image inpainting by patch propagation using patch sparsity. IEEE Trans. Image Process. **19**, 1153–1165 (2010)
25. Studer, C., Kuppinger, P., Pope, G., Bolcskei, H.: Recovery of sparsely corrupted signals. IEEE Trans. Inf. Theor. **58**, 3115–3130 (2012)
26. Guillemot, C., Turkan, M., Meur, O.L., Ebdelli, M.: Object removal and loss concealment using neighbor embedding methods. Sig. Process. Image Commun. **28**, 1405–1419 (2013)
27. Ma, X., Zhang, J., Qi, C.: Position-based face hallucination method. In: Proceedings of the IEEE International Conference on Multimedia and Expo, pp. 290–293 (2009)
28. Jiang, J., Hu, R., Wang, Z., Han, Z.: Face super-resolution via multilayer locality-constrained iterative neighbor embedding and intermediate dictionary learning. IEEE Trans. Image Process. **23**, 4220–4231 (2014)
29. Burgos-Artizzu, X.P., Perona, P., Dollár, P.: Robust face landmark estimation under occlusion. In: IEEE International Conference on Computer Vision, pp. 1513–1520 (2013)
30. Mairal, J., Bach, F., Ponce, J., Sapiro, G.: Online dictionary learning for sparse coding. In: Annual International Conference on Machine Learning, pp. 689–696 (2009)
31. Zhu, X., Ramanan, D.: Face detection, pose estimation, and landmark localization in the wild. In: IEEE Conference on Computer Vision and Pattern Recognition, pp. 2879–2886 (2012)
32. Koestinger, M., Wohlhart, P., Roth, P.M., Bischof, H.: Annotated facial landmarks in the wild: a large-scale, real-world database for facial landmark localization. In: IEEE International Conference on Computer Vision, pp. 2144–2151 (2011)
33. Roweis, S.T., Saul, L.K.: Nonlinear dimensionality reduction by locally linear embedding. Science **290**, 2323–2326 (2000)
34. Belkin, M., Niyogi, P.: Laplacian eigenmaps for dimensionality reduction and data representation. Neural Comput. **15**, 1373–1396 (2003)
35. He, X., Yan, S., Hu, Y., Niyogi, P., Zhang, H.J.: Face recognition using laplacianfaces. IEEE Trans. Pattern Anal. Mach. Intell. **27**, 328–340 (2005)
36. Hu, C., Chang, Y., Feris, R., Turk, M.: Manifold based analysis of facial expression. In: IEEE International Conference on Computer Vision and Pattern Recognition, p. 81 (2004)

37. Lin, M., Chen, L., Wu, Y.M.: The augmented lagrange multiplier method for exact recovery of corrupted low-rank matrices. Technical report, University of Illinois (2009)
38. Phillips, P.J., Wechsler, H., Huang, J., Rauss, P.J.: The FERET database and evaluation procedure for face-recognition algorithms. Image Vis Comput. **16**, 295–306 (1998)
39. Huang, G.B., Ramesh, M., Berg, T., Learned-Miller, E.: Labeled faces in the wild: a database for studying face recognition in unconstrained environments. Technical report 07–49, University of Massachusetts, Amherst (2007)
40. Martinez, A.M., Benavente, R.: The AR Face Database. Technical report, CVC (1998)

Patch Group Based Bayesian Learning for Blind Image Denoising

Jun Xu[1], Dongwei Ren[1,2], Lei Zhang[1($\boxtimes$)], and David Zhang[1]

[1] Department of Computing, The Hong Kong Polytechnic University,
Hong Kong, China
csjunxu@comp.polyu.edu.hk, cslzhang@polyu.edu.hk
[2] School of Computer Science and Technology,
Harbin Institute of Technology, Harbin, China

Abstract. Most existing image denoising methods assume to know the noise distributions, e.g., Gaussian noise, impulse noise, etc. However, in practice the noise distribution is usually unknown and is more complex, making image denoising still a challenging problem. In this paper, we propose a novel blind image denoising method under the Bayesian learning framework, which automatically performs noise inference and reconstructs the latent clean image. By utilizing the patch group (PG) based image nonlocal self-similarity prior, we model the PG variations as Mixture of Gaussians, whose parameters, including the number of components, are automatically inferred by variational Bayesian method. We then employ nonparametric Bayesian dictionary learning to extract the latent clean structures from the PG variations. The dictionaries and coefficients are automatically inferred by Gibbs sampling. The proposed method is evaluated on images with Gaussian noise, images with mixed Gaussian and impulse noise, and real noisy photographed images, in comparison with state-of-the-art denoising methods. Experimental results show that our proposed method performs consistently well on all types of noisy images in terms of both quantitative measure and visual quality, while those competing methods can only work well on the specific type of noisy images they are designed for and perform poorly on other types of noisy images. The proposed method provides a good solution to blind image denoising.

1 Introduction

Image denoising is an important problem in image processing and computer vision. Most existing methods are designed to deal with specific types of noise, e.g., Gaussian noise, mixed Gaussian and impulse noise, etc. Gaussian noise removal is a fundamental problem and has received intensive research interests with many representative work [1–8]. Gaussian noise removal is not only an independent task but also can be incorporated into other tasks, e.g., the removal of mixed Gaussian and impulse noise. The 'first-impulse-then-Gaussian' strategy

L. Zhang—This work is supported by the HK RGC GRF grant (PolyU5313/12E).

C.-S. Chen et al. (Eds.): ACCV 2016 Workshops, Part I, LNCS 10116, pp. 79–95, 2017.
DOI: 10.1007/978-3-319-54407-6_6

is commonly adopted by methods of [9–11], which are designed specifically for mixed Gaussian and impulse noise. However, noise in real images is more complex than simple Gaussian or mixed Gaussian and impulse distribution. Besides, noise is usually unknown for existing methods. This makes image denoising still a challenging problem.

To the best of our knowledge, the study of blind image denoising can be traced back to the BLS-GSM model [12]. In [12], Portilla et al. proposed to use scale mixture of Gaussian in overcomplete oriented pyramids to estimate the latent clean images. In [13], Portilla proposed to use a correlated Gaussian model for noise estimation of each wavelet subband. Based on the robust statistics theory [14], Rabie modeled the noisy pixels as outliers, which could be removed via Lorentzian robust estimator [15]. Liu et al. proposed to use 'noise level function' (NLF) to estimate the noise and then use Gaussian conditional random field to obtain the latent clean image [16]. Recently, Gong et al. proposed an optimization based method [17], which models the data fitting term by weighted sum of ℓ_1 and ℓ_2 norms and the regularization term by sparsity prior in the wavelet transform domain. Later, Lebrun el al. proposed a multiscale denoising algorithm called 'Noise Clinic' [18] for blind image denoising task. This method generalizes the NL-Bayes [19] to deal with signal and frequency dependent noise.

Despite the success of these methods, they have many limitations. On one hand, as suggested in [16,18], Gaussian noise, assumed by [13,15,16], may be inflexible for more complex noise in real images. Hence, better approximation to the noise could bring better image denoising performance [16,18]. On the other hand, the method [17] needs tune specific parameters for different types of noise. This makes the proposed method not a strictly "blind" image denoising. Based on these observations, it is still needed to design an robust and effective model for blind image denoising. Few assumption and no parameter tuning would bring extra points.

In this paper, we propose a new method for blind image denoising task. The key factor of success is to employ the Mixture of Gaussian (MoG) model to fit the patches extracted from the image. Since the noise in real image is unknown, we utilize variational Bayesian inference to determine all the parameters, including the number of components. That is, the noise is modeled by a MoG adapted to the testing image. This data driven property makes our model able to deal with blind noise. Then we employ the nonparametric Bayesian model [20] to reconstruct the latent clean structures in each component. Specifically, the beta-Bernoulli process [21,22] is suitable for this task. In our proposed method, the noise in each component is assumed to be Gaussian. The parameters of the beta-Bernoulli process are automatically determined by nonparametric strategies such as the Gibbs sampling. The proposed model is tested on Gaussian noise, mixed Gaussian and impulse noise, and real noise in photographed images.

To summarize, our paper has the following contributions:

- We proposed a novel framework for blind image denoising problem;
- The proposed model is more robust on image denoising tasks than the competing methods;

- We demonstrated that, the performance of the Beta Process Factor Analysis (BPFA) model can be largely improved by structural clustering strategy and Non-local self similarity property;
- We achieve comparableor even better performance on blind image denoising tasks than the competing methods.

The remainder of this paper is organized as follows. Section 2 introduces the related work. Section 3 introduces the patch group based Mixture of Gaussian model inferred by variational Bayesian method. In Sect. 4, we will formulate the proposed PG based nonparametric Bayesian dictionary learning model. Section 5 summarizes the overall algorithm. In Sect. 6, we will present the experimental results, as well as discussions, on blind image denoising tasks. Section 7 concludes this paper.

2   Related Work

Structural clustering are employed by many image denoising methods. For example, both EPLL [7] and PLE [23] utilized the Mixture of Gaussian (MoG) model [24] for clustering similar patches. NCSR [25] utilized the k-means algorithm. The recently proposed Patch Group Prior based Denoising (PGPD) [26] employ the MoG model to learn the non-local self-similarity (NSS) prior of images. However, all these methods need preset the number of clusters. In [27], the Beta-Bernoulli Factor Analysis (BPFA) model is further extended by nonparametric clustering processes such as Dirichlet process (DP) [28] or probit stick-breaking processes (PSBP) [29]. However, the authors in [27] pointed out that, the resulting 'clustering-aided' models achieved similar performance on image denoising to the original BPFA model [27]. In this paper, we demonstrate that, we could indeed improve the performance of Bayesian methods on image denoising tasks, if we utilize the structural clustering strategy properly.

Dictionary learning is very useful for image denoising tasks. The dictionary can be chosen off-the-shelf (wavelets and curvelets) or can be learned from natural image patches. The seminal work of K-SVD [4] has demonstrated that dictionary learning not only can help achieves promising denoising performance, but also can be used in other image processing applications. In image denoising, dictionary learning is commonly combined with the non-local self-similarity (NSS) prior [2,26], sparse prior [5,6,25], and low rank prior [30], etc. Though work well on Gaussian noise removal, the above methods perform poorly on other types of noises, especially noise in real noisy images.

3   Patch Group Based Bayesian Learning for Structural Clustering

3.1   Patch Group Based Non-local Self Similarity Property

Non-local self similarity (NSS) is a common image prior employed by many image restoration methods [2,5,6,25,26,30]. In PGPD [26], the patch group (PG) based

NSS prior is learned on clean natural images for efficient image denoising. In this paper, however, we apply the PG based NSS prior directly on testing noisy images. Given a noisy image, we firstly extract image patches of size $p \times p$. Then, for each image patch, we find the M most similar patches to it in a large enough local window of size $W \times W$. The similarity measurement is based on Euclidean distance, i.e., ℓ_2 norm, which is commonly used in other methods [5,6,25,26,30]. In this work, we set $p = 8$, $M = 6$, $W = 31$. The PG is a set of similar patches $\{\mathbf{x}_m\}_{m=1}^{M}$, in which the $\mathbf{x}_m \in \mathbb{R}^{p^2 \times 1}$ is the mth patch vector. The group mean of this PG is $\boldsymbol{\mu} = \frac{1}{M} \sum_{m=1}^{M} \mathbf{x}_m$. The mth group mean subtracted patch vector is $\bar{\mathbf{x}}_m = \mathbf{x}_m - \boldsymbol{\mu}$. The $\bar{\mathbf{X}} \triangleq \{\bar{\mathbf{x}}_m\}$, $m = 1, ..., M$ is called the PG variations. In Fig. 1, we show one example of PG, PG mean, and the PG variations after group mean subtraction. The PG mean is the main structure of this PG. Here, we proposed to embed this idea into the BPFA model [27] and compare the performance on the Gaussian noise removal. The results are listed in Fig. 2. When comparing the images (b) and (d), we can see that the introduce of PG based NSS prior can really boost the performance on image denoising of the BPFA model. The improvements on PSNR is nearly 0.8dB and on SSIM is nearly 0.03. In particular, the image quality in (d) is much better than that in (b).

| Noisy Patch Group | Patch Group Mean | Patch Group Variations |

Fig. 1. An example of PG, PG mean (duplicated by five), and PG variations after mean subtraction. The PG is from the image 'House' corrupted by Gaussian noise.

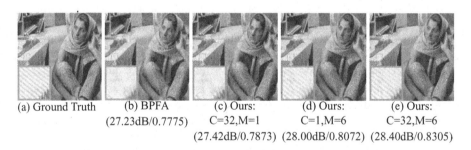

(a) Ground Truth   (b) BPFA   (c) Ours: (d) Ours: (e) Ours:
(27.23dB/0.7775)   C=32,M=1   C=1,M=6   C=32,M=6
(27.42dB/0.7873) (28.00dB/0.8072) (28.40dB/0.8305)

Fig. 2. Denoised images of *Barbara* and PSNR/SSIM results by BPFA and our models (the standard deviation of Gaussian noise is $\sigma = 40$).

3.2   PG Based Bayesian Learning for Structural Clustering

Given a noisy image, we can extracted N PG variations, the m-th variation in the n-th PG is defined as $\bar{\mathbf{x}}_{n,m}$, where $n = 1, ..., N$ and $m = 1, ..., M$. We propose to utilize Mixture of Gaussian (MoG) with C components, to divide the sundry

PG variations into different clusters. The relationship between PG variation $\overline{\mathbf{x}}_{n,m}$ and the MoG model relies on a latent variable defined as $\mathbf{z}_{n,m}$, which is consisted of C elements $z_{n,m,c}$ for $c = 1, ..., C$. If PG variation $\overline{\mathbf{x}}_{n,m}$ belongs to the cth component, $z_{n,m,c} = 1$; and otherwise $z_{n,m,c} = 0$. A natural problem is how to determine C, the number of clusters, as well as other parameters in the MoG model. A reliable solution is to initialize C as a large number and then employ the Dirichlet prior to estimate it [24].

In order to fully estimate the parameters of the MoG model, we can resort to the posterior distribution $p(\{\mathbf{z}_{n,m}\}|\{\overline{\mathbf{x}}_{n,m}\})$ given $\{\mathbf{z}_{n,m}\}$ ($n = 1, ..., N$ and $m = 1, ..., M$). Though its analytical form is not computationally tractable, we could use variational Bayesian inference as an approximational technique. Assuming that the mixing coefficients vector is $\boldsymbol{\pi}$, the conditional distribution of the latent variables $\{\mathbf{z}_{n,m}\}$ is

$$p(\{\mathbf{z}_{n,m}\}|\boldsymbol{\pi}) = \prod_{n=1}^{N} \prod_{m=1}^{M} \prod_{c=1}^{C} \pi_c^{z_{n,m,c}}. \tag{1}$$

The conditional distribution of the PG variations is

$$p(\{\overline{\mathbf{x}}_{n,m}\}|\{\mathbf{z}_{n,m}\}, \{\boldsymbol{\mu}_c\}, \{\boldsymbol{\Sigma}_c\}) = \prod_{n=1}^{N} \prod_{m=1}^{M} \prod_{c=1}^{C} \mathcal{N}(\overline{\mathbf{x}}_{n,m}|\boldsymbol{\mu}_c, \boldsymbol{\Sigma}_c)^{z_{n,m,c}} \tag{2}$$

where $\boldsymbol{\mu}_c$ and $\boldsymbol{\Sigma}_c$ are the corresponding mean vector and covariance matrix of the c-th component of the MoG model. After introducing corresponding priors over the parameters $\{\mathbf{z}_{n,m}\}$, $\{\boldsymbol{\mu}_c\}$ and $\{\boldsymbol{\Sigma}_c\}$, we can write the joint distribution of all the variables by

$$\begin{aligned} p(\{\overline{\mathbf{x}}_{n,m}\}, \{\mathbf{z}_{n,m}\}, \boldsymbol{\pi}, \{\boldsymbol{\mu}_c\}, \{\boldsymbol{\Sigma}_c\}) &= p(\{\overline{\mathbf{x}}_{n,m}\}|\{\mathbf{z}_{n,m}\}, \{\boldsymbol{\mu}_c\}, \\ \{\boldsymbol{\Sigma}_c\}) &\times p(\{\mathbf{z}_{n,m}\}|\boldsymbol{\pi}) \times p(\boldsymbol{\pi}) \times p(\boldsymbol{\mu}_c|\boldsymbol{\Sigma}_c) \times p(\{\boldsymbol{\Sigma}_c\}), \end{aligned} \tag{3}$$

where the mixing coefficients vector $\boldsymbol{\pi}$ is assumed to follow Dirichlet distribution.

Finally, we can perform variational maximization (M step) and expectation estimation of the responsibilities (E step) alternatively. By taking suitable parameters for the Eq. 3, we can re-write the lower bound as a function of the parameters. This lower bound can be used to determine the posterior distribution over the number of components C in the MoG model. Please refer to [24] for more details. We use the Matlab code implemented in Pattern Recognition and Machine Learning Toolbox http://www.mathworks.com/matlabcentral/fileexchange/55826-pattern-recognition-and-machine-learning-toolbox.

3.3 Discussion

Based on above description, we discuss as follows the advantages of the proposed model for PG based modeling directly on noisy images: Firstly, all the parameters of the proposed model are automatically estimated from the noisy image via variational Bayesian inference. This is a major advantage over PGPD

[26]. Secondly, we demonstrated that structural clustering can indeed boost the performance of Bayesian methods such as BPFA [27]. This is demonstrated in the images (b) and (c) of Fig. 2. Thirdly, the Non-local Self Similarity prior can be used to improve the performance of BPFA [27]. This can be demonstrated by comparing the images ((b), (c), (e)) or ((b), (d), (e)).

4   Patch Group Based Bayesian Dictionary Learning

4.1   Truncated Beta-Bernoulli Process for Dictionary Learning

Once we have clustered similar PG variations into different components, we can extract the latent clean PG variations using sparse or low rank image priors. These priors are also frequently employed by many state-of-the-art methods [5–7,25,26,30] for image restoration tasks. We do not fixed the number of atoms in the dictionary learning. This is different from the previous methods, including PGPD. Instead, we set a large number K, which makes our Bayesian model a truncated beta-Bernoulli process. We employ the beta-Bernoulli process [20–22] to seek the sparse priors on infinite feature space.

In this paper, we express noisy PG variations $\overline{\mathbf{X}}$ as

$$\overline{\mathbf{X}} = \mathbf{DW} + \mathbf{V}, \tag{4}$$

where $\overline{\mathbf{X}} \in \mathbb{R}^{p^2 \times M}$, the coefficients vector $\mathbf{W} \in \mathbb{R}^{K \times M}$, and the noise term $\mathbf{V} \in \mathbb{R}^{p^2 \times M}$. The matrix $\mathbf{D} \in \mathbb{R}^{p^2 \times K}$ contains K dictionary atoms. The coefficients matrix $\mathbf{W}$ is represented by $\mathbf{W} = \mathbf{B} \odot \mathbf{S}$. The matrices $\mathbf{B}, \mathbf{S} \in \mathbb{R}^{K \times M}$ are the binary matrix and the coefficients matrix, respectively and $\odot$ is the element-wise product. We denote $\overline{\mathbf{x}}$ as any column of the PG variations $\overline{\mathbf{X}}$ and $\mathbf{w}, \mathbf{b} \in \{0, 1\}^K$, and $\mathbf{s}$ as corresponding columns of the coefficients matrix $\mathbf{W}$, the binary matrix $\mathbf{B}$, and the coefficients matrix $\mathbf{S}$. We denote $\{\mathbf{d}_k\}_{j=1}^{K}$ as columns of the dictionary $\mathbf{D}$, $\{\mathbf{s}_j\}_{j=1}^{n}$ as columns of $\mathbf{S}$, and $\{\mathbf{v}_j\}_{j=1}^{n}$ as columns of the noise matrix $\mathbf{V}$. Each column $\mathbf{w}$ is represented by a binary vector $\mathbf{b} \in \{0, 1\}^K$ and a coefficient vector $\mathbf{s} \in \mathbb{R}^{K \times 1}$, i.e., $\mathbf{w} = \mathbf{b} \odot \mathbf{s}$. Then we can impose suitable priors on these parameters, i.e., $\mathbf{b}$, $\{\mathbf{d}_k\}_{k=1}^{K}$, $\{\mathbf{s}_k\}_{k=1}^{K}$, and $\{\mathbf{v}_j\}_{j=1}^{n}$. The binary vector $\mathbf{b} \in \{0, 1\}^K$ denotes which of the columns (or atoms) in $\mathbf{D}$ are used for the representation of $\overline{\mathbf{x}}$. At beginning, we do not know the suitable K. We can set $K \to \infty$ and impose sparse prior on $\mathbf{b} \in \{0, 1\}^K$ to limit the number of atoms in $\mathbf{D}$ used for representing each PG variation $\overline{\mathbf{x}}$ extracted from noisy images. The beta-Bernoulli process [20–22] provides a convenient way for this purpose. For inference convenience, we impose independent Gaussian priors on $\{\mathbf{d}_k\}_{k=1}^{K}$, $\{\mathbf{s}_k\}_{k=1}^{K}$, and $\{\mathbf{v}_j\}_{j=1}^{n}$.

The dictionary learning model for PG variations is

$$\overline{\mathbf{X}} = \mathbf{DW} + \mathbf{V}, \mathbf{W} = \mathbf{B} \odot \mathbf{S}, \tag{5}$$

where $\odot$ is element-wise product. The beta-Bernoulli process [22] for binary vector is:

$$\mathbf{b} \sim \prod_{k=1}^{K} Bernoulli(\pi_k), \boldsymbol{\pi} \sim \prod_{k=1}^{K} Beta(\frac{a}{K}, \frac{b(K-1)}{K}), \qquad (6)$$

$$\mathbf{d}_k \sim \mathcal{N}(0, P^{-1}\mathbf{I}_P), \mathbf{s}_j \sim \mathcal{N}(0, \gamma_s^{-1}\mathbf{I}_K), \mathbf{v}_j \sim \mathcal{N}(0, \gamma_v^{-1}\mathbf{I}_P), \qquad (7)$$

$$\gamma_s \sim Gamma(c, d), \gamma_v \sim Gamma(e, f), \qquad (8)$$

where $\{a, b\}, \{c, d\}, \{e, f\}$ are the corresponding hyper parameters of the parameters $\boldsymbol{\pi}$, γ_s, and γ_v in the conjugate hyper priors in Eqs. (6), (8). The inference procedures we take for the Bayesian model is Gibbs sampling [31] and we ignore the procedures here. As we can see, the noise estimation is integrated into the overall model in Gibbs sampling process. That is the reason why the proposed algorithm can be used to deal with blind noise. This is also different from the other non-blind image denoising methods [2,5–7,25,26,30], since they do not have the ability to estimate the noise within the noisy images.

We compare the proposed model with the BPFA model on the dictionary elements. With the help of PG based NSS prior and structural clustering, the dictionary learned by the proposed model is more representative and discriminative than that learned by the BPFA model [27]. Take the image "Barbara" for an example, it is corrupted by Gaussian noise with $\sigma = 40$. The initial number of dictionary elements is 512 for both methods. In Fig. 2(a), we demonstrate the dictionary elements learned by the BPFA model [27]. In the (b), (c), and (d) of Fig. 2, we demonstrate the dictionary elements of three components learned by the proposed method. Noted that the BPFA learned from the original patches while the proposed method learned from patch group variations. It can be seen that these dictionary elements express the latent clean structures of the noisy PG variations in this component. For different Gaussian components, we can see that the number of elements are automatically determined by the nonparametric Bayesian inference. The dictionary elements in the subfigures (a) to (d) in Fig. 3 are supplemented by black patches to make sure that these subfigures are square.

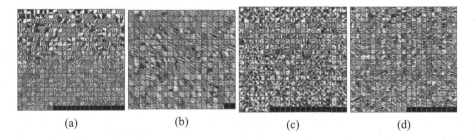

(a)   (b)   (c)   (d)

Fig. 3. The dictionary elements learned by the BPFA model ((a)) and the proposed method ((b), (c), and (d)).

Algorithm 1. The Overall Algorithm

Input: Noisy image $\mathbf{y}$, $a = b = 1, c = d = e = f = 10^{-6}$

1. Extract non-local similar patch groups $\{\mathbf{X}\}$;
2. For each PG, calculate its mean $\boldsymbol{\mu}_x$ and form PG variations $\overline{\mathbf{X}}$;
3. Estimate the parameters in MoG model via variational Bayesian inference;
4. **for** each cluster
5.   **for** $t = 1 : IteNum$ **do** Gibbs Sampling:
6.   Sample $\mathbf{b}$ from equation (6)
7.   Sample $\mathbf{d}_k \sim \mathcal{N}(0, P^{-1}\mathbf{I}_P)$
8.   Sample $\mathbf{s}_j \sim \mathcal{N}(0, \gamma_s^{-1}\mathbf{I}_K), \gamma_s \sim Gamma(c, d)$
9.   Sample $\mathbf{v}_j \sim \mathcal{N}(0, \gamma_v^{-1}\mathbf{I}_P), \gamma_v \sim Gamma(e, f)$
10.   **end for**
11.   Recover each patch group via $\hat{\mathbf{X}} = \mathbf{DW} + \boldsymbol{\mu}_x$;
12. **end for**
13. Average the recovered patch groups to form the recovered image $\hat{\mathbf{y}}$;
14. **Output:** The recovered image $\hat{\mathbf{y}}$

4.2 Discussion

Our method are different from the method of BPFA [27]. Basicly, our model learns flexible number of dictionary atoms on patch groups while BPFA fixes the number of dictionary atoms learned over patches. This modification makes the dictionary more representative and discriminative than those in BPFA [27]. What's more, BPFA is only applied in (non-blind) denoising Gaussian noise while our model is applied to (blind) denoising multiple noises such as Gaussian noise, mixed Gaussian and impulse noise, and real noise. This will be demonstrated in the experimental section.

5 Summary of the Overall Algorithm

The overall algorithm is consisted of three parts. The first is the PG based Bayesian learning for automatically estimating the parameters of the MoG model, including the number of components. After the clustering, we employ a divide-and-conquer strategy. The second part is, for each component, we employ the PG based nonparametric Bayesian dictionary learning to sample the dictionary $\mathbf{D}$ as well as coefficients matrix $\mathbf{W}$. Then, the final denoised PG variations is simply calculated as $\hat{\mathbf{X}} = \mathbf{DW}$. That is the key factor why the proposed algorithm is able to deal with blind image denoising. After this is done for each component, the third part is to recover the denoised image by aggregating the denoised PGs. The overall algorithm is summarized in Algorithm 1. The Matlab source code of our PGPD algorithm can be downloaded at http://www4.comp. polyu.edu.hk/~csjunxu/code/PGBL_BID.zip.

6 Experiments

In this section, we perform image denoising experiments on various types of noise, including synthetic noise and real noise in real world images. The synthetic noise

Fig. 4. The 20 widely used test images.

includes additive white Gaussian noise (AWGN), mixed Gaussian and random value impulse noise (RVIN). The synthetic image denoising experiments are performed on 20 widely used natural images listed in Fig. 4. In the synthetic noise removal experiments, we compare the proposed method with other state-of-the-art methods such as [5,9,11,18,26,27,30]. The BPFA [27] BM3D [5], PGPD [26], and WNNM [30] are designed especially for Gaussian noise removal. The Two Phase [9] and WESNR [11] are designed especially for mixed Gaussian and impulse noise removal. The "Noise Clinic" method [18] is designed especially for real noise removal. It is a state-of-the-art blind denoising method. We evaluate all these methods on PSNR, SSIM [32], and visual quality. On real noise removal, besides the above methods, we also compare with the commercial software Neat Image [33]. This software is embedded in Photoshop CS, a famous commercial software for image processing. The codes or executive package of these methods are provided on the corresponding websites. Since there is no ground truth in real noise removal, we only compare the visual quality of the recovered images by these methods.

6.1   Implementation Details

For the BM3D [5], WNNM [30], and PGPD [26], the input standard derivation of the noise is a key parameter. We employ a robust noise estimation method [34] to estimate the noise standard derivation for these methods. The methods Two Phase [9] and WESNR [11] do not need the noise level as input when performing denoising tasks. But they need some preprocessing for the noise removal tasks. However, our proposed method doesn't need noise estimation nor image preprocessing.

In the proposed method, we set the initial number of components C as 32. The final number of components will be automatically determined by variational Bayesian inference introduced in Sect. 2. The patch size is fixed as 8×8, so the dimension of each patch vector is 64. The number of dictionary atoms K is set to be 256 when number of PG variations in this component is less than 10^4, otherwise the number is set to 512. The hyperparameters are fixed as $a = b = 1$, $c = d = e = f = 10^{-6}$ in all the experiments. We do not tune these parameters in our experiments. The number of patches in a group is set to 6. We set the maximal iteration number as 10. The $IteNum$ is set as 50. The proposed algorithm will be terminated when the noise variance of each component is less than or equal to 1.

6.2   Additive White Gaussian Noise Removal

Here, we compare the proposed method on Gaussian noise removal with other competing methods: BPFA [27], BM3D [5], and PGPD [26], WNNM [30], Two Phase [9], WNSER [11], Noise Clinic [18]. As a common experimental setting, we add additive white Gaussian noise with zero mean and standard deviation σ to the test images. The denoising experiments are performed on multiple noise levels of $\sigma = 30, 40, 50, 75$.

The averaged results on PSNR and SSIM are listed in Tables 1 and 2. We can see that the PSNR and SSIM results of our proposed method are much better than the BPFA, Two Phase, and WESNR, Noise Clinic methods. The results of the proposed method are comparable to BM3D when the noise levels are higher than 30. Though getting inferior results on PSNR and SSIM when compared to WNNM and PGPD, the proposed method can achieve similar or even better performance when compared with other methods. For instance, from Figs. 4 and 5, we can see that the proposed method generates better image quality and less artifacts on the image "House" and "Hill" than the other methods. Considering that the proposed method is fully blind, it is more convincing to see that our proposed method achieves better image quality results than the Noise Clinic [18], which is a state-of-the-art blind image denoising method. We want to mention again that BM3D, PGPD, and WNNM can only work well on the Gaussian noise, which they are designed for, but perform poorly on other types of noise, which will be demonstrated later.

Table 1. Average PSNR(dB) results of different algorithms on 20 natural images corrupted by Gaussian noise.

σ	BPFA	BM3D	WNNM	PGPD	Two Phase	WESNR	Noise Clinic	Ours
30	28.81	29.11	29.35	29.13	18.84	27.44	26.81	28.65
40	27.45	27.68	28.03	27.88	16.52	25.07	24.90	27.59
50	26.38	26.80	27.06	26.89	14.80	22.02	23.47	26.68
75	24.40	25.04	25.25	25.11	11.95	9.02	21.31	24.91

Table 2. Average SSIM results of different algorithms on 20 natural images corrupted by Gaussian noise.

σ	BPFA	BM3D	WNNM	PGPD	Two Phase	WESNR	Noise Clinic	Ours
30	0.7988	0.8108	0.8156	0.8089	0.3137	0.7430	0.6520	0.7971
40	0.7576	0.7706	0.7784	0.7747	0.2361	0.6063	0.5631	0.7668
50	0.7216	0.7430	0.7510	0.7435	0.1855	0.4483	0.4990	0.7382
75	0.6470	0.6803	0.6897	0.6825	0.1147	0.1749	0.4289	0.6766

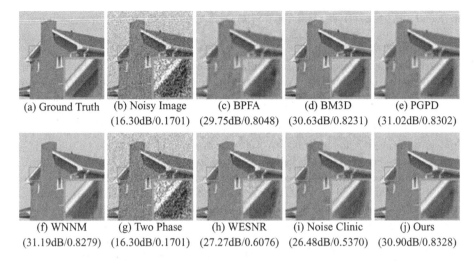

(a) Ground Truth	(b) Noisy Image	(c) BPFA	(d) BM3D	(e) PGPD
	(16.30dB/0.1701)	(29.75dB/0.8048)	(30.63dB/0.8231)	(31.02dB/0.8302)

(f) WNNM	(g) Two Phase	(h) WESNR	(i) Noise Clinic	(j) Ours
(31.19dB/0.8279)	(16.30dB/0.1701)	(27.27dB/0.6076)	(26.48dB/0.5370)	(30.90dB/0.8328)

Fig. 5. Denoised images of *House* and PSNR/SSIM results by different methods (the standard deviation of Gaussian noise is $\sigma = 40$).

(a) Ground Truth	(b) Noisy Image	(c) BPFA	(d) BM3D	(e) PGPD
	(14.68dB/0.1357)	(26.81dB/0.6530)	(27.19dB/0.6745)	(27.22dB/0.6702)

(f) WNNM	(g) Two Phase	(h) WESNR	(i) Noise Clinic	(j) Ours
(27.34dB/0.6772)	(14.68dB/0.1357)	(23.50dB/0.4218)	(25.01dB/0.5128)	(27.02dB/0.6618)

Fig. 6. Denoised images of *Hill* and PSNR/SSIM results by different methods (the standard deviation of Gaussian noise is $\sigma = 50$).

6.3   Mixed Gaussian and Impulse Noise Removal

Here, we compare the proposed method on mixed Gaussian and impulse noise with the compared methods [5,9,11,18,26,27,30]. We consider Random Value Impulse Noise (RVIN) here. The pixels in the testing image corrupted by RVIN is distributed between 0 and 255. This is much harder than salt and pepper noise, which is only 0 or 255 values. In the synthetic noise, the standard derivations of the Gaussian noise are $\sigma = 10, 20$ and the ratios of the impulse noise are 0.15, 0.30, respectively. The RVIN noise is generated by the "impulsenoise" function used by WESNR [11].

For BPFA [27], BM3D [5], and PGPD [26], WNNM [30], they are designed to deal with Gaussian noise. Hence, we still employ the noise estimation method [34] to estimate the noise levels σ. We also compare the proposed method with the methods Two Phase [9] and WESNR [11], which are designed especially for the mixed Gaussian and RVIN noise. Noted that both Two Phase [9] and WESNR [11] employ Adaptive Median Filter (AMF) [35] to preprocess the image before performing image denoising task. We also compare with the Noise Clinic [18].

The results on PSNR and SSIM are listed in Tables 3 and 4. As we can see, the performance of the proposed method is comparable or better than other methods. For the visual quality comparison, the proposed method can generate better results than other methods. Take the image "Couple" for an example, from the Fig. 7, the proposed method removes the noise clearly while all other methods remain some noise or generate some artifacts. From the results on "Barbara" listed in Fig. 8, the proposed method achieves higher SSIM and generates better image quality than all other methods. We have to mention again that the Two Phase and WESNR are two state-of-the-art methods designed especially for the mixed Gaussian and impulse noise. But they perform poorly on other types of noises such as Gaussian.

Table 3. Average PSNR(dB) results of different algorithms on 20 natural images corrupted by mixed of Gaussian and RVIN noise.

σ, Ratio	BPFA	BM3D	WNNM	PGPD	Two Phase	WESNR	Noise Clinic	Ours
10, 0.15	17.12	25.18	22.98	25.41	27.28	27.37	18.66	27.17
10, 0.30	14.19	21.80	21.40	21.74	26.12	21.50	16.44	22.17
20, 0.15	17.62	25.13	23.57	25.33	24.43	27.24	19.66	26.12
20, 0.30	17.61	21.73	21.40	21.64	23.61	22.69	14.46	21.89

Table 4. Average SSIM results of different algorithms on 20 natural images corrupted by mixed of Gaussian and RVIN noise.

σ, Ratio	BPFA	BM3D	WNNM	PGPD	Two Phase	WESNR	Noise Clinic	Ours
10, 0.15	0.2749	0.7037	0.5806	0.7242	0.7091	0.7499	0.3198	0.7459
10, 0.30	0.1576	0.6444	0.6038	0.6405	0.6783	0.4970	0.2042	0.6559
20, 0.15	0.2821	0.7132	0.6190	0.7220	0.5468	0.7568	0.3423	0.7312
20, 0.30	0.2821	0.6414	0.6310	0.6371	0.5183	0.5871	0.1346	0.6470

(a) Ground Truth (b) Noisy Image   (c) BPFA   (d) BM3D   (e) PGPD
                  (17.35dB/0.2852) (17.80dB/0.2871) (25.70dB/0.7126) (25.75dB/0.7208)

(f) WNNM   (g) Two Phase   (h) WESNR   (i) Noise Clinic (j) Ours
(23.81dB/0.6162) (27.47dB/0.7186) (28.40dB/0.7807) (19.48dB/0.3378) (27.26dB/0.7438)

Fig. 7. Denoised images of *Couple* by different methods (the mixed Gaussian and Random Value Impulse Noise is with $\sigma = 10$ and ratio 0.15). The images are better viewed by zooming in on screen.

(a) Ground Truth (b) Noisy Image   (c) BPFA   (d) BM3D   (e) PGPD
                  (16.09dB/0.2644) (17.73dB/0.3199) (25.58dB/0.7552) (25.61dB/0.7633)

(f) WNNM   (g) Two Phase   (h) WESNR   (i) Noise Clinic (j) Ours
(24.03dB/0.6796) (22.66dB/0.5027) (26.74dB/0.7769) (19.87dB/0.3874) (26.58dB/0.7853)

Fig. 8. Denoised images of *Barbara* by different methods (the mixed Gaussian and RVIN noise is with $\sigma = 20$ and ratio 0.15). The images are better viewed by zooming in on screen.

6.4   Real Noisy Image Denoising

In this section, we will test the proposed method on real noise removal. Since the color images are in RGB channels, we firstly transform the color images into YCbCr channels, then perform denoising on the Y channel, and finally transform the denoised YCbCr channel image back into RGB channels. The denoised images are cropped to size of 800 × 600 for better visualization. We do not compare with WNNM [30] and Two Phase [9] here since they achieve worse denoising quality than the other methods such as BM3D [5], WESNR [11], and Noise Clinic [18]. For BM3D and PGPD, the input noise level σ is still estimated by [34]. We also compare with the Neat Image [33], a commercial software embedded in Photoshop CS. In this paper, we take three real noisy images for examples, which are "SolvayConf1927", "Girls", and "Windmill". The image "SolvayConf1927" is an old image provided by the Noise Clinic website on IPOL [36] while the other two images are provided by the Neat Image website [33]. The denoised images are evaluated in Figs. 9, 10 and 11.

From the results listed in Figs. 9, 10, and 11, we can see that the proposed method can remove real noise while preserving details better than other methods.

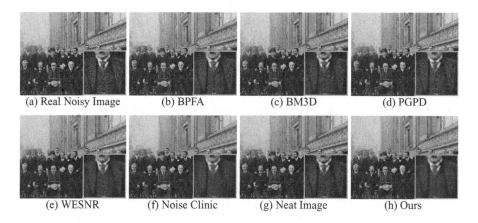

Fig. 9. Denoised images of the old image "SolvayConf1927" by different methods. The images are better viewed by zooming in on screen.

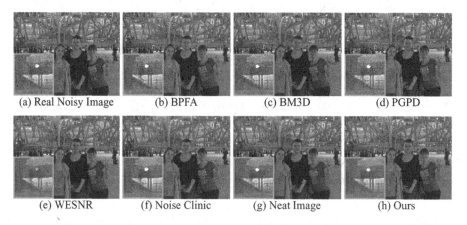

Fig. 10. Denoised images of the image "Girls" by different methods. The images are better viewed by zooming in on screen.

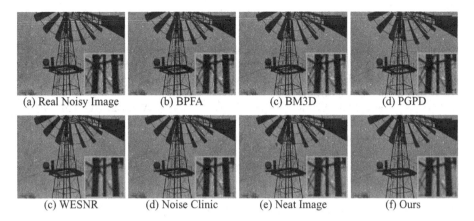

(a) Real Noisy Image   (b) BPFA   (c) BM3D   (d) PGPD

(c) WESNR   (d) Noise Clinic   (e) Neat Image   (f) Ours

Fig. 11. Photo courtesy of Alexander Semenov. Denoised images of the image "Windmill" by different methods. The images are better viewed by zooming in on screen.

For example, in the Fig. 10, the image "Girls" are damaged by night shot noise and hardly recoverable. However, the proposed method can denoise the heavy noise, restore the details under the noise, and make the image looks better. The details of the building in the background is zoomed in to demonstrate the advantages of our proposed method than other compared methods. Another example is the image "Windmill", the noise in night sky is traditionally difficult to remove, since it is very hard to distinct the image noise from the faint stars in the sky. Our method not only reduces the noise while generating much less artifacts, but also preserves details, i.e., the faint stars, better than other methods.

7   Conclusion

Image denoising is a commonly encountered problem in real life. However, most denoising methods [5–7,9,11,26,27,30] need to know the noise distributions, such as Gaussian noise or mixed Gaussian and impulse noise, as well as the noise intensity. In this paper, we developed a novel blind image denoising method by patch group (PG) based nonlocal self-similarity prior modeling. We modeled the PG variations by Mixture of Gaussians [24] whose parameters, including its number of components, are inferred by variational Bayesian method. For each component, we employed nonparametric Bayesian dictionary learning [21,22,27] to reconstruct the latent clean images. The proposed method can deal with unknown or arbitrary noise without knowing the noise distribution. From the experimental results on removing Gaussian noise, mixed Gaussian and random value impulse noise, and the noise in real images, we demonstrated that the proposed method achieves comparable PSNR/SSIM measurements and even better visual quality than those methods [5,9,11,18,26,27,30,33], which are specially designed for specific types of noises.

References

1. Rudin, L.I., Osher, S., Fatemi, E.: Nonlinear total variation based noise removal algorithms. Phys. D: Nonlinear Phenomena **60**, 259–268 (1992)
2. Buades, A., Coll, B., Morel, J.M.: A non-local algorithm for image denoising. In: CVPR, pp. 60–65 (2005)
3. Roth, S., Black, M.J.: Fields of experts. Int. J. Comput. Vis. **82**, 205–229 (2009)
4. Elad, M., Aharon, M.: Image denoising via sparse and redundant representations over learned dictionaries. IEEE Trans. Image Process. **15**, 3736–3745 (2006)
5. Dabov, K., Foi, A., Katkovnik, V., Egiazarian, K.: Image denoising by sparse 3-D transform-domain collaborative filtering. IEEE Trans. Image Process. **16**, 2080–2095 (2007)
6. Mairal, J., Bach, F., Ponce, J., Sapiro, G., Zisserman, A.: Non-local sparse models for image restoration. In: ICCV, pp. 2272–2279 (2009)
7. Zoran, D., Weiss, Y.: From learning models of natural image patches to whole image restoration. In: ICCV, pp. 479–486 (2011)
8. Burger, H.C., Schuler, C.J., Harmeling, S.: Image denoising: can plain neural networks compete with bm3d. In: 2012 IEEE Conference on Computer Vision and Pattern Recognition (CVPR), pp. 2392–2399. IEEE (2012)
9. Cai, J.F., Chan, R.H., Nikolova, M.: Fast two-phase image deblurring under impulse noise. J. Math. Imaging Vis. **36**, 46–53 (2010)
10. Xiao, Y., Zeng, T., Yu, J., Ng, M.K.: Restoration of images corrupted by mixed gaussian-impulse noise via 11–10 minimization. Pattern Recogn. **44**, 1708–1720 (2011)
11. Jiang, J., Zhang, L., Yang, J.: Mixed noise removal by weighted encoding with sparse nonlocal regularization. IEEE Trans. Image Process. **23**, 2651–2662 (2014)
12. Portilla, J., Strela, V., Wainwright, M., Simoncelli, E.: Image denoising using scale mixtures of Gaussians in the wavelet domain. IEEE Trans. Image Process. **12**, 1338–1351 (2003)
13. Portilla, J.: Full blind denoising through noise covariance estimation using gaussian scale mixtures in the wavelet domain. In: 2004 International Conference on Image Processing, ICIP 2004, vol. 2, pp. 1217–1220 (2004)
14. Huber, P.J.: Robust Statistics. Springer (2011)
15. Rabie, T.: Robust estimation approach for blind denoising. IEEE Trans. Image Process. **14**, 1755–1765 (2005)
16. Liu, C., Szeliski, R., Kang, S.B., Zitnick, C.L., Freeman, W.T.: Automatic estimation and removal of noise from a single image. IEEE Trans. Pattern Anal. Mach. Intell. **30**, 299–314 (2008)
17. Gong, Z., Shen, Z., Toh, K.C.: Image restoration with mixed or unknown noises. Multiscale Model. Simul. **12**, 458–487 (2014)
18. Lebrun, M., Colom, M., Morel, J.M.: Multiscale image blind denoising. IEEE Trans. Image Process. **24**, 3149–3161 (2015)
19. Lebrun, M., Buades, A., Morel, J.M.: A nonlocal bayesian image denoising algorithm. SIAM J. Imaging Sci. **6**, 1665–1688 (2013)
20. Hjort, N.L.: Nonparametric bayes estimators based on beta processes in models for life history data. Ann. Stat. **18**, 1259–1294 (1990)
21. Thibaux, R., Jordan, M.I.: Hierarchical beta processes and the Indian buffet process. In: International Conference on Artificial Intelligence and Statistics, pp. 564–571 (2007)

22. Paisley, J., Carin, L.: Nonparametric factor analysis with beta process priors. In: Proceedings of the 26th Annual International Conference on Machine Learning, pp. 777–784. ACM (2009)

23. Yu, G., Sapiro, G., Mallat, S.: Solving inverse problems with piecewise linear estimators: from Gaussian mixture models to structured sparsity. IEEE Trans. Image Process. **21**, 2481–2499 (2012)

24. Bishop, C.M.: Pattern Recognition and Machine Learning. Springer, New York (2006)

25. Dong, W., Zhang, L., Shi, G., Li, X.: Nonlocally centralized sparse representation for image restoration. IEEE Trans. Image Process. **22**, 1620–1630 (2013)

26. Xu, J., Zhang, L., Zuo, W., Zhang, D., Feng, X.: Patch group based nonlocal self-similarity prior learning for image denoising. In: The IEEE International Conference on Computer Vision (ICCV), pp. 244–252 (2015)

27. Zhou, M., Chen, H., Ren, L., Sapiro, G., Carin, L., Paisley, J.W.: Non-parametric bayesian dictionary learning for sparse image representations. In: NIPS, pp. 2295–2303 (2009)

28. Ferguson, T.S.: A bayesian analysis of some nonparametric problems. Ann. Stat. **1**, 209–230 (1973)

29. Ren, L., Du, L., Carin, L., Dunson, D.: Logistic stick-breaking process. J. Mach. Learn. Res. **12**, 203–239 (2011)

30. Gu, S., Zhang, L., Zuo, W., Feng, X.: Weighted nuclear norm minimization with application to image denoising. In: CVPR, pp. 2862–2869 (2014)

31. Zhou, M., Chen, H., Paisley, J., Ren, L., Li, L., Xing, Z., Dunson, D., Sapiro, G., Carin, L.: Nonparametric bayesian dictionary learning for analysis of noisy and incomplete images. IEEE Trans. Image Process. **21**, 130–144 (2012)

32. Wang, Z., Bovik, A.C., Sheikh, H.R., Simoncelli, E.P.: Image quality assessment: from error visibility to structural similarity. IEEE Trans. Image Process. **13**, 600–612 (2004)

33. ABSoft, N.: Neat image. https://ni.neatvideo.com/home

34. Liu, X., Tanaka, M., Okutomi, M.: Single-image noise level estimation for blind denoising. IEEE Trans. Image Process. **22**, 5226–5237 (2013)

35. Hwang, H., Haddad, R.: Adaptive median filters: new algorithms and results. IEEE Trans. Image Process. **4**, 499–502 (1995)

36. Lebrun, M., Colom, M., Morel, J.M.: The noise clinic: a blind image denoising algorithm. http://www.ipol.im/pub/art/2015/125/. Accessed 28 Jan 2015

Low-Rank Tensor Recovery and Alignment Based on ℓ_p Minimization

Kaifei Zhang[1], Di Wang[1(✉)], Xiaoqin Zhang[1], Nannan Gu[2], Hongxing Jiang[1], and Xiuzi Ye[1]

[1] College of Mathematics and Information Science, Wenzhou University, Zhejiang, China
{zhangkaifei,wangdi,xqzhang,hxjiang,yexiuzi}@wzu.edu.cn
[2] Capital University of Economics and Business, Beijing, China
nngu@cueb.edu.cn

Abstract. In this paper, we propose a framework of non-convex low-rank recovery and alignment for arbitrary tensor data. Specially, by using Schatten-p ($0 < p < 1$, the same below) norm and ℓ_p norm to relax the rank function and ℓ_0 norm respectively, the model requires much weaker incoherence conditions to guarantee a successful recovery than the common used nuclear norm and ℓ_1 norm. At the same time, we adopt a set of transformations which acts on the images of the tensor data to compensate the possible misalignments of images. By solving the optimal transformations, the strict alignments of the images are achieved in the low-rank recovery process. Furthermore, we propose an efficient algorithm based on the method of Alternating Direction Method of Multipliers (ADMM) for the non-convex optimization problem. The extensive experiments on the artificial data sets and real image data sets show the superiority of our method in image alignment and denoising.

1 Introduction

With the development of technology and industry, human beings have a large number of imaging devices (such as mobile phones, cameras, surveillance and medical imaging equipment, and so on). This leads to a dramatic increase in the amount of visual data. However, due to the influence of significant variation, partial occlusion, as well as many other degradation factors, the obtained images always have been polluted by noises. The polluted images will bring great difficulty in image segmentation, matching and classification. As a consequence, it is desirable to develop tools that can recover clean images from polluted images.

Recent works have shown that some visual data can be decomposed into a low dimensional structure and a sparse irregular pattern. For example, under the Lambertian assumption, images of a subject with a fixed pose and varying illumination lie close to a linear subspace of dimension 9 [1], and thus the collection of these images has the low-rank structure, and the shadows can be considered as the sparse pattern. In the video denoising, the data constructed by similar

© Springer International Publishing AG 2017
C.-S. Chen et al. (Eds.): ACCV 2016 Workshops, Part I, LNCS 10116, pp. 96–110, 2017.
DOI: 10.1007/978-3-319-54407-6_7

frames is low-rank, and the noise in the frames is sparse. Based on the above idea, John Wright *et al.* [2] propose the model of robust principal component analysis (RPCA), which can recover the principal components of a data matrix even though a fraction of its entries is arbitrarily corrupted. The RPCA model can be expressed as the following optimization problem:

$$\begin{cases} \min_{A,E} \text{rank}(A) + \lambda \|E\|_0 \\ \text{s.t.} \ \ A + E = D \end{cases} \tag{1}$$

where D is an erroneous observation matrix, A and E respectively represent the correspondent structured part and irregular part, $\text{rank}(\cdot)$ is the rank function, $\|\cdot\|_0$ counts the number of nonzero entries in the error matrix E, and $\lambda > 0$ is a parameter that trades off the rank of the solution versus the sparsity of the error. The rank penalty is in fact the ℓ_0 norm of the matrix singular values, then we call this type of problem as ℓ_0 minimizer. Because both rank function and ℓ_0 norm are nonconvex and discontinuous, the optimization problem (1) is NP-hard. A recent convex relaxation technique is to use the nuclear norm and ℓ_1 norm, which are the convex envelops of the rank function and ℓ_0 norm respectively, and thus the principle component pursuit (PCP) model is obtained as following formula:

$$\begin{cases} \min_{A,E} \|A\|_* + \lambda \|E\|_1 \\ \text{s.t.} \ \ A + E = D \end{cases} \tag{2}$$

where $\|\cdot\|_*$ is the nuclear norm and $\|\cdot\|_1$ is ℓ_1 norm. Since the nuclear norm penalty is defined as the ℓ_1 norm of matrix singular values, we call this type of problem as ℓ_1 minimizer. It is shown that the PCP model is able to recover the original low-rank matrix from the unknown random error with the probability of 1 as long as the strong incoherence conditions are satisfied [3,4]. Ganesh *et al.* [5] establish a matrix recovery model with Gauss noise. Zhou *et al.* [6] consider the problem of recovering the low-rank matrix from any high density error. By choosing the proper weighting factor, the low-rank matrix can be successfully recovered with high probability. Ji *et al.* [7] propose a robust video restoration algorithm based on RPCA, which achieves good performance for video denoising and video retrieval. By adopting a set of image transformations in RPCA, Peng *et al.* [8] propose a robust alignment by sparse and low-rank (RASL) decomposition for linearly correlated images. Zhang *et al.* [9] effectively extract a rich class of transform invariant low-rank textures (TILT) from 2D images despite significant distortion and warping. Guo *et al.* [10] consider the low rank of the image sequence and the low-rank texture of each image simultaneously, and propose a method of video editing to deal with temporal alignment, pose rectification, as well as precise recovery of the occlusion. Zhang *et al.* [11] propose a better characterization of the low-rank structure of the matrix, which is called truncated nuclear norm. It minimizes the smallest $N - r$ singular values, where N is the number of singular values and r is the rank of the matrix. In this way, the rank of the matrix can be better approximated than the nuclear norm.

In the two-dimensional case, i.e. the matrix case, the "rank" is a good characterization of sparsity and plays an important part in capturing the global information of visual data. However, most of the existing high-dimensional visual data either has the natural form of tensor, such as videos and multi-channel images. When applied to the data of higher-order tensorial form, they are only able to harness one type of low-dimensional structure at a time, and are not able to exploit the low-dimensional tensorial structures in the data. Liu *et al.* [12] point out that important structures or useful information will very often be lost if we process higher-order tensors as a 2D matrix. Therefore, it is desirable to extend the framework of low-rank matrix recovery to tensor space [12,13]. Inspired by the Tuncker decomposition [14], the tensor rank can be defined as a linear combination of the ranks of a set of matrices unfolded from the tensor. Huang *et al.* [15] propose a class of convex tensor recovery models that can be proved to guarantee exact recovery under certain conditions. Li *et al.* [16] derive a method for the decomposition of an observed tensor into a low-dimensional structure plus unbounded but sparse irregular patterns, and form the following problem:

$$\begin{cases} \min\limits_{\mathcal{A},\mathcal{E}} \sum\limits_{i=1}^{N} \alpha_i \text{rank} \left(\mathcal{A}_{(i)} \right) + \lambda \left\| \mathcal{E} \right\|_0 \\ \text{s.t. } \left\| \mathcal{A} + \mathcal{E} - \mathcal{D} \right\|_F^2 \leq \varepsilon^2 \end{cases} \tag{3}$$

where $\mathcal{A}, \mathcal{D}, \mathcal{E} \in R^{I_1 \times I_2 \times \cdots \times I_N}$ are N-order tensors, $\mathcal{A}_{(i)} \in R^{I_i \times \left(\Pi_{k \neq i} I_k \right)}$ is the mode-i unfolding matrix of $\mathcal{A}$. Applying the same relaxation to problem (3) yields the following convex problem:

$$\begin{cases} \min\limits_{\mathcal{A},\mathcal{E}} \sum\limits_{i=1}^{N} \alpha_i \left\| \mathcal{A}_{(i)} \right\|_* + \lambda \left\| \mathcal{E} \right\|_1 \\ \text{s.t. } \left\| \mathcal{A} + \mathcal{E} - \mathcal{D} \right\|_F^2 \leq \varepsilon^2 \end{cases} \tag{4}$$

They then develop a block coordinate descent (BCD) based algorithm to efficiently solve the above convex problem. Zhang *et al.* [17] propose a general method for recovering low-rank three-order tensors, which aligns and rectifies images with distortion and partial polluted.

Although the ℓ_1 minimization overcomes the computational difficulty of the original model, it needs to impose strong incoherence conditions to guarantee a exact recovery to the sparsest case. In comparison, non-convex minimization problems, such as those with Schatten-p norm and ℓ_p norm, require much weaker incoherence conditions and smaller signal to noise ratio to guarantee a successful recovery [18–22]. Hence, l_p regularizations serve as better alternatives to the popular ℓ_1 ones. Marjanovic and Solo [23] establish a bridge between the rank function and the nuclear norm, and propose the ℓ_p penalized least squares problem for matrix completion. On the other hand, the data constructed by correlated images satisfy low-rank property only in the condition that the images are well aligned. But this is not guaranteed in many different computer vision data. Images of the same object or scene can appear drastically different even under moderate changes in the object's position or pose with respect to the camera. The above low-rank models break down if the images are even slightly misaligned with respect to each other.

In order to solve the above problems, this paper proposes a non-convex approximation model, named low-rank tensor recovery and alignment based on ℓ_p minimization. Specifically, the contributions of this paper are as follows:

- We consider the decomposition of an observed tensor data into a low dimensional structure and a sparse irregular pattern. By seeking optimal dimension and basis for each mode and separating the irregular patterns, the low-dimensional structure of the tensor data could be automatically explored. It is more effective than the low-rank matrix recovery models.
- We use Schatten-p and ℓ_p norm to better approximate the rank function and ℓ_0 norm respectively, and this ensures that the original signal can be recovered successfully under a weaker non-coherence condition.
- We adopt a set of transformations which act on the images to compensate for possible misalignments of the related images.
- We further develop an efficient iterative procedure to solve the proposed non-convex approximation model by using the ADMM method.
- A large number of experiments on artificial data and real data show that our algorithm is effective in image alignment and denoising.

The rest of this paper is organized as follows. In Sect. 2, we propose a low rank tensor recovery and alignment model, and give an efficient optimization algorithm based on ADMM method. In Sect. 3, experiments are conducted on benchmark data sets to illustrate the validity and effectiveness of the proposed method. Conclusion is given in Sect. 4.

2 Low Rank Tensor Recovery and Alignment Based on ℓ_p Minimization

2.1 The Model

We first consider a simple non-convex model, in which the nuclear norm and l_1 norm in problem (4) are replaced by Schatten-p norm and l_p norm respectively, and obtain the following formula of low rank tensor recovery

$$
\begin{cases}
\min_{\mathcal{A}, \mathcal{E}} \sum_i \alpha_i \left\| \mathcal{A}_{(i)} \right\|_p^p + \lambda \left\| \mathcal{E} \right\|_{p,p}^p \\
\text{s.t.} \quad \mathcal{A} + \mathcal{E} = \mathcal{D}
\end{cases}
\tag{5}
$$

Here, $\left\| \mathcal{A}_{(i)} \right\|_p^p = \sum_k \left(\sigma_k^{(i)} \right)^p$ and $\left\| \mathcal{E} \right\|_{p,p}^p = \left\| \mathcal{E}_{(i)} \right\|_{p,p}^p = \sum_{j,k} \left| e_{jk}^{(i)} \right|^p$ for $\forall i = 1, \cdots, N$, where $\sigma_k^{(i)}$ is the k-th singular value of $\mathcal{A}_{(i)}$ and $e_{jk}^{(i)}$ the element of $\mathcal{E}_{(i)}$ in position (j, k).

In model (5), an explicit assumption is that the tensor constructed by images is well aligned. However, for most practical data, such as video and face images, precise alignments are not always guaranteed and even small misalignments will break the low-rank structure of the data. Hence, we need to add some image transformations in model (5). Because most practical data of image sequence we experiment with can be consider as 3-order tensors, we focus on 3-order tensors to study the low-rank recovery problem without loss of generality. Consider an erroneous 3-order tensor $\mathcal{D} \in R^{I_1 \times I_2 \times I_3}$, where I_3 is the number of images and $I_1 \times I_2$ is the image size. To compensate for possible misalignments, we adopt a set of transformations $\Gamma = \{\tau_1, \cdots, \tau_{I_3}\}$ which acts on the images of the tensor data, problem (5) can be changed to

$$
\begin{cases}
\min\limits_{\mathcal{A},\mathcal{E},\Gamma} \sum\limits_{i=1}^{3} \alpha_i \left\| \mathcal{A}_{(i)} \right\|_p^p + \lambda \left\| \mathcal{E} \right\|_{p,p}^p \\
\text{s.t.} \quad \mathcal{A} + \mathcal{E} = \mathcal{D} \circ \Gamma
\end{cases}
\tag{6}
$$

where $\mathcal{D} \circ \Gamma$ represents applying the transformation τ_i to the i-th image data $\mathcal{D}(:,:,i)$ ($i = 1, \cdots, I_3$). However, problem (6) is not directly tractable for the following two reasons: (1) the Schatten-p norm terms of unfolding matrices interdependent; (2) the equality constraint $\mathcal{A} + \mathcal{E} = \mathcal{D} \circ \Gamma$ is highly nonlinear due to the complicated dependence of $\mathcal{D} \circ \Gamma$ on the transformations Γ. To remove the interdependencies of $\mathcal{A}_{(i)}$ ($i = 1, \cdots, 3$), we introduce three auxiliary matrices M_i ($i = 1, 2, 3$) to optimize them independently. For limitation (2), we can approximate the constraint by linearizing about the current estimate of transformations Γ when the change in Γ is small. Accordingly, the first-order approximation to problem (6) is as follows.

$$
\begin{cases}
\min\limits_{\mathcal{A},\mathcal{E},\Delta\Gamma} \sum\limits_{i=1}^{3} \alpha_i \left\| M_i \right\|_p^p + \lambda \left\| \mathcal{E} \right\|_{p,p}^p \\
\text{s.t.} \quad \mathcal{A} + \mathcal{E} = \mathcal{D} \circ \Gamma + \text{fold}_3 \left(\left(\sum_{i=1}^{I_3} J_i \Delta\Gamma \varepsilon_i \varepsilon_i^\top \right)^\top \right) \\
\mathcal{A}_{(i)} = M_i
\end{cases}
\tag{7}
$$

where J_i is the Jacobian of the i-th image with respect to the transformation τ_i, $\Delta\Gamma$ is the change in current Γ, and ε_i denotes the standard basis for R^n. As this linearization is only a local approximation to problem (6), we solve it iteratively in order to converge to a (local) minima of (6).

2.2 Optimization Algorithm

For simplicity, we define $\Delta\tilde{\Gamma} \triangleq \text{fold}_3 \left(\left(\sum_{i=1}^{I_3} J_i \Delta\Gamma \varepsilon_i \varepsilon_i^\top \right)^\top \right)$, then the augmented Lagrangian function for problem (7) can be written as

$$
f_\mu \left(M_i, \mathcal{A}, \mathcal{E}, \Delta\tilde{\Gamma}, \mathcal{Y}, Q_i \right) = \sum_{i=1}^{3} \alpha_i \left\| M_i \right\|_p^p + \lambda \left\| \mathcal{E} \right\|_{p,p}^p + \langle \mathcal{Y}, \mathcal{T} \rangle + \frac{\mu}{2} \left\| \mathcal{T} \right\|_F^2
$$
$$
+ \sum_{i=1}^{3} \left(\langle Q_i, \mathcal{A}_{(i)} - M_i \rangle + \frac{\mu}{2} \left\| \mathcal{A}_{(i)} - M_i \right\|_F^2 \right) \tag{8}
$$

where $\mathcal{T} = \mathcal{A} + \mathcal{E} - \mathcal{D} \circ \Gamma - \Delta\tilde{\Gamma}$, $\mathcal{Y}$ and Q_i are the Lagrange multiplier tensor and matrix respectively, $\langle \cdot, \cdot \rangle$ represents the inner product for matrices or tensors, and μ is a positive scalar. It is difficult to solve the above function directly. A typical iterative process based on the ADMM is to minimize the augmented Lagrange function against only one of the unknown variables at a time, and it can be written explicitly as

$$
\begin{cases}
M_i^{k+1} = \arg\min_{M_i} f_\mu \left(M_i, \mathcal{A}^k, \mathcal{E}^k, \Delta\tilde{\Gamma}^k, \mathcal{Y}^k, Q_i^k \right) \\
\mathcal{A}^{k+1} = \arg\min_{\mathcal{A}} f_\mu \left(M_i^{k+1}, \mathcal{A}, \mathcal{E}^k, \Delta\tilde{\Gamma}^k, \mathcal{Y}^k, Q_i^k \right) \\
\mathcal{E}^{k+1} = \arg\min_{\mathcal{E}} f_\mu \left(M_i^{k+1}, \mathcal{A}^{k+1}, \mathcal{E}, \Delta\tilde{\Gamma}^k, \mathcal{Y}^k, Q_i^k \right) \\
\Delta\tilde{\Gamma}^{k+1} = \arg\min_{\Delta\tilde{\Gamma}} f_\mu \left(M_i^{k+1}, \mathcal{A}^{k+1}, \mathcal{E}^{k+1}, \Delta\tilde{\Gamma}, \mathcal{Y}^k, Q_i^k \right)
\end{cases}
\tag{9}
$$

In detail, the solutions of each term are obtained as follows.

★ For term M_i^{k+1} ($i = 1, 2, 3$):

$$
\begin{aligned}
M_i^{k+1} &= \arg\min_{M_i} \frac{\alpha_i}{\mu} \|M_i\|_p^p + \frac{1}{2} \left\| M_i - \left(\mathcal{A}_{(i)}^k + \frac{1}{\mu} Q_i^k \right) \right\|_F^2 \\
&= U_i T_{\alpha_i/\mu}(\Sigma_i) V_i^\top
\end{aligned}
\tag{10}
$$

where $U_i \Sigma_i V_i^\top = \mathcal{A}_{(i)}^k + Q_i^k/\mu$ and $T_\eta(\cdot)$ is the shrinkage operator:

$$
T_\eta(z) = \begin{cases}
0 & \text{if } |z| < \kappa \\
\{0, \text{sgn}(z)\hat{a}\} & \text{if } |z| = \kappa \\
\text{sgn}(z)\hat{a}^* & \text{if } |z| > \kappa
\end{cases}
\tag{11}
$$

In (11), $\hat{a} = [2\eta(1-p)]^{\frac{1}{2-p}}$, $\kappa = \hat{a} + \eta p \hat{a}^{p-1}$. $\hat{a}^* \in (\hat{a}, |z|)$ is the larger solution of

$$
a + \eta p a^{p-1} = |z|, \quad \text{where } a > 0
\tag{12}
$$

which can be obtained from the iteration $a_{(t+1)} = |z| - \eta p a_{(t)}^{p-1}$ with the initial value $a_{(0)} \in (\hat{a}, |z|)$.

★ For term $\mathcal{E}^{k+1}$:

$$
\begin{aligned}
\mathcal{E}^{k+1} &= \arg\min_{\mathcal{E}} \frac{\lambda}{\mu} \|\mathcal{E}\|_{p,p}^p + \frac{1}{2} \left\| \mathcal{E} - \left(\Delta\tilde{\Gamma}^k - \mathcal{A}^k + \mathcal{D} \circ \Gamma - \mathcal{Y}^k/\mu \right) \right\|_F^2 \\
&= T_{\lambda/\mu} \left(\Delta\tilde{\Gamma}^k - \mathcal{A}^k + \mathcal{D} \circ \Gamma - \mathcal{Y}^k/\mu \right)
\end{aligned}
\tag{13}
$$

★ For term $\mathcal{A}^{k+1}$:

$$
\mathcal{A}^{k+1} = \frac{1}{4} \left(\mathcal{D} \circ \Gamma + \Delta\tilde{\Gamma}^k - \mathcal{E}^{k+1} - \frac{\mathcal{Y}^k}{\mu} + \sum_{i=1}^{3} \text{fold}_3 \left(M_i^{k+1} - \frac{Q_i^k}{\mu}, i \right) \right)
\tag{14}
$$

Algorithm 1. The optimization procedure for problem (6)

Input: Tensor data $\mathcal{D}$ constructed by images, and weights $\lambda, \alpha_i > 0$.
1: Initialize transformations $\Gamma = \{\tau_1, \cdots, \tau_{I_3}\}$.
2: **loop**
3:    Compute Jacobian matrices J_i $(i = 1, \cdots, I_3)$ with respect to current transformation Γ.
4:    Apply transformation Γ to images, and obtain transformed tensor $\mathcal{D} \circ \Gamma$ in which the i-th two-dimensional slice (matrice) is $\mathcal{D}(:,:,i) \circ \tau_i$.
5:    Initialize the variables $\mathcal{A}^0$, $\mathcal{E}^0$ $\Delta\tilde{\Gamma}^0$, $\mathcal{Y}^0$ and Q_i^0. Initialize the parameters $\bar{\mu} \gg \mu^0 > 0$,
      $\rho > 1$, and $k = 0$.
6:    **loop**
7:       $(U_i, \Sigma_i, V_i) = \text{svd} \left(\mathcal{A}_{(i)}^k + Q_i^k / \mu^k \right);$
8:       $M_i^{k+1} = U_i T_{\alpha_i / \mu^k} (\Sigma_i) V_i^\top, i = 1, 2, 3;$
9:       $\mathcal{E}^{k+1} = T_{\lambda/\mu^k} \left(\Delta\tilde{\Gamma}^k - \mathcal{A}^k + \mathcal{D} \circ \Gamma - \mathcal{Y}^k/\mu^k \right);$
10:      $\mathcal{A}^{k+1} = \left(\mathcal{D} \circ \Gamma + \Delta\tilde{\Gamma}^k - \mathcal{E}^{k+1} - \mathcal{Y}^k/\mu^k + \sum_{i=1}^3 \text{fold}_3 \left(M_i^{k+1} - Q_i^k/\mu^k, i \right) \right) / 4;$
11:      $\Delta\tilde{\Gamma}^{k+1} = \mathcal{A}^{k+1} + \mathcal{E}^{k+1} - \mathcal{D} \circ \Gamma + \mathcal{Y}^k/\mu^k;$
12:      $\Delta\Gamma^{k+1} = \sum_{i=1}^{I_3} J_i^\dagger \left(\Delta\tilde{\Gamma}^{k+1} \right)_{(3)}^\top \varepsilon_i \varepsilon_i^\top;$
13:      $\mathcal{Y}^{k+1} = \mathcal{Y}^k + \mu^k \left(\mathcal{A}^{k+1} + \mathcal{E}^{k+1} - \mathcal{A} \circ \Gamma - \Delta\tilde{\Gamma}^{k+1} \right);$
14:      $Q_i^{k+1} = Q_i^k + \mu^k \left(\mathcal{A}_{(i)}^{k+1} - M_i^{k+1} \right);$
15:      $\mu^{k+1} = \max(\rho\mu^k, \bar{\mu}), k = k + 1;$
16:   **end loop**
17:   Let $\Delta\Gamma = \Delta\Gamma^k$, and update transformations $\Gamma = \Gamma + \Delta\Gamma$;
18:   Obtain current optimal solution $\mathcal{A}^* = \mathcal{A}^k$, $\mathcal{E}^* = \mathcal{E}^k$, and $\Gamma^* = \Gamma$.
19: **end loop**

Output: solution $\mathcal{A}^*$, $\mathcal{E}^*$, Γ^* to problem (6).

$\star$ For term $\Delta\tilde{\Gamma}^{k+1}$:

$$\Delta\tilde{\Gamma}^{k+1} = \mathcal{A}^{k+1} + \mathcal{E}^{k+1} - \mathcal{D} \circ \Gamma + \mathcal{Y}^k/\mu \tag{15}$$

As a result, $\Delta\Gamma^{k+1}$ can be computed as follows:

$$\Delta\Gamma^{k+1} = \sum_{i=1}^{I_3} J_i^\dagger \left(\Delta\tilde{\Gamma}^{k+1} \right)_{(3)}^\top \varepsilon_i \varepsilon_i^\top \tag{16}$$

where $J_i^\dagger = (J_i^\top J_i)^{-1} J_i^\top$ is pseudo-inverse of J_i and $\left(\Delta\tilde{\Gamma}^{k+1} \right)_{(3)}$ is the mode-3 unfolding matrix of tensor $\Delta\tilde{\Gamma}^{k+1}$.

$\star$ The Lagrange multipliers are updated by the following formula:

$$\mathcal{Y}^{k+1} = \mathcal{Y}^k + \mu \left(\mathcal{A}^{k+1} + \mathcal{E}^{k+1} - \mathcal{A} \circ \Gamma - \Delta\tilde{\Gamma}^{k+1} \right) \tag{17}$$

$$Q_i^{k+1} = Q_i^k + \mu \left(\mathcal{A}_{(i)}^{k+1} - M_i^{k+1} \right) \tag{18}$$

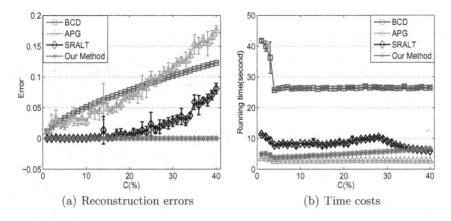

(a) Reconstruction errors   (b) Time costs

Fig. 1. The contrast results of the four methods on the synthetic data sets.

Table 1. The reconstruction errors of the four method. The mean error together with standard deviation is reported.

Noise ratio	RPCA	Li's work	SRALT	Ours (10^{-6})
$\gamma = 5\%$	0.0213 ± 0.0002	0.0335 ± 0.0069	$(4.32 \pm 0.84) \times 10^{-6}$	$\mathbf{1.03 \pm 0.0204}$
$\gamma = 10\%$	0.0266 ± 0.0004	0.0511 ± 0.0058	$(4.87 \pm 0.71) \times 10^{-6}$	$\mathbf{1.08 \pm 0.0246}$
$\gamma = 15\%$	0.0485 ± 0.0004	0.0657 ± 0.0092	0.0013 ± 0.0014	$\mathbf{1.14 \pm 0.0392}$
$\gamma = 20\%$	$0{,}0669 \pm 0.0004$	0.0785 ± 0.0086	0.0042 ± 0.0031	$\mathbf{1.27 \pm 0.1040}$
$\gamma = 25\%$	0.0982 ± 0.0003	0.0905 ± 0.0111	0.0232 ± 0.0193	$\mathbf{1.38 \pm 0.0665}$
$\gamma = 30\%$	0.1127 ± 0.0004	0.1017 ± 0.0091	0.0278 ± 0.0084	$\mathbf{1.61 \pm 0.0363}$
$\gamma = 35\%$	0.1531 ± 0.0005	0.1128 ± 0.0157	0.0589 ± 0.0119	$\mathbf{1.59 \pm 0.0278}$
$\gamma = 40\%$	0.1750 ± 0.0003	0.1228 ± 0.0086	0.0810 ± 0.0085	$\mathbf{1.86 \pm 0.1110}$

The optimization procedure of problem (6) is described in Algorithm 1. In the inner loop, a continuation technique [24], which varies μ, starting from a small initial value μ_0 and increasing it geometrically with each iteration until it reaches the ceil $\bar{\mu}$, can greatly speed up the convergence.

3   Experimental Results

To validate the effectiveness of our work, we first conduct experiments on synthetic data to give a quantitative description of the reconstruction error. And then, we apply the proposed method to three commonly used image sequences. Finally, we evaluate our approach on face recognition data set: YaleB data set [25]. We compare our method with several state-of-the-art algorithms: RPCA [2], Li's work [16] and SRALT [17]. All experiments are performed on a Desktop PC with Intel Core-i5 2.4 GHz CPU, 16G Ram, and Windows7 OS. In all tables,

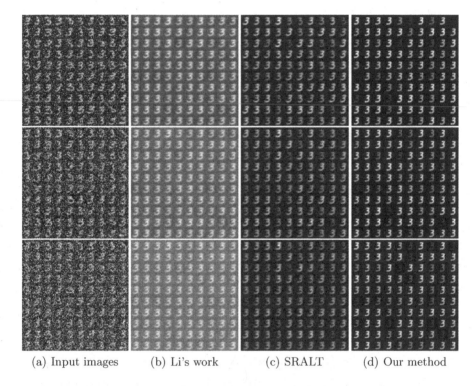

(a) Input images   (b) Li's work   (c) SRALT   (d) Our method

Fig. 2. Recovery results on the data set of handwritten digit "3".

the form of mean value together with standard deviation is used to report the experimental results, and the best results are in boldface.

3.1 Synthetic Data

In this subsection, the four methods are compared on synthetic data. To make a fair comparison, we declare the following implementation details: (1) Because RPCA applies to the recovery of matrix, we apply it to each mode of the tensor and the minimal reconstruction error of the unfolding matrices is used as the result of RPCA. (2) The accelerated proximal gradient algorithm (APG) is selected as the optimization algorithm in RPCA. (3) Because Li's work does not involve the image transformations, here, it is assumed that the synthetic data is well aligned.

We first randomly generate two tensor data: a pure low-rank tensor $\mathcal{A}^* \in R^{50 \times 50 \times 50}$ whose rank is (10,10,10) and an error tensor $\mathcal{E}^* \in R^{50 \times 50 \times 50}$ in which c percent of elements is nonzero (the maximal value of c is set to 40 to guarantee the sparsity of the error). Then $\mathcal{D} = \mathcal{A}^* + \mathcal{E}^*$ is the tensor which we need to recovery. The above four methods are used to recovery the low-rank part of $\mathcal{D}$, denoted by $\hat{\mathcal{A}}$. The relative reconstruction error is defined as $\mathrm{Err} = \frac{\|\hat{\mathcal{A}} - \mathcal{A}^*\|_F}{\|\mathcal{A}^*\|_F}$.

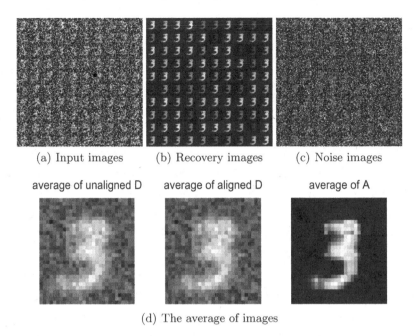

| (a) Input images | (b) Recovery images | (c) Noise images |

| average of unaligned D | average of aligned D | average of A |

(d) The average of images

Fig. 3. The recovery results of our methods on handwritten digit "3" with salt-and-pepper noise ratio 60%.

In order to eliminate the influence of the randomness of the data generation, we repeat each experiment 20 times, and the average of the reconstruction error with the standard deviation is taken as the final result. The reconstruction errors are shown in Fig. 1(a) and Table 1. It can be seen that the reconstruction errors of RPCA, SRALT and Li's work increase sharply with the proportion of corrupted entries increasing. In contrast, the reconstruction errors of our method are maintained at 10^{-6} orders of magnitude. The performance of our algorithm is very robust to the noise ratio. As shown in Fig. 1(b), comparing with Li's work and SRALT, our work can achieve 7 times and 2 times speedup. Our method is slightly slower than RPCA. This is because RPCA is applied to one mode of the tensor each time, and our method considers all modes of the tensor. That is, the optimization variables involved in our method are about three times of those in RPCA. The above results demonstrate the effectiveness and efficiency of our proposed optimization method for low-rank tensor recovery.

3.2 Image Sequence Recovery and Alignment

In this part, to verify the performance of the proposed method on image sequence recovery and alignment, we add different proportions of salt and pepper noises to the original image sequence data.

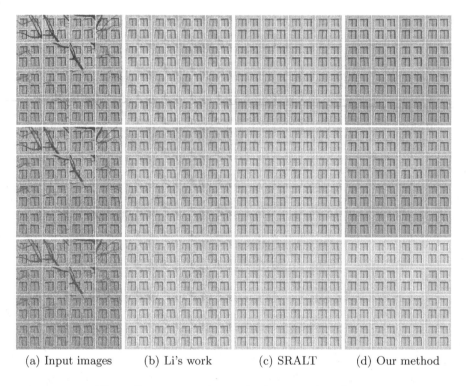

(a) Input images   (b) Li's work   (c) SRALT   (d) Our method

Fig. 4. Recovery results on the data set of windows.

The first data set contains 100 handwritten digits "3". As a result of the individual writing habits, there are significant differences in structures. In Fig. 2, column (a) shows the original images with added salt and pepper noises. From the upper row to bottom row, the percents of added noises are respectively 30%, 40% and 50%. The recovery results of Li'work, SRALT and our method are shown in column (b), (c), (d). We can see that our work has achieved better performance than the other two methods from human's perception, in which the 3's are more clear and their poses are upright. In addition, our method is robust to the proportion of the salt and pepper noises. Even if the proportion of noise is up to 60%, which is very difficult to recognition from human's perception, our algorithm still can efficiently remove the noise and recover the desired low-rank structure, as shown in Fig. 3.

The second data set contains 16 images of architectural surface windows taken from different viewpoints by a perspective camera. Some of the images contain tree branches. Recovery results are shown in Fig. 4. Compared with Li's work and SRALT, our recovery results not only remove the tree branches and salt and pepper noises, but also have the superior performance on image alignment.

The third data set contains 140 frames of Gore Albert talking video. The partial recovery results are shown in Fig. 5. Due to symmetry, the face data is

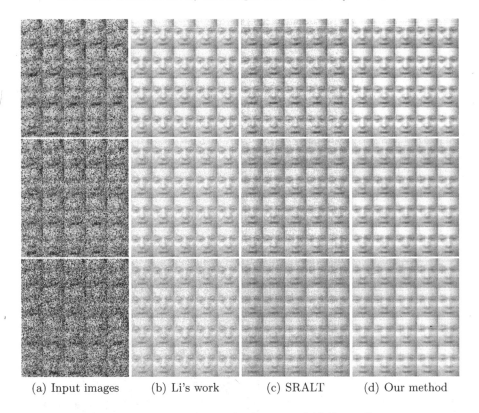

(a) Input images   (b) Li's work   (c) SRALT   (d) Our method

Fig. 5. Recovery results on the data set of AI Gore talking.

rich in low rank structure. In contrast, our recovery results are still significantly better than those of the other two methods, especially in the performance of removal of salt and pepper noise.

3.3   Face Recognition

In this part, experiments are carried out on the YaleB data set. which contains 2414 faces of 38 persons. For each person, there are about 60 gray-scale face images under various lighting conditions. The experiments are designed as following steps:

Step 1. For each person, we randomly select 30 images as training samples, and the rest are left for testing samples.

Step 2. In each image, we add different proportions of salt and pepper noises, which is set to $\gamma = 0.05 : 0.05 : 0.4$.

Step 3. For each proportion of noise, a 3-order tensor is constructed by the polluted images, and denoted by $\mathcal{D}$.

Table 2. The testing accuracies of the four methods. The mean accuracy together with standard deviation is reported.

Noise ratio	Noise images	Li's work	SRALT	Our method
$\gamma = 5\%$	0.5279 ± 0.0103	0.5717 ± 0.0064	0.5738 ± 0.0040	**0.5965 ± 0.0020**
$\gamma = 10\%$	0.4361 ± 0.0161	0.4823 ± 0.0144	0.5653 ± 0.0030	**0.5901 ± 0.0020**
$\gamma = 15\%$	0.3491 ± 0.0164	0.3843 ± 0.0137	0.5570 ± 0.0021	**0.5832 ± 0.0009**
$\gamma = 20\%$	0.2761 ± 0.0118	0.3231 ± 0.0075	0.5338 ± 0.0037	**0.5743 ± 0.0018**
$\gamma = 25\%$	0.2104 ± 0.0122	0.3279 ± 0.0038	0.4843 ± 0.0062	**0.5667 ± 0.0024**
$\gamma = 30\%$	0.1645 ± 0.0128	0.3256 ± 0.0077	0.4113 ± 0.0041	**0.5474 ± 0.0053**
$\gamma = 35\%$	0.1204 ± 0.0034	0.3028 ± 0.0086	0.3926 ± 0.0064	**0.5217 ± 0.0034**
$\gamma = 40\%$	0.0943 ± 0.0087	0.2385 ± 0.0115	0.3630 ± 0.0044	**0.4813 ± 0.0071**

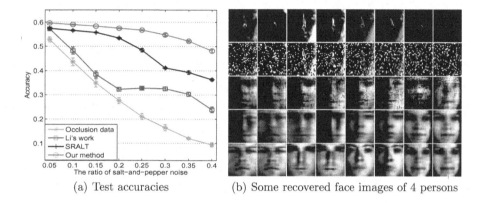

(a) Test accuracies   (b) Some recovered face images of 4 persons

Fig. 6. The contrast results on the YaleB data set.

Step 4. We respectively use Li's work, SRALT and our method to recover the low-rank structure of tensor $\mathcal{D}$.

Step 5. By using the K-nearest neighbor (KNN) algorithm, the recognition accuracies of the testing samples recovered by the three methods are obtained.

The accuracy of the KNN on the data set of original images is 0.6358, which is the baseline. The recognition accuracies of the three methods are shown in Fig. 6(a) and Table 2. We can see that the accuracies of the face recognition decrease with the noise proportion increasing. Specifically, accuracies of Li's work decreased from 0.5717 to 0.2385. SRALT is slightly better than Li's work, but it's worse than our model. With the increase of the noise proportion, accuracies of our model decrease smoothly from 0.5965 to 0.4813. Our method is not only superior to the other two models in terms of accuracy, but also robust to the proportion of noise.

Partial face images of 4 persons are given in Fig. 6(b), and we have selected 2 faces with serious shadows for each person. From top row to bottom row, they are respectively the original face images, the face images with noise proportion 0.3, the face images recovered by Li's work, SRALT and our model. It can be seen that, the images recovered by Li's work still have some shadows and noises. SRALT is better than Li's work in removing shadows, but it can not effectively remove the salt and pepper noises. In contrast, our model can simultaneously remove the shadows and the salt and pepper noises. The good performance on face recognition shows the effectiveness of our model in image denoising.

4 Conclusion

Based on the superior performance of ℓ_p minimization in sparse representation, we propose a framework of non-convex low-rank recovery and alignment for arbitrary tensor data in this paper. Specially, we respectively use Schatten-p norm and ℓ_p norm to relax the rank function and ℓ_0 norm respectively in the low-rank recovery model. They require much weaker incoherence conditions to guarantee a successful recovery than the common used nuclear norm and ℓ_1 norm. To compensate the possible misalignments of images, we adopt a set of transformations which acts on the images of the tensor data. By solving the optimal transformations, the strict alignments of the images are achieved in the denoising process. Furthermore, an efficient algorithm based on the method of ADMM is proposed to solve the non-convex optimization problem. By comparing our work with the three state-of-the-art work through extensive simulations, we demonstrate the effectiveness and efficiency of our method.

Acknowledgement. This work is supported by NSFC (Grants nos. 61305035, 61472285, 61511130084, and 61503263), Zhejiang Provincial Natural Science Foundation (Grants nos. LY17F030004, LR17F030001, LY16F020023, LY12F03016), Project of science and technology plans of Zhejiang Province (Grants nos. 2014C31062, 2015C31168). Project of science and technology plans of Wenzhou (Grants No. G20150017).

References

1. Basri, R., Jacobs, D.W.: Lambertian reflectance and linear subspaces. IEEE Trans. Pattern Anal. Mach. Intell. **25**, 218–233 (2003)
2. Wagner, A., Wright, J., Ganesh, A., Zhou, Z., Mobahi, H., Ma, Y.: Toward a practical face recognition system: robust alignment and illumination by sparse representation. IEEE Trans. Pattern Anal. Mach. Intell. **34**, 372–386 (2012)
3. Donoho, D.L.: Compressed sensing. IEEE Trans. Inf. Theory **52**, 1289–1306 (2006)
4. Candés, E.J., Li, X., Ma, Y., Wright, J.: Robust principal component analysis? J. ACM (JACM) **58**, 1–73 (2011)
5. Ganesh, A., Wright, J., Li, X., Candés, E.J., Ma, Y.: Dense error correction for low-rank matrices via principal component pursuit. In: IEEE International Symposium on Information Theory, pp. 1513–1517 (2010)
6. Zhou, Z., Li, X., Wright, J., Candes, E.J., Ma, Y.: Stable principal component pursuit. In: IEEE International Symposium on Information Theory, pp. 1518–1522 (2010)

7. Ji, H., Huang, S., Shen, Z., Xu, Y.: Robust video restoration by joint sparse and low rank matrix approximation. SIAM J. Imaging Sci. **4**, 1122–1142 (2011)
8. Peng, Y., Ganesh, A., Wright, J., Xu, W., Ma, Y.: RASL: robust alignment by sparse and low-rank decomposition for linearly correlated images. IEEE Trans. Pattern Anal. Mach. Intell. **34**, 2233–2246 (2012)
9. Zhang, Z., Ganesh, A., Liang, X., Ma, Y.: TILT: transform invariant low-rank textures. Int. J. Comput. Vis. **99**, 1–24 (2012)
10. Guo, X., Cao, X., Chen, X., Ma, Y.: Video editing with temporal, spatial and appearance consistency. In: Proceedings of the IEEE Conference on Computer Vision and Pattern Recognition (CVPR), pp. 2283–2290 (2013)
11. Zhang, D., Hu, Y., Ye, J., Li, X., He, X.: Matrix completion by truncated nuclear norm regularization. In: IEEE Conference on Computer Vision and Pattern Recognition (CVPR), pp. 2192–2199 (2012)
12. Liu, J., Musialski, P., Wonka, P., Ye, J.: Tensor completion for estimating missing values in visual data. IEEE Trans. Pattern Anal. Mach. Intell. **35**, 208–220 (2013)
13. Signoretto, M., Plas, R.V.D., Moor, B.D., Suykens, J.A.K.: Tensor versus matrix completion: a comparison with application to spectral data. IEEE Sig. Process. Lett. **18**, 403–406 (2011)
14. Kolda, T.G., Bader, B.W.: Tensor decompositions and applications. SIAM Rev. **66**, 294–310 (2005)
15. Huang, B., Mu, C., Goldfarb, D., Wright, J.: Provable low-rank tensor recovery (2014). http://www.optimization-online.org/DB_HTML/2014/02/4252.html
16. Li, Y., Yan, J., Zhou, Y., Yang, J.: Optimum subspace learning and error correction for tensors. In: Daniilidis, K., Maragos, P., Paragios, N. (eds.) ECCV 2010. LNCS, vol. 6313, pp. 790–803. Springer, Heidelberg (2010). doi:10.1007/978-3-642-15558-1_57
17. Zhang, X., Wang, D., Zhou, Z., Ma, Y.: Simultaneous rectification and alignment via robust recovery of low-rank tensors. In: Advances in Neural Information Processing Systems, pp. 1637–1645 (2013)
18. Mohan, K., Fazel, M.: Iterative reweighted algorithms for matrix rank minimization. J. Mach. Learn. Res. **13**, 3441–3473 (2012)
19. Chen, X., Xu, F., Ye, Y.: Lower bound theory of nonzero entries in solutions of l2-lp Minimization. SIAM J. Sci. Comput. **32**, 2832–2852 (2010)
20. Qin, L., Lin, Z., She, Y., Zhang, C.: A comparison of typical lp minimization algorithms. Neurocomputing **119**, 413–424 (2013)
21. Zuo, W., Meng, D., Zhang, L., Feng, X., Zhang, D.: A generalized iterated shrinkage algorithm for non-convex sparse coding. In: IEEE International Conference on Computer Vision (ICCV), pp. 217–224 (2013)
22. Nie, F., Huang, H., Ding, C.: Low-rank matrix recovery via efficient Schatten p-norm minimization. In: AAAI Conference on Artificial Intelligence (2014)
23. Marjanovic, G., Solo, V.: On lq optimization and matrix completion. IEEE Trans. Sig. Process. **60**, 5714–5724 (2012)
24. Toh, K.C., Yun, S.: An accelerated proximal gradient algorithm for nuclear norm regularized linear least squares problems. Pac. J. Optim. **6**(3), 615–640 (2010)
25. Georghiades, A.S., Belhumeur, P.N., Kriegman, D.J.: From few to many: illumination cone models for face recognition under variable lighting and pose. IEEE Trans. Pattern Anal. Mach. Intell. **23**, 643–660 (2001)

Debluring Low-Resolution Images

Jinshan Pan[1(✉)], Zhe Hu[2], Zhixun Su[1], and Ming-Hsuan Yang[2]

[1] Dalian University of Technology, Dalian, China
sdluran@gmail.com
[2] University of California, Merced, USA

Abstract. The recent years have witnessed significant advances in image deblurring. In general, the success of deblurring methods depends heavily on extraction of salient structures from a blurry image for kernel estimation. Most deblurring methods often operate on high-resolution images where contours or edges can be extracted for kernel estimation. However, recovering reliable structures from low-resolution images becomes extremely challenging. In this paper, we propose a spatially variant deblurring algorithm for low-resolution images based on the exemplars. To exploit the exemplar information, we develop a super-resolution guided method to help the restoration of reliable image structures which can be used for kernel estimation. Experimental evaluations against the state-of-the-art methods show that the proposed algorithm performs favorably in deblurring low-resolution images. Furthermore, we show that the SR results obtained as byproducts in our method are comparable compared to other blind SR methods.

1 Introduction

Image deblurring and super resolution (SR) are two fundamental problems in restoration and enhancement. Image deblurring aims to remove motion blur caused by camera and object movement from a blurry input, whereas super resolution reconstructs a high-resolution (HR) image from a low-resolution (LR) input. These two problems can be formulated with one model,

$$g = H(f) + \varepsilon, \tag{1}$$

where g, f, and ε denote the distorted input, latent/clear image, and noise. In (1), H is the degradation function which represents image convolution in the image deblurring task and downsampling in the super resolution problem.

Numerous approaches have been proposed to specifically address one of these closely related topics. From the perspective of super resolution, most methods assume blur kernels can be modeled well by Gaussians. However, when a blurry image is generated by complex camera and object motion, the assumption with a parametric Gaussian kernel does not hold, and most deblurring methods do not

Electronic supplementary material The online version of this chapter (doi:10. 1007/978-3-319-54407-6_8) contains supplementary material, which is available to authorized users.

© Springer International Publishing AG 2017
C.-S. Chen et al. (Eds.): ACCV 2016 Workshops, Part I, LNCS 10116, pp. 111–127, 2017.
DOI: 10.1007/978-3-319-54407-6_8

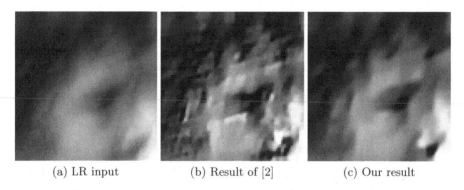

(a) LR input   (b) Result of [2]   (c) Our result

Fig. 1. Deblurring a low-resolution image. State-of-the-art deblurring method [2] fails to generate reliable results.

perform well. There have been a few attempts to understand how blind deconvolution methods help solving the super resolution problem, blind SR specifically, by estimating the blur kernels. In [1], a method for deconvolution and super resolution is proposed to account for unknown blur by using a sequence of LR images. Recently, a nonparametric blind SR method [38] is developed by modeling both internal and external kernels and estimating the kernel with the self-similarity property. From the perspective of image deblurring, few methods explore the potential of blind deconvolution benefitting from SR techniques. Despite significant progress has been made, state-of-the-art deblurring methods [2–10] do not perform well on low-resolution blurred images (see example in Fig. 1). As salient structures cannot be extracted from a low-resolution image for kernel estimation, state-of-the-art methods do not perform well in such scenarios.

In this work, we resort to super resolution methods for recovering high-quality intermediate images in the deblurring framework. We use exemplar information and show how the super resolution benefits kernel estimation in image deblurring. We collect a set of 6,152 natural images as HR exemplars and their downsampled versions. With the known correspondence between LR and HR patches, we develop a deblurring method to jointly estimate the HR latent image and camera motion. The proposed method is able to reconstruct the salient structure and remove camera blur, as shown in Fig. 1(c). Furthermore, we show that by-product SR results in this work are comparable to state-of-the-art blind SR methods.

2   Related Work

In this section, we discuss the most relevant algorithms and put this work in proper context. Since recovering a high-quality image from a blurred LR input involves both deblurring and SR process, a natural solution to this problem is to first apply a deblurring method to obtain a LR image of better quality, and then recover more details with a SR scheme. However, when the degradation model is unknown, generic restoration schemes (e.g., deblurring) do not perform well on real-world images in which salient structures cannot be reliably extracted and thus significant artifacts are introduced.

Image deblurring: Deblurring methods can be broadly categorized into three approaches: Variational Bayesian (VB) formulations, maximum a posteriori (MAP) estimates, and edge predictions [11,12]. Fergus et al. [3] present a method using a mixture of Gaussians to learn the image gradient prior within a variational Bayesian framework. Levin et al. [13] show that the VB based deblurring methods can avoid rendering trivial solutions while naive MAP based approaches cannot. Since the optimization of VB method is computationally expensive, recent methods are formulated based on the MAP framework regularized with priors [2,4,8,14–22] and careful implementations [23] to estimate blur kernels effectively. In addition, numerous methods that explicitly select sharp edges for kernel estimation [5,6,24] have been shown to perform well on the benchmark dataset [25] However, the selection processes often involve heuristics and assume strong edges exist in the latent images. To better reconstruct the sharp edges for kernel estimation, exemplar based methods [26–28] are proposed to exploit information from external datasets.

Real camera motion (with translational as well rotational movements) often results in non-uniform blur that cannot be modeled well with the conventional uniform model. Tai et al. [29] propose a general projective motion model to describe non-uniform motion blur, and Whyte et al. [30] simplify and solve this problem within a VB framework as [3]. In [31], a similar model is proposed in which a motion density function is used to describe the considered camera motion trajectory. To make the non-uniform model efficient, numerous methods [32,33] are developed based on a locally uniform patch-based approach, where the deconvolution step can be efficiently computed by the fast Fourier transform algorithm. On the other hand, numerous approaches have been developed to account for blur caused by object motion [34–36]. However, most of the aforementioned deblurring methods are developed for generic scenes with the assumption that the resolution of input images is sufficiently large.

Super resolution: Existing single-image SR methods often assume that the blur kernel can be modeled well with a Gaussian or bicubic function, which does not hold for all scenarios. To overcome this limitation, a few algorithms use multiple degraded low-resolution frames to recover a clear HR image [1,37]. Recently, Michaeli and Irani [38] propose a blind SR method to estimate the internal/external blur kernels without assuming any parametric formulation of kernels. This method overcomes the limitations of the conventional SR methods based on simple parametric forms of kernels. However, it is still difficult to generate reliable results for complicated scenes involving complex camera motion.

3   Deblurring Low-Resolution Images

Before presenting the proposed algorithm, we first analyze the effect of image resolution on deblurring tasks and discuss why the state-of-the-art deblurring methods fail in deblurring LR images. With the spatially invariant blur assumption, an image is formed by

$$B = I * k + \varepsilon, \tag{2}$$

where I is the latent sharp image, k is the blur kernel (or point spread function), B is the blurred input image, ε is noise, and $*$ is the convolution operator. Solving I and k from a single blurred image B is an ill-posed problem. For low-resolution images, this problem is even more challenging. Assuming that I is given, the commonly used approach for solving k is:

$$k^* = \arg\min_k \|\nabla I * k - \nabla B\|^2 + \varphi(k), \tag{3}$$

where $\varphi(k)$ is a regularization term for the blur kernel k, and $\nabla = (\partial_x, \partial_y)$ represents the derivatives along horizontal and vertical directions.

Considering only the data term $\|\nabla I * k - \nabla B\|^2$, (3) becomes a least squares problem and k^* can be obtained by solving the following linear system:

$$A^\top A k = A^\top b, \tag{4}$$

where A is the Toeplitz matrix of $\partial_x I$ and $\partial_y I$ with a filter k, and b is the vectorization of $\partial_x B$ and $\partial_y B$.

We compute the condition number of matrix $A^\top A$ at the original scale in HR and its downsampled LR image to see how accurate the solution can be. We present the visualization of $A^\top A$ in Fig. 2. The matrix generated with the HR image has a smaller condition number than that of the LR one, which means more stable solutions under the same-level errors in b in HR. The matrix generated with HR image has a stronger diagonal form, and thus more stable results can be computed [5].

Another problem for deblurring an LR image is the missing image edges or textures. When the resolution of an object is low, it is difficult to extract edges or textures, as shown in Fig. 1, that can be used for kernel estimation. As no reliable edges can be extracted, the state-of-the-art deblurring method [2] does not perform well (see Fig. 1(b)) whereas our method is able to estimate kernels well by exploiting exemplar-based SR techniques.

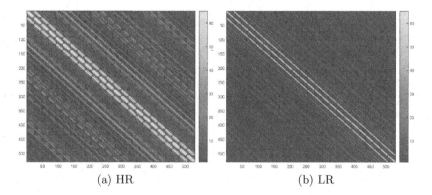

(a) HR   (b) LR

Fig. 2. Visualization of matrix $A^\top A$ of (4) in terms of different image resolutions. (a)-(b) are the visualization matrix $A^\top A$ of HR and the corresponding LR image. The image of (b) is the downsampled version of the image of (a). The matrix $A^\top A$ of HR image has stronger diagonal form and thus the solution in (4) can be computed more stably.

4   Problem Formulation

To deblur LR images, we need to extract reliable edges or textures for kernel estimation. In this work, we solve this problem by exploring exemplar information in a SR formulation to guide kernel estimation and image reconstruction. By exploring exemplar information, the recovered image will contain more useful structures and thus facilitate effective kernel estimation. Previous methods exploit exemplar information via searching and pasting similar patches from exemplars using dense correspondence [27] for deblurring. Due to inherent ambiguity of blurred edges, it is important to develop effective methods to match patches from exemplars for kernel estimation.

In this work, we use an exemplar-based SR approach to estimate latent image at each iteration to reduce ambiguity. The exemplar-based SR technique has been shown to be able to reduce small amounts of blur and to recover sharp images [39]. First, the correspondence between HR and LR patches is constructed using exemplars. In this step, we assume that the LR patches are generated by downsampling the clear patches of exemplars with the same blur kernel as that of the test image, i.e., a Gaussian is used in this work. Given a patch in a test image, we search for the most similar LR patch from exemplars and paste its corresponding HR patch to the output image.

For the SR step, we adopt a similar method as [39] to obtain the relationship between the latent image and the exemplars. We compute the features of each patch in the estimated latent image and search for the closest cluster center in the feature space of the LR exemplars. The updated HR latent image L_h is reconstructed by

$$L_h = Fv, \tag{5}$$

where v represents the features of the estimated image patches and F is the mapping function which indicates the correspondence from the LR exemplar patches to the HR exemplar patches. To learn the mapping function F, we use the following model,

$$F^* = \arg\min_F \|u - Fv\|^2, \tag{6}$$

where u denotes the central regions of the HR exemplar patches.

4.1   Proposed Model

Since the camera motion usually results in spatially variant blur, we consider the non-uniform blur model in this work. Similar to [32], we employ a region-based non-uniform model with a locally uniform kernel at each region

$$B = \sum_r k^{(r)} * (\omega^{(r)} \odot I), \tag{7}$$

where r denotes index of the regions, $\omega^{(r)}$ is the weighting mask for r-th region with the same size as the latent image I, $\odot$ denotes the pixel-wise multiplication, and $k^{(r)}$ represents the kernel at r-th region that is defined with a linear combination of the kernel basis b_θ^r,

$$k^{(r)} = \sum_\theta \mu_\theta b_\theta^{(r)}, \tag{8}$$

where b_θ is the kernel basis induced by the homography of a sampled camera pose θ [32] and μ_θ is the coefficient.

Based on the discussions in Sects. 3 and 4, we incorporate the exemplar information into the deblurring model in a SR manner, where the objective function is written as

$$\min_{I,\mu,v} \| \sum_r \sum_\theta \mu_\theta b_\theta^{(r)} * (\omega^{(r)} \odot I) - B \|^2 + \beta \| I - (Fv) \downarrow_s \|^2$$
$$+ \lambda(\|\nabla I\|_0 + \sigma\|I\|_0) + \gamma\|\mu\|^2, \tag{9}$$

where $\downarrow$ is the downsampling operator, s is the scale factor, and $\mu = (\mu_1, \cdots, \mu_\theta)^\top$. As shown in [40], the ℓ_0 norm on the intensity and gradient of the latent image I is able to restore sharp edges for natural images. We use the ℓ_0 norm to guide the kernel estimation in this work.

4.2 Optimization

We use the alternating minimization method to solve (9) and present the details as below.

Learning mapping function: We note that the mapping function F is pre-learned using the exemplars before solving (9). Using a strategy similar to [39], we first construct the image patches u of HR exemplar patches and v of their down-sampled LR patches, and learn K cluster centers of the extracted LR patches. The learned K cluster centers are used as the index to find the optimal mapping function F. Then the mapping function F is learned by solving the least squares problem (6).

HR image construction: To construct the intermediate HR image, we apply (5) to reconstruct the features of HR patches according to [39]. Given the estimated latent image I, we compute the feature of each patch p_i (LR patch) and search for the closest cluster center from the computed K cluster centers. Then with the closest cluster center, we find the mapping function F^*. Finally, the corresponding HR patch is reconstructed by the F^*p_i.

Intermediate latent image estimation: By fixing the variables μ and v, we can rewrite (9) as,

$$\min_I \| \sum_r \sum_\theta \mu_\theta b_\theta^{(r)} * (\omega^{(r)} \odot I) - B \|^2 + \beta\|I - L_l\|^2 + \lambda(\|\nabla I\|_0 + \sigma\|I\|_0),$$
$$\tag{10}$$

Algorithm 1. Blur kernel estimation algorithm

Input: Blurred image B.
Apply k-means method to find the patch center v and learn the mapping function by minimizing (6).
Initialize I by [2].
for $i = 1 \rightarrow 5$ **do**
    Reconstruct HR image using (5).
    Solve for the latent image I by (10).
    Solve for the coefficients μ by (12).
end for
Compute $k^{(r)}$ using (8).
Output: Blur kernels $\{k^{(r)}\}_r$.

---

where $L_l = (Fv) \downarrow_s$ and each image region $I^{(r)} = \omega^{(r)} \odot I$ can be estimated independently by

$$\min_{I^{(r)}} \| \sum_\theta \mu_\theta b_\theta^{(r)} * I^{(r)} - B^{(r)} \|^2 + \beta \| I^{(r)} - L_l^{(r)} \|^2 + \lambda(\|\nabla I^{(r)}\|_0 + \sigma \|I^{(r)}\|_0), \tag{11}$$

where $L_l^{(r)} = \omega^{(r)} \odot L_l$ and $B^{(r)} = \omega^{(r)} \odot B$. To solve the ℓ_0-regularized optimization problem, we use a half-quadratic optimization method as [2,40].

Kernel estimation: Given the intermediate latent image I and features v, the coefficients μ can be obtained by

$$\min_\mu \| \sum_r \sum_\theta \mu_\theta b_\theta^{(r)} * (\omega^{(r)} \odot \nabla I) - \nabla B\|^2 + \gamma \|\mu\|^2. \tag{12}$$

Here we use the image gradient for better performance as pointed out by [2]. Once the coefficients μ are estimated, we can obtain the kernels at different regions $\{k^{(r)}\}_r$ by computing (8).

We note that the kernel estimation involves iteratively updating v, I and μ. To provide a good initial value I for solving v, we employ the uniform deblurring method [2]. The main steps of the kernel estimation algorithm are summarized in Algorithm 1.

4.3   Image Restoration

For the final latent image I reconstruction, we adopt the $\ell_{0.8}$ regularization [41] to estimate each image region by

$$\min_{I^{(r)}} \|k^{(r)} * I^{(r)} - B^{(r)}\|^2 + \tau \|\nabla I^{(r)}\|_{0.8}, \tag{13}$$

as it is shown to maintain more image textures [41]. For the HR image reconstruction, we use the same SR method [39].

4.4   Analysis

In this section, we provide more insights and analysis on how the proposed algorithm works and performs on image deblurring. There are two main issues to be discussed. One is the benefits of exploiting exemplar information, the other is why we exploit the information in a SR manner. We show some examples here to illustrate those and present comprehensive experiments in Sect. 5.

The SR component (i.e., the second term in (9)) in the proposed method provides the exemplar information. If we set $\beta = 0$, the method naturally falls back to an ℓ_0-regularized spatially variant deblurring method. We provide the comparison of the results with and without the SR component in Fig. 3. There exists information loss when blurring the image and capturing objects in LR, and thus some important salient structures are not extracted when reconstructing the intermediate latent image without using external information. By exploiting the exemplar information, we are able to recover better textures as shown in the intermediate latent image in Fig. 3(b). As a result, the well constructed textures facilitate the following blur estimation process such that the deblurred result in Fig. 3(d) contains fewer ringing artifacts.

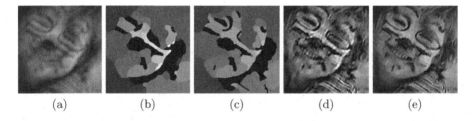

(a)   (b)   (c)   (d)   (e)

Fig. 3. The importance of exploiting the exemplar information in image deblurring. (a) Blurred image. (b) Intermediate result without the exemplar information by setting $\beta = 0$. (c) Intermediate result with the exemplar information. (d) Deblurred image by setting $\beta = 0$. (e) Our result.

We note that the SR component is imposed on the estimated latent image, which is supposed to contain much less blur than the input. The use of SR on a less blurry image helps reduce blur and restore salient structure to some extent. We present an example in Fig. 4. Applying SR technique and downsampling is able to reconstruct sharp edges and structure of a LR object as shown in Fig. 4(b) and (c).

Since the proposed method involves SR step, it does not directly use the information of exemplars due to the effect of scale factor s. Thus, it is different from dictionary-based deblurring methods (e.g., [42,43]). We present some experimental results to clarify this in Sect. 5.3.

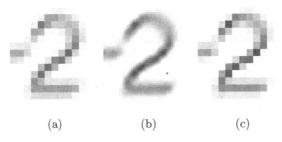

(a)   (b)   (c)

Fig. 4. Deblurring effect of the SR step on an image with small blur. (a) Blurred LR image. (b) SR result with exemplars. (c) Downsampled image of (b). The strokes in (c) are sharper than those of (a).

5   Experimental Results

We conduct the experiments on a machine with a 3.60 GHz Core i7 CPU and 24 GB RAM. Our implementation is in MATLAB and it takes about 9 min to process an image of 255×255 pixels. In all the experiments, we set $\beta = 0.1$, $\lambda = 0.006$, and $\sigma = 1$, respectively. The parameter γ is set according to [2], and τ in (13) is adjusted according to the noise level. In the locally uniform model, we split the image into patches and compute the kernel basis b_θ beforehand according to [32]. We allow 50% overlapping for neighboring patches to reduce the ringing artifact and the patch size is automatically determined by the support size of the kernel. The kernel estimations are carried out on gray-scale images. The final image is reconstructed by applying deconvolution on each channel. We present a few examples in this section and put more results in the supplementary material. For the comparison to other methods, we use the provided codes or the results generated by the authors.

For the SR method, we collect a set of 6,152 natural images as HR exemplars and downsample them with a Gaussian kernel to obtain LR images. We note that a large amount of exemplars is essential for the success of the SR methods. The mapping function is then constructed based on the corresponding LR and HR patches. In the experiments, we use the scale factor $s = 2$ and present the experiment on the scale factor in Sect. 5.3.

5.1   Comparison to Deblurring Methods

We evaluate the proposed algorithm against the state-of-the-art deblurring methods based on edges [2,5,6], exemplars [26,44], and non-uniform approaches [2,30].

Blurred LR images commonly occur in video surveillance. Due to the quality of the sensor, some important information in the captured images is missing, which increases the difficulty for deblurring. Figure 5(a) shows a blurred image captured by a video camera obtained from [45], in which the objects are of small scale and low-resolution. We compare our algorithm with the state-of-the-art deblurring methods [2,5,6,26,30,40]. Due to assumption of uniform blur, the methods [5,6,40] do not recover clean results. Sun et al. [26] make use of exemplars to guide kernel estimation and reconstruct some image regions well. However, this method does not perform well for images with non-uniform blur. The results of the methods [2,30] contain ringing artifacts, which indicate that these two methods do not fully remove the blur. In contrast, the proposed method uses the ℓ_0 prior as [40] and takes a spatial-variant blur model into account. Moreover, our algorithm utilizes the exemplars in a SR manner and the reconstructed result contains less blur, which shows that the camera motion has been accurately estimated. The results show that the proposed algorithm is able to deblur images by exploiting image structured from SR results based on exemplars.

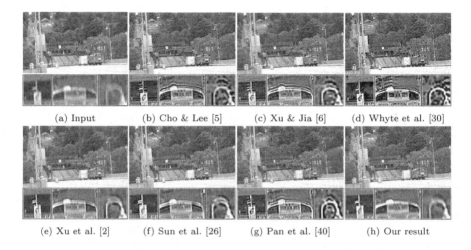

(a) Input   (b) Cho & Lee [5]   (c) Xu & Jia [6]   (d) Whyte et al. [30]

(e) Xu et al. [2]   (f) Sun et al. [26]   (g) Pan et al. [40]   (h) Our result

Fig. 5. A real blurry image which is obtained from the blurred video sequences [45].

We present another real blurry example in Fig. 6. The key success of existing deblurring methods [2,5,6,30,40,44] is to recover salient structure for kernel estimation. We note the methods [5,6,40,44] generate results with ringing artifacts due to the usage of the uniform model. The non-uniform deblurring methods [2,30] recover the regions of arms well, but they still contain ringing artifacts in the face region, which indicates that the camera motion is not well estimated. In contrast, our method is able to restore the image with less ringing artifacts.

Quantitative evaluations on the datasets [13] and [25]: We evaluate the proposed method on the deblurring benchmark datasets [13] and [25] although these two image sets are designed for general purpose.

For the datasets [13], we use the same non-blind deconvolution scheme [7] to generate the results for fair comparisons. The error ratio metric [13] is employed to measure the quality of each algorithm. Overall, the proposed algorithm performs well under the ratio value 2.5 (Fig. 7(a)). Although the proposed method is

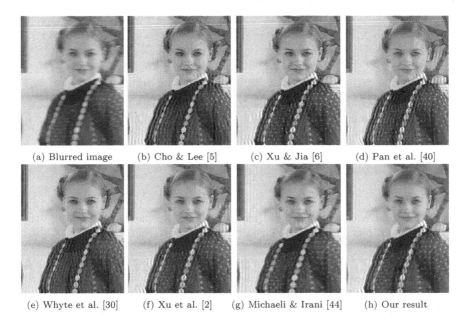

(a) Blurred image   (b) Cho & Lee [5]   (c) Xu & Jia [6]   (d) Pan et al. [40]

(e) Whyte et al. [30]   (f) Xu et al. [2]   (g) Michaeli & Irani [44]   (h) Our result

Fig. 6. Experimental results on a real blurry image.

designed to deblur low-resolution images, it performs well on uniform deblurring with high-resolution images.

We further evaluate the proposed algorithm on the dataset [25], which consists of 4 clear images and 12 camera motions. Figure 8 shows the average PSNR values of different deblurring methods on this dataset. Overall, the proposed method does not work well on the blurry images with large blur kernels. The reason is that the SR step is applied on the estimated latent image in the kernel estimation. If the estimated latent image still contains blur, the reconstructed HR image will be affected by blur effect and thus leads to unreliable kernel estimation results. In this case, the blur would not be reduced in the SR step as it is likely to match the smooth region in the HR image construction. We present the average PSNR values on the same dataset [25] without large blur motions in Fig. 8(b). The results reported in Fig. 8(b) do not contain the blurred examples degraded by the 4 large blur motions in the dataset [25]. The evaluation results demonstrate that the proposed method achieves comparable results to the top-performing methods in this case.

Influence of blur kernel size: We also evaluate the proposed algorithm by using synthetic examples with different level of blur. Figure 7(b) shows that the proposed method performs better than the baseline method [2] when the blur kernel is small. However, compared with the baseline method [2], it does not perform well when the blur kernel is large due to the influence SR step, which indicates that the adopted SR scheme does not perform well with large blur kernels.

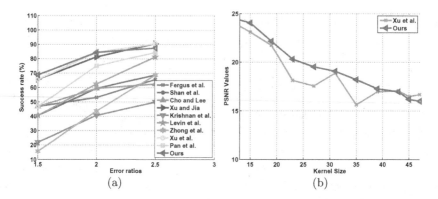

Fig. 7. Quantitative comparisons. (a) Quantitative comparison on the dataset [13]. (b) PSNR plot w.r.t. blur kernel size (image size: 255 × 255 pixels).

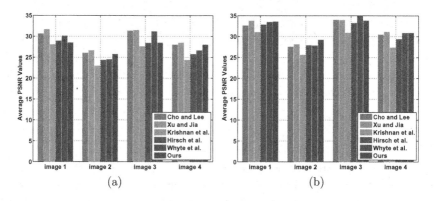

Fig. 8. Quantitative comparison on the dataset [25]. (a) The results on the whole dataset. (b) The results without including large blur examples.

5.2   Blind Super Resolution

The proposed algorithm can generate SR results as byproducts. In this section, we present the results of in which we first apply the state-of-the-art deblurring methods [2] to the inputs and then super resolution schemes [39] sequentially. Furthermore, we compare the proposed algorithm with the exemplar-based and blind super resolution methods [38,39].

In Fig. 9, we apply the non-uniform camera motion (rotation and translation) to a clear image (a) and then downsample it with the scale factor 2 to obtain the LR image (b). Since this image contains motion blur effects, the approaches that simply applies bicubic interpolation or traditional SR method [39] fails to generate clear HR images (Fig. 9(c) and (d)). One straightforward way to solve this problem is to first apply state-of-the-art deblurring methods and then use traditional SR methods to recover HR images. However, estimating reliable blur kernels from LR images is not a trivial task according to our analysis in Sect. 3.

Thus, this simple strategy does not yield a clear image (Fig. 9(e)). The recent method [38] focuses on the blind SR problem by first estimating a blur kernel from a LR input, and then employing SR methods, e.g., [46], with the estimated kernels to restore a HR image. However, this method is less effective for motion blur, especially for the non-uniform scenarios (see Fig. 9(f)). Compared with those methods, our algorithm generates clear images with sharp contrast and less blur effect.

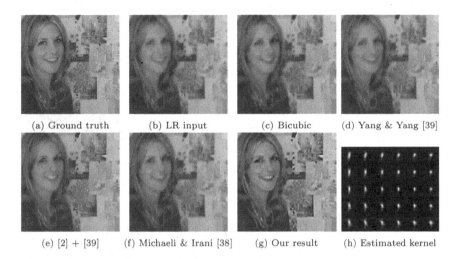

(a) Ground truth   (b) LR input   (c) Bicubic   (d) Yang & Yang [39]

(e) [2] + [39]   (f) Michaeli & Irani [38]   (g) Our result   (h) Estimated kernel

Fig. 9. Comparison on the SR results. The input is of size 188×183 and the scale factor for the SR is set to be 2. The LR input in (b) is shown by nearest neighbor interpolation. (best viewed on high-resolution display).

Figure 10 shows an example which is obtained from [38]. The results from bicubic interpolation and the fast super resolution method [39] are blurry. Although the result generated by [38] is much clearer, the highlighted parts (in boxes) contain some blur compared to our result in (g).

5.3 Effect of Scale Factor

In the proposed algorithm, we make use of the SR technique to recover salient structures for blur kernel estimation. To better understand how the SR technique influences the deblurring process, we show experimental results using different scale factor s. We evaluate the proposed method on a set of 32 blurry images with the scale factor $s = 1, 2, \ldots, 5$. For the case with scale factor $s = 1$, to exploit exemplar information, we simply apply the patch-based reconstruction by searching and pasting the most similar clear patches. As shown in Table 1, the results using SR to exploit exemplar information ($s \geq 2$) are better than those using the direct patch-based reconstruction ($s = 1$) and those not using exemplar information. These results demonstrate that using SR techniques helps extracting salient structures for blur kernel estimation.

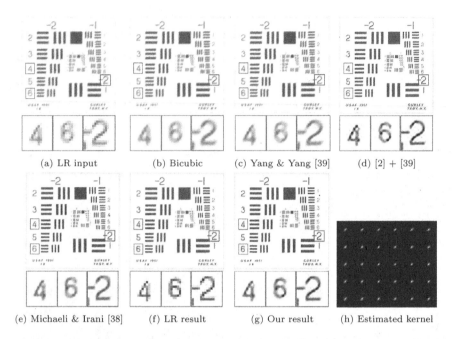

(a) LR input   (b) Bicubic   (c) Yang & Yang [39]   (d) [2] + [39]

(e) Michaeli & Irani [38]   (f) LR result   (g) Our result   (h) Estimated kernel

Fig. 10. A SR example obtained from blind SR method [38]. The input is of size 145×142 and the scale factor for the SR is set to be 2. The LR input in (a) is shown by nearest neighbor interpolation.

Table 1. Quantitative comparisons on different scale factors s.

	Without exemplars	$s = 1$	$s = 2$	$s = 3$	$s = 4$	$s = 5$
Average PSNRs	22.97	25.65	25.73	25.73	25.74	25.73

6   Conclusion and Discussion

In this paper, we propose a spatially variant deblurring method using exemplar-based SR method as a guidance. The proposed method exploits exemplar information to help recover salient structure. The experiments demonstrate its effectiveness of using SR to exploit external information. In addition, the byproduct SR results are comparable to those by the state-of-the-art methods.

The main drawback of the method is that the SR step for restoring salient structure may not always be stable. If the intermediate image still contains significant blur, the results after applying SR would still be blurry as the SR mapping function is learned from clear images. Our future work will focus on how to address these issues with significant motion blurs.

Acknowledgements. This work has been supported in part by NSF CAREER (No. 1149783), NSF IIS (No. 1152576), NSFC (No. 61572099 and 61320106008) and a gift from Adobe.

References

1. Sroubek, F., Cristóbal, G., Flusser, J.: A unified approach to superresolution and multichannel blind deconvolution. IEEE TIP **16**, 2322–2332 (2007)
2. Xu, L., Zheng, S., Jia, J.: Unnatural L_0 sparse representation for natural image deblurring. In: CVPR, pp. 1107–1114 (2013)
3. Fergus, R., Singh, B., Hertzmann, A., Roweis, S.T., Freeman, W.T.: Removing camera shake from a single photograph. ACM TOG **25**, 787–794 (2006)
4. Shan, Q., Jia, J., Agarwala, A.: High-quality motion deblurring from a single image. ACM TOG **27**, 73 (2008)
5. Cho, S., Lee, S.: Fast motion deblurring. ACM TOG **28**, 145 (2009)
6. Xu, L., Jia, J.: Two-phase Kernel estimation for robust motion deblurring. In: Daniilidis, K., Maragos, P., Paragios, N. (eds.) ECCV 2010. LNCS, vol. 6311, pp. 157–170. Springer, Heidelberg (2010). doi:10.1007/978-3-642-15549-9_12
7. Levin, A., Weiss, Y., Durand, F., Freeman, W.T.: Efficient marginal likelihood optimization in blind deconvolution. In: CVPR, pp. 2657–2664 (2011)
8. Krishnan, D., Tay, T., Fergus, R.: Blind deconvolution using a normalized sparsity measure. In: CVPR, pp. 2657–2664 (2011)
9. Goldstein, A., Fattal, R.: Blur-Kernel estimation from spectral irregularities. In: Fitzgibbon, A., Lazebnik, S., Perona, P., Sato, Y., Schmid, C. (eds.) ECCV 2012. LNCS, vol. 7576, pp. 622–635. Springer, Heidelberg (2012). doi:10.1007/978-3-642-33715-4_45
10. Yuan, L., Sun, J., Quan, L., Shum, H.: Image deblurring with blurred/noisy image pairs. ACM TOG **26**, 1 (2007)
11. Jia, J.: Mathematical Models and Practical Solvers for Uniform Motion Deblurring. Cambridge University Press, Cambridge (2014)
12. Lee, S., Cho, S.: Recent advances in image deblurring. In: SIGGRAPH Asia 2013 Course (2013)
13. Levin, A., Weiss, Y., Durand, F., Freeman, W.T.: Understanding and evaluating blind deconvolution algorithms. In: CVPR, pp. 1964–1971 (2009)
14. Lou, Y., Bertozzi, A.L., Soatto, S.: Direct sparse deblurring. J. Math. Imaging Vis. **39**, 1–12 (2011)
15. Zhang, H., Yang, J., Zhang, Y., Huang, T.S.: Sparse representation based blind image deblurring. In: ICME, pp. 1–6 (2011)
16. Cai, J.F., Ji, H., Liu, C., Shen, Z.: Framelet based blind motion deblurring from a single image. IEEE TIP **21**, 562–572 (2012)
17. Zoran, D., Weiss, Y.: From learning models of natural image patches to whole image restoration. In: ICCV, pp. 479–486 (2011)
18. Roth, S., Black, M.J.: Fields of experts: a framework for learning image priors. In: CVPR, pp. 860–867 (2005)
19. Takeda, H., Farsiu, S., Milanfar, P.: Deblurring using regularized locally adaptive Kernel regression. IEEE TIP **17**, 550–563 (2008)
20. Sun, L., Cho, S., Wang, J., Hays, J.: Good image priors for non-blind deconvolution. In: Fleet, D., Pajdla, T., Schiele, B., Tuytelaars, T. (eds.) ECCV 2014. LNCS, vol. 8692, pp. 231–246. Springer, Heidelberg (2014). doi:10.1007/978-3-319-10593-2_16
21. Zhang, H., Wipf, D.P., Zhang, Y.: Multi-image blind deblurring using a coupled adaptive sparse prior. In: CVPR, pp. 1051–1058 (2013)
22. Zhang, H., Wipf, D.P.: Non-uniform camera shake removal using a spatially-adaptive sparse penalty. In: NIPS, pp. 1556–1564 (2013)

23. Perrone, D., Favaro, P.: Total variation blind deconvolution: the devil is in the details. In: CVPR, pp. 2909–2916 (2014)
24. Joshi, N., Szeliski, R., Kriegman, D.J.: PSF estimation using sharp edge prediction. In: CVPR (2008)
25. Köhler, R., Hirsch, M., Mohler, B., Schölkopf, B., Harmeling, S.: Recording and playback of camera shake: benchmarking blind deconvolution with a real-world database. In: Fitzgibbon, A., Lazebnik, S., Perona, P., Sato, Y., Schmid, C. (eds.) ECCV 2012. LNCS, vol. 7578, pp. 27–40. Springer, Heidelberg (2012). doi:10.1007/978-3-642-33786-4_3
26. Sun, L., Cho, S., Wang, J., Hays, J.: Edge-based blur kernel estimation using patch priors. In: ICCP (2013)
27. HaCohen, Y., Shechtman, E., Lischinski, D.: Deblurring by example using dense correspondence. In: ICCV, pp. 2384–2391 (2013)
28. Pan, J., Hu, Z., Su, Z., Yang, M.-H.: Deblurring face images with exemplars. In: Fleet, D., Pajdla, T., Schiele, B., Tuytelaars, T. (eds.) ECCV 2014. LNCS, vol. 8695, pp. 47–62. Springer, Heidelberg (2014). doi:10.1007/978-3-319-10584-0_4
29. Tai, Y., Tan, P., Brown, M.S.: Richardson-Lucy deblurring for scenes under a projective motion path. IEEE PAMI 33, 1603–1618 (2011)
30. Whyte, O., Sivic, J., Zisserman, A., Ponce, J.: Non-uniform deblurring for shaken images. IJCV 98, 168–186 (2012)
31. Gupta, A., Joshi, N., Zitnick, C.L., Cohen, M.F., Curless, B.: Single image deblurring using motion density functions. In: ECCV, pp. 171–184 (2010)
32. Hirsch, M., Schuler, C.J., Harmeling, S., Schölkopf, B.: Fast removal of non-uniform camera shake. In: ICCV, pp. 463–470 (2011)
33. Harmeling, S., Hirsch, M., Schölkopf, B.: Space-variant single-image blind deconvolution for removing camera shake. In: NIPS, pp. 829–837 (2010)
34. Kim, T.H., Ahn, B., Lee, K.M.: Dynamic scene deblurring. In: ICCV, pp. 3160–3167 (2013)
35. Kim, T.H., Lee, K.M.: Segmentation-free dynamic scene deblurring. In: CVPR, pp. 2766–2773 (2014)
36. Chakrabarti, A., Zickler, T., Freeman, W.T.: Analyzing spatially-varying blur. In: CVPR, pp. 2512–2519 (2010)
37. Harmeling, S., Sra, S., Hirsch, M., Schölkopf, B.: Multiframe blind deconvolution, super-resolution, and saturation correction via incremental EM. In: ICIP, pp. 3313–3316 (2010)
38. Michaeli, T., Irani, M.: Nonparametric blind super-resolution. In: ICCV, pp. 945–952 (2013)
39. Yang, C.Y., Yang, M.H.: Fast direct super-resolution by simple functions. In: ICCV, pp. 561–568 (2013)
40. Pan, J., Hu, Z., Su, Z., Yang, M.H.: Deblurring text images via L_0-regularized intensity and gradient prior. In: CVPR, pp. 2901–2908 (2014)
41. Levin, A., Fergus, R., Durand, F., Freeman, W.T.: Image and depth from a conventional camera with a coded aperture. ACM TOG 26, 70 (2007)
42. Zhang, H., Yang, J., Zhang, Y., Huang, T.S.: Close the loop: joint blind image restoration and recognition with sparse representation prior. In: ICCV, pp. 770–777 (2011)
43. Hu, Z., Huang, J.B., Yang, M.H.: Single image deblurring with adaptive dictionary learning. In: ICIP, pp. 1169–1172 (2010)

44. Michaeli, T., Irani, M.: Blind deblurring using internal patch recurrence. In: Fleet, D., Pajdla, T., Schiele, B., Tuytelaars, T. (eds.) ECCV 2014. LNCS, vol. 8691, pp. 783–798. Springer, Heidelberg (2014). doi:10.1007/978-3-319-10578-9_51

45. Cho, S., Wang, J., Lee, S.: Video deblurring for hand-held cameras using patch-based synthesis. ACM TOG **31**, 64:1–64:9 (2012)

46. Glasner, D., Bagon, S., Irani, M.: Super-resolution from a single image. In: ICCV, pp. 349–356 (2009)

Visual Smoke Detection

Abhishek Kumar Tripathi$^{(\boxtimes)}$ and Shanti Swarup

Image Understanding Group, Uurmi Systems Pvt. Ltd., Hyderabad, India
{abhishekt,shanti}@uurmi.com

Abstract. In this paper, we have proposed a novel and efficient visual smoke detection algorithm. Smoke detection in video surveillance is very important for early fire detection. Proposed algorithm uses an unique combination of features to detect smoke efficiently. These features use appearance, energy and motion properties of the smoke. Further analysis of past history of smoke increases the accuracy of the algorithm. These features are less complex and enable the algorithm for real time application. A general assumption is that smoke is a low frequency signal which may smoothen the background. We focused on the nature of the smoke (shape disorder, energy reduction and variability over time) and proposed a novel algorithm which requires no user intervention and prior data training. Due to the large variability in the feature values, we assigned the fuzzy membership to these features instead of hard thresholding to reduce classification errors. Simulation carried out with available dataset, show that smoke is accurately localized both in time and space via proposed approach.

1 Introduction

The ability to provide early and accurate alerts are the critical objective for any smoke detection system and achieving that objective is the primary challenge. The earliest possible warning of any potential smoke increases the time available for responders to take necessary action. This in turn minimizes risks to life and property. Video-based smoke detection techniques detect smoke anywhere within the field of view of the camera. Video based smoke detection techniques have many advantages such as fast response, non-contact, indoor and outdoor detection and ability to record video for smoke progress analysis.

Designing of a smoke detection algorithm is challenging due to following reasons: (1) Variability in density, lighting, diverse background, interfering non-rigid objects etc., (2) None of the image features such as intensity, motion, energy, and obscuration characterizes smoke well, and (3) Visual pattern of smoke is difficult to model.

The complex temporal behavior of the smoke makes the smoke detection very challenging. These behaviors are random shape and motion, change in color and density. Smoke detection algorithm has wide range of applications. It can be used for indoor (Industrial plants, Warehouse, Cold storage, Large boilers & equipments protection, and Laboratories) and outdoor (Forestry services, Mining,

C.-S. Chen et al. (Eds.): ACCV 2016 Workshops, Part I, LNCS 10116, pp. 128–142, 2017.
DOI: 10.1007/978-3-319-54407-6_9

Oil fields, and Gas stations) surveillance. It can be used to protect valuable assets which require continuous monitoring and it can be programmed to release CO_2 when smoke is detected.

2 Background

Many smoke detection algorithms are proposed in past few years. Mainly, these algorithms use appearance, energy, motion features and/or combination of these features (hybrid) for successful detection of smoke.

In the category of appearance features based algorithm; Wang et al. [1] proposed an appearance based real-time smoke detection algorithm. This algorithm uses modified CS-LTP (centre symmetric - local ternary pattern) texture features combined with color information for smoke detection. Difference in the histograms of CS-LTP features of background and foreground detects the smoke regions. Color information is used for further refinement. Relying only on the appearance, restrict the algorithm performance. Maruta et al. [2] proposed a novel smoke detection algorithm based on fractal property. Here, it is assumed that image information of smoke is self affine fractal and local Hurst exponent characterize this features. Value of Hurst exponent decides the presence of smoke, and temporal consistency in exponent gives further confidence in localizing the smoke regions. Chen et al. [3] proposed color and texture features based smoke detection algorithm. Inter frame difference and Local Binary Pattern (LBP) are used to analyze dynamic characteristics of the smoke. Later smoke history image is constructed to reduce false alarms.

In the category of motion features based algorithm; Brovko et al. [4] proposed a motion analysis based smoke detection algorithm. This algorithm uses dynamic and static features of the smoke. Here, optical flow analysis is used to separate out smoke region with the assumption that smoke grows in certain direction only. This assumption restricts the algorithm performance in outdoor scenarios where, smoke can grow in any direction due to wind. Wang et al. [5] proposed an early smoke detection algorithm based on swaying and diffusion features. First, moving regions are separated out using fuzzy integral. Later, smoke regions are determined using analysis of swaying nature of these regions based on centroid. Diffusion features are used for further refinement with the assumption that smoke diffuses over time. For outdoor scenarios, these swaying features generate false alarm when bushes are present in the scene. Li et al. [6] proposed a novel smoke detection algorithm based on fuzzy and mobility characteristics. Modified GMM is used for detection of moving region and characteristics of region mobility are used for final detection of smoke regions. Here, assumption is that, in smoke region displacement in the bottom is slower than that of central of the whole region. However, this assumption fails in many outdoor conditions.

In the category of energy based algorithms; Gonzalez et al. [7] proposed an energy features based smoke detection algorithm. Stationary wavelet transform is used for energy estimation with the assumption that smoke reduces the high frequency details in foreground. To rely on the potential smoke regions, area features are used. Here, it is assumed that the area of the smoke region increases continuously. However, sometimes it happens that one smoke region is partitioned into smaller regions and contradict this assumption.

In the category of hybrid features based algorithms; Li et al. [8] proposed a wavelet features and optical flow based smoke detection algorithm. First, moving areas are detected using Gaussian mixture model. Out of these moving regions, smoke regions are defined where energy reduces in foreground with respect to background. Optical flow is used for further refinement. Chen et al. [9] introduced a new concept of contrast image. On the basis of this contrast image wavelet analysis is performed on the background and foreground images. It is assumed that the smoke is low frequency component, ratio of high frequency and low frequency components is used to decide whether motion region is smoke or not.

Calderara et al. [10] proposed, a smoke detection algorithm by analyzing color and texture features of the moving objects extracted from mixture of Gaussians. Here, wavelet analysis is performed to study the change in energy in smoke regions. Then color analysis based on blending function is used to separate out candidate smoke regions. Avgerinakis et al. [11] proposed appearance and motion features based smoke detection algorithm. Histogram of gradients and optical flow (HOG-HOF) is used to localize smoke in time domain. This information is used along with the color features for classification of smoke region.

Toreyin et al. proposed [12] appearance and color features based spatio-temporal smoke detection algorithm. It is assumed that edges present in image start losing their sharpness over time. This behavior is analyzed using wavelet transform. Periodic behavior of smoke boundaries are analyzed by a Hidden Markov Model. Reduction in energy, periodic behavior in boundaries and reduction in chrominance values, all these clues are used to make final decision of smoke region.

2.1   Contributions

Prior art algorithms are based on the certain assumptions like color of smoke, direction of motion, and energy. It is observed that these features restrict the algorithm performance. However, effective combination of these features in spatial and temporal domain may resolve these issues. Proposed algorithm uses hybrid features for detection of smoke. Novelties in the proposed algorithms are as follows: Unique combination of features (energy, appearance (shape disorder and transparency) and motion (consistency and variability over time)) enables us to detect smoke effectively. Due to the large variability of smoke features, assignment of fuzzy membership helps us to detect smoke effectively. Distance based classification with past history increases the accuracy of the algorithm. No prior data training is required for classification. Algorithm is not using any color features, which makes it robust to detect smoke of any color. Computation of these features are less complex which confirms the real-time application of visual smoke detection. Algorithm analyzes the energy reduction and shape disorder in spatio-temporal domain, which reduces the false alarm significantly. Smoke is transparent in nature, algorithm analyzes the transparency (blending with background) which reduces false alarm. Algorithm analyzes smoke features in block as well as blob level, which reduces miss and false alarms. Proposed algorithm requires a buffer size of 5 frames, which is very low in comparison with prior art algorithms. Existing algorithms require 9 frames to 50 frames.

3   Our Approach

Our smoke detection algorithm analyzes the spatio-temporal behavior of the smoke. Block diagram of the proposed smoke detection algorithm is shown in Fig. 1. First we subtract the background using adaptive Gaussian mixture model (GMM) to detect all foreground (moving) objects present in the scene. Then we estimate the transparency of each moving objects. Semi-transparent moving objects are possible smoke regions. Feature analysis of these possible smoke regions is performed and distance based classifier separate out potential smoke regions and non-smoke regions.

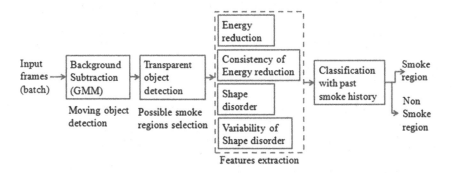

Fig. 1. Block diagram of the proposed smoke detection algorithm.

Adaptive Gaussian mixture model is a reliable method to approximate the background modeling. It models each pixel as a mixture of Gaussians and update the model using an online (Expectation-Maximization) EM algorithm. This model deals with illumination changes and repetitive motion from clutter efficiently. Background modeling by Gaussian mixtures is a pixel based process. Let x is a random process representing intensity value of a pixel at given time. Let $p(x)$ represents a Gaussian mixture having K density components.

$$p(x) = \sum_{k=1}^{K} w_k N(x; \mu_k, \sigma_k) \tag{1}$$

where, w_k are weights and $N(x; \mu_k, \sigma_k)$ is normal distribution with mean μ_k and covariance $\sigma_k I$. GMM estimates these parameters over time to model background. Parameters are initialized with $w_k = w_0$, $\mu_k = \mu_0$ and $\sigma_k = \sigma_0$, if there is a match

$$\frac{\| x - \mu_i \|}{\sigma_i} < \tau \text{ for some } i \in [1, ..., K] \tag{2}$$

where, τ is threshold. These parameters are updated as follows:

$$w_k = (1 - \eta_1)w_k(t-1) + \eta_1 M_k(t)$$
$$\mu_k = (1 - \eta_2)\mu_k(t-1) + \eta_2 x \tag{3}$$
$$\sigma_k^2(t) = (1 - \eta_2)\sigma_k^2(t-1) + \eta_2 \parallel x - \mu_k(t) \parallel^2$$

where, η_1 and η_2 are learning parameters and $M_k(t)$ is 1 for matched components otherwise 0. If there is no match then lowest weight components are reinitialized. The K distributions are sorted in ascending order by $\frac{w_i}{\sigma_i}$. This ordering moves the most probable background and low variance at the top. The first B Gaussians which exceeds the threshold (ρ) is retained for a background defined as

$$B = \underset{x}{\mathrm{argmin}}(\sum_{k=1}^{K} w_k > \rho) \tag{4}$$

If the value of threshold (ρ) is small, the background model is uni-model and is multi-model for high value of threshold. Intuitively, Gaussians with high probability of occurrence and low variability are most likely to be background. Once updated background image is available, we subtract the current frame with the corresponding pixels at the same position on the background image and extract all moving object blobs.

Then for the detection of the possible smoke blobs, transparency of the each moving object is calculated. Here, assumption is that smoke regions are more transparent than non-smoke regions. Current frame I_t is modeled as a combination of foreground FG_t and background BG_t using blending parameter λ ($\lambda \in [0, 1]$) as mentioned in following equation.

$$I_t = \lambda FG_t + (1 - \lambda)BG_t \tag{5}$$

For transparent region λ is 0 and for opaque region λ is 1. For semitransparent region like smoke, value of λ lie within $[0, 1]$. It is claimed [4] that optimum value of λ for smoke is 0.38. Once we have background image BG_t and λ, we can get the foreground FG_t using Eq. (5). Possible smoke candidate region mask for each pixel can be obtained using following condition.

$$Mask^{ij} = \begin{cases} \text{Smoke } (1), & \text{if } FG_t^{ij} > T \\ \text{Non-smoke } (0), & \text{otherwise} \end{cases} \tag{6}$$

where, T is a pre-determined threshold. With this transparent object detection, we separate out smoke regions and other moving objects. For further refinement of the smoke regions, we examined the features of each possible smoke region. We have analyzed the smoke features at block as well as blob level in spatio-temporal domain. It is noted that we have analyzed only those blocks and blobs which support both moving and transparent objects criteria. For block level analysis, we subdivided the image into non overlap blocks and estimated following features for each block.

1. Reduction in energy (α)
2. Consistency of the energy reduction over time (β)

Here, α and β are analyzed in spatial and temporal domain respectively. For blob level analysis, we estimated the following features for each candidate blob.

1. Shape disorder (γ)
2. Variability of the shape disorder over time (δ)

Here, γ and δ are analyzed in spatial and temporal domain respectively. Detail descriptions of these smoke features are as follows:

3.1   Reduction in Energy (α)

It is noted that smoke creates blur in the scene or reduces the sharpness. As edges get blurred, we can assume that smoke reduces the energy of the background. We have computed the energy using wavelet transform. Wavelet transform is typically computed by applying a separable filter bank to the image, given by

$$
\begin{aligned}
cA &= (L_x * (L_y * I) \downarrow_{2,1}) \downarrow_{1,2} \\
cH &= (H_x * (L_y * I) \downarrow_{2,1}) \downarrow_{1,2} \\
cV &= (L_x * (H_y * I) \downarrow_{2,1}) \downarrow_{1,2} \\
cD &= (H_x * (H_y * I) \downarrow_{2,1}) \downarrow_{1,2}
\end{aligned}
\tag{7}
$$

where, I is the original image and $*$ denotes convolution operation in two dimension. $\downarrow_{x,y}$ denotes decimation by x and y in corresponding dimension. L and H are low and high pass filters respectively. Coefficient cA is obtained by low pass filtering in both directions and termed as approximation image. cH, cV and cD are details coefficients obtained via high pass filtering in one or more directions. Due to the sub-sampling involved in this estimation, the total number of coefficients is equal to the pixels present in the original image.

Energy of each block is represented as follows:

$$
E = \frac{cH^2 + cV^2 + cD^2}{cA^2}
\tag{8}
$$

For smoke region, there is a reduction in energy (since details get blurred), If E_{curr} and E_{bg} are the energies of a block in current and background image then reduction in energy (α) is estimated as follows:

$$
\alpha = \frac{E_{curr}}{E_{bg}}
\tag{9}
$$

Reduction in energy (α) for smoke and non smoke region is shown in Fig. 2. It is observed that smoke reduces the sharpness of the scene which result in the reduction of energy. For stationary background or non smoke region, there is no change in energy. The value of α is low for smoke block compared to non-smoke block.

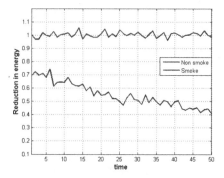

Fig. 2. Reduction of energy with respect to background for smoke and non smoke region.

3.2   Consistency of the Energy Reduction over Time (β)

With time, smoke gets denser and energy at that region keeps on reducing. Since, for smoke block there is a consistent decrease in energy, whereas for rigid moving object energy reduction is not consistent and has fluctuations. As shown in Fig. 2, over time as the smoke density increases, decrease in the reduction in energy continues. Consistency of the energy reduction is estimated by counting number of times α is low (below a certain threshold) for a particular block location over time. Consistency of the energy reduction (β) is estimated as follows:

$$\beta = \left[\sum_{i=0}^{n-1} H[\tau - \alpha_i] \right] \tag{10}$$

where, H is the heaviside step function, whose value is 0 for negative argument and 1 for positive argument. n is the number of frames and τ ($\tau < 1$) is a threshold. Value of β is more for smoke region compared to non smoke region. Maximum value β can attain is n i.e. the number of frames under consideration.

3.3   Shape Disorder (γ)

Disorder or irregularity in shape discriminates the smoke region from other objects [12]. Generally man made objects have simple shape or less disorder in comparison with smoke.

We sampled the shape boundary of each region into N linearly separated points and computed the distance of these points from the centroid of the shape. We got a distance vector $d[n]$ of size $1 \times N$. We performed $1D$ wavelet decomposition of this distance vector. Wavelet decomposition gives low band approximation signal and high band detail signal. Distance vector is decomposed into low band $a[l]$ and high band $b[l]$ sub-signal using following equations:

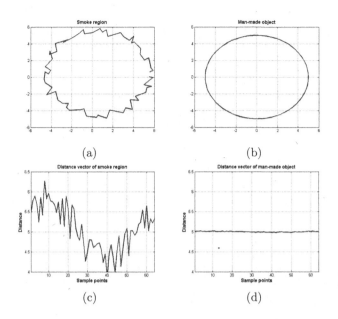

Fig. 3. Shape disorder analysis, shape of a (a) smoke region, and (b) man made object, boundary distance vector of (c) smoke region, and (d) man made object. (axes are adjusted for display).

$$a[l] = \sum_{n} h[2l - n]d[n] \quad \text{where, } h[l] = \{0.25, 0.5, 0.25\}$$

$$b[l] = \sum_{n} g[2l - n]d[n] \quad \text{where, } g[l] = \{-0.25, 0.5, -0.25\} \tag{11}$$

Shape disorder (γ) is calculated as follows:

$$\gamma = \frac{\sum_{l} b[l]}{\sum_{l} a[l]} \tag{12}$$

It is observed that value of γ is more for smoke region compared to non-smoke region. Figure 3 shows the shape and corresponding distance vector of smoke region and man made object. It is evident from the figure that smoke region is having more disorder compared to man made object.

3.4 Variability of the Shape Disorder over Time (δ)

For the estimation of shape disorder variability, we estimated the variation in γ over time. δ is estimated as follows:

$$\delta = \frac{1}{n-1}\left[\sum_{i=0}^{n} x_i^2 - \frac{1}{n}\left(\sum_{i=0}^{n-1} x_i\right)^2\right] \quad \text{where, } x_i = \gamma_i - \frac{1}{n}\left(\sum_{i=0}^{n-1} \gamma_i\right) \tag{13}$$

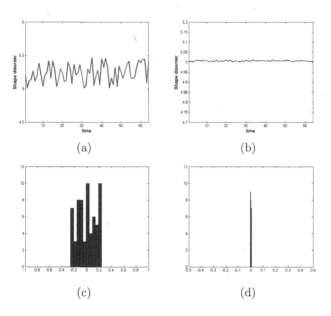

Fig. 4. Variability of the shape disorder over time. Shape disorder over time for (a) smoke region, (b) man made object or non smoke region, Histogram of (c) signal shown in (a), and (d) signal shown in (b). (axes are adjusted for display).

where, γ_i is the shape disorder of a particular block in i^{th} frame and n is the number of frames under observation. Time profile of shape disorder signal and its corresponding histogram (for zero mean signal) is shown in Fig. 4. It is evident from the signal profile and its corresponding histogram that value of δ is high for smoke region compared to non-smoke region.

3.5   Fuzzy Membership Assignment to Features

These estimated features (α, β, γ, and δ) have varying range and very fuzzy in nature. Pre-determined threshold won't be sufficient to discriminate smoke and non-smoke regions. We have assigned a membership to these features. Triangular and S shaped memberships are used for these features.

Triangular membership is denoted as

$$f(x; a, b, c) = \begin{cases} 0; & x \leq a \\ \frac{x-a}{b-a}; & a \leq x \leq b \\ \frac{c-x}{c-b}; & a \leq x \leq b \\ 0; & c \leq x \end{cases} \qquad (14)$$

S Shaped membership is denoted as

$$f(x; a, b) = \begin{cases} 0; & x \leq a \\ 2\left(\frac{x-a}{b-a}\right)^2; & a \leq x \leq \frac{a+b}{2} \\ 1 - 2\left(\frac{x-b}{b-a}\right)^2; & \frac{a+b}{2} \leq x \leq b \\ 1; & x \geq b \end{cases} \tag{15}$$

an example of triangular membership (trimf) for $f(x; 3, 6, 8)$ and S shaped membership (smf) for $f(x; 1, 8)$ is shown in Fig. 5.

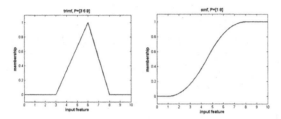

Fig. 5. Membership functions, (left) Triangular membership for $f(x; 3, 6, 8)$, (right) S shaped membership for $f(x; 1, 8)$.

1. **Triangular membership for α:** Due to smoke, there is a reduction in energy at a particular block with respect to the background. However, other moving objects may also increase or decrease the energy. Triangular membership function put a cut-off for abrupt energy reduction and increase. Here, assumption is that change in energy (both reduction and increase) due to rigid moving objects is very large. This fuzzy feature is termed as f_α where, $f_\alpha \in [0, 1]$.

2. **S shaped membership for β:** For a smoke block, energy reduction remains consistent, but other moving objects give fluctuation in energy reduction which reduces the consistency of energy reduction. S shaped membership function put a lower bound cut-off to give low membership to other moving objects. This fuzzy feature is termed as f_β where, $f_\beta \in [0, 1]$.

3. **S shaped membership for γ:** Shape disorder is more in case of smoke in comparison with other moving objects. Value of shape disorder is high for smoke and low for non-smoke or other moving objects. S shaped membership function fulfills this requirement. This fuzzy feature is termed as f_γ where, $f_\gamma \in [0, 1]$.

4. **S shaped membership for δ:** Due to random motion and shape disorder, smoke shape keep on changing over time. Other objects have fixed size and their shape do not change over time. S shaped membership function fulfills this requirement. This fuzzy feature is termed as f_δ where, $f_\delta \in [0, 1]$.

3.6 Past Smoke History (PSH)

We have used one more feature termed as Past Smoke History (PSH). Here, past smoke history is used to increase classification accuracy. Past smoke history is defined based on the previous frame classification decision. This confidence is denoted as the likeliness of the features to represent the assigned class (i.e. smoke and non-smoke).

Classification is performed at block level (non overlapping blocks). We have estimated f_α and f_β at block level and f_γ and f_δ at blob level. For block level classification, for each block f_γ and f_δ are assigned according to the corresponding blob i.e. all blocks fall inside or at the boundary of the blob are given same f_γ and f_δ. Estimated feature vector for b^{th} block of ith frame is denoted as $f_{bi} = [f_{\alpha i}, f_{\beta i}, f_{\gamma i}, f_{\delta i}, PSH_{b(i-1)}]$. For each block, f_α, f_β, f_γ, and f_δ are calculated from the current frame whereas, PSH is calculated from the last frame classification decision (see Eq. (17)). Inclusion of past smoke history as a feature avoid the toggling in classification decision in successive frames. Here, assumption is that a block classified as class t_k ($k \epsilon \{1, 2\}$) in particular frame will more likely to represent same class in next frame.

Target class for smoke feature can be represented as $t_1 = [1, 1, 1, 1, 1]$ (i.e. high fuzzy membership to each feature and ideally 1) and for non-smoke feature can be represented as $t_2 = [0, 0, 0, 0, 0]$ (i.e. low fuzzy membership to each feature and ideally 0).

Steps for class assignment are as follows:

Estimation: For every b^{th} block of ith frame, estimate feature f_{bi}

Matching: Each block is classified by estimating feature f_{bi} and choosing the closet target t_k using following equations:

$$\underset{k}{\operatorname{argmin}} \parallel f_{bi} - t_k \parallel \tag{16}$$

where, $\parallel . \parallel$ represents l^2 norm

Update: update smoke history of each bth block for next frame using following equations:

$$PSH_{bi} = min\left(max\left((PSH_{b(i-1)} + (-1)^{k-1}e^{-d}), 0\right), 1\right) \tag{17}$$

where, PSH_{bi} is the past smoke history of a bth block of ith frame, $d = \parallel f_{bi} - t_k \parallel$, t_k is the assigned target class ($k = 1$ for smoke and $k = 2$ for non smoke).

For every frame, we find out all the blocks supporting smoke target class and these blocks will support some blobs in the mask obtained after transparent object detection. All these supporting blobs are marked as the potential smoke regions. For display, we drawn the boundaries of these blobs on original image. Boundaries of these blobs are obtained by Moore-Neighbor tracing algorithm modified by Jacob's stopping criteria.

4   Experimental Results and Discussion

We implemented algorithm in C using OpenCV library on a system with a 3.29 GHz Intel(R) Core(TM) i3 CPU and 3.40 GB of RAM running on Ubuntu platform. Algorithm is running at 25 frames per second (fps). For quad core processor, CPU utilization is 25%, which ensures system can process 4 camera streams simultaneously. For temporal analysis, we assumed a batch of 5 frames and for spatial analysis we assumed a block of size 8. Frames are resized to 256×256 and each block is of size 8×8, thus, we have 1024 blocks and 1024×5 features (α, β, γ, δ, and PSH) for each frame. For the first frame, we initialized all blocks with PSH equal to 0.5 which means all blocks are equiprobable for smoke and non smoke.

Table 1. Description of the videos used for the experiment.

Video	Description	Challenge in detection
Video 1	Outdoor & moving objects	Person behind the fence with smoke color outfit
Video 2	Outdoor & moving objects	Dense smoke i.e. no reduction in energy over time
Video 3	Indoor	Very sparse smoke and cover entire field of view
Video 4	Outdoor & moving objects	Smoke at far distance
Video 5	Outdoor & moving objects	Illumination variation
Video 6	Indoor & moving objects	Background similar to smoke color
Video 7	Outdoor & moving objects	Very sparse smoke and cover entire field of view
Video 8	Indoor & moving objects	Illumination variation

We performed the qualitative and quantitative analysis of our smoke detection algorithm in outdoor and indoor conditions. Unique combination of features with past history make algorithm robust to all conditions. Videos (used in the experiment) descriptions and challenges in smoke detection are mentioned in Table 1. We performed experiments on all possible scenarios of smoke videos like sparse and dense smoke, illumination variation and background similarity. Results of the intermediate steps of proposed algorithm are shown in Fig. 6. Results show that above mentioned features are sufficient for the detection of all potential smoke regions.

Qualitative results of proposed algorithm are shown in Fig. 7. Results show that proposed algorithm detects smoke regions effectively without producing any significant false alarm. Figure 7a shows smoke in outdoor conditions. Smoke is behind the fence and a person with smoke color shirt is moving. For smoke region, energy reduction and shape disorder are high compared to other moving

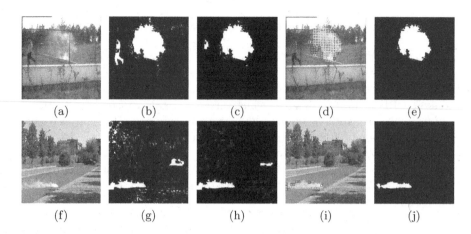

(a)   (b)   (c)   (d)   (e)

(f)   (g)   (h)   (i)   (j)

Fig. 6. Intermediate results of proposed smoke detection algorithm. (a) & (f) Input frame, (b) & (g) detection of all moving objects, (c) & (h) possible smoke region (after transparency criterion), (d) & (i) output of feature extraction and classification (yellow rectangle shows smoke region and blue rectangle shows non-smoke region), (e) & (j) potential smoke region (final mask). (Color figure online)

(a)   (b)   (c)

(d)   (e)   (f)

Fig. 7. Qualitative results of proposed smoke detection algorithm.

object regions which enables us to detect smoke effectively without any false alarm. Figure 7b shows dense smoke in front of a flat surface. As flat surface of background and dense smoke have less energy thus reduction in energy property is less reliable. However, shape disorder and variability in shape disorder properties enable us to detect smoke successfully. Figure 7c shows very sparse smoke which covered entire field of view. Due to coverage of entire field of view, shape

Table 2. Performance evaluation of smoke detection algorithms.

Video	Toreyin et al. [12]		Li et al. [6]		Li et al. [8]		Proposed	
	DR (%)	EDR (%)	DR (%)	EDR (%)	DR (%)	EDR (%)	DR (%)	EDR (%)
Video 1	77.3	7.3	92.8	6.8	91.3	5.4	93.4	1.6
Video 2	76.2	9.5	89.3	7.9	92.7	8.1	93.7	2.3
Video 3	88.2	0	77.4	0	83.6	0	96.3	0
Video 4	81.7	7.1	77.3	5.1	83.4	4.9	87.8	1.8
Video 5	83.4	7.1	87.7	5.2	91.8	5.1	93.4	1.1
Video 6	83.6	5.8	89.3	3.8	92.3	3.5	94.1	0.5
Video 7	77.1	7.1	91.6	6.7	90.2	5.8	92.2	1.7
Video 8	82.3	7.2	86.3	5.6	90.1	5.2	92.6	1.6
Average	**81.23**	**6.39**	**86.46**	**5.14**	**89.43**	**4.75**	**92.94**	**1.33**

disorder and variability in disorder are unreliable features. As smoke is sparse and blurred the background thus reduction in energy and its consistency are strong enough to support smoke detection. Figure 7d shows smoke present at far distance. Shape disorder is easy to detect but due to far distance variation in shape disorder is very low. However, reduction in energy and its consistency over time are supporting features. Transparent object detection is also a plus in this case. Figure 7e shows variation in illumination. As smoke grows up, it becomes less visible in bright sunlight but still algorithm is able to detect smoke. In Fig. 7f background is similar to the smoke. With the adaptive background modeling, we are able to extract smoke regions. As we do not rely on color features thus background similarity is not an issue. Here, smoke and background both are flat which makes energy features unreliable but shape disorder is strong enough to support successful detection. Results verify that proposed algorithm modeled with unique features which make it robust to all real life conditions.

Quantitative analysis is performed in terms of the detection rate (DR) and error detection rate (EDR), where the former indicates the ratio of the number of detected smoke regions to the total number of smoke regions whereas the later stands for the ratio of the number of wrongly detected smoke regions to the total number of smoke regions. The results for smoke localization for the proposed and prior art algorithms are mentioned in Table 2. Proposed algorithm outperformed the prior art algorithms with an average detection and error detection rate of ~93% and 1.33% respectively for all different scenarios. Results confirm that proposed algorithm is effective, robust and has a earlier smoke alarm. Proposed algorithm robustly detects smoke in a video stream in real time at 25 fps.

5   Conclusion

In this paper, we have proposed a novel visual smoke detection algorithm. Proposed algorithm is robust to detect smoke of any color. Algorithm can easily detect smoke and non-smoke regions due to spatio-temporal analysis of energy and shape disorder features. Detection at modular level (first moving objects detection, then transparent objects detection, then feature analysis and finally past history analysis) increases the efficiency of the algorithm. Use of data driven fuzzy thresholds avoids user intervention to select hard thresholds for features discrimination between smoke and non-smoke. Features extraction and classification are less complex in comparison with prior art algorithms which ensures real-time performance. Space-time analysis of unique features combination ensures high efficiency in comparison with other existing algorithms. Proposed algorithm has wide range of surveillance applications in outdoor and indoor conditions.

References

1. Wang, Y., Chua, T.W., Chang, R., Pham, N.T.: Real-time smoke detection using texture and color features. In: ICPR (2012)
2. Maruta, H., Nakamura, A., Yamamichi, T., Kurokawa, F.: Image based smoke detection with local hurst exponent. In: IEEE 17th International Conference on Image Processing (2010)
3. Junzhou, C., Yong, Y., Qiang, P.: Dynamic analysis for video based smoke detection. IJCSI Int. J. Comput. Sci. **102**, 298–304 (2013)
4. Brovko, N., Bogush, R., Ablameyko, S.: Smoke detection algorithm for intelligent video surveillance system. Comput. Sci. J. Moldova **21**, 61 (2013)
5. Wang, S., He, Y., Zou, J.J.: Early smoke detection in video using swaying and diffusion feature. J. Intell. Fuzzy Syst. **26**, 267–275 (2014)
6. Li, J., Yuan, W., Zeng, Y., Zhang, Y.: A modified method of video-based smoke detection for transportation hub complex. In: 9th Asia-Oceania Symposium on Fire Science and Technology (2013)
7. Gonzalez-Gonzalez, R., Alarcon-Aquino, V., Rosas-Romero, R., Starostenko, O., Rodriguez-Asomoza, J., Ramirez-Cortes, J.M.: Wavelet-based smoke detection in outdoor video sequences. In: 53rd IEEE International Midwest Symposium on Circuits and Systems (MWSCAS) (2010)
8. Li, W.H., Fu, B., Xiao, L.C., Wang, Y., Liu, P.X.: A video smoke detection algorithm based on wavelet energy and optical flow eigen-values. J. Softw. **8**, 63–70 (2013)
9. Chen, J., Wang, Y., Tian, Y., Huang, T.: Wavelet based smoke detection method with rgb contrast image and shape constrain. In: Visual Communications and Image Processing (VCIP) (2013)
10. Calderara, S., Piccinini, P., Cucchiara, R.: Smoke detection in video surveillance: a MoG model in the wavelet domain. In: Gasteratos, A., Vincze, M., Tsotsos, J.K. (eds.) ICVS 2008. LNCS, vol. 5008, pp. 119–128. Springer, Heidelberg (2008). doi:10.1007/978-3-540-79547-6_12
11. Avgerinakis, K., Briassouli, A., Kompatsiaris, I.: Smoke detection using temporal hoghof descriptors and energy colour statistics from video. In: International Workshop on Multi-Sensor Systems and Networks for Fire Detection and Management (2012)
12. Toreyin, B.U., Dedeoglu, Y., Cetin, A.E.: Contour based smoke detection in video using wavelets. In: 14th European Signal Processing Conference (2006)

Local Feature-Based Photo Album Compression by Eliminating Redundancy of Human Partition

Chia-Hsin Chan$^{(\boxtimes)}$, Bo-Hsyuan Chen, and Wen-Jiin Tsai

Department of Computer Science, National Chiao Tung University, Hsinchu, Taiwan
terry0201@gmail.com

Abstract. With the explosive growth of photo uploading on the web, traditional photo album compression using individual image coding is needed to be improved to save the storage spaces. Recently, an advance technique of photo album compression via video compression is proposed which utilizes the similarity between photos to improve the compression performance. In this paper, we modify the original scheme to improve the compression performance when photos containing human beings. Experiment results show that the proposed method outperforms the state-of-the-art method by at most 12.7% of bit-rate savings for compressing photo albums with humans. Comparing with traditional JPEG compression, the proposed method achieves 70% to 85% of bit-rate savings.

1 Introduction

Photo capturing devices such as smart phones and digital cameras are now widely used in our daily lives. People can easily build their own online photo albums by taking photos and uploaded to the image hosting websites. It is reported by the image hosting website Flicker that more than 3.5 billion new images are uploaded every day. Since such explosive growth of image production consumes storage spaces rapidly, how to minimize the storage size become an important question. The mostly used image compression format is JPEG and it is already used for decades. Even though individual image compressions enable convenient random access to any picture, the storage space saving is limited to the ordinary single image compression techniques.

In recent years, research works have been proposed to improve the compression performance of photo album containing images with similar contents. One kind of solutions is try to find one representative image (R.I., or average image) from the photo album and compress other images by coding the redundancies between the representative image and the target image [5–7]. However, the limitations of this scheme is that every single image in the photo album needs to be accurately aligned to generate the representative image, thus it only works well when photos in the album are quite similar. Moreover, this one-to-multiple structure does not exploit correlations among images. Another solution is to organize images into a pseudo video and compress it just like a normal video [8–12]. The benefit of this approach is that it does not need accurate alignment

© Springer International Publishing AG 2017
C.-S. Chen et al. (Eds.): ACCV 2016 Workshops, Part I, LNCS 10116, pp. 143–158, 2017.
DOI: 10.1007/978-3-319-54407-6_10

between images. Instead, it deals with images in pseudo sequence as natural video frames and uses inter prediction techniques such as block-based motion compensation (BMC) for efficient compression.

Previous works only focus on compressing similar photos of buildings and scenery, however there are also many pictures taken with people inside. Therefore, in this paper an efficient photo compression algorithm is proposed considering human objects within the photos. The rest of this paper is organized as follows. Section 2 introduces video coding based photo album compression. Section 3 describes the proposed method in detail. Several experimental results are shown in Sect. 4. Finally, concluding remarks are given in Sect. 5.

2   Photo Album Compression via Video Coding

The video coding based photo album compression is to organize images as a pseudo sequence and compress it as a normal video sequence. Generally, the main procedure of this scheme can be divided into two steps. First, clustering images into well separated sets in which images are highly correlated. Second, arranging highly correlated images into a pseudo sequence with a suitable order for efficient compression. There have been many researches working in this field. For example, Chen et al. [9] and Schmieder et al. [10] focus on organizing best coding structure of images in photo album to completely exploit inter correlations. Lu et al. introduces MPEG-like BMC to discover correlations among images [11] and Zou et al. applies the advanced BMC in the latest video coding standard High Efficiency Video Coding (HEVC) [12]. On the other hand, Yeung et al. [8] firstly performs pixel-wise global motion compensation before using BMC.

However, pseudo video is quite different from normal video. In normal videos, consecutive frames are very similar. But every frame in a pseudo video might be taken from different viewpoints, locations, and distances even though they are taken for same scene. In other words, the correlation among images in pseudo video is not as high as that in natural video and the motion disparities among every frame can be complicated. Whats more, since photos in pseudo video might vary in illuminance and shadow, pixel-wise global compensation and BMC will have a hard time to compensate those illumination changes. In this paper, we adopt video coding based scheme since its compression performance is usually better than R.I based coding, and proposes several novel methods including modified predictive cost for prediction structure construction, human transformation index finding for incorporating additional reference picture for human partition, and an adaptive photometric transformation for improving reference picture quality are proposed to deal with the above challenges.

3   Proposed Photo Album Compression Method

The flowchart of the proposed method is depicted in Fig. 1. At the beginning, each image in photo album will be indexed. Then the images are clustered by

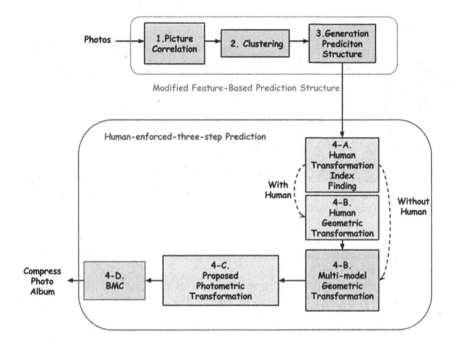

Fig. 1. The flowchart of the proposed photo album compression algorithm.

agglomerative hierarchical clustering with the proposed predictive cost. For each cluster, a minimum spanning tree (MST) of directed graph, with every node in represents an image, is generated to determine the encoding order of images for the pseudo sequence. The first image in the pseudo sequence is intra encoded and the other images are inter encoded. Every inter-encoded image will choose its nearest parent node in the MST as reference picture according to the proposed human-enforced prediction.

3.1   Modified Feature-Based Prediction Structure

Step 1. Picture Correlation

In [1] and [3], SIFT keypoint descriptor [13] matching is proposed to estimate the correlation among images. For both clustering images or building the coding structure, average distances of matched SIFT keypoint descriptors are used to measure the image correlation. However, we observe that this measurement cannot classify images well when they contain structural similar patterns such as square windows, building corners, similar veined lines on the wall, . . . etc. As shown in Fig. 2, when two images have similar window patterns, it will generate "good false match" of SIFT descriptors that misclassifies two unlike photos to be similar. Moreover, we also observe that people usually take several pictures

Fig. 2. Example of "good false match" problem happens in window pattern.

in the same scene with different gestures in personal photo album which can also generate considerable redundancy.

To cope with this problem, first let us recall the original SIFT matching procedure. Let $I_1, I_2, \ldots, I_N$ denote images in the photo album set, with N is the total number of images. Let $F_n = \{f_n(1), f_n(2), \ldots, f_n(K_n)\}$ denotes the SIFT descriptor set of image I_n with K_n is the total number of descriptors of image n. Let $v_n(k)$ be the 128-D vector describing the histogram of gradient directions of $f_n(k)$. Then the distance between $f_i(k_i)$ and $f_j(k_j)$ is defined as

$$d_f(f_i(k), f_j(l)) = \frac{1}{128} \|v_j(l) - v_i(k)\| \tag{1}$$

And they are matched if

$$d_f(f_i(k), f_j(l)) < \alpha \times d_f(f_i(k^{'}), f_j(l)) \tag{2}$$

where $f_i(k)$ and $f_i(k^{'})$ are the closest and second closest descriptors in I_i and α is 0.6. After matched descriptors are found, we can generate the matched descriptor set $F_{i,j} = \{(f_i(k_1), f_j(l_1)), (f_i(k_2), f_j(l_2)), \ldots\}$ between I_i and I_j.

In order to solve the "good false match" problem, our idea is that not only average distance of matched SIFT descriptors should be considered. A modified feature-based prediction structure with a new predictive cost which can correctly measure the correlations between images is proposed as follows. First, each image in the photo album is separated into human partition and background partition through face detection [4]. Then the mean absolute error (MAE) of background partition pixel between image $T_j(I_i)$ and I_j is calculate, where $T_j(I_i)$ is the geometric transformed image aligned with I_j by applying RANSAC algorithm [2] on their matched SIFT descriptors. The reason of using the background partition MAE is that we want to construct the prediction structure mainly according to the background similarities without disturbing by the same foreground object but with large disparities, which is the human part in this paper. Figure 3 shows an example of a transformed image with human partition separation. Moreover,

(a)   (b)

Fig. 3. Example of (a) the target image I_j and (b) the transformed image $T_j(I_i)$. Pixels inside the yellow bounding boxes stand for the foreground region obtained from I_j. (Color figure online)

after RANSAC, part of the descriptors will be identified as inliers and others will be outliers. Generally, the more inliers exist, the more robust the transformation is. Therefore, we also take the number of inliers into considerations.

The complete formulation of the proposed predictive cost from image I_i to image I_j can be written as

$$
Cost_{i,j} = \frac{1}{|F_{i,j}|} \sum_{(f_i(k),f_j(l))F_{i,j}} d_f(f_i(k), f_j(l))
$$
$$
+ \frac{1}{M} \sum_{(x,y)I_j^b,T_j^b(I_i)} |I_j^b(x,y) - T_j^b(I_i)(x,y)| + \frac{TH}{F_{inliers}}
\tag{3}
$$

where $|F_{i,j}|$ is the total number of matched SIFT descriptors between images I_i and I_j, $T_j(I_i)$ is the transformed image from I_i toward I_j, M is the number of pixels of background region in I_j, I_j^b stands for pixels of background region in image I_j, $T_j^b(I_i)$ stands for pixels also bounded by background region area in I_j of the transformed image $T_j(I_i)$, $F_{inliers}$ is the number of inliers after applying RANSAC, TH is a threshold and is set to be 1000 in this paper.

Step 2. Clustering

With the new predictive cost defined in Eq. (3), next we will organize images into correlated sets. The well-known K-means clustering which is also used in the previous work [3] is not adopted in this paper because (i) it needs to specify target number of clusters which may be hard to determined, (ii) its performance is very sensitive to the initial cluster center, and (iii) it cannot find the clusters with small density. Instead, agglomerative hierarchical clustering [15] is adopted, in which two images are combined into one cluster if their predictive cost is smaller than certain threshold. Therefore, each image will iteratively group with another image until all images are grouped into one of the clusters or certain threshold is satisfied. So only one threshold of predictive cost should be manually determined.

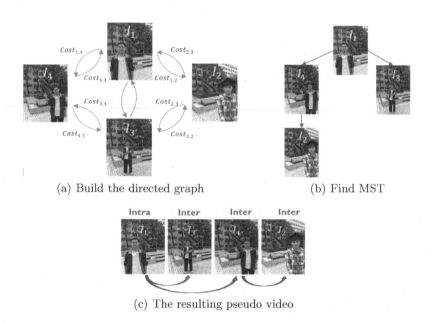

(a) Build the directed graph   (b) Find MST

(c) The resulting pseudo video

Fig. 4. Example of the proposed prediction structure generation.

Step 3. Generation of Prediction Structure

After clustering, the prediction structure need to be constructed for describing the relationships between images for pseudo sequence generation. This problem can be formulated as a graph optimization problem for a directed graph. Constructing a directed graph in which each node represents one image and every edge stands for the predictive cost of encoding the node image with its parent node image as reference The optimal coding structure can be obtained by finding minimum spanning tree (MST) which has the minimum total rate-distortion cost [19]. However, the real rate-distortion costs cannot be obtained because images need to be compressed in advance to calculate the actual bit-rates and distortions. Therefore, the proposed predictive cost in Eq. (3) is used again as the edge definition in the directed graph. The encoding order of the pseudo sequence is determined by depth-first traversing of MST. The root node of MST will be intra-coded and the other nodes will be inter-coded with their parent node serving as the reference picture using the proposed human-enforced prediction described next. Figure 4 illustrates the procedure of this step.

3.2   Human-Enforced Prediction

Unfortunately, disparities in the pseudo video obtained in Step 3 are more complicated than a natural video, and people usually take several pictures in the same scene with different poses/gestures. Therefore, we propose to add human transformation picture(s) as additional reference pictures for better coding performances. We follow the main structure of feature-based three-step prediction proposed by Z. Shi et al. [1,3] to eliminate complicated disparities. Search of human is in Step 4-A. In Step 4-B, Multi-model transformation and the proposed human partition transformation prediction will compensate deformations between two images to further reduce temporal redundancies. In the step 4-C, adaptive photometric transformation comprising the matched feature-based approach and the color based approach is used to determine the best value to compensate illuminance disparities. Finally, we adopt advanced BMC in HEVC to deal with small block disparities. The Fig. 5 shows the whole human-enforced prediction architecture and the details will be described in following subsections.

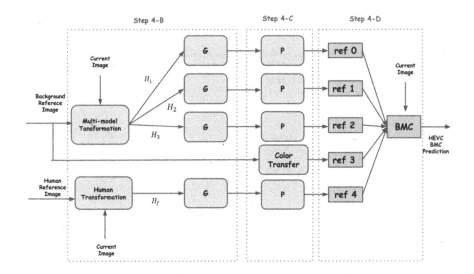

Fig. 5. Human-enforced prediction architecture.

Step 4-A. Human Transformation Index Finding

It is noticed that the picture containing the same people with similar poses or gestures might not be selected as the parent node since its occupation area is too small. Therefore, in order to improve the prediction structure, pictures having human partition will be added as additional reference(s) through the proposed human transformation index finding process, as illustrated in Fig. 6(a). Except for the root node, each human partition of each node image will undergo a search process to try to find the best reference from two preceding pictures according to the encoding order in the pseudo sequence. To make the search process not to be too complicated, we simply compute the MAE value for the human partition between the current picture I_j and the preceding picture I_i, which is formulated

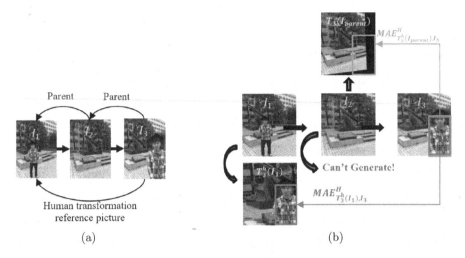

Fig. 6. (a) Additional reference containing human partition. (b) Encoding procedure of human transformation index finding.

as

$$MAE_{i,j} = \sum_{y=0}^{h_{height}} \sum_{x=0}^{h_{width}} |I_i(x,y) - I_j(x,y)| \qquad (4)$$

and the index f is decided by

$$f = \operatorname*{argmax}_{i=parent,j-1,j-2} MAE^H_{T^h_j(I_i),I_j} \qquad (5)$$

If $f = parent$, it means that no suitable picture of human transformation can be used for reference.

Step 4B. Geometric Transformation

Two types of geometric transformations are adopted: multi-model transformation and human partition transformation. For the first part, we use the method in [3] to deduce multi-model geometric transformation matrixes $\{H_1, H_2, \ldots, H_L\}$. In this paper, we limit $L \leq 3$ in order to strike a balance between complexity and performance. After applying geometric transformations, the transformed images are obtained and defined as $\{T_j(I_i)_1, T_j(I_i)_2, \ldots, T_j(I_i)_L\}$ as shown in Fig. 7(c), (d). Moreover, a background reference picture without geometric transform is also added as reference picture. For the human partition transformation, first we search for human faces in I_j and mark it with a bounding box. Then, assuming top-left corner of the face rectangle is at (x_f, y_f), the human partition bounding box is determined by setting the upper left corner at $(x_f - 0.5width_f, y_f)$ and the bottom right corner at $(x_f + 1.5width_f, y_f + 6height_f)$, with $width_f$ and $height_f$ being the width and height of the face bounding box, respectively. Figure 7(g) shows an example of a

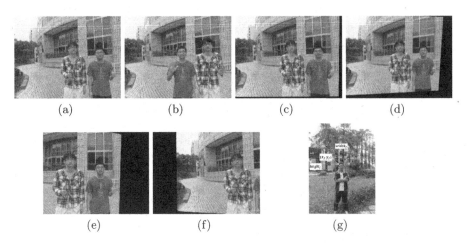

(a)   (b)   (c)   (d)

(e)   (f)   (g)

Fig. 7. (a) Background reference picture I_i. (b) Current image I_j. (c)–(d) Multi-model geometric transformed images $T_j(I_i)_1$, $T_j(I_i)_2$. (e)–(f) Human partition geometric transformed images. (g) Sketch of person partition.

Fig. 8. The bounding box (red rectangle) of geometric transformed image $T_j(I_i)_n$. (Color figure online)

face bounding box and its corresponding human partition bounding box. Once a human partition is obtained, the matched points in this partition are collected to generate human geometric transformations using RANSAC as shown in Fig. 7(e), (f). These geometric transformed images (marked as G in Fig. 5) will also be added as reference pictures.

Step 4C. Photometric Transformation

Although geometric transformation reduces the geometric differences between reference images and the current image, intensity disparities still exists. For the n-th transformed image $T_j(I_i)_n$, we define its photometric transformation as

$$\overline{T_j(I_i)_n} = T_j(I_i)_n + \gamma \tag{6}$$

where γ is intensity offset and $\overline{T_j(I_i)_n}$ is the photometric transformed image in order to minimized the intensity difference between two frames.

$$diff = \sum_{i=1}^{height} \sum_{j=1}^{width} |I_j(i,j) - I_{\overline{T_j(I_i)_n}}(i,j)| \tag{7}$$

For the proposed photometric transformation, we combine two approaches. First, the approach in [3] defines the minimization target value $diff^{Keypoints}$ as

$$diff^{Keypoints} = \sum_{(P^k_{T_j(I_i)_n}, P^k_j) F_l} |I_j(P^k_j) - (I_{T_j(I_i)_n}(P^k_{T_j(I_i)_n}) + \gamma)|^2 \tag{8}$$

where $F_l = \{(P^1_{T_j(I_i)_n}, P^1_j), (P^2_{T_j(I_i)_n}, P^2_j), (P^k_{T_j(I_i)_n}, P^k_j)\}$ is the matched keypoint descriptor set between $T_j(I_i)_n$ and I_j. Second, the approach in [14] tries to alter the colors in one image to resemble that of another image by using the mean and variance values of source and target images which can be written as

$$diff^{Color} = \frac{\sigma^k_T}{\sigma^k_i} \sum_{y=1}^{height} \sum_{x=1}^{width} (I^k_i(x,y) - mean(I^k_i)) + mean(T^k_j(I_i)_n) \tag{9}$$

where $I^k_i(x,y)$ is the source images pixel value of color channel k at location (i,j), σ^k_T is the variance value at channel k in target image, $mean(T^k_j(I_i)_n)$ is the mean value at channel k in target image, with the same meanings applied to σ^k_i and $mean(I^k_i)$. Originally, this measurement is applied on the Lab color space. For fairly comparison with the value of $diff^{Keypoints}$, we only apply this measurement for Y channel under YCrCb color space. Moreover, only pixels within the bounding box confining the geometric transformed image is considered for calculation, as shown in Fig. 8.

In order to find the best offset parameter, we adopt both approaches and dynamically select the one with the smaller difference value. Finally, each photometric transformed image will be added as a reference picture.

Step 4D. Block-Based Motion Compensation

After geometric and photometric transformations, most geometric and illuminate disparities will be eliminated. However, small disparities may still exist between transformed image and target image. Therefore, the HEVC BMC is applied to deal with those local disparities. For BMC, pictures including (i)

Table 1. Description of the test photo album sets.

Photo album name	Number of photos	Photo resolution
CastleEntry [17]	10	768×512
WadhamCollege [18]	5	1024×768
RockBoat [1]	20	784×512
Single Person Album	32	960×1280
Multiple People Album	33	1280×960

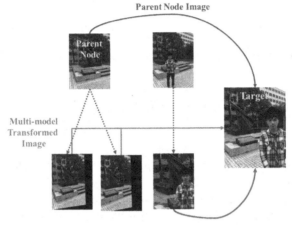

Fig. 9. Example of adding images to reference picture list.

the parent node image in MST (obtained from Step. 3), (ii) the human partition transformed image (obtained from Step. 4-B) after geometric/photometric transformations, and (iii) multi-model transformed images (obtained from Step. 4-B) after geometric/photometric transformations are added to reference picture list (illustrated in Fig. 9).

3.3 Parameters Transmitting

During the encoding process, geometric transformation matrix and photometric transformation parameters will be generated. Each parameter will be scaled and rounded to integers. These parameters will be written in binary files which will be entropy coded by dictionary entropy coding and transmitted to decoder side.

4 Experiment Result

The proposed scheme performance is compared with the feature-based approach in [3] which has currently the best compression efficiency for photo albums.

(a) CastleEntry   (b) WadhamCollege   (c) RockBoat

(d) Single Person Album   (e) Multiple People Album

Fig. 10. Examples images of five photo album test sets.

Five photo albums are used in this paper for evaluation, as described in Table 1. The photo albums *CastleEntry*, *WadhamCollege*, and *RockBoat* are the sets of images with sceneries taken at different viewpoints. The photo albums *Single Person Album* and *Multiple People Album* are the sets of images made by us, containing images with single and two people inside. Figure 10 shows example photos of each album. The HEVC reference software HM16.0 [16] is used for pseudo video compression by following the settings of HEVC common test condition [20]. Three compression ratios are tested by setting the quantization parameters to 22, 27, and 32 to test the performance at different bit-rates.

First we compare the compression performance of the proposed method with the anchor method [3]. Figure 11 shows the rate-distortion curves of the coding results. Clearly, the performance of the proposed method is very close to that of [3] for the photo albums with no human inside. Our approach on average of the three QP settings has 1.4% bit-rate loss in RocketBoat, but obtains 1.4% savings in CastleEntry and 2.8% savings in WadhamCollege. The reasons are that there is no human in the photos and the "good false match" problem does not occur in these photo sets, thus the proposed human-enforced prediction does not take effect.

However, for the result of Single Person Album, the proposed method obtains 12.7% of bit-rate savings which come from solving the wrong clustering caused by the "good false match" problem. For [3], 8 out of 32 photos in the Single Person Album refer to pictures which are not correlate. But for our approach, only 1 photo has this problem. We also conduct detailed analyze the portion of bit-rate savings by selecting between the proposed tool or the original one in [3], as shown in Fig. 12. It is observed that the bit-rate savings mainly comes from the modified predictive cost (92%, 84%), however the human transformation part including the photometric transform still occupies a portion of savings (8%, 16%). We also found that the human transformation part obtains a higher portion of bit-rate savings when there are more human redundancies in the photos. It is worth mentioning that in the Single Person Album there are more different backgrounds

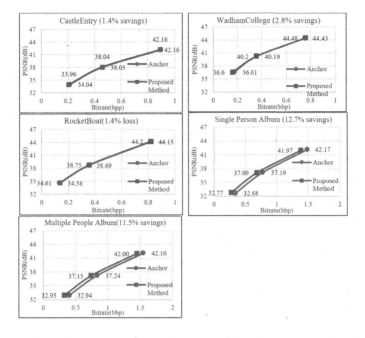

Fig. 11. Rate-distortion curves of the anchor [3] and the proposed method.

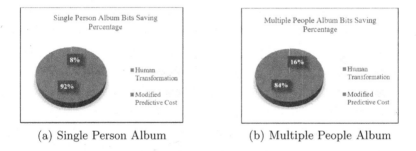

(a) Single Person Album   (b) Multiple People Album

Fig. 12. Bit-rate saving percentages of albums containing human beings.

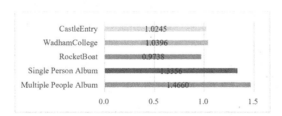

Fig. 13. Time complexity diagram over five album sets. The number of each bar is the running time ratio comparing with the previous work [3].

and different people, but have similar object patterns, consequently the "good false match" problem is easy to happen. That is why the proposed method obtains higher total bit-rate savings than that of Multiple Person Album for avoiding more wrong image clustering.

Figure 13 shows the time complexity analysis of each photo album set compared with [3]. In the albums of scenery photos, there are no human beings inside so the complexity increase only comes from the modified predictive cost the proposed photometric transformation. For the albums containing humans, the proposed method has more time complexity increase because of the computation of the proposed human transformation index finding and the BMC computation for using the additionally incorporated human partition references.

We also compare the proposed method with the traditional JPEG coding for individual photos in the album and the result shows that tremendously 70% to 85% of total bit-rate savings is achieved, as shown in Fig. 14. Although the video coding based compression takes more time comparing with individual image coding, the complex encoding process is only need once for constructing the photo album. When accessing photos, we only need to decode the desired frame from the pseudo video which is much faster than encoding.

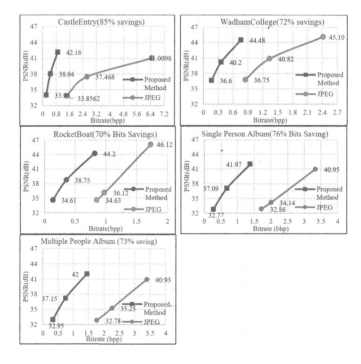

Fig. 14. Rate-distortion curves of the JPEG compression and the proposed method.

5   Conclusion

In this paper, a storage method of photo albums via video compression is proposed. Based on previous works, we further consider efficiently compressing photos containing human beings by proposing a modified predictive cost for feature-based image clustering, a human transformation index finding for incorporating additional reference picture for human partition, and an adaptive photometric transformation for improving reference picture quality. The proposed method solves the "good false match" problem which is unachievable for the previous works and the compression performance are improved for photo albums containing human beings. Our future work will focus on how to dynamically add and delete photos for advanced applications.

References

1. Shi, Z., Sun, X., Wu, F.: Feature-based image set compression. In: IEEE ICME, pp. 1–6 (2013)
2. Fischler, M.A., Bolles, R.C.: Random sample consensus: a paradigm for model tting with applications to image analysis and automated cartography. ACM Commun. **24**, 381–395 (1981)
3. Shi, Z., Sun, X., Wu, F.: Photo album compression for cloud storage using local features. Emerg. Sel. Top. Circuits Syst. **4**, 17–28 (2014)
4. Viola, P., Jones, M.J.: Robust real-time face detection. Int. J. Comput. Vis. **57**, 137–154 (2004)
5. Musatenko, Y.S., Kurashov, V.N.: Correlated image set compression system based on new fast efficient algorithm of Karhunen-Loeve transform, pp. 518–529. International Society for Optics and Photonics (1998)
6. Karadimitriou, K., Tyler, J.M.: The centroid method for compressing sets of similar images. IEEE Pattern Recogn. Lett. **19**, 585–593 (1998)
7. Ait-Aoudia, S., Gabis, A.: A comparison of set redundancy compression techniques. EURASIP J. Adv. Sig. Process. **2006**, 216 (2006)
8. Yeung, C.H., Au, O.C., Tang, K., Yu, Z., Luo, E., Wu, Y., Tu, S.F.: Compressing similar image sets using low frequency template. In: IEEE ICME, pp. 1–6 (2011)
9. Chen, C.P., Chen, C.S., Chung, K.L., Lu, H.I., Tang, G.Y.: Image set compression through minimal-cost prediction structure. In: IEEE ICIP, pp. 1289–1292 (2004)
10. Schmieder, A., Cheng, H., Li, X.: A study of clustering algorithms and validity for lossy image set compression. In: IPCV, pp. 501–506 (2009)
11. Lu, Y., Wong, T.T., Heng, P.A.: Digital photo similarity analysis in frequency domain and photo album compression. In: 3rd International Conference on Mobile and Ubiquitous Multimedia, pp. 237–244 (2004)
12. Zou, R., Au, O.C., Zhou, G., Dai, W., Hu, W., Wan, P.: Personal photo album compression and management. In: IEEE ISCAS, pp. 1428–1431 (2013)
13. Chandrasekhar, V., Takacs, G., Chen, D., Tsai, S.S., Grzeszczuk, R., Girod, B.: CHoG: compressed histogram of gradients a low bit-rate feature descriptor. In: CVPR, pp. 2504–2511 (2009)
14. Reinhard, E., Ashikhmin, M., Gooch, B., Shirley, P.: Color transfer between images. Comput. Graph. Appl. **21**, 34–41 (2001)

15. Day, W.H.E., Edelsbrunner, H.: Efficient algorithms for agglomerative hierarchical clustering methods. J. Classif. **1**, 7–24 (1984)
16. https://hevc.hhi.fraunhofer.de/svn/svn_HEVCSoftware
17. http://cvlabwww.epfl.ch/data/multiview/denseMVS.html
18. http://www.robots.ox.ac.uk/vgg/data2.html
19. Chu, Y.J., Liu, T.H.: On the shortest arborescence of a directed graph. Sci. Sinica **14**, 1396–1400 (1965). Boykov, Y., Veksler, O., Zabih, R.: Fast approximate energy minimization via graph cuts. IEEE PAMI **23**, 1222–1239 (2011)
20. Bossen, F.: Common HM test conditions and software reference configurations. In: JCTVC-L1100 (2013)

Generic 3D Convolutional Fusion
for Image Restoration

Jiqing Wu$^{(\boxtimes)}$, Radu Timofte, and Luc Van Gool

Computer Vision Laboratory, D-ITET, ETH Zurich, Zürich, Switzerland
{jwu,radu.timofte,vangool}@vision.ee.ethz.ch

Abstract. Also recently, exciting strides forward have been made in
the area of image restoration, particularly for image denoising and single
image super-resolution. Deep learning techniques contributed to this significantly. The top methods differ in their formulations and assumptions,
so even if their average performance may be similar, some work better
on certain image types and image regions than others. This complementarity motivated us to propose a novel 3D convolutional fusion (3DCF)
method. Unlike other methods adapted to different tasks, our method
uses the exact same convolutional network architecture to address both
image denoising and single image super-resolution. Our 3DCF method
achieves substantial improvements (0.1 dB–0.4 dB PSNR) over the state-
of-the-art methods that it fuses on standard benchmarks for both tasks.
At the same time, the method still is computationally efficient.

1 Introduction

Image restoration is concerned with the reconstruction/estimation of the uncorrupted image from a corrupted or incomplete one. Typical corruptions include
noise, blur, down-sampling, hardware constraints (*e.g.* Bayer pattern) and combinations of those. After decades of research there is a large literature [1] dedicated
to restoration tasks, whereas the literature studying the fusion of restoration
results is thin [2]. In this paper we tackle such fusion as a means for further
performance improvements. Particularly, we propose a 3D convolutional fusion
(3DCF) method and validate it on image denoising and single image super-
resolution.

1.1 Image Denoising (DN)

Natural image denoising aims at recovering the clean image given a noisy observation. The most often studied case is when the image corruption is caused by
additive white Gaussian (AWG) noise with known variance. Also, the images are
assumed to be natural, capturing every-day scenes, and the quantitative measure
for assessing the recovery result is the peak signal-to-noise ratio (PSNR), which
stands in monotonic relation to the mean squared error (MSE).

© Springer International Publishing AG 2017
C.-S. Chen et al. (Eds.): ACCV 2016 Workshops, Part I, LNCS 10116, pp. 159–176, 2017.
DOI: 10.1007/978-3-319-54407-6_11

The most successful denoising methods employ at least one of the following

Denoising principles as listed in [3]: Bayesian modeling (coupled with Gaussian models for noiseless patches), transform thresholding (assumes sparsity of patches in a fixed basis), sparse coding (sparsity over a learned dictionary), pixel or block averaging (exploits image self-similarity).

Most denoising methods work at a single image scale, the finest one, and often a small image patch is the basic processing unit. The patch captures local image information for a central pixel and a statistical amount of uncorrupted pixels. Zontak *et al.* [4] recently opened up a fresh research direction by proposing a method based on patch recurrence across scales (PRAS). Another partition of the methods is based on whether only the noisy image is used, or also learned priors and/or extra data from other (clean) natural images. This leads to *internal* and *external* methods. Some well known examples of each are:

Internal denoising methods:
NLM (non-local means) [5] reconstructs a noisy patch with a weighted average of similar patches from the same image. It uses the image self-similarity and the fact that the noise is usually uncorrelated.

BM3D (block matching 3D) [6] extends NLM and the DCT denoising method [7]. BM3D groups similar patches into a 3D block, applies 3D linear transform thresholding, and inverses the transform.

WNNM (weighted nuclear norm minimization) [8] follows the self-similarity principle, and applies WNNM to recover the noiseless patch from a matrix of stacked non-local similar patch vectors.

PRAS (patch recurrence across scales) [4] creates (an)isotropic image scale pyramids and extracts the estimated (noiseless) patch from the same corresponding position but at a different scale.

PLE (piecewise linear estimation) [9] is a Bayesian restoration model, including denoising, deblurring, and inpainting. PLE employs a set of 19 Gaussian models obtained from synthetic edge images (as priors) and an estimation-maximization iterative procedure.

External denoising methods:
EPLL (expected patch log likelihood) [10] can be seen as a shotgun extended version of PLE. It learns a Gaussian mixture model with 200 components for 2 million clean patches sampled from external natural images, and tries to maximize the expected log likelihood of any randomly chosen patch in the image.

LSSC (learned simultaneous sparse coding) [11] adapts a sparse dictionary learned over an external database by adding a grouping step to the noise image.

MLP (multi-layer percepton) [12] learns from an external database with clean and noisy images, and was among the first to introduce neural networks to low level image restoration tasks.

CSF (cascade of shrinkage fields) [13] proposes shrinkage fields, combining the image model and the optimization algorithm as a whole. The time complexity is greatly reduced by inherent parallelism.

opt-MRF (Loss-Specific Training of Filter-Based MRFs) [14] revisits loss-specific training and uses bi-level optimization to solve the image restoration problem.

TRD (trained reaction diffusion) [15] extends the solving process of nonlinear reaction diffusion to a deep recurrent neural network, outperforms many of the aforementioned methods, while offering the lowest time complexity for now.

It is quite surprising that most of the recent top denoising methods (such as BM3D, LSSC, EPLL, PRAS, and even WNNM) face a plateau. They perform equally well for a large range of noise, despite that they are quite different in their formulations, assumptions, and information used. This is the reason behind the recent work that fuses them, pushing the limits by combining different approaches [16,17]. We refer the readers to [2] for a study of image fusion algorithms of the past decades. Others investigated the theoretical limits for denoising with natural image patch priors [18], and at least for the lower noise levels, the gap between the most successful methods and the predicted limits seems to rapidly diminish.

Fusion methods:

PatchSNR (patch signal-to-noise ratio). Mosseri *et al.* [19] propose a patch-wise signal-to-noise-ratio to distinguish whether an internal or an external denoising method should be applied. Their fused result slightly improves over the stand-alone methods.

RTF (regression tree fields). Jancsary *et al.* [16] observe that depending on the image content some methods perform better than other. They consider RTFs based on a filterbank (RTF_{plain}), also additional exploitation of BM3D's output (RTF_{BM3D}), or a setting exploiting all the outputs of their benchmarked methods (RTF_{all}). The more methods the better their fusion result. The RTFs are learned on large datasets. It is also worth mentioning that following [16], Schmidt *et al.* [20] propose a cascade of regression tree fields (CRTF) working on deblurring and denoising and obtain good performances in both cases.

NN (neural nets/multi-layer perceptron). Burger *et al.* [17] pursue the success of MLP [12] in denoising, to learn the best fusion. They found the internal denoising methods to suit better images with artificial (human-made) contents, and external ones to work better for natural scenes. They argue against PatchSNR and consider that there is no trivial rule to decide among internal or external method on a patch-by-patch basis, and indeed their NN fusion produces the best denoising results to date. Unfortunately, the learning is quite intensive.

AF (anchored fusion). Timofte [21] clusters the patch space and for each cluster learns an anchored regressor from fused methods' patches to the fusion output.

1.2   Single Image Super-Resolution (SR)

Single image super-resolution (SR) is another active area [22–27] of image restoration aiming at recovering a high-resolution (HR) image from a low-resolution (LR) input image by inferring missing high frequency contents. We can roughly categorize the recent methods in:

Non-neural network methods:

SR (sparse representation) [28] generates a sparse representation/coding of each LR image patch, and then applies the coefficients of this representation to generate the HR image.

A+ (adjusted anchored neighborhood regression) [29], considered to be an advanced version of ANR (anchored neighborhood regression) [30], learns sparse dictionaries and regressors anchored to the dictionary atoms.

RFL (super-resolution Forests) [31] maps low to high-resolution patches using random forests and anchored regressors as in A+.

selfEx (transformed self-exemplars) [32] introduces a self-similarity based image SR algorithm by applying transformed self-exemplars.

Neural network methods:

SRCNN (convolutional neural network) [33] learns an end-to-end mapping between the low/high-resolution images by a deep convolutional neural network.

CSCN (cascade of sparse coding network) [24] combines the key ingredients of deep learning with those of the sparse coding model.

1.3   Contributions

In this paper, we study the patch-by-patch fusion of image restoration methods with particular focus on recent top methods for both DN and SR tasks. To this end, we propose a generic 3D convolutional fusion architecture (3DCF) to learn the best combination of existing methods. Our three main contributions are:

1. We show the complementarity of different methods (*e.g.* internal vs. external).
2. We demonstrate that our method learns sophisticated correlation details from top methods to achieve the best reported results on a wide range of images.
3. The generality of our 3DCF method for both DN and SR.

The paper is organised as follows. Section 2 provides some insights and empirical evidence for the complementarity of the DN/SR methods and analyses oracle bounds for fusion. Section 3 motivates and introduces our novel 3DCF method with the necessary details and mathematical formulations. Section 4 presents the experiments and discusses the results. Section 5 concludes the paper.

2   Insights

Our focus is fusion for improved image restoration results and particularly for denoising in the presence of additive white Gaussian noise (AWG), with validation on single image super-resolution. Here we analyse the complementarity of the restoration methods and fusion strategies.

2.1   Complementarity of Top Methods

Jancsary *et al.* [16], Burger *et al.* [17], and Zontak and Irani *et al.* [4], among others, already observed that each method works best for some particular image contents while being worse than others for other image regions.

First, we pair-wise compare the PSNR performances of BM3D (internal method), and MLP and TRD (external methods) on 68 images from the Berkeley dataset for AWG noise with $\sigma = 50$. The relative improvements (PSNR gain) are reported in Fig. 1. MLP is better than BM3D on all images but is worse than TRD on ~40% of them. Also, BM3D is better than TRD on some images. We conclude there is no absolute winner at image-level.

Second, we compare pixel-wise or patch-wise and see that within the same image there is no absolute winner always getting the best result either. In Fig. 2 for one image altered with AWG noise, $\sigma = 50$, we report pixel-wise selections from BM3D (25.77 dB PSNR) and MLP (26.19 dB) to best match the ground truth image. Despite MLP being significantly better (+0.41 dB) on denoising this image, at pixel-level the results are almost equally divided between the methods. At patch-level (sizes 5×5 and 17×17 pixels) we have a similar pattern.

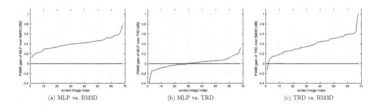

Fig. 1. No absolute winner. Each method is trumped by another on some image.

Fig. 2. An example of oracle pixel and patch-wise selections from BM3D and MLP outputs and the resulting PSNRs for AWG with $\sigma = 50$.

2.2   Average and Selection Fusion and Oracle Bounds

As shown in Fig. 1 for images and in Fig. 2 for patch or pixel regions, the denoising methods are complementary in their performance. Now we study a couple of fusion strategies at image level.

Average fusion directly averages the image results.

Selection of non-overlapping patches assumes that the fusion result contains non-overlapping (equal size) patches with the best image results of the fused methods (see Fig. 2). One needs to learn a patch-wise classifier.

Selection of overlapping patches is similar to the above one in that a patch-wise decision is made, but this time the patches overlap. The final fusion result is obtained by averaging the patches in the overlapped areas (see Fig. 2).

We work with BM3D and MLP, partly because BM3D is an internal while MLP is an external method, and partly because of the result in Fig. 1 where at image level MLP performs better than BM3D. Therefore, the results from fusing BM3D and MLP at patch-level are interesting to see.

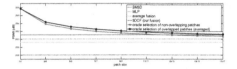 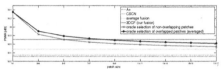

Fig. 3. Average PSNR [dB] comparison of BM3D [6] and MLP [12], average fusion, oracle selection of (overlapping or non-overlapping) patches, and our 3DCF fusion on 68 images, with AWG noise, $\sigma = 50$.

Fig. 4. Average PSNR [dB] comparison of A+ [29] and CSCN [24], average fusion, oracle selection of (overlapping or non-overlapping) patches, and our 3DCF fusion on Set14, upscaling factor $\times 2$.

In Fig. 3 we report how the chosen patch size affects the performance of a selection strategy, on the same Berkeley images corrupted with AWG noise, $\sigma = 50$. We report oracle results, an upper bound for such a strategy. In comparison we report the performance of the fused BM3D and MLP methods, as well as the results of the average fusion and our proposed 3DCF method. We note that (i) overlapping patches lead to better results (while significantly slower) than non-overlapping patches; (ii) the smaller the patch size the better the oracle results become; (iii) the average fusion leads to poorer performance than the fused MLP method; (iv) our 3DCF fusion results are comparable with those from the oracle selection strategies for patch sizes above 9×9.

Complementary, in Fig. 4 we start from the A+ and CSCN methods for the super-resolution (SR) task, where we use the Set14 images and an upscaling factor $\times 2$ (we use the settings described in the experimental section). As in the denoising case, (i) the smaller the patch size is, the better the oracle selection results get; (ii) the overlapped patches lead to better fusion results. However, for SR, (iii) the average fusion improves over both fused methods; (iv) our 3DCF fusion is significantly better than the fused methods, the average fusion, and compares favorably to the oracle selection fusion for patch sizes above 5 × 5.

From these experiments we can conclude that the average and (patch) selection strategies for fusion - while conceptually simple - are either not leading to consistently improved results (case of average fusion) or their oracle upper bounds are quite tight given the difficulty of accurately classifying patches (case of selection strategy). Note that PatchSNR [19] is an example of a selection strategy and that NN [17], a neural network fusion method, reported better results than PatchSNR.

We therefore followed the combination paradigm for image fusion and design and trained an end-to-end 3D convolutional network from the results of two methods to the targeted restored image.

3 Learning Fine Features by 3D Convolution

3.1 Motivation and Related Work

Most of the existing neural network architectures apply spatial filters which address inputs such as 2D images. When it comes to videos, thus 3D inputs, these 2D convolutional neural networks (2DCNN) do not employ crucial information such as the temporal correlation. For example, in human action recognition, the motion information is not captured by 2DCNNs and Ji *et al.* [34] introduced a 3D convolutional neural network (3DCNN) method (see Fig. 5). The 3DCNN architecture has 1 hardwired layer, 3 convolutional layers and 2 subsampling layers. The spatial dimension of inputs 60 × 40 are gradually reduced to 1 × 1 by going through the network, *i.e.* 7 input frames have been converted into a 128-dimensional feature map capturing also the motion information. In the end, each element of the 128-dimensional feature map is fully connected to each unit in the last layer, then the action class is determined.

For performance improvements a brute force approach that proved successful is to deepen the (neural network) architecture [15, 24]. Yet, the improvements decline significantly with the depth while the training time and the demand of hardware (GPU) resources increase. For example, experiments reported in [15] demonstrate that the bulk of the performance is achieved by the first stages in their denoising TRD method while the last 3 stages (from 8) bring merely 0.01 dB to it. In [22] it is shown for SR methods that the first stages are the most

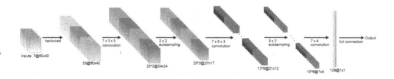

Fig. 5. 3DCNN proposed in [34] for human action recognition.

important and that adding more stages only slightly improves the performance (of A+) further.

On the other hand, for image restoration tasks such as SR it is common to recover the corrupted luminance component instead of the RGB image directly, and to interpolate the chroma. However, exploiting the correlation between corrupted RGB or even extra channels such as depth (D) or near-infrared (NIR) should be beneficial to the restoration task at the price of increased computation. For example, for denoising, Dabov *et al.* [35] apply the same grouping method on chroma channels as on the luminance, and they achieve better PSNR performances than by using BM3D [36] independently on three channels. To sum up, given several highly correlated (corrupted) channels/images, we have a better chance to high quality recovery.

It follows that we can consider the outputs of state-of-the-art methods as highly correlated images, which can be treated as the starting point of our proposed novel 3D convolutional fusion (3DCF) architecture.

3.2   Proposed Generic 3D Convolutional Fusion (3DCF)

General Architecture. As the starting point, we obtain several recovered outputs $\{\mathbf{I}_i\}_{i=1,\dots,n}$ from the same corrupted image, with different methods. We stack those highly correlated images along the channel dimension, which brings us a multichannel image $\mathbf{I}_a = [\mathbf{I}_1, \mathbf{I}_2, \dots, \mathbf{I}_n]$ (see Fig. 6).

Furthermore, since directional gradient filters are sensitive to intensity changes and edges, and our task is about recovering fine image details based on the results of existing methods, hence the correlation between the recovered output image and its gradients can be exploited. To this end, we firstly have the naive average input image $\bar{\mathbf{I}} = \frac{1}{n} \sum_{i=1}^{n} \mathbf{I}_i$, then filter it with the first- and second-order gradients, in both the x and y direction,

$$\mathbf{F}_{1x} = \begin{bmatrix} 1 & -1 \end{bmatrix} = \mathbf{F}_{1y}^T,$$
$$\mathbf{F}_{2x} = \begin{bmatrix} 1 & -2 & 1 \end{bmatrix} / 2 = \mathbf{F}_{2y}^T,$$

(1)

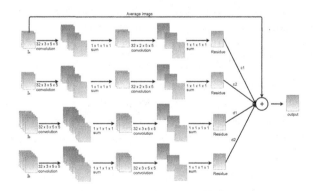

Fig. 6. Proposed 3D convolutional fusion method (3DCF).

followed by stacking those gradient filtered- and average images along the channel dimension, we have another input $\mathbf{I}_b$ as our second starting point,

$$\mathbf{I}_b = [\mathbf{F}_{2x} * \bar{\mathbf{I}}, \mathbf{F}_{1x} * \bar{\mathbf{I}}, \bar{\mathbf{I}}, \mathbf{F}_{1y} * \bar{\mathbf{I}}, \mathbf{F}_{2y} * \bar{\mathbf{I}}]. \tag{2}$$

Next, we intensively explore the correlation within $\mathbf{I}_a, \mathbf{I}_b$ by introducing the 3D convolutional layer. Related recent works such as [15,24,33] mainly exploit deep features with spatial filters. In that case, given the image has multiple channels, they are independently filtered and eventually summed up as the input for the next layer, while the correlations among the channels may not be accurately captured. That is the main reason behind our idea – to fully explore the fine details along the channel dimension. As far as we know, this is the first time that a 3D layer is introduced to address low level image tasks.

Our next step is to update the input images $\mathbf{I}_{a,b}{}^1$ with a 3D hidden layer,

$$\mathbf{H}_1^{a,b}(\mathbf{I}_{a,b}) = \tanh(\mathbf{W}_1^{a,b} * \mathbf{I}_{a,b} + \mathbf{B}_1^{a,b}), \tag{3}$$

where $\mathbf{W}_1^{a,b}$ correspond n 3D filters with $c \times h \times w$ kernel size and $\mathbf{B}_1^{a,b}$ are biases. In our design, due to a tradeoff between the memory constraint and speed, we recommend n and $c \times h \times w$ to be 32 and $3 \times 5 \times 5$ for $\mathbf{I}_b$, along with setting pad to be 0, so that we have the output with same size as input. The default size of filters regarding $\mathbf{I}_a$ showed in Fig. 6 are also determined for the same reason. Besides we use hyperbolic tangent (tanh) as activation function because we allow negative value updates to pass through the network rather than ignore them as ReLU [37] does. In the following step, we use a naive convolutional layer with a single $1 \times 1 \times 1$ filter, which is equivalent to sum up the input $\mathbf{H}_1^{a,b}$

$$\mathbf{H}_2^{a,b}(\mathbf{H}_1^{a,b}(\mathbf{I}_{a,b})) = \tanh(w_2^{a,b} \sum_k \mathbf{H}_{1,k}^{a,b}(\mathbf{I}_{a,b}) + b_2^{a,b}\mathbf{1}), \tag{4}$$

where $w_2^{a,b}$, $b_2^{a,b}$ are the scalar weights and biases, resp. We consider the above two steps as one inference stage. Another important difference between our proposed method and many other neural network methods is that we reconstruct the image residue instead of the image itself (see Fig. 6). Normally, the perturbation on image residues during the optimization is smaller than the one on image values, which increases the odds that the learning process eventually converges. Secondly, residue reconstruction substantiates the robust performance of our general architecture for distinct image restoration tasks. After going through n inference stages, we come to the reconstruction stage,

$$\mathbf{R}_{a,b}(\mathbf{I}_{a,b}) = (w_{2n+2}^{a,b} \sum \mathbf{H}_{2n+1}^{a,b} \circ \mathbf{H}_{2n}^{a,b} \dots \mathbf{H}_2^{a,b} \circ \mathbf{H}_1^{a,b}(\mathbf{I}_{a,b}) + b_{2n+2}^{a,b}\mathbf{1}), \tag{5}$$

where $\mathbf{R}_{a,b}(\mathbf{I}_{a,b})$ are the image residues we want to predict. In order to robustify the performance of our network, we simply duplicate the above mentioned process for each input image array $\mathbf{I}_a$ and $\mathbf{I}_b$ n times, which gives us $2n$ separate

---

[1] Here we abuse of notation, $\mathbf{I}_{a,b}$ indicates two inputs $\mathbf{I}_a, \mathbf{I}_b$.

networks with the same architecture. In the end we sum up the residues and the average image to obtain our output image $F(\mathbf{I}_1, \mathbf{I}_2, \ldots, \mathbf{I}_n)$,

$$F(\mathbf{I}_1, \mathbf{I}_2, \ldots, \mathbf{I}_n) = \frac{1}{n} \sum_k \mathbf{I}_k + \sum_k (c_k \mathbf{R}_a^k(\mathbf{I}_a) + d_k \mathbf{R}_b^k(\mathbf{I}_b)), \tag{6}$$

where c_k, d_k are the coefficients to weight the residues.

Training. Our main task is to learn the parameters $\boldsymbol{\Theta} = (\mathbf{W}, \mathbf{B})$ of the non-linear map F. To this end, we minimize the loss function $l(\boldsymbol{\Theta})$, which computes the Euclidean distance (mean square error (MSE)) between the output image $F(\mathbf{I}_1^i, \mathbf{I}_2^i, \ldots, \mathbf{I}_n^i)$ and ground truth image $\mathbf{I}_g^i$ contained in our training set, *i.e.*,

$$l(\boldsymbol{\Theta}) = \sum_i \| F(\mathbf{I}_1^i, \mathbf{I}_2^i, \ldots, \mathbf{I}_n^i; \boldsymbol{\Theta}) - \mathbf{I}_g^i \|_2^2. \tag{7}$$

The choice of the cost function is appropriate since PSNR is the main evaluation method of image restoration tasks and stands in monotonic relation with MSE. During the training stage, we update the weights/biases with standard back propagation [38,39].

Currently, the optimization of the loss function is dominated by the stochastic gradient descent (SGD) method [40], for example in [15,24,33]. Basically, at the $t+1$-th iteration they update the parameters $\boldsymbol{\Theta}_{t+1}$ with the previous parameter update $\boldsymbol{\Lambda}_t$ and negative gradient $\nabla l(\boldsymbol{\Theta})$,

$$\begin{aligned} \boldsymbol{\Lambda}_{t+1} &= a\boldsymbol{\Lambda}_t - b\nabla l(\boldsymbol{\Theta}_t), \\ \boldsymbol{\Theta}_{t+1} &= \boldsymbol{\Theta}_t + \boldsymbol{\Lambda}_{t+1}, \end{aligned} \tag{8}$$

where a, b are the momentum and learning rate, resp. One weakness of SGD is that the improvements gained from the optimization decrease rapidly with growing iteration steps. In such case, SGD may not be able to recover accurate details from highly corrupted images. This is the main reason why we prefer adaptive moment estimation (Adam) [41] as our optimization method. The Adam method is stated as follows,

$$\begin{aligned} \boldsymbol{\Lambda}_t &= a_1\boldsymbol{\Lambda}_{t-1} + (1 - a_1)\nabla l(\boldsymbol{\Theta}_t), \\ \mathbf{K}_t &= a_2\mathbf{K}_{t-1} + (1 - a_2)\nabla l(\boldsymbol{\Theta}_t)^2, \end{aligned} \tag{9}$$

where a_1, a_2 are moments and $\boldsymbol{\Theta}_{t+1}$ is updated based on $\boldsymbol{\Lambda}_t, \mathbf{K}_t$,

$$\boldsymbol{\Theta}_{t+1} = \boldsymbol{\Theta}_t - b\frac{\sqrt{1 - (a_2)^t}}{1 - (a_1)^t} \frac{\boldsymbol{\Lambda}_t}{\sqrt{\mathbf{K}_t} + \epsilon}, \tag{10}$$

here b is the learning rate and ϵ is used to avoid explosion. At the beginning of the iterations, the cost of $l(\boldsymbol{\Theta})$ converges considerably faster than SGD. Moreover, Eq. (10) shows that the magnitudes of parameter updates are independent of the rescaling of the gradient, therefore it provides a relatively fast convergence speed even after a large amount of iterations.

4   Experiments

In the following we describe the experimental setup and datasets used to validate our 3DCF approach on both the SR and DN tasks, then discuss the results.

4.1   Experimental Setup and Datasets

DN. Like most DN-related papers we add white Gaussian (AWG) noise to ground truth images to create our corrupted images. 3 standard deviations $\sigma \in \{15, 25, 50\}$ are chosen to measure the performance of 3DCF. Under such conditions, we compare our 3DCF with state-of-the-art DN methods as described in the introductory Sect. 1: BM3D [6], LSSC [11], EPLL [10], opt-MRF [14], CRTF [20], WNNM [8], CSF [13], TRD [15], MLP [12], as well as the NN [17] fusion method.

We use the same training data mentioned in [15], *i.e.*, 400 cropped images with 180×180 size from the training part of the Berkeley segmentation dataset (BSD) [42]. We evaluate our method on the 68 test images as in [43], a standard benchmark employed by top methods like [13,15].

SR. For SR we use the same 3DCF architecture as for DN and test it on the standard benchmarks Set5 [44], Set14 [45] (as proposed in [30]) and B100 [29] with 5, 14, 100 images resp., which are widely adopted by the recent literature. To obtain the LR images, according to many of the SR works, we firstly convert the ground truth image into YCbCr color space, then downscale the luminance channel with bicubic interpolation. Our training data is formed by the 200 training BDS images of size 321×481 from which we extract millions of LR-HR image pairs. We report PSNR and SSIM results for the latest methods with top performances: A+ [29], SRCNN(L) [33], RFL [31], SelfEx [32], CSCN [24].

4.2   Implementation Details

We implement our 3DCF method with Caffe [46]. 3DCF is used in the same form for both DN and SR. For clarity and ease of understanding and deployment we prefer stacking two top methods along the channel dimension as our one starting point $\mathbf{I}_a$. For DN we use MLP [12], an external neural network method, and BM3D [36], an internal method. Thus, such combination of two top methods increases our chance to take advantage of the strengths and overcome the weaknesses of both worlds. For SR, the CSCN [24] and A+ [29] are our favorite because of similar reasons – one from CNN and another from non-CNN type of methods. The starting point $\mathbf{I}_b$ is simply obtained by the average image of two methods as well as its corresponding first- and second order gradients along x/y direction. To enable 3DCF to recover more accurate details, we use two networks for each starting point $\mathbf{I}_a, \mathbf{I}_b$ (See Fig. 6), while slightly perturbing the value as the input of each activation, by multiplying -1. For the same reason we fix the coefficients c_1, c_2 to be 1 and 0.1. So are the coefficients d_1, d_2. Now Eq. (6) looks as follows:

$$F(\mathbf{I}_1, \mathbf{I}_2) = \frac{1}{2}(\mathbf{I}_1 + \mathbf{I}_2) + \mathbf{R}_a^1(\mathbf{I}_a) + 0.1\mathbf{R}_a^2(\mathbf{I}_a) + \mathbf{R}_b^1(\mathbf{I}_b) + 0.1\mathbf{R}_b^2(\mathbf{I}_b). \qquad (11)$$

For the sake of time complexity and memory saving, each network showed in Fig. 6 has 4 layers, and the filter size $n \times c \times h \times w$ is set to be ($32 \times 3 \times 5 \times 5, 1 \times 1 \times 1 \times 1, 32 \times 3 \times 5 \times 5, 1 \times 1 \times 1 \times 1$) for $\mathbf{I}_a$, while $\mathbf{I}_b$ has the almost same settings except for the 3rd layer with $32 \times 2 \times 5 \times 5$. We also set the channel-, height- and width stride to be 1 for all layers. It is expected that our output is a single image with the same spatial size as the input image. To this end, the channel-, height- and width padding size are determined to be ($1 \times 2 \times 2, 0 \times 0 \times 0, 0 \times 2 \times 2, 0 \times 0 \times 0$) for $\mathbf{I}_a$, and for $\mathbf{I}_b$ we follow the same setup except the first layer parameters are determined to be $0 \times 2 \times 2$. We also initialize the weights by a Gaussian distribution with standard deviation 0.05 for convolutional layers, and put the weight to 1 for sum layers, and the bias to 0 for all cases.

Meanwhile, we simply use the default learning- and decay rate 1 when learning the weights/biases for each layer. In the end, for Eq. 10 the learning rate b for the whole network is considered to be 0.001, the moments a_1, a_2 have the default value 0.9, 0.999, and ϵ is also set to the default 10^{-8}. It is worth mentioning that all the parameters are exactly the same for the two tasks, DN and SR.

4.3 Denoising Results

We demonstrate our 3DCF method on 68 standard images [43] from BSD [42]. We apply the best setup for the compared methods, already described in the introductory Sect. 1. CRTF [20] has 5 cascades, CSF [13] employs the 7×7 filter, the same as TRD [15] with 8 stages. Table 1 shows that our 3DCF method achieves top performances compared to other methods for 3 different standard deviations. For example, if we start our method with BM3D [6] and MLP [12], we are 0.11 dB and 0.1 dB better than the top standalone method MLP for $\sigma \in \{25, 50\}$. Due to the lack of an MLP model trained for $\sigma = 15$, we use BM3D+TRD instead. Still, the performance of our 3DCF is consistent with the other cases, 0.09 dB higher than TRD, the currently best method. Interestingly, if we compare 3DCF with the NN fusion method under the same conditions, that is, with the same starting methods BM3D and MLP, the proposed method outperforms NN with 0.15 and 0.07 dB for $\sigma \in \{15, 25\}$. Such observation confirms the non-trivial improvements achieved by 3DCF. Moreover, Fig. 7 indicates that the naive average of MLP and BM3D is even worse than MLP. Besides, it is also notable from Fig. 7 that the PSNR gradually increases with the growth of back propagation. 3DCF is robust to the fused methods, TRD + MLP leads to relative improvements comparable with those achieved starting from BM3D + MLP or BM3D + TRD.

Table 1. Average PSNR values [dB] on 68 images from BSD dataset as in [43] for $\sigma \in \{15, 25, 50\}$. The best is with bold. The results with (*) are obtained from [15].

Method	σ		
	15	25	50
BM3D [6]	31.08	28.57	25.61
*LSSC [11]	31.27	28.70	25.72
*EPLL [10]	31.19	28.68	25.67
*opt-MRF [14]	31.18	28.66	25.70
*CRTF$_5$ [20]		28.75	
*WNNM [8]	31.37	28.83	25.83
CSF$_{7\times7}$ [13]	31.24	28.71	
TRD$^8_{7\times7}$ [15]	31.42	28.93	25.99
MLP [12]		28.96	26.01
NN (BM3D+MLP) [17]		28.92	26.04
3DCF (BM3D+TRD)	**31.51**	29.03	26.10
3DCF (BM3D+MLP)		**29.07**	26.11
3DCF (TRD+MLP)		**29.07**	**26.12**

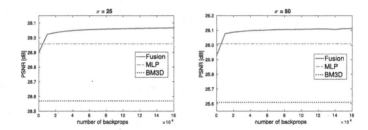

Fig. 7. PSNR versus backprops on 68 images for $\sigma \in \{25, 50\}$.

4.4   Super Resolution Results

The PSNR and SSIM results are listed in Table 2. Here our 3DCF fuses A+ [29] with CSCN [24]. Note that we modify the steps of downscaling the image for CSCN to be consistent with other methods including A+ and SRCNN(L). That is the reason why we obtain different PSNR results for CSCN than in the original work [24]. As in the case of DN, our 3DCF shows significant improvements over the starting methods. The PSNR improvements vary from 0.11 dB on (B100, ×3) to 0.35 dB on (Set5, ×2) over the best result from SRCNN (L,with largest model). The SSIM improvements follow the same trend. Note that for SR, the naive average fusion of A+ and CSCN results improves over both fused methods. However, our 3DCF results are on average 0.2 dB higher than the average fusion, as shown in Fig. 8.

Table 2. Average PSNR/SSIMs for upscaling factors ×2, ×3, and ×4 on datasets Set5, Set14, and B100. The best results are with bold.

Dataset	Scale	A+ [29]	SRCNN(L) [33]	RFL [31]	SelfEx [32]	CSCN [24]	3DCF (CSCN+A+)
		PSNR/SSIM	PSNR/SSIM	PSNR/SSIM	PSNR/SSIM	PSNR/SSIM	PSNR/SSIM
Set5	x2	36.56/0.9612	36.68/0.9609	36.52/0.9589	36.50/0.9577	36.55/0.9605	**37.03/0.9631**
	x3	32.67/0.9199	32.83/0.9198	32.50/0.9164	32.63/0.9190	32.68/0.9197	**33.11/0.9255**
	x4	30.33/0.8749	30.52/0.8774	30.17/0.8715	30.32/0.8728	30.44/0.8779	**30.82/0.8865**
Set14	x2	32.32/0.9607	32.52/0.9612	32.30/0.9599	32.27/0.9584	32.36/0.9593	**32.71/0.9623**
	x3	29.16/0.8869	29.35/0.8886	29.07/0.8842	29.19/0.8873	29.19/0.8850	**29.48/0.8907**
	x4	27.33/0.8277	27.53/0.8285	27.23/0.8251	27.43/0.8279	27.41/0.8256	**27.69/0.8334**
B100	x2	31.16/0.8857	31.32/0.8874	31.13/0.8842	31.15/0.8860	31.20/0.8836	**31.48/0.8899**
	x3	28.25/0.7824	28.37/0.7853	28.20/0.7814	28.25/0.7821	28.28/0.7804	**28.48/0.7881**
	x4	26.76/0.7073	26.86/0.7089	26.70/0.7068	26.81/0.7078	26.83/0.7072	**26.99/0.7147**

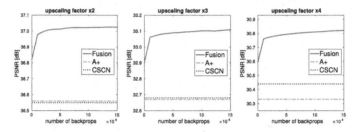

Fig. 8. PSNR versus backprops on Set5 dataset for upscaling factors ×2, ×3, ×4.

4.5 Other Aspects

Visual assessment. In general, the visual results are consistent with PSNR results. Some image results are shown in Fig. 9 for DN and in Fig. 10 for SR. We can observe that the 3DCF results have generally fewer artifacts and sharper edges in comparison with the other methods.

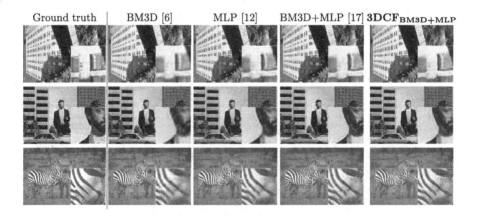

Fig. 9. Denosing results for $\sigma = 50$. Best zoomed on screen.

| Ground truth | A+ [29] | CSCN [24] | SRCNN(L) [33] | **3DCF**CSCN+A+ |

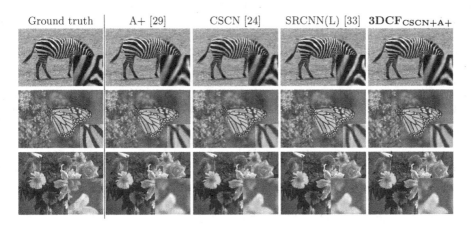

Fig. 10. Super-resolution results (×4). Best zoomed on screen.

Running time. 3DCF runs on roughly 0.04 second per 321 × 480 image on nVidia TitanX GPU, which is quite competitive and shows that at the price of slight increase in processing time one could fuse available image restoration results. 3DCF needs about 5 h training time to obtain meaningful improvements over the fused methods, and this is mainly due to the Adam method.

General. To summarize, our 3DCF method shows wide adaptability for two important image restoration tasks, DN and SR, with non-trivial improvements. Also, the training and running times of 3DCF are competitive in comparison with other neural network architectures. For certain combinations of existing methods our proposed fusion method only shows mild progress, for example for the case of TRD+MLP (see Table 1). This sensitivity to the starting point drives us to be careful of the choice of starting methods.

5   Conclusions

We propose a novel 3D convolutional fusion (3DCF) network for image restoration. With the same settings, for both single image super resolution and image denoising, we achieve significant improvements over the fused methods and other fusion methods on several standard benchmarks. For speeding up the training, we apply an adaptive moment estimation method. The testing and training times are also competitive to other recent deep neural networks.

Acknowledgments. This work was supported by the ERC project *VarCity* (*#273940*), the ETH General Fund (OK) and by an Nvidia GPU grant.

References

1. Katsaggelos, A.K.: Digital Image Restoration. Springer Publishing Company, Incorporated, Heidelberg (2012)
2. Stathaki, T.: Image Fusion: Algorithms and Applications. Academic Press, Amsterdam (2011)
3. Lebrun, M., Colom, M., Buades, A., Morel, J.: Secrets of image denoising cuisine. Acta Numerica **21**, 475–576 (2012)
4. Zontak, M., Mosseri, I., Irani, M.: Separating signal from noise using patch recurrence across scales. In: CVPR (2013)
5. Buades, A., Coll, B., Morel, J.M.: A non-local algorithm for image denoising. In: IEEE Computer Society Conference on Computer Vision and Pattern Recognition, CVPR 2005, vol. 2, pp. 60–65. IEEE (2005)
6. Dabov, K., Foi, A., Katkovnik, V., Egiazarian, K.: Image denoising by sparse 3d transform-domain collaborative filtering. IEEE Trans. Image Process. **16**, 2080–2095 (2007)
7. Yu, G., Sapiro, G.: DCT image denoising: a simple and effective image denoising algorithm (2011)
8. Gu, S., Zhang, L., Zuo, W., Feng, X.: Weighted nuclear norm minimization with application to image denoising. In: CVPR (2014)
9. Yu, G., Sapiro, G., Mallat, S.: Solving inverse problems with piecewise linear estimators: from gaussian mixture models to structured sparsity. IEEE Trans. Image Process. **21**(5), 2481–2499 (2012)
10. Zoran, D., Weiss, Y.: From learning models of natural image patches to whole image restoration. In: IEEE International Conference on Computer Vision, pp. 479–486 (2011)
11. Mairal, J., Bach, F., Ponce, J., Sapiro, G., Zisserman, A.: Non-local sparse models for image restoration. In: IEEE 12th International Conference on Computer Vision, pp. 2272–2279 (2009)
12. Burger, H., Schuler, C., Harmeling, S.: Image denoising: can plain neural networks compete with bm3d? In: IEEE Computer Vision and Pattern Recognition, pp. 2392–2399 (2012)
13. Schmidt, U., Roth, S.: Shrinkage fields for effective image restoration. In: Proceedings of the IEEE Conference on Computer Vision and Pattern Recognition, pp. 2774–2781 (2014)
14. Chen, Y., Pock, T., Ranftl, R., Bischof, H.: Revisiting loss-specific training of filter-based MRFs for image restoration. In: Weickert, J., Hein, M., Schiele, B. (eds.) GCPR 2013. LNCS, vol. 8142, pp. 271–281. Springer, Heidelberg (2013). doi:10.1007/978-3-642-40602-7_30
15. Chen, Y., Yu, W., Pock, T.: On learning optimized reaction diffusion processes for effective image restoration. In: Proceedings of the IEEE Conference on Computer Vision and Pattern Recognition, pp. 5261–5269 (2015)
16. Jancsary, J., Nowozin, S., Rother, C.: Loss-specific training of non-parametric image restoration models: a new state of the art. In: IEEE European Conference of Computer Vision (2012)
17. Burger, H.C., Schuler, C., Harmeling, S.: Learning how to combine internal and external denoising methods. In: Weickert, J., Hein, M., Schiele, B. (eds.) GCPR 2013. LNCS, vol. 8142, pp. 121–130. Springer, Heidelberg (2013). doi:10.1007/978-3-642-40602-7_13

18. Levin, A., Nadler, B., Durand, F., Freeman, W.T.: Patch complexity, finite pixel correlations and optimal denoising. In: Fitzgibbon, A., Lazebnik, S., Perona, P., Sato, Y., Schmid, C. (eds.) ECCV 2012. LNCS, vol. 7576, pp. 73–86. Springer, Heidelberg (2012). doi:10.1007/978-3-642-33715-4_6

19. Mosseri, I., Zontak, M., Irani, M.: Combining the power of internal and external denoising. In: IEEE International Conference on Computational Photography (ICCP), pp. 1–9 (2013)

20. Schmidt, U., Jancsary, J., Nowozin, S., Roth, S., Rother, C.: Cascades of regression tree fields for image restoration (2014)

21. Timofte, R.: Anchored fusion for image restoration. In: ICPR (2016)

22. Timofte, R., Rothe, R., Van Gool, L.: Seven ways to improve example-based single image super resolution. In: The IEEE Conference on Computer Vision and Pattern Recognition (CVPR) (2016)

23. Kim, J., Kwon Lee, J., Mu Lee, K.: Accurate image super-resolution using very deep convolutional networks. In: The IEEE Conference on Computer Vision and Pattern Recognition (CVPR) (2016)

24. Wang, Z., Liu, D., Yang, J., Han, W., Huang, T.: Deep networks for image super-resolution with sparse prior. In: Proceedings of the IEEE International Conference on Computer Vision, pp. 370–378 (2015)

25. Agustsson, E., Timofte, R., Van Gool, L.: Regressor basis learning for anchored super-resolution. In: ICPR (2016)

26. Dai, D., Timofte, R., Van Gool, L.: Jointly optimized regressors for image super-resolution. Comput. Graph. Forum **34**, 95–104 (2015)

27. Timofte, R., De Smet, V., Van Gool, L.: Semantic super-resolution: when and where is it useful? Comput. Vis. Image Underst. **142**, 1–12 (2016)

28. Yang, J., Wright, J., Huang, T.S., Ma, Y.: Image super-resolution via sparse representation. IEEE Trans. Image Process. **19**, 2861–2873 (2010)

29. Timofte, R., De Smet, V., Van Gool, L.: A+: adjusted anchored neighborhood regression for fast super-resolution. In: Cremers, D., Reid, I., Saito, H., Yang, M.-H. (eds.) ACCV 2014. LNCS, vol. 9006, pp. 111–126. Springer, Cham (2015). doi:10.1007/978-3-319-16817-3_8

30. Timofte, R., Smet, V., Gool, L.: Anchored neighborhood regression for fast example-based super-resolution. In: Proceedings of the IEEE International Conference on Computer Vision, pp. 1920–1927 (2013)

31. Schulter, S., Leistner, C., Bischof, H.: Fast and accurate image upscaling with super-resolution forests. In: Proceedings of the IEEE Conference on Computer Vision and Pattern Recognition, pp. 3791–3799 (2015)

32. Huang, J.B., Singh, A., Ahuja, N.: Single image super-resolution from transformed self-exemplars. In: 2015 IEEE Conference on Computer Vision and Pattern Recognition (CVPR), pp. 5197–5206. IEEE (2015)

33. Dong, C., Loy, C.C., He, K., Tang, X.: Image super-resolution using deep convolutional networks (2015)

34. Ji, S., Xu, W., Yang, M., Yu, K.: 3d convolutional neural networks for human action recognition. IEEE Trans. Pattern Anal. Mach. Intell. **35**, 221–231 (2013)

35. Dabov, K., Foi, A., Katkovnik, V., Egiazarian, K.: Color image denoising via sparse 3d collaborative filtering with grouping constraint in luminance-chrominance space. In: IEEE International Conference on Image Processing, ICIP 2007, vol. 1, p. I-313. IEEE (2007)

36. Dabov, K., Foi, A., Katkovnik, V., Egiazarian, K.: Image denoising by sparse 3-d transform-domain collaborative filtering. IEEE Trans. Image Process. **16**, 2080–2095 (2007)

37. Nair, V., Hinton, G.E.: Rectified linear units improve restricted Boltzmann machines. In: Proceedings of the 27th International Conference on Machine Learning (ICML 2010), pp. 807–814 (2010)
38. Rumelhart, D.E., Hinton, G.E., Williams, R.J.: Learning representations by back-propagating errors. Cogn. Model. **5**, 1 (1988)
39. LeCun, Y., Bottou, L., Bengio, Y., Haffner, P.: Gradient-based learning applied to document recognition. Proc. IEEE **86**, 2278–2324 (1998)
40. Bottou, L.: Large-scale machine learning with stochastic gradient descent. In: Lechevallier, Y., Saporta, G. (eds.) Proceedings of COMPSTAT 2010, pp. 177–186. Springer, Heidelberg (2010)
41. Kingma, D., Ba, J.: Adam: A method for stochastic optimization. arXiv preprint arXiv:1412.6980 (2014)
42. Martin, D., Fowlkes, C., Tal, D., Malik, J.: A database of human segmented natural images and its application to evaluating segmentation algorithms and measuring ecological statistics. In: Proceedings of the 8th International Conference on Computer Vision, vol. 2, pp. 416–423 (2001)
43. Roth, S., Black, M.J.: Fields of experts. Int. J. Comput. Vis. **82**, 205–229 (2009)
44. Bevilacqua, M., Roumy, A., Guillemot, C., Alberi-Morel, M.L.: Low-complexity single-image super-resolution based on nonnegative neighbor embedding (2012)
45. Zeyde, R., Elad, M., Protter, M.: On single image scale-up using sparse-representations. In: Boissonnat, J.-D., Chenin, P., Cohen, A., Gout, C., Lyche, T., Mazure, M.-L., Schumaker, L. (eds.) Curves and Surfaces 2010. LNCS, vol. 6920, pp. 711–730. Springer, Heidelberg (2012). doi:10.1007/978-3-642-27413-8_47
46. Jia, Y., Shelhamer, E., Donahue, J., Karayev, S., Long, J., Girshick, R., Guadarrama, S., Darrell, T.: Caffe: convolutional architecture for fast feature embedding. In: Proceedings of the ACM International Conference on Multimedia, pp. 675–678. ACM (2014)

Video Super Resolution Using Non-Local Means with Adaptive Decaying Factor and Searching Window

Yawei Li$^{(\boxtimes)}$, Xiaofeng Li, Cui Yao, Zhizhong Fu, and Xiuxia Yin

University of Electronic Science and Technology of China, No. 2006, Xiyuan Avenue,
West Hi-Tech District, Chengdu 611731, China
lywuestc@163.com

Abstract. Power consumption, transmission bandwidth, and spatial-temporal resolution tradeoff are among the most important factors that affect video processing on mobile devices. Scalable video coding (SVC) provides a possible solution to overcome these problems. Every video in SVC format consists of high-resolution (HR) frames and low-resolution (LR) frames. For a better viewing experience, super resolution (SR) is introduced to refine the LR frames. Non-local means (NLM) based SR has a promising prospect. Since NLM originates from image denoising, the fixed parameters of NLM is not fit for SR tasks. A fixed decaying factor tends to blur the details in flat regions and a fixed searching window results in mismatches among pixels. Thus, we propose two adaptive parameters to address these problems. We generalize key features that affect SR methods' applicability of implementation on hardware and show NLM is fit for hardware implementation. The experimental results validate the proposed algorithm.

1 Introduction

There has been a great demand for high definition video in mobile devices recently. However, some practical constrains and considerations hinder the industry to meet that consumer demand. Among those constrains, transmission bandwidth is the first one under consideration [1]. Since wireless bandwidth is limited especially for mobile devices, devising efficient video compression methods seems to be urgent. Meanwhile, a large array of sensors is necessary to capture high-resolution (HR) video sequences. Thus, high power consumption becomes an unavoidable requirement which aggravates the problem of the short battery life of mobile devices [2,3]. Finally, it is well-known that there is a tradeoff between spatial resolution (pixel intensity) and temporal resolution (frame rate) [4–6]. It is anything but easy to design a device that can acquire HR videos with high frame rate.

Fortunately, one way to overcome the aforementioned problems is the video sequence with HR guide frames (scalable video coding, SVC) shown in Fig. 1. A video sequence consists of HR frames (key frames) and low-resolution (LR)

© Springer International Publishing AG 2017
C.-S. Chen et al. (Eds.): ACCV 2016 Workshops, Part I, LNCS 10116, pp. 177–190, 2017.
DOI: 10.1007/978-3-319-54407-6_12

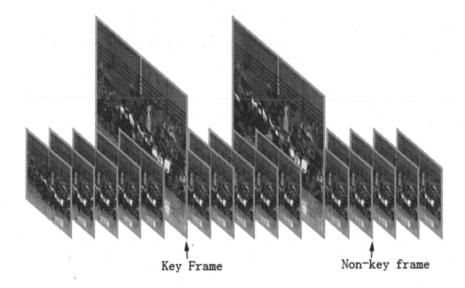

Key Frame   Non-key frame

Fig. 1. Video sequence with key frames and non-key frames.

frames (non-key frames) [7]. During the capture process, a non-key frame is obtained from an original full resolution frame by blurring and decimation, i.e.

$$\mathbf{f_n} = DB\mathbf{F}_n \tag{1}$$

where B is the blurring filter, D is the decimation operator, $\mathbf{F}_n$ is the original full resolution frame, and $\mathbf{f}_n$ is the non-key frame.

In this mechanism, one can capture, store, and transmit non-key frames in low resolution so that the problems of spatial-temporal resolution tradeoff, power consumption, and limited bandwidth are all addressed. The whole video sequence is then refined using signal processing techniques such as interpolation and super resolution (SR).

Several state-of-the-art image interpolation methods have been proposed. These approaches belong to a class of visually oriented interpolation techniques, including edge directed, content adaptive, and wavelet-based methods [8–11]. However, the video sequences refined by interpolation suffer from perceived loss of detail in texture regions because they are unable to estimate the high-frequency information. Thus, SR is of necessity for a better viewing experience.

Compared with the traditional interpolation approaches, SR algorithms can recover the missing high-frequency details in one image. Since the breakthrough made by Huang and Tsai [12], SR algorithms have been improved a lot. The existing SR algorithms are mainly divided into restoration-based algorithms and exampled-based algorithms. Restoration-based SR algorithms use subpixel shifts among several LR images to reconstruct a HR image [13–17]. When the magnification factor is large, restoration-based algorithms usually fail to create an image with satisfactory textures. On the other hand, example-based methods incorporate the concept of learning. Freeman *et al.* exploited a series of HR images to learn the high-frequency content for the LR images [18,19]. They first used these known HR images to build a database in a training phase. The database consisted of pairs of low-frequency and high-frequency information. Then the authors harnessed the established database to recover the high-frequency details for given blocks of LR images.

To refine the SVC video sequence, Ancuti *et al.* proposed a maximum a posteriori probability (MAP) estimator [17]. During the optimization phase of the MAP estimator, the authors added two priors that constrained the high-frequency information from HR frames and the edge information, respectively. Contrary to the MAP treatment [17], Brandi *et al.* directly separated a key frame into a low-frequency part and a high-frequency part [7]. For every anchor block in the non-key frame, a matching block was found in the key frame with minimum mean square error (MMSE). The high-frequency component in the matching block was then added to the anchor block. In the light of learning concept, Chang *et al.* proposed a SR method through neighbor embedding [20].

To avoid the block artifacts of the aforementioned block-wise SR methods, one must introduce some special treatment, for example, overlapping the blocks by one pixel. On the contrary, non-local means (NLM) directly computed the value of each pixel by averaging all the pixels within its neighborhood [21]. Weights that reflect the similarities between pixels are computed and used to average the neighbor pixels. Since oriented towards image denoising, NLM does not provide SR results. Therefore, Protter *et al.* generalized NLM to adapt to the SR tasks by minimizing the error energy [22]. Basavaraja *et al.* combined the separation treatment [23] and NLM to create a novel NLM-based SR algorithm. Lenyel *et al.* incorporated illuminance and gradient information into the similarity comparison and reduced the averaging pixels by thresholding [24]. However, there are two problems of those NLM methods. First, the decaying factor used to calculate the weights is a constant, which is not suitable for different regions such as flat regions and texture regions. Second, there are object or scene motions between frames. Thus, an adaptive size of neighborhood (searching window) is better to compensate for the motions. In this paper, considering those two modifications, we propose a NLM-based algorithm.

The rest of the paper is organized as the follows. Section 2 explains the video acquisition as inputs of the SR algorithm. Section 3 introduces the traditional NLM method and discusses two problems of it. Section 4 explains our proposed framework in detail. The proposed method can overcome the shortcomings of the traditional NLM and provide good SR results. Section 5 explores the possibility of implementing the proposed method on hardware. Section 6 shows the experiments and results. Section 7 concludes the paper.

2   Video Acquisition

The SVC format can solve the bandwidth, power consumption, and tradeoff problems that hinder the development of HR video. In this section, we introduce the various ways of video acquisition out of these constrains.

2.1   Hybrid Imaging System

An imaging system needs a minimum time interval to accumulate light energy. Assume that the sensor size remains constant. If the spatial resolution increases, then by definition this means that the pixel footprints would get smaller and they would receive less irradiance. As a result, the exposure time needs to be increased and the temporal resolution goes down. Thus, capturing a HR image requires a longer exposure time. On the other hand, to capture HR video sequences, the transfer rate of data from sensors to memories is too large for the traditional imaging system to achieve. Therefore, one is faced with a fundamental tradeoff between spatial resolution and temporal resolution of an imaging system [4–6].

To solve this problem, Ben-Ezra and Nayar proposed a hybrid imaging system which could capture HR frames with low frame rates (key frames) and LR frames with high frame rates (non-key frames) simultaneously [4]. The authors then estimated the point spread function (PSF) using the motion information obtained by the cameras. Then this function was used to deblur the captured image. Tai *et al.* introduced the use of hybrid imaging system to address spatially varying blur and video sequence [5]. They further developed this scheme to obtain HR frames with high frame rate [6].

2.2   Low-Power Acquisition

Low-power video acquisition has attracted a lot of attention. The variable acuity sensor imager (VASI) has different spatial resolution at locations with different interests [25]. It simulates the fovea of human and can capture high-frequency details within the region of interest. With the VASI framework, a content-based video capture scheme was proposed [2]. This scheme could reduce the power consumption by conditionally turning off some sensors in flat regions and naturally provide SR results without extensive post-processing.

2.3   Online Transmission

Apart from the hardware systems, a mobile device can also download video sequences online. In this scenario, bandwidth and transmission rate are among the most important factors for a better consumer experience. To transmit HR video, the high transmission rate imposes great pressure on the bandwidth. With the SVC format, the viewer can only get a degraded video sequence on mobile devices. However, if SR algorithms are used in the display end, viewers can obtain a HR video sequence without requiring a large bandwidth. For example, if every fifth frame is decimated by half, the total data size will be reduced by forty percent.

3   The Non-Local Means Algorithm

NLM is originally proposed as an image denoising method and results in very successful denoised images. The basic assumption of NLM is that a patch pattern may repeat within the range of an image. NLM exploits that spatial redundancy of an image to denoise it. A pixel is replaced by the weighted average of its neighbors. Every neighbor of the pixel is assigned with a weight that reflects the similarity between pixels, namely

$$\hat{\mathbf{x}}(k,l) = \frac{\displaystyle\sum_{(i,j)\in N(k,l)} \omega_{k,l}(i,j) \cdot \mathbf{y}(i,j)}{\displaystyle\sum_{(i,j)\in N(k,l)} \omega_{k,l}(i,j)} \tag{2}$$

where $N(k,l)$ is the neighborhood (or the searching window in this paper) of pixel (k,l), $\omega_{k,l}(i,j)$ is the weight of every pixel (i,j) in the neighborhood of (k,l), $\mathbf{y}$ is the noisy image, and $\hat{\mathbf{x}}$ is the denoised image. The weights of pixels are calculated by

$$\omega_{k,l}(i,j) = \exp\left(\frac{-\left\| (R_{k,l}^S \mathbf{y} - R_{i,j}^S \mathbf{y})G_{\sigma_s} \right\|_2^2}{2\sigma^2} \right) \tag{3}$$

where σ is a fixed decaying factor, $R_{i,j}^S$ is an operator that extracts a patch of size $S \times S$ centered at (i,j), G_{σ_s} is a Gaussian window with zero mean and standard deviation σ_s. The patch difference of each pixel (i,j) is given by

$$E_{k,l}^2(i,j) = \left\| (R_{k,l}^S \mathbf{y} - R_{i,j}^S \mathbf{y})G_{\sigma_s} \right\|_2^2 . \tag{4}$$

The above two Eqs. (2) and (3) formulate the basic 2-D functions and are able to denoise one single image. By incorporating time domain as a third dimension, the NLM algorithm can denoise video sequences, i.e.

$$\hat{\mathbf{x}}(k,l,t) = \frac{\displaystyle\sum_{s\in T_t} \sum_{(i,j)\in N(k,l,s)} \omega_{k,l,s}(i,j) \cdot \mathbf{y}_s(i,j)}{\displaystyle\sum_{s\in T_t} \sum_{(i,j)\in N(k,l,s)} \omega_{k,l,s}(i,j)} \tag{5}$$

where $\omega_{k,l,s}(i,j)$ is the 3-D counterpart of $\omega_{k,l}(i,j)$, t is the current frame, and T_t is the set containing all the successive adjacent frames of the current frame (the current frame is also included in T_t). Equation (5) uses the input frames to calculate the desired pixel. Based on (5), researchers have designed video denoising method without motion estimation [21,22].

One important difference between the 2-D formula and the 3-D formula is brought by motion. For a video sequence, there are object or scene motions between two successive frames. In order to find a similar pair between two frames, the NLM algorithm must compensate for the motion. The common trick here is using a larger search window. Thus, the noise can be removed without explicit motion estimation. This way, one would consider the 3-D functions as a coarse motion estimator.

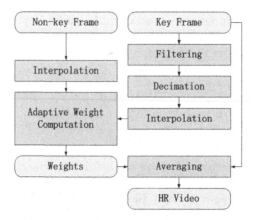

Fig. 2. Framework of the proposed algorithm.

4   The Proposed Algorithm

When applied to SR task, NLM does not work out well. The main reason is that the essence of image denoising is distinct from that of SR. For denoising problem, NLM searches the neighbors of a pixel to find those with similar grey level information and geometric configuration to the pixel. All of the similar neighbors are averaged with respect to the calculated weights. From the stochastic point of view, it is proved that NLM can reduce the noise level [21]. So for denoising problem, NLM can be seen as a mean filter with different weights.

On the other hand, the essence of NLM-based SR method is quite different. Although NLM still searches pixels with similar information for every pixel, it should only keep the pixels that highly resemble the target pixel. If the algorithm selects too many pixels, it will tend to blur the resulting image. Thus, when applied as a SR algorithm, NLM is a high-frequency information extractor rather than a mean filter.

When applied to SR, traditional NLM has two drawbacks. The first one is that the decaying factor σ is a constant. This is not suitable for regions with different characteristics. The second drawback is the huge computation complexity caused by large searching window. In order to overcome the two drawbacks of NLM-based SR algorithms, two adaptive parameters including the decaying factor and the size of the search window are introduced. The overall framework is shown in Fig. 2.

4.1   Input Video Sequence

The input video sequences consist of non-key frames and key-frames, namely

$$\{\mathbf{f}_n | n = Rz + r, z \in N, r = 1, 2, ..., R - 1\} \tag{6a}$$

$$\{\mathbf{F}_k | k = Rz, z \in N\} \tag{6b}$$

where R is the period of key frames. For every non-key frame, its bilateral key frame counterparts are

$$\{\mathbf{F}_b | b \in \Phi\} \tag{7a}$$

$$\Phi = \{\lfloor n/R \rfloor \times R, \lceil n/R \rceil \times R\} \tag{7b}$$

where $\lfloor \cdot \rfloor$ is the round-down operator and $\lceil \cdot \rceil$ the round-up operator. The two key-frames are used to super-resolve one non-key frame. The formation of the non-key frames includes blurring with PSF and decimation. Additive noise is omitted in the SR task. The non-key frames are first scaled up to the same full resolution as key frames, i.e.

$$\tilde{\mathbf{f}}_n = U\mathbf{f}_n \tag{8}$$

where the U operator is interpolation. The key frames go through the same operations with non-key frames, including filtering, decimation, and interpolation, namely

$$\tilde{\mathbf{f}}_k = UDB\mathbf{F}_k \tag{9}$$

where the operators B and D are the same as those in (1), the U as that in (9). This arrangement guarantees the key and non-key frames undergo the same processing. $\tilde{\mathbf{f}}_n$ and $\tilde{\mathbf{f}}_k$ have the same spatial resolution as key frames but they don't contain high-frequency information. NLM uses the processed frames $\tilde{\mathbf{f}}_n$ and $\tilde{\mathbf{f}}_k$ to calculate the weights of every pixel in the neighborhood around a center pixel. This processing incorporates the proposed adaptive parameters. After that one pixel is replaced by the weighted average of its neighbors.

4.2 Adaptive Searching Window

In NLM, The neighborhood (searching window) size of a pixel should be adaptive to different regions to not only maintain the excellent performance of NLM but also reduce the computational complexity. The determination of the searching window size P should be efficient and simple to implement. Thus, the searching begins with the minimum window size $P_{\min}$. The algorithm first finds the minimum patch difference of its neighbors, i.e.,

$$E_{k,l}^2 = \min_{b \in \Phi} \min_{(i,j) \in N^P(k,l)} E_{k,l}^2(i,j,b) . \tag{10}$$

If $E_{k,l}^2$ is larger than a predefined value ξ, then enlarge the window size P by a step Δ and repeat the above process. If $E_{k,l}^2$ is no larger than ξ, then choose the current window size as the desired one. The searching processing is stopped if $E_{k,l}^2$ is no larger than ξ or P exceeds a predefined value $P_{\max}$. Note that during the iterations, one need only to compute the patch difference of the new added pixels, which can save lots of computation. This processing is summarized in Table 1.

Table 1. Determination of the adaptive searching window.

Objective: Determine the window size.
Initialization: $E_{k,l}^2 = \xi + 1$, $P = P_{\min}$
Pseudocode: While $E_{k,l}^2 > \xi$ and $P \le P_{\max}$
Compute patch error for new added pixels.
Compute $E_{k,l}^2$ using (10).
$P = P + \Delta$
end

4.3 Adaptive Decaying Factor

In fact, the adaptive decaying factor is a by-product of the adaptive searching window. In the former stage, one obtains the minimum patch difference. To determine the decaying factor, we force the minimum patch difference decays to a predefined value, namely

$$\frac{E_{k,l}^2}{2\sigma_A^2} = \alpha \tag{11}$$

where the subscript of σ means that the parameter is chosen uniquely per pixel. If α is set to 2, the computed weight is 0.1353 for $E_{k,l}^2$ and 0.0183 for $2E_{k,l}^2$. The former is almost 6.5 times larger than the latter, which is suitable for decaying large values of the patch error. Solving the above equation results in

$$\sigma_A = \sqrt{\frac{E_{k,l}^2}{2\alpha}} \ . \tag{12}$$

4.4 Weighted Average

Having derived the two adaptive parameters, the following is to calculate the pixels by averaging. Incorporating the adaptive searching window and the adaptive decaying factor, the value of a pixel in a non-key frame is given by

$$\tilde{\mathbf{f}}_n^H(k,l) = \frac{\displaystyle\sum_{b \in \Phi} \sum_{(i,j) \in N^P(k,l)} \omega_{k,l}(i,j,b) \cdot \mathbf{F}_b(k,l)}{\displaystyle\sum_{b \in \Phi} \sum_{(i,j) \in N^P(k,l)} \omega_{k,l}(i,j,b)} \ . \tag{13}$$

Considering the adaptive decaying factor, the weights are given by

$$\omega_{k,l}(i,j,b) = \exp\left(\frac{-\left\|(R_{k,l}^S \tilde{\mathbf{f}}_n - R_{i,j}^S \tilde{\mathbf{f}}_b)G_{\sigma_s}\right\|_2^2}{2\sigma_A^2}\right) \ . \tag{14}$$

5   Discussion About Hardware Implementation

In this section, we discuss the potential of SR algorithms to be implemented on hardware systems. Instead of designing a new hardware system, we generalize some factors that affect the applicability of SR algorithms on those systems. We also apply these factors to justify the applicability of the adaptive NLM and conclude that it is fit for hardware implementation. Hardware implementation is important since it is more efficient than software implementation. For video display on mobile devices, one would prefer real-time display to waiting for a long time. The typical frame rate for a video sequence is 30 Hz. If the period of key frame is 2, then one SR algorithm needs to process 15 frames per second, with 66.7 ms per frame. This requirement may be beyond the capacity of most SR algorithms. However, the high efficiency of hardware provides a possibility for real-time display.

Vast literature discusses the hardware implementation of various SR algorithms. Those proposed implementations are based on different hardware devices like FPGA, DSP, and GPU. We refer to the papers [26–30] as the representatives. Factors that affect the applicability of SR methods on hardware include non-iteration, symmetric and independent computation, and memory considerations.

5.1   Non-iteration

Some SR methods require iteration in order to improve the image quality during each time of the iterations. Iterations are common for restoration-based SR methods [16,17]. Some exampled-based approaches also need iteration [3]. For real-time display, computation without iteration is important because it affects the efficiency of SR algorithms.

5.2   Symmetric and Independent Computation

Another advantage of NLM is that the computation of each pixel is symmetric and independent. The symmetry means the computation for each pixel is the same. The independent computation means the computation for one pixel does not affects that of another pixel. Thus, one can calculate two pixels independently at the same time. These two features indicate that one can design a system with several similar modules. Meanwhile, parallel calculation is possible, which can increase the efficiency of the NLM algorithm.

5.3   Memory Consideration

For mobile devices, memory is usually an important constrain that controls the applicability of an algorithm. If a SR method occupies too much memory, one should have sufficient reason to doubt its applicability. Fortunately, for NLM algorithms, one can save memory because every pixel is calculated independently.

Table 2. PSNR (dB) results of all considered test sequences.

Sequence	Ballroom	Exit	Crowd	Vassar	Foreman	Mobile	Container
Bilinear	30.7105	33.3100	26.5038	32.7461	31.7585	20.0480	25.0366
TNLM	29.3321	36.3067	28.4654	35.6520	36.6919	22.2506	31.0955
NLMAT	34.0473	37.8545	27.8664	35.7561	33.0152	20.1810	27.5071
NLMAD	34.7160	38.4935	31.4407	35.5461	36.7604	23.1547	31.4924
NLMAW	34.3213	38.6036	30.6330	**36.6687**	36.9561	22.8397	31.1410
Proposed	**36.5129**	**38.7558**	**32.7694**	35.8126	**36.9572**	**23.6688**	**31.5191**

Table 3. SSIM results of all considered test sequences.

Sequence	Ballroom	Exit	Crowd	Vassar	Foreman	Mobile	Container
Bilinear	0.7717	0.7827	0.8018	0.7727	0.6083	0.4484	0.5427
TNLM	0.9726	0.9811	0.9799	0.9755	0.9610	0.8913	0.9574
NLMAT	0.9761	0.9749	0.9733	0.9671	0.9479	0.7724	0.9359
NLMAD	0.9798	**0.9818**	0.9877	0.9657	0.9582	0.9007	0.9541
NLMAW	0.9804	0.9799	0.9880	**0.9746**	**0.9606**	0.8981	**0.9594**
Proposed	**0.9822**	0.9789	**0.9901**	0.9648	0.9580	**0.9072**	0.9545

6   Experimental Results

We validates the proposed NLM-based SR algorithm with three experiments in this section. We test the proposed algorithm on 7 video sequences, including *ballroom, exit, crowd, vassar, foreman, mobile, and container*. The algorithms under comparison include bilinear interpolation (Bilinear), traditional NLM (TNLM) [21], NLM with adaptive thresholding (NLMAT) [24], NLM with the proposed adaptive decaying factor (NLMAD), NLM with the proposed searching window (NLMAW), and the proposed NLM. Bilinear interpolation is the benchmark of those comparisons.

In order to simulate the imaging process of non-key frames, Lanczos filter serves as a PSF to filter the original full resolution frames [31]. Then the blurred frames are decimated, resulting in the non-key frames. The original full frames corresponding to the key frames remain unchanged.

The searching window for non-adaptive NLM methods is set to 15×15, while for adaptive NLM methods, the searching window is initialized by 5×5. To compensate for object or scene motion, the largest adaptive searching window is 53×53. The parameter ξ is initialized by 1. With these setups, the proposed algorithm can find good matches with high similarity for every pixel within a small neighborhood. Only a few pixels with large motion need a large searching window to find good matches. The Peak Signal-to-Noise Ratio (PSNR) and Structural Similarity (SSIM) are used to compare different methods [32].

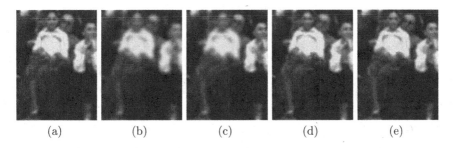

(a)   (b)   (c)   (d)   (e)

Fig. 3. Comparison of different methods for *ballroom* (a) original, (b) bilinear interpolation, (c) NLMAT, (d) TNLM, and (e) the proposed.

In the first experiment, the aforementioned 6 methods are tested on all of the 7 video sequences. The period of key frames is 6. The decimation factor is 2. From the results in Tables 2 and 3, we find that both the adaptive decaying factor and the adaptive searching window can improve the performance of NLM algorithms. By combining the two adaptive parameters, the PSNR results are further improved except *vassar*. Comparing TNLM and NLMAD, we find that TNLM outperforms bilinear interpolation with only a small margin. For *ballroom*, TNLM is even worse than bilinear interpolation. This is because the object in *ballroom* undergoes severe motion. On the other hand, NLMAD improved the results more. Note that the only difference between the two algorithms is the adaptive decaying parameter. This phenomenon shows the necessity of adaptive decaying factor.

The details of the refined images of *ballroom, foreman,* and *mobile* are shown in Figs. 3, 4, and 5, respectively. Note that the proposed algorithm can recover the details in an image. Take *foreman* for example. The tongue of the man is blurred by its surrounding pixels in Fig. 4(d) while Fig. 4(e) removes this artifact. Since different authors may choose different parameter setups, the data may be different somehow. For example, the period of key frame in [24] is 5, while ours is 6. By carefully selecting the parameters, the results of the methods can change and improve. However, the trends of comparisons between methods remain valid and our algorithm is still effective.

Table 4. PSNR (dB) and SSIM results of different key frame periods for *ballroom*.

Period of key frames	Bilinear PSNR/SSIM	TNLM PSNR/SSIM	NLMAT PSNR/SSIM	Proposed PSNR/SSIM
2	30.9896/0.7732	33.3189/0.9766	35.1958/0.9792	**37.3953/0.9832**
4	30.9896/0.7732	28.4757/0.9594	34.4199/0.9777	**36.5979/0.9891**
8	30.9896/0.7732	24.1655/0.9370	33.8053/0.9762	**35.6061/0.9879**
10	30.9896/0.7732	22.5685/0.9271	33.7870/0.9761	**34.2527/0.9878**

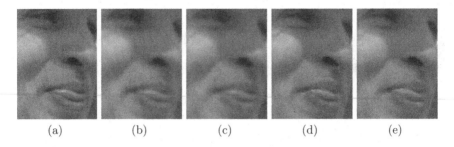

    (a)   (b)   (c)   (d)   (e)

Fig. 4. Comparison of different methods for *foreman* (a) original, (b) bilinear interpolation, (c) NLMAT, (d) TNLM, and (e) the proposed.

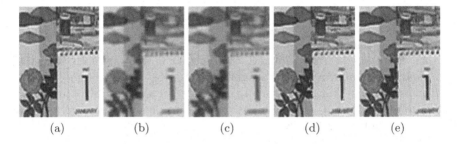

    (a)   (b)   (c)   (d)   (e)

Fig. 5. Comparison of different methods for *mobile* (a) original, (b) bilinear interpolation, (c) NLMAT, (d) TNLM, and (e) the proposed.

We also test the robustness of the proposed algorithm by changing the period of key frames and the decimation factor. These results are shown in Tables 4 and 5, respectively. We only show the results of the sequence *ballroom*. In those circumstances, both the PSNR values and SSIM values drop quickly with the increase of the two parameters. This is because some low-frequency components are lost as those parameters increase. Thus, the similarity comparison by weights is no longer accurate. However, the proposed still performs best in those circumstances.

Table 5. PSNR (dB) and SSIM results of different decimation factors for *ballroom*.

Decimation factor	Bilinear PSNR/SSIM	TNLM PSNR/SSIM	NLMAT PSNR/SSIM	Proposed PSNR/SSIM
2	30.7105/0.7717	29.3321/0.9726	33.8002/0.9760	**36.5159/0.9822**
4	24.6768/0.6168	27.8290/0.9389	28.7866/0.9359	**31.6728/0.9494**
8	20.6398/0.4074	24.4099/0.8495	24.9692/0.8642	**25.5540/0.8693**
10	19.6661/0.3562	23.0464/0.8042	24.0643/0.8288	**24.3542/0.8375**

7   Conclusion

The NLM algorithm has a very promising application in SR tasks. However, the traditional NLM algorithms suffer from two main artifacts, i.e., the fixed decaying factor and searching window. The fixed decaying parameter is unfit for regions with different characteristics. It tends to blur the relatively flat regions in the image, resulting in perceived loss of detail. On the other hand, the fixed searching window leads to mismatches between pixels, causing unbearable degradation of the video. In this paper, we propose two adaptive correspondences of those parameters to address the problems. The proposed algorithm behaves more like a high-frequency extractor rather than a mean filter. Due to the symmetric and independent computation, non-iteration, and memory consideration, the NLM is very applicable to hardware implementation, making real-time display possible. The experimental results validate the effectiveness and robustness of the proposed algorithm.

Acknowledgement. This work was supported by the Natural Science Foundation of China (61671126).

References

1. Healy, D., Mitchell, O.R.: Digital video bandwidth compression using block truncation coding. IEEE Trans. Commun. **29**, 1809–1817 (1982)
2. Banerjee, S.: Low-power content-based video acquisition for super-resolution enhancement. IEEE Trans. Multimedia **11**, 455–464 (2009)
3. Shen, M., Xue, P.: Low-power video acquisition with super-resolution reconstruction for mobile devices. IEEE Trans. Consum. Electron. **56**, 2520–2528 (2010)
4. Ben-Ezra, M., Nayar, S.K.: Motion deblurring using hybrid imaging. In: Proceedings of the CVPR, pp. 657–664 (2003)
5. Tai, Y., Du, H., Brown, M., Lin, S.: Image/video deblurring using a hybrid camera. In: Proceedings of the CVPR, pp. 1–8 (2008)
6. Tai, Y., Du, H., Brown, M., Lin, S.: Correction of spatially varying image and video motion blur using a hybrid camera. IEEE Trans. Pattern Anal. Mach. Intell **32**, 1012–1028 (2010)
7. Brandi, F., de Queiroz, R., Mukherjee, D.: Super-resolution of video using key frames and motion estimation. In: Proceedings of the ICIP, pp. 321–324 (2010)
8. Zhang, L., Wu, X.: An edge-guided image interpolation algorithm via directional filtering and data fusion. IEEE Trans. Image Process. **15**, 2226–2238 (2006)
9. Allebach, J.P., Wong, P.W.: Edge-directed interpolation. In: Proceedings of the ICIP, pp. 707–710 (1996)
10. Li, X., Orchard, M.T.: New edge-directed interpolation. IEEE Trans. Image Process. **10**, 1521–1527 (2001)
11. Wang, Q., Ward, R.K.: A new orientation-adaptive interpolation method. IEEE Trans. Image Process. **16**, 889–900 (2007)
12. Tsai, R.Y., Huang, T.S.: Multiple frame restoration and registration. In: Advances in Computer Vision and Image Processing, pp. 317–339 (1984)
13. Capel, D., Zisserman, A.: Computer vision applied to super-resolution. IEEE Sig. Process. Mag. **20**, 75–86 (2003)

14. Park, S.C., Park, M.K., Kang, M.G.: Super-resolution image reconstruction: a technical overview. IEEE Sig. Process. Mag. **20**, 21–36 (2003)
15. Farsiu, S., Robinson, M.D., Elad, M., Milanfar, P.: Fast and robust multiframe super resolution. IEEE Trans. Image Process. **13**, 1327–1344 (2004)
16. Fu, Z., Li, Z., Ding, L., Nguyen, T.: Translation invariance-based super resolution method for mixed resolution multiview video. In: Proceedings of the ICIP, pp. 5457–5461 (2014)
17. Ancuti, C., Ancuti, C.O., Bekaert, P.: Video super-resolution using high quality photographs. In: Proceedings of the ICASSP, pp. 862–865 (2010)
18. Capel, D., Zisserman, A.: Super-resolution from multiple views using learnt image models. In: Proceedings of the CVPR, pp. 627–634 (2001)
19. Freeman, W.T., Jones, T.R., Pasztor, E.C.: Example-based super-resolution. IEEE Comput. Graph. Appl. **22**, 56–65 (2002)
20. Chang, H., Yeung, D.Y., Xiong, Y.: Super-resolution through neighbor embedding. In: Proceedings of the CVPR, pp. 275–282 (2004)
21. Buades, A., Coll, B., Morel, J.M.: Nonlocal image and movie denoising. Int. J. Comput. Vis. **76**, 123–139 (2008)
22. Protter, M., Elad, M., Takeda, H., Milanfar, P.: Generalizing the nonlocal-means to super-resolution reconstruction. IEEE Trans. Image Process **18**, 36–51 (2009)
23. Basavaraja, S.V., Bopardikar, A.S., Velusamy, S.: Detail warping based video super-resolution using image guides. In: Proceedings of the ICIP, pp. 2009–2012 (2010)
24. Lengyel, R., Soroushmehr, S.M.R., Shirani, S.: Multi-view video super-resolution for hybrid cameras using modified NLM and adaptive thresholding. In: Proceedings of the ICIP, pp. 5437–5441 (2014)
25. Wong, A., Fieguth, P., Clausi, D.: A perceptually adaptive approach to image denoising using anisotropic non-local means. In: Proceedings of the ICIP, pp. 537–540 (2008)
26. Amanatiadis, A., Andreadis, I., Konstantinidis, K.: Design and implementation of a fuzzy area-based image-scaling technique. IEEE Trans. Instrum. Meas. **57**, 1504–1513 (2008)
27. Lopez, S., Callico, G.M., Tobajas, F., Lopez, J.F., Sarmiento, R.: A novel real-time DSP-based video super-resolution system. IEEE Trans. Consum. Electron. **55**, 2264–2270 (2009)
28. Singhal, N., Park, I.K., Cho, S.: Implementation and optimization of image processing algorithms on handheld gpu. In: Proceedings of the ICIP, pp. 4481–4484 (2010)
29. Szydzik, T., Callico, G.M., Nunez, A.: Efficient FPGA implementation of a high-quality super-resolution algorithm with real-time performance. IEEE Trans. Consum. Electron. **57**, 664–672 (2011)
30. Wang, C., Chan, S.C.: A new bandwidth adaptive non-local kernel regression algorithm for image/video restoration and its GPU realization. In: Proceedings of the ISCAS, pp. 1388–1391 (2013)
31. Duchon, C.E.: Lanczos filtering in one and two dimensions. J. Appl. Meteorol. **18**, 1016–1022 (1979)
32. Wang, Z., Bovik, A.C., Sheikh, H.R., Simoncelli, E.P.: Image quality assessment: from error visibility to structural similarity. IEEE Trans. Image Process. **13**, 600–612 (2004)

Single Image Super-Resolution Reconstruction Based on Edge-Preserving with External and Internal Gradient Prior Knowledge

Ruxin Wang, Congying Han$^{(\boxtimes)}$, Mingqiang Li, and Tiande Guo

Key Laboratory of Big Data Mining and Knowledge Management,
School of Mathematical Science, University of Chinese Academy of Sciences (UCAS),
Beijing, China
hancy@ucas.ac.cn

Abstract. Single image super-resolution (SISR) reconstruction is currently a very fundamental and significant task in image processing. Instead of upscaling the image in spatial domain, we propose a novel SISR method based on edge preserving integrating the external gradient priors by deep learning method (auto-encoder network) and internal gradient priors using non-local total variation (NLTV). The gradient domain effectively reflects the high frequency details and edge information of nature image to some extent. The joint perspective exploits the complementary advantages of external and internal gradient prior knowledge for reconstructing the HR image. The experimental results demonstrate the effectiveness of our approach over several state-of-art SISR methods.

1 Introduction

Single image super-resolution (SISR) reconstruction is currently a very fundamental and significant task in image processing, which aims at predicting a high-resolution (HR) image from one given low-resolution (LR) image. Usually, there is an unknown degradation process from the ground truth to the observed LR image. For the HR image X and the LR image Y, it can be typically formulated by

$$Y = DBX + N \tag{1}$$

where D is the downsampling matrix, B is a blurring operator and N is the additive noise. However, SISR is an essentially ill-posed problem in that the ground truth is unknown. Over the past decades, extensive studies have been proposed which can be broadly categorized into three groups: interpolation-based [1,2], reconstruction-based [3–5] and example learning-based methods [6–8]. Interpolation-based approaches are usually fast, which estimate the unknown pixels using the neighbors information. But a notable limitation is that the interpolation results are lack of fine details. Currently, Reconstruction-based and example-based methods are the most popular techniques for SISR. The reconstruction-based methods are based on a common model. From a probability based perspective, it could be treated as a Bayesian posterior estimation

© Springer International Publishing AG 2017
C.-S. Chen et al. (Eds.): ACCV 2016 Workshops, Part I, LNCS 10116, pp. 191–205, 2017.
DOI: 10.1007/978-3-319-54407-6_13

problem. Different representative priors have been adopted for SISR. For example, the total variation (TV) [9], non-local means (NLM) [10], local structure similarity and some edge priors [11] and so on. These added prior knowledge according to the different properties of image can effectively improve the recovery results. The example learning-based methods attempt to learn reconstruction prior knowledge from a large number of external LR-HR patch pairs for predicting the missing high frequency information. There are two main techniques. One is learning the mapping relations between LR-HR pairs directly [12,13], the other is coding the LR and HR feature spaces for each other [14,15]. A kind of very classic algorithm is based on the method of dictionary learning using sparse coding. In particular, Yang et al. [6] proposed jointly learning LR-HR dictionary pair for SISR. Timofte et al. [16] proposed a fast SISR method called Anchored Neighborhood Regression (ANR) by precalculating and storing projection matrices for improving the execution speed without sacrificing the quality of the reconstruction image. An improved variant of ANR named A+ is proposed with improved quality better than ANR and excellent time complexity [8]. In recent years, deep learning has achieved great success in many computer vision tasks. Dong et al. [7] introduce a convolutional neural network (SRCNN) with three convolutional layers for end-to-end LR to HR learning. Kim et al. [17] present a highly accurate SISR method using a very deep convolutional network. The deep learning method improves the quality of image super-resolution significantly. The example learning-based methods are propitious for predicting the missing details. However, only using the external priors may produce some unexpected artifacts. In addition, Timofte et al. [18] sum up and put forward seven generic techniques to improve example-based single image super-resolution without any changes to the core baseline method.

The gradient domain effectively reflects the high frequency details and edge information of nature image to some extent. Instead of upscaling the image in spatial domain, some edge-directed methods have been proposed [19–22], which reconstruct the HR image by inferring the HR gradients (profile). In this paper, we propose a novel SISR method based on edge preserving integrating the external gradient priors by deep learning method (auto-encoder network) [23] and internal gradient priors using non-local total variation (NLTV) [24]. Different from many approaches sharpening the edge using predefined transformation operator, we learn the mapping relation between the LR and HR gradient patch by directly training numerous external data using the auto-encoder (AE) network. At the same time, the non-local similarity of gradient domain is considered for improving the reconstruction results. The joint perspective exploits the complementary advantages of external and internal gradient priors for reconstructing the HR image. At last, a 2D complex shock filter is employed for enhancing the weak recovered details and edges in the HR image.

The remainder of the paper is organized as follows: Sect. 2 gives an overview of the proposed SISR algorithm. The specific algorithm is detailed in Sect. 3. Section 4 presents experimental results and compare it to other state-of-the-art methods. Finally, we conclude our paper in Sect. 5.

2   Overview of the Proposed SISR Algorithm

In this paper, we use the reconstruction-based framework and integrate the example learning-based method for SISR, which can roughly expressed as follows:

$$\min_{X} ||Y \uparrow_{bicubic} -X||_2^2 + \lambda P_{external}(\nabla X) + \beta P_{internal}(\nabla X) \qquad (2)$$

where $Y \uparrow_{bicubic}$ denotes the upsampling image with same size of the target image X by bicubic interpolation. $P_{external}(\nabla X)$ indicates the gradient priors based on external examples learning and $P_{internal}(\nabla X)$ represents the gradient priors from the image itself.

There are two main stages of our method (Fig. 1). For the training stage, numerous external LR-HR gradient patch pairs are extracted from the training data firstly. Then the auto-encoder (AE) network is adopted for learning the mapping relation of between the LR and HR gradient patches. In testing stage, bicubic interpolated image' gradient is calculated firstly. Then the HR gradient is estimated by the AE with the well trained parameters and improved by non-local total variation priors. At last, a 2D complex shock filter is adopted for edge enhancement.

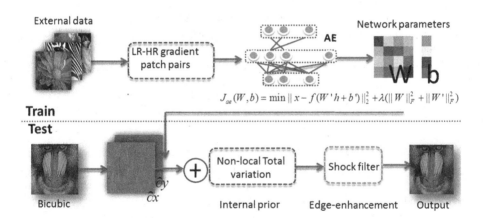

Fig. 1. Overview of the proposed method.

3   Proposed Edge-Preserving Method

In this section, we will introduce our proposed method in detail and give the corresponding mathematical model.

3.1 External Gradient Prior Knowledge Learning with Auto-Encoder

For the external gradient priors, we hope to have a good estimation of HR gradient. The external priors can be formulated by:

$$P_{external}(\nabla X) = ||\nabla X - \nabla \widehat{X}_{estimate}||_2^2 \tag{3}$$

where $\nabla \widehat{X}_{estimate} = (\partial_x \widehat{X}_{estimate}, \partial_y \widehat{X}_{estimate})$ denotes the estimated gradient vector of HR and $\partial_x \widehat{X}_{estimate}$ ($\partial_y \widehat{X}_{estimate}$) denotes the horizontal (vertical) component of $\nabla \widehat{X}_{estimate}$. Different from previous approaches, a non-linear auto-encoder network is employed for learning the mapping relation from numerous external LR gradient samples to its corresponding HR gradient samples directly. In practice, two gradient components are learned for each other. Therefore, formula (3) can be equivalently written as follows:

$$P_{external}(\nabla X) = \lambda_1 ||\partial_x X - \partial_x \widehat{X}_{estimate}||_2^2 + \lambda_2 ||\partial_y X - \partial_y \widehat{X}_{estimate}||_2^2 \tag{4}$$

The auto-encoder (AE) is one approach to automatically learn features from unlabeled data, and it can effectively learn a intrinsic representation between the input and output. This kind of network structure has strong nonlinear representation ability which has been applied for many computer vision tasks. In this paper, we use the AE as a nonlinear regression method. It is a fully connected networks using some nonlinear function as the activation function of the hidden layer neurons. For each gradient component patch pairs $[\{y_i^k\}_1^m, \{x_i^k\}_1^m]$, where $y_i^k, x_i^k \in \mathbb{R}^d$, m denotes the number of patch pairs and $k = \{h, v\}$ indicates the type of component, the cost function of AE is as follows:

$$J_{AE}(W, b) = \min_{W,b} \sum_i ||x_i^k - f(W'^k h_i^k + b'^k)||_2^2 + \lambda(||W^k||_F^2 + ||W'^k||_F^2) \tag{5}$$

where $h_i^k = f(W^k y_i^k + b^k)$ represents the output of the network's hidden layer, $f(.)$ is a group of non-linear mapping functions with parameters $W = (W^k, W'^k)$ and $b = (b^k, b'^k)$. The parameters can be updated by the classical back propagation (BP) algorithm iteratively.

The learned mapping relation by AE network can effectively guide the reconstruction as the external gradient prior knowledge.

3.2 Internal Gradient Prior Knowledge Learning with Non-local Total Variation

Benefiting from the development of non-local means and cross-scale self-similarity methods, many approaches have incorporated the internal information for improving the reconstruction quality in recent years and they have achieved very good performances. Inspired by these methods, we consider the non-local similarity in gradient domain. According to the statistical properties of natural image, many approaches for image restoration simulate the gradient profile of the natural image as a heavy-tailed distribution [25]. Specifically, a generalized Gaussian distribution is used as follows:

$$P(\nabla X) \propto \frac{\gamma \lambda^{\frac{1}{\gamma}}}{2\Gamma(\frac{1}{\gamma})} exp(-\lambda ||\nabla X||^{\gamma}) \qquad (6)$$

Γ is a Gamma function and γ, λ are the shape parameters which determine the shape of the distribution. Based on the above viewpoint, prior knowledge for constraining the gradient profile is a feasible and effective method. The total variation (TV) is a general prior knowledge for many image processing problem. It has achieved good results in SISR. In TV, the gradient profile is supposed to obey i.i.d. Laplacian distribution with zero-mean for all pixels $(TV_X : ||\nabla X||_1)$. However, this constrain condition is strict and it has weaker adaptive ability for gradient profile.

Different from constraining the gradient magnitude obeying $Laplace(0, \sigma^2)$ strictly, the non-local property is added by considering the spatial similarity of the gradient patches. The non-local total variation (NLTV) provides a better constrain effect. Mathematically, let $X \in \mathbb{R}^N$ denote the original image vector and R_i is an image patch extracting operator from X at location i, $x_i = R_i X \in \mathbb{R}^{n \times n}$. The non-local total variation is defined as follows:

$$NLTV_X : ||\nabla_\omega X||_1 = \sum_{i \in \Omega(X)} \sqrt{\sum_{j \in \Omega(i)} ((X_j - X_i)^2 \omega_{ij})} \qquad (7)$$

where the $\Omega(i)$ denotes the index set for similar pixels of X_i, the $\Omega(X)$ denotes the index set for all the pixels of X and the similarity between the patch $R_i X$ and $R_j X$ is defined by:

$$\omega_{ij} = exp(-\frac{||R_i X - R_j X||^2}{2\eta^2}) \qquad (8)$$

In order to effectively utilize the information within the image to guide the reconstruction, so the internal gradient prior in this paper is formulated by:

$$P_{internal}(\nabla X) = ||\nabla_\omega X||_1 \qquad (9)$$

3.3   Edge Enhancement Using 2D Complex Shock Filter

Based on the LR upsampling result $\widehat{X_0}$ using the external and internal gradient priors, a 2D complex shock filter [26] is employed for further enhancing the weak recovered details and edges in the high-resolution image. Shock filter is a blind deconvolution technique which has been widely applied in many image processing problem as an effective supplementary approach [27,28]. General 2D complex shock filter is defined as follows:

$$filter(\widehat{X_0}) = -\frac{2}{\pi}(aIm(\frac{\widehat{X_0}}{\theta}))||\widehat{X_0}|| + \lambda \widehat{X_0}_{\eta\eta} + \tilde{\lambda} \widehat{X_0}_{\xi\xi} \qquad (10)$$

where a controls the sharpness of the slope near zero, θ is the phase angle of the complex part $\frac{\widehat{X_0}}{\theta}$, which is defined by $Im(\frac{\widehat{X_0}}{\theta})$. $\widehat{X}_{0\eta\eta}$ and $\widehat{X}_{0\xi\xi}$ are the second derivative with direction η and ξ respectively. λ is a complex scalar and $\tilde{\lambda}$ is a real scalsr. More details can be found from [26].

Therefore, $\widehat{X_0}$ can be updated by the following:

$$\widehat{X_0}^{k+1} := \widehat{X_0}^{k} + d * filter(\widehat{X_0}^{k}) \tag{11}$$

3.4 Summary of the Algorithm

The proposed SISR algorithm has been described, and now the mathematical model for SISR can be formulated by:

$$\min_{X} ||Y\uparrow_{bicubic} -X||_2^2 + \lambda||\nabla X - \nabla\widehat{X}_{estimate}||_2^2 + \beta||\nabla_\omega X||_1. \tag{12}$$

The entire process is outlined in Algorithm 1.

Algorithm 1. The proposed method

Input: training dataset: LR-HR image pairs, test LR image Y.
1: Extract training patch pairs $[\{y_i^k\}_1^m, \{x_i^k\}_1^m]$.
2: For each component, use (5) to learn the network parameters.
3: Let $Y_0 = Y\uparrow_{bicubic}$ and extract its gradient component.
4: Estimate X gradient component $(\partial_x\widehat{X}_{estimate}, \partial_y\widehat{X}_{estimate})$ using the trained parameters.
5: Update X using the external priors by gradient descent method:
   $X_0 = argmin_X||Y_0 - X||_2^2 + \lambda_1||\partial_x X - \partial_x\widehat{X}_{estimate}||_2^2 + \lambda_2||\partial_y X - \partial_y\widehat{X}_{estimate}||_2^2$
6: Update X using the internal priors by split bregman iteration:
   $\widehat{X}_0 = argmin_X||X_0 - X||_2^2 + \beta||\nabla_\omega X||_1$
7: Update the X using the 2D complex shock filter for edge enhancement.
8: Update the X using the back-projection.
Output: High-resolution image X.

---

4   Experimental Results

In our experiments, the RGB images are first transformed into the YCbCr space and the SISR algorithm are only applied on the Y channel (namely, the illuminance channel). 9×9 patches are selected as the basic processing unit. All experiments in this section were performed on a workstation installed with an Intel core i7 CPU with 16 GB of RAM.

4.1 Training

For a fair comparison, we conducted experiments on the same dataset in [29], which consists of 91 natural images. For our AE network with one hidden layer, the number of neurons in the input layer is set to 81, and the hidden layer contains 3 times of the input layer neurons. The $tanh(.)$ function is adopted as the network's activation function in that we normalize the gradient patch to $[-1, 1]$. The number of training patch pairs is 1000000.

4.2 Testing

To verify the robustness of our SISR method, Set5, Set14 [29] and five images from Berkeley Segmentation Datasets (BSDS500)[1] are selected as the testing sets (Fig. 2). We test the algorithm on Set5 and BSD500 for two upscaling factors ($\times 3$, $\times 4$).

Fig. 2. The testing images.

In our experiment, seven state-of-the-art SISR methods are considered, including: (1) Bicubic, (2) ScSR [6], (3) NE+NNLS (Neighbor Embedding with Non-Negative Least Squares [16], similar to Bevilacqua *et al.* [30]), (4) NE + LLE (Neighbor Embedding with Locally Linear Embedding [16], similar to Chang *et al.* [31]), (5) ANR [16] and (6) BPJDL [32], (7) SRCNN [7]. In order to objectively assess the quality of SISR reconstruction, the peak signal to noise ratio (PSNR) and the structural similarity (SSIM) [33] are employed.

4.3 Results

Table 1 shows the contribution of each of the used components (external vs. internal priors vs. edge enhancement). From which we can find that each of the priors is effective for the improvement of the reconstruction result. Tables 2, 3 and 4 compare the different upscaling results on the testing database: Set5, Set14 and

---

[1] http://www.eecs.berkeley.edu/Research/Projects/CS/vision/grouping/segbench/.

Table 1. The average PSNR improvement of each components

Datasets	Scale	Bicubic	External priors	Internal priors	Edge enhancement
Set5	×3	30.39	31.88↑ +1.49	32.11↑ +0.23	32.38↑ +0.27
	×4	28.42	29.58↑ +1.16	29.78↑ +0.20	30.04↑ +0.26
Set14	×3	27.54	28.74↑ +1.20	28.87↑ +0.13	28.99↑ +0.12
	×4	26.13	26.93↑ +0.80	27.06↑ +0.13	27.22↑ +0.16

Table 2. The PSNR and SSIM of seven methods in Set5

Images	Scale	Bicubic	ScSR	NNLS	LLE	ANR	BPJDL	SRCNN	Ours
baby	×3	33.9	34.3	34.8	35.1	35.1	35.0	35.0	**35.3**
	–		0.9967	0.9971	0.9973	0.9973	0.9972	0.9973	**0.9974**
Bird	×3	32.6	34.1	34.3	34.6	34.6	34.3	34.9	**34.9**
	–		0.9931	0.9933	0.9937	0.9938	0.9933	**0.9942**	0.9941
Butterfly	×3	24.0	25.6	25.6	25.8	25.9	25.8	**27.6**	27.2
	–		0.9711	0.9699	0.9708	0.9718	0.9707	**0.9809**	0.9786
Head	×3	32.9	33.6	33.5	33.6	33.6	33.5	33.5	**33.7**
	–		0.9943	0.9947	0.9949	0.9949	0.9948	0.9948	**0.9950**
Woman	×3	28.6	30.4	29.9	30.2	30.3	30.1	**30.9**	30.8
	–		0.9917	0.9915	0.9921	0.9923	0.9916	**0.9933**	0.9929
Average	×3	30.39	31.42	31.60	31.84	31.92	31.74	32.38	**32.38**
	–		0.9894	0.9893	0.9898	0.9900	0.9896	**0.9921**	0.9916
Baby	×4	31.8	–	32.8	33.0	33.0	33.0	33.0	**33.3**
	–		–	0.9954	0.9956	0.9956	0.9953	0.9956	**0.9958**
Bird	×4	30.2	–	31.5	31.7	31.8	31.9	32.0	**32.0**
	–		–	0.9874	0.9880	0.9883	0.9869	0.9986	**0.9986**
Butterfly	×4	22.1	–	23.3	23.4	23.5	23.9	**25.1**	24.4
	–		–	0.9477	0.9486	0.9503	0.9446	**0.9653**	0.9572
Head	×4	31.6	–	32.1	32.2	32.3	32.2	32.2	**32.3**
	–		–	0.9923	0.9930	0.9930	0.9928	0.9929	**0.9930**
Woman	×4	26.5	–	27.6	27.7	27.8	28.0	28.2	28.2
	–		–	0.9854	0.9856	0.9860	0.9843	0.9873	**0.9874**
Average	×4	28.42	–	29.47	29.61	29.69	29.80	**30.09**	30.04
	–		–	0.9818	0.9822	0.9827	0.9808	**0.9859**	0.9844

partial BSDS500 database with different SISR methods. Some reconstruction examples are shown in Figs. 3, 4, 5, 6 and 7. As shown in the results, our SISR method significantly outperforms the other state-of-art methods and has better visual quality.

Table 3. The PSNR and SSIM of seven methods in Set14

Images	Scale	Bicubic	ScSR	NNLS	LLE	ANR	BPJDL	SACNN	Ours
Baboon	×3	23.2	23.5	23.5	23.6	23.6	23.6	23.6	**23.6**
	–		0.8871	0.8866	0.8886	0.8890	0.8894	**0.8902**	0.8899
Barbara	×3	26.2	26.4	26.7	26.7	26.7	26.8	26.7	**26.9**
	–		0.9566	0.9594	0.9600	0.9600	**0.9600**	0.9593	0.9543
Bridge	×3	24.4	24.8	24.9	25.0	25.0	25.3	25.1	**25.4**
	–		0.9629	0.9630	0.9641	09643	0.9643	0.9650	**0.9651**
Coastguard	×3	26.6	27.0	27.0	27.1	27.1	26.8	**27.2**	27.1
	–		0.9675	0.9672	0.9677	0.9678	0.9660	**0.9687**	0.9666
Comic	×3	23.1	23.9	23.8	24.0	24.0	24.0	**24.4**	24.3
	–		0.9506	0.9490	0.9507	0.9516	0.9510	**0.9555**	0.9537
Face	×3	32.8	33.1	33.5	33.6	33.6	33.5	33.6	**33.7**
	–		0.9942	0.9947	0.9948	0.9949	0.9947	0.9948	**0.9949**
Flowers	×3	27.2	28.2	28.2	28.4	28.5	28.5	28.9	**28.9**
	–		0.9791	0.9787	0.9795	0.9800	0.9800	**0.9821**	0.9814
Foreman	×3	31.2	32.0	32.9	33.2	33.2	32.9	33.3	**33.6**
	–		0.9927	0.9940	0.9945	0.9945	0.9939	0.9946	**0.9948**
Lenna	×3	31.7	32.6	32.8	33.0	33.1	33.0	**33.4**	33.3
	–		0.9896	0.9901	0.9905	0.9906	0.9903	**0.9912**	0.9911
Man	×3	27.0	27.8	27.7	27.9	27.9	27.9	28.2	**28.2**
	–		0.9778	0.9775	0.9783	0.9786	0.9782	0.9798	**0.9800**
Monarch	×3	29.4	30.7	30.8	30.9	31.1	31.2	**32.4**	32.1
	–		0.9788	0.9785	0.9793	0.9800	0.9800	**0.9851**	0.9840
Pepper	×3	32.4	33.3	33.6	33.8	33.8	33.8	34.4	**34.4**
	–		0.9930	0.9934	0.9937	0.9938	0.9935	0.9944	**0.9944**
ppt3	×3	23.7	25.0	24.8	24.9	25.0	25.0	**26.0**	25.4
	–		0.9787	0.9776	0.9782	0.9787	0.9771	**0.9832**	0.9804
Zebra	×3	26.6	28.0	28.1	28.3	28.4	28.4	28.9	**28.9**
	–		0.9798	0.9800	0.9807	0.9813	0.9803	0.9828	**0.9828**
Average	×3	27.54	28.31	28.44	28.60	28.65	28.62	**29.01**	28.99
	–		0.9706	0.9707	0.9715	0.9718	0.9713	**0.9733**	0.9722

Table 4. The PSNR and SSIM of seven methods in partial BSDS500 database

Images	Scale	Bicubic	ScSR	NNLS	LLE	ANR	BPJDL	SACNN	Ours
Corn	×3	26.0	26.6	26.7	27.0	27.0	27.0	27.1	**27.2**
		–	0.9758	0.9759	0.9772	0.9773	0.9772	0.9781	**0.9782**
Parthenon	×3	26.3	26.8	26.7	26.8	26.9	26.9	27.1	**27.1**
		–	0.9583	0.9575	0.9589	0.9591	0.9594	0.9612	**0.9612**
Portrait-1	×3	26.6	27.9	27.9	28.1	28.2	28.5	**28.7**	28.6
		–	0.9887	0.9888	0.9893	0.9894	0.9891	**0.9911**	0.9903
Portrait-2	×3	30.9	31.3	31.4	31.6	31.6	31.6	31.7	**31.7**
		–	0.9932	0.9934	0.9936	0.9936	0.9936	**0.9938**	0.9937
Starfish	×3	28.1	28.8	28.8	29.0	29.0	29.2	**29.4**	29.2
		–	0.9809	0.9807	0.9816	0.9818	0.9820	**0.9835**	0.9825
Average	×3	27.58	28.28	28.31	28.49	28.52	28.54	**28.80**	28.76
		–	0.9794	0.9793	0.9801	0.9802	0.9803	**0.9815**	0.9812
Corn	×4	24.1	–	24.9	25.1	25.2	25.1	25.3	**25.4**
		–	–	0.9626	0.9643	0.9647	0.9639	0.9657	**0.9664**
Parthenon	×4	25.1	–	25.5	25.5	25.6	25.6	25.8	**25.8**
		–	–	0.9446	0.9449	0.9450	0.9448	**0.9476**	0.9471
Portrait-1	×4	24.8	–	25.7	25.8	25.9	25.8	**26.5**	26.4
		–	–	0.9804	0.9810	0.9811	0.9798	**0.9836**	0.9823
Portrait-2	×4	29.6	–	30.0	30.1	30.1	30.2	30.3	**30.3**
		–	–	0.9907	0.9909	0.9910	0.9910	0.9912	**0.9914**
Starfish	×4	26.5	–	27.0	27.2	27.1	27.3	**27.6**	27.5
		–	–	0.9712	0.9720	0.9724	0.9718	**0.9742**	0.9731
Average	×4	26.03	–	26.62	26.74	26.78	26.80	27.08	**27.08**
		–	–	0.9699	0.9707	0.9709	0.9703	**0.9725**	0.9721

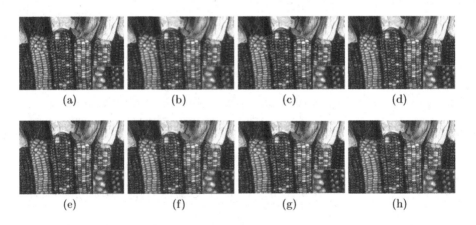

(a)   (b)   (c)   (d)

(e)   (f)   (g)   (h)

Fig. 3. Visual comparison of different SISR methods on 'corn' with upscaling ×3: (a) the ground truth, (b) Bicubic, (c) NE+NNLS, (d) NE+LLE, (e) ANR, (f) BPJDL, (g) SRCNN, (h) the proposed method.

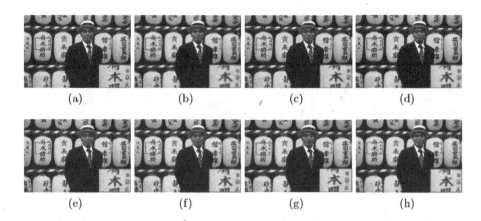

Fig. 4. Visual comparison of different SISR methods on 'portrait-1' with upscaling ×3: (a) the ground truth, (b) Bicubic, (c) NE+NNLS, (d) NE+LLE, (e) ANR, (f) BPJDL, (g) SRCNN, (h) the proposed method.

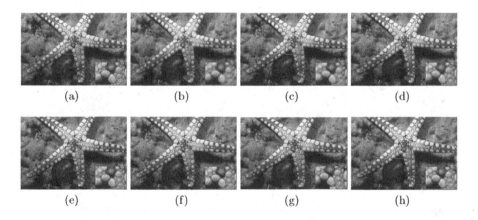

Fig. 5. Visual comparison of different SISR methods on 'starfish' with upscaling ×3: (a) the ground truth, (b) Bicubic, (c) NE+NNLS, (d) NE+LLE, (e) ANR, (f) BPJDL, (g) SRCNN, (h) the proposed method.

Fig. 6. Visual comparison of different SISR methods on 'portrait-2' with upscaling ×4: (a) the ground truth, (b) Bicubic, (c) NE+NNLS, (d) NE+LLE, (e) ANR, (f) BPJDL, (g) SRCNN, (h) the proposed method.

Fig. 7. Visual comparison of different SISR methods on 'flowers' with upscaling ×4: (a) the ground truth, (b) Bicubic, (c) NE+NNLS, (d) NE+LLE, (e) ANR, (f) BPJDL, (g) SRCNN, (h) the proposed method.

5   Conclusion

This paper presents a novel edge-preserving image super-resolution reconstruction method using the external and internal gradient prior knowledge. In the training stage, the auto-encoder network is adopted as a non-linear regression method for learning the intrinsic mapping relation between the LR and HR gradient patches directly. The learned external gradient priors is utilized for the initial SISR reconstruction. Then, the reconstruction image's gradient magnitude is constrained by the non-local total variation using the internal spatial similarity of gradient patches. At last, a 2D complex shock filter is adopted for the further edge enhancement. The experimental results demonstrate the effectiveness of our approach which can preserve the sharpness over several state-of-art SISR methods.

Acknowledgement. The authors would like to sincerely thank J. C. Yang, L. He, R. Timofte and C. Dong *et al.* for sharing the source codes of the ScSR, BP-JDL, NE+NNLS, NE+LLE, ANR and SRCNN methods. This work was funded by the Chinese National Natural Science Foundation (11331012, 71271204, 11571014).

References

1. Li, X., Orchard, M.T.: New edge-directed interpolation. IEEE Trans. Image Process. **10**, 1521–1527 (2001)
2. Zhang, L., Wu, X.: An edge-guided image interpolation algorithm via directional filtering and data fusion. IEEE Trans. Image Process. **15**, 2226–2238 (2006)
3. Marquina, A., Osher, S.J.: Image super-resolution by TV regularization and bregman iteration. J. Sci. Comput. **37**, 367–382 (2008)
4. Zhang, K., Gao, X., Tao, D., Li, X.: Single image super-resolution with non-local means and steering kernel regression. IEEE Trans. Image Process. **21**, 4544–4556 (2012)
5. Lin, Z., Shum, H.Y.: Fundamental limits of reconstruction-based super resolution algorithms under local translation. IEEE Trans. Pattern Anal. Mach. Intell. **26**, 83–97 (2004)
6. Yang, J., Wright, J., Huang, T.S., Ma, Y.: Image super-resolution via sparse representation. IEEE Trans. Image Process. **19**, 2861–2873 (2010)
7. Dong, C., Loy, C.C., He, K., Tang, X.: Learning a deep convolutional network for image super-resolution. In: Fleet, D., Pajdla, T., Schiele, B., Tuytelaars, T. (eds.) ECCV 2014. LNCS, vol. 8692, pp. 184–199. Springer, Cham (2014). doi:10.1007/978-3-319-10593-2_13
8. Timofte, R., De Smet, V., Van Gool, L.: A+: adjusted anchored neighborhood regression for fast super-resolution. In: Cremers, D., Reid, I., Saito, H., Yang, M.-H. (eds.) ACCV 2014. LNCS, vol. 9006, pp. 111–126. Springer, Cham (2015). doi:10.1007/978-3-319-16817-3_8
9. Osher, S., Burger, M., Goldfarb, D., Xu, J., Yin, W.: An iterative regularization method for total variation-based image restoration. Multiscale Model. Simul. **4**, 460–489 (2005)
10. Buades, A., Coll, B., Morel, J.M.: A non-local algorithm for image denoising. In: IEEE Computer Society Conference on Computer Vision and Pattern Recognition, CVPR 2005, vol. 2, pp. 60–65. IEEE (2005)

11. Tai, Y.W., Liu, S., Brown, M.S., Lin, S.: Super resolution using edge prior and single image detail synthesis. In: 2010 IEEE Conference on Computer Vision and Pattern Recognition (CVPR), pp. 2400–2407. IEEE (2010)

12. Yang, C.Y., Yang, M.H.: Fast direct super-resolution by simple functions. In: Proceedings of the IEEE International Conference on Computer Vision, pp. 561–568 (2013)

13. Kim, K.I., Kwon, Y.: Single-image super-resolution using sparse regression and natural image prior. IEEE Trans. Pattern Anal. Mach. Intell. 32, 1127–1133 (2010)

14. Zhang, K., Gao, X., Li, X., Tao, D.: Partially supervised neighbor embedding for example-based image super-resolution. IEEE J. Sel. Top. Sign. Proces. 5, 230–239 (2011)

15. Zeyde, R., Elad, M., Protter, M.: On single image scale-up using sparse-representations. In: Boissonnat, J.-D., Chenin, P., Cohen, A., Gout, C., Lyche, T., Mazure, M.-L., Schumaker, L. (eds.) Curves and Surfaces 2010. LNCS, vol. 6920, pp. 711–730. Springer, Heidelberg (2012). doi:10.1007/978-3-642-27413-8_47

16. Timofte, R., Smet, V., Gool, L.: Anchored neighborhood regression for fast example-based super-resolution. In: Proceedings of the IEEE International Conference on Computer Vision, pp. 1920–1927 (2013)

17. Kim, J., Lee, J.K., Lee, K.M.: Accurate image super-resolution using very deep convolutional networks. arXiv preprint (2015). arXiv:1511.04587

18. Timofte, R., Rothe, R., Van Gool, L.: Seven ways to improve example-based single image super resolution. arXiv preprint (2015). arXiv:1511.02228

19. Sun, J., Xu, Z., Shum, H.Y.: Gradient profile prior and its applications in image super-resolution and enhancement. IEEE Trans. Image Process. 20, 1529–1542 (2011)

20. Wang, L., Xiang, S., Meng, G., Wu, H., Pan, C.: Edge-directed single-image super-resolution via adaptive gradient magnitude self-interpolation. IEEE Trans. Circuits Syst. Video Technol. 23, 1289–1299 (2013)

21. Li, Y., Liu, J., Yang, W., Guo, Z.: Neighborhood regression for edge-preserving image super-resolution. In: 2015 IEEE International Conference on Acoustics, Speech and Signal Processing (ICASSP), pp. 1201–1205. IEEE (2015)

22. Si, D., Hu, Y., Gan, Z., Cui, Z., Liu, F.: Edge directed single image super resolution through the learning based gradient regression estimation. In: Zhang, Y.-J. (ed.) ICIG 2015. LNCS, vol. 9218, pp. 226–239. Springer, Cham (2015). doi:10.1007/978-3-319-21963-9_21

23. Bengio, Y.: Learning deep architectures for AI. Found. Trends Mach. Learn. 2, 1–127 (2009)

24. Zhang, X., Burger, M., Bresson, X., Osher, S.: Bregmanized nonlocal regularization for deconvolution and sparse reconstruction. SIAM J. Imaging Sci. 3, 253–276 (2010)

25. Cho, T.S., Zitnick, C.L., Joshi, N., Kang, S.B., Szeliski, R., Freeman, W.T.: Image restoration by matching gradient distributions. IEEE Trans. Pattern Anal. Mach. Intell. 34, 683–694 (2012)

26. Gilboa, G., Sochen, N.A., Zeevi, Y.Y.: Regularized shock filters and complex diffusion. In: Heyden, A., Sparr, G., Nielsen, M., Johansen, P. (eds.) ECCV 2002. LNCS, vol. 2350, pp. 399–413. Springer, Heidelberg (2002). doi:10.1007/3-540-47969-4_27

27. Liu, W., Li, S.: Sparse representation with morphologic regularizations for single image super-resolution. Sig. Process. 98, 410–422 (2014)

28. Zhou, Q., Chen, S., Liu, J., Tang, X.: Edge-preserving single image super-resolution. In: Proceedings of the 19th ACM International Conference on Multimedia, pp. 1037–1040. ACM (2011)

29. Dong, C., Loy, C., He, K., Tang, X.: Image super-resolution using deep convolutional networks. IEEE Trans. Pattern Anal. Mach. Intell. **38**, 295–307 (2016)
30. Bevilacqua, M., Roumy, A., Guillemot, C., Alberi-Morel, M.L.: Low-complexity single-image super-resolution based on nonnegative neighbor embedding (2012)
31. Chang, H., Yeung, D.Y., Xiong, Y.: Super-resolution through neighbor embedding. In: Proceedings of the 2004 IEEE Computer Society Conference on Computer Vision and Pattern Recognition, CVPR 2004, vol. 1, p. 1. IEEE (2004)
32. He, L., Qi, H., Zaretzki, R.: Beta process joint dictionary learning for coupled feature spaces with application to single image super-resolution. In: Proceedings of the IEEE Conference on Computer Vision and Pattern Recognition, pp. 345–352 (2013)
33. Wang, Z., Bovik, A.C., Sheikh, H.R., Simoncelli, E.P.: Image quality assessment: from error visibility to structural similarity. IEEE Trans. Image Process. **13**, 600–612 (2004)

A Dual Adaptive Regularization Method to Remove Mixed Gaussian-Poisson Noise

Ziling Wu[1,2], Hongxia Gao[1,2], Ge Ma[1,2(✉)], and Yanying Wan[1,2]

[1] School of Automation Science and Engineering,
South China University of Technology,
Guangzhou 510641, Guangdong, People's Republic of China
m.ge.tina@gmail.com
[2] Engineering Research Centre for Manufacturing Equipment
of Ministry of Education, South China University of Technology,
Guangzhou 510641, Guangdong, People's Republic of China

Abstract. The noise in low photon-counting imaging system can often be described as mixed Gaussian-Poisson noise. Regularization methods are required to replace the ill-posed image denoising problems with an approximate well-posed one. However, the sole constraint in non-adaptive regularization methods is harmful to a good balance between the noise-removing and detail-preserving. Meanwhile, most existing adaptive regularization methods were aimed at unitary noise model and dual adaptive regularization scheme remained scarce. Thus, we propose a dual adaptive regularization method based on local variance to remove the mixed Gaussian-Poisson noise in micro focus X-ray images. Firstly, we raise a new 3-step image segmentation scheme based on local variance. Then, a self-adaptive p-Laplace variation function is used as the regularization operator while the regularization parameter is adaptively obtained via a barrier function. Finally, experimental results demonstrate the superiority of the proposed method in suppressing noise and preserving fine details.

1 Introduction

Image denoising is still a challenging topic in many imaging systems, such as micro focus X-ray imaging in electronic industry, low dose medical computed tomography (CT), astronomic imaging, microscopic imaging and etc. These special imaging systems are constrained to tiny focal spot, in which only a few photons are collected by the photon sensors in one shot [23]. In these situations, the common Gaussian noise model isn't sufficient to describe the imaging process because that Poisson noise exists definitely in this kind of imaging systems related to low photon-counting [17]. Consequently, it is more reasonable to concern about the mixed Gaussian-Poisson noise with the observation model (1)

$$b = Poisson(u) + r \tag{1}$$

where $b \in \Re^n$ is the observed image, u is the unknown true image and $r \in \Re^n$, represents Gaussian noise. $Poisson(u)$ means that the image is corrupted by

© Springer International Publishing AG 2017
C.-S. Chen et al. (Eds.): ACCV 2016 Workshops, Part I, LNCS 10116, pp. 206–221, 2017.
DOI: 10.1007/978-3-319-54407-6_14

Poisson noise. In this mixed noise model, the quantization error of photovoltaic conversion in digital imaging device is taken into consideration together with thermal noise.

The problem of estimating u from the observed image b is often ill-posed, regularization methods are required to stabilize the solution. The basic idea of regularization is to impose a priori knowledge constraint and replace the original ill-posed problem with a 'nearby' well-posed problem. The general expression of regularization methods for Gaussian-Poisson noise removing is

$$minT(u) = F(b|u) + \lambda R(u) \qquad (2)$$

where $\lambda > 0$ is the regularization parameter which provides a tradeoff between fidelity to the measurements and noise sensitivity. $F : \Re^n \to \Re$ is data fidelity term derived from a MAP probability estimation, $R : \Re^n \to \Re$ is a regularization operator.

In general, the regularization methods to solve the optimization problem (2) can be classified as non-adaptive methods and adaptive methods. Non-adaptive methods often choose the invariant regularization operator and regularization parameter. One of the popular non-adaptive regularization techniques is Tikhonov regularization [33] which added a quadratic penalty to the objective function, resulting in isotropic diffusion and over-smooth. To overcome this difficulty, total variation (TV) regularization and the wavelet l_1 regularization have attracted a revived interest. Considering the mixed Gaussian-Poisson noise removal, Jezierska et al. [20] took advantage of the convexity and Lipschitz differentiability properties of the Poisson-Gaussian negative log-likelihood and developed a primal-dual proximal algorithm to the TV-based image restoration problem. And Li et al. [22] studied the weighted l_2 fidelity term derived from Gaussian approximation of Poisson statistics and extended it to the mixed Poisson-Gaussian noise case. Combined with the sparse framelets regularization, a Split Bregman iteration based algorithm was proposed. However, non-adaptive methods impose a sole constraint to the whole image, resulting in a same penalty on the edges and noise which is harmful to a good balance between the noise-removal and detail-preserving.

Compared to the non-adaptive methods, adaptive regularization methods adopt different regularization operator or regularization parameter according to different image characteristics. As a result, adaptive regularization methods can solve the contradiction between the noise-removing and detail-preserving. In this paper, we propose a dual adaptive regularization method for mixed Gaussian-Poisson noise removal, in which both the regularization operator and regularization parameter are self-adaptive according to image information.

Our main contributions in this paper are as follows: firstly, we raise a new 3-step image segmentation scheme based on local variance. Secondly, a p-Laplace variation function is used as the regularization operator, in which p is self-adaptive. Finally, a novel design scheme of the adaptive regularization parameter is put forward based on a barrier function, which can preserve more useful details. Experimental results indicate that proposed dual adaptive regularization method is superior to the contrast methods for both micro focus X-ray images and natural images.

2   Related Work

In this section, we will introduce the related present work on adaptive regularization methods, including adaptive regularization operator and adaptive regularization parameter.

The adaptive methods adjusting regularization operator depend on how to design regularization operator according to the neighbour's feature. Jeon [19] developed an adaptive regularization algorithm which performed different regularization operators for flat and edge regions respectively. As a result, this method reduced amplification of noise in flat regions and avoided edge regions' ringing artefacts. Bing Song [4] method to eliminate Gaussian noise, which calculated the image gradient and imposed a l_2 norm regularization operator on pixels with small value gradient, while a TV term on other pixels. [15,38] described an l_p-TV algorithm adaptively selecting the regularization operator to preserve edges in detail areas and remove Gaussian noise in smooth areas. [8] adjusted the regularization term based on a new edge indicator named difference curvature. If the difference curvature was large, the regularization term was approximate to TV norm; else it was approximate to H^1 norm. According to the image gradients, [35] applied TV functional ($\int_\Omega |\nabla u| d\Omega$) on edges regions and H^1 functional ($\int_\Omega |\nabla u|^2 d\Omega$) on ramps regions.

Apart from the above mentioned methods, most researchers currently concentrate on spatially adaptive regularization parameter selection [3,30,31] for eliminating a particular type of noises, such as Gaussian or Poisson noise. Lagendijk et al. [21] gave a classic adaptive regularization cost function to remove white Gaussian noise, in which the regularization parameter is not a scalar value but a spatially dependent coefficient matrix. [13] automatically adjusted the regularization parameter by utilizing local variance estimators of the residual image to remove Gaussian noise adaptively. [28] modified the method in [13] to remove salt-and-pepper noise. Unlike [13,28], [27] proposed an adaptive regularization algorithm for vector-value image with Poisson noise, which estimated the restored image's local constant regions by the observed images' local variance and set larger regularization parameter for them. Meanwhile, this regularization parameter for local constant regions was increased in the following iterations step by step. On the contrary, a reverse operation was done for detail regions to preserve edges. In [14], if the gradient was larger than the threshold, the regularization parameter was smaller than that in last iteration, otherwise it remained unchangeable. [37] used difference curvature defined in [8] to detect edges. Then the weighted parameter was in inverse proportion to difference curvature. [12,36] adaptively selected the regularization parameter according to the idea that regularization parameter was in proportion to the data fidelity term, meanwhile inversely proportional to the regularization item. [6,34] selected regularization parameter by matching the residual' norm to a certain upper bound relying on the noise level, so called Morozov's discrepancy principle [25]. Based on this discrepancy principle, a new method was proposed to remove multiplicative noise [9]. It utilized the noise's local statistical characteristics to distinguish the texture-rich regions and decrease this region's regularization parameter. To avoid the high computational costs in [9], [11] presented two fast linearized alternating direction algorithms. The Regularization parameter in [10] was automatically selected based on the local discrepancy function for images with Poisson noise. Clearly, all these methods require a specific discrepancy function for a particular noise.

However, almost all of the existing relevant adaptive regularization methods were aimed at unitary noise but not mixed noise. And dual adaptive regularization methods like [8] remained scarce. Moreover, images obtained from low photon-counting machines like micro focus X-ray images are usually low-contrast and contain lots of weak structures, methods mentioned above usually failed to save weak details which are important for diagnosis or detection. Consequently, it's necessary to study a self-adaptive regularization method for mixed Gaussian-Poisson noise removal.

3   Dual Adaptive Regularization Method

3.1   Model Formulation

The degraded model has been shown as formula (1). Under the MAP criterion, minimization problem (2) can be formulated as a quadratic optimization problem by the following steps:

Firstly, the probability density function (pdf) of the observing y corrupted by Poisson noise is

$$P_{poi}(y|u) = \prod_{k=1}^{n} \frac{(u)_k^{y_k} e^{-(u)_k}}{y_k!} \tag{3}$$

where $(u)_k$ is the k_{th} component of u.

The probability density function (pdf) of the observing y corrupted by Gaussian noise (zero mean) is formulated as

$$P_{gau}(y|u) = \frac{1}{\sqrt{2\pi\sigma^2}} e^{\frac{-(u-y)^2}{2\sigma^2}} \tag{4}$$

Although the mixed noise is not separated in imaging process, independence hypothesis is still common in regularization method on account of the complexity of cost function based on joint probability distribution, which should be approximated to make the solution feasible. For instance, [2] transformed the cost function via asymptotic approximation in order to get a feasible numerical algorithm. Therefore, the mixed Gaussian-Poisson noise model is expressed as:

$$P_{mixed}(b|u) = \prod_{k=1}^{n} \frac{(u)_k^{b_k} e^{-(u)_k}}{b_k!} \frac{1}{\sqrt{2\pi\sigma^2}} e^{\frac{-\|u-b\|_2^2}{2\sigma^2}} \tag{5}$$

Then, under MAP criterion, we can write the log-likelihood as

$$F(b|u) = -log^{P_{mixed}(b|u)} = \sum (u)_k - b_k log[(u)_k] + \|u - b\|_2^2, s.t.u \geq 0 \quad (6)$$

It is obviously that $F(b|u)$ is convex. And a second-order Taylor expansion [18] of $P(u) = \sum (u)_k - b_k log[(u)_k]$ around u^i is used, where u^i is the current estimate of u at the i_{th} iteration. Then, we obtain the following quadratic approximation $P_i(u)$ of $P(u)$:

$$P_i(u) = P(u^i) + (u - u^i)^T \nabla P(u^i) + \frac{\eta_i}{2}\|u - u^i\|_2^2 \quad (7)$$

where ∇ is the gradient operator. With the Barzilai-Borwein [1] method, we denote $d^i = u^i - u^{i-1}$ and $r^i = \nabla P(u^i) - \nabla P(u^{i-1})$, and η_i is calculated as follows:

$$\eta_i = \frac{(d^i)^T r^i}{(d^i)^T d^i} \quad (8)$$

Replacing $P(u)$ in (6) by $P_i(u)$, then

$$argmin_u F(b|u)$$
$$= argmin_u \sum(u)_k - b_k log[(u)_k] + \|u - b\|_2^2$$
$$= argmin_u P(u^i) + (u - u^i)^T \nabla P(u^i) + \frac{\eta_i}{2}\|u - u^i\|_2^2 + \|u - b\|_2^2$$
$$= argmin_u \frac{1}{\eta_i}(u - u^i)^T \nabla P(u^i) + \frac{1}{2}\|u - u^i\|_2^2 + \frac{1}{\eta_i}\|u - b\|_2^2$$
$$= argmin_u \|u - [u^{i-1} - \frac{1}{\eta_i}\nabla P(u^i)]\|_2^2 + \frac{1}{\eta_i}\|u - b\|_2^2$$
$$= argmin_u \|u - s^i\|_2^2 + \frac{2}{\eta_i}\|u - b\|_2^2 \quad (9)$$

where $s^i = u^i - \frac{1}{\eta_i}\nabla P(u^i)$.

Finally, the minimization problem (2) is formulated as:

$$minT(u) = \|u - s^i\|_2^2 + \frac{2}{\eta_i}\|u - b\|_2^2 + \lambda R(u), s.t.u \geq 0 \quad (10)$$

where $s^i = u^i - \frac{1}{\eta_i}\nabla P(u^i)$.

3.2   Adaptive Regularization Parameter and Operator

Here, we will consider the regularization term including the regularization operator and regularization parameter. A new dual adaptive scheme is put forward based on local variance, which can ensure a better effectiveness on noise-removing and detail-preserving.

Firstly, the basis of the regularization operator and parameter selection will affect the denoising performance greatly. As the value of local variance can reflect the image features directly, we raise a 3-step scheme to segment the image based on local variance:

Step 1: Compute the local variance

$$\sigma_u^2(x,y) = \frac{1}{(2K+1)(2L+1)} \sum_{k=-K}^{k=K} \sum_{l=-L}^{l=L} [u(i+k,j+l) - u(i,j)]^2 \qquad (11)$$

$$\overline{u}(x,y) = \frac{1}{(2K+1)(2L+1)} \sum_{k=-K}^{k=K} \sum_{l=-L}^{l=L} [u(i+k,j+l)] \qquad (12)$$

where K and L are the window size, usually $K = L = 2$. For no-reference noisy image, u refers to the filtered image.

Step 2: Compute the related binary image u_{BW} of the local variance image via Ostu, and denote the background-pixel sets as $\phi_{bg} = \{(x,y)|u_{BW}(x,y) = 0\}$.

Step 3: Define $\Omega = \{\sigma_\phi^2(x,y)|(x,y) \in \phi_{bg}\}$ in which $\sigma_\phi^2(x,y)$ is the corresponding local variance of the pixels of ϕ_{bg}. Calculate the mean value M_{σ^2} of Ω, and the image is segmented into two types:

$$\begin{cases} (x,y) \in detail \quad regions & \text{if } \sigma_u^2(x,y) > M_{\sigma^2} \\ (x,y) \in smooth \quad regions & \text{if } \sigma_u^2(x,y) \leq M_{\sigma^2} \end{cases} \qquad (13)$$

Secondly, a p-Laplace variation functional is used as the regularization operator:

$$R(u) = \|\nabla u\|_{p(s)} \qquad (14)$$

where $p(s)$ is a potential function that $p(s) \in [1,2]$ with the variable $s \in [0,\infty)$.

The key idea of the adaptive $R(u)$ is: in detail regions, the TV norm ($p = 1$) can keep more edges, while in smooth regions, a H^1 norm ($p = 2$) can suppress the noise better [5]. Therefore, $p(x,y)$ is chosen as

$$\begin{cases} p(x,y) = 1 & \text{if } \sigma_u^2(x,y) > M_{\sigma^2} \\ p(x,y) = 2 & \text{if } \sigma_u^2(x,y) \leq M_{\sigma^2} \end{cases} \qquad (15)$$

Thirdly, a novel scheme is designed to adaptively select the regularization parameter for each pixel. The present selection criteria of regularization parameters were mainly about applying a small parameter in detail regions and a larger one in smooth regions, such as the inverse proportion to gradient. However, weak details with small gradient and relatively large parameter may be over-smoothened. Hence, we utilize a barrier function, and $\lambda(x,y)$ is calculated as

$$\begin{cases} \lambda_d(x,y) = ln[\frac{\sigma_u^2(x,y)}{M_{\sigma^2}}] + \frac{M_{\sigma^2}}{\sigma_u^2(x,y)} & \text{if } \sigma_u^2(x,y) > M_{\sigma^2} \\ \lambda_s(x,y) = 1.5 * max\{\lambda_d\} & \text{if } \sigma_u^2(x,y) \leq M_{\sigma^2} \end{cases} \qquad (16)$$

Generally, in detailed regions, the regularization parameter is an increasing function so that weak details with small local variance correspond to small parameters. And it tends to mild smoothness. With local variance approaching the infinite, the regularization parameter is nearly fixed, which avoids over-smoothing strong edges and can moderately denoise. In smooth regions, we revise it to be fixed and largest, so that it can better remove noise (Table 1).

Table 1. The proposed algorithm.

Algorithm 1
(1) Calculate the local variance $\sigma_u^2(x, y)$ via (11);
(2) Divide the local variance image via Otsu;
(3) Compute segmentation threshold M_{σ^2};
(4) Determine $p(x, y)$ and $\lambda(x, y)$ based on (15) and (16);
(5) Initialize: choose u^0 (usually an constant matrix), $\eta_0 > 0, \alpha > 0, \rho > 0, \varepsilon > 0, i \to 0$;
(6) Repeat:
A. Compute: $s^i = u^i - \frac{1}{\eta_i}\nabla P(u^i)$, $\eta_i = \frac{(d^i)^T r^i}{(d^i)^T d^i}$;
B. $u = argmin_u \|u - s^i\|_2^2 + \frac{2}{\eta_i}\|u - b\|_2^2 + \frac{2\lambda(x,y)}{\eta_i}\|\nabla u\|_{p(s)}$;
C. if
$$T(u) \leq max_{m=max(m-M,0),\cdots,i}T(u^m) - \frac{\rho}{2}\eta_i\|u - u^i\|^2,$$
then: $u^i = u$, turn to (7);
else: $\eta_i = \alpha\eta_i$, turn to B;
(7) Stop criterion:
if
$$\frac{\|u - u^{i-1}\|_2^2}{\|u^{i-1}\|_2^2} \leq \varepsilon,$$
then: $u^* = u^i$;
else: turn to (6).

3.3 The Proposed Dual Adaptive Regularization Method

According to the analysis carried out in Sects. 3.1 and 3.2, the objective function has the following form:

$$u^{i+1} = argmin_u \|u - s^i\|_2^2 + \frac{2}{\eta_i}\|u - b\|_2^2 + \frac{2\lambda(x,y)}{\eta_i}\|\nabla u\|_{p(x,y)}, s.t.u \geq 0 \quad (17)$$

where $s^i = u^i - \frac{1}{\eta_i}\nabla P(u^i)$.

It must be pointed out that this adaptive regularization model is convex and can be solved by various methods, such as gradient descent method [7]. We just present a feasible solution, and other efficient algorithm will be proposed in future research. The general procedure of the proposed dual adaptive regularization method as follows.

4   Experimental Results

In this section, two sets of experiments are presented to demonstrate the superiority of the proposed dual adaptive regularization and the effectiveness of the proposed method for micro focus X-ray images. Parameters of the contrast methods were set to obtain the best performance.

4.1   Superiority of Dual Adaptive Regularization

In this experiment, we will demonstrate the superiority of the proposed dual adaptive regularization in comparison with the non-adaptive TV [29], sole-adaptive l_p-TV [38] and dual-adaptive ATV [8]. This set of experiments are performed on standard 'Cameraman' (256*256) and 'Barbara' (512*512), which are corrupted by mixed Poisson and Gaussian noise ($\mu = 0, \sigma^2 = 30$). The test images and noisy images are shown in Fig. 1. Full-reference (FR) image quality assessments Mean Structural Similarity Index Measurement (MSSIM) and Peak Signal to Noise Ratio (PSNR) are employed to evaluate the objective quality of the denoising images.

    (a)   (b)   (c)   (d)

Fig. 1. Cameraman and Barbara. From left to right: (a) Cameraman; (b) Noisy 'Cameraman'; (c) Barbara; (d) Noisy 'Barbara'

Figures 2 and 3 show the restored results of 'Cameraman' and 'Barbara' respectively. The related objective quality assessments are shown in Table 2. It is clearly that TV, l_p-TV and ATV fail to reach a great balance between noise-removing and detail-preserving in Figs. 2 and 3(a)–(c). TV and l_p-TV suffer from staircase effects and the textures regions are over-smooth. Although ATV can avoid the staircase effects, it is trapped with blurred vision and vague edges. The proposed method is superior to other methods in suppressing noise and preserving details. Moreover, the 'spot' weak features in the smooth regions are also restored in Fig. 2(d). Similarly to the 'Cameraman', the proposed method can also obtain the same effects in 'Barbara' in Fig. 3. The objective quality assessments also demonstrate the superiority of the proposed method.

To compare more clearly, the enlarged parts of the restored 'Barbara' are given in Fig. 4. The proposed method can suppress noise better in smooth regions free from staircase effects in Fig. 4(d). Considering detail-preserving quality, TV and l_p-TV can't restore weak edges. ATV is slightly better but still blur. The proposed method attains a pleasant denoising result for both strong and weak edges.

4.2   Effectiveness for Micro Focus X-Ray Images

In this section, we present two experiments to verify the validity mixed Gaussian-Poisson noise assumption for micro focus X-ray images and the proposed

(a)   (b)   (c)   (d)

Fig. 2. Denoising results of 'Cameraman'. From left to right: the restoration results of: (a) TV; (b) l_p-TV; (c) ATV; (d) The proposed

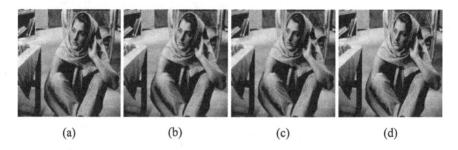

(a)   (b)   (c)   (d)

Fig. 3. Denoising results of 'Barbara'. From left to right: the restoration results of: (a) TV; (b) l_p-TV; (c) ATV; (d) The proposed

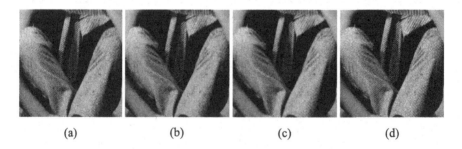

(a)   (b)   (c)   (d)

Fig. 4. Enlarged part of the denoising results of 'Barbara'. From left to right: the restoration results of: (a) TV; (b) l_p-TV; (c) ATV; (d) The proposed

Table 2. The comparison of MSSIM and PSNR.

		Noisy	TV	l_p-TV	ATV	The proposed
Cameraman	MSSIM	0.59	0.88	0.88	0.88	0.90
Cameraman	PSNR	26.42	30.23	30.48	30.54	31.12
Barbara	MSSIM	0.87	0.92	0.92	0.93	0.94
Barbara	PSNR	26.59	28.72	28.31	29.20	29.61

method's superiority in denoising. The test image in Fig. 5 is micro focus X-ray images of nano chips which generated by micro focus X-ray detector. Focal spot of the detector is smaller than $5\mu m$. To have a better comparison, we also show the enlarged part of the micro focus X-ray images. Because it is unable to obtain the noise-free micro focus X-ray images in actual industry, no-reference (NR) image quality assessments including Mean to Standard Deviation Ratio (MSR) [16] and Laplacian Sum (LS) [32] are employed to evaluate the objective quality. The higher MSR and LS are, the better performance is.

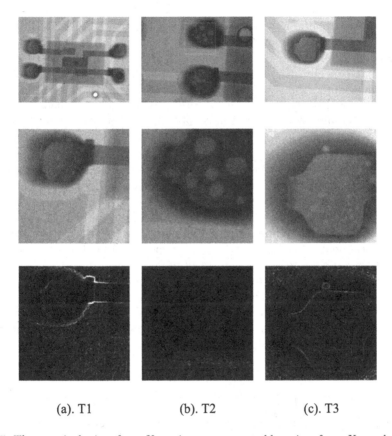

(a). T1   (b). T2   (c). T3

Fig. 5. Three typical micro focus X-ray images generated by micro focus X-ray detector whose focal spot is smaller than 5 m for detecting nano chip. We name them as T1, T2 and T3. From top to bottom: original micro focus X-ray images, enlarged part of the original images, local variance of the enlarged part

Firstly, we compare TV for Gaussian noise (GTV) [29], TV for Poisson noise (PTV) [17], TV for mixed Gaussian-Poisson noise (GPTV) and the proposed method to verify the mixed Gaussian-Poisson noise assumption for micro focus X-ray images.

From the denoised images Fig. 6, it is obvious that the results of GTV are over-smooth. Oppositely, PTV doesn't suppress enough noises. PGTV has a better result in noise-removing and detail-preserving. And the proposed method performs best.

Secondly, under the mixed Gaussian-Poisson noise assumption of micro focus X-ray images, the state-of-the-art mixed denosing methods are employed to compare with our proposed method, such as PURE-LET [23] (All extensions of PURE-LET, namely UWT PURE-LET, BDCT PURE-LET, and UWT/BDCT PURE-LET are conducted on every micro focus X-ray images, we choose the best result.), GAT+BM3D [24] and GAT+BLS-GSM [26]. Figure 7 gives the details of the denoised results. It is obvious that all of PURE-LET, GAT+BM3D and GAT+BLS-GSM suffer from over-smooth.

Moreover, artefacts exist in the results of PURE-LET and GAT+BLE-GSM, see Fig. 7(a)–(c). For the results of the proposed method Fig. 7(d), noise are removed in smooth regions with little staircase effects and artefacts. Meanwhile, most details are kept with more evident transitions between different types of features. MSR and LS in Table 3 also verify the superiority of our proposed method.

To compare more clearly, the corresponding local variance images are also displayed in Fig. 8. From Fig. 8(a)–(c), we can observe that strong edges can be saved by PURE-LET, GAT+BM3D and GAT+BLS-GSM, but weak edges may be lost. It is more clearly in the restoration of bubble defects on the micro focus X-ray images. Differently, our method preserves most weak details, even if the small and dense cluster of bubbles or the weak contours, Fig. 8(d). The objective quality assessments also indicate that the proposed method can take a better balance between noise-removing and detail-preserving.

In order to have a better comparison, we test anther 6 micro focus X-ray images, and their objective quality assessments are displayed in Table 4. It can be seen that the proposed method attains best results for both MSR and LS, which indicates that the proposed method can achieve a better balance between noise-smoothing and detail-preserving.

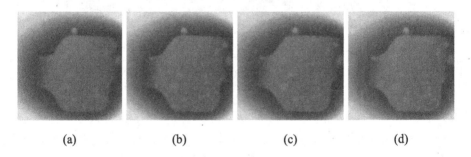

(a)   (b)   (c)   (d)

Fig. 6. Denoising results of the noise model assumption experiment. From left to right: (a) GTV; (b) PTV; (c) PGTV; (d) The proposed

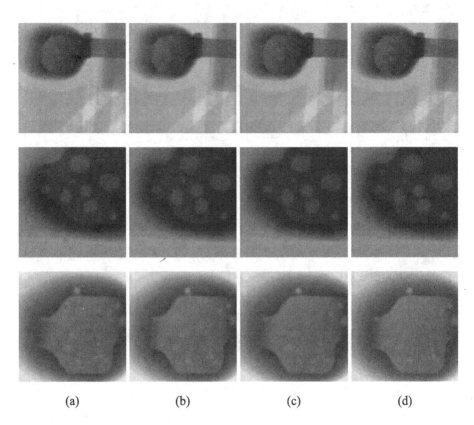

(a)   (b)   (c)   (d)

Fig. 7. Denoising details of three typical micro focus X-ray images. From top to bottom: the first row: T1; the second row: T2; the third row: T3. From left to right: (a) GAT+BLS-GSM; (b) GAT+BM3D; (c) PURE-LET; (d) The proposed

Table 3. The comparison of MSR and LS of T1–T3.

Image		GAT+BLS+GSM	GAT+BM3D	PURE-LET	The proposed
T1	MSR	27.92	54.13	54.77	54.98
T1	LS	2.55	2.47	2.55	4.92
T2	MSR	67.96	68.14	68.06	69.667
T2	LS	2.41	2.50	2.58	4.12
T3	MSR	29.33	46.86	47.29	48.42
T3	LS	2.44	2.10	2.73	4.27

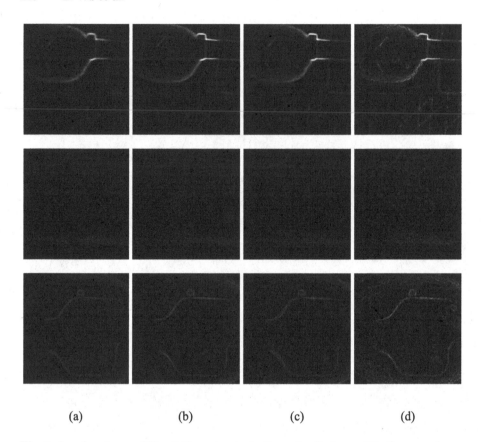

(a)   (b)   (c)   (d)

Fig. 8. Local variance of Fig. 7. From top to bottom: the first row: T1; the second row: T2; the third row: T3. From left to right: (a) GAT+BLS-GSM; (b) GAT+BM3D; (c) PURE-LET; (d) The proposed

Table 4. The comparison of MSR and LS of anther 6 micro focus X-ray images.

Image		GAT+BLS+GSM	GAT+BM3D	PURE-LET	The proposed
E1	MSR	15.53	23.56	24.34	28.92
E1	LS	2.47	2.35	2.70	3.78
E2	MSR	11.39	38.20	40.64	47.77
E2	LS	2.14	2.08	2.31	2.99
E3	MSR	30.39	31.65	33.21	35.55
E3	LS	2.75	2.36	2.56	3.63
E4	MSR	9.60	19.72	19.62	20.20
E4	LS	2.00	1.97	2.12	2.80
E5	MSR	12.14	20.06	20.52	24.10
E5	LS	5.95	7.80	6.79	7.87
E6	MSR	16.35	58.16	58.05	62.64
E6	LS	3.44	5.41	3.34	5.74

5   Conclusions

In this paper, we focus on the mixed Gaussian-Poisson noise model of low photon-counting mechanism such as micro focus X-ray imaging system. The emphases of present regularization methods are on single noise model or sole constraint for the whole image, resulting in difficulties in inner defect detection for precise integrated circuit. To overcome this difficulty, a new 3-step image segmentation scheme based on local variance is firstly raised, and then we propose a novel spatially dual adaptive MAP-based regularization method to reduce the mixed Gaussian-Poison noise. In this paper, both regularization operator and regularization parameter are self-adjusted. On the one hand, a p-Laplace function with adaptive p is used as regularization operator. On the other hand, the regularization parameter is designed based on a barrier function. Experimental results show that the performance of proposed method is superior to the competitive methods in mixed Gaussian-Poisson removal, both in reducing noise and keeping details. Moreover, more fine structures are saved in the view of the local variance images which is significantly important for defect detection. Further works will focus on better image segmentation method and improving the time efficiency.

Acknowledgement. This work was supported by the National Natural Science Foundation of China under Grant 61403146 and the Fundamental Research Funds for the Central Universities (x2zd-D2155120).

References

1. Barzilai, J., Borwein, J.M.: Two-point step size gradient methods. IMA J. Numer. Anal. **8**(1), 141–148 (1988)
2. Benvenuto, F., Camera, A.L., Theys, C., et al.: The study of an iterative method for the reconstruction of images corrupted by Poisson and Gaussian noise. Inverse Prob. **24**(3), 35016–35020 (2008)
3. Bertalmío, M., Caselles, V., Rougé, B., et al.: TV based image restoration with local constraints. J. Sci. Comput. **19**(1–3), 95–122 (2003)
4. Bing, S.: Topics in variational PDE image segmentation, inpainting and denoising. PhD Thesis. University of California Los Angeles, USA (2003)
5. Blomgren, P., Chan, T.F., Mulet, P., et al.: Total variation image restoration: numerical methods and extensions. In: IEEE International Conference on Image Processing (ICIP), vol. 3, pp. 384–387 (1997)
6. Blomgren, P., Chan, T.F.: Modular solvers for image restoration problems using the discrepancy principle. Numer. Linear Algebra Appl. **9**(5), 347–358 (2002)
7. Boyd, S., Vandenberghe, L.: Convex Optimization. Cambridge University Press, Cambridge (2004)
8. Chen, Q., Montesinos, P., Sun, Q.S., et al.: Adaptive total variation denoising based on difference curvature. Image Vis. Comput. **28**(3), 298–306 (2010)
9. Chen, D.Q., Cheng, L.Z.: Spatially adapted total variation model to remove multiplicative noise. IEEE Trans. Image Process. **21**(4), 1650–1662 (2012)
10. Chen, D.Q., Cheng, L.Z.: Spatially adapted regularization parameter selection based on the local discrepancy function for Poissonian image deblurring. Inverse Prob. **28**(1), 15004–15027 (2012)

11. Chen, D.Q., Cheng, L.Z.: Fast linearized alternating direction minimization algorithm with adaptive parameter selection for multiplicative noise removal. J. Comput. Appl. Math. **257**(2), 29–45 (2014)

12. Dai, S., Xiang, H., Du, Z., et al.: Adaptive regularization of infrared image super-resolution reconstruction. In: International Conference on Computing, Communication and Networking Technologies (ICCCNT), pp. 1–4 (2014)

13. Dong, Y., Hintermüller, M., Rincon-Camacho, M.M.: Automated regularization parameter selection in multi-scale total variation models for image restoration. J. Math. Imaging Vis. **40**(1), 82–104 (2011)

14. Fan, Z., Gao, R.X.: An adaptive total variation regularization method for electrical capacitance tomography. In: IEEE International Instrumentation and Measurement Technology Conference (I2MTC), pp. 2230–2235 (2012)

15. Gao, L., Li, C.: An adaptive TV model for image denoising. In: International Conference on Natural Computation (ICNC), pp. 766–770 (2013)

16. Geronimo, J.S., Hardin, D.P., Massopust, P.R.: Fractal functions and wavelet expansions based on several scaling functions. J. Approximation Theo. **78**(3), 373–401 (1994)

17. Harmany, Z.T., Marcia, R.F., Willett, R.M.: This is SPIRAL-TAP: sparse poisson intensity reconstruction algorithms–theory and practice. IEEE Trans. Image Process. **21**(3), 1084–1096 (2010)

18. Harmany, Z.T., Marcia, R.F., Willett, R.M.: SPIRAL out of convexity: sparsity-regularized algorithms for photon-limited imaging. In: SPIE - The International Society for Optical Engineering, vol. 7533, pp. 75330R–75330R-12 (2010)

19. Jeon, W., Yi, T.: Image restoration using dual adaptive regularization operators. In: IEEE International Conference on Pattern Recognition (ICPR), vol. 3, pp. 45–48 (2000)

20. Jezierska, A., Chouzenoux, E., Pesquet, J.C., et al.: A primal-dual proximal splitting approach for restoring data corrupted with Poisson-Gaussian noise. In: IEEE International Conference on Acoustics, Speech and Signal Processing (ICASSP), pp. 1085–1088 (2012)

21. Lagendijk, R.L., Biemond, J., Boekee, D.E.: Regularized iterative image restoration with ring reduction. IEEE Trans. Acoust. Speech Signal Process. **36**(12), 1874–1887 (1988)

22. Li, J., Shen, Z., Yin, R., et al.: A reweighted l_2 method for image restoration with Poisson and mixed Poisson-Gaussian noise. Inverse Prob. Imaging **9**(3), 875–894 (2015)

23. Luisier, F., Blu, T., Unser, M.: Image denoising in mixed Poisson-Gaussian noise. IEEE Trans. Image Process. **20**(3), 696–708 (2011)

24. Makitalo, M., Foi, A.: Poisson-gaussian denoising using the exact unbiased inverse of the generalized anscombe transformation. In: IEEE International Conference on Acoustics, Speech, and Signal Processing (ICASSP), pp. 1081–1084 (2012)

25. Morozov, V.A., Nashed, Z.: Methods for solving incorrectly posed problems. Springer, New York (2012)

26. Portilla, J., Strela, V., Wainwright, M.J.: Image denoising using scale mixtures of Gaussians in the wavelet domain. IEEE Trans. Image Process. **12**(11), 1338–1351 (2003). IEEE Signal Processing Society

27. Rodriguez, P.: Total variation regularization for poisson vector-valued image restoration with a spatially adaptive regularization parameter selection. In: International Symposium on Image and Signal Processing and Analysis (ISPA), pp. 402–407 (2011)

28. Rojas, R., Rodriguez, P.: Spatially adaptive total variation image denoising under Salt and Pepper noise. European Signal Processing Conference (EUSIPCO) 314–318(2011)
29. Rudin, L.I., Osher, S., Fatemi, E.: Nonlinear total variation based noise removal algorithms. Phys. D, Physica D: Nonlinear Phenomena **60**(1–4), 259–268 (1992)
30. Strong, D.M., Chan, T.F.: Spatially and scale adaptive total variation based regularization and anisotropic diffusion in image processing. Diffusion in Image Processing UCLA Math Department CAM Report (1996)
31. Strong, D.M., Aujol, J.F., Chan, T.F.: Scale recognition, regularization parameter selection, and Meyer's G norm in total variation regularization. Multiscale Model. Simul. **5**(1), 273–303 (2006)
32. Tang, X.: Research of regularization method in image restoration. Huazhong University of Science and Technology (2006)
33. Tikhonov, A.N., Arsenin, V.Y.: Solution of ill-posed problems. Math. Comput. **32**(144), 491–491 (1978)
34. Wen, Y., Chan, R.: Parameter selection for total-variation-based image restoration using discrepancy principle. IEEE Trans. Image Process. **21**(4), 1770–1781 (2012)
35. Xu, Y., Song, X., Dong, F., et al.: An adaptive total variation regularization method for elec-trical resistance tomography. In: IEEE International Conference on Imaging Systems and Techniques (IST), pp. 127–131 (2013)
36. Yong, X., Fang, Y., Gao, X., et al.: An adaptive regularization image super-resolution reconstruction algorithm. In: Chinese Control Conference (CCC), pp. 7258–7262 (2014)
37. Yuan, Q., Yan, L., Li, J., et al.: Remote sensing image super-resolution via regional spatially adaptive total variation model. In: IEEE International Geoscience and Remote Sensing Symposium (IGARSS), pp. 3073–3076 (2014)
38. Zhang, H., Peng, Q.: Adaptive image denoising model based on total variation. Opt. Eng. **33**(3), 50–53 (2006)

Workshop on Assistive Vision

A Study of Combining Re-coloring and Adding Patterns to Images for Dichromats

Wei-Ta Chu$^{(\boxtimes)}$ and Tsung-Han Yang$^{(\boxtimes)}$

National Chung Cheng University, Chiayi, Taiwan
wtchu@ccu.edu.tw

Abstract. Color is one of the most important modality to convey information. However, around the world about 200 million people are with color vision deficiency (CVD). Some works have been developed to improve viewing experience for people with CVD, such as simulating colorblind vision, re-coloring images, and using patterns to encode images. In this work, we advocate that combining re-coloring and adding patterns to image might be more helpful to colorblind people and is worth deep research. We propose a framework to combine patterns and re-coloring. We first simulate colorblind vision, determine how to add patterns according to the degree of deformation, and then re-color images overlaid with patterns. In the evaluation, we verify effectiveness of combining adding patterns and re-coloring, and demonstrate content-dependent characteristics through the studies based on different types of images and different types of patterns.

1 Introduction

Many people have problems in identifying colors (color vision deficiency, CVD) due to malfunction of some photoreceptors. Human trichromatic vision relies on three unequal kinds of photoreceptors, which are ingenious to long (L-cones), medium (M-cones), and short (S-cones) visible wavelengths. The most commonly seen CVDs include anomalous trichromacy, dichromacy, and monochromacy. *Anomalous trichromats* are the major proportion of CVDs, and they are different from normal people for that they see colors with distortion. This distortion happens because at least one kind of photoreceptors does not work correctly. *Dichromats* are the people with an injured color channel, so that one kind of cones entirely does not work. If the color space to be identified by normal human vision is three-dimensional, the one to be identified by dichromats is two-dimensional. For *monochromats*, only one kind of cones can work correctly, and the color space to be identified by them is reduced to only one dimension.

Figure 1 shows some poster or magazine images that were not specially designed to people with CVD. In the images of the first row, mixed red and green regions disturb users with red-green colorblindness. In the images of the second row, mixed blue and yellow regions cause problems to blue-yellow colorblindness.

© Springer International Publishing AG 2017
C.-S. Chen et al. (Eds.): ACCV 2016 Workshops, Part I, LNCS 10116, pp. 225–238, 2017.
DOI: 10.1007/978-3-319-54407-6_15

Fig. 1. Poster and magazine images that were not designed for colorblind (Color figure online).

Several works have been proposed to improve viewing experience of people with CVD (mostly for anomalous trichromats and dichromats), such as finding the area of image that people with CVD can't recognize [1], simulation of CVD [2–4], re-recoloring images [5–9], and using patterns to enhance images [10]. In this work, we advocate that combining re-coloring and adding patterns to image might be more helpful to colorblind people. We are also wondering if such enhancement is consistently effective for all types of images.

We propose a framework to combine adding patterns and re-coloring to images. We first simulate colorblind vision, and according to the degree of color deformation, the patterns to be overlaid are determined. To further enhance viewing experience, we adjust image's contrast and shift patterns for continuity [10]. Finally, we re-color images with overlaid patterns to combine two types of enhancements. In the evaluation, we verify effectiveness of the combination approach, and demonstrate that different patterns have varied influences on different types of images.

The major contributions of this paper are listed as follows.

- We verify the effectiveness of combining re-coloring techniques and the pattern encoding scheme to enhance images for CVDs. We also design different patterns and investigate performance variations when different patterns are overlaid on images.
- We unveil content-dependent characteristics of enhancement performance. Three types of images are collected and evaluated, and performance variations of the proposed method on different types of images are comprehensively studied.

The rest of this paper is organized as follows. Related works are surveyed in Sect. 2. In Sect. 3, we present our method to combine re-recoloring with patterns for dichromats. Experimental design and results are shown in Sect. 4. Finally, conclusion and future work will be given in Sect. 5.

2   Related Works

Wang et al. [1] proposed an in-image accessibility detection method that automatically detects regions that cannot be identified by colorblind people. This work can not only be used to check an image's readability but also helps raising visual feasibility of an image. Machado et al. [2] presented a model for simulating the colorblind's vision, derived from electrophysiological studies. This work is the first one to jointly consider normal color vision, anomalous trichromacy, and dichromacy in a unified approach. This model was verified to be able to improve visualization experiences for people with CVD[1], through an experiment to normal individuals and color vision deficiency ones.

Rasche et al. [9] presented a re-coloring technique that maintains visual details by preserving contrasts and luminance consistency. Preserving contrasts can facilitate color discrimination, while maintaining luminance consistency facilitate distinguishing color variance. Kuhn et al. [5] presented an efficient and automatic image re-coloring technique based on the mass-spring optimization framework. The global consistency property guarantees that the same color in the original image will be transformed to the same color in the re-coloring image. Brightness of the original image is retained that makes people use color gradients to distinguish colors easily. Ribeiro et al. [6] proposed an algorithm to adjust colors for colorblind people (especially protanopia and deuteranopia). Their algorithm was more efficient and easier than previous studies by adjusting the HSV color model and processing every pixel just once.

In contrast to re-coloring approaches to enhance images, Sajadi et al. [10] presented a content-independent method to overlay patterns on images that not only minimizes ambiguities but also enables color identification. Overlaid patterns does not affect the underlying original color, and thus don't obstruct people with normal vision. Their experiments show that overlaid patterns effectively improve dichromats' recognition performance in several tests.

In our work, we are wondering if the re-coloring approach and the pattern embedding approach can be combined so that recognition performance of colorblind people can be further improved. We thus design a framework to combine them, and verify feasibility of this idea through several experiments.

3   Combining Re-coloring with Adding Patterns

3.1   Framework

Figure 2 shows the flowchart and step-by-step results of the proposed framework. Given an image, the model proposed in [2] is used to simulate colorblind people's

---

[1] In this paper, we interchangeably describe them as "colorblind people", "people with CVD", or "dichromats".

vision. This method maps RGB colors onto a color plane P, which is the set of colors colorblind people can see. We then measure the distance between an original color and its orthogonal projection on the plane P to measure the degree of color distortion. At the second step, we use the method proposed in [10] to add patterns, which encode color information, to enhance images. With color adjustment and pattern shift for continuity [10], we overlay smooth patterns to images. Finally, we re-color the image based on the method proposed in [6], and get the enhanced image for dichromats. In the following, we will first briefly describe the principle of colorblindness, and then describe details the enhancement process.

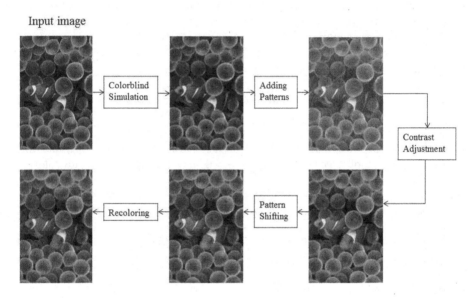

Fig. 2. The flowchart of combining re-coloring and adding patterns (better viewed in colorful version).

3.2 Principle of Colorblindness

Human trichromatic vision needs joint functions of three kinds of photoreceptors, which are sensitive to long (L-cones), medium (M-cones), and short (S-cones) visible wavelengths. The colors normal people can identify collectively form a color space like a 3D cube. However, because one or more kinds of cones do not work correctly, the set of colors dichromats can identify just forms a plane inside the aforementioned 3D cube. Figure 3 shows the planes that three types of dichromats can perceive, i.e., protanope, deuteranope, and tritanope, respectively. We can see that the planes perceived by protanope and deuteranope are similar, and this is why protanope and deuteranope are jointly known as red-green colorblindness. Because more than 99% of colorblind people are red-green colorblindness, we focus on this type of colorblindness.

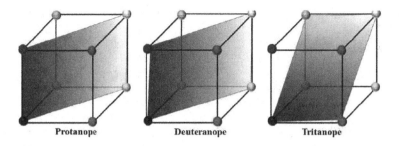

Fig. 3. The planes constituted by the colors identified by protanope, deuteranope, and tritanope, respectively [10].

3.3   Combining Re-Coloring with Adding Patterns

Overlaying Patterns. The idea of encoding color information as patterns is that, according to different degrees of color distortion, different patterns are used to replace pixels. We use the process proposed in [10] to add patterns to images. Given an image I, for each pixel with color c, an 4×4 patch C_S is used to replace this pixel, where the patch C_S is colored with c. We then use the method proposed in [2] to simulate the vision of dichromats. The color c is mapped to the color c_p, which is on the plane P that consist of colors colorblind people can identify (Fig. 3). A pixel's color distortion is then measured by the distance d between c and c_p. Note that pixels with different colors have different distortions d's. On the other hand, all the colors along $c_p + dN$ will map onto the same color c_p, where N is the normal vector of the plane P, d is a real number indicating the distance between c and c_p. According to d, the patch C_S is overlaid with a pre-defined pattern, and the patch with overlaid patterns is denoted as C_T. After patch replacement and pattern overlaying, the image I is enlarged and consists of a set of textured patched C_T's.

There are several ways to design the overlaid patterns. In [10], they designed various line patterns by varying the orientation of a white line inside a 4×4 patch. Different line patterns correspond to different ranges of color distortion d's. In this work, we further design different types of patterns, i.e., dot patterns, geometry patterns, and wave patterns, and investigate performance variation yielded by adding different types of patterns. In the following, we will take line patterns as the main instance. Samples of different patterns can be seen in Fig. 8. Figure 4 shows a sample image with overlaid line patterns as defined in [10].

Contrast Adjustment. To enhance visibility of images with patterns, we decrease pattern contrast based on the absolute value of d. Let's denote the absolute maximum value of d as d_{MAX}. We use the weighted average of C_S and C_T by using $\alpha = \frac{|d|}{d_{MAX}}$ as the weighting coefficient to produce the final patch C_F:

$$C_F = \alpha C_T + (1 - \alpha)C_S. \tag{1}$$

The adjusted patches C_F's are actually used to replace pixels in the image I.

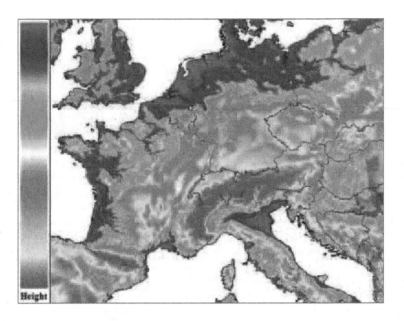

Fig. 4. A sample image overlaid with line patterns.

Pattern Shifting. The steps mentioned above replace pixels of the same color with patches of the same pattern, i.e., orientation of lines in these patches is the same. However, adjacent patches overlaid with lines of the same orientation would form a region with fragmented line segments, as shown in Fig. 5. To enhance continuity of lines, we adopt the method in [10] to shift line patterns according to their locations in images.

We first compute vertical and horizontal shifts of each patch. According to slope L of the line in a patch, the horizontal and vertical shifts are $s_x = \lfloor \frac{m}{L} \rfloor$ and $s_y = \lfloor m \times L \rfloor$ pixels, respectively. Shifting is done only in one direction (horizontal or vertical). If the absolute value of the pattern slope is more than 1, only horizontal shift is conducted; if the absolute value of pattern slope is less than 1, only vertical shift is conducted.

For the line pattern at pixel position (x, y), we conduct horizontal shift by $((y \bmod m)\, s_x \bmod m)$, or conduct vertical shift by $((x \bmod m)\, s_y \bmod m)$. This transformation ensures that adjacent line segments can be continuous. Figure 5 shows examples of two types of shifts.

Re-coloring. After adding patterns, we use the method proposed in [6] to re-color images. This algorithm was designed for deuteranopes and protanopes because they are the most common dichromats. We represent each pixel in the HSV color space. In this space, we have to transform the hue value prone to be confused by dichromats:

$$H' = H + \Delta H, \quad S' = S, \quad V' = V \tag{2}$$

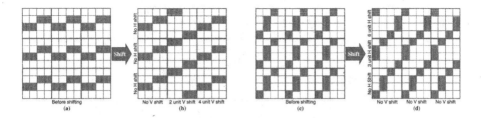

Fig. 5. We show 4×4 patterns in two different orientations before shifting (a)(c) and after shifting (b)(d). The shifts are implemented in only one direction [10].

where the variation of hue ΔH is defined as

$$\Delta H = \begin{cases} 0, & \text{if } H \in R \\ \rho(r,g,b), & \text{if } H \notin R \end{cases} \tag{3}$$

where $R \approx [60°, 280°]$ represents the range of blues and greens.

Deuteranopes perceive blues just like normal people, and thus $\Delta H = 0$ for the hue values in R. On the other hand, deuteranopes have problems in identifying red, pink, orange, and magenta. Therefore, for the colors outside the region R, we apply the hue remapping function:

$$\rho(r,g,b) = (H - 130) \cdot \mu(r,g,b) \tag{4}$$

where μ is the magenta function $\mu(r,g,b) = 1 - \frac{g}{255}$ presented in [3]. The values (r,g,b) are the colors obtained by transferring (H, S_{MAX}, V_{MAX}) to (r,g,b), where S_{MAX} is the maximum value of the saturation channel, and V_{MAX} is the maximum value of the value channel in the HSV color space.

Fig. 6. Sample images from three different types of images in the collected dataset.

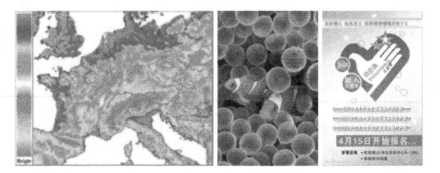

Fig. 7. Sample results of adding *line patterns* to enhance images. Left to right: map image, photography, and poster image (before re-coloring).

4   Experimental Results

4.1   Dataset

To investigate whether performance of enhancement is content-dependent, we collect three different types of images and enhance them with different types of patterns, in addition to the line patterns mentioned previously. We performed image search with the queries "map", "photo", or "poster" on Google Image Search. We limit our selection to images with resolution higher than 640×480 or 480×640, and with prominent colors red or green, because we focus on red-green colorblindness. We collect 100 images for each image category. Figure 6 shows some sample images.

4.2   Effectiveness of Combination

We first evaluate the effectiveness of combining re-coloring and pattern embedding to enhance images. We take the line patterns designed in [10] as the main investigation target. According to the distance d between an original color and the projection plane, different line patterns with different orientations are selected to replace original pixels, followed by the re-coloring process proposed in [6]. Figure 7 shows three sample results of adding patterns to enhance map, photo, and poster images, respectively.

We asked 21 people with CVD to see 15 sets of images (there are five sets of map images, five sets of photos, and five sets of poster images). In each set, the original image, the image with adding patterns, the image with re-coloring, and the image with both re-coloring and adding patterns are randomly juxtaposed to the subject, who was asked to judge how easy he/she can distinguish colors by giving a score from 1 to 4 (larger means better, and each score can be given for just once). This mechanism enforces subjects to clearly discriminate different images, and facilitates us to quantitatively study the influence of adding patterns.

The first sub-table of Table 1 shows average overall scores based on all types of images. We can see the original images get the lowest score, which shows that

Table 1. Average scores obtained based on different settings. The scores from top to down are average scores obtained based on all images, map images only, photos only, and poster images only, respectively.

Overall	Original	Adding patterns (P)	Re-coloring (R)	P + R
	1.5	2.21	3	3.12
Map	Original	Adding patterns (P)	Re-coloring (R)	P + R
	1	3	1.8	3.8
Photo	Original	Adding patterns (P)	Re-coloring (R)	P + R
	1	2	3.1	3.8
Poster	Original	Adding patterns (P)	Re-coloring (R)	P + R
	2	2	3.5	2

colors in them are hard to be identified by people with CVD. Images with re-coloring get a higher score, and images with both re-coloring and adding patterns get the highest score. This means re-coloring is an effective enhancement method, and it would be more helpful if images are further overlaid with appropriate patterns.

The second sub-table shows that map images with re-coloring receive lower average score than that with adding patterns (1.8 vs. 3). Map images are often with diverse and widely distributed color regions, and cause difficulty for people with CVD to identify colors. On the other hand, adding patterns to images facilitates users to distinguish colors in this case, and again the highest average score can be obtained by combining re-coloring and adding patterns.

Characteristics of the third sub-table are similar to that of the first sub-table. From the fourth sub-table, combining re-coloring and adding patterns doesn't yield the best performance for posters. Poster images have much more text information, and the overlaid line patterns make images blurred. People do not like blurred effects especially when they want to clearly see the text information inside posters. Therefore, in this case, images with re-coloring receive a much higher average score than that with adding patterns.

Here we verify effectiveness of the combination approach. But more interestingly, we also uncover the content-dependent characteristics of re-coloring techniques and pattern embedding techniques.

4.3   Influence of Different Patterns

We have verified effectiveness of the combination approach and the performance variations depending on image content. In this section, we want to further investigate if different patterns, other than line patterns, have different effects in enhancing images for dichromats. We therefore design various 4 × 4 patterns, which are roughly categorized into four types, i.e., line patterns, dot patterns, geometry patterns, and wave patterns. Figure 8 shows eight designed samples of each type of patterns.

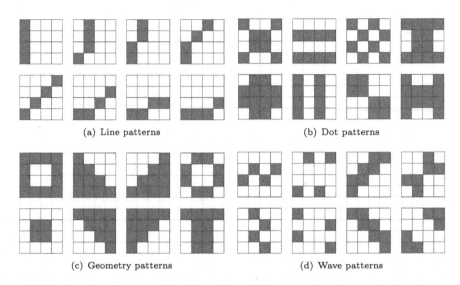

Fig. 8. The designed line patterns, dot patterns, geometry patterns, and wave patterns.

Figure 9 shows three sample images, respectively map, photo, and poster, overlaid by the dot patterns. We can see visual effects obviously different from that with line patterns shown in Fig. 7. Geometry patterns and wave patterns can also be overlaid on images, according to the distance between original colors and distorted colors. Sample images with overlaid geometry patterns and wave patterns are omitted here due to space limitation. Note that for adding dot patterns, geometry patterns, or wave patterns, the pattern shifting process is not employed. It is not required to maintain orientation continuity between adjacent patterns when such patterns are applied.

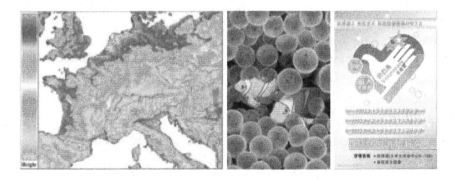

Fig. 9. Sample results of adding *dot patterns* to enhance images. Left to right: map image, photography, and poster image.

We also design a questionnaire to evaluate whether different patterns yield different identification performance. We asked 22 people with CVD to see 15 sets of images (there are five sets of map images, five sets of photos, and five sets of poster images). Each set has four images, which are images enhanced with line patterns plus re-coloring, with dot patterns plus re-coloring, with geometry patterns plus re-coloring, and with wave patterns plus re-coloring. These images are randomly juxtaposed to the subjects, and then each subject is asked to give· a score from 1 to 4 (larger means easier to identify different colors). Similarly, each score can only be given once for us to more clearly quantize performance variation of different patterns.

Table 2. Average scores of images enhanced with different types of patterns. Scores obtained based on all images, map images, photos, and poster images, respectively.

	Line	Dot	Geometry	Wave
Overall	3.00 (±1.00)	2.20 (±1.00)	2.40 (±1.00)	2.40 (±1.00)
Map	3.40 (±0.80)	1.20 (±0.40)	2.80 (±0.75)	2.60 (±1.00)
Photo	3.80 (±0.40)	1.80 (±0.75)	1.60 (±0.50)	2.80 (±1.00)
Poster	2.20 (±0.75)	2.80 (±0.50)	2.20 (±1.00)	2.80 (±0.50)

The first row of Table 2 shows the overall average scores of images overlaid with four different types of patterns. We can see that images with line patterns get the highest average score, which verifies that line patterns are more useful, and conforms to the selection of [10]. Images with dot patterns get the lowest average score, showing that dot patterns are weaker to help people with CVD to distinguish colors.

The last three rows of Table 2 show detailed results of three different types of images, respectively. We can see that map images and photo images with line patterns receive the highest scores (the second and the third rows), while poster images with dot patterns receive the highest scores (the fourth row). Map images have high color variations, and line patterns provide higher discrimination than other patterns. This trend can also be seen from photos. On the other hand, poster images with dot patterns can be most identified. Overlaying dot patterns yields weaker blurred effects on poster images, and thus poster images with dot patterns obtain a higher average score.

4.4 Number of Identified Colors

In this section we try to more quantitatively evaluate how dichromats can identify colors through the images with enhancement. In this experiment, we construct a questionnaire and ask 22 people with CVD to see five sets of colorful images. Each set has four images, which are the original image, the images with overlaid patterns, the image with re-coloring, and the image with both re-coloring and overlaid patterns. Each subject was asked to indicate how many different colors he/she can identify. The number of different colors in these test images is actually five. We study the number of colors being able to be identified when images are enhanced with different settings. Figure 10 shows sample images used in this experiment.

Fig. 10. Sample images with different enhancement settings. (a) original image, (b) image with re-coloring, (c) image with adding patterns, and (d) image with both re-coloring and adding patterns.

Table 3. Average numbers of colors able to be identified from images enhanced with different settings.

	Original	Adding patterns (P)	Re-coloring (R)	P + R
Avg. number	3.2	3.51	3.9	4.5

Table 3 shows average numbers of colors able to be identified from images enhanced with different settings. We can see that, from original images, least number of colors can be identified, and most colors can be identified from images enhanced with both re-coloring and adding patterns. This again verifies the effectiveness of combining re-coloring and adding patterns to enhance images for dichromats.

5 Conclusion

We have presented a work for combining re-coloring and adding patterns to people with color vision deficiency. The effectiveness of combining adding patterns and re-coloring to facilitate color identification is first verified. We then study the influence of different types of patterns on color identification, i.e., dot patterns, geometry patterns, and wave patterns, in addition to line patterns. We also found the content-dependent characteristics of color identification based on different types of images and different types of overlaid patterns. Overall, this would be the first study on content-dependent performance variation on image enhancement for colorblindness.

In this work, globally overlaying patterns and re-coloring to images achieves good enhancement results for people with CVD. However, we find that the proposed approach makes image blurred after adding patterns. This annoys users especially there is text in images. Therefore, especially considering text information when adding patterns would be an interesting future work.

Acknowledgement. The work was partially supported by the Ministry of Science and Technology in Taiwan under the grant MOST103-2221-E-194-027-MY3, MOST104-2221-E-194-014, and MOST105-2628-E-194-001-MY2.

References

1. Wang, M., Sheng, Y., Liu, B., Hua, X.S.: In-image accessibility indication. IEEE Trans. Multimedia **12**, 330–336 (2010)
2. Machado, G., Oliveira, M., Fernandes, L.: A physiologically-based model for simulation of color vision deficiency. IEEE Trans. Visual. Comput. Graph. **15**, 1291–1298 (2009)
3. Yang, S., Ro, Y.: Visual contents adaptation for color vision deficiency. In: Proceedings of IEEE International Conference on Image Processing, pp. 453–456 (2003)
4. Vienot, F., Brettel, H., Mollon, J.D.: Digital video colourmaps for checking the legibility of displays by dichromats. Color Res. Appl. **24**, 243–252 (1999)
5. Kuhn, G., Oliveira, M., Fernandes, L.: An efficient naturalness-preserving image-recoloring method for dichromats. IEEE Trans. Visual. Comput. Graph. **14**, 1747–1754 (2008)
6. Ribeiro, M., Gomes., A.: A skillet-based recoloring algorithm for dichromats. In: Proceedings of IEEE International Conference on e-Health Networking, Applications and Services, pp. 702–706 (2013)

7. Liu, B., Wang, M., Yang, L., Wu, X., Hua, X.: Efficient image and video recoloring for colorblindness. In: Proceedings of IEEE international Conference on Multimedia and Expo, pp. 906–909 (2009)
8. Huang, J., Chen, C., Wang, S.J., Jen, T.C.: Image recolorization for the colorblind. In: Proceedings of IEEE International Conference on Acoustics, Speech and Signal Processing, pp. 1161–1164 (2009)
9. Rasche, K., Geist, R., Westall, J.: Re-coloring images for gamuts of lower dimension. ACM Trans. Graph. **25**, 423–432 (2005)
10. Sajadi, B., Majumder, A., Oliveira, M., Schneider, R.: Using patterns to encode color information for dichromats. IEEE Trans. Visual. Comput. Graph. **19**, 118–129 (2013)

Calorie Counter: RGB-Depth Visual Estimation of Energy Expenditure at Home

Lili Tao$^{(\boxtimes)}$, Tilo Burghardt, Majid Mirmehdi, Dima Damen, Ashley Cooper, Sion Hannuna, Massimo Camplani, Adeline Paiement, and Ian Craddock

SPHERE, Faculty of Engineering, University of Bristol, Bristol, UK
lili.tao@bristol.ac.uk

Abstract. We present a new framework for vision-based estimation of calorific expenditure from RGB-D data - the first that is validated on physical gas exchange measurements and applied to daily living scenarios. Deriving a person's energy expenditure from sensors is an important tool in tracking physical activity levels for health and lifestyle monitoring. Most existing methods use metabolic lookup tables (METs) for a manual estimate or systems with inertial sensors which ultimately require users to wear devices. In contrast, the proposed pose-invariant and individual-independent vision framework allows for a remote estimation of calorific expenditure. We introduce, and evaluate our approach on, a new dataset called *SPHERE-calorie*, for which visual estimates can be compared against simultaneously obtained, indirect calorimetry measures based on gas exchange. We conclude from our experiments that the proposed vision pipeline is suitable for home monitoring in a controlled environment, with calorific expenditure estimates above accuracy levels of commonly used manual estimations via METs. With the dataset released, our work establishes a baseline for future research for this little-explored area of computer vision.

1 Introduction

The large majority of research into the physical activity levels of people with, or at risk of, chronic disease have measured either total physical activity or physical activity acquired in specific activities, such as walking, that generally occur outside the home. Very little is known about how activities of normal daily living in the home environment may contribute to prevention of or recovery from/management of chronic disease, such as obesity and diabetes. An accurate assessment of physical activity within the home is thus important to understand recovery progress and long term health monitoring [1].

Energy expenditure, also referred to as 'calorific expenditure', is one commonly used single metric to quantify physical activity levels over time. It provides a key tool for the assessment of physical activity; be that for the long term monitoring of health and lifestyle aspects associated to chronic conditions or for recovery medicine. Calorific expenditure is traditionally measured either using direct methods, such as a sealed respiratory chamber [2], or indirect calorimetry,

© Springer International Publishing AG 2017
C.-S. Chen et al. (Eds.): ACCV 2016 Workshops, Part I, LNCS 10116, pp. 239–251, 2017.
DOI: 10.1007/978-3-319-54407-6_16

which requires carrying gas sensors and wearing a breathing mask [3]. The latter is often based on the respiratory differences of oxygen and carbon dioxide in the inhaled and exhaled air. It forms the measurement standard for non-stationary scenarios where the person can move freely. Recently, the use of wearable devices – with a focus on coarse categorisations of activity levels by wrist-worn inertial sensors [4] – has become a popular monitoring choice due to its low cost, low energy consumption, and data simplicity. Among these, tri-axial accelerometers are the most broadly used inertial sensors [5].

Visual systems, on the other hand, are already a key part of home entertainment systems today, and their RGB-D sensors [6] allow for a rich and fine-grained analysis of human activity within the field of view. Recent advances in computer vision have now opened up the possibility of integrating these devices seamlessly into home monitoring systems [7, 8].

With this in mind, we propose a framework for estimating energy expenditure from RGB-D data in a living room environment. Figure 1 shows in bold a flowchart of the proposed method – mapping visual flow and depth features to calorie estimates using activity-specific models (AS in short). The method implements a cascaded and recurrent approach, which explicitly detects activities as an intermediate to select type specific mapping functions for final calorific estimation. We compare this proposed method against a ground truth of gas-exchange measurements (GT in short) and two off-the-shelf alternatives: (1) mapping features directly to calorie estimates via a monolithic classifier (DM in short), and (2) manual mapping from activity classes to calorie estimates via metabolic equivalent task lookup tables [9] (MET in short) as most often applied in practice today.

This is a new application in computer vision where no existing datasets are available. In order to quantify the performance, we introduce a new dataset, *SPHERE-calorie*[1], for calorific expenditure estimation collected within a home environment. The dataset contains 11 common household activities performed over up to 20 sessions, lasting up to 30 min for each session, in each of which the activities are performed continuously. The setup consists of an RGB-D Asus Xtion camera mounted at the corner of a living room and a COSMED K4b2 [3] indirect calorimeter for ground truth measurement. The *SPHERE-calorie* dataset will be publicly released. In summary, the major contributions of this paper are (a) a first-ever framework for a vision-based estimation of calorific expenditure from RGB-D data only, applicable to daily living scenarios, and (b) a novel dataset linking more than 10 h of RGB-D video data to ground truth calorie readings from indirect calorimetry based on gas exchange.

2   Related Work

Applying computer vision techniques to help with the diagnosis and management of health and wellbeing conditions has gained significant momentum over the

---

[1] The dataset is released on SPHERE website http://www.ircsphere.ac.uk/work-package-2/calorie.

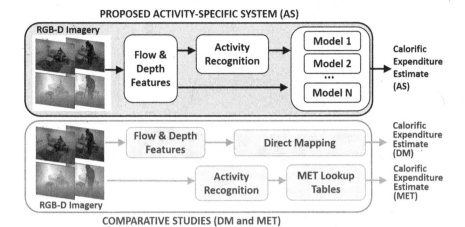

Fig. 1. Framework Overview. RGB-D videos are represented by a combination of flow and depth features. The proposed recurrent method AS (top) then selects activity-specific models which map to energy expenditure estimates. We compare this method to a direct mapping method DM and a manual estimate via lookup tables MET (bottom).

last years. However, studies on energy expenditure using visual sensors have been relatively limited. Our work explores this field further and builds on several relevant subject areas in vision.

Feature Representation - The visual trace of human activity in video forms a spatio-temporal pattern. To extract relevant properties from this for the task at hand, one aims at compactly capturing this pattern and highlighting important aspects related to the properties of interest. Assuming that both body configuration and body motion [10] are relevant to infer calorific uptake, the pool of potential features is large - ranging from local interest point configurations [11], over holistic approaches like histograms of oriented gradients and histograms of motion information [12], to convolutional neural network features [13].

Motion information in the first place could also be recovered in various ways, e.g. from RGB data using optical flow *or* from depth data using 4D surface normals [14]. Whilst a composition of these features via concatenation of per-frame descriptors is straight forward, this approach suffers from the curse of dimensionality and unaffordable computational cost. Sliding window methods [15], on the other hand, can limit this by predicting current values only from nearby data within a temporal window. Further compaction may be achieved by converting large feature arrays into a single, smaller vector with a more tractable dimension count via, for instance, bags of visual words [16], Fisher vectors [17] or time series pooling [18]. In summary, the challenge of feature representation will require capturing visual aspects relevant to calorific expenditure, whilst limiting the dimensionality of the descriptor.

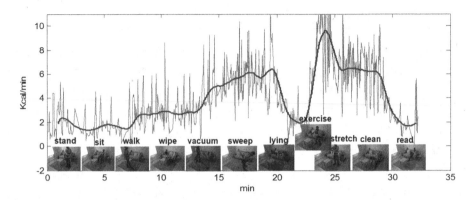

Fig. 2. Ground truth example sequence. Raw per breath data (red) and smoothed COSMED-K4b2 calorimeter readings (blue) and sample colour images corresponding to the activities performed by the subject. (Color figure online)

Activity Recognition - There exists a significant body of literature describing the inference of activities from 2D colour intensity imagery [19], RGB-D data [6], and skeleton-based data [20]. Knowledge about the type of activity undertaken has been shown to correlate with the calorific expenditure incurred [9]. In alignment with Fig. 1, we will argue in this work that an explicit activity recognition step in the vision pipeline can, as an intermediate component, aid the visual estimate of energy uptake.

Visual Energy Expenditure Estimation - 2D video has recently been used by Edgcomb and Vahid [21] coarsely to estimate daily energy expenditure. In their work a subject is segmented from the scene background. Changes in height and width of the subject's motion bounding box, together with vertical and horizontal velocities and accelerations, are then used to estimate calorific uptake. Tsou and Wu [22] take this idea further and estimate calorie consumption using full 3D joint movements tracked as skeleton models by a Microsoft Kinect. Both of the above methods use wearable accelerometry as the target ground truth, which in fact does not provide an accurate benchmark; and skeleton data is commonly noisy and currently only operates reliably when the subject is facing the camera. This limits applicability in more complex visual settings as contained in the *SPHERE-calorie* dataset.

As outlined in the following section, our work attempts to remedy these shortcomings by using skeleton-independent, RGB-D based vision to estimate calorific expenditure against a standardised calorimetry sensor COSMED-K4b2 based on gas exchange (see Fig. 2).

3   Proposed Method

We propose an activity-specific pipeline to estimate energy expenditure utilising both depth and motion features as input. Importantly, our setup as shown in Fig. 1 is designed to reason about activities *first*, before estimating calorie expenditure via a set of models which are each separately trained for particular activities.

3.1   Features

We first simultaneously collect RGB and depth imagery using an Asus Xmotion. For each frame t, appearance and motion features are extracted, with the latter being computed with respect to the previous frame (level 0). A set of temporal filters is then applied to form higher level motion features (level 1). We extract the features within the bounding box returned by the OpenNI SDK [23] person detector and tracker. To normalise the utilised image region due to varying heights of the subjects and their distance to the camera, the bounding box is scaled by fixing its longer side to $M = 60$ pixels, a size recognised as optimal for human action recognition [24], while maintaining aspect ratio. The scaled bounding box is then centred in a $M \times M$ square box and horizontally padded.

Motion feature encoding - Inspired by [24], optical flow measurements are taken over the bounding box area and split into horizontal and vertical components. These are re-sampled to fit the normalised box and a median filter with kernel size 5×5 is applied to smooth the data. A spatial pyramid structure is used to form hierarchical features from this. Such partitioning of the image into an iteratively growing number of sub-regions increases discriminative power. The normalised bounding box is divided into a $n_g \times n_g$ non-overlapping grid, where n_g depends on the pyramid level, and the orientations of each grid cell are quantised into n_b bins. The parameters for our experiments are empirically

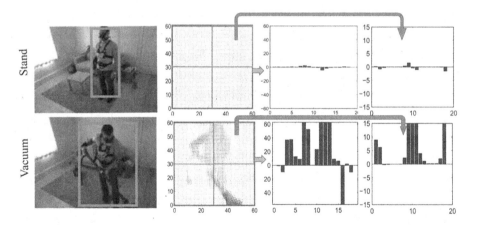

Fig. 3. Flow feature encoding via spatial pyramids. *Top row:* limited motion while standing still. *Bottom row:* significant motion features when moving during vacuuming. *First column:* colour images with detected person. *Second column:* optical flow patterns. *Third column:* motion features at level 0. *Last column:* motion features from the top-right quadrants of the image at level 1 (at which the image is subdivided into four quadrants).

determined as $n_b = 9$ and $n_g = 1$ and 2 for levels 0 and 1 respectively. Figure 3 exemplifies optical flow patterns and their encoding for different activities.

Appearance feature encoding - We extract depth features by applying the histogram of oriented gradients (HOG) feature on raw depth images [25] within the normalised bounding box. We then apply Principal Component Analysis (PCA) and keep the first 150 dimensions of this high-dimensional descriptor which retains 95% of the total variance.

Pyramidal temporal pooling - Given the motion and appearance features extracted from each frame in a sequence of images, it is important to capture both short and long term temporal changes and summarise them to represent the motion in the video. Pooled motion features were first presented in [18], designed for egocentric video analysis. We modify the pooling operator to make it more suitable for our data as follows.

An illustration of the temporal pyramid structure and the process for pooling operations are shown in Fig. 4. The time series data $\mathbf{S}$ can be represented as a set of time segments at level i as $\mathbf{S} = [\mathbf{S}_i^1, \ldots, \mathbf{S}_i^{2^i}]$. The final feature representation is a concatenation of multiple pooling operators applied to each time segments at each level. The time series data can also be explained as T number of per-frame feature vector, such that $\mathbf{S} = \{S_1, \ldots, S_N\}, \mathbf{S} \in \mathbb{R}^{N \times T}$ for a video in matrix form, where N is the length of the per-frame feature vector, and T is the number of frames. A time series $S_n = [s_n(1), \ldots, s_n(T)]$ is the nth feature across $1, \ldots, T$ frames, where $s_n(t)$ denotes n^{th} feature at frame t. A set of temporal filters with multiple pooling operators is applied to each time segment $[t_{min}, t_{max}]$ and produces a single feature vector for each segment via concatenation. We use two conventional pooling operators, max pooling and sum pooling, as well as frequency domain pooling. They are defined respectively as:

$$\mathcal{O}_{\max}(S_n) = \max_{t=t_{\min}\cdots t_{\max}} s_n(t) \quad \text{and} \quad \mathcal{O}_{\text{sum}}(S_n) = \sum_{t=t_{\min}}^{t_{max}} s_n(t) \tag{1}$$

Frequency domain pooling is used to represent the time series S_n in the frequency domain by the discrete cosine transform, where the pooling operator takes the absolute value of the j lowest frequency components of the frequency coefficients D,

$$\mathcal{O}_{\text{dct}}(S_n) = |M_{1:j} S_n| \tag{2}$$

where M is the discrete cosine transformation matrix.

3.2   Learning and Recurrency

Energy expenditure estimation can be formulated as a sequential and supervised regression problem. We train a support vector regressor to predict calorie values from given features over a training set. The sliding window method is used to map each input window of width w to an individual output value y_t. The window contains the current and the previous $w - 1$ observations. The window feature is represented by temporal pooling from the time series $\mathbf{S} = \{S_{t-w+1}, \ldots, S_t\}$.

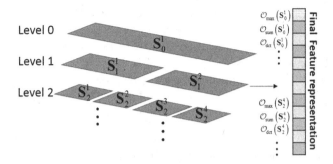

Fig. 4. Temporal pyramid pooling and its feature representation. This schematic shows the temporal subdivision of data into various pyramidal levels (left) and the concatenation of resulting feature (e.g. max, sum and dict) into a descriptor vector (right).

We note that energy values for a particular time are highly dependent on the energy expenditure history. In our system, these are most directly expressed by previous calorific predictions during operation. Thus, employing recurrent sliding windows offers an option to not only use the features within a window, but also take the most recent d predictions $\{\hat{y}_{t-d}, \ldots, \hat{y}_{t-1}\}$ into consideration to help predict y_t. During learning, as suggested in [26], the ground truth labels in the training set are used in place of recurrent values.

4   Experimental Results

We introduce the RGB-D *SPHERE-calorie dataset* captured in a real living environment. The ground truth was captured by the COSMED K4b2 portable metabolic measurement system. The dataset was generated over 20 sessions by 10 subjects with varying anthropometric measurements containing up to 11 activity categories per session, and totalling around 10 h recording time. The categories and their associated MET values (in brackets) are: light intensity activities = $\{$*sit still (1.3), stand still (1.3), lying down (1.3), reading (1.5)*$\}$; light+ intensity activities = $\{$*walking (2.0), wiping table (2.3), cleaning floor stain (3.0)*$\}$; moderate vigorous intensity activities = $\{$*vacuuming (3.3), sweeping floor (3.3), squatting (5.0), upper body exercise (4.0)*$\}$.

Colour and depth images were acquired at a rate of 30 Hz. The calorimeter gives readings per breath, which occurs approximately every 3 s. To better model transitions between activity levels, we consider 9 different combinations of the above three activity intensities in each session. Figure 2 shows a detailed example of calorimeter readings and associated sample RGB images from the dataset. The raw breath data is noisy (in red), and so we apply an average filter with a span of approximately 20 breaths (in blue). The participants were asked to perform the activities based on their own living habits without any extra instructions.

We compare the proposed method AS to the direct mapping method DM and the metabolic equivalent table method MET. DM is formalised as $Y_t = f(X_t)$, where Y_t is the target calorie value regardless of activity at time t, and X_t contains the associated feature vector over a window. The goal is to find a function $f(.)$ that best predicts Y_t from training data X_t. MET, widely used by clinicians and physiotherapists, assumes N clusters of activity $\mathbf{A} = \{A_1, A_2, \ldots, A_N\}$ are known. A MET value is assigned to each cluster, together with anthropometric characteristics of individuals. The amount of activity-specific energy expended can then be estimated as $energy = 0.0175(kcal/kg/min) \times weight(kg) \times$ MET values [9].

4.1 Evaluation and Parameter Settings

In our experiments, we use non-linear SVMs with Radial Basis Function (RBF) kernels for activity classification and a linear support vector regressor for energy expenditure prediction. The libsvm [27] implementation was used in the experiments. We perform a grid search algorithm to estimate the hyper-parameters of the SVM. For testing, we implement leave-one-subject-out cross validation on the dataset. This process iterates through all subjects, and the average testing error and standard deviation of all iterations are reported. We use the root-mean-squared error (RMSE) as a standard evaluation metric for the deviation of estimated calorie values from the ground truth.

4.2 Quantitative Evaluation

Temporal Window Size - The accuracy of predicted calorie values is linked to the number of previous frames utilised for making the prediction. The test described in this section looks at the relation between window length on the one hand, and activity recognition and calorie prediction errors on the other. All the sequences are tested with various window frame lengths $w = \{450, 900, 1800\}$, corresponding to a 15, 30 and 60 s time slot. Table 1 illustrates the activity recognition rates and the average RMSEs for calorie prediction of different window length w.

In general, the best performance for recognising activities is achieved when a 15 s window is applied. This is particularly prominent for individually highly variable activity types. For example, the recognition rates for exercise and stretch are significantly lower when $w = 1800$. In these cases, data are likely to be better explained at a relatively small temporal interval, for which local temporal information are more descriptive. On the contrary, calorie values are better predicted using larger window sizes. Here, human body adaptation causes an exponential increase/decrease to a plateau in oxygen consumption until a steady state corresponding to the current activity is attained [28].

So far, we have applied the same window length for the prediction of calories and the detection of activities. In order to test how the estimated calorie value is influenced by the performance of action recognition, the proposed method is also tested with fixed window length $w = 1800$ for predicting calorific expenditure, whilst different window lengths are applied to achieve different activity recognition rates. We also compare this to an idealised case by assuming all the activities

Table 1. Activity recognition rate (%) and calorific expenditure prediction error (RMSE) with different window length. The best results in each activity are in bold.

	w	stand	sit	walk	wipe	vacuum	sweep	lying	exercise	stretch	clean	read	overall
Activity	450	**86.5**	77.6	88.3	69.4	79.0	76.5	**62.3**	39.2	61.1	**91.4**	38.9	**73.7**
	900	85.0	79.1	**89.4**	**71.9**	**81.1**	75.2	54.3	**40.3**	57.8	90.4	36.8	71.1
	1800	81.1	**79.7**	85.1	66.0	77.2	72.9	33.0	29.3	52.7	90.0	35.9	68.2
Calorie	450	1.41	1.12	0.76	1.23	1.19	1.63	1.95	3.37	2.91	1.57	1.68	1.55
	900	1.25	0.87	**0.76**	**1.09**	1.26	1.47	1.75	2.82	2.91	**1.46**	1.42	1.41
	1800	**0.92**	**0.76**	0.82	1.17	**1.19**	**1.40**	**1.54**	**2.16**	**2.81**	1.49	**1.32**	**1.31**

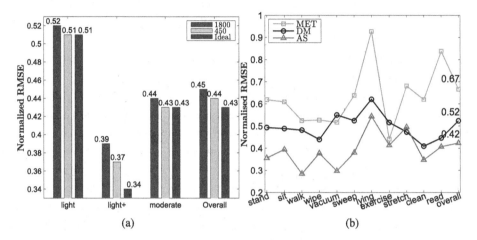

(a)   (b)

Fig. 5. Prediction Accuracy of Calorific Expenditure. (a) Average calorie prediction errors (Normalised RMSE) for different intensities and action recognition rates; (b) Average calorie prediction errors (Normalised RMSE) of three models.

are correctly recognised. For better visualisation, the 11 actions are grouped into three clusters based on their intensity level in Fig. 5(a) which summarises the calorie prediction error for different intensities and action recognition rates. We use normalised RMSE to facilitate the comparison between data with different scales.

Evaluation of Recurrent System Layout - To evaluate the use of recurrency, we set the activity-specific model using the sliding window technique above as our baseline method. We now introduce two methods, which are based on recurrent sliding window approaches. The first method (Recurrent1) uses the most recent predictions of the baseline method as input together with both visual features to predict current calorie value. Thus, it implements indirect recurrency utilising the predicted values from the baseline as recent predictions. The second method (Recurrent2) implements full recurrency, i.e. it uses its own output as recurrent input together with visual features.

Table 2 shows the effect of using recurrent information, with the best results for each activity highlighted. In general, indirect recurrency, Recurrent1,

Table 2. Average calorific expenditure prediction errors (RMSE) for each activity with different learning approaches. The best results in each activity are in bold.

	stand	sit	walk	wipe	vacuum	sweep	lying	exercise	stretch	clean	read	overall
Baseline	0.70	0.72	0.76	1.09	1.11	1.40	1.51	2.27	2.92	1.43	1.17	1.30
Recurrent1	0.61	**0.67**	**0.72**	**1.01**	**1.04**	**1.37**	**1.44**	2.11	2.69	**1.37**	**1.07**	**1.24**
Recurrent2	**0.60**	0.82	0.80	1.38	2.05	1.91	1.48	**1.95**	**2.48**	1.63	1.20	1.50

outperforms the other approaches at an average RMSE of 1.24. We note that the full recurrency, Recurrent2, suffers from drift and produces the worst results for half of the activities and also overall.

Model Comparison - We select the indirect recurrency model with the best window configurations as AS and analyse the performance: we compare AS with

Table 3. Ground truth and predicted calorie values in total per sequence and its accuracy and correlation. The best results for each sequence are in bold.

Sequence	GT	Prediction (Calories)			Accuracy %			Correlation		
		AS	DM	MET	AS	DM	MET	AS	DM	MET
1	59	71	83	76	**80.2**	61.8	71.3	**0.83**	0.80	0.66
2	89	80	82	78	**90.3**	90.2	88.2	**0.85**	0.81	0.57
3	74	81	90	69	90.1	80.7	**92.7**	**0.84**	0.78	0.63
4	79	48	46	43	**60.4**	54.9	55.0	**0.87**	0.79	0.78
5	37	39	41	28	**98.6**	87.9	77.6	**0.90**	0.88	0.77
6	89	86	83	107	**94.3**	93.4	79.9	**0.82**	0.73	0.63
7	101	96	94	114	**95.3**	90.1	87.6	**0.61**	0.57	0.61
8	39	42	41	35	91.9	**94.0**	84.4	**0.93**	0.42	0.57
9	82	76	85	94	92.8	**96.0**	85.3	**0.87**	0.78	0.71
10	49	68	77	76	**61.5**	45.1	45.2	0.54	**0.61**	0.42
11	28	38	40	38	65.5	62.9	**66.8**	**0.64**	0.62	0.56
12	98	88	91	79	90.3	**91.4**	80.8	0.56	0.61	**0.66**
13	56	66	82	77	**82.3**	55.3	62.7	**0.78**	0.76	0.62
14	141	84	77	74	**57.4**	54.9	52.8	0.86	**0.90**	0.60
15	40	41	40	30	**98.9**	98.9	74.5	**0.94**	0.93	0.94
16	29	31	30	38	**97.3**	94.3	69.1	**0.88**	0.84	0.81
17	81	85	93	100	**94.2**	88.8	76.0	**0.74**	0.70	0.70
18	65	86	87	94	**69.3**	66.1	54.7	**0.83**	0.70	0.48
19	92	89	84	101	**94.2**	90.6	90.6	0.75	**0.81**	0.72
20	63	83	82	86	66.9	**69.5**	64.4	**0.81**	0.72	0.41
Average	−	−	−	−	**82.9**	78.2	73.7	**0.80**	0.74	0.64

DM and MET against the ground truth (GT). For MET, we use the ground truth labels to select activities to keep this procedure identical to the commonly used manual estimate. Figure 5(b) shows the average normalised RMSE for each activity estimated by each of the models. Lower values indicate a lower residual. AS gives an average of 0.42 for normalised RMSE, which is 19% less than DM at 0.52 for normalised RMSE, and 36% less than MET at 0.67. The overall improvements are similar across all activities except for upper-body stretching, where DM is slightly better than AS.

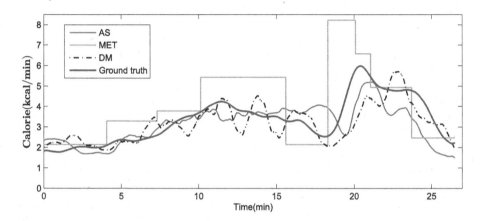

Fig. 6. Example of Calorie Uptake Prediction. In comparison to DM and MET, AS shows its ability to better predict calories and model the transition between activities.

Table 3 presents the detailed results for each sequence. The accuracy is calculated over the total calorie expended in each recording session. We also measure the correlation between the ground truth and the observed values. Note that the total calorie value for sequence 5, 8, 11, 15 and 16 are relatively low due to shorter sequences[2]. The proposed AS achieves higher accuracy and correlation in more sequences than DM and MET model based methods, and obtains better rates on average. Figure 6 illustrates an example (corresponding to sequence 6 in Table 3) of a visual trace of calorie values.

The proposed AS matches most closely to the ground truth while DM and MET diverge widely for some time periods. MET, as one of the most commonly used models in many fields, fails by its nature to capture the transition periods between activities, while the proposed AS model has the ability to capture transitions fairly well.

---

[2] This was caused by camera errors where only the first half of a sequence was saved.

5   Conclusion

This paper presented a system for estimating calorific expenditure from an RGB-D sensor. We demonstrated the effectiveness of the method through a comparative study of different models. The proposed activity-specific method used pooled spatial and temporal pyramids of visual features for activity recognition. Subsequently, we utilised a model for each activity built on a recurrent sliding window approach. To test the methodology, we introduced the challenging *SPHERE-calorie* dataset which covers a wide variety of home-based human activities comprising 20 sequences over 10 subjects. The proposed method demonstrates its ability to outperform the widely used METs based estimation approach. Possible future work includes taking into account anthropometric features. We hope this work, and the new dataset, will establish a baseline for future research in the area.

Acknowledgment. This work was performed under the SPHERE IRC project funded by the UK Engineering and Physical Sciences Research Council (EPSRC), Grant EP/K031910/1.

References

1. Samitz, G., Egger, M., Zwahlen, M.: Domains of physical activity and all-cause mortality: systematic review and dose-response meta-analysis of cohort studies. Int. J. Epidemiol. **40**, 1382–1400 (2011)
2. Ravussin, E., Lillioja, S., Anderson, T., Christin, L., Bogardus, C.: Determinants of 24-hour energy expenditure in man. Methods and results using a respiratory chamber. J. Clin. Investig. **78**, 1568 (1986)
3. Cosmed K4b2. http://www.cosmed.com/
4. Altini, M., Penders, J., Vullers, R., Amft, O.: Estimating energy expenditure using body-worn accelerometers: a comparison of methods, sensors number and positioning. IEEE J. Biomed. Health Inform. **19**, 219–226 (2015)
5. Chen, C., Jafari, R., Kehtarnavaz, N.: Improving human action recognition using fusion of depth camera and inertial sensors. IEEE Trans. Hum. Mach. Syst. **45**, 51–61 (2015)
6. Aggarwal, J., Xia, L.: Human activity recognition from 3D data: a review. Pattern Recogn. Lett. **48**, 70–80 (2014)
7. Zhu, N., Diethe, T., Camplani, M., Tao, L., Burrows, A., Twomey, N., Kaleshi, D., Mirmehdi, M., Flach, P., Craddock, I.: Bridging eHealth and the internet of things: the SPHERE project. IEEE Intell. Syst. **30**(4), 39–46 (2015)
8. Woznowski, P., et al.: A multi-modal sensor infrastructure for healthcare in a residential environment (2015)
9. Ainsworth, B., et al.: Compendium of physical activities: an update of activity codes and met intensities. Med. Sci. Sports Exerc. **32**, 498–504 (2000)
10. Guo, G., Lai, A.: A survey on still image based human action recognition. Pattern Recogn. **47**, 3343–3361 (2014)
11. Laptev, I.: On space-time interest points. Int. J. Comput. Vision **64**, 107–123 (2005)

12. Tao, L., Burghardt, T., Hannuna, S., Camplani, M., Paiement, A., Damen, D., Mirmehdi, M., Craddock, I.: A comparative home activity monitoring study using visual and inertial sensors. In: IEEE International Conference on E-Health Networking, Application and Services (2015)
13. Jia, Y., Shelhamer, E., Donahue, J., Karayev, S., Long, J., Girshick, R., Guadarrama, S., Darrell, T.: Caffe: Convolutional architecture for fast feature embedding, pp. 675–678 (2014)
14. Oreifej, O., Liu, Z.: Hon4d: Histogram of oriented 4D normals for activity recognition from depth sequences, pp. 716–723 (2013)
15. Tao, L., Paiement, A., Damen, D., Mirmehdi, M., Hannuna, S., Camplani, M., Burghardt, T., Craddock, I.: A comparative study of pose representation and dynamics modelling for online motion quality assessment. Comput. Vis. Image Underst. **148**, 136–152 (2016)
16. Laptev, I., Marszałek, M., Schmid, C., Rozenfeld, B.: Learning realistic human actions from movies, pp. 1–8 (2008)
17. Perronnin, F., Sánchez, J., Mensink, T.: Improving the fisher kernel for large-scale image classification. In: Daniilidis, K., Maragos, P., Paragios, N. (eds.) ECCV 2010. LNCS, vol. 6314, pp. 143–156. Springer, Heidelberg (2010). doi:10.1007/978-3-642-15561-1_11
18. Ryoo, M., Rothrock, B., Matthies, L.: Pooled motion features for first-person videos, pp. 896–904 (2015)
19. Aggarwal, J., Ryoo, M.: Human activity analysis: a review. ACM Comput. Surv. **43**, 16 (2011)
20. Presti, L.L., La Cascia, M.: 3D skeleton-based human action classification: a survey. Pattern Recogn. **53**, 130–147 (2016)
21. Edgcomb, A., Vahid, F.: Estimating daily energy expenditure from video for assistive monitoring, pp. 184–191 (2013)
22. Tsou, P.F., Wu, C.C.: Estimation of calories consumption for aerobics using kinect based skeleton tracking, pp. 1221–1226 (2015)
23. OpenNI organization: OpenNI User Guide (2010)
24. Tran, D., Sorokin, A.: Human activity recognition with metric learning. In: Forsyth, D., Torr, P., Zisserman, A. (eds.) ECCV 2008. LNCS, vol. 5302, pp. 548–561. Springer, Heidelberg (2008). doi:10.1007/978-3-540-88682-2_42
25. Dalal, N., Triggs, B.: Histograms of oriented gradients for human detection, vol. 1, pp. 886–893 (2005)
26. Dietterich, T.G.: Machine learning for sequential data: a review. In: Caelli, T., Amin, A., Duin, R.P.W., Ridder, D., Kamel, M. (eds.) SSPR /SPR 2002. LNCS, vol. 2396, pp. 15–30. Springer, Heidelberg (2002). doi:10.1007/3-540-70659-3_2
27. Chang, C., Lin, C.: Libsvm: a library for support vector machines. ACM Trans. Intell. Syst. Technol. **2**, 27 (2011)
28. McArdle, W., Katch, F., Katch, V.: Exercise physiology: energy, nutrition, and human performance. Med. Sci. Sports Exerc. **23**, 1403 (1991)

Emotion Understanding Using Multimodal Information Based on Autobiographical Memories for Alzheimer's Patients

Juan Manuel Fernandez Montenegro[1]([✉]), Athanasios Gkelias[2], and Vasileios Argyriou[1]

[1] Kingston University, Kingston upon Thames, England
{J.Fernandezmontenegro,Vasileios.Argyriou}@kingston.ac.uk
[2] Imperial College, London, England
a.gkelias@imperial.ac.uk

Abstract. Alzheimer Disease (AD) early detection is considered of high importance for improving the quality of life of patients and their families. Amongst all the different approaches for AD detection, significant work has been focused on emotion analysis through facial expressions, body language or speech. Many studies also use the electroencephalogram in order to capture emotions that patients cannot physically express. Our work introduces an emotion recognition approach using facial expression and EEG signal analysis. A novel dataset was created specifically to remark the autobiographical memory deficits of AD patients. This work uses novel EEG features based on quaternions, facial landmarks and the combination of them. Their performance was evaluated in a comparative study with a state of the art methods that demonstrates the proposed approach.

1 Introduction

Alzheimer Disease (AD) is a dementia characterized by the decline of various cognitive domains such as memory and learning ability, language expression difficulties or social cognition problems. Based on these symptoms and the importance of AD early detection, many research works are focused on the detection of those cognitive handicaps that characterise Alzheimer disease. This work is focused on the social cognition problems and memory related problems, in particular, these related with emotion expressions.

Many works are focused on studying dementia patients' capability to recognise emotions [10,11] whereas a minority tries to analyse patients' facial expressions to specific stimulus. In contrast to other dementias such as Lewy Body dementia where there is lack of facial expression, AD patients' facial expression

Electronic supplementary material The online version of this chapter (doi:10.1007/978-3-319-54407-6_17) contains supplementary material, which is available to authorized users.

C.-S. Chen et al. (Eds.): ACCV 2016 Workshops, Part I, LNCS 10116, pp. 252–268, 2017.
DOI: 10.1007/978-3-319-54407-6_17

is increased [14]. This work focuses on the automatic detection of emotions to certain stimulus for AD early detection.

Different approaches for automatic emotion recognition are focused on the variety of human interaction capabilities or biological data. For example, the study of speech and other acoustic cues in [18], body movements in [19], Electroencephalogram (EEG) in [25], facial expressions or combinations of previous ones such as speech and facial expressions in [22] or EEG and facial expressions in [20]. Our approach will focus on EEG and facial emotion detection.

The study of facial expression was part of various disciplines since Aristotelian era but it was in 1978 when the first automatic recognition study appeared [8,16]. Several techniques have been proposed for facial expressions interpretation. The most well known system is the Facial Action Coding System(FACS) [17]. FACS describes facial expressions as action units (AU), where each AU corresponds to a facial configuration. When it comes to the computational side of face analyses the known approaches can be classified as spatial or spatio-temporal and appearance or shape based. The first approach differentiates between methodologies that work with single images or with groups of successive frames. The second approach groups methods that use the appearance features of the face, such as pixel intensity and methods that use a description of the face shape. All of them face the same challenges, such as head-pose and illumination variations, registration errors, occlusions and identity bias. Some of these problems are not included in most of the available databases therefor some of them may not work properly on real conditions.

Several datasets, focusing on different applications, are available for emotion recognition. For example, DEAP dataset provides EEG and face recordings of participants while they watch musical videos just for the analysis of human affective states [9]; SEMAINE database aims to provide voice and facial information to study the behaviour of subjects interacting with virtual avatars [23]; MAHNOB-HCI database was created for the study of emotions while humans are watching multimedia, supplying several data such as audio, an RGB video and five monochrome videos of the face, EEG, ECG, respiration amplitude, skin temperature and eye-gaze data [21]; or CASMEII dataset which studies facial micro-expressions for security and medical applications, requiring cameras of higher frame rate and spatial resolution [24].

It has been proved that for AD patients, semantic, autobiographical and implicit memory are more preserved than recent memory; therefore our work is based on the subjects' autobiographical memory [12,13,15]. Thus a novel dataset was created based on these symptoms providing RGB, IR and Depth video data of the participants' faces, EEG and eye-gaze data.

The purpose of this work is to introduce human behaviour and face expression recognition techniques for the detection of early dementia symptoms. Our novel dataset contains recordings of the participants' reactions when specific images, related and unrelated with their personal life stories, are shown. The classification of different reactions related to the images displayed is performed using different data features included in our dataset, such as facial landmarks and EEG signals, as input to supervised learning approaches. Our study analyses expected emotions. Thus our classification is based on the expected emotions according to the images displayed during the test instead of classifying accordingly to the emotions felt (represented on the captured video). This work investigates healthy people to analyse the differences and level of the emotional inputs generated from the available image classes and generate a model that describes the reactions associated to the healthy group of people. Thus, every reaction detected out of this model could be considered as a possible sign of dementia.

The remainder of this paper is organized as follows: Sect. 2 describes previous related work on behaviour and face expression recognition. Section 3 analyses the proposed methodology and in Sect. 4 details on the evaluation process and the obtained results are presented. Section 5 gives some conclusion remarks.

2   Previous Work

In this section current state of the art facial and EEG based emotion recognition approaches are analysed.

2.1   Facial Emotion Recognition Approaches

Images and video sequences of faces are highly utilised as source for emotion recognition. There are several models to represent emotions and they define emotions according to the number of dimensions, such as the three dimension Schlosberg Model: pleasantness-unpleasantness, attention-rejection and sleep-tension [40]. Most of the facial recognition approaches use the Facial Action Coding System (FACS) [17] to describe facial human emotions such as happiness, sadness, surprise, fear, anger or disgust; where each of these emotions is described as a combination of AUs. Other approaches abandon the path of specific emotions recognition and focus on emotions' dimensions, measuring their valence, arousal or intensity [22,41,46].

The methods for facial emotion recognition can be classified according to the approaches used during the recognition stages: registration, features selection, dimensionality reduction or classification/recognition [8,16].

Three different approaches can be used for face registration: whole face, parts or points registration. These registration approaches usually are based on Active Appearance Models (AAM) [27,28]; a method that matches a statistical model of the face to the images to extract face landmarks and specific face areas. Whole face approaches get the features from the whole face. Littlewort et al. [42] get image based features of the whole face, such as Gavor Wavelets, in order to detect AUs for pain recognition. Face parts approaches use face areas that contain the maximum amount of information related to face expressions, such as the eyebrows and the mouth. Nicolle et al. [41] propose a method for emotion recognition (valence, arousal, expectancy and power) using a combination of whole face, face parts, points and audio features. This approach gets patches of the face on regions of interest and they use the log-magnitude Fourier spectra and other measures as features. Points based approaches use fiducial points for shape representation. Michel et al. [33] use a tracker to get 22 fiducial points and calculate the distance of each point between a neutral and a peak frame. These distances are used as features of an Support Vector Machine (SVM) algorithm in order classify the emotions. Neutral and peak frames are automatically detected when the motion of the points is almost zero. Valstar et al. uses Particle Filtering Likelihoods [39] in order to extract 20 fiducial points, but they still have to select the initial position of these points manually. These points are normalised by respecting a neutral point(tip of the nose) and a scale transformation is also applied. The distances between certain points are used as features to recognise specific AUs using SVM.

When it comes to feature representation, methods can be divided in spatial and spatio-temporal approaches. Spatial approaches include shape representations, low-level histograms or Gabor representations amongst others. For example, Huang et al. [43] proposed a spatial shape representation using groups of three fiducial points (triangular features) as input to a neural network classifier; and Sariyanidi et al. presented in [44] a low-level histogram representation using local Zernike moments for emotion recognition based on kNN and SVM classifiers. On the other hand, spatio-temporal approaches get the features from a range of frames within a temporal window, detecting more efficiently emotions that cannot be easily differentiated in spatial approaches. Zhao et al. [45] proposed a method that uses spatio-temporal local binary patterns as features and SVM for classifying facial expressions.

Once the features are selected, dimensionality reduction techniques such as PCA are used before classification in order to reduce challenges such as illumination variation, registration errors and identity bias.

The results from most of the approaches are not always reliable since many of them are tested on posed datasets such as CK [37] and MMI [38]. Therefore, the results are not reliable on naturalistic conditions regarding illumination, head-pose variations and nature of expressions. Nevertheless, there are non-posed datasets to test naturalistic expressions such as SEMAINE [23], MAHNOB-HCI [21] or DECAF [52]. In these cases the illumination and head-pose variation problems are taken into account depending on the aim of the study.

2.2 EEG Emotion Recognition Approaches

EEG based emotion recognition is a less common approach since the majority uses facial or speech data as source for emotion detection. Considering that these sources are easy to fake [25] amongst other problems, EEG provides an extra source that solves problems such as falseness, illumination or speech impaired subjects. On the other hand, EEG signal deals with other challenges such as noise and biological and non-biological artifacts [20,30], such as electrooculogram (EOG), electromyogram (EMG) and electrocardiogram (ECG). Nevertheless, these biological artifacts are also affected by emotions and are expected to provide extra information to EEG signal for emotion recognition [20].

Two types of descriptors can be used for EEG signal analysis: simple descriptors such as frequency and amplitude; and more complex ones such as asymmetry metrics, time/frequency analysis, topographic mapping, coherence analysis or covariation measures. These descriptors are used depending on the area of study; for example, asymmetry metrics are usually applied in cognitive neuroscience [30]. In particular, asymmetric hemispheric differences are used for emotion recognition [29,31]. Furthermore, state of the art methods use techniques such as Independent Component Analysis (ICA) for removing some artifacts, then they extract different features such as amplitude or spectral power and use them in classifiers such as k-Nearest Neighbour (kNN) or Support Vector Machine (SVM). For example, Vijayan et al. [26] use DEAP data (data captured using 32 sensors) and filter 50 Hz frequency to remove noise. Afterwards, they get the Gamma band from the signal and use auto-regressive modeling to obtain the features that are passed to an SVM classifier.

2.3   Facial and EEG Emotion Recognition Approaches

Few approaches utilise a combination of EEG and facial information to recognise emotions. The work in [20] considers both types of data using the MAHNOB-HCI database [21]. The EEG signal was captured using 32 sensors and the power spectral density was extracted from overlapping one second windows. The facial approach extracts 49 fiducial points and calculates the distance from 38 of these points to a reference point. Finally, they use regression models for emotion detection. As a result, they have obtained better results using the facial data and conclude that the good performance of the EEG results are due to the facial artifacts present in the EEG signal.

In this work a novel multimodal non-posed database is introduced. Due to the nature of our study, the environment where the RGB video is recorded is controlled avoiding illumination variations and occlusions. In addition, the head-pose variations are minimal since the video sequences are recorded while the participants are looking at the screen in front of them. Using this novel multimodal database a method based on expected emotions is presented. These emotions are not defined as specific standard emotions, therefore our approach does not use FACS or any other emotion coding system. The facial modality presented uses geometric based spatio-temporal features. For the EEG data a new feature is introduced based on quaternion principal component analysis using only four channels. Both modalities individually and combined are studied and compared with with state of the art methods.

3   Proposed Dataset and Methodology

This section describes the approach utilised to recognize the spontaneous reactions to specific visual stimulus. Next subsections describe our novel multimodal database and the proposed features used for emotion recognition.

3.1   Spontaneous Emotion Multimodal Database (SEM-db)

SEM database is a multimodal dataset for spontaneous emotional reaction recognition that contains multimodal information of nine participants aged between 30 to 60 years old with different educational background taken while completing cognitive/visual tests. Ten repetitions have been recorded per participant providing a total of 90 instances.

The novelty of SEM dataset is the non-posed reactions to autobiographical and non-autobiographical visual stimulus data. The main contribution of SEM database is the use of personalized images for each participant. These images are photos of themselves or their relatives and friends both from the recent and distant past. Moreover, the participants did not know that those images were used so the reactions were genuine. Additionally, images of famous and unknown to the subjects persons or places were shown. In more details we had the following classes of images with the corresponding expected spontaneous emotions or reactions.

(a) 10 images of distant past faces of the subjects and their relatives.
(b) 10 images of recent past faces of the subjects and their relatives.
(c) 10 images of distant past group of relatives, including themselves.
(d) 10 images of recent past group of relatives, including themselves.
(e) 10 images of famous people.
(f) 10 images of unknown to the subject persons.
(g) 10 images of famous places/objects.
(h) 10 images of unknown to the subject places/objects.

The recorded data is provided in different data modalities: HD RGB, depth and IR frames of the face, EEG signal and eye gaze data; which were recorded

Fig. 1. Data modalities contained in the database and the related classes analysed in our approach (see Table 1 for the emotion definitions). The left figure shows from top to bottom, images of people from distant vs recent past; and famous vs unknown people. The right figure shows from top to bottom images of group of people from distant vs recent past; and famous vs unknown places. The different modalities from left to right in each case are EEG, gaze tracked heat map, RGB, facial landmarks, depth and IR (Color figure online).

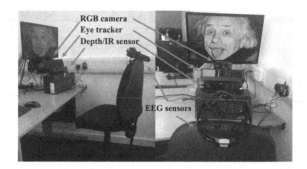

Fig. 2. Location of the devices during the recording of the database.

using 4 different devices: a 30fps HD RGB camera, IR/Depth sensors (Kinect), an eye tracker (Tobii eye tracker) and EEG sensors (Emotiv headset)(see Fig. 1). The recording of the data has been done in done in a controlled environment e.g. an office. The participants were asked to put on the EEG headset and they were seated in a comfortable chair in front of the test screen, the RGB camera, the Kinect sensor and the eye tracker. The height of their chair was adjusted in order the eye tracker to detect their eye movements (see Fig. 2). Once the eye tracker is detecting the participants' eyes and all the EEG sensors are receiving good quality signal the test begins. The instructions of the test are provided before they start and at the beginning of each test, a red image is displayed for synchronization.

3.2 Emotion Recognition Using Fiducial Points and EEG Quaternion Based Supervised Learning

Our approach intends to classify the reaction of the participants using two data modalities: the EEG data and the fiducial points obtained from the RGB face images. Using each modality and combining them (see Fig. 3), two binary classifications have been performed trying to recognise spontaneous reactions from distant and recent memories that were triggered during our experiments (see Table 1). The main reaction to be detected is the 'positive recognition' reaction versus the 'indifference' reaction. Additionally, it is expected a stronger recognition reaction when the participant watches images from the distant past.

Our approach extracts features from both data modalities: EEG and Facial points. The facial fiducial points were obtained using Baltru et al. approach [28] from a 30 frame rate video. This approach obtains 68 fiducial points per frame. The coordinates of the fiducial points have been preprocessed normalizing them according to a neutral face point (i.e. nose) [32] to obtain rigid head motions invariant features. The EEG data was recorded using a EEG headset (Emotiv Epoc) which collects EEG from 14 sensors at 128 Hz.

Once the data was collected and preprocessed, the features were extracted. The spatio-temporal facial features studied were based on Michel et al. work [33].

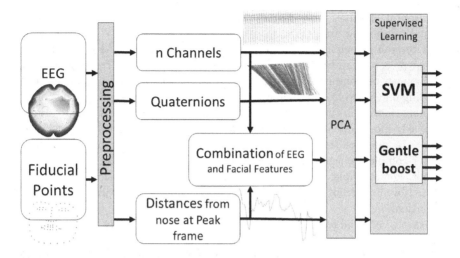

Fig. 3. Diagram demonstrating our approach. The two modalities of data used (Fiducial Points, EEG) go through the process independently. The combination of the features concatenates the features extracted from the fiducial points with the EEG features, including all combinations.

Table 1. Classes chosen for recognition and the expected reaction.

Id	Class 1	Class 2	Expected emotion
1	Famous faces	Unknown faces	Recognition vs Neutral reaction
2	Distant past images of the participant family and friends faces	Recent past images of the participant family and friends faces	Long term memory recognition vs short term memory recognition
3	Distant past images of group of people including the participant family and friends faces	Recent past images of group of people including the participant family and friends faces	Long term memory recognition vs short Term memory recognition
4	Famous places, objects/brands	Unknown places and objects	Recognition vs Neutral reaction

The distance of each coordinate to the nose point was measured. The first frame of each subject data was considered as neutral face since at the beginning of the test a neutral pose is expected. Each frame was compared to the neutral face, calculating the frame that varies most from the neutral face. That frame was selected as the peak frame and used as a p points long feature vector. For EEG, three variations of features were analysed: (i) a combination of the 14 channels, (ii) a combination of the four frontal channels and (iii) novel features combining the 4 frontal channels into a quaternion representation based on quaternion principal component analysis.

Quaternion principal component analysis (Quaternion PCA) is based on the fact that a vector can be decomposed in linearly independent components, such that they can be combined linearly to reconstruct the original vector. However, depending on the event that changes the vector, correlation between the components may exist from the statistical point of view (i.e. two uncorrelated variables are linearly independent but two linearly independent variables are not uncorrelated). In most of the cases during the feature extraction process complex or hyper-complex features are generated but decomposed to be computed by a classifier. For example, normals and gradients in 2D/3D are features that are consisted by more than one element and this decomposition can imply a loss of information.

To do so, vectorial features can be represented more precisely using a complex or hyper-complex representation [47,48]. Since, in our case and many similar scenarios, vectorial features such as a location, speed, gradients or angles, are the primary source of information, a hyper-complex representation of these features is more efficient allowing better correlation between these channels [47–49]. The proposed method exploits the hyper-complex (quaternion) representation capturing the dependencies within the EEG sensors located on the sides of the head and the ones over the eyes, [50,51].

Quaternion PCA is applied in order to reduce the number of the selected hyper-complex features without increasing the complexity. In more details, the quaternion representation was introduced in [35,36] as a generalization of the complex numbers. A quaternion $q \in \mathcal{H}$ has four components:

$$q = q_r + q_i i + q_j j + q_k k \tag{1}$$

where $q_r, q_i, q_j, q_k \in \Re$ and i, j, and k satisfy

$$
\begin{aligned}
i^2 = j^2 = k^2 = -1, \; ij = -ji = k \\
jk = -kj = i, ki = -ik = j
\end{aligned}
\tag{2}
$$

Conjugation of quaternions denoted by H is analogous to conjugation of complex numbers elements and is defined as:

$$q^H = q_r - q_i i - q_j j - q_k k. \tag{3}$$

The square of the norm of a quartenion is defined as

$$||q||^2 \doteq q_r^2 + q_i^2 + q_j^2 + q_k^2 = q^H q. \tag{4}$$

with $(q_1 q_2)^H = q_2^H q_1^H$ and the four components (q_r, q_i, q_j, q_k) to correspond to the available four frontal EEG channels (AF3, AF4, F7 and F8).

Let quaternion column vector $\mathbf{q} = [q_1, \ldots, q_F]^T \in \mathcal{H}^F$ where T denotes simple transposition be the EEG values over time. The conjugate transpose of vector $\mathbf{q}$ is denoted by $\mathbf{q}^H$. There is an isomorphy between a quaternion and a complex 2×2 matrix defined as

$$\mathbf{Q} = \begin{bmatrix} q_r + q_i i & q_j + q_k i \\ -q_j + q_k i & q_r - q_i i \end{bmatrix} \tag{5}$$

Let $\mathbf{x}_l$ be the F-dimensional vector obtained by writing in lexico- graphic ordering and form $\mathbf{X} = [\mathbf{x}_1|\cdots|\mathbf{x}_N] \in \mathcal{H}^{F\times N}$. Also we denote by $\bar{\mathbf{x}} = \frac{1}{N}\sum_{i=1}^{N}\mathbf{x}_i$ and $\overline{\mathbf{X}}$ the sample mean and the centralized sample matrix $\mathbf{X}$, respectively. A projection vector is denoted by $\mathbf{u} \in \mathcal{H}^F$ and by $y_i = \mathbf{u}^H\mathbf{x}_i$ the projection of $\mathbf{x}_i$ onto $\mathbf{u}$. We want to maximize the (sum of the) variances of the data assigned to a particular class

$$\begin{aligned} E(\mathbf{u}) &= \sum_{l=1}^{N}||y_l - \tilde{m}||^2 = \sum_{l=1}^{N}||\mathbf{u}^H(\mathbf{x}_i - \mathbf{m})||^2 \\ &= \mathbf{u}^H\sum_{l=1}^{N}(\mathbf{x}_l - \mathbf{m})(\mathbf{x}_l - \mathbf{m})^H\mathbf{u} \\ &= \mathbf{u}^H\mathbf{S}\mathbf{u} \end{aligned} \tag{6}$$

where $\mathbf{S} = \overline{\mathbf{X}}\overline{\mathbf{X}}^H$. It can be easily proven that matrix $\mathbf{S}$ is a quaternion Hermitian matrix i.e., $S_{ij} = S_{ji}^H$.

In order to find K projections $\mathbf{U} = [\mathbf{u}_1|\ldots|\mathbf{u}_k] \in \mathcal{H}^{F\times K}$ we may generalize $E(\mathbf{U})$:

$$\begin{aligned} \mathbf{U}_o &= \arg\max_{\mathbf{U}\in\mathbb{H}^{F\times p}} E(\mathbf{U}) \\ &= \arg\max_{\mathbf{U}\in\mathbb{H}^{F\times p}} \text{tr}[\mathbf{U}^H\mathbf{S}\mathbf{U}] \\ \text{s.t. } \mathbf{U}^H\mathbf{U} &= \mathbf{I}. \end{aligned} \tag{7}$$

We aim at solving the above noted problem by using the isomorphic complex form that can be reformulated as

$$\begin{aligned} \tilde{\mathbf{U}}_o &= \arg\max_{\tilde{\mathbf{U}}} \text{tr}[\tilde{\mathbf{U}}^H\tilde{\mathbf{S}}\tilde{\mathbf{U}}] \\ \text{s.t. } \tilde{\mathbf{U}}^H\tilde{\mathbf{U}} &= \mathbf{I}. \end{aligned} \tag{8}$$

Since $\mathbf{S}$ is a quaternion Hermitian matrix, $\tilde{\mathbf{S}}$ is a complex Hermitian. Also, given that $\tilde{\mathbf{S}}$ is a positive semidefinite Hermitian matrix (i.e., it has only non-negative eigenvalues) the solution $\tilde{\mathbf{U}}_0$ is given by the p eigenvectors of $\tilde{\mathbf{S}}$ that correspond to p largest eigenvalues. We want an efficient algorithm for performing eigen-analysis to $\tilde{\mathbf{S}}$, which is a complex $2F\times2F$ matrix and can be written as $\tilde{\mathbf{S}} = \tilde{\mathbf{X}}\tilde{\mathbf{X}}^H$ where $\tilde{\mathbf{X}} \in \mathbb{C}^{2n\times F}$ and needs $O((2F)^3)$ time.

In general, given a quaternion Hermitian matrix $\mathbf{A}$ then it has n nonnegative real eigenvalues (due to the non-commutative multiplication property of quaternions, there exist two kinds of eigenvalue; in this paper we are interested only on the left eigenvalues) $\mathbf{l} = [\sigma_1,\ldots,\sigma_n]$. Let $\tilde{\mathbf{A}}$ be its complex form

$$\tilde{\mathbf{A}} = \begin{bmatrix} \mathbf{A}_r + i\mathbf{A}_i & \mathbf{A}_j + i\mathbf{A}_k \\ -\mathbf{A}_j + i\mathbf{A}_k & \mathbf{A}_r - i\mathbf{A}_i. \end{bmatrix}$$

then the eigenvalues of $\mathbf{l}_{2n} = [\sigma_1,\sigma_1,\ldots,\sigma_n,\sigma_n]$. Representing $\mathbf{A} = \mathbf{B}\mathbf{B}^H$, where $\mathbf{B}$ is a quaternion matrix, and considering $\tilde{\mathbf{A}}$ and $\tilde{\mathbf{B}}$ to be the complex forms of matrices $\mathbf{A}$ and $\mathbf{B}$, respectively, then, $\tilde{\mathbf{A}}$ will be given by $\tilde{\mathbf{A}} = \tilde{\mathbf{B}}\tilde{\mathbf{B}}^H$. So, based on this analysis, we can write $\tilde{\mathbf{S}} = \bar{\mathbf{X}}\bar{\mathbf{X}}^H$. Also by defining matrices $\mathbf{A}$ and $\mathbf{B}$ such that $\mathbf{A} = \mathbf{\Gamma}\mathbf{\Gamma}^H$ and $\mathbf{B} = \mathbf{\Gamma}^H\mathbf{\Gamma}$ with $\mathbf{\Gamma} \in \mathcal{C}^{m \times r}$, and considering $\mathbf{U}_A$ and $\mathbf{U}_B$ to be the eigenvectors corresponding to the non-zero eigenvalues $\mathbf{\Lambda}_A$ and $\mathbf{\Lambda}_B$ of $\mathbf{A}$ and $\mathbf{B}$, respectively, we finally obtain $\mathbf{\Lambda}_A = \mathbf{\Lambda}_B$ and $\mathbf{U}_A = \mathbf{\Gamma}\mathbf{U}_B\mathbf{\Lambda}_A^{-\frac{1}{2}}$.

Thus, according to the above, in a classification problem, we may represent the quaternion Hermitian matrix (descriptor) providing a subspace analysis method in the quaternion domain. Assuming that we have a quaternion matrix P with dimension $m \times n$, we consider n to be the total number of the captured data and m the number of the actual hyper-complex features. A quaternion PCA of P, as it was analysed above, seeks a solution that contains r $(r < m, n)$ linearly independent quaternion eigenvectors in the columns of Q $(m \times r)$ such that $P = QA$; where the rows of A $(r \times n)$ contain the r quaternion principal component (QPC) series. As a result, a solid representation of the selected quaternion features is obtained, while the computational complexity is low.

Besides the individual features modalities, a combination of the aforementioned EEG and facial features has been also analysed. This combination comprises the attachment of the EEG features vector to the facial one. Once the features are structured properly, dimensionality reduction is applied using PCA and the reduced features are used as input to two supervised learning algorithms: SVM and GentleBoost. A leave one out approach is used so the features obtained from N-1 of the participants, being N the number of participants, are used for training and the remaining participant data are used for testing. Moreover, k-fold cross-validation has been applied so the final results are the average of all the folds.

4   Results

This section shows and analyses the classification results obtained using the EEG and Facial approaches presented in the previous section using SVM and gentleboost classifiers. The results are represented by the F1 score which is a measure of accuracy that takes into account the precision and recall. A leave one out approach and a k-fold cross validation is applied for all the participants in our database (http://staffnet.kingston.ac.uk/~ku43576/?page_id=414). These results are compared with the ones obtained using as features the suggested in [20].

Tables 2 and 3 show the F1 scores for all the modalities and both classifiers, SVM and gentleboost, respectively. Also, the precision and recall values are shown in Table 4, while an overview of the best outcomes is presented in Fig. 4. Furthermore, the ROC curves of the proposed method based on facial features in comparison to the ones proposed by Soleymani is shown in Fig. 5. The results of both individual modalities (EEG and facial) are coherent and adequate for the detection of emotions with overall F1 values around 70%. Comparing both data modalities, facial fiducial landmarks provide slightly better results than EEG signal for both classifiers; and the combination of both modalities only improves slightly the results using the gentleboost classifier. These results are in alignment

Table 2. F1 scores obtained using SVM. See Table 1 for id definitions.

SVM		id 1	id 2	id 3	id 4	Overall
EEG						
Soleymani [20]		0.6002	0.5677	0.6194	0.7122	0.6249
Proposed	14 Ch	0.5972	0.6882	0.6507	0.6704	0.6516
	4 Ch	0.6637	0.6965	0.7177	0.6725	0.6876
	Quaternion	**0.7105**	**0.7043**	**0.7225**	**0.7553**	**0.7232**
Face						
Soleymani [20]		0.6235	0.6699	0.6722	**0.6942**	0.6650
Proposed	Dist	**0.6987**	**0.8028**	**0.7750**	0.6438	**0.7301**
EEGFace						
Soleymani [20]		0.6429	0.7090	0.6502	0.6461	0.6620
Proposed	Dist + 14 Ch	**0.6950**	0.6825	0.7452	0.7313	0.7135
	Dist + 4 Ch	0.6887	0.7470	**0.7843**	0.7319	0.7380
	Dist + Quaternion	**0.6945**	**0.7699**	0.7774	**0.7372**	**0.7448**

Table 3. F1 scores obtained using Gentleboost. See Table 1 for id definitions.

Boost		id 1	id 2	id 3	id 4	Overall
EEG						
Soleymani [20]		0.6891	0.6843	0.6515	0.7540	0.6947
Proposed	14 Ch	0.7030	0.6508	0.6901	0.6670	0.6777
	4 Ch	0.7061	0.6622	0.7792	0.7035	0.7128
	Quaternion	**0.7297**	**0.6861**	**0.7439**	**0.7565**	**0.7291**
Face						
Soleymani [20]		**0.7068**	0.7362	0.7295	0.6579	0.7076
Proposed	Dist	0.6200	**0.7934**	**0.8024**	**0.7481**	**0.7410**
EEGFace						
Soleymani [20]		**0.7146**	0.7296	0.6711	0.7412	0.7141
Proposed	Dist + 14 Ch	0.6444	0.7084	0.7841	0.7148	0.7129
	Dist + 4 Ch	0.6327	**0.7694**	0.7871	**0.7680**	**0.7393**
	Dist + Quaternion	0.6753	0.6911	**0.8145**	0.7077	0.7222

with the results obtained by [8]. On the other hand, the emotions related with unknown and known people or places have been recognised with higher accuracy using EEG features. We assume that this is due to a minimal difference on facial expressions during the recognition of famous, but not personally related, versus the unknown people or places.

In EEG, the use of 4 channels provides similar results with the 14 ones. The proposed quaternion based features improves the overall results by more than 1%.

Table 4. Best precision and recall values of our approach. corresponding to EEG Quaternion. Facial distance and Distance plus Quaternion features. in comparison with the ones obtained by [20] features.

		SVM					Boost				
		id 1	id 2	id 3	id 4	OA	id 1	id 2	id 3	id 4	OA
		EEG					EEG				
[20]	Prec	0.664	0.687	0.733	0.735	0.705	0.749	0.721	0.714	0.787	0.743
	Rec	0.644	0.622	0.655	0.716	0.659	0.705	0.694	0.672	0.761	0.708
Quat	Prec	0.758	0.737	0.746	0.784	**0.757**	0.751	0.723	0.783	0.780	**0.759**
	Rec	0.722	0.711	0.727	0.761	**0.730**	0.733	0.694	0.750	0.761	**0.734**
		Face					Face				
[20]	Prec	0.716	0.713	0.727	0.807	0.741	0.820	0.749	0.806	0.762	0.784
	Rec	0.661	0.722	0.705	0.722	0.702	0.738	0.777	0.750	0.705	0.743
Dist	Prec	0.749	0.849	0.838	0.710	**0.787**	0.684	0.823	0.860	0.786	**0.788**
	Rec	0.711	0.816	0.794	0.683	**0.751**	0.644	0.800	0.816	0.755	**0.754**
		EEGFace					EEGFace				
[20]	Prec	0.797	0.730	0.708	0.711	0.737	0.751	0.742	0.790	0.808	**0.773**
	Rec	0.688	0.733	0.688	0.683	0.698	0.744	0.772	0.705	0.755	**0.744**
Dist +	Prec	0.764	0.806	0.806	0.767	**0.786**	0.722	0.736	0.847	0.735	0.760
Quat	Rec	0.711	0.777	0.783	0.744	**0.754**	0.688	0.705	0.822	0.716	0.733

The proposed facial features also provide better F1 scores than the ones used in [20] in most of the classification scenarios. On the other hand, the results of the combined features are not always consistent in terms of which combination is the best one.

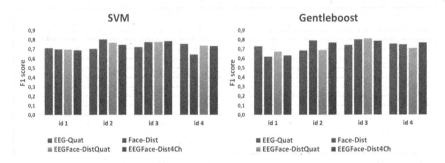

Fig. 4. Results of the features that provide the best results for each classification using SVM and Gentleboost classifier: EEG quaternion, Face distance and EEGFace distances plus quaternion.

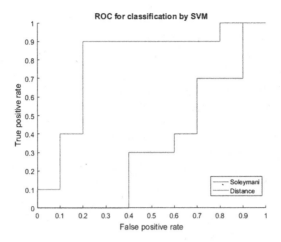

Fig. 5. ROC curve of the proposed method based on facial features in comparison to the ones proposed by Soleymani.

5   Conclusion

A novel database (SEM-db) has been created focusing on natural reactions to specific autobiographical and non-autobiographical stimulus that intend to elicit different emotions. This database provides facial videos and EEG signals, amongst other information, that can be used for emotion recognition. Using this database this work presents an approach for expected emotion recognition based on novel feature descriptors. The novel quaternion EEG and facial features result accurate classification rates. The overall results demonstrate that facial features outperform the EEG ones for emotion recognition.

References

1. Alpher, A.: Advances in frobnication. J. Foo **12**, 234–778 (2002)
2. Alpher, A., Fotheringham-Smythe, J.P.N.: Frobnication revisited. J. Foo **13**, 234–778 (2003)
3. Herman, S., Fotheringham-Smythe, J.P.N., Gamow, G.: Can a machine frobnicate? J. Foo **14**, 234–778 (2004)
4. Smith, F.: The Frobnicatable Foo Filter. GreatBooks, Atown (2009)
5. Wills, H.: Frobnication tutorial. Technical report CS-1204, XYZ University, Btown (1999)
6. Rosler, A., Mapstone, M.E., Hays, A.K., Mesulam, M., Rademaker, A., Gitelman, D.R., Weintraub, S.: Alterations of visual search strategy in Alzheimer's disease and aging. Neuropsychology **14**(3), 398–408 (2000)
7. Pereira, M.L., Camargo, M.V.Z.A., Aprahamian, I., Forlenza, O.V.: Eye movement analysis and cognitive processing: detecting indicators of conversion to Alzheimer's disease. Neuropsychiatric Dis. Treat. **10**, 1273–1285 (2014)

8. Alpher, A.: Automatic analysis of facial affect: a survey of registration, representation, and recognition. IEEE Trans. Pattern Anal. Mach. Intell. **37**(6), 1113–1133 (2015)
9. Koelstra, S., Muehl, C., Soleymani, M., Lee, J.S., Yazdani, A., Ebrahimi, T., Pun, T., Nijholt, A., Patras, I.: Deap: a database for emotion analysis; using physiological signals. IEEE Trans. Affect. Comput. **3**(1), 18–31 (2012)
10. Sapey-Triomphe, L.A., Heckemann, R.A., Boublay, N., Dorey, J.M., Hnaff, M.A., Rouch, I., Padovan, C.: Neuroanatomical correlates of recognizing face expressions in mild stages of Alzheimers disease. PLoS ONE **10**(12) (2015)
11. Van den Stock, J., De Winter, F.L., de Gelder, B., Rangarajan, J.R., Cypers, G., Maes, F., Sunaert, S., Goffin, K., Vandenberghe, R., Vandenbulcke, M.: Impaired recognition of body expressions in the behavioral variant of frontotemporal dementia. Neuropsychologia **75**, 496–504 (2015)
12. Han, K.H., Zaytseva, Y., Bao, Y., Pppel, E., Chung, S.Y., Kim, J.W., Kim, H.T.: Impairment of vocal expression of negative emotions in patients with Alzheimers disease. Front. Aging Neurosci. **6**(101), 1–6 (2014)
13. Irish, M., Hornberger, M., Lah, S., Miller, L., Pengas, G., Nestor, P.J., Hodges, J.R., Piguet, O.: Profiles of recent autobiographical memory retrieval in semantic dementia, behavioural-variant frontotemporal dementia, and Alzheimers disease. Neuropsychologia **49**(9), 2694–2702 (2011)
14. Seidl, U., Lueken, U., Thomann, P.A., Kruse, A., Schrder, J.: Facial expression in Alzheimers disease impact of cognitive deficits and neuropsychiatric symptoms. Am. J. Alzheimer's Dis. Other Dementias **27**(2), 100–106 (2012)
15. American Psychiatric Association: Diagnostic and statistical manual of mental disorders (DSM-5). American Psychiatric Association Publishing (2013)
16. Bettadapura, V.: Face expression recognition and analysis: the state of the art. Tech. Report, pp. 1–27. arXiv:1203.6722 (2012)
17. Ekman, P., Friesen, W.V.: The Facial Action Coding System: A Technique for The Measurement of Facial Movement. Consulting Psychologists Press, San Francisco (1978)
18. Weninger, F., Wllmer, M., Schuller, B.: Emotion recognition in naturalistic speech and language a survey. In: Emotion Recognition: A Pattern Analysis Approach, pp. 237–267 (2015)
19. Chowdhuri, M.A.D., Bojewar, S.: Emotion detection analysis through tone of user: a survey. Emotion **5**(5), 859–861 (2016)
20. Soleymani, M., Asghari-Esfeden, S., Fu, Y., Pantic, M.: Analysis of EEG signals and facial expressions for continuous emotion detection. IEEE Trans. Affect. Comput. **7**(1), 17–28 (2016)
21. Soleymani, M., Lichtenauer, J., Pun, T., Pantic, M.: A multimodal database for affect recognition and implicit tagging. IEEE Trans. Affect. Comput. **3**(1), 42–55 (2012)
22. Nicolaou, M.A., Gunes, H., Pantic, M.: Continuous prediction of spontaneous affect from multiple cues and modalities in valence-arousal space. IEEE Trans. Affect. Comput. **2**(2), 92–105 (2011)
23. McKeown, G., Valstar, M., Cowie, R., Pantic, M., Schroder, M.: The semaine database: annotated multimodal records of emotionally colored conversations between a person and a limited agent. IEEE Trans. Affect. Comput. **3**(1), 5–17 (2012)
24. Yan, W.J., Li, X., Wang, S.J., Zhao, G., Liu, Y.J., Chen, Y.H., Fu, X.: CASME II: an improved spontaneous micro-expression database and the baseline evaluation. PLoS ONE **9**(1), e86041 (2014)

25. Lokannavar, S., Lahane, P., Gangurde, A., Chidre, P.: Emotion recognition using EEG signals. Emotion **4**(5), 54–56 (2015)
26. Vijayan, A.E., Sen, D., Sudheer, A.P.: EEG-based emotion recognition using statistical measures and auto-regressive modeling. In: IEEE International Conference on Computational Intelligence and Communication Technology (CICT), vol. 14(1), pp. 587–591 (2015)
27. Cootes, T.F., Edwards, G.J., Taylor, C.J.: Active appearance models. IEEE Trans. Pattern Anal. Mach. Intell. **23**(6), 681–685 (2001)
28. Baltru, T., Robinson, P., Morency, L.P.: OpenFace: an open source facial behavior analysis toolkit. In: IEEE Winter Conference on Applications of Computer Vision (WACV), pp. 1–10 (2016)
29. Sohaib, A.T., Qureshi, S., Hagelbäck, J., Hilborn, O., Jerčić, P.: Evaluating classifiers for emotion recognition using EEG. In: Schmorrow, D.D., Fidopiastis, C.M. (eds.) AC 2013. LNCS (LNAI), vol. 8027, pp. 492–501. Springer, Heidelberg (2013). doi:10.1007/978-3-642-39454-6_53
30. Müller-Putz, G.R., Riedl, R., Wriessnegger, S.C.: Electroencephalography (EEG) as a research tool in the information systems discipline: foundations, measurement, and applications. Commun. Assoc. Inform. Syst. **37**(46), 911–948 (2015)
31. Petrantonakis, P.C., Hadjileontiadis, L.J.: Emotion recognition from brain signals using hybrid adaptive filtering and higher order crossings analysis. IEEE Trans. Affect. Comput. **1**, 81–97 (2010)
32. Valstar, M.F., Patras, I., Pantic, M.: Facial action unit detection using probabilistic actively learned support vector machines on tracked facial point data. In: IEEE Computer Society Conference on Computer Vision and Pattern Recognition (CVPR 2005), vol. 3, pp. 76–84 (2005)
33. Michel, P., El Kaliouby, R.: Real time facial expression recognition in video using support vector machines. In: Proceedings of the 5th International Conference on Multimodal Interfaces, pp. 258–264 (2003)
34. Hamilton, W.R.: On quaternions, or on a new system of imaginaries in algebra. Philos. Mag. **25**(3), 489–495 (1844)
35. Chen, M., Meng, X., Wang, Z.: Quaternion fisher discriminant analysis for bimodal multi-feature fusion. In: Abraham, A., Jiang, X.H., Snášel, V., Pan, J.-S. (eds.) Intelligent Data Analysis and Applications. AISC, vol. 370, pp. 479–487. Springer, Cham (2015). doi:10.1007/978-3-319-21206-7_41
36. Le Bihan, N., Sangwine, S.J.: Quaternion principal component analysis of color images. In: International Conference on Image Processing (ICIP), vol. 1, pp. I-809–I-812 (2003)
37. Kanade, T., Cohn, J.F., Tian, Y.: Comprehensive database for facial expression analysis. In: Fourth IEEE International Conference on Automatic Face and Gesture Recognition, pp. 46–53 (2000)
38. Pantic, M., Valstar, M., Rademaker, R., Maat, L.: Web-based database for facial expression analysis. In: IEEE International Conference on Multimedia and Expo, pp. 317–321 (2005)
39. Patras, I., Pantic, M.: Particle filtering with factorized likelihoods for tracking facial features. In: Sixth IEEE International Conference on Automatic Face and Gesture Recognition, pp. 97–102 (2004)
40. Izard, C.E.: Human Emotions. Springer, New York (2013)
41. Nicolle, J., Rapp, V., Bailly, K., Prevost, L., Chetouani, M.: Robust continuous prediction of human emotions using multiscale dynamic cues. In: 14th ACM International Conference on Multimodal Interaction, pp. 501–508 (2012)

42. Littlewort, G.C., Bartlett, M.S., Lee, K.: Automatic coding of facial expressions displayed during posed and genuine pain. Image Vis. Comput. **27**(12), 1797–1803 (2009)
43. Huang, K.C., Huang, S.Y., Kuo, Y.H.: Emotion recognition based on a novel triangular facial feature extraction method. In: 2010 International Joint Conference on Neural Networks (IJCNN), pp. 1–6(2010)
44. Sariyanidi, E., Gunes, H., Gkmen, M., Cavallaro, A.: Local zernike moment representation for facial affect recognition. In: British Machine Vision Conference (2013)
45. Zhao, G., Pietikinen, M.: Boosted multi-resolution spatiotemporal descriptors for facial expression recognition. Pattern Recogn. Lett. **30**(12), 1117–1127 (2009)
46. Wllmer, M., Eyben, F., Reiter, S., Schuller, B., Cox, C., Douglas-Cowie, E., Cowie, R.: Abandoning emotion classes-towards continuous emotion recognition with modelling of long-range dependencies. Interspeech **12**, 597–600 (2008)
47. Adali, T., Schreier, P.J., Scharf, L.L.: Complex-valued signal processing: the proper way to deal with impropriety. IEEE Trans. Sig. Process. (Overview Pap.) **59**(11), 5101–5123 (2011)
48. Li, X.-L., Adali, T., Anderson, M.: Noncircular principal component analysis and its application to model selection. IEEE Sig. Proc. **59**(10), pp. 4516i–4528i (2011)
49. Chai, Z., Ma, K.K., Liu, Z.: Complex wavelet-based face recognition using independent component analysis. In: Fifth International Conference on Intelligent Information Hiding and Multimedia Signal Proceedings, pp. 832–835 (2009)
50. Bonita, J.D., Ambolode, II., L.C.C., Rosenberg, B.M., Cellucci, C.J., Watanabe, T.A.A., Rapp, P.E., Albano, A.M.: Time domain measures of inter-channel EEG correlations: a comparison of linear, nonparametric and nonlinear measures. Cogn. Neurodyn. **8**(1), 1–15 (2014)
51. Li, K., Sun, G., Zhang, B., Wu, S., Wu, G.: Correlation between forehead EEG and sensorimotor area EEG in motor imagery task. In: Eighth IEEE International Conference on Dependable, Autonomic and Secure Computing, DASC 2009, pp. 430–435 (2009)
52. Abadi, M.K., Subramanian, R., Kia, S.M., Avesani, P., Patras, I., Sebe, N.: DECAF: MEG-based multimodal database for decoding affective physiological responses. IEEE Trans. Affect. Comput. **6**(3), 209–222 (2015)

Video Captioning via Sentence Augmentation and Spatio-Temporal Attention

Tseng-Hung Chen$^{(\boxtimes)}$, Kuo-Hao Zeng, Wan-Ting Hsu, and Min Sun

Department of Electrical Engineering,
National Tsing Hua University, Hsinchu, Taiwan
s104061544@m104.nthu.edu.tw, s103061614@m103.nthu.edu.tw
s105061525@m105.nthu.edu.tw, sunmin@ee.nthu.edu.tw

Abstract. Generating video descriptions has many important applications such as human-robot interaction, video indexing, video summarization and assisting for the visually impaired. Many significant breakthroughs in deep learning and releases of large-scale open-domain video description datasets allow us to explore this task more effectively. Recently, Venugopalan et al. (S2VT) propose to caption a video via the technique on machine translation. We propose tracklet attention method to capture spatio-temporal information in the decoding phase and reserve the encoding phase similar to S2VT to retain the technique on machine translation. On the other hand, labels for video captioning are expensive and scarce, and training corpus is hard to completely cover rare words presenting in testing set. Hence, we propose to use sentence augmentation method to enrich our training corpus. Finally, we conduct experiments to demonstrate that tracklet attention and sentence augmentation improve the performance of S2VT on the validation set of Microsoft Research Video to Text dataset (MSR-VTT). In addition, we also achieve the state-of-the-art performance on Video Titles in the Wild dataset (VTW) by applying tracklet attention.

1 Introduction

Generating natural language descriptions of visual contents has received lots of interests lately in the computer vision community. It has many important applications such as human-robot interaction, video indexing, video summarization and assisting for the visually impaired. Hence, the task of generating description from visual contents are highly relevant to the development of assisitve vision system in the future. Thanks to the recent advances in deep learning [13,25,38] and Recurrent Neural Networks (RNN), many attempts [44,45,53,56] have been made to jointly model videos and their corresponding sentence descriptions or titles. This task is often referred to as video captioning.

S2VT [44] proposes to caption a video via the technique on machine translation and SA [53] utilizes soft attention mechanism to describe a video. We aim to caption a video by combining their ideas together. Thanks to the great success on similar sequence-to-sequence tasks such as speech recognition [11] and

© Springer International Publishing AG 2017
C.-S. Chen et al. (Eds.): ACCV 2016 Workshops, Part I, LNCS 10116, pp. 269–286, 2017.
DOI: 10.1007/978-3-319-54407-6_18

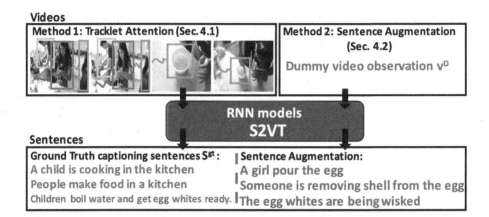

Fig. 1. Our model is built on S2VT [44] with two proposed methods. Method 1 is tracklet attention: We track bounding boxes (brown, blue, green color in the presented frames) related to object through time and utilize soft attention mechanism on those tracklets. Method 2 is sentence augmentation: In addition to the ground truth captioning sentences, we augment training corpus via retrieving similar sentences from other large video captioning corpus (Color figure online).

machine translation [40], we divide video captioning problem as two parts based on S2VT. The first part is the encoding of the visual observations for a given video and the second part is the decoding of the descriptions.

In general, a good description not only describes the visual objects but also expresses how objects interact with each other (activities) as well as their attributes. Thus we propose to utilize tracklets to capture spatio-temporal information in the decoding phase (bounding boxes with wavy line for each frame in Fig. 1). We let the model itself decide what the information it needs to generate the next word from the given previous words and the encoded visual observations. The tracklet is produced by an object detector using Faster-RCNN [33] trained on MSCOCO [27] object detection dataset in the first. Then, a naïve tracker is built on Forward-Backward error estimation [17] and a bipartite graph matching. Inspired by many successful works [29,50,53] using attention model on computer vision, we further use the soft attention mechanism on the tracklets for each video. Since soft attention mechanism is an unsupervised learning approach, we do not need to label the supporting evidence on tracklets for each video.

Secondly, we have discovered that the two latest movie description datasets [35,42] share the challenge of diverse sentences. On these datasets, state-of-the-art methods [44,53] have reported fairly low performances. One of the reasons behind the low performances is that the training corpus is hard to completely cover rare words presenting in testing set. Hence, it is important to "increase the number of sentences" for training a more robust language model. We further use sentence augmentation method introduced in [56] to enrich our training corpus (blue color captioning sentences in Fig. 1). We use Word2Vec [28] as feature for each caption and retrieve the sentences having similar meaning from other large video captioning data corpus. The sentence augmentation method not only increases the opportunity for completely coverage of testing corpus during training, but also enhances the model capacity for understanding the visual observation from different viewpoints. For application, it also provides an opportunity to enrich and interest the video summarization via augmented dictionary corpus.

Our Contributions. We have four main contributions summarized as follows.

- Fuse machine translation technique and soft attention mechanism together on video captioning and video title generation.
- Utilize soft attention mechanism on tracklets.
- Apply sentence augmentation approach introduced in [56] on Microsoft Research Video to Text dataset.
- Achieve the state-of-the-art performance on title generation in Video Titles in the Wild dataset.

Finally, on Microsoft Research Video to Text dataset (MSR-VTT) [49], we conduct experiments to demonstrate that tracklet attention and sentence augmentation approaches improve the performance of S2VT. In addition, the results of ablation experiments show the effectiveness both of the tracklet attention and the sentence augmentation approaches. Furthermore, we achieve the state-of-the-art performance on Video Titles in the Wild dataset (VTW) [56] dataset by applying tracklet attention.

2   Related Work

2.1   Image Captioning

Multiple works [9,18,21,46] have successfully addressed the image captioning problem. Many of them leverage the power of deep neural networks such as Recurrent Neural Networks (RNN) and in particular Long-Short Term Memory networks (LSTM). Concurrently, image description datasets such as Flickr30k [54], MSCOCO [27] and Visual Genome [23] have been released. [27] further presents several algorithms to evaluate the performance for automatic caption generation.

2.2   Video Captioning

In the past, early work on video captioning [4,8,12,22,24,36,41] typically perform a two-stage procedure which first identifies the semantic content (subject, verb, object) and then generates a sentence according to a template-based or statistical machine translation approaches. In the first stage, classifiers are used to detect objects, actions, and scenes. In the second stage, a model that integrates visual scores with a language model is used to estimate the combination of subject, verb, object, and scene. Then, a sentence is generated based on a predefined template. However, these methods require some hand-crafted components such as the template and the content to be classified. Therefore, the generated sentences are often have less diversity structure-wise compared to natural human description.

Recently, video captioning methods incorporate the Convolutional Neural Networks (CNN) and Recurrent Neural Networks (RNN) approaches. Generally, the models are directly learned from a large number of video-sentence pairs. Rohrbach et al. [34] propose to combine different RNN architectures with multiple CNN classifiers for classifying verbs (actions), objects, and places. Venugopalan et al. [45] use a fix dimension feature to represent a video by average-pooling CNN features of many frames, then take the feature as input of the RNN to generate a sentence. However, the temporal information of the video is discarded.

Venugopalan et al. [44] propose to use a recurrent encoder to model a ordered sequence of CNN features extracted from video frames. This sequence encoding approach outperforms their previous work [45] significantly. In the meanwhile, Yao et al. [53] propose to model the temporal structure of visual features in two approaches. First, they design a 3-D CNN based on dense trajectory-like features (HoG, HoF, MBH) [47] to capture the short temporal dynamics. Then, it integrates with a temporal attention mechanism which selectively focus on subsets of video observations for generating each word. Besides the above works that use RNN for encoding or decoding, Xu et al. [51] propose to embed both video and sentence to a joint space. It uses CNN to embed a video, and uses a dependency-tree structure model to embed sentences.

Most recently, several works have further tackled more sophisticated tasks and tried to improve the state of the art of video captioning. First of all, Pan et al. [32] put forward a framework to jointly perform visual-semantic embedding and learn a RNN model for video captioning. Then Pan et al. [31] proposed a Hierarchical RNN to take advantage of temporal structure in a long-range video. In addition, Yu et al. [55] present a novel hierarchical framework, including a sentence generator and a paragraph generator.

2.3   Attention Mechanism

In the computer vision and natural language processing community, there are quite some works using attention mechanism to tackle their problems. Typically, attention mechanism can be divided into hard attention and soft attention mechanisms.

For hard attention mechanism, Xu et al. [50] and Zeng et al. [56] use hard attention mechanism on image captioning and video title generation. There are multiple works [2,29] that apply reinforcement learning to decide attention bounding box to recognize handwriting digit numbers [26] and Street View House Numbers (SVHN) [30]. Hu et al. [16] utilize proposed bounding box to retrieve object for given query sentence in natural language.

For soft attention mechanism, Bahdanau et al. [3] use soft attention mechanism and machine translation to align and translate English to French. Many works [50,53,56] leverage soft attention mechanism on image captioning, video captioning and video title generation. Furthermore, there are several works [1,48,52] that apply soft attention mechanism on either image or question sentence to do visual question answering. Sukhbaatar et al. [39] incorporates attention mechanism to memory cells iteratively to different values stored in the memory for question answering in the text domain. In addition, Chan et al. [5] utilize soft attention mechanism to focus on vehicles potentially involved in accident to anticipate accident.

2.4   Data Augmentation

In order to bridge between computer vision and natural language, video captioning models typically need rich information both of visual part and linguistic part.

On visual part, Krizhevsky et al. [25] propose several data augmentation methods proposed for image classification such as image randomly sampled or horizontal flip and standard color augmentation. On linguistic part, Hendricks et al. [14] and Zeng et al. [56] propose sentence augmentation methods to enrich the information in linguistic domain for image captioning, video captioning, and video title generation.

Our work is most similar to [44,53,56]. We use sequence to sequence technique and soft attention mechanism to caption a video. However, our soft attention mechanism is applied on tracklets, not on the regions in one frame or a single image. We also apply the sentence augmentation approach introduced by [56] to enrich our training corpus.

3   Preliminaries

We first give preliminaries of the standard RNN and LSTM before we describe the S2VT machine translation model.

3.1   Recurrent Neural Network (RNN)

The idea behind RNN is to make use of sequential observations $(\mathbf{x}_1, \mathbf{x}_2, \ldots, \mathbf{x}_T)$. It outputs a sequence of learned hidden representations $(\mathbf{h}_1, \mathbf{h}_2, \ldots, \mathbf{h}_T)$, where $\mathbf{h}_t$ encodes the sequence observations $(\mathbf{x}_1, \mathbf{x}_2, \ldots, \mathbf{x}_t)$. The hidden representation is generated by a recursive equation below,

$$\mathbf{h}_t = g(\mathbf{W}\mathbf{x}_t + \mathbf{H}\mathbf{h}_{t-1} + \mathbf{b}) , \tag{1}$$

where $g(\cdot)$ is a non-linear function applied element-wise (e.g., sigmoid), $\mathbf{W}, \mathbf{H}, \mathbf{b}$ are the model parameters to be learned. The hidden representation $\mathbf{h}_t$ is use to predict a target output. For example, if we wants to predict the next word in a sentence, the target output $\mathbf{o}_t$ will be a vector of probabilities across our vocabulary.

$$\mathbf{o}_t = \text{softmax}(\mathbf{W}_s \mathbf{h}_t + \mathbf{b}_s) , \tag{2}$$

The softmax function computes the probability of each word (i.e., $\sum_i o_t^i = 1$), and $\mathbf{W}_s, \mathbf{b}_s$ are the model parameters to be learned.

3.2   Long-Short Term Memory (LSTM)

LSTM introduces a memory cell $\mathbf{c}$ to maintain information over time. It can be considered as the state of the recurrent system. LSTM extends the standard RNN by replacing the recursive equation in Eq. 1 with

$$(\mathbf{h}_t, \mathbf{c}_t) = \text{LSTM}(\mathbf{x}_t, \mathbf{h}_{t-1}, \mathbf{c}_{t-1}) \,, \qquad (3)$$

where the memory cell $\mathbf{c}$ allows RNN to model long-term contextual dependencies. The LSTM has the ability to remove or add information to the state, carefully regulated by three gates: input gate $\mathbf{i}$, forget gate $\mathbf{f}$, and output gate $\mathbf{o}$ (see Fig. 2). Gates are designed to optionally let information through and to control the interaction among the input, memory cell, and output. At each frame t, LSTM first computes gate activations: $\mathbf{i}_t, \mathbf{f}_t$ (Eqs. 4 and 5) and updates its memory cell from $\mathbf{c}_{t-1}$ to $\mathbf{c}_t$ (Eq. 6). Then it computes the output gate activation $\mathbf{o}_t$ (Eq. 7), and outputs a hidden representation $\mathbf{h}_t$ (Eq. 8). The hidden state of the cell is recurrently connected back to the input and three gates.

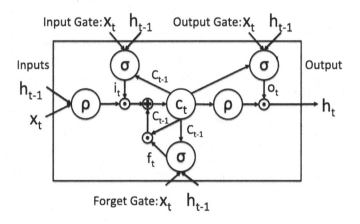

Fig. 2. Workflow of LSTM [15].

$$\mathbf{i}_t = \sigma(\mathbf{W}_i \mathbf{x}_t + \mathbf{U}_i \mathbf{h}_{t-1} + \mathbf{V}_i \mathbf{c}_{t-1} + \mathbf{b}_i) \qquad (4)$$

$$\mathbf{f}_t = \sigma(\mathbf{W}_f \mathbf{x}_t + \mathbf{U}_f \mathbf{h}_{t-1} + \mathbf{V}_f \mathbf{c}_{t-1} + \mathbf{b}_f) \qquad (5)$$

$$\mathbf{c}_t = \mathbf{f}_t \odot \mathbf{c}_{t-1} + \mathbf{i}_t \odot \rho(\mathbf{W}_c \mathbf{x}_t + \mathbf{U}_c \mathbf{h}_{t-1} + \mathbf{b}_c) \qquad (6)$$

$$\mathbf{o}_t = \sigma(\mathbf{W}_o \mathbf{x}_t + \mathbf{U}_o \mathbf{h}_{t-1} + \mathbf{V}_o \mathbf{c}_t + \mathbf{b}_o) \qquad (7)$$

$$\mathbf{h}_t = \mathbf{o}_t \odot \rho(\mathbf{c}_t) \qquad (8)$$

We now define the common notations in Eqs. 4–8. $\odot$ is an element-wise product, and the logistic function σ and the hyperbolic tangent function ρ are both applied element-wise. $\mathbf{W}_*, \mathbf{V}_*, \mathbf{U}_*, \mathbf{b}_*$, and $\mathbf{V}_*$[1] are the parameters. Please note that the input and forget gates participate in updating the memory cell (Eq. 6). More specifically, forget gate controls how much of memory to forget, and the input gate allows newly computed values (based on the current observation and previous hidden representation) to update the memory cell. The output gate determines which parts of the memory cell flow into the output. The output gate together with the memory cell computes the hidden representation (Eq. 8). Since the current memory cell only goes through a binary operation (i.e., forget gate) and a summation (Eq. 6), the gradient with respect to the memory cell does not vanish as fast as standard RNN when back propagating through time.

3.3   Sequence to Sequence Video to Text (S2VT) [44]

Figure 3 represents an overview of the S2VT model. The S2VT model is a Sequence2Sequence translation model [7,40] which directly translates sequences of video frames to sequences of words. A stack of two LSTMs are used in this model. The top LSTM layer models the encoding of visual inputs while the bottom LSTM layer models the decoding of language. $< BOS >$ indicates the Beginning of Sentence and $< EOS >$ for the End of Sentence. The sequences are padded with zero vectors $< pad >$ to fill up the empty channels in different stages. In detail, during the encoding stage, the top LSTM layer is fed the video frames while the bottom LSTM layer receives the concatenation of zero-padded input words and the hidden representation of the video sequences. In the decoding stage, the visual frame representation of the top LSTM layer is replaced by zero-padded input frames. While training in the decoding stage, we use the bottom LSTM cell output z_t to predict the next word s_{t+1} in each time step.

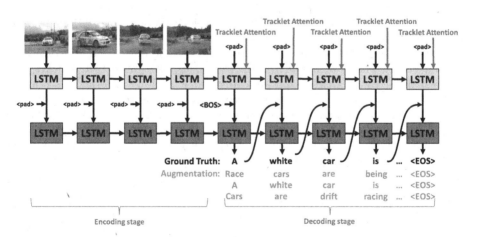

Fig. 3. An overview of our proposed methods: (red color) tracklet soft attention and (brown color) sentence augmentation (Color figure online).

---

[1] The subscript $*$ denotes any symbol.

Here a softmax function is applied after the LSTM layer to get the probability distribution over all the D words in the vocabulary V:

$$p(s_{t+1}|z_t) = \frac{\exp(W_s z_t)}{\sum\limits_{s' \in V} \exp(W_{s'} z_t)} \ , \tag{9}$$

where W_s is the embedding matrix in the softmax layer. At test time, the predicted word will be fed as the LSTM input for next timestep. While during training, the next input word is provided by the annotated sentence. Each word is chosen with the maximum probability until the $< EOS >$ token.

4 Proposed Methods

In order to effectively combine S2VT and soft attention mechanism and enrich training corpus, we introduce tracklet attention method (red color in Fig. 3) and apply sentence augmentation method (brown color in Fig. 3) as follows.

4.1 Tracklet Attention

In order to have more precise descriptions for videos, we propose a spatial-attention mechanism on tracklets via soft attention mechanism.

Bounding Boxes Detection. We use the Faster R-CNN [33] model pre-trained on the 2014 COCO detection dataset [27] to first detect top 10 bounding boxes for each frame.

Track Selection from Video. First, we detect 10 objects in the first frame as initial points for the tracks. Following the Forward-Backward error estimation [17], we forward each track in the current frame to the next frame to form the tracked boxes. Then, we match the tracked boxes with the detected boxes in the next frame by bipartite graph matching under the IoU^2 criteria. It is worth noting that we only compute the IoU score between the tracked box and the detected box belonging to the same object category. If there is no detected box having high enough IoU score, we treat that the track is broken and we let the box be the new initial point to form another track. Thus, we can obtain several tracks for each video and denote each track as tracklet **T** in the following paragraphs. Both *detection by tracking*[3] and *tracking by detection* are applied for this task. For each tracklet, we mean pool the feature of bounding boxes across time as the representation. For each video, we extract the top 10 tracklets based on both of the average confidence score over time and the length of the tracklet.

---

[2] Intersection over Union: an evaluation method for object detection. We set 0.5 overlap as threshold for IoU criteria.

[3] We add tracked boxes into region proposals for object detection in the next frame.

Spatial-Attention Mechanism. In order to focus on visual tracklets in different time intervals and video regions, the technique on soft attention is employed. In the decoding phase of S2VT, we utilize three embedding functions and a hyperbolic tangent function depicted in Eq. (10) to calculate the score for each tracklet $\mathbf{T}_i$ given the hidden state $\mathbf{h}_{t-1}$ in the previous LSTM unit.

$$e_i^{(t)} = \mathbf{w}^T tanh(\mathbf{W}_h \mathbf{h}_{t-1} + \mathbf{W}_T \mathbf{T}_i + \mathbf{b}), \tag{10}$$

where i denotes which tracklet is computed and t denotes which word s would be generated. $\mathbf{w}$, $\mathbf{W}_h$, $\mathbf{W}_T$, and $\mathbf{b}$ are all the model parameters in our spatial-attention mechanism. Moreover, $\mathbf{W}_h$ and $\mathbf{W}_T$ embed the hidden representation $\mathbf{h}_{t-1}$ and each tracklet $\mathbf{T}_i$ to the same dimensional subspace. The dot-product of $\mathbf{w}$ and the output from the non-linear function computes a single scalar value $e_i^{(t)}$. Once we have the scores $e_i^{(t)}$ for all the tracklets $i = \{1, ..., n\}$, we normalize them to form the soft-attention weight α_i for each tracklet $\mathbf{T}_i$,

$$\alpha_i^{(t)} = \frac{\exp(e_i^{(t)})}{\sum_{j=1}^{n} \exp(e_j^{(t)})}. \tag{11}$$

Finally, we weighted-sum over all of them to acquire the attentive visual observation to generate next word,

$$\mathbf{T}_{att}^{(t)} = \sum_{i=1}^{n} \alpha_i \mathbf{T}_i. \tag{12}$$

By employing spatial-attention mechanism, each generated word pays different attention to the tracklets and thus decides how spatio-temporal visual observations are represented. As a result, we can follow [44] to formulate the video captioning task as the probability of a generated sentence given a video and the tracklets in this video as follows,

$$p(S|\mathbf{V}, \mathbf{T}_{att}; \theta) = \prod_{t=1}^{m} p(s_t|S_{1:(t-1)}, \mathbf{V}, \mathbf{T}_{att}^{(t)}). \tag{13}$$

Here, S denotes a generated sentence, m denotes the length of S, $\mathbf{V}$ is encoded visual observation, and θ denotes all of our model parameters. We add the spatial-attention feature $\mathbf{T}_{att}$ as input based on the tracklets for each video in the decoding phase. Accordingly, we have established the model combining machine translation and soft attention mechanism on spatial-temporal feature such like tracklets for video captioning.

4.2   Sentence Augmentation

We train a RNN model with both video-sentence pairs and sentence-only examples, where sentence-only examples are additional sentences retrieved from other corpora. We use the method introduced in [56] to apply the technique on sentence augmentation as described below.

Sentence Retrieval. Given a large corpus, we retrieve additional sentences for sentence augmentation as follows. Each training sentence is used as a query to retrieve similar sentences in the corpus. In detail, we use the mean of Word2Vec [28] feature extracted from non-stop words in each sentence as the sentence-based feature. Then, we use cosine similarity to measure sentence-wise similarity between query sentence and the sentences in the corpus. Among sentences in the corpus with similarity above 0.75, we sample without replacement to retrieve a target number of sentences. In total, we retrieve 39116 sentences from the Microsoft Youtube dataset corpus (MSVD) [6] for augmentation.

Dummy Video Observation. Since retrieved sentences may not be completely relevant to visual observation, video captioning model probably obtains poor performance if directly training on these sentences. In order to avoid such problem, we simply design visual observation feature $\mathbf{V}$ and spatial-attention feature $\mathbf{T}_{att}$ as all-zero vectors so that $\mathbf{h}_t$ is not related to visual observations. This $\mathbf{h}_t$ is a function of only $\mathbf{W}_{h*}$, $\mathbf{b}_*$ and other parameters of LSTM unit. Intuitively, this design allows simplicity of handling additional sentence-only examples with dummy video observations. For the original video-captioning pairs, all visual observations are sequentially encoded by RNN as well.

5   Experiments

In this section we first describe the datasets used and the details of our models. Then we analyze our approach on the datasets and explore different design choices. We also compare our best system to prior work.

5.1   Datasets

We conduct our experiments on both Microsoft Research Video to Text dataset (MSR-VTT) [49] and Video Titles in the Wild dataset (VTW) [56].

Microsoft Research Video to Text Dataset. MSR-VTT is a large-scale video benchmark for translating video to text. It is collected by 257 popular queries from a commercial video search engine, with 118 videos for each query. In its current version, MSR-VTT provides 10K web video clips with 41.2 h and 200K clip-sentence pairs in total. Each clip is annotated with about 20 natural sentences by 1,327 AMT workers. We follow the setting in [49] to split dataset into 6.5K clips for training, 3K clips for testing, 0.5K clips for validation. Because the testing annotated natural sentences have not been released yet, we evaluate all the ablation experiments on the validation set and report the best model on the testing set evaluated by ACM MM 2016 Multimedia Grand Challenge Solutions on MSR Video to Language Challenge.

Video Titles in the Wild Dataset. VTW is a large-scale video benchmark for video title generation. It is a dataset of 18100 automatically crawled user-generated videos and titles. Each video is on an average 1.5 min (45 s median duration) and contains a highlight event which makes this video interesting associated with one title sentence. We follow the setting in [56] to split dataset into 14100 videos for training, 2000 videos for testing, 2000 videos for validation. We use validation set to determine model hyper-parameters. Also, the performance reported on testing set is evaluated by best model selected from validation set.

5.2   Details

Features. We utilize both appearance and local motion features. For video object appearance, we extract features for each frame using VGG [38] and ResNet [13] pre-trained on the ImageNet dataset [37]. The VGG feature is the output of fc7 layer from the 16-layer VGG model. The ResNet feature is extracted from the global average pooling layer of the 152-layer ResNet model. For video motion and activities, we extract a fixed-length feature vector every 16 frames from fc7 layer of the pre-trained C3D [43] (on the Sports-1M dataset [19]).

Video Observation. We divide a video into maximum 45 clips. Then, we average-pool all the VGG, C3D and ResNet features in each clip to obtain a video observation. We concat these features then linearly embed to a lower 500 dimensional space. The weights of the linear embedding are learned jointly with the LSTM layers during training. It is worth noting that we only use VGG and C3D on VTW dataset due to the limited computing resource, but we still achieve the state-of-the-art performance on the testing set of VTW dataset.

TensorFlow Implementation. We reimplement S2VT using TensorFlow [10]. We implement sequence to sequence translation between video observations and corresponding video captionings. For simplification, each LSTM block in the figure is implemented by the standard module in TensorFlow. In order to separate visual embedding feature and language embedding feature, we use the all zero vector to implicitly show the absence of each feature. While only a visual feature channel is shown in Fig. 3, we follow [44] to replicate the same architecture with different learnable weights and bias for VGG [38], C3D [43] and ResNet [13]. Then, a late-fusion layer is used to combine the results from different feature sources.

Training Details. We implement our model using TensorFlow with the batch size of 100 and selected the final model according to the best validation accuracy. In all experiments, we use Adam optimizer [20] with an initial learning rate of 0.0001. Training runs up to 150 maximum epochs.

5.3 Ablation Experiments

We have proposed a tracklet attention method to combine machine translation and soft attention mechanism together. For alleviating the problem that training corpus is too scarce to completely cover rare words presenting in testing set, we have proposed a sentence augmentation methods to enrich our training corpus. In order to realize the effectiveness of these two methods, we plan to experiment on them separately, and the combination of them. Besides, thanks to the 152-layer residual net proposed by Kaiming He et al. [13] last year, many tasks in computer vision have been addressed more effectively. Thus, we also experiment on S2VT model only using VGG and C3D features and using all of VGG, C3D and ResNet features. The overall results are shown in Table 1 and the discussion is in the next section.

Table 1. The performance for different variants on the val. set of MSR-VTT [49]: B@4 denotes BLEU at 4-gram. MET. stands for METEOR, ROU. stands for ROUGE-L, and CID. stands for CIDEr-D. C3D, VGG and Res denote which feature type is used. Aug. and Tra. denote that sentence augmentation and/or tracklet attention method is used.

	Model	B@4	MET.	ROU.	CID.
1	**SA** [53] **(C3D + VGG)**	30.3	24.9	55.6	34.6
2	**S2VT (C3D + VGG)**	32.3	25.8	56.7	29.6
3	**S2VT w/ Aug. (C3D + VGG)**	33.3	26.6	57.2	32.5
4	**S2VT (C3D + Res.)**	35.2	26.6	58.6	37.8
5	**S2VT w/ Tra. (C3D + Res.)**	36.0	26.8	58.9	38.8
6	**S2VT (C3D + VGG + Res.)**	34.6	26.9	58.3	37.9
7	**S2VT w/ Aug. (C3D + VGG + Res.)**	35.3	**27.4**	58.9	38.3
8	**S2VT w/ Tra. (C3D + VGG + Res.)**	**36.5**	27.1	**59.2**	**40.3**
9	**S2VT w/ Aug. + Tra. (C3D + VGG + Res.)**	35.6	27.0	58.9	38.1

5.4 Results

We use the evaluation code provided by MSCOCO [27] to evaluate all of our experiments on the validation set of MSR-VTT[4] [49] and the testing set of VTW [56].

MSR-VTT. We show the result in Table 1. According to Table 1, both row-3 comparing to row-2 and row-7 comparing row-6 show that sentence augmentation improves the performance based on S2VT model. Although in some cases the additional sentences might not correlate with the corresponding videos, the

---

[4] http://ms-multimedia-challenge.com/challenge.

model still can learn from grammatical and semantically correct sentences. The sentence augmentation approach consistently improves the performance as it leads to a more robust language model. Row-5 comparing to row-4 and row-8 comparing to row-6 show that tracklets attention improves performance significantly. This indicates that focusing on different object tracklets helps. It uses details to enhance understanding and improves visual comprehension. The performance of the combination of both proposed methods are shown in row-9. We also show that concatenating all of C3D, VGG and ResNet has the highest performance compared to the other two combinations. In addition, we apply another state-of-the-art model SA [53] as a baseline. However, the performance (row-1) is lower than S2VT model (row-2). In order to gain more insights on our model, we show a few typical examples with tracklets for both S2VT model trained on C3D + VGG + Res. and C3D + VGG + Res. + Tra. (having the highest performance) in Fig. 4.

In Table 2, we show the result for our method (selected according to the best validation performance) on the testing set of [49] evaluated by ACM MM 2016 MSR Video to Language Challenge. Our testing results can potentially be improved by applying beam search and/or combining a set of independently trained models using standard ensemble methods.

VTW. We show the result in Table 3. According to Table 3, both of row-3 and row-4 are done by [56] and they show that sentence augmentation approach improves both of SA and S2VT models. Most importantly, row-5 shows that tracklet attention mechanism leads S2VT to outperform the state-of-the-art performance in [56] with sentence augmentation. It demonstrates that tracklet attention mechanism let the captioning model generate title more effective.

Table 2. The performance for our method on the testing set of MSR-VTT [49]. C1, C2 and C3 are Human evaluation of the captions in terms of Coherence, Relevance, and helpful for blind on a scale of 1–5 (The higher the better).

	B@4	MET.	ROU.	CID.	C1	C2	C3
Ours	34.4	26.0	58.4	36.7	3.19	2.75	2.80

Table 3. The performance for different variants on the testing set of VTW [56].

Model	B@1	B@2	B@3	B@4	MET.	CID.
1 **SA** [53]	11.6	5.3	2.9	1.8	5.6	24.9
2 **S2VT** [44]	11.4	4.9	2.5	1.6	6.2	24.9
3 **SA w/ Aug.** [56]	11.8	5.5	2.9	1.9	5.7	25.1
4 **S2VT w/ Aug.** [56]	11.7	5.1	2.6	1.6	6.2	25.4
5 **S2VT w/ Aug. + Tra.**	**12.5**	**5.8**	**3.2**	**2.0**	**6.4**	**27.7**

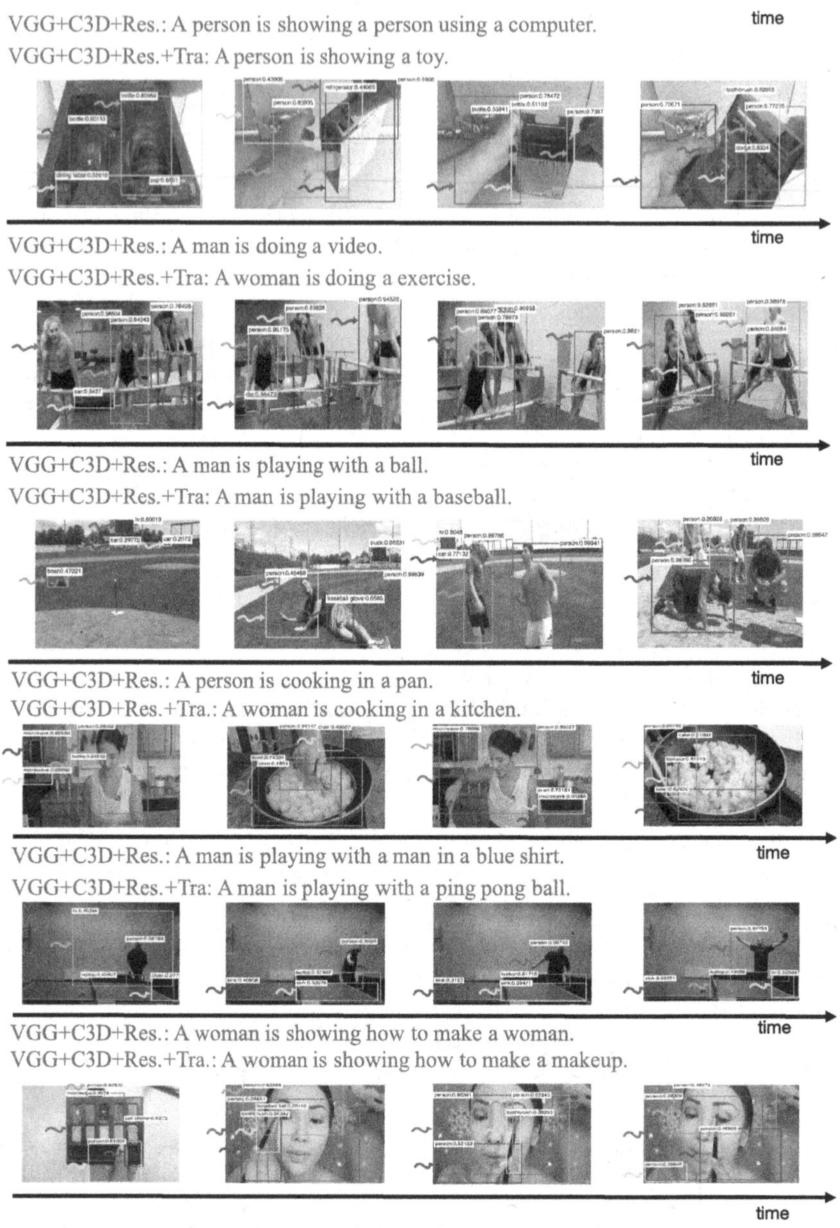

VGG+C3D+Res.: A person is showing a person using a computer.   time
VGG+C3D+Res.+Tra: A person is showing a toy.

VGG+C3D+Res.: A man is doing a video.   time
VGG+C3D+Res.+Tra: A woman is doing a exercise.

VGG+C3D+Res.: A man is playing with a ball.   time
VGG+C3D+Res.+Tra: A man is playing with a baseball.

VGG+C3D+Res.: A person is cooking in a pan.   time
VGG+C3D+Res.+Tra.: A woman is cooking in a kitchen.

VGG+C3D+Res.: A man is playing with a man in a blue shirt.   time
VGG+C3D+Res.+Tra: A man is playing with a ping pong ball.

VGG+C3D+Res.: A woman is showing how to make a woman.   time
VGG+C3D+Res.+Tra.: A woman is showing how to make a makeup.

time

Fig. 4. Typical examples on the testing set of MSR-VTT [49]. The S2VT model using C3D+VGG+Res. as baseline method is in blue color. Our model is S2VT model using C3D+VGG+Res.+Tra. in red color. Every bounding box with wave line denotes the tracklet in the local spatial-temporal segment. Please note that we select top 4 tracklet in the local spatial-temporal segment according to the detection score. The final top 10 tracklets for each video need to take the length of each tracklet into account (Color figure online).

6   Conclusion

To sum up, we propose the method to combine the techniques on both machine translation and soft attention. Moreover, we extend single-frame-based soft attention to spatial-temporal attention on tracklets. In addition, we use sentence augmentation to increase presenting frequency of rare words of testing set in training corpus. The experiments show that both of the methods improve the performance of the state-of-the-art video captioning model. Most importantly, we achieve the state-of-the-art performance on title generation task by combining the two methods. In the future, we have two directions to improve the video captioning model. The one is to employ more sophisticated tracking algorithm so that we could acquire more reliable and more robust tracklet. The other one is to utilize more excellent multimodal approach to integrate sentences augmentation and original video-captioning pairs well.

Acknowledgements. We thank Microsoft Research Asia (MSRA) project grants and MOST 103-2218-E-007-025 and MOST 104-3115-E-007-005 in Taiwan for their support. Kuo-Hao Zeng was supported by NOVATEK Fellowship.

References

1. Andreas, J., Rohrbach, M., Darrell, T., Klein, D.: Deep compositional question answering with neural module networks. In: Proceedings of the IEEE Conference on Computer Vision and Pattern Recognition (2016)
2. Ba, J., Mnih, V., Kavukcuoglu, K.: Multiple object recognition with visual attention. arXiv preprint arXiv:1412.7755 (2014)
3. Bahdanau, D., Cho, K., Bengio, Y.: Neural machine translation by jointly learning to align and translate. arXiv preprint arXiv:1409.0473 (2014)
4. Barbu, A., Bridge, E., Burchill, Z., Coroian, D., Dickinson, S., Fidler, S., Michaux, A., Mussman, S., Narayanaswamy, S., Salvi, D., Schmidt, L., Shangguan, J., Siskind, J.M., Waggoner, J., Wang, S., Wei, J., Yin, Y., Zhang, Z.: Video in sentences out. In: UAI (2012)
5. Chan, F.-H., Chen, Y.-T., Xiang, Y., Sun, M.: Anticipating accidents in dashcam videos. In: ACCV (2016)
6. Chen, D.L., Dolan, W.B.: Collecting highly parallel data for paraphrase evaluation. In: NAACL-HLT, pp. 190–200. Association for Computational Linguistics (2011)
7. Cho, K., Van Merriënboer, B., Gulcehre, C., Bahdanau, D., Bougares, F., Schwenk, H., Bengio, Y.: Learning phrase representations using RNN encoder-decoder for statistical machine translation. arXiv preprint arXiv:1406.1078 (2014)
8. Das, P., Xu, C., Doell, R., Corso, J.: A thousand frames in just a few words: lingual description of videos through latent topics and sparse object stitching. In: CVPR (2013)
9. Donahue, J., Anne Hendricks, L., Guadarrama, S., Rohrbach, M., Venugopalan, S., Saenko, K., Darrell, T.: Long-term recurrent convolutional networks for visual recognition and description. In: Proceedings of the IEEE Conference on Computer Vision and Pattern Recognition, pp. 2625–2634 (2015)
10. Abadi, M., et al.: TensorFlow: Large-scale machine learning on heterogeneous systems (2015). www.tensorflow.org

11. Graves, A., Jaitly, N.: Towards end-to-end speech recognition with recurrent neural networks. In: ICML, vol. 14, pp. 1764–1772 (2014)
12. Guadarrama, S., Krishnamoorthy, N., Malkarnenkar, G., Venugopalan, S., Mooney, R., Darrell, T., Saenko, K.: Youtube2text: recognizing and describing arbitrary activities using semantic hierarchies and zero-shot recognition. In: ICCV (2013)
13. He, K., Zhang, X., Ren, S., Sun, J.: Deep residual learning for image recognition. arXiv preprint arXiv:1512.03385 (2015)
14. Hendricks, L.A., Venugopalan, S., Rohrbach, M., Mooney, R., Saenko, K., Darrell, T.: Deep compositional captioning: describing novel object categories without paired training data. In: Proceedings of the IEEE Conference on Computer Vision and Pattern Recognition (2016)
15. Hochreiter, S., Schmidhuber, J.: Long short-term memory. Neural Comput. **9**, 1735–1780 (1997)
16. Hu, R., Xu, H., Rohrbach, M., Feng, J., Saenko, K., Darrell, T.: Natural language object retrieval. In: Proceedings of the IEEE Conference on Computer Vision and Pattern Recognition (2016)
17. Kalal, Z., Mikolajczyk, K., Matas, J.: Forward-backward error: automatic detection of tracking failures. In: 2010 20th International Conference on Pattern recognition (ICPR), pp. 2756–2759. IEEE (2010)
18. Karpathy, A., Fei-Fei, L.: Deep visual-semantic alignments for generating image descriptions. In: Proceedings of the IEEE Conference on Computer Vision and Pattern Recognition, pp. 3128–3137 (2015)
19. Karpathy, A., Toderici, G., Shetty, S., Leung, T., Sukthankar, R., Fei-Fei, L.: Large-scale video classification with convolutional neural networks. In: CVPR (2014)
20. Kingma, D.P., Ba, J.: Adam: A method for stochastic optimization. arXiv:1412.6980 (2014)
21. Kiros, R., Salakhutdinov, R., Zemel, R.S.: Unifying visual-semantic embeddings with multimodal neural language models. arXiv preprint arXiv:1411.2539 (2014)
22. Kojima, A., Tamura, T., Fukunaga, K.: Natural language description of human activities from video images based on concept hierarchy of actions. IJCV **50**(2), 171–184 (2002)
23. Krishna, R., Zhu, Y., Groth, O., Johnson, J., Hata, K., Kravitz, J., Chen, S., Kalantidis, Y., Li, L.-J., Shamma, D.A., et al.: Visual genome: Connecting language and vision using crowdsourced dense image annotations. arXiv preprint arXiv:1602.07332 (2016)
24. Krishnamoorthy, N., Malkarnenkar, G., Mooney, R.J., Saenko, K., Guadarrama, S.: Generating natural-language video descriptions using text-mined knowledge. In: AAAI (2013)
25. Krizhevsky, A., Sutskever, I., Hinton, G.E.: Imagenet classification with deep convolutional neural networks. In: Pereira, F., Burges, C., Bottou, L., Weinberger, K. (eds.) NIPS (2012)
26. LeCun, Y., Bottou, L., Bengio, Y., Haffner, P.: Gradient-based learning applied to document recognition. Proc. IEEE **86**(11), 2278–2324 (1998)
27. Lin, T.-Y., Maire, M., Belongie, S., Hays, J., Perona, P., Ramanan, D., Dollár, P., Zitnick, C.L.: Microsoft COCO: common objects in context. In: Fleet, D., Pajdla, T., Schiele, B., Tuytelaars, T. (eds.) ECCV 2014. LNCS, vol. 8693, pp. 740–755. Springer, Cham (2014). doi:10.1007/978-3-319-10602-1_48
28. Mikolov, T., Chen, K., Corrado, G., Dean, J.: Efficient estimation of word representations in vector space. arXiv:1301.3781 (2013)
29. Mnih, V., Heess, N., Graves, A., et al.: Recurrent models of visual attention. In: Advances in Neural Information Processing Systems, pp. 2204–2212 (2014)

30. Netzer, Y., Wang, T., Coates, A., Bissacco, A., Wu, B., Ng, A.Y.: Reading digits in natural images with unsupervised feature learning (2011)
31. Pan, P., Xu, Z., Yang, Y., Wu, F., Zhuang, Y.: Hierarchical recurrent neural encoder for video representation with application to captioning. arXiv:1511.03476 (2015)
32. Pan, Y., Mei, T., Yao, T., Li, H., Rui, Y.: Jointly modeling embedding and translation to bridge video and language. arXiv:1505.01861 (2015)
33. Ren, S., He, K., Girshick, R., Sun, J.: Faster R-CNN: towards real-time object detection with region proposal networks. In: NIPS (2015)
34. Rohrbach, A., Rohrbach, M., Schiele, B.: The long-short story of movie description. In: Gall, J., Gehler, P., Leibe, B. (eds.) GCPR 2015. LNCS, vol. 9358, pp. 209–221. Springer, Cham (2015). doi:10.1007/978-3-319-24947-6_17
35. Rohrbach, A., Rohrbach, M., Tandon, N., Schiele, B.: A dataset for movie description. In: CVPR (2015)
36. Rohrbach, M., Qiu, W., Titov, I., Thater, S., Pinkal, M., Schiele, B.: Translating video content to natural language descriptions. In: ICCV (2013)
37. Russakovsky, O., Deng, J., Su, H., Krause, J., Satheesh, S., Ma, S., Huang, Z., Karpathy, A., Khosla, A., Bernstein, M., et al.: Imagenet large scale visual recognition challenge. IJCV 115(3), 211–252 (2015)
38. Simonyan, K., Zisserman, A.: Very deep convolutional networks for large-scale image recognition. In: ICLR (2015)
39. Sukhbaatar, S., Weston, J., Fergus, R.: End-to-end memory networks. In: NIPS (2015)
40. Sutskever, I., Vinyals, O., Le, Q.V.: Sequence to sequence learning with neural networks. In: NIPS, pp. 3104–3112 (2014)
41. Thomason, J., Venugopalan, S., Guadarrama, S., Saenko, K., Mooney, R.: Integrating language and vision to generate natural language descriptions of videos in the wild. In: COLING (2014)
42. Torabi, A., Pal, C.J., Larochelle, H., Courville, A.C.: Using descriptive video services to create a large data source for video annotation research. arXiv:1503.01070 (2015)
43. Tran, D., Bourdev, L., Fergus, R., Torresani, L., Paluri, M.: Learning spatiotemporal features with 3D convolutional networks. In: ICCV (2015)
44. Venugopalan, S., Rohrbach, M., Donahue, J., Mooney, R., Darrell, T., Saenko, K.: Sequence to sequence - video to text. In: ICCV (2015)
45. Venugopalan, S., Xu, H., Donahue, J., Rohrbach, M., Mooney, R., Saenko, K.: Translating videos to natural language using deep recurrent neural networks. In: NAACL (2015)
46. Vinyals, O., Toshev, A., Bengio, S., Erhan, D.: Show and tell: a neural image caption generator. In: Proceedings of the IEEE Conference on Computer Vision and Pattern Recognition, pp. 3156–3164 (2015)
47. Wang, H., Schmid, C.: Action recognition with improved trajectories. In: ICCV (2013)
48. Xu, H., Saenko, K.: Ask, attend and answer: Exploring question-guided spatial attention for visual question answering. arXiv preprint arXiv:1511.05234 (2015)
49. Xu, J., Mei, T., Yao, T., Rui, Y.: MSR-VTT: a large video description dataset for bridging video and language. In: CVPR (2016)
50. Xu, K., Ba, J., Kiros, R., Cho, K., Courville, A., Salakhutdinov, R., Zemel, R.S., Bengio, Y.: Show, attend and tell: Neural image caption generation with visual attention. arXiv preprint arXiv:1502.03044 2(3): 5 (2015)

51. Xu, R., Xiong, C., Chen, W., Corso, J.J.: Jointly modeling deep video and compositional text to bridge vision and language in a unified framework. In: AAAI, pp. 2346–2352 (2015)
52. Yang, Z., He, X., Gao, J., Deng, L., Smola, A.: Stacked attention networks for image question answering. arXiv preprint arXiv:1511.02274 (2015)
53. Yao, L., Torabi, A., Cho, K., Ballas, N., Pal, C., Larochelle, H., Courville, A.: Describing videos by exploiting temporal structure. In: ICCV (2015)
54. Young, P., Lai, A., Hodosh, M., Hockenmaier, J.: From image descriptions to visual denotations: new similarity metrics for semantic inference over event descriptions. Trans. Assoc. Comput. Linguist. 2, 67–78 (2014)
55. Yu, H., Wang, J., Huang, Z., Yang, Y., Xu, W.: Video paragraph captioning using hierarchical recurrent neural networks. arXiv:1510.07712 (2015)
56. Zeng, K.-H., Chen, T.-H., Niebles, J.C., Sun, M.: Title generation for user generated videos. In: Leibe, B., Matas, J., Sebe, N., Welling, M. (eds.) ECCV 2016. LNCS, vol. 9906, pp. 609–625. Springer, Cham (2016). doi:10.1007/978-3-319-46475-6_38

Bottom-Up Fixation Prediction
Using Unsupervised Hierarchical Models

Hamed R. Tavakoli$^{(\boxtimes)}$ and Jorma Laaksonen

Department of Computer Science, Aalto University, Espoo, Finland
hamed.r-tavakoli@aalto.fi

Abstract. Fixation prediction, also known as saliency modelling, has been a subject undergoing intense study in various contexts. In the context of assistive vision technologies, saliency modelling can be used for development of simulated prosthetic vision as part of the saliency-based cueing algorithms. In this paper, we present an unsupervised multi-scale hierarchical saliency model, which utilizes both local and global saliency pipelines. Motivated by bio-inspired vision findings, we employ features from image statistics. Contrary to previous research, which utilizes one-layer equivalent networks such as independent component analysis (ICA) or principle component analysis (PCA), we adopt independent subspace analysis (ISA), which is equivalent to a two-layer neural architecture. The advantage of ISA over ICA and PCA is robustness towards translation meanwhile being selective to frequency and rotation. We extended the ISA networks by stacking them together, as done in deep models, in order to obtain a hierarchical representation. Making a long story short, (1) we define a framework for unsupervised fixation prediction, exploiting local and global saliency concept which easily generalizes to a hierarchy of any depth. (2) we assess the usefulness of the hierarchical unsupervised features, (3) we adapt the framework for exploiting the features provided by pre-trained deep neural networks, (4) we compare the performance of different features and existing fixation prediction models on MIT1003, (5) we provide the benchmark results of our model on MIT300.

1 Introduction

Visual saliency estimation is a well-established problem in computer vision. It deals with identifying (1) the prominent areas of an image where a salient object is present, or (2) the locations humans look at in a free-viewing task. The first is known as salient object detection and the latter is recognized as fixation prediction or saliency modeling. The two practices are following the same concept, though they produce different results. Figure 1 depicts the result of each approach. Both paradigms are useful for development of assistive vision technologies as part of an efficient computer vision solution, e.g., [1] applied saliency modelling for retinal prostheses as part of a fast image processing pipeline. Similarly, [2] applied salient object detection for a fast indoor obstacle avoidance system. Here, we address the problem of fixation prediction.

© Springer International Publishing AG 2017
C.-S. Chen et al. (Eds.): ACCV 2016 Workshops, Part I, LNCS 10116, pp. 287–302, 2017.
DOI: 10.1007/978-3-319-54407-6_19

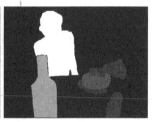

Fig. 1. Saliency estimation task in computer vision, left: the original image, middle: the fixation density map overlaid on the image, used in fixation prediction, right: the salient object masks, used in salient object detection. As depicted, the two tasks are highly correlated, though the ground truth and expected result are different. This paper deals with the fixation prediction. The images are from [3].

The fixation prediction is mostly influenced by the models of visual attention. In this paradigm, the feature integration theory has been playing an important role, particularly in the early models of saliency estimation such as [4]. Over the past decade, many computational models have been developed for fixation prediction. According to [5,6], these models fall into several categories such as information theoretic methods [7,8], spectral analysis models [9–11], Bayesian models [12,13], etc.

The rise of deep convolutional neural networks (CNNs) has also influenced the saliency models. The CNN-based models can be studied under the span of learning-based approaches [14]. The CNN-based techniques can be categorized into two types: (1) end-to-end models and (2) semi end-to-end models. The first group tries to learn an end-to-end mapping, e.g. [15], while the latter category employs the responses of various layers as features for a classification/regression technique such as SVM, e.g., [16], where the error is not back-propagated. The CNN-based models are fully supervised and rely on huge amount of data in order to learn the saliency. In this paper, we seek an unsupervised approach for establishing a hierarchical model for fixation prediction. The advantage of unsupervised techniques is that they do not require huge amount of annotated data, which is expensive to obtain for eye tracking-based databases. Nonetheless, the unsupervised models are less prone to overfit and to learn the properties of a specific dataset, i.e., the unsupervised models are more generic, albeit less accurate, and easily apply to any data meanwhile the deep models often require fine-tuning in order to achieve maximum performance on a data.

To summarize, the contributions are, (1) we define a framework for unsupervised fixation prediction, which exploits local and global saliency concepts and easily generalizes to a hierarchical model of any depth, (2) we assess the usefulness of the hierarchical unsupervised features, (3) we adapt the framework for exploiting the features provided by pre-trained deep neural networks, (4) we compare the performance of different features and existing fixation prediction models, particularly, some of the publicly available deep models of fixation prediction, on MIT1003 database [17], (5) we provide the benchmark results for our framework on MIT300 [18] benchmark.

1.1 Related Work

The saliency estimation literature is replete with numerous models, which going through each is beyond the scope and extent of this paper. Instead, we summarize some of the most relevant works here by presenting some of the recent deep models and reminding some of the relevant works.

The Ensembles of Deep Networks (eDN) method [16] adopts the neural filters learned during image classification task by deep neural networks and learns a classifier to perform fixation prediction. eDN can be considered an extension to [17] in which the features are obtained from layers of deep neural network. For each layer of the deep neural network, eDN first learns the optimal blend of the neural responses of all the previous layers and the current layer by a guided hyperparameter search. Then, it concatenates the optimal blend of all the layers to form a feature vector for learning a linear SVM classifier. It is worth noting that the final saliency map is explicitly center-biased by adding a Gaussian.

Deepgaze [19] utilizes the pre-trained model of Krizhevsky (AlexNet) [20] for the fixation prediction task by treating the problem as a point processing one. Despite this model is justified differently than [16,17], in practice, it boils down to the same framework. Nonetheless, the objective function to be minimized is slightly different due to the explicit incorporation of the center-bias factor and the imposed sparsity constraint in the framework.

Multiresolution CNN (Mr-CNN) is another technique for fixation prediction [21], in which the convolutional neural network is designed to discriminate image patches centered on fixations from non-fixated image patches at multiple resolutions. To this end, the authors train a convolutional neural network at each scale, which results in three parallel networks. The outputs of these networks are connected together through a common classification layer in order to learn the best resolution combination. To infer the saliency map, they perform patch sampling over the input and evaluate each patch using the network.

Deepfix [22] is an end-to-end deep model of fixation prediction which tries to overcome the shortcoming of Mr-CNN by considering the whole image rather than patches. It introduces a location-sensitive convolutional layer in order to incorporate the center-bias in the architecture. To avoid overfitting, caused by the architecture design and data limitations, this model incorporates drop-out in the architecture.

Pan et al. [15] introduce two architectures, a deep network and a shallow one, for end-to-end fixation prediction. SalNet is the deep convolutional network consisting of 13 layers, where the last layer is a deconvolution layer. The network is initialized by the VGG-16 architecture [23] and fine-tuned to perform fixation prediction using a large dataset of fixation data. The shallow network, a.k.a shallowNet, has only five layers where the last two layers are fully connected performing saliency inference. The shallow network is, however, more prone to overfitting compared to the deep one and requires more nuance training tricks.

SALICON [24] develops a model by fine-tuning the convolutional layers of networks, trained on ImageNet, using saliency evaluation metrics as objective functions. SALICON, however, utilizes only the last convolution layer for fixation prediction, albeit in multiple scales. This model feeds the image into the network at two resolutions, coarse and fine. The response of the last convolution layer is obtained for each scale. These responses are then concatenated together to build a feature vector, which is fed into a linear integration scheme, optimizing the Kullback-Leibler divergence between the network output and the ground-truth fixation maps. The error is back-propagated to the convolution layers for fine-tuning the network.

The idea of fixation prediction by neural responses is indeed older than the deep models, though the early models are shallow. We shortly address two of the most relevant models that influence this paper. Bruce and Tsotsos [7] sets the bases of a model based on information maximization. They employ one layer of neurons modelled by the Independent Component Analysis (ICA) and measured the entropy of features to compute saliency. In a similar vain, Zhang et al. [12] define a Bayesian approach for maximizing the local self-information, which has more flexibility for top-down information integration. They also demonstrate that ICA features are better means of fixation prediction than Gabor features.

In the rest of this paper, we first introduce the proposed saliency estimation framework. We then review the unsupervised hierarchical features employed in this work. Next, the adaptation of pre-trained deep neural networks to the saliency framework is explained. We continue with the experiments and results, followed by conclusions.

2   Saliency Estimation

Figure 2 visualizes an overall overview of the proposed model. As depicted, an image pyramid is built, where each resolution is treated by a stack of neural networks, i.e., a hierarchical feature extractor. The features are used by local and global saliency estimators which are eventually integrated into one saliency map. It is worth noting that we have left the normalization steps out for the sake of visualization.

We define the total saliency in terms of global and local saliency over scales. On a single scale, we define the global saliency using a global rarity scheme defined by random walks. The local saliency is then measured using the self-information criteria for each scale. The integrated saliency over each scale is defined as $S^\sigma = S_l^\sigma \times S_g^\sigma$, where σ indicates the scale, and S_l and S_g are the local saliency and the global saliency, respectively. The final saliency is defined by integrating over the scales as follows:

$$S = \sum_\sigma \hat{S}^\sigma, \tag{1}$$

where $\hat{S}$ is the saliency value normalized by softmax. Borrowing the scale-space concept, we implement the scales by employing an image pyramid architecture.

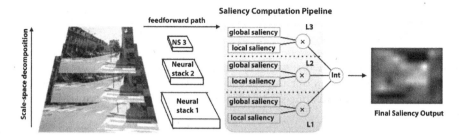

Fig. 2. An overview of the proposed model with three layers, **Int** refers to the integration scheme, i.e. sum in computing (1). In theory, there is no limit for the number layers and **Int** can be replaced by other integration schemes.

2.1 Local Saliency

The local saliency is a well exploited concept as a measure of visual feature conspicuity, examples can be found in [13, 25, 26]. Here, we define the local saliency of an image patch, defined by its feature vector $\mathbf{f}$, in terms of self-information as follows:

$$S_l = -\log(P(\mathbf{f})) = -\sum_{i=1}^{n} \log(P(f_i)), \tag{2}$$

where f_i is the i-th feature of n-dimensional feature vector $\mathbf{f}$, corresponding to the image patch. Assuming $f_i \sim \mathcal{N}(0, 1)$, the local saliency boils down to

$$S_l = \alpha + \frac{1}{2} \sum_{i=1}^{n} f_i^2, \tag{3}$$

where α is a constant.

2.2 Global Saliency

The global saliency can be considered as the rarity of features in the whole image [26]. It can be measured using the likelihood of feature occurrence, e.g. [27]. The rarity of features can be interpreted in terms of their reachability in a random walk, where starting from a node, one visits all the possible nodes and ends in the starting node. To estimate the global saliency, we choose a random walk approach as in [28, 29].

Considering each pixel as a state, we formulate the problem as an ergodic chain where each pixel is reachable from another one. Thus, the probability of reaching a particular state is the same at equilibrium, and we can write:

$$\boldsymbol{\pi} = \mathbf{P}\boldsymbol{\pi}, \tag{4}$$

where $\mathbf{P}$ is a stochastic transition matrix and $\boldsymbol{\pi}$ is the equilibrium probability of $\mathbf{P}$, corresponding to the left eigenvector $\lambda = 1$. It is worth noting that each component of π corresponds to one state or pixel. Under the equilibrium, the saliency is defined as

$$S_g = \exp(-1/(k\pi)), \tag{5}$$

where k is a smoothing factor, which is kept constant and equal to the maximum dimension of the image in our implementation. We define the transition matrix $\mathbf{P}$ in terms of feature distances, where p_{ij}, the probability of transition from state i to j, is

$$p_{ij} = \frac{e^{-D(\mathbf{f}_i, \mathbf{f}_j)}}{\sum_z e^{-D(\mathbf{f}_i, \mathbf{f}_z)}}, \tag{6}$$

$$D(\mathbf{x}, \mathbf{y}) = \sum_{k=1}^d \frac{(x_k - y_k)^2}{x_k + y_k}, \tag{7}$$

where $\mathbf{f}_i$ and $\mathbf{f}_j$ are the feature vectors corresponding to states/pixels i and j, respectively. In principle, $D(\cdot, \cdot)$ can be replaced by any distance function. During experiments, we empirically observed that the Chi-square distance (7) performs better than some common distances such as the Euclidean distance.

3   Unsupervised Hierarchical Feature

The proposed unsupervised hierarchical feature is based on the natural image statistics. To be more specific, it relies on the Independent Subspace Analysis (ISA) [30] as the basis of each level and extends it to extract a hierarchy of features. In this section, we first explain the standard ISA algorithm. We then explain how it is extended to multiple layers.

3.1   Independent Subspace Analysis

The ISA can be seen as a nonlinear extension to Independent Component Analysis (ICA) [31], in which the components of several independent subspaces are estimated. It is worth noting that while the subspaces are independent, the components in the subspace are not necessarily independent. The features learned by ISA are robust towards local translation while being selective to frequency and rotation. Hence, they are providing a good approximation to V1 neural responses in the cortex.

Having a set of unlabeled, whitened data, $\{\mathbf{x}^t\}_{t=1}^T$, where $\mathbf{x}^t = [x_1^t, \cdots, x_n^t]$, $x_i^t \in \mathbb{R}^d$, the ISA can be defined in terms of the following minimization:

$$\underset{W}{\text{minimize}} \quad \sum_{t=1}^T \sum_{i=1}^m h_i(\mathbf{x}^t; W, V), \tag{8}$$

$$\text{s.t.} \qquad WW^T = \mathbf{I},$$

$$h_i(\mathbf{x}^t; W, V) = \left(\sum_{z=1}^k V_{iz} \left(\sum_{j=1}^n W_{zj} x_j^t\right)^2\right)^{0.5}, \tag{9}$$

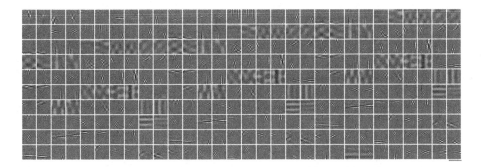

Fig. 3. The learned ISA filters from the first layer of the network. These are the first 300 filters that are used in the training of the second layer.

where $W \in \mathbb{R}^{k \times n}$ is the weights of the inputs, and $V \in \mathbb{R}^{m \times k}$ are the output weights of the ISA neural unit, n is the number of inputs, k is the number of subspaces or poolings, and m is the number of outputs. Figure 3 visualizes the filters learned using ISA algorithm from color images. As depicted, these features seems different than PCA/ICA features in the sense they are grouped as being selective to frequency and rotation. A detailed comparison between properties of ISA features and PCA/ICA features and their superiority is provided in [32], including, a side-by-side visual comparison.

3.2 Hierarchy of ISA

The general approach from natural image statistics towards hierarchy of features is to stack multiple neural architectures together, where the output of one is pooled together to be fed to the next layer, e.g. [33–38]. The stacking operation is an essential part for scaling the architecture towards being deep [39]. Thus, our architecture resembles a deep convolutional neural network, in which each module, group of multi-layer neural networks, performs convolution operations and feed-forwards the result. To learn the ISA filters through the hierarchy, we first learn the first layer of the architecture on several image patches. Then, we compute the responses of the learned ISA filters over some input images and learn the next level of ISA filters over these responses. The process can be implemented efficiently by employing larger patches for the successor layers and avoiding convolution over the whole image [36]. Figure 4 shows the architecture design used in this paper.

To train the ISA over a large number of patches, in our case one million image patches, we employ the stochastic gradient descent over batches of 60,000 image patches. The algorithm converges on average after 300 iterations and the whole training process takes less than 2 h of time, which is relatively fast.

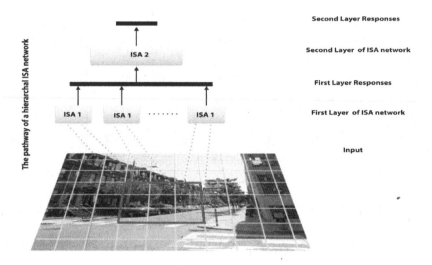

Fig. 4. The design of the ISA hierarchy. The yellow grid represents the fovea corresponding to the first layer. The red rectangle shows the fovea for the higher level. The first layer fovea can have overlap. We used a non-overlapping grid for demonstration purposes. (Color figure online)

4   Adapting the Framework for Pre-trained CNNs

An alternative to the unsupervised features, learned from natural image statistics, are the pre-trained filters of deep CNNs. In essence, they are sharing the common grounds of image statistics, though the procedure for learning them is different. To employ the pre-trained CNNs, we rely on the convolutional layers of networks trained on ImageNet. In other words, we discard the fully connected layers and compute the saliency using (1). There is, however, one blockade for computing the global saliency, that is the dimension of responses on the early layers of the very deep models that exceeds the memory constraints on a workstation. To keep the computation easy and memory-wise compact, we utilize only the local saliency. In other words, the saliency would be $S = \sum_\sigma \hat{S}_l^\sigma$.

5   Experiments

To utilize the described framework with unsupervised features, we train a series of unsupervised hierarchical ISA features. We learn the filters by setting the fovea size, i.e. the image patch size, to be 16×16 and 20×20 for the first and second layers of the stacked ISA, respectively. The ISA filters are then trained using one million image patches sampled from the McGill Calibrated Colour Image Database [40] using stochastic gradient descent and batches of 60,000 in a single precision operation. To compute the first layer responses, we use a somewhat dense overlapping criteria, where the fovea is shifted by two pixels during the convolution operation. We employ the proposed saliency framework also to the features obtained by pre-trained deep neural networks. For this purpose, we utilize the VGG-16 [23], which is reported of having good performance for this task by [24].

The performance evaluation is carried out on the MIT1003 dataset [17], which has the ground truth publicly available for easier model evaluation. To evaluate the model performance, there exists numerous metrics [41] of which, we use shuffled AUC (sAUC) and Normalized Scanpath Score (NSS), following the suggestion of [42], that sAUC and NSS are good complementary measures. The sAUC has chance level of 0.5 and is robust towards center-bias. The NSS score is center-bias sensitive. Therefore, we prioritize the performance of sAUC over NSS. For both metrics, the higher the value is, the better the performance is.

5.1   Scale Effect

We assess the hypothesis that inference over multiple scales improves the fixation prediction results. We also measure the computation time since the number of scales affects the number of convolutions to be performed. For this purpose, we start with a single scale scenario and change the scale by 100 pixels in each step. The results are summarized in Fig. 5. The results motivate that the proposed framework benefits from multiple scales, albeit at a quadratic computational cost. Thus, in the rest of this paper, to keep the computations feasible we employ a four-scale model with the on average computation time of 2.1 s per image.

5.2   Deep Network Layers

Based on the argument of [43], that the neural responses at higher layers in the hierarchy encode more meaningful semantic representations than at lower layers, [24] relies on the last convolutional layer for saliency prediction. To get an insight about the performance of the layers of the VGG-16 adapted to the proposed framework, we divided the network into three almost equal sequences of layers and utilized the responses of these parts to estimate saliency. These sequences of layers are located in the beginning (s1), middle (s2), and the end (s3) of the network architecture. We also test the features from only the last

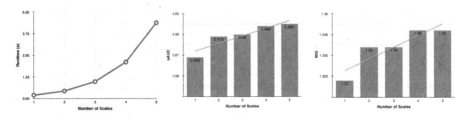

Fig. 5. The scale effect. On the left: the effect of scales on the computation time of the proposed framework. On the middle: sAUC behavior. On the right: NSS behavior against various scales.

Table 1. Performance of VGG-16 layers in the unsupervised architecture.

Metric	Models					
	s1	s1–s2	s2–s3	s3	s1–s2–s3 (full architecture)	last layer
sAUC	0.62	0.68	0.72	0.72	0.70	0.63
NSS	0.72	0.88	1.15	1.13	1.06	0.83

layer. Table 1 summarizes the results for this experiment. It seems that taking the responses from s2 and s3 together produces the best performance, followed by the features from s3. However, the last layer does not contribute significantly to the proposed framework by itself. In the rest of the paper, we will use all the responses from all the layers of VGG-16.

5.3 Performance Analysis

We compare the proposed model and features with several models on the MIT1003 dataset. We compare with SalNet [15], eDN [16], GBVS [28], SUN [12], AIM [7], FES [13], and Itti [44]. The models are chosen to be among the best performing models according to [18] and share some common grounds with the proposed framework. Thus, we can interpret the results easily without possible confusion in model differences. The Itti's model was selected as the minimum baseline of fixation prediction. We choose AIM, SUN, and GBVS as they have similarity with the components of the proposed framework. The SalNet [15] and eDN [16] are both among the top performing ones. Unfortunately, we could not access the code/maps from other deep models due to unavailability to public at the time. The saliency maps were computed using default parameters, except in the case of eDN, for which the precomputed maps are provided.

The results are summarized in Table 2. The proposed model, compared to the deep models by sAUC metric, outperforms eDN, though falls short of SalNet. Knowing that eDN is learning the best linear combination of pre-learned filters with no fine-tuning, we can infer that a simple uniform integration, performed for pre-trained CNN filters, or weighting the filters from image statistics, here ISA filters, with their global saliency before integration as for the unsupervised features is better than a classification/regression operation on the deep CNN filters without any fine-tuning of filters.

To interpret the NSS score, we first remind that NSS is center-bias sensitive. For example, eDN has a good NSS score because it biases the maps towards the center (see Fig. 6). The GBVS algorithm outperforms the proposed framework since its maps are slightly boosted in the center compared to ours. Nonetheless, by considering the sAUC metric, we infer that the proposed framework is the second best in the list of current models since its inefficiency considering NSS can be compensated by exploiting the center-bias. To demonstrate this phenom-

Table 2. Performance comparison on MIT1003 dataset. On this data, the upper-bound, human performance, vs. chance level, Gaussian blob, are the following: sAUC: 0.75 vs. 0.5 and NSS: 3.1 vs. 1.3.

Metric	Models								
	SalNet	eDN	GBVS	SUN	AIM	FES	Itti	Proposed VGG-16	Proposed Unsupervised
sAUC	0.72	0.67	0.67	0.64	0.67	0.65	0.62	0.70	0.68
NSS	1.88	1.38	1.51	0.92	0.93	1.10	0.89	1.06	1.35

ena, we biased the saliency map values with respect to the center of the image resulting in NSS = 1.5 and sAUC = 0.68 for the case of unsupervised features.

We utilized two types of features, unsupervised hierarchical features and features from VGG-16. Looking closely to the performance and fixation maps, we learn that the VGG-16 features produce a very fine-grained saliency map, explaining its low score on NSS. To compensate, we also evaluated a blurred version of the maps ($\sigma = 0.4$) for VGG-16, resulting in a slight performance improvement of sAUC = 0.71 and NSS = 1.11. Despite the gap between proposed model with VGG-16 and the SalNet is not significant considering sAUC, we observe a huge gap in terms of NSS.

To summarize, there is a theoretical similarity between the proposed model and GBVS, SUN and AIM. The local pipeline is based on self-information theory akin to SUN and the global pipeline is similar to GBVS. In the proposed framework the features are, however, different and the global pipeline is used as weights to the local pipeline. The proposed model improves over SUN on both metrics and GBVS in terms of sAUC, not to forget that GBVS slightly exploits the border-effect that boosts NSS. In overall, the results on MIT1003 motivate that the proposed features and pipeline have improved over SUN, AIM, and GBVS for MIT1003 data.

We visualize the top five saliency maps and the worst five ones of the proposed model along with others in Fig. 6. The visual inspection of the fixation maps backs the interpretation of the scores on MIT1003.

6   Benchmark Results

To get a better insight, we further benchmark the model on MIT300 [18]. The results are summarized in Table 3, sorted based on the sAUC metric. We report all the available metrics on the benchmark (please check [18] for a detailed description).

While the proposed model improves over SUN model significantly on all the metrics, interpreting the results in comparison with AIM and GBVS is a bit more difficult. The proposed model using the VGG-16 features improves over AIM on all metrics, while the unsupervised features fall short of AIM only on sAUC metric. Both features improve over GBVS in terms of sAUC in the

Fig. 6. Visualization of the predicted fixation maps, from left to right: original image, human fixation map, SUN [12], AIM [7], GBVS [28], eDN [16], SalNet [15], Proposed VGG-16, Proposed Unsupervised. The first five rows are the first five top performances for the proposed model, and the last five rows are among the worst performances.

proposed model. They, however, fall short on some of the other metrics, which are center-bias sensitive, making a definite conclusion difficult since GBVS exploits border-effect in its implementation. The border-effect has a similar influence as center-bias on model performance and can boost some metrics significantly, e.g., AUC-Judd can be boosted up to 24% with only four pixel zero borders on a saliency map [12].

It is interesting that the proposed model improves over eDN in terms of several metrics, including sAUC, NSS and KL. Knowing that eDN adds a strong center-bias to its maps, we can speculate that the proposed framework outperforms the regression-based CNN combination framework of eDN, where its AUC performance can be associated to center-bias. Nonetheless, the benchmark

Table 3. Performance comparison on MIT300 Benchmark. Note: the ground-truth is not publicly available and we use Itti model implemented by GBVS.

Models	sAUC	AUC-Judd	EMD	AUC-Borji	CC	NSS	KL
SALICON [24]	0.74	0.87	2.62	0.85	0.74	2.12	0.54
SalNet [15]	0.69	0.83	3.31	0.82	0.58	1.51	0.81
Proposed VGG-16	0.67	0.78	3.96	0.76	0.45	1.24	1.02
AIM [7]	0.66	0.77	4.73	0.75	0.31	0.79	1.18
BMS [45]	0.65	0.83	3.35	0.82	0.55	1.41	0.81
Proposed Unsup.	0.64	0.80	4.11	0.79	0.47	1.21	1.00
GBVS [28]	0.63	0.81	3.51	0.80	0.48	1.24	0.87
ShallowNet [15]	0.63	0.80	3.99	0.79	0.56	1.47	0.95
Itti [44]	0.63	0.75	4.26	0.74	0.37	0.97	1.03
eDN [16]	0.62	0.82	4.56	0.81	0.45	1.14	1.14
SUN [12]	0.61	0.67	5.10	0.66	0.25	0.68	1.27
Judd [17]	0.60	0.81	4.45	0.80	0.47	1.18	1.12
FES [13]	0.59	0.80	3.36	0.73	0.48	1.27	1.20

setup of eDN maps on MIT300 is not disclosed and making such a conclusion is difficult without concrete evidence. We can, however, infer that the proposed framework adapted for pre-trained CNN features can be a good baseline evaluator for learning-based deep models. Also, the proposed model can be used for choosing an appropriate deep architecture for saliency modeling prior to any computational process, by evaluating the CNN features for fixation prediction. It is worth noting that due to the lack of enough data, most of the deep saliency models exploit pre-trained networks and fine-tune.

7   Discussion and Conclusions

The proposed framework has some commonalities with the deep CNN architectures such as convolution operation, module stacking and multi-layer neural networks in each module. The parameters of neural layers of each module can be learned. Furthermore, in theory, there is no limitation for stacking indefinite number of ISA neural architectures to build a very deep network.

Despite the accuracy of the unsupervised models may not be as high as some deep-learning based supervised models, we believe that assisstive vision methods benefit from the unsupervised bottom-up fixation prediction more because their saliency maps seem more appropriate for exploration of unknown visual environment compared to those of the deep models. It would be the next step in our research to validate the performance of unsupervised models in task-specific assistive vision scenarios.

To conclude, this paper presented a framework for saliency estimation based on unsupervised hierarchical features learned from natural image statistics, more specifically independent subspace analysis (ISA). It stacked several ISA networks to build the hierarchy of features. The framework was then adapted to utilize pre-trained features from deep convolutional neural networks. The experiments showed the usefulness of the proposed framework and features. The proposed model provides a performance comparable with some deep models, though it is less accurate. The results motivate the benefits of multiscale feature learning and natural image statistics in saliency estimation. Furthermore, for the task of saliency prediction, the proposed framework can be used for choosing an appropriate deep architecture prior to any computational process like fine-tunning by evaluating the CNN features for saliency modelling.

The model's code is available at: http://github.com/hrtavakoli/UHM.

Acknowledgement. The authors would like to acknowledge the Finnish Center of Excellence in Computational Inference Research (COIN) and the computational resources provided by the Aalto Science-IT project.

References

1. Parikh, N., Itti, L., Weiland, J.: Saliency-based image processing for retinal prostheses. J. Neural Eng. **7**, 016006 (2010)
2. Huang, H.C., Hsieh, C.T., Yeh, C.H.: An indoor obstacle detection system using depth information and region growth. Sensors **15**, 27116–27141 (2015)
3. Li, Y., Hou, X., Koch, C., Rehg, J.M., Yuille, A.L.: The secrets of salient object segmentation. In: CVPR (2014)
4. Itti, L., Koch, C.: A saliency-based search mechanism for overt and covert shifts of visual attention. Vis. Res. **40**, 1489–1506 (2000)
5. Borji, A., Sihite, D., Itti, L.: Quantitative analysis of human-model agreement in visual saliency modeling: a comparative study. TIP **22**, 55–69 (2013)
6. Borji, A., Itti, L.: State-of-the-art in visual attention modeling. PAMI **35**, 185–207 (2013)
7. Bruce, N.D.B., Tsotsos, J.K.: Saliency based on information maximization. In: NIPS (2006)
8. Hou, X., Zhang, L.: Dynamic visual attention: searching for coding length increments. In: NIPS (2008)
9. Guo, C., Ma, Q., Zhang, L.: Spatio-temporal saliency detection using phase spectrum of quaternion fourier transform. In: CVPR (2008)
10. Hou, X., Harel, J., Koch, C.: Image signature: highlighting sparse salient regions. PAMI **34**, 194–201 (2012)
11. Vig, E., Dorr, M., Martinetz, T., Barth, E.: Intrinsic dimensionality predicts the saliency of natural dynamic scenes. PAMI **34**, 1080–1091 (2012)
12. Zhang, L., Tong, M.H., Marks, T.K., Shan, H., Cottrell, G.W.: Sun: a Bayesian framework for saliency using natural statistics. J. Vis. **8**, 32 (2008)
13. Rezazadegan Tavakoli, H., Rahtu, E., Heikkilä, J.: Fast and efficient saliency detection using sparse sampling and kernel density estimation. In: Heyden, A., Kahl, F. (eds.) SCIA 2011. LNCS, vol. 6688, pp. 666–675. Springer, Heidelberg (2011). doi:10.1007/978-3-642-21227-7_62
14. Zhao, Q., Koch, C.: Learning saliency-based visual attention: a review. Signal Process. **93**, 1401–1407 (2013)

15. Pan, J., McGuinness, K., Sayrol, E., O'Connor, N., Giro-i Nieto, X.: Shallow and deep convolutional networks for saliency prediction. In: CVPR (2016)
16. Vig, E., Dorr, M., Cox, D.: Large-scale optimization of hierarchical features for saliency prediction in natural images. In: CVPR (2014)
17. Judd, T., Ehinger, K., Durand, F., Torralba, A.: Learning to predict where humans look. In: ICCV (2009)
18. Bylinskii, Z., Judd, T., Borji, A., Itti, L., Durand, F., Oliva, A., Torralba, A.: Mit saliency benchmark (2016)
19. Kmmerer, M., Theis, L., Bethge, M.: Deep gaze i: Boosting saliency prediction with feature maps trained on imagenet. In: ICLR Workshop (2015)
20. Krizhevsky, A., Sutskever, I., Hinton, G.E.: Imagenet classification with deep convolutional neural networks. In: NIPS (2012)
21. Liu, N., Han, J., Zhang, D., Wen, S., Liu, T.: Predicting eye fixations using convolutional neural networks. In: CVPR (2015)
22. Kruthiventi, S.S., Ayush, K., Babu, R.V.: Deepfix: A fully convolutional neural network for predicting human eye fixations (2015)
23. Simonyan, K., Zisserman, A.: Very deep convolutional networks for large-scale image recognition. In: ICLR (2015)
24. Huang, X., Shen, C., Boix, X., Zhao, Q.: Salicon: Reducing the semantic gap in saliency prediction by adapting deep neural networks. In: ICCV (2015)
25. Bruce, N.D.B., Tsotsos, J.K.: Saliency, attention, and visual search: an information theoretic approach. J. Vis. 9, 5 (2009)
26. Borji, A., Itti, L.: Exploiting local and global patch rarities for saliency detection. In: CVPR (2012)
27. Mancas, M.: Computational attention: towards attentive computers. PhD thesis, CIACO University (2007)
28. Harel, J., Koch, C., Perona, P.: Graph-based visual saliency. In: NIPS (2007)
29. Gopalakrishnan, V., Hu, Y., Rajan, D.: Random walks on graphs to model saliency in images. In: CVPR (2009)
30. Hyvarinen, A., Hoyer, P.: Emergence of phase- and shift-invariant features by decomposition of natural images into independent feature subspaces. Neural Comput. 12, 1705–1720 (2000)
31. Comon, P.: Independent component analysis - a new concept? Signal Process. 36, 287–314 (1994)
32. Hyvärinen, A., Hurri, J., Hoyer, P.O.: Natural Image Statistics - A Probabilistic Approach to Early Computational Vision. Springer, London (2009)
33. Matsuda, Y., Yamaguchi, K.: Linear multilayer independent component analysis for large natural scenes. In: NIPS (2004)
34. Matsuda, Y., Yamaguchi, K.: Linear multilayer ICA generating hierarchical edge detectors. Neural Comput. 19, 218–230 (2007)
35. Matsuda, Y., Yamaguchi, K.: Linear multilayer ICA using adaptive PCA. Neural Process. Lett. 30, 133–144 (2009)
36. Le, Q.V., Zou, W.Y., Yeung, S.Y., Ng, A.Y.: Learning hierarchical invariant spatio-temporal features for action recognition with independent subspace analysis. In: CVPR (2011)
37. Gutmann, M.U., Hyvärinen, A.: A three-layer model of natural image statistics. J. Physiol. Paris 107, 369–398 (2013)
38. Hosoya, H., Hyvarinen, A.: A hierarchical statistical model of natural images explains tuning properties in V2. J. NEUROSCI. 35(29), 10412–10428 (2015)
39. Bengio, Y., LeCun, Y.: Scaling learning algorithms towards AI. Large-Scale Kernel Mach. 34(5), 1–41 (2007)

40. Olmos, A., Kingdom, F.A.: A biologically inspired algorithm for the recovery of shading and reflectance images. Perception **33**, 1463–1473 (2004)
41. Borji, A., R.-Tavakoli, H., Sihite, D.N., Itti, L.: Analysis of scores, datasets, and models in visual saliency prediction. In: ICCV (2013)
42. Riche, N., Duvinage, M., Mancas, M., Gosselin, B., Dutoit, T.: Saliency and human fixations: state-of-the-art and study of comparison metrics. In: ICCV (2013)
43. Zeiler, M.D., Fergus, R.: Visualizing and understanding convolutional networks. In: Fleet, D., Pajdla, T., Schiele, B., Tuytelaars, T. (eds.) ECCV 2014. LNCS, vol. 8689, pp. 818–833. Springer, Cham (2014). doi:10.1007/978-3-319-10590-1_53
44. Itti, L., Koch, C., Niebur, E.: A model of saliency-based visual attention for rapid scene analysis. PAMI **20**, 1254–1259 (1998)
45. Huang, L., Pashler, H.: A boolean map theory of visual attention. Psychol. Rev. **114**, 599 (2007)

Face Detection and Object Recognition for a Retinal Prosthesis

Derek Rollend, Paul Rosendall, Seth Billings, Philippe Burlina, Kevin Wolfe, and Kapil Katyal(✉)

The Johns Hopkins University Applied Physics Laboratory,
11100 Johns Hopkins Road, Laurel, MD, USA
kapil.katyal@jhuapl.edu

Abstract. We describe the recent development of assistive computer vision algorithms for use with the Argus II retinal prosthesis system. While users of the prosthetic system can learn and adapt to the limited stimulation resolution, there exists great potential for computer vision algorithms to augment the experience and significantly increase the utility of the system for the user. To this end, our recent work has focused on helping with two different challenges encountered by the visually impaired: face detection and object recognition. In this paper, we describe algorithm implementations in both of these areas that make use of the retinal prosthesis for visual feedback to the user, and discuss the unique challenges faced in this domain.

1 Introduction

The ability for humans to visually sense features within the environment is a fundamental capability that enables effortless interaction with everyday objects. Individuals with limited or severely impaired vision find mobility, interaction with other individuals, and interaction with objects to be arduous tasks. Such visual impairment may arise from degenerative disorders, trauma, or other diseases. Several commercial systems have been developed over the last several decades that provide at least a subset of the visually impaired population with the ability to perceive information about the environment. One such system is the Argus II retinal prosthesis developed by Second Sight Medical Products (SSMP). The Argus II is an FDA-approved medical device that consists of wearable glasses (with an integrated monocular camera), a 60-channel (10×6) electrode array and a video processing unit (VPU) used to host computer vision algorithms that ultimately determine which electrodes to stimulate and the intensity of that stimulation.

Electronic supplementary material The online version of this chapter (doi:10. 1007/978-3-319-54407-6_20) contains supplementary material, which is available to authorized users.

C.-S. Chen et al. (Eds.): ACCV 2016 Workshops, Part I, LNCS 10116, pp. 303–313, 2017.
DOI: 10.1007/978-3-319-54407-6_20

One of the major challenges of the system is to distill the numerous and varying visual features present in a typical world scene down to a 60-channel electrode array. This problem becomes exacerbated when trying to locate an object or a face in the presence of clutter. In this paper, we describe several algorithms that were developed to provide a means of highlighting objects of interest to the blind individual. In addition, we provide a description of the optimization strategies necessary to allow some of these algorithms to run on an embedded platform in real time at 30 frames per second (fps), and the prototype implementations that currently run on a laptop.

This paper is organized as follows: Sect. 2 provides an overview of the Argus II system as well as the vision algorithm framework, Sect. 3 provides a high-level description of the face detection and object recognition algorithms, Sect. 4 describes optimization strategies and example implementations, and Sect. 5 provides a summary and the planned future research efforts.

2 System Overview

Figure 1 highlights the main components of the Argus II retinal prosthetic system. This device consists of wearable glasses, a 60-channel electrode array and a VPU. The wearable glasses contain an integrated 640 × 480 resolution monocular RGB outward facing camera capable of capturing live video feeds from the patient's field of view. The VPU runs computer vision algorithms with the goal of distilling the world scene down to a 5-bit 10 × 6 resolution image containing

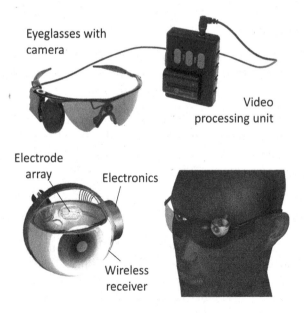

Fig. 1. Argus II system diagram

information to be presented to the individual. An RF coil on the glasses wirelessly transmits this data to the retinal implant. These signals are then sent to the 60-channel electrode array which emits small pulses of electricity along the optic nerve to the brain.

The VPU consists of a Texas Instruments OMAP DM3730 1 GHz System on Module (SOM) chip with 512 MB of NAND Flash. This power-efficient processor was chosen in order to extend battery life for everyday use. One of the significant challenges of using this processor is the limited CPU resources available to run computationally intensive computer vision algorithms in parallel with the stimulation and safety monitoring software. At present, the face detection algorithm described below has been implemented on the VPU hardware, using optimization strategies discussed in Sect. 4.1. For prototype algorithm testing, a laptop or external processing platform may be used in conjunction with the existing VPU by feeding in a different video signal than the video coming from the patient's glasses. This setup enables exploratory testing while maintaining device safety protocols, and was used in this work for testing the object recognition algorithm (Sect. 3.2).

3 Methods: Vision Algorithms

There are a variety of areas in which computer vision can increase the quality of life of the visually impaired and retinal prosthesis users. The initial focus on face detection and object recognition during this collaboration was born out of patient feedback on the desire to find and recognize faces around them, and the basic ability to know what is in front of them in unfamiliar surroundings.

3.1 Face Detection

One feature often requested by blind individuals is to have a mode where only faces in the scene are highlighted by the retinal implant. This allows the patient to clearly identify and recognize people without having to rely on audio cues. Previous work has demonstrated the feasibility of using the Argus II for this task [18].

The face detection portion of our module uses the classical Viola-Jones cascaded classifier detection method available in OpenCV, operating on local binary pattern (LBP) features [2,10,19]. When a face is first detected using LBP features, a second cascaded classifier based on Haar features is used to verify the face and reduce false positives. Despite the fact that LBP features are considered very efficient, the face detection portion is the most computationally expensive part of the entire face detection module and was the subject of various optimizations during development.

When a detected face intersects the implant field of view (FOV), an oval region around the face is segmented, contrast-enhanced via histogram equalization, and then added back to the reduced intensity background region around

Fig. 2. Example face detection outputs, where the rectangle in the middle of the frames represents the implant FOV image that is further downsampled and sent to the implant. Left: Sample output of the face enhancement phase. Right: Sample output of the spatial cueing phase. The black square around the face indicates a face has been detected and is only used for visualization.

the face, as shown in Fig. 2 (left) (note that the output shown does not include the downsampling to the required 10×6 electrode input).

Spatial cueing is used when a detected face lies outside the implant FOV to indicate to the user where the face is relative to where they are currently looking. This concept is illustrated in Fig. 2 (right) wherein a bright stimulation rectangle is presented on the edge of the electrode array closest to where the face is present in the global forward-facing camera FOV. When a certain amount of intersection between the detected face and the implant FOV region occurs as the user turns their head toward the face, the algorithm will switch to face enhancement output (Fig. 2, left). For further algorithm details, please see [16].

3.2 Object Recognition

Another feature frequently requested by visually impaired individuals is the ability to perform object recognition. Recently, tremendous advancements have been made in this area due to the success of Convolutional Neural Networks (CNNs), even approaching/surpassing the level of human performance in some cases [5]. Most of the initial success of CNNs occurred in the single-image classification task, wherein the algorithm must identify the class of the primary object in the image. However, recently more research has focused on using CNNs to perform object detection and localization, both in still images and in videos [3,4,8,9,13–15,17]. In this work, we utilize the Single Shot MultiBox Detector (SSD) CNN from [13], designed for use with still images, but adapt it to operate on the video from a camera in the Argus II user's glasses. The CNN model used in this work was trained on the 80 classes in the Microsoft Common Objects in Context (COCO) dataset [11], a subset of which has been selected for initial (indoor) system testing.

System Description. An overview of the object recognition scenario is shown in Fig. 3. In the current work, the user sits in front of a static scene and the camera imagery is fed to the object detection CNN. Only CNN detections above a certain confidence threshold are considered valid and kept for the filtering/tracking phase. The tracking step (further described below) is critical as it filters out both spurious false positives and false negatives, thereby stabilizing the system's representation of the scene.

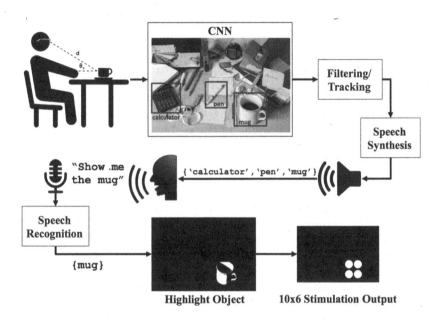

Fig. 3. Object recognition system overview. Video from the scene camera on the user's glasses is fed to the CNN, and the detected objects are tracked over time. Speech recognition is used to select an object of interest, after which the object is highlighted in the frame and converted to the correct 10×6 stimulation output.

Upon user request, speech synthesis is used to describe the scene by announcing any recognized and currently tracked objects. Additionally, speech recognition is used to listen for an object selection from the list of detected objects that have been aurally presented to the user. When a selected object has been successfully recognized, the system informs the user that it is currently looking for that object via speech synthesis, highlights the desired object in the camera input frame, and downsamples the stimulation image before sending it to the retinal implant.

Object Filtering/Tracking. By nature, the SSD CNN from [13] produces both false positives and false negatives when run on consecutive frames from a video stream. These are primarily due to the fact that the network is trained on

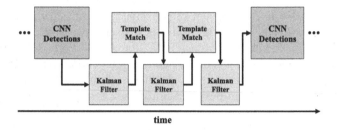

Fig. 4. Object tracking high-level concept.

single images and lacks any temporal information that might help account for changes in perspective as either the camera moves around the scene or objects move within the scene.

In order to smooth out the noisy detections, we developed a basic tracking scheme that employs both a Kalman filter and template matching in order to keep track of detected objects over time (see Fig. 4). We note that elements of this approach are similar to ideas from the object tracking literature, primarily the Tracking-Learning-Detection (TLD) algorithm from [7]. When an object has been detected by SSD, a track is only "soft"-initialized for that object, *i.e.* it is not considered an actual track. If the object has been observed by SSD a minimum number of times, then it is "hard"-initialized and added to the list of known objects in the scene.

A Kalman filter is used at each timestep in order to predict and measure a tracked object's bounding box centroid (x, y), width, and height. Therefore, the estimated track state, x, contains six elements: $x = [x_c, y_c, \dot{x}_c, \dot{y}_c, w, h]$, where (x_c, y_c) is the bounding box centroid, $\dot{x}_c$ and $\dot{y}_c$ are the associated centroid velocity estimates, and w, h are the bounding box width and height, respectively. The Kalman filter system is defined by the following equations:

Prediction:

$$x_{k+1} = \mathbf{A}x + \mathbf{B}u \tag{1}$$

$$\mathbf{P}_{k+1} = \mathbf{A}\mathbf{P}_k\mathbf{A}^T + \mathbf{Q} \tag{2}$$

Update:

$$\mathbf{K} = \mathbf{P}_k\mathbf{H}^T(\mathbf{H}\mathbf{P}_k\mathbf{H}^T + \mathbf{R})^{-1} \tag{3}$$

$$x_{k+1} = x_k + \mathbf{K}(z_k - \mathbf{H}x_k) \tag{4}$$

$$\mathbf{P}_{k+1} = (\mathbf{I} - \mathbf{K}\mathbf{H})\mathbf{P}_k \tag{5}$$

where:

$\mathbf{A}$ is the state matrix
$\mathbf{B}$ is the control matrix
$\mathbf{P}$ is the state covariance matrix
$\mathbf{Q}$ is the process noise covariance matrix
$\mathbf{H}$ is the observation matrix

R is the measurement noise covariance matrix
K is the Kalman gain
z_k is the measurement at time k

For our use case, we use a zero control matrix and a basic linear motion model with the following noise covariance values:

$$\mathbf{Q} = 0.005 \times \mathbf{I}_6 \qquad \mathbf{R} = (3 \times 10^{-5}) \times \mathbf{I}_4$$

To associate SSD detections with the tracked objects at each iteration, three criteria are used: matched class label, minimum intersection over union (IOU), and a maximum Euclidean distance between centroids. The minimum IOU threshold, T_{IOU}, and maximum distance threshold, T_D, were determined empirically to be $T_{IOU} = 0.1$ and $T_D = 2 * H_{track}$ where H_{track} is the length of the hypotenuse formed by the bounding box width and height (w, h).

In the event that no track association is found for an SSD detection, a new track is soft-initialized as discussed above. Conversely, if no matching SSD detection is found for an existing track, then template matching is employed between the most recent SSD-detected image patch for that track and an expanded search region centered at the current track centroid (x_c, y_c). If the maximum normalized cross-correlation metric exceeds a certain threshold, T_{TM}, a match is declared and the Kalman filter is corrected with this measured location in the input frame. The stored image patch for a given track is not updated until a new associated SSD detection has been found. At each iteration we also increase the "age" of every track (both soft and hard-initialized) and reset the age of tracks that have been either SSD-associated or template matched. All tracks that exceed a certain age are removed, ensuring any objects that were false positives or objects that have exited the camera's field of view are no longer processed by the algorithm.

4 Results

In this section we describe implementations of both algorithms discussed in Sect. 3. Due to the fact that the algorithms are in different stages of development, we discuss the results for each in a different manner. In Sect. 4.1 we focus on the various optimization strategies used to enable the face detection and cueing algorithm to run in real-time on the current Argus II embedded, ARM-based hardware platform described in Sect. 2. In Sect. 4.2 we show a prototype implementation of the object recognition system, currently designed to run on a modern laptop with a graphics processing unit (GPU). It is important to note that while both algorithms are at different maturity levels, they have both been tested with and received favorable feedback from current Argus II users.

4.1 Face Detection

In order to reduce the execution time of the face detection algorithm, we consider the effects of two algorithm parameters: the input image resolution and the required detectable face size. The input image resolution is first downsampled from 640 × 480 to 320 × 240 before face detection. After downsampling, we only consider an expanded implant FOV region during initial detection, where the amount of expansion is configurable and is set based on the maximum detectable face size. After a face has been detected and verified, the tracking phase begins, wherein detection is only performed on a slightly expanded region of interest (ROI) around the current face location (see Fig. 5). In both the initial detection and tracking phases, the amount of expansion is dictated by the detectable face size.

The maximum and minimum detectable face sizes are determined by the camera FOV, face anthropometry, and by the sizes of faces used to train the cascaded classifier. For initial detection, we expand the downsampled implant FOV size by 30 pixels on all sides for a total operating area of 140 × 108 pixels. This enables detection down to 0.25 m from the camera. The minimum face size during initial detection, 24 × 24 pixels, is determined by the trained cascaded classifier and allows for detection up to 2 m from the camera with the current system.

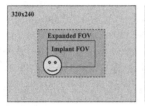

Fig. 5. Left: Expanded implant FOV during initial detection and verification. Right: Subsequent operation only on an expanded region of interest around the detected face during the tracking phase.

During the tracking phase (after initial face detection and verification), the maximum and minimum face sizes are set dynamically based on the current bounding box around the face. The maximum size is based on padding added to the current bounding box, nominally one third of the current width and height. The minimum size is set to half of the current face bounding box size or the minimum detectable face size parameter, whichever is larger. This adaptive search ROI enables faster than real-time operation on the embedded hardware platform after a face has been detected.

These optimizations helped improve the qualitative user experience and reduced the computational strain on the VPU system as a whole. The high latency induced in the non-optimized case makes it difficult for the algorithm to keep pace with the changing input as users move their head to scan the environment, causing spurious detections and no temporal consistency. Additionally, having the nominal runtime consistently higher than the frame rate (30 frames/s or 33 ms/frame) would drive the system's CPU utilization towards 100%, creating an unrealistic scenario for a real-time embedded system. Table 1 provides

Table 1. Optimization summary for the face detection algorithm.

Algorithm phase	Non-optimized (ms)	Optimized (ms)	Speedup
Initial detection	402.9	43.4	9.3×
Verification	558.3	164.9	3.4×
Tracking	402.9	18.9	21.3×

a summary of the effects of the various optimization strategies used in the face detection algorithm.

4.2   Object Recognition

The object tracking algorithm described in Sect. 3.2 was implemented in Python using the OpenCV 3.0 Kalman filter implementation [2]. An example output frame from the CNN detection and tracking module can be seen in Fig. 6. Currently tracked objects are represented with bounding boxes drawn around the objects on the camera input frame, with the current confidence score displayed next to the object labels. This confidence is output from the SSD CNN, and averaged over all detections of a given object so that low confidence false alarms may be ignored. The desired object to be identified is selected via speech recognition (in this case the cup) and highlighted, as seen in Fig. 7. A video showing the raw SSD detections and filtered detections side-by-side can be found in the supplemental material.

The core object detection CNN uses a variant of the Caffe library [6] available from [12], and is built with CUDA 8.0 and cuDNN v5.0 support. A prototype application was developed in Python to combine the object detection/tracking module, the speech synthesis module (using Festival [1]), and the speech

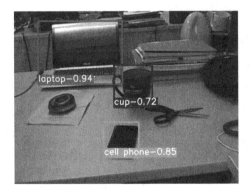

Fig. 6. Detected and tracked objects from a subset of classes from the Microsoft COCO dataset [11] used during initial testing.

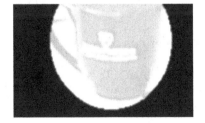

Fig. 7. The selected object ("cup") is emphasized, and the background pixels are darkened before sending the implant FOV image to the Argus II system.

recognition module (using CMU Sphinx [20]). The entire vision algorithm runs
in 99.1 ms (≈10 fps) on a laptop with a quad-core Core i7 2.80 GHz CPU and an
NVIDIA Quadro M1000M GPU, where 98% of the processing time is spent in the
forward pass of the image through the CNN.[1] A video demonstrating successful
use of the system by an Argus II user can be found in the supplemental material.

5 Conclusions

In this paper, we present our approach to address two fundamental needs for
blind individuals using existing retinal prosthetic devices, namely the ability to
identify and locate faces in the scene, as well as identify and locate objects in
front of them. We describe optimization strategies necessary to allow the face
detection algorithm to run in real-time on low power, embedded devices as found
in the existing Argus II System. We also describe a filtered object recognition
algorithm that enables a single-image object recognition CNN to be used with
live video input. At the time of this writing, we have tested these algorithms on
a limited number of blind individuals with positive feedback. A more comprehensive publication focusing on formal patient performance is forthcoming after
additional patient testing.

Future work will include further patient testing of the two algorithms presented here, with potential integration into the next generation Argus system.
We also plan to refine our object detection in video approach detailed here by
testing it on the ILSVRC Object Detection in Video dataset [17]. The recent
and significant advances in image captioning might be leveraged as well in order
to provide a more descriptive depiction of the scene using speech synthesis. In
addition, we plan on examining Simultaneous Localization and Mapping (SLAM)
algorithms in order to assist users with guided navigation in unknown environments. These algorithms will all be used together to further improve the quality
of life for current and future retinal prosthesis users.

Acknowledgement. This work was supported by an Alfred E. Mann collaboration
grant. We would also like to thank Arup Roy, Avi Caspi, and Robert Greenberg, our
collaborators from Second Sight Medical Products.

References

1. Black, A.W., Taylor, P.A.: The festival speech synthesis system: system documen-
   tation. Technical report HCRC/TR-83, Human Communciation Research Centre,
   University of Edinburgh, Scotland, UK (1997). http://www.cstr.ed.ac.uk/projects/
   festival.html
2. Bradski, G.: The OpenCV Library. Dr. Dobb's J. Softw. Tools (2000). http://code.
   opencv.org/projects/opencv/wiki/CiteOpenCV

---

[1] The speech synthesis and recognition modules run asynchronously from the vision
algorithm and their computational demands are minimal compared to the CNN detec-
tion/tracking portion.

3. Burlina, P.: MR-CNN: a stateful fast R-CNN. In: International Conference on Pattern Recognition (2016)
4. Girshick, R., Donahue, J., Darrell, T., Malik, J.: Rich feature hierarchies for accurate object detection and semantic segmentation. In: Computer Vision and Pattern Recognition (2014)
5. He, K., Zhang, X., Ren, S., Sun, J.: Delving deep into rectifiers: surpassing human-level performance on imagenet classification. CoRR abs/1502.01852 (2015)
6. Jia, Y., Shelhamer, E., Donahue, J., Karayev, S., Long, J., Girshick, R., Guadarrama, S., Darrell, T.: Caffe: convolutional architecture for fast feature embedding. arXiv preprint (2014). arXiv:1408.5093
7. Kalal, Z., Mikolajczyk, K., Matas, J.: Tracking-learning-detection. IEEE Trans. Pattern Anal. Mach. Intell. 34(7), 1409–1422 (2012)
8. Kang, K., Li, H., Yan, J., Zeng, X., Yang, B., Xiao, T., Zhang, C., Wang, Z., Wang, R., Wang, X., Ouyang, W.: T-CNN: tubelets with convolutional neural networks for object detection from videos. CoRR abs/1604.02532 (2016)
9. Kang, K., Ouyang, W., Li, H., Wang, X.: Object detection from video tubelets with convolutional neural networks. CoRR abs/1604.04053 (2016)
10. Liao, S., Zhu, X., Lei, Z., Zhang, L., Li, S.Z.: Learning Multi-scale Block Local Binary Patterns for Face Recognition. In: Lee, S.-W., Li, S.Z. (eds.) ICB 2007. LNCS, vol. 4642, pp. 828–837. Springer, Heidelberg (2007). doi:10.1007/978-3-540-74549-5_87
11. Lin, T., Maire, M., Belongie, S.J., Bourdev, L.D., Girshick, R.B., Hays, J., Perona, P., Ramanan, D., Dollár, P., Zitnick, C.L.: Microsoft COCO: common objects in context. CoRR abs/1405.0312 (2014)
12. Liu, W.: SSD Caffe (2015). https://github.com/weiliu89/caffe/tree/ssd
13. Liu, W., Anguelov, D., Erhan, D., Szegedy, C., Reed, S.E.: SSD: single shot multibox detector. CoRR abs/1512.02325 (2015)
14. Redmon, J., Divvala, S.K., Girshick, R.B., Farhadi, A.: You only look once: unified, real-time object detection. CoRR abs/1506.02640 (2015)
15. Ren, S., He, K., Girshick, R., Sun, J.: Faster R-CNN: towards real-time object detection with region proposal networks. In: Advances in Neural Information Processing Systems (NIPS) (2015)
16. Rollend, D., Rosendall, P., Wolfe, K., Kleissas, D., Billings, S., Oben, J., Helder, J., Tenore, F., Burlina, P., Roy, A., Greenberg, R., Katyal, K.: Embedded clutter reduction and face detection algorithms for a visual prosthesis. In: 2016 38th Annual International Conference of the IEEE Engineering in Medicine and Biology Society (EMBC), August 2016
17. Russakovsky, O., Deng, J., Su, H., Krause, J., Satheesh, S., Ma, S., Huang, Z., Karpathy, A., Khosla, A., Bernstein, M., Berg, A.C., Fei-Fei, L.: ImageNet large scale visual recognition challenge. Int. J. Comput. Vis. (IJCV) 115(3), 211–252 (2015)
18. Stanga, P., Sahel, J., Mohand-Said, S., daCruz, L., Caspi, A., Merlini, F., Greenberg, R.: Face detection using the argus II retinal prosthesis system. Invest. Ophthalmol. Vis. Sci. 54, 1766 (2013)
19. Viola, P., Jones, M.: Rapid object detection using a boosted cascade of simple features. In: Proceedings of the 2001 IEEE Computer Society Conference on Computer Vision and Pattern Recognition (CVPR 2001), vol. 1, pp. I-511-I-518 (2001)
20. Walker, W., Lamere, P., Kwok, P., Raj, B., Singh, R., Gouvea, E., Wolf, P., Woelfel, J.: Sphinx-4: a flexible open source framework for speech recognition. Technical report, Mountain View, CA, USA (2004)

Hyperspectral Image and Signal Processing

Spectral Dichromatic Parameter Recovery from Two Views via Total Variation Hyper-priors

Filippo Bergamasco[1], Andrea Torsello[1], and Antonio Robles-Kelly[2,3(✉)]

[1] Dipart. di Sci. Ambientali, Informatica e Statistica,
Università Ca' Foscari Venezia, Venice, Italy
[2] CSIRO, DATA61, Tower A, 7 London Cct., Canberra, ACT 2601, Australia
antonio.robles-kelly@data61.csiro.au
[3] College of Engineering and Computer Science, Australian National University,
Canberra, Australia

Abstract. In this paper, we propose an approach for the recovery of the dichromatic model from two hyperspectral or multispectral images, *i.e.*, the joint estimation of illuminant, reflectance, and shading of each pixel, as well as the optical flow between the two views. The approach is based on the minimization of an energy functional linking the dichromatic model to the image appearances and the flow between the images to the factorized reflectance component. In order to minimize the resulting under-constrained problem, we apply vectorial total variation regularizers both to the scene reflectance, and to the flow hyper-parameters. We do this by enforcing the physical priors for the reflectance of the materials in the scene and assuming the flow varies smoothly within rigid objects in the image. We show the effectiveness of the approach compared with single view model recovery both in terms of model constancy and of closeness to the ground truth.

1 Introduction

In computer vision, the modelling and recovery of photometric parameters is a topic of pivotal importance for purposes of surface analysis and image understanding. Since the estimation of illuminant and material reflectance are mutually interdependent, the problem of recovering physically meaningful parameters that govern the image formation process is closely related to the ability to resolve the intrinsic material reflectance from their trichromatic colour images captured under varying illumination conditions. Existing methods often rely upon the use of statistics of illuminant and material reflectance or draw upon the physics-based analysis of local shading and specularity of the objects in the scene.

Statistics-based approaches often employ Bayes's rule [3] to compute the best estimate from a posterior distribution. The illuminant and surface reflectance spectra typically take the form of a finite linear model with a Gaussian basis [9], where a correlation matrix is built for a set of known plausible illuminants to characterise all the possible image colours (chromaticities) that can be observed. Contrary to these statistics-based approaches, physics-based colour constancy

© Springer International Publishing AG 2017
C.-S. Chen et al. (Eds.): ACCV 2016 Workshops, Part I, LNCS 10116, pp. 317–333, 2017.
DOI: 10.1007/978-3-319-54407-6_21

analyses the physical processes by which light interacts with the object surface [16, 26].

Regarding specularities and shading, there have been several attempts to remove specular highlights from images of non-Lambertian objects. For instance, Brelstaff and Blake [4] used a thresholding strategy to identify specularities on moving curved objects. Narasimhan *et al.* [21] have formulated a scene radiance model for the class of "separable" Bidirectional Reflectance Distribution Functions (BRDFs). More recently, Zickler *et al.* [32] introduced a method for transforming the original RGB colour space into an illuminant-dependent colour space to obtain photometric invariants. Other alternatives elsewhere in the literature aiming at detecting and removing specularities either make use of additional hardware [22], impose constraints on the input images [18] or require colour segmentation [14].

Here, we depart from the dichromatic model so as to describe the image radiance as a combination of shading, specular highlights, surface reflectance and the illuminant power spectrum. Our multi-view dichromatic parameter recovery method separates the scene illuminant, shading and object surface reflectance by linking the reflectance of objects present in two images to the flow between them. This is achieved by minimising the total variation of reflectance and flow, subject to the notion that the object reflectance and illuminant power spectrum across the two images should not change. This also imposes further constraints on the optical flow which are akin to those imposed upon brightness in trichromatic imagery [2]. This leads to an optimisation problem where a total variation regularization approach is used to enforce the consistency of the scene photometric parameters over an image sequence. This contrasts with previous approaches where the intersection of dichromatic planes [9, 30], assumed chromaticities of common light sources [9], or structural optimisation [13] are used on single images. In [15], Kong et al. use polarised images to separate the background and reflection layers from each of the input images.

2   Contributions

The contributions of this paper are the following:

- To the best of our knowledge, this is the first approach that uses reflectance constancy across multiple images to improve the recovery of the dichromatic parameters, relating the reflectance to the optical flow between multiple images.
- We introduce a novel homographic hyper-prior for the flow similar in spirit to the affine formulation presented in [17]. This, in combination with a total variation regularization provides a natural modelling of the scene resulting in an improvement of the optical flow estimation in parallel with the improvement to the photometric parameters.
- Our method is quite general in nature and can be modified in a straightforward manner to regularise any vectorial field whose total variation under consideration is to be minimised simultaneously with other terms in an energy

functional. Indeed, regularization methods have been reported in the contexts of optical flow computation [20], curvature-based surface shape representation [29] and the smoothing of stereo disparity fields [19]. We would like to stress, however, that the focus of the work presented here is the recovery of illuminant and reflectance in multispectral and hyperspectral images.

– We employ homographic hyperpriors in the total variation regularizer so as to impose a physically sound set of constraints on the solution. This not only improves the reflectance recovery results, but also delivers better localisation of the specular highlights.

– Experiments show that the approach is capable of providing a more stable recovery of illuminant, reflectance, shading and specular parameters with respect to the sate of the art. This is as our approach can achieve both, better photometric accuracy and can naturally handle the computationally challenging task of simultaneously processing a large number of wavelength indexed bands.

3   Multi-view Dichromatic Model Recovery

In this section, we present the multi-view dichromatic model energy functional which permits us, later on, to enforce consistency upon the reflectance across corresponding scene points in multiple images.

By assuming a uniform illuminant power spectrum across the scene, the dichromatic model [26] expresses the image radiance $I(\mathbf{u}, \lambda)$ at pixel location $\mathbf{u} = (u_1, u_2)$ and wavelength λ as follows:

$$I(\mathbf{u}, \lambda) = g(\mathbf{u})L(\lambda)S(\mathbf{u}, \lambda) + k(\mathbf{u})L(\lambda) \tag{1}$$

where $L(\lambda)$ and $S(\mathbf{u}, \lambda)$ are the illuminant power spectrum and surface reflectance at wavelength λ, respectively, $g(\mathbf{u})$ is the shading factor governing the proportion of diffuse light reflected from the object and $k(\mathbf{u})$ is the specular coefficient at pixel $\mathbf{u}$.

Note that the dichromatic model above assumes a single "global" illuminant power spectrum while allowing the intensity of the light to vary across the scene. This is not an overly restrictive assumption. In fact, the dichromatic model has been used extensively in colour constancy [9]. Here, we also make the assumption that $\sum_\lambda S(\mathbf{u}, \lambda)^2 = 1$. Note that this can be done without any loss of generality since the illuminant power spectrum can be normalised such that the shading factor and specular coefficients are rescaled accordingly.

3.1   Optical Flow and Reflectance Coherence

One of the main features of the dichromatic model is that the reflectance $S(\mathbf{u}, \lambda)$ is a characteristic of the object's material, being invariant to the geometry of the object and its relative position with respect to the light source and the viewer. As a consequence, it is preserved across multiple images. We model this correspondence in a two image setting by maintaining one single reflectance function on one image and relating it to the reflectance on a second image through an optical flow function $f(\mathbf{u}) = \mathbf{u}' : \Omega_1 \to \Omega_2$ which maps points from the first to the second image. This results in an energy term per each image comparing the measured irradiance $I(\mathbf{u}, \lambda)$ with the irradiance reconstructed from the model parameters using Eq. (1).

Note that, for highly specular pixels, the reflectance information is effectively lost at capture, *i.e.* $g(\mathbf{u})L(\lambda) S(\mathbf{u}, \lambda) \approx 0$. For this reason, throughout the paper, we make use of the multiplicative gating function

$$W(\mathbf{u}) = \exp\left(-\tau||I(\mathbf{u}, \lambda) - \mathcal{P}(I(\mathbf{u}, \lambda))||\right) \tag{2}$$

where $\mathcal{P}(I(\mathbf{u}, \lambda))$ is the projection of the image radiance $I(\mathbf{u}, \lambda)$ onto the dichromatic plane [8] spanned by the radiance over the neighbourhood about pixel location $\mathbf{u}$. The dichromatic plane can computed using SVD [24].

The gating function above reflects the observation that, as the deviation of the image radiance from the dichromatic plane increases, the diffuse reflection decreases in importance [8]. Therefore, the function $W(\mathbf{u})$ can be viewed as a weight in the illuminant and reflectance recovery error. Further, $W(\mathbf{u})$ decreases in value for increasingly specular pixels. This is in accordance with the dichromatic plane formalism used to define $W(\mathbf{u})$, which implies that, for specular highlights, the gating function tends to zero, *i.e.* the gating function and the specular coefficient are nearly orthogonal with respect to each other. Hence, using this weighting function, the contribution of the specular pixels to the energy functional is negligible. As a result, we remove the specular coefficient $k(\mathbf{u})$ from further consideration for purposes of our optimisation approach and, instead, compute it analytically at the end of the process, once the reflectance, illuminant power spectrum, and shading are in hand.

Under these assumptions, we obtain the following energy terms comparing measured and reconstructed irradiance:

$$E_{\mathrm{DI}_1} = \int_{\Omega_1} W_1(\mathbf{u})^2 \sum_{\lambda} \left(I_1(\mathbf{u}, \lambda) - L(\lambda)g_1(\mathbf{u})S(\mathbf{u}, \lambda)\right)^2 d\mathbf{u}$$

$$E_{\mathrm{DI}_2} = \int_{\Omega_1} W_2(\mathbf{u}')^2 \sum_{\lambda} \left(I_2(\mathbf{u}', \lambda) - L(\lambda)g_2(\mathbf{u}')S(\mathbf{u}, \lambda)\right)^2 d\mathbf{u}$$

where the subscript indicate the index for either of the two images. Note that, even for the term related to the second image, the integration is performed over the domain Ω_1 of the first image whereby the relations with Ω_2 is always mediated through the flow f.

3.2   Total Variation Regularization

Our goal is to minimize the energy terms over the flow f and the dichromatic model parameters. We tackle the under-determination of the problem we by adding a regularization term to the energy functional above. The *Total Variation* (TV) of a function $\phi : \mathbb{R}^m \supseteq \Omega \to \mathbb{R}^n$ is an operator defined as

$$\mathrm{TV}(\phi) = \sup_{p_1,\ldots,p_m} \left\{ \int_\Omega \sum_{i=1}^n \phi_i(\mathbf{x}) \nabla \cdot p_i(\mathbf{x}) \, d\mathbf{x} \; : \; p_1,\ldots,p_m \in \mathcal{C}^1(\Omega,\mathbb{R}^n) \right\} \quad (3)$$

where $\mathcal{C}^1(\Omega,\mathbb{R}^n)$ is the set of continuously differentiable functions from Ω to $\mathbb{R}^n$, and $p_1,\ldots,p_m$ satisfy $\sum_{i=1}^m \|p_i(x)\|^2 \leq 1$ everywhere except at most in a subset of measure 0. Further, if ϕ is a differentiable function, the TV assumes the equivalent form

$$\mathrm{TV}(\phi) = \int_\Omega \|D\phi(x)\|_2 \, dx, \quad (4)$$

where $D\phi$ is the differential or Jacobian matrix of ϕ and $\|\cdot\|_2$ denotes the Frobenius norm.

Used as a regularizer, TV privileges piecewise constant solutions. For this property, it has found a multitude of applications ranging from image processing restoration [25], to segmentation [23], to the estimation of the optical flow [7,31]. In our proposed formulation we adopt TV to impose smoothness priors both on the reflectance and flow estimates. The reflectance component is assumed to be constant over image patches of uniform material, thus TV is naturally applicable to S, seen as a function from Ω_1 to $\mathbb{R}^\ell$ where ℓ is the number of spectral bands.

For the flow, however, there is no reason to assume a piecewise constant model. Most approaches in the literature opt to express the flow as a displacement $f(\mathbf{u}) = \mathbf{u} + T(\mathbf{u})$ where the displacement is regularized, resulting in a piecewise uniform translation. Here we opt for a higher order smoothness priors, and we compare the use of an affine prior, similar to one proposed in [17], and an homographic prior, which to the best of our knowledge has never been used before.

In the affine model we assume the displacement to be locally affine: $f(\mathbf{u}) = \mathbf{u} + A(\mathbf{u})\mathbf{u}$, where

$$A(\mathbf{u}) = \begin{pmatrix} a_1(\mathbf{u}) & a_2(\mathbf{u}) & a_3(\mathbf{u}) \\ a_4(\mathbf{u}) & a_5(\mathbf{u}) & a_6(\mathbf{u}) \\ 0 & 0 & 0 \end{pmatrix} \quad (5)$$

while for the homographic model we assume the full coordinate transformation to be projective: $\lambda f(\mathbf{u}) = H(\mathbf{u})\mathbf{u}$ where λ is a scaling factor and $H = (h_{ij})$ is in the special linear group SGL(3), *i.e.*, the group of 3×3 real matrices with unit determinant. Both models can be seen as capturing view transformation of locally planar patches. The homographic model does so exactly assuming the image to follow the pinhole model, while the affine model approximates it assuming a so called "weak camera model". Under these assumptions, the hyperparameters defining the entries of A and H can be assumed to be piecewise constant within such patches. In the rest of the paper we will use Θ and $\mathrm{Dom}(\Theta)$ to refer to the flow hyperparameters and their domain when these can be indifferently the affine of the homographic hyperpriors, and A or H when we want to specify models used. Note that $\mathrm{Dom}(\Theta)$ is $\mathbb{R}^6$ for the affine model and $SGL(3)$ for the homographic one.

Finally, we perform a convex relaxation of the total variation functional [5] transforming the TV regularized optimization problem $\min_\phi E(\phi) + TV(\phi)$ into the relaxed problem

$$\min_{\phi, \phi_{\mathrm{TV}}} E(\phi) + \int \frac{||\phi - \phi_{\mathrm{TV}}||^2}{\delta} + TV(\phi_{\mathrm{TV}}). \tag{6}$$

While the size increases with the addition of the auxiliary function ϕ_{TV}, assuming $E(\phi)$ convex, the formulation becomes convex for $\delta > 0$ and converges to the original variational problem as $\delta \to 0$.

3.3 Multi-view Dichromatic Functional

Assembling the data fidelity terms and the regularizers, we obtain the energy Multi-view dichromatic functional

$$E = \alpha \left(\mu E_{\mathrm{DI}_1} + (1 - \mu) E_{\mathrm{DI}_2} \right) + \rho_S \int_{\Omega_1} \frac{||S(\mathbf{u}) - S_{\mathrm{TV}}(\mathbf{u})||^2}{\delta_S} d\mathbf{u}$$
$$+ \rho_S \int_{\Omega_1} ||DS_{\mathrm{TV}}(\mathbf{u})||_2 \, d\mathbf{u} + \rho_f \int_{\Omega_1} \frac{||\Theta(\mathbf{u}) - \Theta_{\mathrm{TV}}(\mathbf{u})||_2^2}{\delta_f} d\mathbf{u} \tag{7}$$
$$+ \rho_f \int_{\Omega_1} ||D\Theta_{\mathrm{TV}}(\mathbf{u})||_2 \, d\mathbf{u}$$

which is then minimized over S, Θ, L, g_1, g_2, S_{TV}, and Θ_{TV} subject to $\Theta \in \mathrm{Dom}(\Theta)$, to obtain simultaneous flow estimation and joint factorization of the dichromatic model over the two spectral images. Here α, ρ_S, and ρ_f are constants balancing the data fidelity and regularization terms, while $\mu \in [0; 1]$ is used to limit the effect that errors in the estimation of the flow can have in the dichromatic factorization originating form the second image. Note that, as mentioned earlier, due to the $W(\mathbf{u})k(\mathbf{u})$ orthogonality we can eliminate the minimization over k, and recover the specular coefficient after the optimization from the optimal illuminant, reflectance, and shading with the relation $k(\mathbf{u}) = \frac{1}{\ell} \sum_\lambda \frac{I(\mathbf{u},\lambda)}{L(\lambda)} - g(\mathbf{u})S(\mathbf{u},\lambda)$.

4 Implementation and Discussion

4.1 Minimization Process

To optimize E we adopt an alternating minimization procedure, rotating trough the following steps:

1. Minimize with respect to $L(\lambda)$, $g_1(\mathbf{u})$, and $g_2(f(\mathbf{u}))$, keeping $S(\mathbf{u}, \lambda)$, Θ, $S_{\mathrm{TV}}(\mathbf{u}, \lambda)$ and $\Theta_{\mathrm{TV}}(\mathbf{u})$ fixed;
2. Update $S(\mathbf{u}, \lambda)$ and Θ through a projected gradient descent step, keeping all other variables fixed;

3. Minimize the total variation terms to obtain a new estimate of $\Theta_{TV}(\mathbf{u})$ and $S_{TV}(\mathbf{u})$.

For the first step, we differentiate E with respect to $g_1(\mathbf{u})$ and $g_2(f(\mathbf{u}))$ and set both equations to zero so as to obtain

$$g_1(\mathbf{u}) = \frac{\sum_\lambda I_1(\mathbf{u},\lambda)S(\mathbf{u},\lambda)L(\lambda)}{\sum_\lambda S(\mathbf{u},\lambda)^2 L(\lambda)^2} \qquad g_2(f(\mathbf{u})) = \frac{\sum_\lambda I_2(f(\mathbf{u}),\lambda)S(\mathbf{u},\lambda)L(\lambda)}{\sum_\lambda S(\mathbf{u},\lambda)^2 L(\lambda)^2}.$$

$$(8)$$

Similarly, we differentiate $E = \mu E_{DI_1} + (1-\mu)E_{DI_2} + \text{const.}$ with respect to $L(\lambda)$ and set the derivative to zero, which yields

$$L(\lambda) = C_1 \frac{\int_{\Omega_1} S(\mathbf{u},\lambda)\Delta_I(\mathbf{u},\lambda)\, d\mathbf{u}}{\int_{\Omega_1} S(\mathbf{u},\lambda)^2 \Delta_S(\mathbf{u},\lambda)\, d\mathbf{u}}, \qquad (9)$$

where

$$\Delta_I(\mathbf{u},\lambda) = \mu W_1(\mathbf{u})^2 I_1(\mathbf{u},\lambda)g_1(\mathbf{u}) + (1-\mu)W_2(f(\mathbf{u}))^2 I_2(f(\mathbf{u}),\lambda)g_2(f(\mathbf{u}))$$
$$\Delta_S(\mathbf{u},\lambda) = \mu W_1(\mathbf{u})^2 g_1(\mathbf{u})^2 + (1-\mu)W_2(f(\mathbf{u}))^2 g_2(f(\mathbf{u}))^2$$

and C_1 is a normalizing constant satisfying $\sum_\lambda L(\lambda)^2 = 1$.

Hence, for the first step of the optimization process, we find the global optimum of E with respect to $g_1(\mathbf{u})$, $g_2(f(\mathbf{u}))$, and $L(\lambda)$ by alternating Eqs. (8) and (9). Note that, while we are estimating $g_1(\mathbf{u})$ in the regular lattice of the first image, we also do so for $g_2(f(\mathbf{u}))$ through f. This means that, in a discrete image setting, the estimated values of the second image's shading factor are not aligned with that image's regular lattice, but are shifted according to the flow f.

For the second step, we compute the gradient of E with respect to the reflectance $S(\mathbf{u},\lambda)$ and the hyper-parameter $\Theta(\mathbf{u})$. Further, in the case of the homographic hyperparameter H, we project the gradient onto the tangent plane of SGL(3) before taking the gradient step and then reproject the updated H onto SGL(3). The constraint defining SGL(3) is $C = \det(H) - 1 = 0$, from which we get the projection of the gradient on the tangent space as

$$\partial_H^{\parallel} E = \partial_H E - \frac{\partial_H C^T \partial_H E}{\partial_H C^T \partial_H C}\partial_H C. \qquad (10)$$

The reprojection of the updated hyperparameter H is obtained by dividing it by the cubic root of its determinant.

Note that the data fidelity term only depends of $f(\mathbf{u})$, thus, using the chain rule for the data fidelity term only, we can write

$$\partial_{\Theta(\mathbf{u})} E = (\partial_{\mathbf{u}'} E)(\partial_{\Theta(\mathbf{u})}\mathbf{u}') + \rho_f \frac{\Theta(\mathbf{u}) - \Theta_{TV}(\mathbf{u})}{\delta_f}. \qquad (11)$$

The gradient with respect to the flow can be computed easily in terms of the dichromatic parameters

$$
\begin{aligned}
\partial_{\mathbf{u}'} E = \alpha(1-\mu) \Big[& E_{\mathrm{DI}_2}(\mathbf{u}')\partial_{f(\mathbf{u})} W_2(\mathbf{u}') + 2W_2(\mathbf{u}') \sum_{\lambda} \Big(I_2(\mathbf{u}',\lambda) \\
& - L(\lambda)g_2(\mathbf{u}')S(\mathbf{u},\lambda) \Big) \cdot \big(\partial_{\mathbf{u}'} I_2(\mathbf{u}',\lambda) - L(\lambda)S(\mathbf{u},\lambda)\partial_{f(\mathbf{u})} g_2(\mathbf{u}') \big) \Big].
\end{aligned}
\tag{12}
$$

For the affine model, the derivative of the flow with respect to the hyperparameters is a linear function:

$$
\partial_A \mathbf{u}' = \partial_{(a_1,\dots,a_6)} f(\mathbf{u}) = \begin{pmatrix} 1 & 0 \\ 0 & 1 \end{pmatrix} \otimes (u\ v\ 1)
\tag{13}
$$

where $A \otimes B$ is the Kronecker product of matrices A and B and $\mathbf{u} = (u,v)^T$. For the homographic model we have:

$$
\begin{aligned}
\partial_H \mathbf{u}' &= \big(\frac{\partial \mathbf{u}'}{\partial h_{11}}, \dots, \frac{\partial \mathbf{u}'}{\partial h_{13}}, \frac{\partial \mathbf{u}'}{\partial h_{21}}, \dots, \frac{\partial \mathbf{u}'}{\partial h_{33}} \big) \\
&= \frac{1}{\zeta} \begin{pmatrix} 1 & 0 & -(h_{11}u + h_{12}v + h_{13}) \\ 0 & 1 & -(h_{21}u + h_{22}v + h_{23}) \end{pmatrix} \otimes (u\ v\ 1)
\end{aligned}
$$

where $\zeta = h_{31}u + h_{32}v + h_{33}$.

Furthermore, the energy gradient with respect to reflectance can be expressed as:

$$
\begin{aligned}
\partial_{S(\mathbf{u},\lambda)} E = & - 2\alpha\mu g_1(\mathbf{u})W_1(\mathbf{u})L(\lambda)\,(I_1(\mathbf{u},\lambda) - g_1(\mathbf{u})L(\lambda)S(\mathbf{u},\lambda)) \\
& - 2\alpha(1-\mu)g_2(\mathbf{u}')W_2(\mathbf{u}')L(\lambda) \Big(I_2(\mathbf{u}',i) - L(\lambda)g_2(\mathbf{u}')S(\mathbf{u},\lambda) \Big) \\
& + 2\rho_S \frac{S(\mathbf{u},\lambda) - S_{\mathrm{TV}}(\mathbf{u},\lambda)}{\delta_S}.
\end{aligned}
$$

We approximate $\partial_{\mathbf{u}'} W_2(\mathbf{u}')$ and $\partial_{\mathbf{u}'} I_2(\mathbf{u}')$ with central finite differences that are pre-computed at the beginning of the optimization process. As we mentioned before, obtaining $\partial_{f(\mathbf{u})} g_2(\mathbf{u}')$ is not straightforward since we never optimize $g_2(\mathbf{u})$ in the regular lattice of the second image but only its representation warped to the first image through the flow $f(\mathbf{u})$. However, by the chain rule, we have:

$$
\frac{\partial}{\partial \mathbf{u}} g_2\big(f(\mathbf{u})\big) = \frac{\partial g_2\big(f(\mathbf{u})\big)}{\partial f(\mathbf{u})} \cdot \frac{\partial f(\mathbf{u})}{\partial \mathbf{u}},
\tag{14}
$$

from which we have

$$
\frac{\partial g_2\big(f(\mathbf{u})\big)}{\partial f(\mathbf{u})} = \frac{\partial}{\partial \mathbf{u}} g_2\big(f(\mathbf{u})\big) \left(\frac{\partial f(\mathbf{u})}{\partial \mathbf{u}} \right)^{-1}
\tag{15}
$$

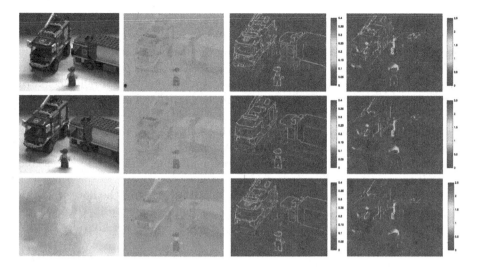

Fig. 1. Sample image pair showing the effect of the priors on the regularized parameters. First column: input image pair and optical flow. Second column: Reflectance value computed by H&RK (top) and by our approach with affine and homographic hyper-priors (center and bottom). Third column: Reflectance norm magnitude computed by H&RK (top) and by our approach with affine and homographic hyper-priors (center and bottom). Last column: Frobenius norm of the differential of the initial (top) affine and homographic hyper-parameters (center and bottom).

or, equivalently

$$\nabla_{f(\mathbf{u})} g_2\big(f(\mathbf{u})\big) = \big(D_{\mathbf{u}} f(\mathbf{u})^T\big)^{-1} \nabla_{\mathbf{u}} g_2\big(f(\mathbf{u})\big), \tag{16}$$

where both terms $\nabla_{\mathbf{u}} g_2\big(f(\mathbf{u})\big)$ and $D_{\mathbf{u}} f(\mathbf{u})$ are computed with standard central differences from $g_2\big(f(\mathbf{u})\big)$ and $f(\mathbf{u})$ respectively.

Finally, for the third optimization step, we follow the fast iterative method proposed by Bresson and Chan [5].

4.2   Initialization

Note that our approach relies on an initial estimate of the flow $f(\mathbf{u})$, illuminant power spectrum and specular highlights. This is since, if the illuminant power spectrum and the specular coefficient is known, the reflectance and the shading factor can be obtained via algebraic manipulation and normalisation operations [12]. Indeed, there are a number of methods elsewhere in the literature that can be used to obtain these initial estimates. Here, we use the method in [28] to recover the image highlights and that in [9] for the recovery of the initial estimate of the illuminant power spectrum.

For the optical flow, we avoid the common coarse-to-fine-approaches, proposing to rather exploit a small set of initial sparse matches as a starting point for the flow optimisation. This is a similar approach to that used in recent works by Leordeanu et al. [17] or Brox and Malik [6] which are proven to deal with very large displacements. To this end, we compute a small set of reliable sparse matches from an image pair following the method in [1] and making use of SURF features extracted from the initial shading factor. We modified the original payoff function to include a similarity term that weights the angular error of the reflectance spectra among two matches. As a consequence, we are better able to select a good set of inliers without epipolar filtering, which is not a feasible option if the scene objects are allowed to move.

We use these sparse matches to get an initial estimate of the flow around a limited set of points in our optimization domain, where we have designed an energy functional composed by a data term and a simple L2 regularizer given by

$$E_f = \alpha \int_\Omega D(\mathbf{u}) H(\mathbf{u}) \left[\mu_1 Es(\mathbf{u}) + \mu_2 Er(\mathbf{u}) \right] + ||\partial_\mathbf{u} T(\mathbf{u})||_2^2 d\mathbf{u} \qquad (17)$$

with

$$Es(\mathbf{u}) = \left[g_1(\mathbf{u}) - g_2 \big(\mathbf{u} + T(\mathbf{u}) \big) \right]^2 \qquad (18)$$

$$Er(\mathbf{u}) = e^{-\sum_\lambda S_1(\mathbf{u}, \lambda) S_2 \big(\mathbf{u} + T(\mathbf{u}), \lambda \big)} \qquad (19)$$

$$D(\mathbf{u}) = e^{-\frac{1}{\sigma} \min_{m \in M} ||\mathbf{u} - m||} \qquad (20)$$

$$H(\mathbf{u}) = \gamma \frac{\sum_\lambda ||\partial_\mathbf{u} S_1(\mathbf{u}, \lambda)||^2}{\max_{\mathbf{u}'} \sum_\lambda ||\partial_{\mathbf{u}'} S_1(\mathbf{u}', \lambda)||^2} + 1 \qquad (21)$$

where γ, μ_1 and μ_2 are constants, σ is the radius of the spatial weighting term and we have written $S_i(\mathbf{u}, \lambda)$ to imply that the reflectance corresponds to the ith view under consideration, i.e. $i = \{1, 2\}$.

In the expression above, the data term accounts for both the photometric Es and material Er consistency between the two images trough an L2-norm penalty function. The spatial weighting term D moderates the effect of the L2-regularizer with respect to the data term as a function of the distance from the closest match in the initial set M whereas (H) is used to allow discontinuities in the proximity of edges. Here, we tackle the minimization of the functional above as a standard variational problem by solving the set of associated Euler-Lagrange equations [10] and have set all constants to unity.

4.3 Effect of the Regularization Terms

Recall that the total variation hyper-prior regularization term was introduced to enforce the patch-wise uniform material assumption and the locally uniform flow assumption formalized in terms of locally-homographic transformation. Figure 1 shows the effect of the priors on the regularized parameters on a sample hyperspectral image pair. The left-most column shows the input color image pair (we

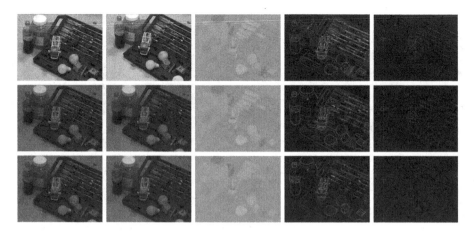

Fig. 2. Qualitative example of the results obtained for Scene 4. Top row: Initial input pairs, reflectance value, reflectance gradient magnitude and specular factors computed by H&RK. Central and Bottom rows: Shading, reflectance value, reflectance gradient magnitude and specular factors computed by the proposed method with the affine and homographic hyper-priors respectively.

show the pseudocolour obtained using the colour matching functions of Stiles and Burch [27] over the 30 bands in the visible range as delivered by the camera) and the initial flow. The second column shows the reflectance (again, in pseudo-colour) as returned by [12] (H&RK) and as optimized by our process with the affine and homographic priors respectively in the first, second and third row. The third column shows the gradient magnitude of the reflectance for the same three methods, while the last column shows the Frobenius norm of the differential of the hyper-parameters A and H at initialization and after optimization.

From the figure, it is clear that the algorithm is capable of clustering together regions of uniform material that had significant variation in the estimated reflectance with H&RK. For example, look at the gradient magnitude of the reflectance in areas like the roof of the truck on the right or the wheels of the trick on the left. In both cases the materials are uniform and thus should exhibit uniform reflectance, but the wide variation in shading leaks into variations in the reflectance estimated with H&RK, on the other hand, our approach strongly reduces the variation in reflectance, while maintaining sharp variations across different materials.

Also note that the regularization of the hyper-parameters significantly improve the details captured by the flow. For instance, the flow around the logos on the two trucks correspond to a pure change in material and should not have any effect on the flow. However, the edges of the logos are clearly visible in the gradient magnitude of the flow hyper-parameters at initialization, which indicates a leakage of information from the estimated reflectance to the estimated flow. After optimization, not only is the flow generally more uniform, with high gradient mostly in correspondence with depth discontinuities or occluded

Fig. 3. Qualitative example of the results obtained for Scene 5. Top row: Initial input pair, reflectance value, reflectance gradient magnitude and specular factors computed by H&RK. Central and Bottom rows: Shading, reflectance value, reflectance gradient magnitude and specular factors computed by the proposed method with the affine and homographic hyper-priors respectively.

pixels, but the boundaries of the logos vanish almost completely. Indeed, from the reflectance gradient in the 3rd column, we can see many more shading arte-facts and edge ghosting effects in the results yielded by the H&RK alternative than in those obtained with our approach. As expected, the homographic hyperprior performs better than the affine one as it captures the assumption of locally planar patches. This is clear by observing the yellow logo on the truck. The affine case (central row) suffers a loss of contrast which is almost negligible for the homographic case while still effectively clustering uniform patches of similar material.

5   Experiments

For purposes of comparison, we have used the method in [12]. Our choice hinges in the fact that the alternative is aimed at processing imaging spectroscopy data based upon the dichromatic model. Moreover, the method in [12] is an optimisation approach. Both our method and the alternative have been initialised using the same estimates of the illuminant power spectrum and specular highlights.

For the experiments shown in this section, we have used four image sequences acquired using an uncalibrated multispectral camera delivering six channels in the visible spectrum *i.e.* six wavelength indexed bands in the range 430–680 nm with 50 nm steps, and one in the near-infrared at 950 nm. It is worth noting in passing that our method can be easily applied to data comprising any number of wavelength bands, as shown in Fig. 1, where the images comprised 30 bands. Each of our image sequences here comprises 10 frames, depicting scenes containing a wide variety of objects made of different materials and depicting a wide variety of shapes. Each of these scenes is illuminated by different lights, spanning artificial sunlights, tungsten and incandescent lamps. For each of these, the ground truth illuminant power spectrum has been acquired using a LabSphere Spectralon calibration target. For our dataset, we have computed reflectance images for groundtruthing purposes following the procedure in [11]. All our pseudocolour images have been obtained using the colour matching functions of Stiles and Burch [27].

In Figs. 2 and 3 we present some qualitative results comparing the proposed method against H&RK for two sample scenes in our dataset. In the first row, a pseudo-colour image pair obtained in the same way as in [12] is shown together with the reflectance, reflectance gradient magnitude and specular factor estimated by H&RK. In the second and the third row the output of our method for affine and homographic hyper priors respectively is shown together with the estimated shading factor. Note how the total variation regularisation process in our approach has improved the reflectance estimate by removing artefacts arising from the surface geometry. The specular highlights delivered by our method are in better accordance with the input imagery as can be appreciated by observing the screwdriver box, the jars and the cups present in the scenes. Overall, both the regularizer performs better than H&RK producing similar results. However, a better estimation of the flow given by the homographic hyperprior improves the reflectance contrast and specular highlights localization.

In Figs. 4 and 5 we illustrate the results yielded by our method and the alternative regarding the recovery of the reflectance and the illuminant power spectrum. To this end, the left-hand side of Fig. 4, we plot the illuminant delivered by H&RK and our method superimposed over the ground-truth (red-line). In the panel, the top trace depicts the spectrum whereas the bottom bar plot corresponds to the standard deviation of the illuminant per band over the corresponding image sequence. Note that the standard deviation for the illuminant power spectrum is much lower for our method. This is also the case for the reflectance. In Fig. 5 we show the reflectance for four colour tiles on an XRite colour checker placed in one of our scenes. For the sake of clarity of presentation, the figure shows a close up of the color checker and, in a fashion similar to Fig. 4, the spectrum as a trace at the top of the plots with the standard deviation at the bottom on a bar plot.

In Table 2 we show the RMS and Euclidean angular error for the illuminant power spectrum recovered by our approach and the H&RK method across all the images for the four scenes in our dataset. Note that, for both measures, our method exhibits a lower error. This is consistent with our qualitative results showed earlier. Finally, in Table 1, we show the RMS and Euclidean angular error of the recovered reflectance averaged among each coloured tile shown in Fig. 5.

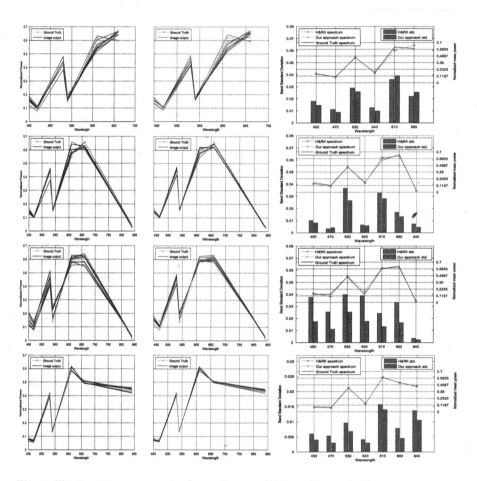

Fig. 4. Illuminant power spectra for each scene. First and second column: power spectrum for each scene image as computed by H&RK and our approach respectively. Third column: average spectrum with the standard deviation for each band. (Color figure online)

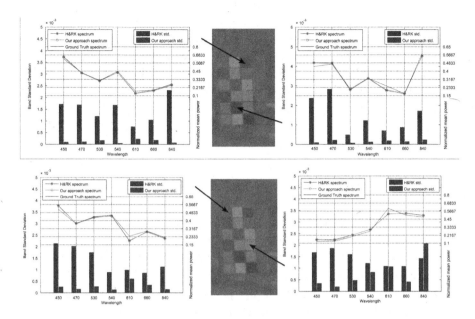

Fig. 5. Reflectance spectra and the standard deviation for each band for the pixels inside the respective color tiles. (Color figure online)

Table 1. Average and standard deviation of the RMS and Euclidean angular error (θ) of the estimated reflectance inside the coloured tiles shown in Fig. 5.

	Mean	St. dev.
H&RK θ	0.075242	0.01740
Our θ	0.073063	0.02032
H&RK RMS	0.028431	0.00657
Our RMS	0.027607	0.00767

Table 2. RMS and Euclidean angular error (θ) for the illuminant recovered by our approach and the H&RK method for all the four scenes of our dataset.

Scene	H&RK θ	Our θ	H&RK RMS	Our RMS
1	0.080354	0.080045	0.026777	0.026675
2	0.066695	0.055120	0.023576	0.019485
3	0.076167	0.074565	0.025383	0.024849
4	0.021691	0.020638	0.007669	0.007296

6 Conclusions

In this paper, we proposed a novel method for dichromatic model recovery from a spectral image pair by means of an energy minimization that simultaneously take into account the model parameters and the flow between the images. We introduced a novel affine hyper-prior for the flow that, in combination with a Total Variation regularization, provides a natural piecewise-planar assumption of the scene under the pinhole camera model. The same kind of regularizer is used for the reflectance imposing the assumption that objects are composed by local patches of uniform materials. As a result, we are able to obtain a better reflectance estimation with respect to the current single-image state of the art approaches. Moreover, our approach has shown a significant lower variance while computing the illuminant spectrum over a sequence of images of the same scene. This behaviour is crucial for many applications for which a coherence of the dichromatic parameters is advisable when analysing multiple instances of the same objects involved in a sequence for which the illuminant is constant across the scene. Furthermore, qualitative results shows that the method discriminates better between the shading (i.e. the geometrical features of a surface) and the texture an object.

References

1. Albarelli, A., Rodolà, E., Torsello, A.: Imposing semi-local geometric constraints for accurate correspondences selection in structure from motion: a game-theoretic perspective. IJCV 97(1), 36–53 (2012)
2. Barron, J.L., Fleet, D.J., Beauchemin, S.S.: Performance of optical flow techniques. Int. J. Comput. Vis. 12(1), 43–77 (1994)
3. Brainard, D.H., Delahunt, P.B., Freeman, W.T., Kraft, J.M., Xiao, B.: Bayesian model of human color constancy. J. Vis. 6(11), 1267–1281 (2006)
4. Brelstaff, G., Blake, A.: Detecting specular reflection using Lambertian constraints. In: International Conference on Computer Vision, pp. 297–302 (1988)
5. Bresson, X., Chan, T.F.: Fast dual minimization of the vectorial total variation norm and applications to color image processing (2008)
6. Brox, T., Malik, J.: Large displacement optical flow: descriptor matching in variational motion estimation. IEEE TPAMI 33(3), 500–513 (2011)
7. Drulea, M., Nedevschi, S.: Total variation regularization of local-global optical flow. In: 2011 14th International IEEE Conference on Intelligent Transportation Systems (ITSC), pp. 318–323, October 2011
8. Finlayson, G.D., Schaefer, G.: Convex and non-convex illuminant constraints for dichromatic colour constancy. In: IEEE CVPR, vol. 1, pp. 598–604 (2001)
9. Finlayson, G.D., Schaefer, G.: Solving for colour constancy using a constrained dichromatic reflection model. IJCV 42(3), 127–144 (2001)
10. Forsyth, A.: Calculus of Variations. Dover Books on Advanced Mathematics. Dover Publications, New York (1960)
11. Foster, D.H., Amano, K., Nascimento, S.M.C., Foster, M.J.: Frequency of metamerism in natural scenes. J. Opt. Soc. Am. A 23(10), 2359–2372 (2006)
12. Huynh, C.P., Robles-Kelly, A.: A solution of the dichromatic model for multispectral photometric invariance. IJCV 90(1), 1–27 (2010)
13. Huynh, C.P., Robles-Kelly, A., Hancock, E.R.: Shape and refractive index recovery from single-view polarisation images. In: IEEE CVPR (2010)
14. Klinker, G., Shafer, S., Kanade, T.: A physical approach to color image understanding. Int. J. Comput. Vis. 4(1), 7–38 (1990)
15. Kong, N., Tai, Y., Shin, J.S.: A physically-based approach to reflection separation: from physical modeling to constrained optimization. IEEE Trans. Pattern Anal. Mach. Intell. 36(2), 209–221 (2014)
16. Land, E.H., Mccann, J.J.: Lightness and retinex theory. J. Opt. Soc. Am. 61, 1–11 (1971)
17. Leordeanu, M., Zanfir, A., Sminchisescu, C.: Locally affine sparse-to-dense matching for motion and occlusion estimation. In: IEEE ICCV, December 2013

18. Lin, S., Shum, H.: Separation of diffuse and specular reflection in color images. In: International Conference on Computer Vision and Pattern Recognition (2001)
19. Marr, D., Poggio, T.: A computational theory of human stereo vision. Proc. R. Soc. Lond. Ser. B Biol. Sci. **204**, 301–328 (1979)
20. Nagel, H., Enkelmann, W.: An investigation of smoothness constraints for the estimation of displacement vector fields from image sequences. IEEE Trans. Pattern Anal. Mach. Intell. **8**, 565–593 (1986)
21. Narasimhan, S.G., Nayar, S.K.: Contrast restoration of weather degraded images. IEEE TPAMI **25**, 713–724 (2003)
22. Nayar, S., Bolle, R.: Reflectance based object recognition. Int. J. Comput. Vis. **17**(3), 219–240 (1996)
23. Pock, T., Cremers, D., Bischof, H., Chambolle, A.: An algorithm for minimizing the Mumford-Shah functional. In: ICCV, pp. 1133–1140. IEEE (2009)
24. Robles-Kelly, A., Huynh, C.P.: Imaging Spectroscopy for Scene Analysis. Springer, London (2013)
25. Rudin, L.I., Osher, S., Fatemi, E.: Nonlinear total variation based noise removal algorithms. Phys. D **60**(1–4), 259–268 (1992)
26. Shafer, S.A.: Using color to separate reflection components. Color Res. Appl. **10**(4), 210–218 (1985)
27. Stiles, W.S., Burch, J.M.: Interim report to the Commission Internationale de l'Éclairage Zurich, 1955, on the National Physical Laboratory's investigation of colour-matching. Optica Acta **2**, 168–181 (1955)
28. Tan, R.T., Nishino, K., Ikeuchi, K.: Separating reflection components based on chromaticity and noise analysis. IEEE TPAMI **26**(10), 1373–1379 (2004)
29. Terzopoulos, D.: Multilevel computational processes for visual surface reconstruction. Comput. Vis. Graph. Image Underst. **24**, 52–96 (1983)
30. Tominanga, S., Wandell, B.A.: Standard surface-reflectance model and illuminant estimation. J. Opt. Soc. Am. A **6**, 576–584 (1989)
31. Werlberger, M., Pock, T., Bischof, H.: Motion estimation with non-local total variation regularization. In: CVPR, pp. 2464–2471. IEEE (2010)
32. Zickler, T., Mallick, S.P., Kriegman, D.J., Belhumeur, P.N.: Color subspaces as photometric invariants. IJCVs **79**(1), 13–30 (2008)

A Combinatorial Approach for Hyperspectral Image Segmentation

José Antonio Valero Medina[1]([⊠]), Pablo Andrés Arbeláez Escalante[2],
and Iván Alberto Lizarazo Salcedo[3]

[1] Universidad Distrital, Cra 7 No 40B–53, Bogotá, Colombia
jvalero@udistrital.edu.co
[2] Universidad de los Andes, Cra 1 No 18A–12, Bogotá, Colombia
pa.arbelaez@uniandes.edu.co
[3] Universidad Nacional, Cra 45 No 26–85, Edif. Uriel Gutiérrez, Bogotá, Colombia
ializarazos@unal.edu.co

Abstract. A common strategy in high spatial resolution image analysis
is to define coarser geometric space elements, *i.e.* superpixels, by grouping near pixels based on (a, b)–connected graphs as neighborhood definitions. Such an approach, however, cannot meet some topological axioms
needed to ensure a correct representation of connectedness relationships.
Superpixel boundaries may present ambiguities because one-dimensional
contours are represented by pixels, which are two-dimensional. Additionally, the high spatial resolution available today has increased the volume
of data that must be processed during image segmentation even after
data reduction phases such as principal component analysis. The inherent complexity of segmentation algorithms, including texture analysis,
along with the aforementioned volume of data, demands considerable
computing resources. In this paper, we propose a novel way for segmenting hyperspectral imagery data by defining a computational framework
based on Axiomatic Locally Finite Spaces (ALFS) provided by Cartesian
complexes, which provide a geometric space that complies with the T_0
digital topology. Our approach links also oriented matroids to geometric
space representations and is implemented on a parallel computational
framework. We evaluated quantitatively our approach on a subset of
hyperspectral remote sensing scenes. Our results show that, by depart-
ing from the conventional pixel representation, it is possible to segment
an image based on a topologically correct digital space, while simultaneously taking advantage of combinatorial features of their associated
oriented matroids.

1 Introduction

Digital imagery analysis largely relies in some nearby pixel grouping [1–3] scheme
to form elements of coarser grain (superpixels). In doing this, a common strategy
is to define coarser geometric space elements by grouping near pixels based on (a,
b)–connected graphs as neighbourhood relationships. However, some ambiguities
could appear when using such a space representation as (a, b)–connected graphs

© Springer International Publishing AG 2017
C.-S. Chen et al. (Eds.): ACCV 2016 Workshops, Part I, LNCS 10116, pp. 334–348, 2017.
DOI: 10.1007/978-3-319-54407-6_22

could not meet some topological axioms needed [4]. Specifically digital images segmentation is heavily dependent on proper boundary definitions which are very hard to obtain in a 2D space from 2D–elements as (a, b)–connected graphs [5].

In this paper we propose a computational framework that allows representing digital images in a correct way based on the Axiomatic Locally Finite Space (ALFS) definition provided by Cartesian complexes [5–9]. Cartesian complexes rely on Abstract Cell Complexes (ACC) [10] supplemented with irreflexive, asymmetric and transitive bounding relationships. This digital space definition allows having all nearest pairs of elements formed by elements with different dimension value between a minimum and a maximum. So it is possible to have a digital topology, which meets the T_0 separation property, for defining properly the boundary of higher dimension space elements from its nearest lower dimension pairs. Figure 1 shows how in an ALFS, in contrast with a conventional pixel space, two one–dimensional and one zero–dimensional additional interpixel space elements are included in the digital representation.

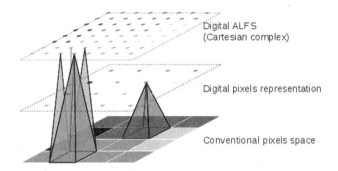

Fig. 1. Conventional pixels (*bottom*) are digitally (middle) represented (*The small connecting pyramid*) as scalars or vectors allocated by row and column on a grid. In contrast, in a Cartesian complex besides pixels (*The tall connecting pyramid*), interpixel space elements are also digitally represented (*two connecting triangles for two interpixel one–dimensional elements (edges) and a connecting line for one interpixel zero–dimensional element*).

Having an adequate space representation is not the only obstacle to overcome. Hyperspectral images have a considerably large number of spectral bands and nowadays it is common to have increasing spatial and radiometric discernibility. While this means that the ambiguity is decreasing, it also means that for a given geographic area the volume of data that must be processed during segmentation of multi and hyperspectral imaging is considerably large even after data reduction stages such as principal component analysis. This ambiguity reduction suggests that an analysis performed by segmentation algorithms, *e.g.* texture analysis, achieving better results but requiring intensive usage of computing resources. Even with the computing power available today, the algorithmic complexity of segmentation analysis and the large volume of data to be processed can cause problems of lack of memory or delay in obtaining results.

In order to overcome the aforementioned performance obstacle, we propose two strategies. The first one is to codify topological and geometrical characteristics using oriented matroids [11,12] linked with hyperplanes defined by Standard Separating Form (SSF) [4] available in the Cartesian complex geometric space. Figure 2 shows a hyperplane central arrangement [13] formed by two SSFs and two subspace examples produced by using a Chirotope transformation [14]. Each SSF is defined based on interpixel space elements available in the Cartesian complex. Specifically, a hyperplane central arrangement was used in establishing the valid orientation set for both filter bank [15] production and oriented gradient calculation.

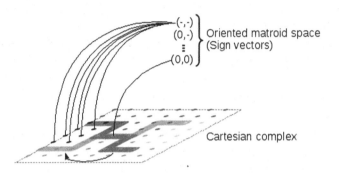

Fig. 2. Combinatorial approach includes Cartesian complex geometry to oriented matroid position vector transformation.

The second strategy is to design and implement all algorithms by using a parallel computational framework. In the segmentation procedural framework [16] based on filtering, texture and oriented gradient analysis, except watershed transform, the process outcome calculated in a space element is independent of that calculated in another. Therefore each pipeline process can be conducted simultaneously for all space elements being limited only by the number of available computing processors.

This paper's aim is to reformulate the widely used segmentation framework of [16] in terms of Cartesian complexes and to encode some geometric features present in a Cartesian complex using oriented matroids in order to improve segmentation processes. We ran an experiment including the oversegmented image production and a later segmentation assessment using the ground truth of a small subset of Hyperspectral Remote Sensing Scenes website. We show that it is possible to segment an image based on a topologically correct digital space representation and to take advantage of combinatorial features by using associated oriented matroids as an alternative to conventional pixel representation.

This article is organized as follows. In Sect. 2 we describe the dataset used. In Sect. 3 we discuss our current proposed framework. Results and discussion are presented in Sects. 4 and 5 respectively and finally, main conclusions are drawn in Sect. 6.

2 Data

For this work we used a small subset of Hyperspectral Remote Sensing Scenes website (http://goo.gl/0zKFTx). We used Pavia Centre scene acquired by the ROSIS sensor over Pavia, northern Italy. The chosen scene is made up 102 spectral bands for Pavia Centre. After taking off without data area, it is a 1096×715 pixels image with spatial resolution of 1.3 m and radiometric of 14 bits. Ground truth image (see Fig. 3 right) differentiates 9 classes: Water, Trees, Asphalt, Self–Blocking Bricks, Bitumen, Tiles, Shadows, Meadows and Bare Soil.

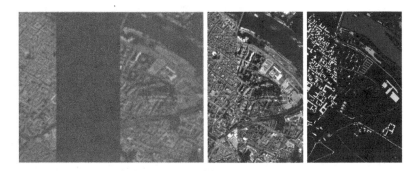

Fig. 3. Used dataset covers Pavia Centre (Italy) (*complete area on the left, used dataset without missing data area in the middle, and ground truth image on the right*).

3 Methods

We conducted experiments including the oversegmented image production based on underlying Cartesian complex implementation and a later segmentation assessment. Figure 4 shows current Java implementation components we used in this work.

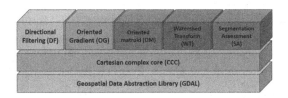

Fig. 4. Framework components relies on GDAL and CCC layers.

We implemented the Cartesian complex support on top of Geoespatial Data Abstraction Library (GDAL) [17] as a multi band image depending on the Cartesian complex space element types used. A GDAL band was created for each used type. The geographic projection and affine transformation coefficients definitions are supported via GDAL.

The model for hyper and multi spectral imaging covers all conceptual elements that participate in the segmentation process. Those elements support the functionalities shown in Fig. 5.

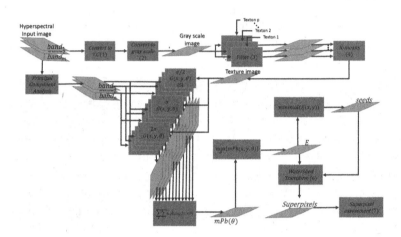

Fig. 5. Segmentation pipeline includes filtering, oriented gradient and watershed transform processes.

First we conduct contour detection and then a watershed transform based superpixel production following [16] but using a Cartesian complex space rather than the conventional image space. We used the weights $\alpha_{i,s}$ that they learned by gradient ascent on the F-measure using the training images and corresponding ground-truth of the Berkeley Segmentation Dataset (BSDS). In the following there is an overview for each component.

3.1 Underlying Cartesian Complex

We used our current Underlying Cartesian complex (CC) framework implementation to compose the pipeline shown on Fig. 5 as a simplified version of that proposed in [16] but using our underlying Cartesian complex. Specifically in filtering, oriented gradient and watershed transform we used only one–dimensional space elements available.

The employment of Cartesian complexes for representing digital images would entail to use four more times storage space than conventionally is required (see Fig. 1) because of the additional inter pixel elements (0– and 1–cells) for each pixel (2–cell) [4]. Now, it was not always necessary to use and save cells of all dimensions, but if a type of cell became necessary it was applied EquNaLi set membership rule [5] on fly the first time and then saved in memory the result for any posterior usage.

Current Cartesian complex implementation includes support for necessary conceptual elements. The defined space elements are *point* for zero–dimensional, *crack* for one–dimensional and *pixel* for two–dimensional with both a combinatorial coordinate and a vector of values but if additional features are necessary they should be added via extended subclasses which must be announced when creating the respective Cartesian complex. Smallest Open Neighbourhood (SON) and Smallest Closed Neighbourhood (SCN) topological and standard separating form (SSF) and digital straight segment (DSS) geometrical [4] concepts are also in our framework core.

3.2   Gray Scale Conversion

The first process is to produce a gray scale image from multi/hyper spectral input image as a Cartesian complex with only pixel space elements. We used a parallel implementation of average image from input bands.

The value calculated in a pixel element is independent of that calculated in another. Therefore each output pixel value can be calculated simultaneously for all pixels.

3.3   Filtering

Image segmentation focus on boundary arrangements that are not areal but linear, so the directional filtering should focus on that structures. One–dimensional linear structures are not available in a conventional image analysis context. In contrast, a Cartesian complex provides cracks (1–cells) which we used as the filtering building block.

As an example, Fig. 6 shows sobel [18] filter kernel definition in a conventional pixel space compared to that in a CC. Conventional sobel kernel definition uses a 3×3 pixel arrangement to produce the convolution for the central pixel. In contrast we propose to use a 2×3 pixel arrangement to produce the convolution for the central crack. In a CC the one–dimensional pattern (the border) found by the kernel is kept in a one–dimensional space element, the central crack.

The filtering stage is based on bank filters for texture recognition [15] but having a crack as central space element. A bank filter is composed by a number of Gaussian derivative filter kernels depending on a number of orientations, a radius around the central crack and a number of granularity scales. Each orientation is based on a SSF of the SSF central arrangement (see Sect. 3.5) for attempting to have only geometrical valid ones. A filter bank spatial kernel has $(2n + 1) \times 2n$ pixels where n is the number of cracks both before and after the central one.

On our underlying Cartesian complex we implemented Gaussian filter kernels production as Berkeley Computer Vision Group proposal available at http:// goo.gl/WrC7gT. As in the gray scale conversion, the value calculated at a crack element is independent of that calculated at another, thus the whole filter bank is applied simultaneously over as many cracks as the number of available processors.

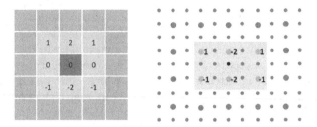

Fig. 6. Conventional sobel filter kernel definition needs 3 × 3 pixel arrangement (*left*) while in a CC this needs a 2 × 3 (*right*). (*Conventional pixels are represented as squares, those used in the filter kernel appear clearer, while in the CC they are represented as little bit bigger dots. Smallest dots correspond to interpixel elements. Central element is the darker one.*)

3.4   K–means Clustering

After filter bank stage each crack has a response vector with as many values as the number of filter kernels in the filter bank. The next stage is to cluster response vectors using a similarity measure. For that we implemented a parallel version of k–means clustering technique using minimum euclidean distance as similarity measurement. Our initial cluster centres are obtained from the mean diagonal, between the minimum and maximum coordinates in the filter bank response space. Iterative regrouping goes until convergence or until a invariability threshold is met to produce a texture image.

3.5   Oriented Gradient

An oriented gradient stage [16] is conducted on both multi/hyper spectral input and texture images. We used principal component analysis (PCA) for reducing the spectral resolution from 102 to 5. The oriented gradient signal on each image is computed for each crack placing at it a disk split in two half discs by one of a number of oriented diameter (see Fig. 7). If a gradient calculation algorithm histograms each disk halve for each oriented diameter, he would really recalculate the common disk sector histograms as many times as he number of orientations.

Instead of that, we propose a combinatorial procedure for this calculation. A disk is split in as many subsets as that produced by the SSF central arrangement formed overlaying all SSFs passing by a central crack [13] and then combining appropriately the histograms calculated for each so produced subset to calculate the gradient for each orientation. We implemented those ideas using oriented matroids [11,12] to produce a combinatorial solution as described in the following.

SSF central arrangement. Orientation numbers define a hyperplane central arrangement [13] being the central crack the common origin. In a CC can have

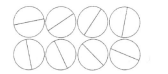

Fig. 7. Each oriented diameter splits a disk in two halves. Oriented gradient can be independently calculated for each orientation (*let*) or, taking into account the common sectors, once for each subspace and then suitably to combine them (*right*).

only a restricted number of SSFs as they need mutual prime numbers as their parameters [4]. To the extent that the number of cracks, before and after the central one, also does the number of SSFs that can be defined (see Fig. 8) so the maximum number of SSFs can be combinatorially obtained from Eq. 1. When the number of orientations is odd the respective central arrangement will contain the horizontal one (see Fig. 8) but when it is even, the central arrangement must be suitably rotated as in that case it is impossible to have a vertical SSF containing the central crack (see Fig. 7 right).

$$maxSSFs = cn(cn + 1) + 1. \tag{1}$$

where cn = radius in number of additional cracks besides the central one.

When a SSF is used the space is subdivided in three subsets: (1) the interpixel elements on that SSF, (2) space elements on one side of the SSF, and (3) space elements on the other side. Each time a SSF is added sharing the same central crack divides a sub space on each side in three new subspaces, besides that containing the central crack itself.

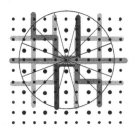

Fig. 8. Each orientation must be approximated to the nearest SSF available in a CC.

Oriented matroid. The SSF central arrangement geometry to associated combinatoric is transferred through its oriented matroid by using the concept of position vector [14]. The position vector takes a geometric coordinate in combinatorial coordinates and put it in the sign vector space of a oriented matroid associated with the respective SSF central arrangement (see Fig. 1). Space elements belonging to a same subspace must to produce identical sign vector. We proposed to use Eq. 2 in order to establish the sign for each space element regarding to a specific SSF i.

$$\delta(c)_i = \begin{cases} -, & H_i(c) < 0 \\ 0, & H_i(c) = 0, \\ +, & H_i(c) > 0 \end{cases} \tag{2}$$

where $c = $ a CC space element,
$H_i(c) = i$th SSF value at combinatorial coordinate of c.

So the position vector $\delta(x)$ of each CC space element is obtained applying Eq. 2 for each SSF in the central arrangement from the most horizontal (rotated if orientation number is even) one in clockwise (see Fig. 2). It should be notice that an image is scanned from top to down and from left to right.

Oriented gradient. For indexing each subspace we propose to encode a position vector as though each sign value were a 3 base one according to the Eq. 3. Then, a position vector value is given by Eq. 4.

$$val(\delta(c)_i) = \begin{cases} 0, & \delta(c)_i = - \\ 1, & \delta(c)_i = 0. \\ 2, & \delta(c)_i = + \end{cases} \tag{3}$$

$$val(\delta(c)) = \sum_{i=0}^{n-1} val(\delta(c)_i) \times 3^i, \tag{4}$$

where $n = $ number of SSFs in the central arrangement.

A subspace index ss, of which a space element c is member, is established by Eq. 5. When c's ss has been established, the respective c's radiometric and texture values are used for updating the subspace histogram.

$$ss(c) = \begin{cases} 2\log_3(\frac{1}{2}(val(\delta(c)) + 1) + 1, & val(\delta(c)) \text{ is even} \\ 2\log_3(val(\delta(c)) + 1), & val(\delta(c)) \text{ is odd.} \end{cases} \tag{5}$$

The histograms of each channel (spectral band or texture) get calculated only once for each subspace and then they are combined to form both disk halves in agreement with an orientation θ to calculate the oriented gradient $G(c_x, c_y, \theta)$ at a cell c (see Fig. 5). The maximum response and the respective orientation are searched for in the oriented gradient value set at each c to produce maximum mPb image E (see Fig. 5).

3.6   Watershed Transform

The maximum gradient image E is used as input of our watershed transform implementation to produce the final superpixels. We implemented Vincent – Soille algorithm [19] but on one–dimensional space elements. For exploring a eventual basin expansion we propose to process cracks in the union of SON of

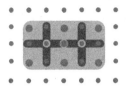

Fig. 9. Cracks (*dots in darker area*) in the union of SON (*clearer area*) of elements in the central crack closure (*all circled dots*), except the central one, are used in exploring basin expansion from the central one (*central circled dot*).

elements in the central crack closure, except the central one (Fig. 9), instead of 8– or 4–connected conventional pixel neighbourhood.

The watershed transform output can be the over segmentation image or a multilevel segmentation with a given number of levels. A multilevel segmentation is got from a given number n of ranges each one of them with $2^{i+1}+1$ slots where i is between 1 and n.

3.7 Superpixel Geometry Assessment

As last stage we conducted two superpixel geometry assessments using (1) superpixel and ground truth images intersection, and (2) BSDS segmentation assessment benchmark [16,20]. In the first evaluation, we compared the ground truth image boundaries against superpixels intersecting them using Eq. 6.

$$spAss = \frac{1}{\sum_{sp \in Bint} area(sp)} \sum_{sp \in Bint} \max_{gt}(area(int(sp, gt))). \tag{6}$$

where
$Bint$ = ground truth boundary intersecting superpixel set,
gt = a connected ground truth area,
$int(sp, gt)$ = set of intersections between a superpixel sp and any gt.

For the second assessment, we evaluated the suitability of our superpixels for downstream recognition appliances. For this purpose, we supposed all our superpixels were correctly classified by assigning to each segmentation superpixel the ground truth region label more frequent (mode) inside it. Then we assessed each so produced segmentation using the BSDS benchmarks taking as ground truth the semantic annotations of the hyperspectral images.

4 Results

A filter bank with a Gaussian second derivative and its Hilbert transform filters in 8 orientations and one center surround filter all of them at 3 scales produced a texton image with 51 channels using the gray scale image. Figure 10 (left) shows filtering outcomes for more detailed scale, *i.e.* 3 cracks before and 3 cracks after the central one on each of 8 respective SSFs. Texture image (see Fig. 10 right) was produced with our k–means implementation in as many iterations as that necessary for guaranteeing at least 90% of no changes.

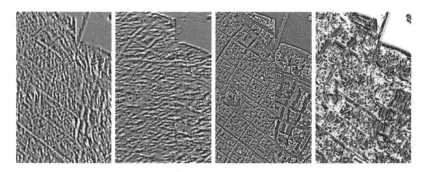

Fig. 10. Texton images were produced for eight orientations and three scales (*first, fifth and center surround ones for the finest scale are on the left*) from Gray scale image. A texture image (*right*) was also produced from textons ones applying k–means grouping.

Our oriented gradient implementation was run on both PCA and texture images for 8 orientations using one–scale gradient disk. For a smaller image window Fig. 11 shows maximum oriented gradient horizontal crack images for first, second, third principal components and then for texture images.

Fig. 11. Oriented gradient images were got for both PCA and the texture images for eight orientation and a crack radius of 3 (*first three components, on the left, and texture, on the right, maximum oriented gradients horizontal crack images*).

Table 1 shows an oriented gradient execution time comparative between Berkeley Computer Vision Group Matlab and our implementations using three radio measurements in pixels for Matlab's and cracks for ours.

In agreement with our pipeline (see Fig. 5), oriented gradient images were weighted summed to produce the boundary strength image mPb for each orientation using our own implementation of that proposed by Berkeley Computer

Table 1. Execution time was measured for three radius (*in pixels for Matlab and cracks for ours*) and several orientations (*the maximum always*).

Radius	1		2		3	
Orientations	3	3	7	3	8	15
Matlab	23.94	23.54	44.35	23.12	56.81	99.75
Our	23.33	27.33	35.33	26.33	43.33	47.67

Vision Group. The maximum boundary strength image values were used to produce a multilevel segmentation from the over segmented one to the most general, *i.e.* that produced using 5 boundary strength ranges (see Fig. 12).

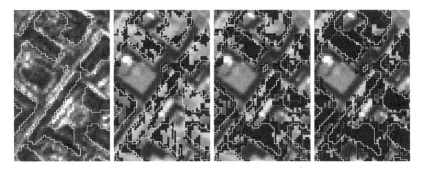

Fig. 12. Boundary strength image (*left with ground truth on top*) was produced from both PCA and texture oriented gradient signals. A multilevel segmentation image (*on the right: over, 9 height ranges and 5 height ranges segmentations*) was also produced by watershed transform.

Table 2 shows assessment outcomes of superpixels intersecting ground truth boundaries. The superpixel geometry assessment was conducted on each of 9 levels in the multilevel segmentation produced by our watershed transform implementation. Both an over segmentation image and a 8 levels multilevel segmentation images were produced. First row in Table 2 corresponds to the over segmentation image superpixel geometry assessment, the other rows assess superpixel geometry for seventh and eighth segmentation levels, *i.e.* using 9 and 5 boundary strength values respectively.

Table 3 shows the Optimal Image Scale (OIS) boundary and region quality measures for our multilevel segmentation.

Table 2. Superpixel intersecting boundary ground truth geometry assessment includes several segmentations (*one by row*). The first column is the number of boundary strength ranges used, the second is the total number of superpixels, the third is the number of intersecting superpixels, the fourth is total intersecting area (*in number of pixels*), the fifth is total maximum intersecting superpixel area, and the last one is the superpixel geometry assessment.

Ranges	Superpixels	Intersecting	$\sum area(sp)$	$\sum \max(area(int(sp, gt)))$	spAss
All	145638	13157	161865	128577	0.7943
9	102694	9433	209932	165522	0.7885
5	74956	5321	286630	224402	0.7829

Table 3. Boundary performance in terms of precision and recall reports the precision, recall and F-measure while region quality reports covering of the ground-truth (*GT covering*).

Boundary			Region
Precision	Recall	F-measure	GT covering
0.99	0.42–0.60	F(0.60, 1.00) = 0.75	0.92

5   Discussion

In boundary detection it was possible to link filtering outcomes to one–dimensional space elements because they were available as interpixel ones. The use of SSF central arrangements made possible to constraint the filter and gradient orientations to only those available in the digital space.

The calculation of oriented gradients signals not only is benefited by the halve discs combinatorial formation from the subspaces, this also does not need to search subspace belonging to a particular crack but the crack subspace belonging is based on subspace indexing from the position vector value we introduced. Since a gradient disk is created once along with its combinatorial structure, *i.e.* space element location to subspace mapping, insofar as the radius increases the improvement over the conventional implementation becomes more evident (see Table 1).

The use of ALFSs for representing the underlying space which meets a topological space T_0 allowed us to define connectivity relationships based on unambiguous topological concepts. This was the case of eventual basin expansion explorations in the watershed transform. In that case we managed to evaluate each crack in the subset of those that belonged to the SON union of the elements included in the closure of the central crack being evaluated.

Our current implementation allowed having boundary strengths that managed to constrain superpixel growth as it passed coarser scales. Table 2 shows how the border precision kept almost the same at coarser scales of multilevel segmentation. Figure 12 shows how highest boundary strengths tends to follow ground truth borders. Ground truth covering report (see Table 3) means that if we managed to classify correctly all of superpixels we would got 0.92 semantic segmentation [21] metric due to a good boundary definition provided by our segmentation.

It should be noticed that despite having used the ground truth image background as a class and therefore source of additional connected regions, superpixels geometry assessment always was above 78% for superpixels intersecting ground truth connected region boundaries independently of the segmentation level.

6   Conclusions

With this work we managed to reformulate the segmentation framework of [16] in terms of Cartesian complexes. This allowed us to define most of the procedures using interpixel space elements there available and to take advantage of the underlying topological model to unambiguously represent geometric and topological features. We also managed to link oriented matroids to geometric features to improve the performance.

Specifically, the interpixel space element availability allowed the filtering stage evaluating filter kernel responses at one–dimensional elements, which agree with the border features dimensionality. In addition, topological features available in underlying ALF provided by Cartesian complexes let us implement an alternative watershed transform procedure which could unambiguously make connectedness tests, *e.g.* basin expansions.

The combinatorial implementation of oriented gradients improved the computational performance due to oriented matroids we linked to the SSF central arrangement by using position vector values to index subspaces in which the respective histogram had to be calculated only once.

While the geometric evaluation of superpixels showed an accuracy close to 80%, the benefits of having a topologically correct representation of space and using combinatorial features through association of oriented matroids highlight the interest of extending our framework in the future to other parts of the processing pipeline.

Acknowledgement. Centro de Computo de Alto Desempeño of Universidad Distrital (CECAD) provided the computing environment.

References

1. Grady, L.: Targeted image segmentation using graph methods (2012)
2. Lizarazo, I., Elsner, P.: Fuzzy segmentation for object-based image classification. Int. J. Remote Sens. **30**, 1643–1649 (2009)
3. Brun, L., Domenger, J.P., Mokhtari, M.: Incremental modifications of segmented image defined by discrete maps. J. Vis. Commun. Image Represent. **14**, 251–290 (2003)
4. Kovalevsky, V.A.: Geometry of locally finite spaces. Int. J. Shape Model. **14**, 231–232 (2008)

5. Kovalevsky, V.A.: Finite topology as applied to image analysis. Comput. Vis. Graph. Image Process. **46**(2), 141–161 (1989). doi:10.1016/0734-189X(89)90165-5
6. Kovalevsky, V.A.: Discrete topology and contour definition. Pattern Recogn. Lett. **2**, 281–288 (1984)
7. Kovalevsky, V.: Algorithms and data structures for computer topology. In: Bertrand, G., Imiya, A., Klette, R. (eds.) Digital and Image Geometry. LNCS, vol. 2243, pp. 38–58. Springer, Heidelberg (2001). doi:10.1007/3-540-45576-0_3
8. Kovalevsky, V.: Algorithms in digital geometry based on cellular topology. In: Klette, R., Žunić, J. (eds.) IWCIA 2004. LNCS, vol. 3322, pp. 366–393. Springer, Heidelberg (2004). doi:10.1007/978-3-540-30503-3_27
9. Kovalevsky, V.: Axiomatic digital topology. J. Math. Imaging Vis. **26**, 41–58 (2006)
10. Listing, J.B.: Der Census räumlicher Complexe: oder Verallgemeinerung des euler'schen Satzes von den Polyädern. Abhandlungen der Königlichen Gesellschaft der Wissenschaften in Göttingen **10**, 97–182 (1862)
11. Whitney, H.: On the abstract properties of linear dependence. Am. J. Math. **57**, 509–533 (1935)
12. Oxley, J.G.: Matroid Theory. Oxford Graduate Texts in Mathematics. Oxford University Press, Inc., New York (2006)
13. Fukuda, K.: Lecture notes on oriented matroids and geometric computation. Technical report RO-2004.0621, course of Doctoral school in Discrete System Optimization, EPFL 2004 (2004)
14. De Loera, J.A., Rambau, J., Santos, F.: Triangulations: Structures for Algorithms and Applications, 1st edn. Springer, Heidelberg (2010)
15. Leung, T., Malik, J.: Detecting, localizing and grouping repeated scene elements from an image. In: Buxton, B., Cipolla, R. (eds.) ECCV 1996. LNCS, vol. 1064, pp. 546–555. Springer, Heidelberg (1996). doi:10.1007/BFb0015565
16. Arbelaez, P., Maire, M., Fowlkes, C., Malik, J.: Contour detection and hierarchical image segmentation. IEEE Trans. Pattern Anal. Mach. Intell. **33**, 898–916 (2011)
17. GDAL Development Team: GDAL - Geospatial Data Abstraction Library, Version 2.1.0. Open Source Geospatial Foundation (2016)
18. Szeliski, R.: Computer Vision: Algorithms and Applications, 1st edn. Springer, New York (2010)
19. Vincent, L., Soille, P.: Watersheds in digital spaces: an efficient algorithm based on immersion simulations. IEEE Trans. Pattern Anal. Mach. Intell. **13**, 583–598 (1991)
20. Martin, D.R., Fowlkes, C.C., Malik, J.: Learning to detect natural image boundaries using local brightness, color, and texture cues. IEEE Trans. Pattern Anal. Mach. Intell. **26**, 530–549 (2004)
21. Everingham, M., Van Gool, L., Williams, C.K.I., Winn, J., Zisserman, A.: The PASCAL Visual Object Classes Challenge 2008 (VOC2008) Results (2008). http://www.pascal-network.org/challenges/VOC/voc2008/workshop/index.html

Hyperspectral Image Classification via a Joint Weighted K-Nearest Neighbour Approach

Chunjuan Bo[1,2], Dong Wang[1(✉)], and Huchuan Lu[1]

[1] School of Information and Communication Engineering,
Dalian University of Technology, Dalian, China
wdice@dlut.edu.cn
[2] College of Electromechanical Engineering,
Dalian Nationalities University, Dalian, China

Abstract. In this paper, we propose a simple yet effective classification framework to conduct hyperspectral image (HSI) classification based on K-nearest neighbour (KNN) and joint model. First, we extend the traditional KNN method to deal with the HSI classification problem by introducing its domain knowledge in HSI data. To be specific, we develop a joint KNN approach to solve the HSI classification problem by considering the distances between all neighbouring pixels of a given test pixel and training samples. Second, we exploit a set-to-point distance between neighbouring pixels and each training sample, and introduce this distance into the joint KNN framework. In addition, a weighted KNN method is adopted to achieve stable performance based on our empirical observations. Both qualitative and quantitative results illustrate that our method achieves better performance than other classic and popular methods.

1 Introduction

Recently, the research of hyperspectral imagery (HSI) [1] has been a very active topic in many application fields including agriculture, military, surveillance, environmental protection and so on. One of the most important directions in HSI is image classification, in which the pixels in the HSI data are categorized into one of several land-cover classes by exploiting their spectral and spatial characteristics. Although much progress has been made in this topic [2–8], it is still a difficult problem to develop an effective HSI classification method due to many challenging factors (such as the limited number of available training samples, the high dimensionality of spectral bands and so on).

For the HSI classification problem, the spatial consistency among the test pixels is a very important prior, which should be carefully considered in designing the classification method. Broadly speaking, the neighbouring pixels within a certain region usually consist of the same type of materials and are of similar spectral characteristics, which makes that these pixels tend to share the same label. Thus, the research of HSI classification should take both the spectral features and spatial information of adjacent pixels into consideration (usually

© Springer International Publishing AG 2017
C.-S. Chen et al. (Eds.): ACCV 2016 Workshops, Part I, LNCS 10116, pp. 349–360, 2017.
DOI: 10.1007/978-3-319-54407-6_23

called spectral-spatial classification). The existing spectral-spatial classification algorithms can be divided into four categories, which are introduced as follows.

The first category is to extract the spectral-spatial features, from the perspective of which morphological profiles are founded to be effective in fusing spectral and spatial information [9,10]. Morphological profiles are extracted on different image layers by using geodesic (e.g., opening and closing) operations, and these profiles are stacked together to depict the spectral-spatial characteristics. By using several attitude filters, the extended morphological profiles [11,12] are proposed to extract a richer description of the HSI regions based on multiple attributes (e.g., area, width, moment of inertia and deviation).

The second category is to use the kernel combination (or fusion) manner for conducting spectral-spatial classification. In 2006, Camps-Valls et al. [13] propose a series of composite kernels to combine spectral information and spatial information directly in the support vector machine kernels. After that, a generalized composite kernel framework is developed, which uses the extend multi-attribute morphological profiles [14] to extract the spectral-spatial features and exploits the multinomial logistic regression method [15] to combine different kernels in a flexible manner.

Another effective way of mining the spectral-spatial characteristics is to integrate classic classification algorithms with various graphic models. Moser et al. [16] develop a unified contextual classification framework by combining support vector machines and Markov random filed models. Srinivas et al. [17] present a probabilistic graphical model framework to fuse sparse representation classifiers and probabilistic graphical models. In this framework, different neighbouring pixels are modeled by multiple sparse representations and their class conditional correlations are exploited by using probabilistic graphical models. In [18], a semi-local spatial graph regularization term is proposed and added into the kernel sparse representation method [19] for considering the spectral-spatial information and improving the classification performance.

The fourth category of considering both spectral and spatial information is the joint representation model, which has also drawn many researchers' attention due to the recent trend of research on sparse representation [20]. Chen et al. [21] adopt a joint sparse model to incorporate the spatial contextual information by simultaneously representing neighbouring pixels by all training pixels and infer the representation coefficient matrix by using the simultaneous orthogonal matching pursuit (SOMP) method. As shown in [21], the joint sparse model performs better than many traditional HSI classification algorithms and sparse models without joint representation. Inspired by the success of the joint sparse model [21,22], much work has been done to combine different sparse models and joint representation manners by focusing on various research lines, including kernel-based joint sparse model [18,19], structured joint sparse model [23,24], local model [25], feature selection [26,27], dictionary learning [28,29] and so on.

In this work, we present a simple but effective HSI classification method motivated by the K-nearest neighbours (KNN) algorithm and the joint representation model. To consider both spectral and spatial information, we extend the traditional KNN method to a joint manner, the basic distance of which is to measure the differences between all neighbouring pixels of a given test pixel and training samples. Specially, we exploit a set-to-point distance between neighbouring pixels and each training sample as the basic distance, which can be effectively calculated by the least squares regression manner. Finally, this set-to-point distance is introduced into the proposed joint KNN framework, in which a weighted KNN method is adopted to achieve stable performance. To evaluate different classification algorithms, we adopt two HSI classification data sets (including AVIRIS Indian Pines and University of Pavia) to compare the proposed weighted joint KNN method with other classic and popular methods. Experimental results demonstrate that our method is able to achieve better performance than other competing ones in these challenging data sets.

2   Joint Weighted K-Nearest Neighbours for HSI Classification

To begin with, we define some notions to make this section be self-contained. The matrix $\mathbf{A} = \begin{bmatrix} \mathbf{A}^1, \mathbf{A}^2, ..., \mathbf{A}^M \end{bmatrix}$ denotes the sample set that consists of all training pixels with M classes totally, where the sub-matrix $\mathbf{A}^m = \begin{bmatrix} \mathbf{a}_1^m, \mathbf{a}_2^m, ..., \mathbf{a}_{n_m}^m \end{bmatrix}$ stands for the training pixels belonging to the m-th class (n_m is the number of sample vectors in class m). The sample vector $\mathbf{a}_i^m \in \mathbb{R}^{d \times 1}$ is a column vector, which denotes the i-th column of the subset $\mathbf{A}^m$ (d is the dimension of the spectral feature for each pixel). For a given image pixel l, we use $\mathbf{y}_l$ to denote its spectral feature. The matrix $\mathbf{Y}_l = \begin{bmatrix} \mathbf{y}_1, \mathbf{y}_2, ..., \mathbf{y}_l, ..., \mathbf{y}_{|\mathbb{N}_l|} \end{bmatrix}$ denotes the set of all neighbouring pixels of pixel l (including pixel l itself), where $\mathbb{N}_l$ stands for the neighbourhood system and $|\mathbb{N}_l|$ is the number of pixels within $\mathbb{N}_l$.

2.1   K-Nearest Neighbours (KNN) and Weighted K-Nearest Neighbours (WKNN)

NN: The nearest neighbour (NN) method is one of the basic algorithms in pattern recognition and machine learning, which usually acts as the baseline algorithm in many domains (e.g., face recognition, image search and visual tracking). If the NN method is applied for classifying the l-th pixel, it requires to determine a distance between the spectral feature vector $\mathbf{y}_l$ and any training sample $\mathbf{a}_i^m$ (i.e., a point-to-point distance) at first. Here we adopt the squared euclidean distance due to its effectiveness and simplicity, which is defined as

$$d\left(\mathbf{y}_l; \mathbf{a}_i^m\right) = \left\| \mathbf{y}_l - \mathbf{a}_i^m \right\|_2^2. \tag{1}$$

Based on the point-to-point distance (1), the NN method is to achieve the classification process by assigning the label z of pixel l as the index that corresponds to the smallest distance $d\left(\mathbf{y}_l; \mathbf{a}_i^m\right)$, i.e.,

$$z = \arg\min_m \left[\min_{i=1,2,...,n_m} d\left(\mathbf{y}_l; \mathbf{a}_i^m\right) \right]. \tag{2}$$

KNN: The KNN algorithm has been one of the most well-known classification methods in pattern recognition, where a test pixel l is labelled by a majority voting of its k nearest neighbours in the training set. For the KNN method, the k nearest neighbours with small distances are searched, which are denoted as $\mathbf{a}_1^{NN}, \mathbf{a}_2^{NN}, ..., \mathbf{a}_k^{NN}$. In this work, we assume that $\mathbf{a}_1^{NN}, \mathbf{a}_2^{NN}, ..., \mathbf{a}_k^{NN}$ are arranged in an increasing order, i.e., $d\left(\mathbf{y}_l; \mathbf{a}_1^{NN}\right) \leq d\left(\mathbf{y}_l; \mathbf{a}_2^{NN}\right) \leq ... \leq d\left(\mathbf{y}_l; \mathbf{a}_k^{NN}\right)$. The corresponding class labels are denoted as $z_1^{NN}, z_2^{NN}, ..., z_k^{NN}$.

$$z = \arg\max_{\widetilde{z}} \sum_{i=1}^{k} \delta\left(\widetilde{z} = z_i^{NN}\right). \tag{3}$$

WKNN: In order to overcome the unstability of the traditional KNN method, Dudani [30] introduces a weighted voting method for the KNN method, in which the closer neighbours are weighted more heavily than the farther ones. By defining the weights of k neighbours as

$$w_i = \begin{cases} \frac{d\left(\mathbf{y}_l; \mathbf{a}_k^{NN}\right) - d\left(\mathbf{y}_l; \mathbf{a}_i^{NN}\right)}{d\left(\mathbf{y}_l; \mathbf{a}_k^{NN}\right) - d\left(\mathbf{y}_l; \mathbf{a}_1^{NN}\right)} & , \; if \; d\left(\mathbf{y}_l; \mathbf{a}_k^{NN}\right) \neq d\left(\mathbf{y}_l; \mathbf{a}_1^{NN}\right) \\ 1 & , \; if \; d\left(\mathbf{y}_l; \mathbf{a}_k^{NN}\right) = d\left(\mathbf{y}_l; \mathbf{a}_1^{NN}\right) \end{cases} \tag{4}$$

We can see that a neighbour with smaller distance is weighted more heavily than one with greater distance.

Then, the classification rule of the WKNN method can be conducted by the majority weighted voting.

$$z = \arg\max_{\widetilde{z}} \sum_{i=1}^{k} w_i \times \delta\left(\widetilde{z} = z_i^{NN}\right). \tag{5}$$

2.2   Joint Weighted K-Nearest Neighbours (JWKNN)

We note that the NN, KNN and WKNN methods merely consider the feature information of the test pixel itself and completely ignore the contributions of its surrounding pixels to the final classification. Inspired by the idea of the joint representation model, we first present a joint nearest neighbour (JNN) method and then extend it into a WKNN manner. This method considers the spatial information by using a point-to-set distance $d\left(\mathbf{Y}_l; \mathbf{a}_i^m\right)$, which depicts a distance from the neighbouring sample set $\mathbf{Y}_l$ to any training sample $\mathbf{a}_i^m$,

$$d\left(\mathbf{Y}_l; \mathbf{a}_i^m\right) = \|\mathbf{Y}_l\widehat{\mathbf{x}} - \mathbf{a}_i^m\|_2^2, \tag{6}$$

where the coefficient vector with respect to each training sample is calculated by using a least squares manner, i.e.,

$$\widehat{\mathbf{x}} = \arg\min_{\mathbf{x}} \|\mathbf{Y}_l\mathbf{x} - \mathbf{a}_i^m\|_2^2. \tag{7}$$

This problem can be analytically solved by $\widehat{\mathbf{x}} = \left(\mathbf{Y}_l^\top\mathbf{Y}_l\right)^{-1}\mathbf{Y}_l^\top\mathbf{a}_i^m$ if $\mathbf{Y}_l^\top\mathbf{Y}_l$ is non-singular. In practice, it should better use ridge regression to obtain a stable

The joint weighted KNN method for hyperspectral image classification.

Input: The query matrix $\mathbf{Y}_l$, and the training matrix $\mathbf{A}$.

1: Calculate all joint distances of nearest neighbours of the query $\mathbf{Y}_l$, denoted as $d\left(\mathbf{Y}_l; \mathbf{a}_i^m\right)$, for all i, m.

2: Sort the distances in an ascending order and search k nearest neighbours of the query $\mathbf{Y}_l$. The k nearest neighbours are denoted as $\mathbf{a}_i^{NN}, i = 1, 2, ..., k$. The corresponding class labels are recorded as $z_i, i = 1, 2, ..., k$.

3: Calculate the weights of k nearest neighbours, i.e., $w_i = 1, 2, ..., k$.

$$w_i = \begin{cases} \frac{d\left(\mathbf{Y}_l; \mathbf{a}_k^{NN}\right) - d\left(\mathbf{Y}_l; \mathbf{a}_i^{NN}\right)}{d\left(\mathbf{Y}_l; \mathbf{a}_k^{NN}\right) - d\left(\mathbf{Y}_l; \mathbf{a}_1^{NN}\right)} & , if\ d\left(\mathbf{Y}_l; \mathbf{a}_k^{NN}\right) \neq d\left(\mathbf{Y}_l; \mathbf{a}_1^{NN}\right) \\ 1 & , if\ d\left(\mathbf{Y}_l; \mathbf{a}_k^{NN}\right) = d\left(\mathbf{Y}_l; \mathbf{a}_1^{NN}\right) \end{cases}$$

4: Assign a majority weighted voting class label to the query $\mathbf{Y}_l$.

$$z = \arg\max_{\tilde{z}} \sum_{i=1}^{k} w_i \times \delta\left(\tilde{z} = z_i^{NN}\right).$$

Output: The class label z.

solution, i.e., $\hat{\mathbf{x}} = \left(\mathbf{Y}_l^{\top}\mathbf{Y}_l + \lambda\mathbf{I}\right)^{-1}\mathbf{Y}_l^{\top}\mathbf{a}_i^m$, where λ is a small positive constant to avoid model degradation.

By adopting the joint distance (6) as the basic distance, we propose a novel joint weighted KNN (JWKNN) method to achieve HSI classification, the fundamental idea of which is illustrated in Fig. 1. It is obvious that the JWKNN method effectively combines the ideas of both joint distance and WKNN for developing a stable HSI classification algorithm.

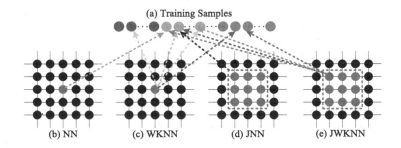

Fig. 1. The differences between the traditional NN, WKNN and the proposed JNN, JWKNN methods. Figure (a) provides an illustration of training samples, where different classes are described by different color circle dots. Figure (b) shows the basic idea of the NN method, in which the red circle dot denotes the pixel to be classified and the black circle dot stands for other pixels. Figure (d) illustrates the motivation of the JNN method, in which the 3×3 neighbourhood system is chosen as an example. Figure (c) denotes the basic idea of the WKNN method, in which different colors of arrows represents different weights in the WKNN scheme. Finally, the proposed JWKNN method is illustrated in Figure (e), which can be viewed as a combination of the JNN method and WKNN method. (Color figure online)

The overall classification framework is presented in Table 1. We note that the JWKNN method is different from the WKNN method from two aspects: (a) the joint distance (6) is adopted as a basic distance and is used for searching k nearest neighbours; and (b) the weights for the WKNN method are calculated based on the joint distances (6) rather than the euclidean distances.

3   Experiments and Discussions

In this experiment, we compare our JWKNN method with eight competing algorithms, including nearest neighbour (NN), support vector machine (SVM) with a polynomial kernel [2], SVM with the composite kernel framework (SVMCK) [13], the fusion of SVM and MRF (SVMMRF) [31], automatic standard deviation attribute profiles (ASDAP) [12], simultaneous orthogonal matching pursuit (SOMP) [21], joint sparse representation-based nearest neighbour (JSRNN) [32], and the implemented joint nearest neighbours (JNN). For the SVMCK method, we adopt a composite kernel that combines the spectral and spatial information via a weighted kernel summation, where a polynomial kernel is adopted to describe the spectral information and a Gaussian kernel is exploited to depict the spatial contextual information. All parameters in the SVM-based algorithms (SVM and SVMCK) are estimated by cross-validation. For the sparsity-based methods (SOMP and JSRNN), the sparsity levels of them are set to be 5 and the neighbourhood size is chosen as 5×5 or 7×7 based on its better performance.

We note that all methods are evaluated on the AVIRIS Indian Pines and University of Pavia datasets by using three quantitative criteria, including overall accuracy (OA), average accuracy (AA) and κ coefficient[1]. For the proposed JWKNN algorithm, the default parameters are set as follows: the 7×7 window size is chosen for describing spatial neighbourhood information; and the number of neighbours for the WKNN scheme is set to be 6.

3.1   Experiments on AVIRIS Indian Pines

First, we evaluate different HSI classification algorithms on the AVIRIS Indian Pines dataset, which was collected by the AVIRIS (Airborne Visible/Infrared Imaging Spectrometer) sensor over the Indian Pines region in 2012 [33]. This dataset consists of 145×145 spatial pixels and 224 spectral bands in the wavelength range 0.4–2.5 μm, and include 16 ground-truth classes with different kinds of plants (such as corns, grasses, soybeans, wheats and so on). The number of spectral bands is reduced to 200 by removing noisy bands that cover the region of water absorption.

---

[1] The overall accuracy is defined as the ratio of correctly classified samples to all test samples, the average accuracy is calculated by simply averaging the accuracies for each class, and the κ coefficient is computed based on the confusion matrix of different classes.

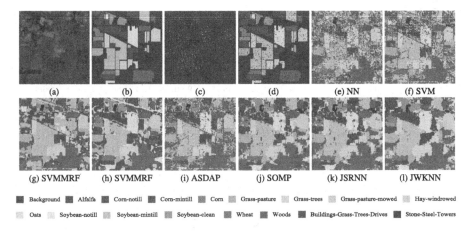

Fig. 2. Qualitative results on the AVIRIS Indian Pines data set. (a) Three-band color composite image (bands 1, 20 and 50), (b) Ground truth, (c) Training set, (d) Test set. The classification maps of different methods are shown in (e–l) respectively.

Figure 2 illustrates the classification maps of different algorithms. From this figure, we can see that the proposed method achieves better classification performance and more smoother visual effect than other competing methods. Table 1 compares different classification methods quantitatively, which reports the classification accuracy for each class, overall accuracy (OA), κ coefficient measure (κ), and average accuracy (AA). From this table, we can also conclude that the proposed JWKNN method achieves better performance than other competing ones.

3.2 Effects of Parameters and Components

In this subsection, we investigate the effects of parameters and components for the proposed JWKNN method by using the AVIRIS Indian Pines data set (5% for training and 95% for test). Figure 3 (a) demonstrates the effect of the window size for describing the spatial neighbourhood information. For one thing, if the window size is too small (e.g., 3×3), the proposed JWKNN model does not capture sufficient spatial information. Thus, it cannot achieve satisfying classification results in these cases. For another, a very large window size will introduce unexpected noises and degrade the classification model. It can be seen from Fig. 3 (a) that the stable performance is achieved when the window size is chosen as 7×7. Figure 3 (b) shows the overall classification results with different numbers of neighbours, from which we can see that the best performance is achieved when k is around 6. In Fig. 3 (c), we conduct a comparison of related models with different basic distances and different KNN scheme (i.e., the traditional KNN method and the weighted KNN method). From this figure, we have the following observations: (1) the results of algorithms with the traditional KNN technique are not stable with different choices of the neighbourhood size k. In contrast, the

Table 1. Classification accuracy on the Indian Pines data set for different methods (5% samples for training).

Class No.	No. of Samples		NN	SVM	SVMMRF	SVMCK	ASDAP	SOMP	JSRNN	JNN	JWKNN
	Train	Test									
1	2	44	43.9	40.9	50.0	53.2	93.6	91.6	86.6	94.4	96.4
2	71	1357	51.9	71.3	80.5	89.4	87.5	84.4	90.2	91.8	93.6
3	42	788	48.7	58.6	87.9	91.8	89.9	86.0	87.6	88.3	90.8
4	12	255	41.7	39.4	60.0	81.2	73.1	89.8	85.8	89.6	92.9
5	34	459	75.7	89.2	94.7	92.9	94.2	94.0	92.5	93.7	94.9
6	37	693	81.5	93.8	92.6	98.2	98.9	96.5	93.9	93.1	94.4
7	1	27	59.8	54.4	50.0	70.4	71.5	94.0	76.3	88.5	95.6
8	24	454	91.1	95.4	90.6	99.6	99.9	97.0	99.2	98.9	98.8
9	1	19	23.4	18.9	50.0	64.7	24.7	73.6	33.7	48.0	48.8
10	49	923	52.9	62.8	84.8	84.9	87.1	86.0	89.4	89.4	90.1
11	123	2332	65.8	74.0	61.9	91.5	93.6	86.4	94.4	93.1	93.6
12	30	563	41.9	52.3	85.7	81.8	89.7	78.6	86.8	90.0	91.8
13	10	195	81.3	94.6	99.6	98.1	98.9	96.4	88.2	88.6	92.0
14	63	1202	89.0	90.1	93.6	97.0	98.5	94.9	97.1	97.8	97.9
15	19	367	43.9	44.5	93.0	88.7	94.4	90.7	85.3	92.5	94.7
16	5	88	98.9	88.9	95.0	92.0	90.8	79.9	85.1	85.0	91.5
OA (%)	-	-	65.1	73.5	78.6	91.1	92.3	88.4	91.8	92.5	93.7
κ	-	-	0.602	0.696	0.749	0.899	0.913	0.868	0.907	0.914	0.928
AA (%)	-	-	62.0	66.8	79.4	86.0	86.6	88.7	85.8	88.9	91.1

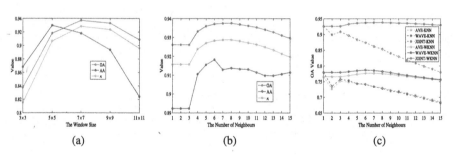

(a)   (b)   (c)

Fig. 3. The effects of different parameters and components. (a) The effects of the neighbouring window size. (b) The effects of the number of neighbours. (c) The comparisons of different basic distances with KNN and WKNN methods.

results of algorithms with WKNN [34] achieve more robust performance; (2) the adopted basic distance based on least squares performs better than other basic ones.

3.3 Experiments on University of Pavia

The second hyperspectral image, University of Pavia, was acquired by the ROSIS (Reflective Optics System Imaging Spectormeter) sensor. The ROSIS

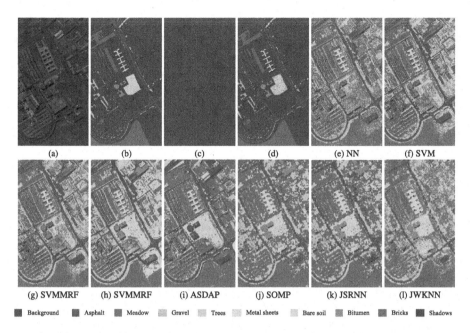

(a)　　　　(b)　　　　(c)　　　　(d)　　　　(e) NN　　　(f) SVM

(g) SVMMRF　　(h) SVMMRF　　(i) ASDAP　　(j) SOMP　　(k) JSRNN　　(l) JWKNN

■ Background　■ Asphalt　■ Meadow　■ Gravel　■ Trees　■ Metal sheets　■ Bare soil　■ Bitumen　■ Bricks　■ Shadows

Fig. 4. Qualitative results on the University of Pavia data set. (a) Three-band color composite image (bands 102, 56 and 31), (b) Ground truth, (c) Training set, (d) Test set. The classification maps of different methods are shown in (e–l) respectively.

sensor captures 115 spectral bands totally, which are in the wavelength range $0.43 - 0.86\,\mu\mathrm{m}$ and the spatial resolution of $1.3\,\mathrm{m}$ per pixel [35]. This dataset includes nine ground-truth classes of interests and consists of 610×340 pixels,

Table 2. Classification accuracy on the University of Pavia data set for different methods (5% samples for training).

Class No.	No. of Samples Train	Test	NN	SVM	SVMMRF	SVMCK	ASDAP	SOMP	JSRNN	JNN	JWKNN
1	332	6299	90.9	90.2	98.8	94.8	93.9	98.5	86.1	99.2	99.4
2	932	17717	86.1	97.0	100.0	99.5	98.3	94.9	99.6	97.9	98.0
3	105	1994	62.5	68.2	64.2	68.1	86.2	83.5	91.6	88.9	93.0
4	153	2911	93.6	91.9	95.0	94.7	97.7	98.8	89.6	97.9	98.1
5	67	1278	99.0	99.4	99.7	99.7	98.1	99.3	100.0	93.9	93.8
6	251	4778	68.4	68.5	92.6	86.2	99.1	94.8	94.5	95.8	97.0
7	67	1263	65.9	76.4	75.2	88.0	98.0	84.1	97.5	92.5	93.3
8	184	3498	63.0	85.7	90.2	90.4	84.7	78.9	94.7	81.1	80.5
9	47	900	99.8	96.8	97.3	99.5	100.0	100.0	77.3	98.2	99.0
OA (%)	-	-	82.0	89.3	95.1	94.2	95.9	93.3	94.8	95.4	95.7
κ	-	-	0.762	0.854	0.935	0.922	0.946	0.911	0.931	0.939	0.943
AA (%)	-	-	81.0	86.0	90.3	91.2	95.1	92.5	92.3	93.9	94.7

which are of 103 bands with 12 noisy bands removed. The classification results of different algorithms are summarized in Fig. 4 and Table 2, in which 5% samples are randomly selected for training and the remaining 95% are adopted for testing (this process is repeated 10 times and the average values and standard deviations are reported in Table 2). The experimental results on the University of Pavia dataset also demonstrate that the proposed JWKNN method yields much better overall performance.

4   Conclusions and Further Works

This paper presents a simple yet effective HSI classification method called joint weighted K-nearest neighbour (JWKNN). The proposed JWKNN algorithm extends the traditional KNN method into a joint manner, in which the a set-to-point distance is adopted as a basic distance to measure the difference between neighbouring pixels and each training sample. Based on empirical observations, a weighted KNN method is adopted to achieve stable HSI classification performance. By using two standard HSI benchmarks, the proposed methods are compared with other classic and popular algorithms. The experimental results demonstrate that our method performs better than other competing methods. In the further, we plan to introduce the kernel trick and metric learning techniques into the JWKNN framework for conducting HSI classification.

Acknowledgement. This work was supported in part by the Natural Science Foundation of China under Grant No. 61502070, and in part by Fundamental Research Funds for Central Universities under Grant No. DUT16RC(4)16.

References

1. Landgrebe, D.A., Serpico, S.B., Crawford, M.M., Singhroy, V.: Introduction to the special issue on analysis of hyperspectral image data. IEEE Trans. Geosci. Remote Sens. **39**, 1343–1345 (2001)
2. Gualtieri, J.A., Chettri, S.R., Cromp, R.F., Johnson, L.F.: Support vector machine classifiers as applied to aviris data. In: Airborne Geoscience Workshop (1999)
3. Melgani, F., Serpico, S.B.: A statistical approach to the fusion of spectral and spatio-temporal contextual information for the classification of remote-sensing images. Pattern Recogn. Lett. **23**, 1053–1061 (2002)
4. Bruzzone, L., Cossu, R.: A multiple-cascade-classifier system for a robust and partially unsupervised updating of land-cover maps. IEEE Trans. Geosci. Remote Sens. **40**, 1984–1996 (2002)
5. Palmason, J.A., Benediktsson, J.A., Sveinsson, J.R., Chanussot, J.: Classification of hyperspectral data from urban areas using morphological preprocessing and independent component analysis. In: IGARSS, pp. 176–179 (2005)
6. Ham, J., Chen, Y., Crawford, M.M., Ghosh, J.: Investigation of the random forest framework for classification of hyperspectral data. IEEE Trans. Geosci. Remote Sens. **43**, 492–501 (2005)

7. Bruzzone, L., Chi, M., Marconcini, M.: A novel transductive svm for semisupervised classification of remote-sensing images. IEEE Trans. Geosci. Remote Sens. **44**, 3363–3373 (2006)

8. Gao, L., Li, J., Khodadadzadeh, M., Plaza, A., Zhang, B., He, Z., Yan, H.: Subspace-based support vector machines for hyperspectral image classification. IEEE Geosci. Remote Sens. Lett. **12**, 349–353 (2014)

9. Pesaresi, M., Benediktsson, J.A.: A new approach for the morphological segmentation of high-resolution satellite imagery. IEEE Trans. Geosci. Remote Sens. **39**, 309–320 (2001)

10. Benediktsson, J.A., Pesaresi, M., Amason, K.: Classification and feature extraction for remote sensing images from urban areas based on morphological transformations. IEEE Trans. Geosci. Remote Sens. **41**, 1940–1949 (2003)

11. Benediktsson, J.A., Palmason, J.A., Sveinsson, J.R.: Classification of hyperspectral data from urban areas based on extended morphological profiles. IEEE Trans. Geosci. Remote Sens. **43**, 480–491 (2005)

12. Marpu, P.R., Pedergnana, M., Mura, M.D., Benediktsson, J.A., Bruzzone, L.: Automatic generation of standard deviation attribute profiles for spectral-spatial classification of remote sensing data. IEEE Geosci. Remote Sens. Lett. **10**, 293–297 (2013)

13. Camps-Valls, G., Gomez-Chova, L., Muñoz-Marí, J., Vila-Francés, J., Calpe-Maravilla, J.: Composite kernels for hyperspectral image classification. IEEE Geosci. Remote Sens. Lett. **3**, 93–97 (2006)

14. Mura, M.D., Benediktsson, J.A., Waske, B., Bruzzone, L.: Morphological attribute profiles for the analysis of very high resolution images. IEEE Trans. Geosci. Remote Sens. **48**, 3747–3762 (2010)

15. Krishnapuram, B., Carin, L., Figueiredo, M.A.T., Hartemink, A.J.: Sparse multinomial logistic regression: fast algorithms and generalization bounds. IEEE Trans. Pattern Anal. Mach. Intell. **27**, 957–968 (2005)

16. Moser, G., Serpico, S.B.: Combining support vector machines and markov random fields in an integrated framework for contextual image classification. IEEE Trans. Geosci. Remote Sens. **51**, 2734–2752 (2013)

17. Srinivas, U., Chen, Y., Monga, V., Nasrabadi, N.M., Tran, T.D.: Exploiting sparsity in hyperspectral image classification via graphical models. IEEE Geosci. Remote Sens. Lett. **10**, 505–509 (2013)

18. Liu, J., Wu, Z., Sun, L., Wei, Z., Xiao, L.: Hyperspectral image classification using kernel sparse representation and semilocal spatial graph regularization. IEEE Geosci. Remote Sens. Lett. **11**, 1320–1324 (2014)

19. Chen, Y., Nasrabadi, N.M., Tran, T.D.: Hyperspectral image classification via kernel sparse representation. IEEE Trans. Geosci. Remote Sens. **51**, 217–231 (2013)

20. Wright, J., Yang, A.Y., Ganesh, A., Sastry, S.S., Ma, Y.: Robust face recognition via sparse representation. IEEE Trans. Pattern Anal. Mach. Intell. **31**, 210–227 (2009)

21. Chen, Y., Nasrabadi, N.M., Tran, T.D.: Hyperspectral image classification using dictionary-based sparse representation. IEEE Trans. Geosci. Remote Sens. **49**, 3973–3985 (2011)

22. Chen, Y., Nasrabadi, N.M., Tran, T.D.: Sparse representation for target detection in hyperspectral imagery. IEEE J. Sel. Topics Sig. Process. **5**, 629–640 (2011)

23. Qian, Y., Ye, M., Zhou, J.: Hyperspectral image classification based on structured sparse logistic regression and three-dimensional wavelet texture features. IEEE Trans. Geosci. Remote Sens. **51**, 2276–2291 (2013)

24. Sun, X., Qu, Q., Nasrabadi, N.M., Tran, T.D.: Structured priors for sparse-representation-based hyperspectral image classification. IEEE Geosci. Remote Sens. Lett. **11**, 1235–1239 (2014)
25. Li, J., Zhang, H., Huang, Y., Zhang, L.: Hyperspectral image classification by nonlocal joint collaborative representation with a locally adaptive dictionary. IEEE Trans. Geosci. Remote Sens. **52**, 3707–3719 (2014)
26. Li, S., Qi, H.: Sparse representation based band selection for hyperspectral images. In: ICIP, pp. 2693–2696 (2011)
27. Yang, S., Jin, H., Wang, M., Ren, Y., Jiao, L.: Data-driven compressive sampling and learning sparse coding for hyperspectral image classification. IEEE Geosci. Remote Sens. Lett. **11**, 479–483 (2014)
28. Wang, Z., Nasrabadi, N.M., Huang, T.S.: Discriminative and compact dictionary design for hyperspectral image classification using learning vq framework. In: ICASSP, pp. 3427–3431 (2013)
29. Soltani-Farani, A.A., Rabiee, H.R., Hosseini, S.A.: Spatial-aware dictionary learning for hyperspectral image classification. IEEE Trans. Geosci. Remote Sens. **53**, 527–541 (2015)
30. Dudani, S.A.: The distance-weighted k-nearest-neighbor rule. IEEE Trans. Syst. Man Cybern. **6**, 325–327 (1976)
31. Fauvel, Y.T.M., Chanussot, J., Benediktsson, J.A.: Svm-and mrf-based method for accurate classification of hyperspectral images. IEEE Geosci. Remote Sens. Lett. **7**, 736–740 (2010)
32. Zou, J., Li, W., Du, Q.: Sparse representation-based nearest neighbor classifiers for hyperspectral imagery. IEEE Geosci. Remote Sens. Lett. **12**, 2418–2422 (2015)
33. The indian pines data set. http://www.ehu.es/ccwintco/index.php/Hyperspectral_Remote_Sensing_Scenes
34. Dudani, S.A.: The distance-weighted k-nearest neighbor rule. IEEE Trans. Syst. Man Cybern. **6**, 325–327 (1976)
35. Plaza, A., et al.: Recent advances in techniques for hyperspectral image processing. Remote Sens. Environ. **113**, S110–S122 (2009)

Perceptual Color Classification Based on Lightning Environment with Hyperspectral Data

Yuko Ozasa[1]([⊠]), Kenji Iwata[2], Naoko Enami[3], and Yutaka Satou[2]

[1] Faculty of Science and Technology, Keio University, 3-14-1, Hiyoshi, Kohoku-ku, Yokohama, Kanagawa 223-8522, Japan
yuko.ozasa@keio.jp
[2] National Institute of Advanced Industrial Science and Technology (AIST), Central 1, 1-1-1 Umezono, Tsukuba, Ibaraki 305-8560, Japan
{kenji.iwata,yu.satou}@aist.go.jp
[3] Organization of Advanced Science and Technology, Kobe University, 1-1 Rokkodai, Nada, Kobe 657-8501, Japan
naoko.enami@port.kobe-u.ac.jp

Abstract. Identifying objects by its color name is an essential capability for a service robot to understand and interact with the physical world and be of use in everyday life scenarios. The robot uses an image for the identification, but the image is strongly affected by lightning environment while human does not affected by the environment when they decide its color name. We present lightning environment estimation and perceptual color classification based on the estimation using hyperspectral data. Support Vector Machine (SVM) or Multiple Kernel Learning SVM is used for the estimation and classification. Originality of our paper is that the hyperspectral data is used for both lightning environment estimation and perceptual color classification of an object. Suppose that the lightning environment is given, the perceptual color of the object is classified in each environment. Additionally, perceptual color is estimated by using result of the lightning environment estimation. A novel dataset which consists of the hyperspectral image of 15 objects taken in 4 different lightning environments is constructed in our experiment. Experimental result of the hyperspectral data are compared by those of common color spaces, such as RGB and L*a*b. The estimation and classification with SVM and MKL SVM are compared in the experiments. From the experimental results, hyperspectral data enabled us to present a separated scheme which consists of the lightning environment estimation and perceptual color classification while previous color spaces could not.

1 Introduction and Related Work

Identification of task-relevant objects is an essential problem for advanced service robots. This paper deals with a task where a robot identifies an object that a human asks it to bring by its color name. The achievement of the task requires a method for color recognition based on perceptual color of human. The perceptual

© Springer International Publishing AG 2017
C.-S. Chen et al. (Eds.): ACCV 2016 Workshops, Part I, LNCS 10116, pp. 361–373, 2017.
DOI: 10.1007/978-3-319-54407-6_24

Fig. 1. Color perception of human and robot vision.

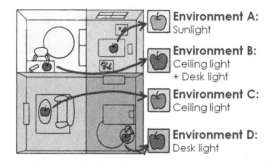

Fig. 2. Lightning environments in home environment.

color is a color which is classified with categorical color perception, such as red, blue, and yellow etc. [1]. Human see an object with human eyes and determines a perceptual color of the object. The robot captures an image of the object with camera and recognizes its color of the image. (See Fig. 1).

Human vision perceives the same color even under various lighting environments. This is referred to as a color consistency [2]. The color of images captured by camera built in the robot is affected by lightning environment. An example of home environment is shown in Fig. 2. As illustrated, environment (a) receives sunlight, environment (b) receives ceiling light and desk light, environment (c) receives ceiling light, and environment (d) receives desk light. Even if the same object is located in each environment, sensing data captured by camera differ from each other.

Should not be pessimistic for this fact. There are some constrained conditions of lightning environments in home environment. The conditions are (1) the light is different in each room: (2) the light is fixed in place: (3) the number of the room in home environment is finite. Focus on the facts, the lightning environments may be estimated in home environment. Additionally, if the lightning environments are estimated, the accuracy of the color classification can be improved by using classifiers optimized for each environments. This paper proposes a perceptual color classification based on estimation of lightning environment. For the first step of the color recognition of each object, the classification is operated with each pixels of an image in this paper.

Over the past years, several research groups reported the technique of object identification for service robot whose eyes are RGB camera [3–5]. Although some work introduced the methods for object identification based on color name, the accuracy of them are not high enough [4,5]. The cameras which can capture various data were developed in recent years, for example, 3D camera and hyperspectral camera. Not far equipped with the cameras is the day for robot. The hyperspectral camera can capture a spectrum that has been split into more than a few tens of bands while the RGB camera can capture a spectrum whose number of bands are only three [6]. Invisible and detailed information can obtained by the hyperspectral camera, it is possible to get more information that human eyes and the RGB cameras cannot obtain [7]. Aims to mount the hyperspectral camera on robot in the near future, the perceptual color classification with hyperspectral data. The color recognition is the first step for the perceptual color recognition. If the classification goes well, then the recognition goes well.

The camera used in this paper is the hyperspectral camera $NH-7$ [8] which can capture 151 bands of spectrum from 350 nm to 1100 nm. 85 bands of 151 bands are visible light. RGB and L*a*b color spaces are consists of 3 band of these 85 bands. Our method uses not only the 3 bands but also the 85 bands of spectrum for the classification.

Support Vector Machine (SVM) is a supervised approach, for the classification of hyperspectral images [9]. SVMs are good candidates for a supervised non-parametric classification of hyperspectral images [10]. SVMs belong to the elite group of machine learning algorithms that utilize optimization tools, which seek to identify an optimal separating hyperplane to discriminate between two classes of interest [11]. In this paper, SVM is used for the classification.

While the dimensions of the hyperspectral data is high, the feature selection is effective. A possible answer to the needs of the SVM-based feature selection can be found in the framework of multiple kernel learning (MKL) [12]. In order to reduce the effect of the high dimensional data, MKL SVM is used in this paper, too. In MKL, the SVM kernel function is defined as a weighted linear combination of kernels built using subsets of features. MKL works iteratively, optimizing both the weights and the kernel parameter. This way, the SVM model is optimized at the same time as the combination of base kernels. When constructing the base kernels with single features or physically inspired groups of features, the optimization of the SVM kernel works as a feature selector providing a weighted ranking of the importance of its components [13].

2 Proposed Method

The perceptual color classification and recognition is affected by lightning environment, so it may be effective that a classifier of each lightning environment is used for perceptual color classification. Even if the lightning environment is not given, the lightning environment is estimated in this paper. Hyperspectral data is used for both lightning environment estimation and perceptual color classification.

2.1 Lightning Environment Estimation and Perceptual Color Classification Using Hyperspectral Data

A hyperspectral data of each pixel is classified into its lightning environment and perceptual color in this paper. An input of the classification is a hyperspectral data of a pixel, and output is lightning environment and perceptual color name. A number of dimensions of the data is a number of bands of the data. The classification is the first step for lightning environment and perceptual color recognition of an object. An input of the recognition is hyperspectral data of whole image, and output are the environment and its color name of the object. If the classification success, the recognition becomes possible.

For the classifier, SVM or Multiple kernel learning SVM (MKL SVM) is used in this paper. Multiple Kernel Learning methods aim to construct a kernel model where the kernel is a linear combination of fixed base kernels. Learning the kernel then consists of learning the weighting coefficients for each base kernel, rather than optimizing the kernel parameters of a single kernel [14]. Common kernel functions, like Gaussian (RBF) or polynomial, are rigid representations of the data, that may be replaced by more flexible and data-adapted kernels [13]. In the multiple-kernel framework, the optimal kernel is learned from data by building a weighted linear combination of M base kernels. The use of multiple kernels can enhance the performance of the model and, more importantly, the interpretability of the results. Let β be a vector of weights for the mixture of kernels. A multiple kernel is the combination of the M basis kernels k_d, and the representation is

$$k(\mathbf{H_i}, \mathbf{H_j}) = \sum_{d=1}^{M} \beta_d k_d(\mathbf{H_i}(d), \mathbf{H_j}(d)) \qquad (1)$$

where $\mathbf{H_i}$ and $\mathbf{H_j}$ are samples of the hyperspectral data i and j. β_l denotes a weights of l-th subkernel k_i. The subkernels are defined at each dimension of the hyperspectral data. d is a hyperspectral data of d-th dimension. The weights of MKL β_d is learned with margin maximization as SVM. Dual problem for optimization in MKL using margin maximization is as follows,

$$\max \alpha, \beta \sum_i \alpha_i - \frac{1}{2} \sum_{i,j} \alpha_i \alpha_j y_i y_j \sum_d \beta_d k_d(\mathbf{H_i}, \mathbf{H_j})$$

$$s.t. \begin{cases} \sum_i y_i \alpha_i = 0, & 0 \le \alpha_i \le C \\ \sum_d \beta_d = 1, & \beta_d \ge 0 \end{cases} \qquad (2)$$

where α_i is Lagrange coefficient, and y_i is a variable which denotes a class$(-1, 1)$, and C is a variable which determines the trade-off between the margin and error rate of training data. Both α_i and β_d are obtained by two steps of iteration. In the first step, α_i is updated by the same way to SVM while β_d is fixed. Next, β_d is updated while α_i is fixed. These two steps are repeated until it reaches the end condition. Then, the subkernels are linear-combined. The integrated kernels are applied to SVM and the estimation or classification is operated using the SVM.

2.2 Perceptual Color Classification Based on Lightning Ennvironment

The perceptual color classification is affected by lightning environment. It may be effective that a classifier of each lightning environment is used for perceptual color classification if the lightning environment is given.

Even if the lightning environment is not given, the lightning environment is estimated by using the hyperspectral data. The lightning environment is estimated for the preprocessing of the classification. Using the result of the lightning environment estimation, input data is classified into the color class.

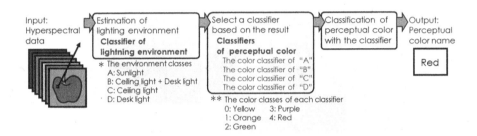

Fig. 3. A rough sketch of architecture for perceptual color classification based on lightning environment estimation.

The architecture of our method is shown in Fig. 3. First, a hyperspectral data of a pixel is input, then the lightning environment is estimated by the data. For the estimation, a classifier of the lightning environment is necessary to be learned. Second, a classifier of perceptual color is selected by the result of the lightning environment estimation. A number of the color classifier required for this method is a number of variations of the lightning environment prepared in the preprocessing. The input data is classified by a classifier selected from these classifiers. For example, when the lightning environment is estimated as "Sunlight", the color classifier of "Sunlight" is selected and the classification is executed by the classifier. Then, the output is a perceptual color name. For the estimation and the classification, SVM or MKL SVM is used in this paper.

3 Experiments

Our method consists of two parts which are lightning environment estimation and perceptual color classification. We evaluated the lightning environment and perceptual color classification respectively, and then we evaluated the perceptual color classification based on the lightning environment estimation.

ID	1	2	3	4	5
Annotation	Green	Red	Green	Orange	Yellow
RGB image	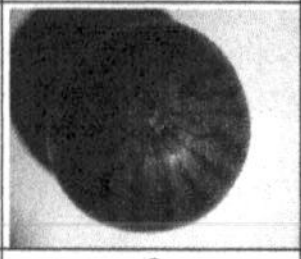	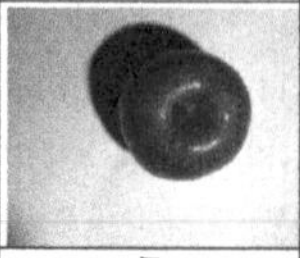	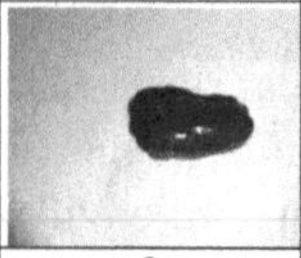	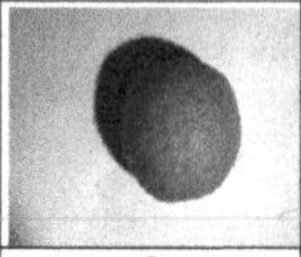	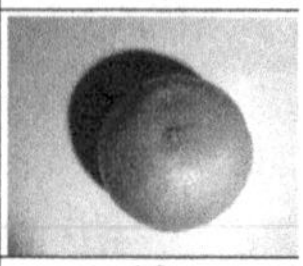
ID	6	7	8	9	10
Annotation	Yellow	Green	Red	Yellow	Red
RGB image		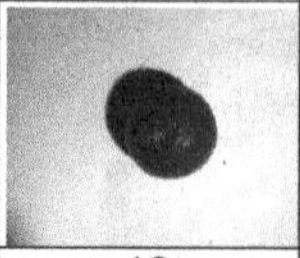	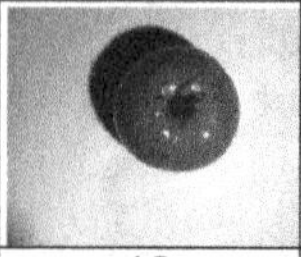		
ID	11	12	13	14	15
Annotation	Orange	Orange	Green	Purple	Purple
RGB image	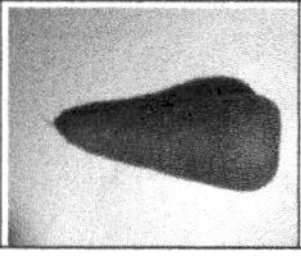	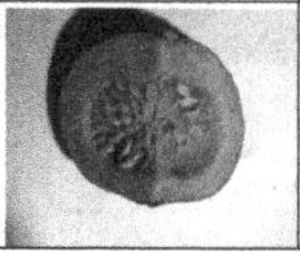	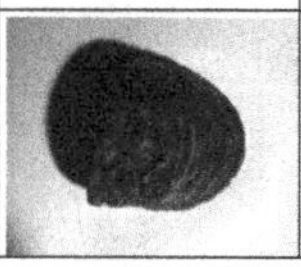		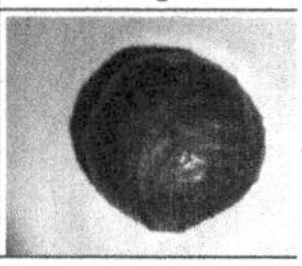

Fig. 4. Objects and its annotations used in experiment (Color figure online).

3.1 Hyperspectral Dataset

In our experiments 15 objects were used, and the objects were fruit and vegetables, such as a watermelon, an apple, and a green pepper. A participant annotated color labels of the objects. The participant saw the real objects and annotated their colors from categorical color names (11 colors) [1] such as blue, brown, grey, green, orange, pink, purple, red, white, and yellow. Number of the color names selected by the participant was 5 colors among 11 colors in this experiment. The objects and its annotation were shown in Fig. 4.

The objects were captured under 4 lighting environments by the hyperspectral camera $NH - 7$ [8]. The camera captures 151 bands of hyper-spectrum and its image size is 1280×1024 pixels. 85 bands of visible light were used from 151 bands in our experiments. Its wavelength is 350 nm to 1100 nm, and spectral resolution is 5 nm.

4 lighting environments were prepared: (A) a halogen light, (B) a ceiling light and a desk light, (C) a ceiling light, and (D) a desk light. The environments were shown in Fig. 5. Although the effect of sunlight cannot be ignored, the sunlight is not stable, so the halogen light is substituted as sunlight in the experiment. Object region was manually extracted from the image and the extracted objet region is used for an experiment.

Figure 6 shows sample spectral signature of each object in each lightning environments. The spectral signature was taken from a pixel data of each object data. A, B, C, D in Fig. 6 denote the lightning environments in our experiments. There are some differences between 15 objects. The spectral signature of lightning environment A is broad and others are not.

| (A) | (B) | (C) | (D) |
| Halogen light | Ceiling light and desk light | Ceiling light | Desk light |

Fig. 5. Variations of Lightning environmrnts.

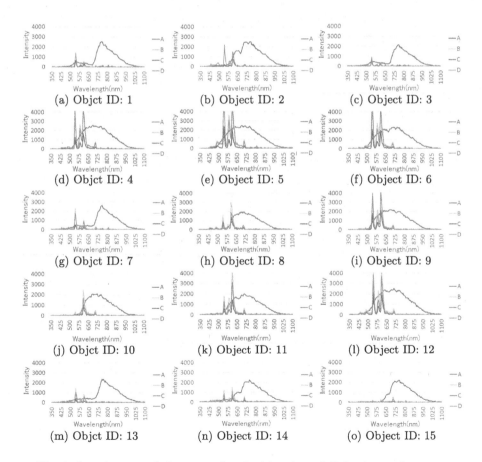

Fig. 6. Sample spectral signature of each object in each lightning environment.

3.2   Lightning Environment Estimation Using Hyperspectral Data

3000 pixel values of object region were sampled by random from the images in each environment. 1000 samples were used as test samples and other 2000 samples were used as learning sample in each environment.

4 types of samples were used in the experiment: 3 bands of the hyperspectral raw data, 85 bands of the raw data, RGB color spaces data converted from the 3 bands of the raw data [15], and L*a*b color spaces data converted from the RGB color spaces data. The 85 bands is the bands of a visible light region.

The classifiers used in experiment were a support vector machine (SVM) and multiple kernel learning support vector machine (MKL SVM), and we adopt the one-vs-rest strategy for multi-class classification. 3 types of kernel SVM were used, such as linear, polynomial, and Gaussian kernels. 2types of MKL SVM kernel were used, such as polynomial and Gaussian kernels.

The estimation results were shown in the Table 1. The accuracy of hyperspectral data were almost twice higher than that of color spaces. The estimation using RGB with SVM based on Gaussian kernel was the most effective in all the estimation using color spaces. The condition whose accuracy was the highest of all conditions was using 85 bands of hyperspectral data with MKL based on polynomial kernel. Figure 7 shows the accuracy of the lightning environment estimation at each environment which conditions are RGB with SVM based on Gaussian kernel and 85 bands of hyperspectral data with MKL based on polynomial kernel. Both result shows that the accuracy of the estimation in environment A and C were higher than that of the estimation in environment B and D. The environment A is the halogen light which lightning characteristic is broad, and the estimation in environment A with hyperspectral data went well. Due to the characteristic of the florescent light, the estimation in environment B and D were more difficult than that in environment A.

Table 1. Accuracy of lightning environment estimation (%).

	Color space		Hyperspectral raw data	
	RGB	L*a*b	3 bands	85 bands
SVM				
Linear Kernel	25.70	25.65	52.78	63.45
Polynomial kernel	25.68	26.60	46.83	64.93
Gaussian kernel	35.40	34.18	52.75	74.88
MKL SVM				
Polynomial kernel	27.48	27.13	52.35	**76.20**
Gaussian kernel	25.65	25.65	48.18	67.10

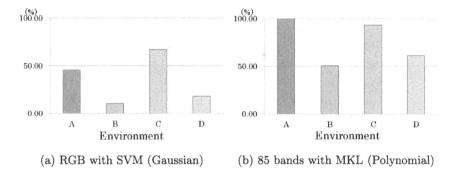

(a) RGB with SVM (Gaussian)   (b) 85 bands with MKL (Polynomial)

Fig. 7. Accuracy of lightning environment estimation at each environment. (a) RGB with SVM (Gaussian) (b) 85 bands with MKL (Polynomial) (Color figure online)

3.3 Perceptual Color Classification Using Hyperspectral Data

We evaluate the perceptual color recognition without lightning environment estimation in this subsection.

5 type of data were used in the experiment: data of environment A, B, C and D, and data sampled by random from the environments. 2000 pixel values of object region were sampled by random from each color in each environment. Number of the colors was 5. 1000 samples were used as test sample, and 1000 samples were used as learning sample. Types of classifiers and samples used in the experiment were as same as the experiments in subsection B.

The result of the experiment is shown in Table. 2. "Not given" in the table denotes that the lightning environment of test data was not given and a classifier of color classification was prepared for each sample, such as RGB, L*a*b, 3 bands, and 85 bands. "Given" denotes that the lightning environment of test sample was given and the classifiers were prepared for each lightning environment, so 4 types of classifiers were prepared for each sample type. The most effective condition in color space was the condition which uses RGB color space with SVM based on Gaussian kernel even if the lightning environment was given or not. The most effective condition in hyper spectral data was using 85 bands of the hyper spectral data with SVM based on Gaussian kernel even if the environment is given or not. The condition which uses 85 bands of the hyperspectral data with Gaussian kernel SVM is the most effective in all conditions. Although the accuracy of MKL SVM is known to be higher than that of SVM, the accuracy in condition using MKL SVM is lower than using SVM with Gaussian kernel. It is considered as the reason that the learning sample is too small for the learning of the MKL SVM. Compared between the accuracy of "Not given "and "Given", the accuracy of "Given" is higher than that of "Not given". This result shows that the lightning environment estimation is effective for the color classification. Figure 8 shows the accuracy of the perceptual color classification with Gaussian kernel SVM in each color at environment A and C. The types of the samples are RGB and 85 bands of hyper spectral data. Using 85 bands of hyper spectral

Table 2. Accuracy of perceptual color classification (%)

Not given	Color space		Hyperspectral raw data	
	RGB	L*a*b	3 bands	85 bands
SVM				
Linear Kernel	36.64	35.57	38.01	60.47
Polynomial kernel	36.05	36.50	50.63	50.63
Gaussian kernel	40.63	40.19	39.71	**76.56**
MKL SVM				
Polynomial kernel	17.89	21.04	24.80	65.34
Gaussian kernel	21.08	21.64	33.33	60.97
Given	Color space		Hyperspectral raw data	
	RGB	L*a*b	3 bands	85 bands
SVM				
Linear Kernel	37.64	36.28	41.55	71.05
Polynomial kernel	34.42	35.07	27.21	72.45
Gaussian kernel	40.67	39.55	45.52	**78.41**
MKL SVM				
Polynomial kernel	38.68	39.33	40.14	72.18
Gaussian kernel	39.84	38.57	42.98	65.06

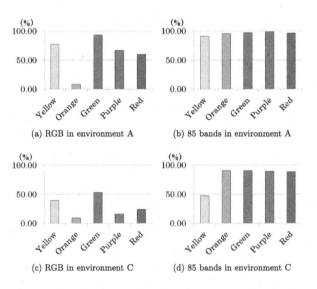

(a) RGB in environment A   (b) 85 bands in environment A

(c) RGB in environment C   (d) 85 bands in environment C

Fig. 8. Rate of the number of test samples that the perceptual color classification with SVM (Gaussian) at each color was suceed or not. (a) RGB in environment A (b) 85 bands in environment A (c) RGB in environment C (d) 85 bands in environment C (Color figure online)

data is more effective than using RGB. Using the RGB, the accuracy of orange is the lowest in both environments. Using the 85 bands data, the accuracy of yellow is lowest in both environments. There is a difference in what these data are good at.

3.4 Perceptual Color Classification Based on Lightning Environment Estimation

We evaluated the perceptual color recognition based on the estimation result of lighting environment. The number of the test samples was 4000 samples. The internal division of these is 1000 samples from 4 lightning environments. The method used for the estimation and classification was SVM based on Gaussian kernel. The classification results is shown in the Table 3. "True" in the table denotes that the estimation or classification succeed, and "False" denotes that the estimation or classification did not succeed. The values in the table are the rate of the number of test samples. The value that both estimation and classification are "True" denotes the accuracy of the perceptual color classification based on the lightning environment estimation. The accuracy of the method using 85 bands of hyperspectral data

Table 3. Accuracy of perceptual color classification based on lightning environment estimation (%).

RGB	Perceptual color classification	
	True	False
Environment estimation		
True	13.15	22.25
False	24.88	39.73
L*a*b	Perceptual color classification	
	True	False
Environment estimation		
True	12.25	21.93
False	23.78	42.05
3 bands	Perceptual color classification	
	True	False
Environment estimation		
True	20.73	36.53
False	15.78	26.98
85 bands	Perceptual color classification	
	True	False
Environment estimation		
True	54.78	17.40
False	19.08	6.05

was the highest in the table. The result shows that the proposed method is the most effective for perceptual color classification.

4 Conclusion

The paper introduced the perceptual color classification based on lightning environment estimation using hyperspectral data. MKL SVM is used for both estimation and classification.

Originality of our paper is as folows.

- The hyperspectral data is used for both lightning environment estimation and perceptual color classification.
- Suppose that the lightning environment is given, the perceptual color is classified in each environment.
- Perceptual color is estimated by using result of the lightning environment estimation.

Contribution is as follows.

- A novel dataset which consists of the hyperspectral image of 15 objects taken in 4 different lightning environments is constructed in our experiment.
- Experimental result of the hyperspectral data are compared by those of common color spaces, such as RGB and L*a*b.
- The estimation and classification with SVM and MKL SVM are compared in the experiments.

From the experimental results, the accuracy of the color recognition became higher when the lightning environment was given. When the lightning environment can be estimated with sufficiently high accuracy, the estimation becomes the effective preprocessing. Only the hyperspectral data enable to estimate the lightning environment estimation with enough accuracy. Furthermore, the accuracy of the perceptual color classification using hyperspectral data is higher than that of the classification using color space data. The hyperspectral data enable us to have a separated scheme which is consists of the lightning environment estimation and perceptual color classification. This paper shows the possibility that the proposed method is useful for the perceptual color recognition of objects. In our future work, we apply the proposed method for the perceptual color and object recognition.

References

1. Özgen, E., Davies, I.R.: Acquisition of categorical color perception: a perceptual learning approach to the linguistic relativity hypothesis. J. Exp. Psychol. Gen. **131**, 477 (2002)
2. HaCohen, Y., Shechtman, E., Goldman, D.B., Lischinski, D.: Optimizing color consistency in photo collections. ACM Trans. Graph. (TOG) **32**, 38 (2013)

3. Graf, B., Hans, M., Schraft, R.D.: Care-o-bot II: development of a next generation robotic home assistant. Auton. Robots **16**, 193–205 (2004)
4. Sun, Y., Bo, L., Fox, D.: Attribute based object identification. In: Proceedings of 2013 International Conference on the Robotics and Automation (ICRA), pp. 2096–2103. IEEE (2013)
5. Fang, R., Liu, C., Chai, J.Y.: Integrating word acquisition and referential grounding towards physical world interaction. In: Proceedings of 2012 International Conference on Multimodal Interaction (ICMI), pp. 109–116. ACM (2012)
6. Brelstaff, G.J., Párraga, A., Troscianko, T., Carr, D.: Hyperspectral camera system: acquisition and analysis. In: International Society for Optics and Photonics, Satellite Remote Sensing, vol. II, pp. 150–159 (1995)
7. Mehl, P.M., Chen, Y.R., Kim, M.S., Chan, D.E.: Development of hyperspectral imaging technique for the detection of apple surface defects and contaminations. J. Food Eng. **61**, 67–81 (2004)
8. EBA Japan Co., Ltd.: (Hyper spectrum camera nh-7). http://www.ebajapan.jp/NH-7.html
9. Camps-Valls, G., Bruzzone, L.: Kernel-based methods for hyperspectral image classification. IEEE Trans. Geosci. Remote Sens. **43**, 1351–1362 (2005)
10. Shah, C., Watanachaturaporn, P., Varshney, P., Arora, M.: Some recent results on hyperspectral image classification. In: 2003 IEEE Workshop on Advances in Techniques for Analysis of Remotely Sensed Data, pp. 346–353. IEEE (2003)
11. Mourao-Miranda, J., Friston, K.J., Brammer, M.: Dynamic discrimination analysis: a spatial-temporal SVM. NeuroImage **36**, 88–99 (2007)
12. Bach, F.R., Lanckriet, G.R., Jordan, M.I.: Multiple kernel learning, conic duality, and the SMO algorithm. In: Proceedings of the Twenty-First International Conference on Machine Learning, vol. 6. ACM (2004)
13. Tuia, D., Camps-Valls, G., Matasci, G., Kanevski, M.: Learning relevant image features with multiple-kernel classification. IEEE Trans. Geosci. Remote Sens. **48**, 3780–3791 (2010)
14. Sonnenburg, S., Rätsch, G., Schäfer, C., Schölkopf, B.: Large scale multiple kernel learning. J. Mach. Learn. Res. **7**, 1531–1565 (2006)
15. Reinhard, E., Ashikhmin, M., Gooch, B., Shirley, P.: Color transfer between images. IEEE Comput. Graph. Appl. **21**, 34–41 (2001)

Joint Multiview Fused ELM Learning with Propagation Filter for Hyperspectral Image Classification

Yu Shen[(✉)], Liang Xiao[(✉)], and Mohsen Molaei[(✉)]

Nanjing University of Science and Technology, Nanjing, China
shenyu0305@126.com, xiaoliang@mail.njust.edu.cn, mohsenmolaei@yahoo.com

Abstract. In this paper, we propose a fused extreme learning machine (ELM) method with multiview learning for hyperspectral imagery. The proposed approach consists of the following aspects. First, multiple views of spectral-spatial features are generated from the hyperspectral image by using a multiscale spectral-spatial context aware propagation filter. We next apply the weighted-based probabilistic ELM to these multiple feature views to obtain a robust supervised classification results with high accuracy. The advantages of the proposed method are twofold: (1) the multiscale local spectral-spatial contexts of the image are able to be exploited to improve the classification performance significantly; and (2) the algorithm is simple but very robust to the small size of training labeled samples. The experimental results suggest that the proposed algorithm obtains a competitive performance and outperforms other state-of-the-art ELM-based classifiers and the classical SVM classifier.

1 Introduction

Recent developments in remote sensing technologies allow us to collect enormous of high spectral and resolution hyperspectral images (HSIs) via airborne and spaceborne platforms. Due to the abundant spectral information among these bands, HSIs can be used to realize accurate and robust characterization, identification, and classification of the land-covers. Among all of these aforementioned applications of HSI, classification has attracted increasing interests and become a hot researching area over the past decades.

Recently, many machine leaning methods have been developed for HSI classification. Among these methods, discriminative approaches have shown success in HSI classification due to their advantages in dealing with high dimensionality and small-size training samples. For instance, neural networks (NNs) and support vector ma-chine (SVM) have shown remarkable success for supervised HSI classification task [1,2]. The main learning feature of SVM is that the standard optimization method is used to find the solution of maximizing the separating margin of two different classes while minimizing the training errors. As a single-hidden layer feed-ward neural net-works, extreme learning machine (ELM) [3] has a comparable classification accuracies but high efficiency learning speed compared with SVM. Through randomly generating the hidden node parameters and

© Springer International Publishing AG 2017
C.-S. Chen et al. (Eds.): ACCV 2016 Workshops, Part I, LNCS 10116, pp. 374–388, 2017.
DOI: 10.1007/978-3-319-54407-6_25

analytically computing the networks weights in a close form solution, ELM can be trained extremely fast. Owing to the aforementioned advantages of ELM, a family of the ELM based algorithms has been widely used for HSI classification [4,5]. The algorithms can be divided into two categories: spectral-only classifier and spectral-spatial context based classifier.

In general, spectral-only HSI classifier usually results in many classification errors or outliers in the homogenous area due to the lack of spatial contextual constraint. This phenomena will also be caused in ELM based and SVM based spectral-only HSI classifiers. To cope this drawback, one solution is to use the kernel tricks to improve the linear separability of the data. For instance, in [6], kernel-based ELM was considered for HSI classification and provided comparable performance to kernel-SVM. Another solution is to adopt the spectral-spatial context based method. Specifically, spatial coherence and clustering is the main assumption for spectral-spatial classification of HSI, which assumes that the nearest neighboring pixels consist of the same type of materials (same class) and have similar spectral characteristics. Moreover, some geometrical structures (e.g., edges) and texture patterns can also be exploited to further improve the performance of HSI classification. In a word, the combination of spectral information and spatial context can improve the result of HSI classification.

There are two major solutions to integrate the spectral-spatial information: one family of algorithms usually integrate a spectral-spatial feature extraction procedure before the classifiers; and another family algorithms usually adopt a post-processing after the classifiers. For the first categories, different type of spatial features such as morphological profiles [7], wavelet [8] features, Gabor features [9] as well as local binary pattern (LBP) [10] have been successfully used for HSI classification. An algorithm based on multi-hypothesis (MH) [11] prediction was also proposed for noise-robust HSI classification. For the second categories, four groups of classification postprocessing strategies were considered [12]: (1) filtering; (2) random field; (3) object-based voting; and (4) relearning. In this paper, we will mainly focus on the spectral-spatial extraction based classification.

The aforementioned spatial feature extraction strategies can be viewed as a form of preprocessing prior to classification, which generates spectral-spatial features from the original images. In this paper, we will present a novel preprocessing prior to classification, and we call it a multiscale context aware propagation filter (MCPF), which is derived from the propagation filter which is recently proposed in traditional image processing [13]. By using the MCPF, we propose a fused ELM method with multiview learning for HSI classification. We first separate a cube data of HSI into different band-subsets [14], and each subset whose spectral bands will has a high structure similarity. Then we introduce multiscale spectral-spatial context aware propagation filter as a tool to extract multiviews of spectral-spatial features. Finally, the weighted-based probabilistic ELM [15] use the multiple features to obtain a refined classification results. The advantages of the proposed method are twofold: (1) the multiscale local spectral-spatial contexts of the image are able to be exploited to improve the classification performance significantly; and (2) the algorithm is simple but very robust to the small size of training labeled samples.

To demonstrate the performance of the proposed method, many experiments on popular HSI datasets will be conducted to declare the superiority owing to the preprocessing prior produce by MCPF as well as the multiview learning with weighted-based probabilistic ELM scheme. The experimental results suggest that the proposed algorithm obtains a competitive performance and outperforms other state-of-the-art ELM-based classifiers and the classical SVM classifier.

2   Related Work

In this section, we briefly review ELM and propagation filter.

2.1   Extreme Learning Machine

ELM [3] was originally developed from feed-forward neural networks. The weights between input layer and hidden layer are randomly assigned, and the weights of out-put layer are computed using a close form solution. Compared with traditional classifiers, the computational cost of ELM is much lower.

For C classes, let the class labels be defined as $y_c \in \{0, 1\}$, $1 \leq c \leq C$. Thus, a constructed row vector $\mathbf{y} = [y_1, ..., y_c, ...y_C]$ indicates the class to which a sample belongs to. For example, if $y_c = 1$ and other elements in $\mathbf{y}$ are zero, then the sample belongs to the cth class. Thus, the training samples and corresponding labels are represented as $\{\mathbf{x}_i, \mathbf{y}_i\}_{i=1}^{G}$, where and ($B$ is the number of the spectral bands or dimensionality). The output function of an ELM having L hidden neurons can be represented as

$$f(\mathbf{x}_i) = \sum_{j=1}^{L} \beta_j h(\boldsymbol{\omega}_j \cdot \mathbf{x}_i + e_j) = \mathbf{y}_i, \tag{1}$$

where $h(\cdot)$ is a nonlinear activation function (e.g. Sigmoid function), $\beta_j \in \mathbb{R}^c$ is the weight vector connecting the jth hidden neuron and the output neurons, $\boldsymbol{\omega}_j \in \mathbb{R}^B$ is the weight vector connecting the jth hidden neuron and input neurons, and is the bias e_j of the jth hidden neuron. $\boldsymbol{\omega}_j \cdot \mathbf{x}_i$ denotes the inner product of $\boldsymbol{\omega}_j$ and $\mathbf{x}_i$. With G equations, Eq. (1) can be written compactly as

$$\mathbf{H}\beta = \mathbf{Y}, \tag{2}$$

where $\mathbf{Y} = [\mathbf{y}_1^T \mathbf{y}_2^T \mathbf{y}_3^T ... \mathbf{y}_G^T]^T \in \mathbb{R}^{G \times C}$, $\beta = [\beta_1^T \beta_2^T \beta_3^T ... \beta_L^T]^T \in \mathbb{R}^{L \times C}$, and $\mathbf{H}$ is the hidden layer out-put matrix of the neural network:

$$\mathbf{H} = \begin{bmatrix} \mathbf{h}(\mathbf{x}_1) \\ \vdots \\ \mathbf{h}(\mathbf{x}_G) \end{bmatrix}_{G \times L}, \tag{3}$$

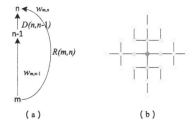

Fig. 1. Illustration of propagation filter. (a) the definition of filtering weight $w_{m,n}$, (b) the pattern of performing 2D propagation filtering with $d = 3$ pixels.

$\mathbf{H}(\mathbf{x}_i)$ is the output of the hidden neurons with respect to the input, which maps the data from the B-dimensional input space to the L-dimensional feature space. In most cases, the number of hidden neurons is much smaller than the number of training samples, the smallest norm least-squares solution of Eq. (2) proposed in [3] is defined as

$$\beta' = \mathbf{H}^\dagger \mathbf{Y}, \tag{4}$$

where $\mathbf{H}^\dagger$ is the *Moore-Penrose* generalized inverse of matrix $\mathbf{H}$. The *Moore-Penrose* generalized inverse of $\mathbf{H}$ can be calculated as $\mathbf{H}^\dagger = \mathbf{H}^T(\mathbf{H}\mathbf{H}^T)^{-1}$. For better stability and generali-zation, a positive value ρ^{-1} is added to the diagonal elements of $\mathbf{H}\mathbf{H}^T$. Therefore, we have the output function of ELM classifier

$$f(\mathbf{x}_i) = \mathbf{h}(\mathbf{x}_i)\beta = \mathbf{h}(\mathbf{x}_i)\mathbf{H}^T(\frac{\mathbf{I}}{\rho} + \mathbf{H}\mathbf{H}^T)^{-1}\mathbf{Y}. \tag{5}$$

In ELM, a feature mapping $\mathbf{h}(\mathbf{x}_i)$ is usually known to users. If a feature mapping is unknown to users, a kernel matrix for ELM can be defined as follows:

$$\Omega_{ELM} = \mathbf{H}\mathbf{H}^T : \Omega_{ELM} = \mathbf{h}(\mathbf{x}_p) \cdot \mathbf{h}(\mathbf{x}_q) = \mathbf{K}(\mathbf{x}_p, \mathbf{x}_q). \tag{6}$$

Thus, the output function of KELM can be written as

$$f(\mathbf{x}_i) = \begin{bmatrix} \mathbf{K}(\mathbf{x}_i, \mathbf{x}_1) \\ \vdots \\ \mathbf{K}(\mathbf{x}_i, \mathbf{x}_G) \end{bmatrix} (\frac{\mathbf{I}}{\rho} + \Omega_{ELM})^{-1}\mathbf{Y}. \tag{7}$$

The label of the input data is determined by the index of the output node with the largest value.

2.2 Propagation Filter

Propagation filter [12] is a novel image filtering operator, with the goal of smoothing over neighboring image pixels while preserving image context like edges or textural regions. In particular, this filter does not utilize explicit spatial kernel

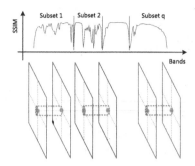

Fig. 2. Context aware propagation filter on band-subsets.

function as bi-lateral and guided filters do. It has been verified it performs favorably against existing bilateral and guided filter on processing characteristics and details of adjacent image regions.

Given an input image U, the filtered output $\widetilde{U}(m)$ produced by propagation filter is calculated by

$$\widetilde{U}(m) = \frac{1}{Z(m)} \sum_{n \in \mathcal{N}} w_{m,n} U(m), \qquad (8)$$

where $U(m)$ denotes the value at pixel m, and $\mathcal{N}(m)$ indicates the set of neighboring pixels centered at m. We have $w_{m,n}$ as the weight for each pixel n to performing the filtering of $U(m)$, while $Z(m)$ as the normalization factor for ensuring the sum of all weights equal to 1.

The algorithm is driven by the idea that, for the pixel m being related to pixel n, the intermediate pixels between m and n not only need to be photometrically related to m, they are also required to be adjacent-photo-metrically related to their predecessors.

As a result, we derive the filter weight $w_{m,n}$ by the follow definition:

$$w_{m,n} = w_{m,n\text{-}1} D(n, n\text{-}1) R(m, n), \qquad (9)$$

where $D(n, n\text{-}1)$ represents the adjacent photometric relationship between pixel n and $n\text{-}1$, which can be defined their value difference by a Gaussian function:

$$D(n, n\text{-}1) = \exp(\frac{-\|U(n) - U(n\text{-}1)\|^2}{2\sigma_a^2}). \qquad (10)$$

$R(m, n)$ measures the adjacent-photometric relationship between m and n. We define

$$R(m, n) = \exp(\frac{-\|U(m) - U(n)\|^2}{2\sigma_r^2}), \qquad (11)$$

where σ_a and σ_r are the filter parameters. For simplicity, we choose $\sigma_a = \sigma_r$. We use Fig. 1a and b to illustrate how to calculate $w_{m,n}$.

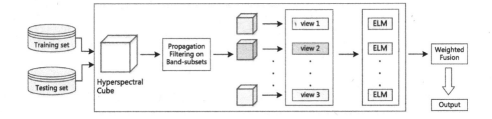

Fig. 3. Flowchart of the proposed method.

3 Proposed Framework

3.1 Multiview Generation via Multiscale Context Aware Propagation Filter

Let $\mathbf{I} = [\mathbf{I}_1, \mathbf{I}_2, \cdots, \mathbf{I}_B]^T$ represents one HSI, B is the number of the spectral bands or dimensionality and each $\mathbf{I}_i \in \mathbb{R}^{M \times N}$ is ith band of a HSI. The structural similarity between spectral bands $\mathbf{I}_i$ and $\mathbf{I}_{i+1}$ can be generated as

$$SSIM(\mathbf{I}_i, \mathbf{I}_{i+1}) = \frac{(2\mu_{\mathbf{I}_i}\mu_{\mathbf{I}_{i+1}} + C_1)(2\sigma_{\mathbf{I}_i,\mathbf{I}_{i+1}} + C_2)}{(\mu_{\mathbf{I}_i}^2 + \mu_{\mathbf{I}_{i+1}}^2 + C_1)(\sigma_{\mathbf{I}_i}^2 \sigma_{\mathbf{I}_{i+1}}^2 + C_2)}, \tag{12}$$

where $\mu_{\mathbf{I}_i}(\mu_{\mathbf{I}_{i+1}})$ is the mean value of the spectral band $\mathbf{I}_i(\mathbf{I}_{i+1})$, $\sigma_{\mathbf{I}_i}(\sigma_{\mathbf{I}_{i+1}})$ is the standard deviation of $\mu_{\mathbf{I}_i}(\mu_{\mathbf{I}_{i+1}})$, $\sigma_{\mathbf{I}_i,\mathbf{I}_{i+1}}$ represents the covariance of $\mathbf{I}_i$ and $\mathbf{I}_{i+1}$, and C_1 and C_2 are two constant numbers.

By applying (12) on adjacent spectral bands, a correlation curve C can be generated, where $C(i) = SSIM(\mathbf{I}_i, \mathbf{I}_{i+1})$. From the curve in Fig. 2, continuous spectral bands with high structural similarity correspond to relative stable trend, while obvious drops existing in curve C demonstrate the adjacent spectral bands have much lower structural similarity. Therefore, the band-subset partition can be achieved based on detecting the sharp drops in curve C. Let one band-subset as S_q, where S_q contains B_q spectral bands with size of $M \times N$.

In our work, we apply the propagation filter for HSI band-subsets to extract the local contextual spectral-spatial feature, as shown in Fig. 2. Let $S_q(m)$ represents a pixel of S_q, the filtered output $\widetilde{S}_q$ produced by propagation filter is calculated by

$$\widetilde{S}(m) = \frac{1}{Z(m)} \sum_{n \in \mathcal{N}} w_{m,n} S(m), \tag{13}$$

The weights can be defined in the same way as Eq. (9), where every pixel $S_q(m)$ is a vector rather than a single value. After applying the propagation filter in every S_q, we merge them into a new data cube, which will be served as a robust spectral signatures for the input of ELM classifiers.

To further improve the performance of HSI classification, we propose MCPF by using the propagation filter with variable window size (d=1, 2, 3, 4, 5, etc.). By applying the filter at different window size in preprocessing stage, we can

Table 1. Class labels and train-test distribution of samples for Indian Pines.

#	Class	Train	Test
1	Alfalfa	6	48
2	Corn-no till	144	1290
3	Corn-min till	84	750
4	Corn	24	210
5	Grass/pasture	50	447
6	Grass/tree	75	627
7	Grass/pasture-mowed	3	23
8	Hay-windrowed	49	440
9	Oats	2	18
10	Soybeans-no till	97	871
11	Soybeans-min till	247	2221
12	Soybeans-clean till	62	552
13	Wheat	22	190
14	Woods	130	1164
15	Bldg-grass-tree-drives	38	342
16	Stone-steel towers	10	85
	Total	1043	9323

obtain multiple scale feature cubes which contain local spectral-spatial contexts at different scales. Each feature cubes of a single scale will be regarded as a feature view. Thus, the multiple feature views can be generated by using MCPF.

As illustrated in Fig. 2, we give some interpretations on the feature extraction of context aware propagation filter. For a current pixel in a HSI, the context aware propagation filter will integrate the local adaptive spatial-spectral contextual information. To achieve a better spectral band adaptiveness, the local adjacent spectral bands with high structure similarity will be used to guide the filtering in the direction of spectral dimension. For the spatial domain, an edge and texture information in a local neighborhood will be integrated in the filtering by using a contextual adaptive weighted scheme.

3.2   Decision Leval Fusion via Weighted ELM

In this paper, to combine multiple feature classification results, we employ decision level fusion via weighted ELM [15].

First, we transform output of ELM into a probabilistic type, defined as

$$P(f_c(\mathbf{x})) = \frac{1}{1 + \exp(-(f_c(\mathbf{x})))}, \tag{14}$$

Table 2. Class labels and train-test distribution of samples for University of Pavia.

#	Class	Train	Test
1	Alfalfa	30	6601
2	Meadows	30	18619
3	Gravel	30	2069
4	Trees	30	3034
5	Painted metal sheets	30	1315
6	Bare soil	30	4999
7	Bitumen	30	1300
8	Self-blocking bricks	30	3652
9	Shadows	30	917
Total		270	42506

we can obtain the probability of each class c by normalizing $P(f_c(\mathbf{x}))$, denoted as $p_c(\mathbf{x})$. In our work, several ELM classifiers are applied on a test sample. In a J classifiers and C classes problem, we can compute the probability matrix as

$$P(\mathbf{x}) = \begin{bmatrix} p_{11}(\mathbf{x}) & \cdots & p_{1J}(\mathbf{x}) \\ \vdots & \ddots & \vdots \\ p_{C1}(\mathbf{x}) & \cdots & p_{CJ}(\mathbf{x}) \end{bmatrix}_{C \times J}. \tag{15}$$

For any sample, the maximum probability output of jth classifier is considered as the weight of the jth ELM, which can be defined as

$$p_j = \max_{c=1,2,\cdots,C}[p_{cj}((x)], \tag{16}$$

$$\mu_j = \frac{p_j}{\sum_{j=1}^{J} p_j} \tag{17}$$

where $p_{cj}(\mathbf{x})$ is the probability of cth class in jth classifier, μ_j is the fusion weight of every classifier. Thus, the class label can be calculated as

$$\text{label}(\mathbf{x}) = \arg\max_{c=1,2,\cdots,C}[\sum_{j=1}^{J} \mu_j p_{cj}(\mathbf{x})]. \tag{18}$$

Decision level fusion is to merge results from every single classifier. The mechanism combines classification results obtained from multiview ELM into a final decision, improving the accuracy of a single classifier that uses a certain scale of propagation filtered feature.

The flowchart of the proposed method is shown in Fig. 3. As we can see, in the first stage, we apply MCPF to obtain multi-features, and then ELM classifiers are applied on these feature subsets. In the second stage, the ELM classification results are combined by fusion weights. In this way, a robust ELM learning classification result is obtained.

Table 3. Class labels and train-test distribution of samples for Center of Pavia.

#	Class	Train	Test
1	Water	745	664533
2	Trees	785	5722
3	Meadow	797	2094
4	Bricks	485	1667
5	Soil	820	5729
6	Asphalt	678	6847
7	Bitumen	808	6479
8	Stile	223	2899
9	Shadows	195	1970
Total		5536	97940

Table 4. Overall classification accuracy, average classification accuracy (in percent) and Kappa coefficient of different classification methods for the Indian Pines.

Class	Spec-SVM	Bilat-SVM	Prop-SVM	MF-SVM	Spec-ELM	Bilat-ELM	Prop-ELM	MF-ELM
1	45.00	18.54	20.62	43.13	68.75	95.83	91.67	97.92
2	82.37	91.22	92.91	90.01	83.10	96.97	97.60	99.07
3	75.13	86.52	84.11	89.37	64.27	91.60	97.33	98.40
4	65.33	88.24	88.24	73.76	62.86	99.52	96.19	98.57
5	92.34	87.61	89.26	95.28	91.28	96.42	97.54	95.30
6	94.94	86.53	89.00	97.38	94.20	98.21	99.40	99.26
7	38.70	3.04	18.26	50.43	30.43	65.21	100	100
8	98.84	93.82	94.70	99.16	98.86	97.27	93.86	99.77
9	0	0	6.47	1.67	55.56	66.67	50.00	100
10	77.01	86.50	87.30	86.18	78.42	95.75	97.36	97.36
11	89.66	94.11	94.45	92.99	81.31	98.33	99.23	98.87
12	86.21	88.08	85.11	88.41	87.32	93.84	95.11	97.64
13	97.63	72.79	67.47	98.84	99.47	91.57	94.21	99.47
14	96.80	95.82	95.30	97.04	97.25	99.39	96.99	99.23
15	56.90	83.92	85.56	71.43	59.94	95.32	99.12	98.83
16	89.76	66.24	49.29	93.76	67.06	76.47	88.24	77.65
OA	79.92	89.39	89.50	90.97	83.30	96.50	97.43	98.40
AA	67.79	71.44	71.75	79.30	76.25	91.15	93.37	97.73
k	0.770	0.879	0.880	0.897	0.809	0.960	0.971	0.982

4   Experimental Results

4.1   Experimetal Data

We apply the proposed method to three hyperspectral airborne images described as follows:

The first data set is Indian Pines, which is generated by the Airborne/Visible Infra-red Imaging Spectrometer (AVIRIS) sensor over the Indian Pines test site in north-western Indiana. The AVIRIS sensor generates 220 spectral bands rang-

Table 5. Overall classification accuracy, average classification accuracy (in percent) and Kappa coefficient of different classification methods for the Indian Pines.

Class	Spec-SVM	Bilat-SVM	Prop-SVM	MF-SVM	Spec-ELM	Bilat-ELM	Prop-ELM	MF-ELM
1	77.31	95.24	89.55	87.49	63.16	92.86	84.65	86.15
2	64.67	86.99	92.21	93.52	84.14	91.39	97.17	98.58
3	74.00	95.65	87.00	79.85	77.28	88.88	94.83	93.72
4	97.23	90.38	95.52	95.25	90.01	91.86	93.67	99.64
5	99.70	99.62	96.27	99.70	98.86	99.54	99.35	99.77
6	82.44	70.11	89.64	98.26	76.32	97.82	99.16	98.90
7	92.54	97.69	87.31	93.23	75.15	99.85	99.46	100
8	83.46	96.55	81.82	85.27	54.03	94.61	86.86	95.35
9	99.02	99.89	100	98.47	99.67	97.82	99.35	99.56
OA	75.79	88.77	90.73	92.18	79.97	93.24	94.04	96.35
AA	85.59	92.46	91.03	92.34	77.85	94.96	94.61	96.85
k	0.698	0.853	0.878	0.898	0.715	0.911	0.921	0.952

ing from 0.2 to $2.4\,\mu$m and has a spatial resolution of $20\,$m per pixel. The hyperspectral image consists of a $145 \times 145 \times 22$ data cube reduced to 200 bands after removing the water absorption and noisy bands. The ground truth contains 16 land cover classes and 10366 labeled pixels. In our experiment, 10% of samples are used for training and the rest are testing set, as is shown is Table 1.

The second data named The University of Pavia is an urban image collected by Reflective Optics System Imaging Spectrometer (ROSIS), which has 115 spectral bands ranging from 0.43 to 0.86μm. It contains 9 classes spread over the 610×340 pixels. The 12 noisiest bands are removed. For this dataset, the training and test split is used where the training set consists of 30 samples per class and the rest 42506 pixels are used for testing which are listed in Table 2.

The third data set named Center of Pavia image is the other urban image collected by the ROSIS sensor over the center of Pavia city. This image consists of 1096×429 pixels, each having 102 spectral bands after 13 noisy bands are removed. The nine ground-truth classes and the number of training and test samples for each class as are shown in Table 3. 5536 labeled samples (about 5.35% percentage of all the labeled samples) are used for training purpose and the rest are for testing.

4.2   Measurements and Parameter Setting

Before doing the experiments, we first describe the parameters and the measurements of classification. In our proposed method, Gaussian kernel parameter σ_a and σ_r in filtering stage are selected by cross validation. The maximum window size of propagation filter is set to $d = 5$ while the window size of bilateral filter based classification method is set to $d = 4$.

We show the classification results by overall accuracy (OA), average accuracy (AA) as well as kappa (k) statistic. The OA is computed by the ratio between correctly classified test samples and the total number of test samples, and the

Table 6. Overall classification accuracy, average classification accuracy (in percent) and Kappa coefficient of different classification methods for the Indian Pines.

Class	Spec-SVM	Bilat-SVM	Prop-SVM	MF-SVM	Spec-ELM	Bilat-ELM	Prop-ELM	MF-ELM
1	99.23	99.10	99.34	99.25	99.83	98.97	99.95	99.81
2	92.64	94.25	97.39	96.14	94.68	98.36	97.94	98.17
3	95.51	97.42	95.65	96.13	96.26	99.48	99.31	98.24
4	81.64	85.84	84.70	95.80	85.42	99.49	98.93	100
5	96.04	97.54	99.04	98.34	95.60	97.63	97.22	98.66
6	94.99	96.44	96.39	94.74	98.80	99.43	99.39	99.40
7	94.61	97.04	98.40	97.96	90.33	97.09	96.16	98.23
8	99.72	99.86	99.97	100	95.96	99.74	99.74	99.87
9	99.95	99.80	100	98.43	99.77	99.95	99.74	98.15
OA	97.70	98.18	98.47	98.50	97.98	98.82	99.30	99.42
AA	894.93	96.36	96.43	97.42	95.18	98.90	98.73	98.95
k	0.958	0.967	0.972	0.930	0.965	0.980	0.988	0.990

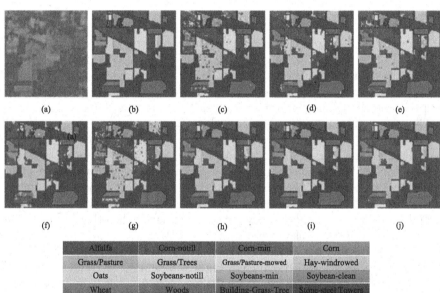

Fig. 4. Thematic maps resulting from classification for the Indian Pines data. (a) Pseudocolor image; (b) Ground truth image; (c) SVM; (d) Bilat-SVM; (e) Prop-SVM; (f) MF-SVM; (g) ELM; (h) Bilat-ELM; (i)Prop-ELM; (j) MF-ELM

AA is the mean of the 16 (or 9) classes accuracies. The k statistic is computed by weighting the measured by weighting the measured accuracies.

4.3   Classification Results

The performance of the proposed classification method is shown in Tables 4, 5 and 6, where the MF-ELM and MF-SVM indicate joint multiview fused ELM (SVM)

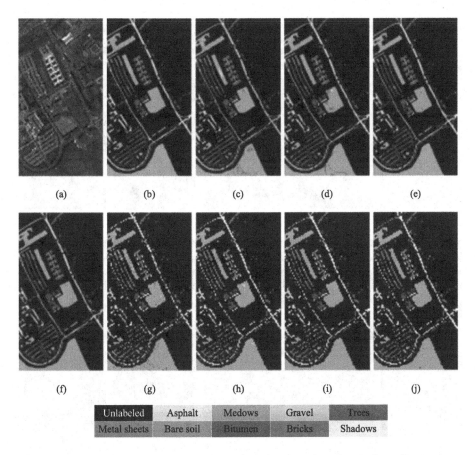

(a)   (b)   (c)   (d)   (e)

(f)   (g)   (h)   (i)   (j)

Unlabeled	Asphalt	Medows	Gravel	Trees
Metal sheets	Bare soil	Bitumen	Bricks	Shadows

Fig. 5. Thematic maps resulting from classification for the Indian Pines data. (a) Pseudocolor image; (b) Ground truth image; (c) SVM; (d) Bilat-SVM; (e) Prop-SVM; (f) MF-SVM; (g) ELM; (h) Bilat-ELM; (i)Prop-ELM; (j) MF-ELM

learning with MCPF. All experiments are carried out using MATLAB 2014a on an Intel Xeon 3.00-GHz machine with 8 GB of RAM.

From the results of each single classifier, the performance of propagation filter (Prop-SVM and Prop-ELM) is much better than spectral signature only (Spec-SVM and Spec-ELM) or bilateral filter based classification method (Bilat-SVM and Bilat-ELM); for instance, in Table 4, Prop-SVM gives over 9% higher accuracy than Spec-SVM, and Prop-ELM yields 10% higher accuracy than Spec-ELM. Moreover, only propagation filter based classification method (i.e. Prop-SVM and Prop-ELM) achieves higher classification accuracies than bilateral filter based classification method (i.e. Bilat-SVM and Bilat-ELM) for all three experimental data. This clearly demonstrates that propagation filter based method is a highly efficient local textural preserving operator. On the other hand, from the Tables 4, 5 and 6, ELM has a better performance than SVM in general.

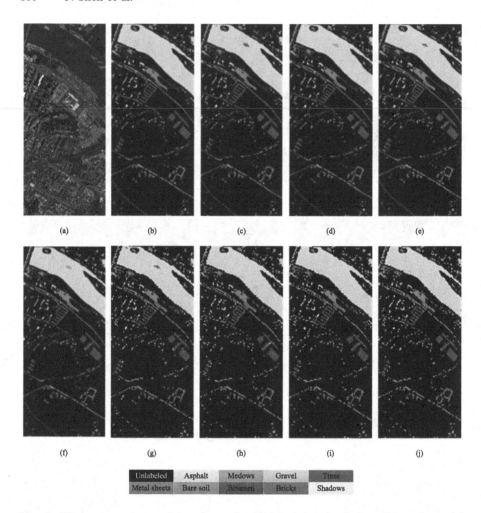

Fig. 6. Thematic maps resulting from classification for the Indian Pines data. (a) Pseudocolor image; (b) Ground truth image; (c) SVM; (d) Bilat-SVM; (e) Prop-SVM; (f) MF-SVM; (g) ELM; (h) Bilat-ELM; (i)Prop-ELM; (j) MF-ELM

Therefore, ELM-based classification methods turn out to be very effective for HSI classification under the small sample size condition.

As is shown in the Tables 4, 5 and 6, our proposed method, joint multiview ELM learning with propagation filter achieves the best classification accuracy.

Figures 4, 5 and 6 show the pseudocolor image, ground truth and classification maps of all the mentioned classifiers on three HSI data sets. These maps provide the consistent results shown in Tables 4, 5, 6. As we can see, bilateral or propagation filter based classifier contain much less noise than spectral signature only classification method. Furthermore, the map of our proposed method shows a vision which is close to the ground truth map.

5   Conclusion

In this paper, a fused ELM method with multiview learning is proposed to improve the performance of HSI classification. In multiview generation stage, context aware propagation filter is employed on band-subsets to extract multi-scale local spectral-spatial contextual information. In decision level fusion stage, weighted-based probabilistic ELM is applied to acquire robust supervised classification results. Compared with bilateral-filter-based classification method, the context aware propagation filter is more efficient in HSI spectral-spatial feature extraction. Moreover, experiments demonstrate that the fused ELM method with multiview learning classification outperforms other ELM-based classifiers and SVMs under small training samples.

Acknowledgement. This work was supported in part by the National Key Research and Development Program of China under Grant No. 2016YFF0103604, by the National Natural Science Foundation of China under Grant No. 61571230, Grant No. 11431015 and Grant No. 61171165; by the National Scientific Equipment Developing Project of China under Grant No. 2012YQ050250; and by the Natural Science Foundation of Jiangsu Province under Grant No. BK20161500.

References

1. Ratle, F., Camps-Valls, G., Weston, J.: Semisupervised neural networks for efficient hyperspectral image classification. IEEE Trans. Geosci. Remote Sens. **48**(5), 2271–2282 (2010)
2. Camps-Valls, G., Rodrigo-González, A., Muñoz-Marí, J., Gómez-Chova, L., Calpe-Maravilla, J: Hyperspectral image classification with mahalanobis relevance vector machines. In: IEEE International Geoscience and Remote Sensing Symposium (IGARSS 2007), Barcelona, Spain (2007)
3. Huang, G.-B., Zhu, Q.-Y., Siew, C.-K.: Extreme learning machine: theory and applications. Neurocomputing **70**(1), 489–501 (2006)
4. Pal, M.: Extreme-learning-machine-based land cover classification. Int. J. Remote Sens. **30**(14), 3835–3841 (2009)
5. Moreno, R., et al.: Extreme learning machines for soybean classification in remote sensing hyperspectral images. Neurocomputing **128**, 207–216 (2014)
6. Pal, M., Maxwell, A.E., Warner, T.A.: Kernel-based extreme learning machine for remote-sensing image classification. Remote Sens. Lett. **4**(9), 853–862 (2013)
7. Benediktsson, J., Palmason, J.A., Sveinsson, J.R.: Classification of hyperspectral data from urban areas based on extended morphological profiles. IEEE Trans. Geosci. Remote Sens. **43**(3), 480–491 (2005)
8. Qian, Y., Ye, M., Zhou, J.: Hyperspectral image classification based on structured sparse logistic regression and three-dimensional wavelet texture features. IEEE Trans. Geosci. Remote Sens. **51**(4), 2276–2291 (2013)
9. Bau, T.C., Sarkar, S., Healey, G.: Hyperspectral region classification using a three-dimensional Gabor filterbank. IEEE Trans. Geosci. Remote Sens. **48**(9), 3457–3464 (2010)
10. Li, W., et al.: Local binary patterns and extreme learning machine for hyperspectral imagery classification. IEEE Trans. Geosci. Remote Sens. **53**(7), 3681–3693 (2015)

11. Chen, C., et al.: Spectral-spatial classification of hyperspectral image based on kernel extreme learning machine. Remote Sens. **6**(6), 5795–5814 (2014)
12. Huang, X., et al.: New postprocessing methods for remote sensing image classification: a systematic study. IEEE Trans. Geosci. Remote Sens. **52**(11), 7140–7159 (2014)
13. Chang, J.-H.R., Wang, Y.-C.F.: Propagated image filtering. In: IEEE Conference on Computer Vision and Pattern Recognition (CVPR 2015), Boston, USA (2015)
14. Lu, T., Li, S.: Gradient-guided sparse representation for hyperspectral image denoising. In: IEEE International Geoscience and Remote Sensing Symposium (IGARSS 2015), Italy, Milan (2015)
15. Wen-Bo, Z., Hong-Bing, J.: Fusion of extreme learning machines. J. Electron. Inform. Technol. **35**(11), 2278–2279 (2013)

Unsupervised Band Selection Based on Group-Based Sparse Representation

Hung-Chang Chien, Chih-Hung Lai, and Keng-Hao Liu[✉]

Department of Mechanical and Electro-Mechanical Engineering,
National Sun Yat-sen University, Kaohsiung, Taiwan
{m043020080,m033020087}@student.nsysu.edu.tw, keng3@mail.nsysu.edu.tw

Abstract. Band selection (BS) is one of the important topics in hyperspectral image data analysis. How to search the representative bands that can effectively represent the image with lower inter-band redundancy is an long-term issue. Recently, the sparse representation (SR) was used to solve BS problem, called SR-BS. It aimed to find a set of representative bands that can represent the whole bands based on the minimization of reconstructed error in SR. However, those SR-BS methods suffer from an issue about higher complexity in the optimization process, even though the greedy strategy, such as orthogonal matching pursuit (OMP) algorithm, is used to accelerate them. Another issue is that the bands selected by SR-BS may not be really complementary since the homogeneity of adjacent spectral bands is not considered. To make SR-BS more efficient, in this paper, we model the BS problem as group sparse representation (GSR) problem where the dictionary matrix (i.e., all spectral bands) are pre-clustered to several non-overlapping groups based on the spectral similarity. Later, we adopt group orthogonal matching pursuit (GOMP) algorithm to solve the optimization problem. We named the proposed approach as GOMP-BS. Since GOMP-BS picks bands under group structure, it not only reaches higher computational efficiency but also makes the selected bands more less redundant. The experiments show that GOMP-BS achieves higher classification performance when the number of selected bands is low and requires less computation time than OMP-based methods.

1 Introduction

Due to use of hundreds of spectral bands, hyperspectral imagery generally has enormous data volume and contains vast amount of information. Using full spectral bands for data analysis may cause extremely high computation issue and even deteriorate the overall performance. Thus, how to reduce the inter-band redundancy in a hyperspectral image has been a popular topic in remote sensing community. One common way is data dimensionality reduction (DR) [1] which implements a transform to reduce data dimensions in accordance with a certain criterion. Another way is called band selection (BS) [2] which selects a set of representative bands from original band set that can well represent the data.

© Springer International Publishing AG 2017
C.-S. Chen et al. (Eds.): ACCV 2016 Workshops, Part I, LNCS 10116, pp. 389–401, 2017.
DOI: 10.1007/978-3-319-54407-6_26

Compared to DR, BS has an advantage of preserving original information from the data so it is more widely used in the application.

Many methods have been proposed to achieve BS. One common rule is to find a subset of spectral bands that can maximize the spectral information. However, determining the number of bands required to be selected (p), and designing an effective criterion that can evaluate amount of the containing information, are two major issues for BS. It is necessary to design a way that can resolve both in a single framework.

Sparse representation (SR) was received much attention in computer vision and pattern recognition [3]. The SR can sparsely represent any input vector as a linear combination of some basis from the dictionary with the non-zero coefficients. The SR framework has been utilized to solve many problems of hyperspectral image processing such as spectral unmixing [4], endmember extraction [5], and also band selection [6–8] in the recent years. However, most of those existing SR-based BS(SR-BS) methods could not resolve the above-mentioned issues efficiently because they usually need empirical parameter setting or require extra support of optimization tools or other algorithms. Based on this trend, a new SR-BS method was proposed in [9]. It follows the self-sparse model (SSR) in [8,10] to formulate the BS, and use OMP's principle to find the representative bands by forward greedy search without any empirical parameters and theoretical guarantees of convex optimization. This method, called OMP-BS, selects band in a sequential manner so that the results can be used for data transmission or progressive band processing [11]. It also can be used to estimate the p as long as the reconstructed error (residual) is pre-set.

However, there are still two issues existed in OMP-BS. The first one is computation time. Since it needs to find the most appropriate basis band by solving a least square (LS) problem one time in each iteration, the overall number of LS problems required to deal with would be $\frac{L!}{(L-p)!}$. One thing we can expect is that the computation time would be high. The second one is about the information complementarity. As the above-mentioned, the OMP-BS selects bands based on the minimization of reconstructed residual without imposing any terms related to spectral dissimilarity constraints. Therefore, doing this way may still result in that some selected bands are located in a narrow spectrum range, and some bands in certain spectrum ranges cannot be selected. Finally, the information spanned by the OMP-BS selected bands will not reach the maximal.

The most simple way to resolve the dilemma is using group sparse representation (GSR) structure to re-formulate the BS. In this paper, we propose a GSR-based BS method. In GSR, the basis signals (bands) with similar spectral properties are pre-clustered to g non-overlapping group, $g \leq L$. Then we can use group orthogonal matching pursuit (GOMP) [12] algorithm to perform stagewise group variable selection. The main difference between OMP and GOMP is that the former selects one base at each iteration but the latter selects an entire group of bases. The number of LS problems encountered in the optimization is reduced to $\frac{g!}{(g-p)!}$. It can be expected that the convergence will be faster. Since one group of bands is selected in each stage, we can further designate the centroid base of the group as the output band, to avoid selecting informatively duplicate bands.

The remainder of this paper is organized as follows. Before we introduce the proposed GOMP-BS method, the self-sparse model for band selection and OMP-BS algorithm are introduced in Sect. 2. Section 3 introduces the details of the proposed GOMP-BS, where the several variants of GOMP-BS methods are presented. The experiments discussion will be shown in Sect. 4. Finally, the conclusion is drawn in Sect. 5.

2   The OMP-BS Method

2.1   The Self-sparse Model for BS

The self-sparse (SR) can sparsely represent any input vector as a linear combination of some basis from the dictionary. The SR framework has been utilized to solve the BS. In general, the self-sparse representation (SSR) [8,10] model is used to formulate BS. Let L and N be the number of bands and pixels in a hyperspectral image, respectively $(N \gg L)$. The objective is to find p bands that can represent whole bands as good as possible. The SSR assumes that the observation matrix could be sparsely described by the matrix itself. Suppose $\mathbf{b}_i \in R^{N \times 1}$ presents the i-th band vector. In SSR model, the observation matrix as well as the dictionary matrix are both comprised of all band vectors denoted by $\mathbf{B} = [\mathbf{b}_1, \mathbf{b}_2, ..., \mathbf{b}_L]$. The coefficient matrix is denoted by $\mathbf{C}$. Now, the goal is to find the coefficient matrix $\mathbf{C}$ that minimizes the reconstructed error $||\mathbf{B} - \mathbf{BC}||_F^2$. Since there are only p bands required to be selected, we could impose the sparsity constraint $||\mathbf{C}||_{0,2} \leq p$ where mixed norm $||\mathbf{C}||_{0,2}$ counts the number of the non-zero rows in $\mathbf{C}$. Thus, the SR-BS optimization problem is presented by

$$\hat{\mathbf{C}} = \arg \min_{\mathbf{C}} ||\mathbf{B} - \mathbf{BC}||_F^2, s.t. ||\mathbf{C}||_{0,2} \leq p \qquad (1)$$

However, solving the l_0-norm is an NP hard problem. There are two approached to approximate it, convex optimization and greedy algorithm. In the former, the l_0-norm is replaced by its nearest convex l_1 surrogate the norm and the resulting form is solved by quadratic programming techniques. A seminal study [13] studies the conditions under which the l_1-norm minimization is equivalent to the l_0-norm minimization. However solving the l_1-norm minimization problem is computationally expensive, whereas greedy approximate algorithms for solving Eq. (1) are computationally fast. Greedy (sub-optimal) algorithms try to find an approximate solution for Eg. (1) directly. Although they do not have the same theoretical guarantees of convex optimization, it has been found that using greedy algorithms can produce quite accurate results. One of famous greedy algorithm, namely orthogonal matching pursuit (OMP) [14], is a sequential forward search method. It has been used to solve many sparse problems. In [9], the OMP was first adopted to solve the SSR-based BS problem. We name this method as OMP-BS. By virtue of greedy operation, it selects band in a sequence manner. It implies that changing p does not require repeated computation. Based on the results shown in [9], the OMP-BS method has excellent performance in land cover classification for real hyperspectral images.

2.2 The OMP-BS Algorithm

Given an observation signal $\mathbf{v}$, a measurement (dictionary) matrix $\mathbf{\Phi}$, and a pre-given sparsity level p, OMP recovers the signal as the linear combination of m column bases in $\mathbf{\Phi}$ accompanied with a coefficient vector which has p nonzero elements. At each iteration, the next column basis is chosen by searching the most strongly correlated one with the remaining part (residual) of $\mathbf{v}$. By virtue of the spirit, it is also possible to find the OMP solution for problems Eg. (1). It has been proposed in [9]. The principle of finding the next basis is equivalent to finding a new one that can most significantly reduce the remaining unconstructed error. Based on the principle, we could use the following steps to select the representative bands:

1. At first, it searches all L band vectors in $\mathbf{B}$ and pick the one, denoted as $\mathbf{b}_{b_1}$, which can produce the most less reconstructed error in $||\mathbf{B} - \mathbf{b}_{b_1}\mathbf{Q}||_F^2$. We denote b_1^* as the corresponding band index, and set $\Omega_1 = b_1^*$. In this case, the coefficient matrix $\mathbf{Q}$ will have nonzero coefficients only in the b_1^*-th row.
2. At j-th iteration ($j \leq p \leq L$), we search all the remaining band vectors and find the one $\mathbf{b}_j$, that produces the smallest residual $||\mathbf{B} - \mathbf{PQ}||_F^2$. Here $\mathbf{P} = [\mathbf{B}_{\Omega_{j-1}}\mathbf{b}_{b_j}]$ is the temporary basis set and Ω_{j-1} indicates the selected band indices obtained in previous step. Once $\mathbf{b}_j^*$ is found, update $\Omega_j \leftarrow \Omega_{j-1} \cup b_j^*$.
3. Repeat Step 2 until j reaches p. The final band indices set Ω_p is the OMP-BS result.

In each step, solving the following least-square optimization problem is required:

$$\hat{\mathbf{Q}} = \arg \min_{\mathbf{Q} \in R^{j \times N}} ||\mathbf{B} - \mathbf{PQ}||_F^2 \qquad (2)$$

Solving Eq. (2) is analogous to solve spectral unmixing problem [1]. If no constraint is set, the least square (LS) solution would be $\hat{\mathbf{Q}} = \mathbf{P}^+\mathbf{B} = (\mathbf{P}^T\mathbf{P})^{-1}\mathbf{P}^T\mathbf{B}$. If $\mathbf{1}^T\mathbf{C} = \mathbf{1}^T$ is imposed, we can use SCLS [1] to solve $\hat{\mathbf{Q}} = [\mathbf{q}_1, \mathbf{q}_2, \ldots, \mathbf{q}_L]$ by finding each column $\mathbf{q}_l, l \in [1, L]$ individually. Similarly, if $\mathbf{C} \geq 0$ is imposed, we use NCLS [1]. At last, if both $\mathbf{C} \geq 0$ and $\mathbf{1}^T\mathbf{C} = \mathbf{1}^T$ are imposed, we use FCLS [1] to estimate $\hat{\mathbf{Q}}$. More details can be found in [9].

There is one thing worthy to note. In general OMP applications (e.g. signal recovery, signal compression), since the observation is just a vector, it only iteratively selects the member that maximizes the sum of absolute correlation with the residual obtained in previous step. However, in SSR-BS the observation is a matrix. At each j, it further uses an inner loop to search the band that can best reconstruct (i.e. least square problem Eq. (2)) $\mathbf{B}$ with the previously selected bands. Such a two-fold nested loop structure will make OMP-BS computationally expensive.

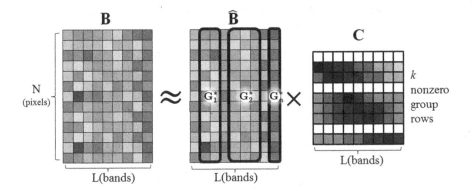

Fig. 1. Group self-sparse model for BS.

3   The Proposed GOMP-BS Method

The problem of group sparse representation (GSR) is relatively new concept. It aims to find the coefficient vector (or matrix) where the nonzero coefficients form a group structure. The group sparsity optimization is regarded as an extension of LASSO method [15]. One of the famous algorithms is Group LASSO [16] which performs variable selection on predefined groups of variables in linear regression model. By integrating group sparsity concept into the SSR-BS, the new optimization problem is shown in Eq. (3). In this case, $\mathbf{C}$ should be in the form of group sparsity, where g means the number of group ($p \leq g \leq L$).

$$\hat{\mathbf{C}} = \arg\min_{\mathbf{C}} ||\mathbf{B} - \mathbf{BC}||_F^2, s.t.||\mathbf{C}||_{0,2} \leq k \tag{3}$$

where $\widehat{\mathbf{C}} = [\underbrace{c_{(1,1)}, \ldots, c_{(1,n_1)}}_{\mathbf{G}_1}, \underbrace{c_{(2,1)}, \ldots, c_{(2,n_2)}}_{\mathbf{G}_2}, \cdots, \underbrace{c_{(g,1)}, \ldots, c_{(g,n_g)}}_{\mathbf{G}_g}]$

In order to achieve that, the predefined group label for $\mathbf{B}$ is needed. Fortunately, many existing band grouping (or band clustering) methods [17,18], or common spectral similarity methods (e.g., SID) can be used to produce band groups. In this paper, we simply choose K-mean as the band grouping algorithm to pre-group $\mathbf{B}$ into g disjoint groups. In order to stabilize the results, we set the bands selected by uniform BS (UBS) as the initial conditions for K-mean. Now we mark $\widehat{\mathbf{B}}$ as grouped hyperspectral band vectors. Similarly, we can adopt the OMP strategy to pick the most informatively complement group at new iteration until the number of selected bands reaches p. Such an algorithm is called group orthogonal matching pursuit (GOMP) [12]. We name the group-based SSR-BS solved by GOMP as GOMP-BS. The model of GOMP-BS is shown is Fig. 1. In later subsections, we will introduce two different searching strategy of GOMP.

$$\widehat{\mathbf{B}} = [\underbrace{\mathbf{b}_{(1,1)}, \ldots, \mathbf{b}_{(1,n_1)}}_{\mathbf{G}_1}, \underbrace{\mathbf{b}_{(2,1)}, \ldots, \mathbf{b}_{(2,n_2)}}_{\mathbf{G}_2}, \cdots, \underbrace{\mathbf{b}_{(g,1)}, \ldots, \mathbf{b}_{(g,n_g)}}_{\mathbf{G}_g}] \qquad (4)$$

3.1   GOMP-BS Algorithm

The standard GOMP algorithm is described in Algorithm 1. Compared to OMP-BS [9], the main feature of GOMP-BS is that it selects only one group index but entire bands in the group will be selected. However, if we directly apply GOMP-BS results, denoted by, $\mathbf{B}_{BS} = [\mathbf{G}_1^*, \mathbf{G}_2^*, \ldots, \mathbf{G}_p^*]$ ($\mathbf{G}_i^*$ presents the selected group at i-th iteration), to perform data analysis, it would be expected that the analysis performance may not be high because the inter-band redundancy within each group is generally high. Therefore, we further provide an alternative output $\mathbf{b}_{(i,centroid)}^*$, which presents the centroid of the selected the band group $\mathbf{G}_i^*$. In this case, the BS-results will be $\mathbf{B}_{BS} = [\mathbf{b}_{(1,centroid)}^*, \mathbf{b}_{(2,centroid)}^*, \cdots, \mathbf{b}_{(p,centroid)}^*]$.

The SVM analysis results of both will be shown and compared in Sect. 4. They are named as GOMP-BS and GOMP-BS (centroid), respectively. Obviously, if we use GOMP-BS, the algorithm will be terminated faster since it obtains multiple bands at each iteration. If we use GOMP-BS (centroid), it will perform better in the aspect of information complement but require more computing time since it only picks the representative band (i.e., centroid) from the group.

It is worth mentioning that OMP-BS [9] is a special case of GOMP-BS if we set $g = p$. Note that both GOMP-BS and GOMP-BS (centroid) use full group elements to solve regression problem, but they output the results by full elements and centroid element, respectively.

3.2   Simplified GOMP-BS (sGOMP-BS) Algorithm

GOMP-BS can be simplified to a more efficient version by reduced the size of temporary basis matrix $\mathbf{P}$, we name it as simplified GOMP-BS (sGOMP-BS) method. In sGMOP-BS, the $\mathbf{P}$ is formed by only considering the average of each group $\mathbf{G}_j$. So it has size of $N \times j$. The advantage of sGOMP is $\mathbf{P}$ will significantly increase as the selected bands p increased so that it can save computation time. The sGOMP-BS algorithm is described in Algorithm 2 below.

In the same way, we could choose full group elements or centroid as the output. We mark these two methods as sGOMP-BS and sGOMP-BS (centroid). In conclusion, sGMOP use mean elements to solve regression problem. The output can be designated as full group elements or centroid element. It can be expected that sGOMP-BS methods will be calculated faster than GOMP-BS methods.

In Algorithms 1 and 2, we use LS formula $\hat{\mathbf{Q}} = \mathbf{P}^+\mathbf{B} = (\mathbf{P}^T\mathbf{P})^{-1}\mathbf{P}^T\mathbf{B}$ to solve non-constrain regression problem in (*1) and (*2), and use SCLS formula [1] for sum-to-one constrain problem.

Algorithm 1 GOMP-BS

Input: Group structure $\widehat{\mathbf{B}} = [\mathbf{G}_1, \mathbf{G}_2, \ldots, \mathbf{G}_g]$, number of required bands p , output *option* (=1 or 2)

Output: selected bands matrix $\mathbf{B}_{BS}$, coefficient matrix $\mathbf{C}$

1: Initialization:$\mathbf{G}_{BS} = \phi$, $\mathbf{B}_{BS} = \phi$ (empty matrix)
2: **for** $j = 1, \ldots, p$ **do**
3:     Set $s = \infty$
4:     **for all** $(i \in [1, g]) \cap (\mathbf{G}_i \notin \mathbf{G}_{BS})$ **do**
5:         Let $\mathbf{P} = [\ \mathbf{G}_{BS} \ \ \mathbf{G}_i \ ]$
6:         Solve $e = \min_{\mathbf{Q}} \| \ \mathbf{B} - \mathbf{PQ} \|_F^2$——————————— (*1)
7:         **if** $e < s$ **then**
8:             $s = e$
9:             $\mathbf{G}_{new} = \mathbf{G}_i$
10:        **end if**
11:    **end for**
12:    Update $\mathbf{G}_{BS} = [\mathbf{G}_{BS} \ \ \mathbf{G}_{new}]$
13:    **if** $(option = 1) \cap (size(\mathbf{G}_{BS}, 2) \geq p)$ **then**
14:        $\mathbf{B}_{BS} = \mathbf{G}_{BS}$
15:        break
16:    **else if** $(option = 2))$ **then**
17:        $\mathbf{B}_{BS} = [\mathbf{B}_{BS} \ \ \mathbf{b}_{(new, centroid)}]$
18:    **end if**
19: **end for**
20: Finally, solve $\mathbf{C} = \arg\min_{\mathbf{C}' \in R^{p \times L}} \| \mathbf{B} - \mathbf{B}_{BS} \mathbf{C}' \|_F^2$——————— (*2)

Algorithm 2 sGOMP-BS

Input: Group structure $\widehat{\mathbf{B}} = [\mathbf{G}_1, \mathbf{G}_2, \ldots, \mathbf{G}_g]$, number of required bands p , output *option* (=1 or 2)

Output: selected bands matrix $\mathbf{B}_{BS}$, coefficient matrix $\mathbf{C}$

1: Initialization:$\mathbf{G}_{BS} = \phi$, $\mathbf{B}_{BS} = \phi$, $\mathbf{G}_{BSmean} = \phi$ (empty matrix)
2: **for** $j = 1, \ldots, p$ **do**
3:     Set $s = \infty$
4:     **for all** $(i \in [1, g]) \wedge (\mathbf{G}_{(i, mean)} \notin \mathbf{G}_{BSmean})$ **do**
5:         Let $\mathbf{P} = [\ \mathbf{G}_{BSmean} \ \ \mathbf{G}_{(i, mean)} \ ]$
6:         Solve $e = \min_{\mathbf{Q}} \| \ \mathbf{B} - \mathbf{PQ} \|_F^2$——————————— (*1)
7:         **if** $e < s$ **then**
8:             $s = e$
9:             $\mathbf{G}_{(new, mean)} = \mathbf{G}_{(i, mean)}$
10:        **end if**
11:    **end for**
12:    Update $\mathbf{G}_{BS} = [\mathbf{G}_{BS} \ \ \mathbf{G}_{new}]$
13:    Update $\mathbf{G}_{BSmean} = \mathbf{G}_{(new, mean)}$
14:    **if** $(option = 1) \wedge (size(\mathbf{G}_{BS}, 2) \geq p)$ **then**
15:        $\mathbf{B}_{BS} = \mathbf{G}_{BS}$
16:        break
17:    **else if** $(option = 2))$ **then**
18:        $\mathbf{B}_{BS} = [\mathbf{B}_{BS} \ \ \mathbf{b}_{(centroid, new)}]$
19:    **end if**
20: **end for**
21: Finally, solve $\mathbf{C} = \arg\min_{\mathbf{C}' \in R^{p \times L}} \| \mathbf{B} - \mathbf{B}_{BS} \mathbf{C}' \|_F^2$——————— (*2)

---

4   Experiments

The dataset used for experiments are two real hyperspectral image, called Pavia and Purdue. The former has very high spatial resolution but weak spectral resolution. In contrast, Purdue has less spatial resolutions but high spectral resolution. We use these two data to verify our proposed method. When BS is done, the progressive land cover classification will be implemented. Theoretically, the proposed GOMP-BS/sGOMP-BS could perform as good as OMP-BS but require less computing time.

4.1   Pavia Data

The Pavia dataset was collected by the ROSIS optical sensor over an urban area of the University of Pavia. The size of the Pavia image is 610×340 with very high spatial resolution about 1.3 m per ground pixel. The original data contains 115 spectral bands with spectral range from $0.43 \mu m$ to $0.86 \mu m$. After removing the noisy bands, the remaining 103 bands are used for the experiments. Figure 2(a–c) respectively show the image scene of 80-th band, the geometric locations of all target classes, and the corresponding spectral signatures. According to the ground truth in Fig. 2(b), there are 9 classes in this image scene, consisting of several urban targets such as vegetation, soil, and roads.

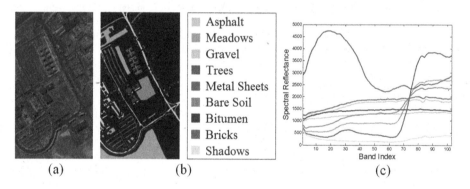

Fig. 2. ROSIS image scene: University of Pavia (a) Band 80-th. (b) Ground truth map for 9 classes. (c) Spectral signatures of ten classes.

Table 1 tabulates the first 50 selected bands selected by 8 different methods, the PBS-Variance [11], CEM-CBS [19], LP [20], OMP-BS [9] and the proposed GOMP-BS/sGOMP-BS. The time on 3rd column lists the corresponding computing time under the environment: CPU Intel(R) Core(TM) i7-4770 at 3.4 Ghz with 16 GB RAM. The K-mean clustering pre-process for Pavia data is 2.01 s. According the table, the proposed GOMP/sGOMP methods can select bands more complementarily in the spectrum range.

Table 1. 30 selected bands of using different BS methods from Pavia data.

	Pavia data 30 selected bands	Time
PBS-Variance [11]	91/90/92/88/89/87/95/94/93/96/82/86/83/97/103/102/ 98/81/100/101/99/85/84/80/79/63/62/61/64/60	0.04 s
CEM-BCC [19]	93/4/6/5/7/8/9/10/11/12/2/13/14/15/16/17/18/19/20/ 21/22/1/23/24/25/26/27/28/29/30	228.20 s
LP [20]	95/63/15/1/73/83/85/3/33/47/5/102/2/7/78/9/100/11/ 68/18/55/98/27/40/4/13/88/22/103/76	8.49 s
OMP-BS-LS [9]	74/42/94/24/66/2/6/81/1/13/54/4/100/33/86/9/83/18/ 102/7/48/11/3/28/61/15/97/70/89/21	44.55 s
GOMP-BS-LS	[72, 73]/[35, 36, 37]/[81, 82, 83, 84]/[11, 12, 13, 14, 15, 16]/ [53, 54, 55, 56, 57, 58]/[100, 101, 102, 103]/[1, 2]/[5, 6]/ [59, 60, 61, 62, 63, 64, 65]	23.77 s
GOMP-BS-LS (centroid)	72/36/82/13/56/101/2/5/62/23/93/45/3/8/30/76/88 [/19/69/98//39/50/79/26/85/33/42/74/71/66	44.81 s
sGOMP-BS-LS	[74, 75]/[41, 42, 43]/[11, 12, 13, 14, 15, 16]/[87, 88, 89, 90]/ [53, 54, 55, 56, 57, 58]/[1, 2]/[100, 101, 102, 103]/ [28, 29, 30, 31]	16.65 s
sGOMP-BS-LS (centroid)	74/42/13/88/56/2/101/30/85/66/3/82/5/98/62/8/19/ 79/50/36/71/93/23/26/45/33/72/39/76/69	28.29 s

4.2 Purdue Data

The Purdue dataset was gathered by AVIRIS sensor over the Indian Pines test site in North-western Indiana and consists of 145×145 pixels and 224 spectral reflectance bands in the wavelength range $0.4\,\mu m$ to $2.5\,\mu m$. This scene is a subset of a larger one. Figure 3(a–c) respectively show the image scene of 30-th band, the geometric locations of all target classes, and the corresponding spectral signatures. In Purdue dataset, there are 16 classes in this image scene, consisting two-thirds agriculture, and one-third forest or other natural perennial

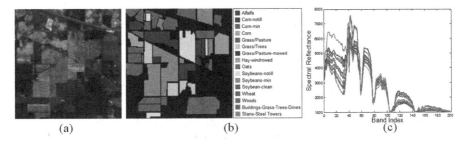

(a)   (b)   (c)

Fig. 3. AVIRIS image scene: Purdue University Agronomy farm northwest (a) Band 30-th. (b) Ground truth map for 16 classes. (c) Spectral signatures of seventeen classes.

Table 2. 50 selected bands of using different BS methods from Purdue data.

	Purdue data 50 selected bands	Time
PBS-Variance [11]	29/28/27/26/25/30/42/32/41/24/33/23/31/43/22/44/ 39/21/48/49/50/20/45/51/52/34/19/53/38/47/18/46/17/ 16/15/14/12/13/11/54/40/10/9/35/55/37/56/69/70/68	0.23 s
CEM-BCC [19]	86/28/33/55/95/64/124/9/32/39/118/101/31/172/135/ 162/90/72/82/1/76/40/60/149/81/113/97/141/116/187/ 178/103/184/18/159/138/26/192/150/127/99/200/176/ 185/71/174/190/188/23/74	966.47 s
LP [20]	42/29/1/35/2/70/3/4/6/117/5/38/32/7/41/8/39/96/37/ 34/45/9/44/43/10/33/46/36/40/47/11/48/17/49/12/86/ 50/13/51/14/52/63/31/15/94/53/16/54/93/55	80.11 s
OMP-BS-LS [9]	81/113/19/50/6/38/189/29/1/68/34/2/97/3/131/42/9/ 4/32/5/89/41/36/10/49/39/7/74/8/35/45/44/37/63/43/ 46/33/20/40/47/11/48/12/90/15/31/51/13/58/14	468.20 s
GOMP-BS-LS	[67, 68, 69, 70, 71, 72, 73, 74, 75, 76]/[25, 26, 27, 28, 29, 30, 31, 32, 33]/[121, 122, 135, 136, 137, 138, 139, 140, 141]/[39, 40, 41, 42, 43, 44, 45]/[4, 5, 6, 7, 8, 9, 10, 11, 12]/[81, 82, 83, 84 , 85, 86, 87, 88, 89, 90, 91, 92, 93, 94]	23.39 s
GOMP-BS-LS (centroid)	36/34/58/9/50/126/26/64/1/88/179/2/43/3/74/15/97/ 117/37/22/35/38/55/165/155/78/79/80/77/101/168/138 /103/201/112/104/144/105/202/106/107/143/145/142 /146/147/200/152/199/197	138.22 s
sGOMP-BS-LS	36/34/[58, 59]/[4, 5, 6, 7, 8, 9, 10, 11, 12]/[46, 47, 48, 49, 50, 51, 52]/[123, 124, 125, 126, 127, 128, 129, 130, 131, 132, 133, 134]/[25, 26, 27, 28, 29, 30, 31, 32, 33]/[60, 61, 62, 63, 64, 65, 66]/1	26.54 s
sGOMP-BS-LS (centroid)	36/34/58/9/50/126/26/64/1/88/179/2/43/3/74/15/97/ 117/37/22/35/38/55/165/155/78/79/80/77/101/168 /138/103/201/112/104/144/105/202/106/107/143/145 /142/146/147/200/152/199/197	64.72 s

vegetation. There are two major dual lane highways, a rail line, as well as some low density housing, other built structures, and smaller roads. Table 2 tabulates the first 50 selected bands selected by 8 different methods, and the corresponding computing time.

4.3 Classification Results and Comparison

In order to evaluate the effectiveness of the proposed methods, we perform progressive SVM classification on the p-selected bands obtained by different BS methods shown in Tables 1 and 2. Theoretically, higher classification rate means that the band set has lower redundancy or higher complement information. The p value is set from 2 to 30 to conduct progressive analysis in Pavia dataset while in

Purdue dataset the p value is set from 2 to 50. For the proposed GOMP-BS, two abundance estimators, LS, SCLS, are used to solve the sub optimization problem (*1)(*2), respectively. To maximize the effectiveness of band de-correlation in each group, the K-mean is implemented by $K = 30$ and 50 for Pavia and Purdue data, respectively. The SVM classifier uses linear kernel with default kernel parameters and regularization value. The training data is formed by 10% randomly selected samples from each class. For the quantitative analysis, we adopt overall accuracy (OA) as the measure to evaluate the classification performance. The random part is repeated 10 times and the average results are reported.

Figure 4 shows the corresponding progressive classification results of PBS-Variance [11], CEM-CBS [19], LP [20], OMP-BS [9] and the proposed GOMP-BS (centroid) and sGOMP-BS (centroid) in OA curve, respectively, where x-axis denotes the number of the selected bands (p) and y-axis denotes the corresponding OA values. It can be observed that the LP and CEM-BCC seem to produce the best classification results in Pavia and Purdue data, respectively, in some p values. However, the proposed GOMP-BS/sGOMP-BS and OMP-BS [9] can perform overall better and more stably on both dataset. Most importantly, GOMP-BS resolves the issue of high computational complexity caused by the greedy search in OMP-BS. According to Tables 1 and 2, GOMP-BS takes half of time in Pavia data and even one-tenth of time in Purdue than OMP-BS.

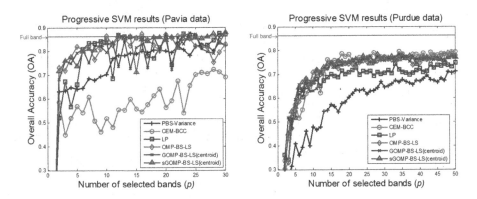

Fig. 4. The classification results of all compatred BS methods in Pavia and Purdue dataset.

To analyze the influence about the imposed constraints, and the difference between group output and centroid output, Fig. 5 further shows the OA results of all four proposed methods, with using no constrain (LS) and sum-to-one constraint (SCLS) optimizers, respectively. It can be observed that imposing sum-to-one constraint or not seems to provide almost the same classification performance. It also can be found that using group output provides lower OA values as we expected. The reasons that the bands in the same group have very high spectral similarity so that those duplicate information could not effectively contribute to classification.

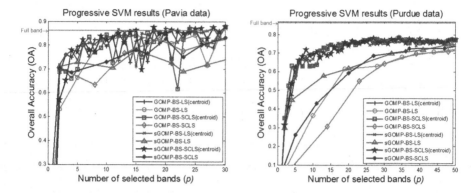

Fig. 5. The classification results of all GOMP/sGOMP methods in Pavia and Purdue dataset.

5   Conclusion

This paper develops a new sparse-based band selection method based on group sparse structure, called group orthogonal matching pursuit-based band selection (GOMP-BS), which can select the bands in a sequential manner and resolves the issue of high complexity caused by the greedy search of traditional OMP algorithm. We assume that the all spectral bands can be sparsely represented by a set of representative bands, and use the self-sparse model to formulate the problem. To find the representative bands in an efficient way, we utilize the idea of group orthogonal matching pursuit (GOMP) to pick the next representative band group based on the minimization of the model residual. The experimental results show that the proposed GOMPs can effectively select the bands that are highly informative for land cover classification of real hyperspectral images with appropriate computing time. The future work would be extending the methodology to nonlinear group sparse regression, and further reducing the computational complexity in the regression process.

References

1. Chang, C.I.: Hyperspectral Imaging: Techniques for Spectral Detection and Classification, vol. 1. Springer, New York (2003)
2. Chang, C.I.: Hyperspectral Data Processing: Algorithm Design and Analysis. Wiley, New York (2013)
3. Wright, J., Ma, Y., Mairal, J., Sapiro, G., Huang, T.S., Yan, S.: Sparse representation for computer vision and pattern recognition. Proc. IEEE **98**, 1031–1044 (2010)
4. Iordache, M.D., Bioucas-Dias, J.M., Plaza, A.: Sparse unmixing of hyperspectral data. IEEE Trans. Geosci. Remote Sens. **49**, 2014–2039 (2011)
5. Ehler, M., Hirn, M.: Sparse endmember extraction and demixing. In: IGARSS, pp. 1385–1388 (2012)

6. Li, S., Qi, H.: Sparse representation based band selection for hyperspectral images. In: 2011 18th IEEE International Conference on Image Processing, pp. 2693–2696. IEEE (2011)

7. Du, Q., Bioucas-Dias, J.M., Plaza, A.: Hyperspectral band selection using a collaborative sparse model. In: 2012 IEEE International Geoscience and Remote Sensing Symposium, pp. 3054–3057. IEEE (2012)

8. Elhamifar, E., Sapiro, G., Vidal, R.: See all by looking at a few: sparse modeling for finding representative objects. In: 2012 IEEE Conference on Computer Vision and Pattern Recognition (CVPR), pp. 1600–1607. IEEE (2012)

9. Lai, C.H., Chen, C.S., Chen, S.Y., Liu, K.H.: Sequential band selection method based on group orthogonal matching pursuit. In: 8th Workshop on Hyperspectral Image and Signal Processing (Whispers 2016). IEEE (2016)

10. Sun, W., Zhang, L., Du, B.: A sparse self-representation method for band selection in hyperspectral. In: 7th Workshop on Hyperspectral Image and Signal Processing (Whispers 2015). IEEE (2015)

11. Chang, C.I., Liu, K.H.: Progressive band selection of spectral unmixing for hyperspectral imagery. IEEE Trans. Geosci. Remote Sens. **52**, 2002–2017 (2014)

12. Swirszcz, G., Abe, N., Lozano, A.C.: Grouped orthogonal matching pursuit for variable selection and prediction. In: Advances in Neural Information Processing Systems, pp. 1150–1158 (2009)

13. Donoho, D.L.: For most large underdetermined systems of linear equations the minimal ℓ1-norm solution is also the sparsest solution. Commun. Pure Appl. Math. **59**, 797–829 (2006)

14. Tropp, J.A., Gilbert, A.C.: Signal recovery from random measurements via orthogonal matching pursuit. IEEE Trans. Inf. Theory **53**, 4655–4666 (2007)

15. Blumensath, T., Davies, M.E.: On the difference between orthogonal matching pursuit and orthogonal least squares (2007)

16. Yuan, M., Lin, Y.: Model selection and estimation in regression with grouped variables. J. Roy. Stat. Soc.: Ser. B (Stat. Methodol.) **68**, 49–67 (2006)

17. Mojaradi, B., Emami, H., Varshosaz, M., Jamali, S.: A novel band selection method for hyperspectral data analysis. Int. Arch. Photogramm Remote Sens. Spat. Inf. Sci. **37**, 447–454 (2008)

18. Su, H., Yang, H., Du, Q., Sheng, Y.: Semisupervised band clustering for dimensionality reduction of hyperspectral imagery. IEEE Geosci. Remote Sens. Lett. **8**, 1135–1139 (2011)

19. Chang, C.I., Wang, S.: Constrained band selection for hyperspectral imagery. IEEE Trans. Geosci. Remote Sens. **44**, 1575–1585 (2006)

20. Du, Q., Yang, H.: Similarity-based unsupervised band selection for hyperspectral image analysis. IEEE Geosci. Remote Sens. Lett. **5**, 564–568 (2008)

Computer Vision Technologies for Smart Vehicle

Cost-Based Feature Transfer for Vehicle Occupant Classification

Toby Perrett[✉] and Majid Mirmehdi

Visual Information Laboratory, University of Bristol, Bristol, UK
TP8961@bristol.ac.uk

Abstract. Knowledge of human presence and interaction in a vehicle is of growing interest to vehicle manufacturers for design and safety purposes. We present a framework to perform the tasks of occupant detection and occupant classification for automatic child locks and airbag suppression. It operates for all passenger seats using a single overhead camera. A transfer learning technique is introduced to make full use of training data from all seats, whilst still maintaining some control over the bias necessary for a system designed to penalize certain misclassifications more than others. An evaluation is performed on a challenging dataset with both weighted and unweighted classifiers that demonstrates the effectiveness of the transfer process.

1 Introduction

One of the main areas in which computer vision can aid vehicle design is improving safety for occupants. Examples of this include lane tracking [1,2], collision prediction systems [3–5], and alertness monitoring for the driver [6,7]. One such area that has received particular research interest is automatic occupant classification [8–11]. This has traditionally been for the purpose of airbag suppression, where it is sometimes safer to not deploy an airbag in a collision if the occupant is a small child [12,13].

A variety of non-vision sensors have been used for the task of occupant detection. Weight sensors [14] are the most widely used, while ultrasound [15] and inductive sensors [16] have also been proposed, and for classification, arrays of weight sensors have been suggested [17]. However, there are a number of incentives to replace all of these with a camera based system. These include the possibility of introducing additional functionality without any additional hardware (for example, determining if an occupant is using a control surface [18]), more robustness to tall and heavy child seats (some weigh more than 10 kg) and reducing the cost to the vehicle manufacturer.

Designing a computer vision system to work reliably in a safety-critical automotive environment presents a number of challenges. These include a large variety of lighting conditions (including over exposure in some cases), skin tones, clothing, postures and occupant behaviours. The bias of the system should also be considered - is it better to incorrectly identify an adult as a child or vice versa, for example?

© Springer International Publishing AG 2017
C.-S. Chen et al. (Eds.): ACCV 2016 Workshops, Part I, LNCS 10116, pp. 405–419, 2017.
DOI: 10.1007/978-3-319-54407-6_27

In this paper, a method is proposed that can handle the tasks of occupant detection and occupant classification in all passenger seats using a single camera. An occupant detection system would allow a vehicle to automatically turn on infotainment systems and seatbelt warnings, for example. The two use cases of occupant classification we address here are airbag suppression and automatic child locks. Currently, these two functionalities, which have different age cutoffs and intra-class costs, have to be switched on and off manually, leading to the possibility of human error. To address the issues raised above, it is important to make full use of the available training data from every seat. Thus, a transfer learning technique is introduced to allow a classifier for one seat to use training information from others whilst still maintaining some control over the bias generated by the transfer process. More specifically, Histogram of Oriented Gradient (HOG) and binned motion statistics are extracted from training images and used as features. Joint Discriminant Analysis (JDA) is applied to find a space of reduced dimensionality shared by both the source and target features. Source features are then transferred to bring them closer to the target feature distributions whilst being attracted to classes with a lower misclassification cost. Finally, these features are used to train a weighted Support Vector Machine (SVM) classifier, which can then classify unseen images from the target seat.

To evaluate the proposed method, a dataset was collected featuring 40 adults and 60 children. To the best of our knowledge this is the first time such a large dataset, including children, has been used to evaluate an occupant classification method using computer vision.

Our contributions in this work are threefold. Perhaps the most important is the proposed cost-based transfer learning method. This is the first time intra-class costs have been addressed from a transfer learning perspective, and applied to a weighted classification problem. Secondly, computer vision has not previously been used to investigate automatic child locks, which is presented here in the same framework as an automatic airbag suppression method. Finally, this is also the first time a single sensor has been used to perform the occupant classification task for multiple seats simultaneously.

2 Related Work

Although some works have used multiple cameras or time-of-flight cameras [11, 19–21] to obtain depth and segment the foreground for occupancy classification, they often have a limited depth range and also a field of view that makes observing more than one occupant difficult. We found a cheap, single camera with a wide field of view lens (e.g. see Fig. 1) would be the ideal choice of sensor from a manufacturer's perspective[1] as one could be used to observe the whole cabin.

Most vision-based occupant classification systems, such as [9–11,19,21–24], tend to follow a standard classification pipeline, usually consisting of a feature extraction stage, dimensionality reduction stage, and classification stage. However, there are issues that are yet to be addressed in the literature. These include

---

[1] The steer for this work comes from Jaguar Land Rover Research.

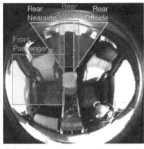

(a) An empty cabin with the seat regions that will be classified labelled.

(b) 3 children. The front passenger is not seated correctly. The rear passengers are putting seatbelts on.

(c) 2 children and 1 adult. Some overexposure is present, particularly in the front passenger region.

(d) 3 children and 1 adult driver.

(e) 4 adults. Better exposure level for the rear of the cabin.

(f) 2 children. Very dark rear seats.

Fig. 1. Some example images from our dataset with child faces blurred.

how a system that should work on more than one seating position at a time can most effectively use all the training data available.

Intuitively, occupancy detection and classification seems like an ideal problem for improvement via some transfer of knowledge - there should be some kind of consistent difference, for example, in the features extracted from a child and adult in a front seat and the features extracted from a child and adult in a rear seat. Transfer learning, or domain adaptation, is the process of taking information learned in one domain (the *source*), and applying it to a different domain (the *target*). In this paper, we treat each seat region as a different domain - Fig. 1(a) shows these seat regions labelled. The proposed method will be a form of inductive transfer as both source and target labels are known [25, 26]. Transductive techniques [27–29] are used when no source labels are known, but could influence our task because they tend to rely on knowledge of the overall source and target distributions [30]. This information will be available as we have a large number of both source and target labels. However, none of these methods are able to handle the potential costs of misclassification when performing a feature transfer. Some works have investigated giving weightings to the transferred features' contributions to a classification [31], but these have been based on how sparse the feature space is at certain locations and again do not adjust the transferred features' locations or account for relative misclassification costs. Other works have foregone a transfer stage, instead assuming the source and target distributions are similar enough to apply an additional penalty to source features when training a classifier [32].

3 Dataset

Previous evaluations have used blankets [21] and dolls [33] to simulate children or experimented on only a small dataset (for example, 6 children and 9 adults in [8] or 3 adults and 1 child in [10]). These evaluations and datasets cannot provide the necessary variability to adequately reflect a system that is critical in nature. As such, for this study we recruited 60 children between the ages of 3 and 9, along with 40 adults. There are a large number of clothing types, skin tones, lighting conditions (including over- and under-exposure), postures and behaviours in the footage.

A single near infra-red (NIR) camera with a 190° field of view lens was mounted just above the rear view mirror in two different cars, so that all the seats were fully visible apart from the rear nearside and offside seats, which were mostly, but sufficiently, visible. An NIR camera was chosen as it is more robust to the wide variety of illumination conditions found in vehicles, including the ability to operate at night [33]. The first 30–60 s were filmed after the occupants entered the vehicle. Participants were seated in different configurations (e.g. adult in front passenger seat, child in booster seat in rear nearside, empty rear middle and rear offside, etc.), with each participant taking part in around three different sequences. In total, there are 122 sequences each containing footage of 4 seats, some with occupants and some empty. Examples are given in Fig. 1.

4 Proposed Method

In the proposed approach, after fisheye lens distortion removal, features are extracted from each frame and put through a dimensionality reduction stage. The dimensionality reduction process works on source and target features simultaneously to give a reduced, shared, feature space. This is followed by a procedure to adjust the reduced source features based on the cost of inter-class misclassification. Next, a classifier is trained on these source and target features, which allows it to classify unseen features from the target domain.

4.1 Data Preparation, Feature Extraction and Dimensionality Reduction

The first stage of processing is to correct the fisheye distortion of the 190° FOV lens by following the calibration procedure of Scaramuzza *et al.* [34]. Various features have been used in the literature for occupancy detection, including Haar responses [9], edge features [23], and shape statistics [33], amongst others. We trialled a number of these, but found using HOG features, concatenated with optical flow [35] based motion descriptors (the mean flow magnitude and orientation from the previous 5 frames taken from each bin), produced the best results. We note here that the main contribution in this paper is the cost-based feature transfer, which is explained in detail in Sects. 4.2 and 4.3. The exact type of feature used is of less concern as long as it provides a suitable means of evaluating the transfer process.

Once image features are extracted, dimensionality reduction is performed before they are passed to a classifier. A commonly used technique when using features from different distributions is to exploit the Maximum Mean Discrepancy (MMD) measure [36], which gives a measurement of how different two distributions are, for example see [37, 38] amongst others. Long $et\ al.$ [39] introduced Joint Discriminant Analysis (JDA), where they expanded on the MMD measure by allowing it to determine the differences in marginal and conditional distributions of the features. Figure 2 shows example reduced source and target feature distributions generated by JDA. In practice 50 dimensions are retained after the dimensionality reduction stage, compared to 7752 for the original feature vectors given as input.

4.2 · Feature Transfer via Weighted Gaussians

Once the initial embeddings of the source features in the shared reduced space have been found, we would like to take into account the way classes are arranged in this shared space and adjust each feature's location accordingly. Briefly, separate class distributions for the source and target features are obtained. Each source feature's location relative to the source class distributions is taken, and a new location is found that has the most similar location relative to the target class distributions. For this process to take into account inter-class misclassification costs, the distances used when finding the new location relative to the target class distributions are modified according to these intra-class costs.

The first step is to model the location and distribution of each source class, which is done by fitting a mixture of Gaussians to each class separately. Here, we take C classes with G Gaussians per class, and say the jth Gaussian belonging to the ith class has the following mean, covariance and weighting parameters

$$\lambda_{ij}^{src} = \left\{ \mu_{ij}^{src}, \Sigma_{ij}^{src}, w_{ij}^{src} \right\}. \tag{1}$$

Any feature f can then be represented by a matrix $F^{src}(f)$ containing its relative likelihood with respect to each source Gaussian on a class by class basis. We write the ith row and jth column as

$$F^{src}(f)_{ij} = \frac{w_{ij}^{src} P(f|\lambda_{ij}^{src})}{\sum_{j'=1}^{G} w_{ij'}^{src} P(f|\lambda_{ij'}^{src})}. \tag{2}$$

Here j' is a dummy variable that sums over every column, and is chosen to avoid overloading j. The same process can also be performed with respect to the target classes, which gives

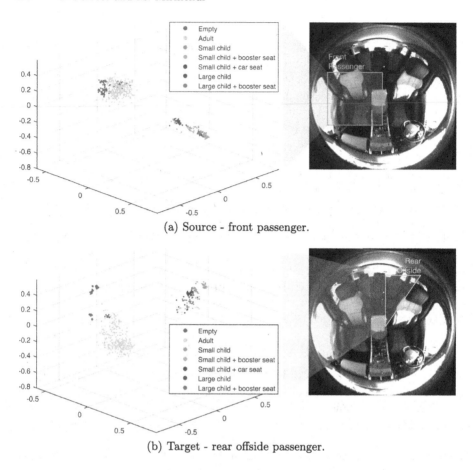

(a) Source - front passenger.

(b) Target - rear offside passenger.

Fig. 2. Example source and target distributions in the shared reduced space. Note, only the three most significant dimensions from the JDA process are shown, while our classifier uses 50 dimensions (where better separation between the classes are achieved).

$$\lambda_{ij}^{tar} = \left\{ \mu_{ij}^{tar}, \Sigma_{ij}^{tar}, w_{ij}^{tar} \right\} \tag{3}$$

and

$$F^{tar}(f)_{ij} = \frac{w_{ij}^{tar} P(f|\lambda_{ij}^{tar})}{\sum_{j'=1}^{G} w_{ij'}^{tar} P(f|\lambda_{ij'}^{tar})}. \tag{4}$$

Given a set of all N source features $S = \{f_1, ..., f_N\}$ with class labels $\{l_1, ..., l_N\}$, a corresponding set of new features $T = \left\{ \tilde{f}_1, ..., \tilde{f}_N \right\}$ is introduced with the same set of labels. We would like the locations of members of T with respect to the target class distributions to be similar to the locations of members of S with respect to the source class distributions. For each pair f_k and $\tilde{f}_k$ this is achieved by fixing f_k and calculating $\tilde{f}_k$ as

$$\tilde{f}_k = \arg \min_{f^\star} \mathcal{D}(F^{src}(f_k), F^{tar}(f^\star)), \tag{5}$$

where $\mathcal{D}$ is a similarity measure (e.g. the norm function) and f^* is an unfixed dummy feature. This can be thought of as allowing the new feature $\tilde{f}_k$ to vary until its location with respect to target class distributions is as similar as possible to the location of the original source feature f_k with respect to the source class distributions. A high penalty should be imposed for large differences between $F^{src}(f_k)$ and $F^{tar}(\tilde{f}_k)$, whereas smaller differences will be less important to correct. It will also be desirable for a large number of non-zero differences to contribute, so the $L_{3,1}$ norm is chosen such that $\mathcal{D}(A,B) = \sum_j \left(\sum_i |A_{ij} - B_{ij}|^3 \right)^{1/3}$. Note how $\mathcal{D}$ may have to be chosen differently if applied to different tasks.

Minimising (5) is necessary in the case where all classes are treated equally, i.e. where the metric used to evaluate the system is the true positive rate. However, in our occupant detection and classification system, it is desirable to have some control over the bias of this feature transfer procedure. For example, when deciding whether to activate a child lock, it does not matter if the 'empty' class is classified as 'adult' or 'child', but a manufacturer would be concerned about confusion of the 'adult' and 'child' classes. Moreover, they may wish to bias the system, for example by penalising a misclassification of a child as an adult more than a misclassification of an adult as a child for safety reasons.

Equation (5) can also take into account this desired bias if (4) is modified. First, a C by C cost matrix Φ is defined, where the entry $\Phi_{u,v}$ denotes the cost of misclassifying class u as class v. The conditions $\forall u, \forall v, 0 \leq \Phi(u,v) \leq 1$ and $u = v \implies \Phi(u,v) = 0$ are enforced. Next, a monotonically increasing function Ψ is chosen, such that $\forall x \in [0,1], \Psi(x) > 0$. Then (4) becomes

$$F^{tar}(f)_{ij} = \frac{\Psi(\Phi_{il}) w^{tar}_{ij} P(f|\lambda^{tar}_{ij})}{\sum_{j'=1}^{G} w^{tar}_{ij'} P(f|\lambda^{tar}_{ij'})}. \tag{6}$$

Here, l is the class label of f, and Φ_{il} is the cost of misclassifying class i as class l. The more "aggressive" Ψ is (i.e. the harder it pulls features towards the centre of their target class distribution and pushes them away from target class distributions which have a higher cost), the lower the probability of the transferred features causing an undesirable misclassification. However, increasing this aggression means that the transferred features are less likely to provide useful additional information to the classifier. Conversely, a weaker Ψ can influence a classifier's predictions more, but at the possible cost of more undesirable misclassifications. We set Ψ as an exponential function and determine Φ via cross validation using the classifier and task-specific cost matrices both presented in Sect. 5. Note that Eq. 6 simplifies to Eq. 4 if the matrix Φ has a 0 diagonal with all other entries being 1, i.e. an unweighted classification problem.

4.3 Gradient Descent Procedure

Now that we have an objective function to minimize (Eq. (5)), we can proceed by gradient descent (we use the method from Lagarias et al. [40]). As we are dealing with high dimensional data with the potential for a large number of local minima, the final state of a gradient descent process is likely to be sensitive to its initial state. Farajidavar *et al.* [27] proposed a method for transferring features when target labels are not known. Their method relied on an initial embedding based on the overall shapes of the source and target data as the target labels were not available. This initial embedding could be used here, but as we have access to target labels it is appropriate to incorporate this additional information to enhance the initial embedding. We thus proceed by adopting the approach in [27], but using it one class at a time. For each D (reduced) dimensional source feature f_k, an initial embedding $\tilde{f}_k$ is taken, where $\tilde{f}_k = f_k + \tau v_k$, and τ is a large scaling factor. To calculate v, first Gaussians are fitted to the target classes with parameters $\lambda_{l_k,j}^{tar}$ for $j = 1, ..., G$. Next, a quantity ν is defined as

$$\nu_{jd} = \sum_{k'=1}^{N} \frac{1}{N\sqrt{w_{l_{k'}j}^{tar}}} P(\lambda_{l_{k'}j}^{tar}|f_{k'}) \frac{f_{k'd} - \mu_{l_{k'}jd}^{tar}}{\sigma_{l_{k'}jd}^{tar}}. \tag{7}$$

where $d = 1, ..., D$. Then each d'th dimension of v_k is set as

$$v_{kd} = \frac{1}{G} \sum_{j'=1}^{G} \nu_{j'd}. \tag{8}$$

4.4 Classification

For the classification stage, an SVM is used with a radial basis function (RBF) kernel. When training a classifier with no weightings provided (as used to evaluate the general performance over all classes in Sect. 5), a grid search over the parameters for the LIBSVM implementation is conducted [41]. When training and testing for specific tasks for which weightings are provided (as used for the main weighted evaluation in Sect. 5), the cost-sensitive one-versus-one (CSOVO) SVM modification by Lin is used [42].

5   Experiments

We now present an evaluation of the proposed method using the dataset introduced in Sect. 2. For comparative analysis, we show our results against two other approaches. The first is the method of Zhang *et al.* [23][2] which relies on a dense Haar feature response as the feature vector and an RBF SVM for classification. Many works use edge-based features instead of Haar responses, such as [10,19,43], with variations on the classification stage. As such, the second method we compare against deploys a standard edge feature extractor (as used in [43]) followed by PCA and an RBF SVM, and hereafter we refer to this as the "Classic Approach." Some works have used techniques such as finite

---

[2] We developed our own implementation of [23] as faithfully as possible.

state machines to combine results from multiple classifications done at different times [43], whereas here we are focussing on maximising the performance of a single classifier. This aim has obvious benefits for any system that uses multiple predictions, as they would benefit from improved single classification accuracy. We chose the approaches we compare against because clear implementation details were available, they use similar camera positions for the front passenger which indicated the features used would work well with our footage, and the amount of training and testing data we use suggests an SVM is both an appropriate choice of classifier and allows more direct comparison.

To investigate how the use of transferred information and cost-based feature transfer are fundamental parts of our approach, we analyse our proposed method's performance when these two aspects are 'switched off'. When the system is just trained and tested on data from a single seat, this gives a non-transfer baseline which we call 'Proposed (non trans)'. When we take data from multiple seats and apply JDA, but do not use the cost-based feature adjustment, this gives a standard transfer baseline which we call 'Proposed (JDA)'.

Results are given for the front passenger (FP), rear nearside (RN), rear middle (RM) and rear offside (RO) seats. The driver's seat is not relevant here, as detection and classification of the driver are unnecessary if the car is being driven. To reduce the possibility of overfitting to certain occupants, no occupant appears in both the training and testing sets. To prevent very similar images appearing in either the training or testing sets, images from the same video sequence are chosen that are more than one second apart. The following results are for training and testing sets of 1000 randomly selected images per seat (subject to the above conditions), and the accuracy presented is taken as the mean over 10 runs. Each run is performed on a new training and testing set selection.

First, results will be presented for weighted classifiers (RBF CSOVO SMVs [42] introduced in Sect. 4) tuned to the tasks of occupant detection, classification for child locks, and classification for airbag suppression, as would be necessary in a production vehicle. These will be followed by results for unweighted classifiers (unweighted RBF SVMs [41] also introduced in Sect. 4) to investigate standard classification performance over all labelled classes.

5.1 Weighted Classification Performance

Data was annotated with the following eight class labels: empty seat, small child in booster seat, small child in child seat, small child with no additional seat, large child in booster seat, large child in child seat, large child with no additional seat, and adult. Children up to Year 2 of school were assigned to the small child classes, and those in Year 3 and above were assigned to the large child classes, making the cutoff around 6 years old. For the three use cases of occupant detection, classification for childlocks, and classification for airbag suppression, it is possible to reduce the number of labelled classes to four - Empty (E), Adult (A), Small child (S) and Large child (L).

Figure 4 shows the weighted classification accuracies, using the cost matrix in Fig. 3(a) for the occupant detection task. The transferred features increase

	E	A	S	L
E	0	1	1	1
A	1	0	0	0
S	1	0	0	0
L	1	0	0	0

(a) Occupant detection.

	E	A	S	L
E	0	0	0	0
A	0	0	1	1
S	1	1	0	0
L	1	1	0	0

(b) Child locks.

	E	A	S	L
E	0	0	0	0
A	1	0	1	0
S	0	1	0	1
L	1	0	1	0

(c) Airbag suppression.

Fig. 3. Cost matrices used by weighted SVMs. The classes denoted are Empty seat (E), Adult (A), Small child (S) and Large child (L).

Method	FP	RN	RM	RO	Avg.
Proposed (non tans)	96.1	93.2	93.3	92.5	93.8
Proposed (JDA)	97.6	93.4	94.1	92.6	94.4
Proposed	**98.2**	**95.3**	**97.3**	**95.9**	**96.7**
Zhang [23]	90.9	88.1	89.9	74.3	85.8
Classic Approach	96.2	90.0	90.5	90.1	91.7

Fig. 4. Weighted classification % accuracy for occupant detection using the cost matrix in Fig. 3(a). FP denotes the front passenger seat, RN the rear nearside seat, RM the rear middle seat and RO the rear offside seat.

Method	FP	RN	RM	RO	Avg.
Proposed (non tans)	94.2	95.1	98.2	89.2	94.1
Proposed (JDA)	94.8	95.5	98.2	89.0	94.4
Proposed	**98.9**	**97.1**	**98.5**	**94.0**	**97.1**
Zhang [23]	90.1	88.0	87.2	70.4	83.9
Classic Approach	94.7	91.1	92.0	90.8	92.2

Fig. 5. Weighted classification % accuracy for automatic child locks using the cost matrix in Fig. 3(b).

the average classification accuracy in this case from 93.8% to 94.4% when using JDA and 96.7% when using the full proposed method, compared to 91.7% for the classic approach and 85.8% for [23]. An example decision outcome is shown in Fig. 7(a).

Figure 5 shows the results for the child lock task, using the cost matrix in Fig. 3(b). This time, the transferred features increase the average classification accuracy from 94.1% to 94.4% when using JDA and 97.1% when using the full proposed method, compared to 92.2% for the classic approach and 83.9% for [23]. An example decision outcome is shown in Fig. 7(b).

Finally, Fig. 6 shows the results for the weighted classification for automatic airbag suppression, and uses the cost matrix in Fig. 3(c). Again, the transferred features increase the average classification accuracy, from 92.0% to 93.0% when using JDA and 94.5% when using the full proposed method, compared to 89.1%

Method	FP	RN	RM	RO	Avg.
Proposed (non tans)	92.8	93.6	92.1	89.3	92.0
Proposed (JDA)	94.3	94.2	93.3	90.1	93.0
Proposed	**96.3**	**96.1**	**94.3**	**91.1**	**94.5**
Zhang [23]	90.4	85.4	85.7	75.6	84.3
Classic Approach	92.8	88.4	87.9	87.2	89.1

Fig. 6. Weighted classification % accuracy for automatic airbag suppression using the cost matrix in Fig. 3(c).

for the classic approach and 84.3% for [23]. An example decision outcome is shown in Fig. 7(c).

5.2 Unweighted Classification Performance

To evaluate the proposed method on a normal unweighted problem, tests were run using all eight labelled classes, as well as the four superclasses introduced above. Results are shown in Table 1. For the eight class problem, using transferred information with JDA averages 79.1%, and the full proposed method averages 78.9%, an improvement over 76.5% when not using transferred information, 74.0% when using the classic approach and 65.9% when using [23].

For the four class problem, the transferred features increase the average classification accuracy from 87.1% to 90.0% when using JDA and 90.2% when using the full proposed method (the classic approach scores 82.7% and [23] 72.2%).

5.3 Discussion

The most important result to take away from the experiments is that the inclusion of transferred information does indeed increase the success rate for the weighted classification tasks. The proposed method delivers improvements over the non-transfer and JDA transfer baselines, as well as the other methods compared against. These improvements can be observed in the successful decisions for occupancy detection, classification for automatic child locks, and classification for automatic airbag suppression.

Another trend that is visible across all our weighted classification experiments is that the front passenger seat obtains higher scores than the rear seats. This is not necessarily surprising, as more detail is visible due to proximity to the camera. The front seats and occupants can also partially occlude the view of the rear of the vehicle.

An improvement is also observed when including transferred features for unweighted classification. However, in the four and eight class classification experiments, there is little difference between the JDA baseline and the full proposed method (JDA outscores the full method by 0.2% on average for eight classes, and the full method outscores JDA by 0.2% on average for four classes). An interesting failure case worth noting can be found in the results for the eight

 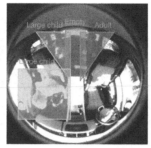 

(a) Occupant detection task. A blue region indicates an occupant is present and a purple region indicates an occupant is not present.

(b) Automatic childlocks. A blue region indicates a childlock should be engaged, and a purple region indicates a childlock should not be engaged.

(c) Automatic airbag suppression. A blue region indicates an airbag should be suppressed, and a purple region indicates an airbag should not be suppressed.

(d) Occupant detection task. Rear nearside small child is misclassified as a large child, but all occupants are still detected.

(e) Automatic childlocks. Rear nearside small child is misclassified as a large child, but the childlocks are still activated correctly.

(f) Automatic airbag suppression. The small child in the front passenger seat is misclassified as an adult, so the airbag is not suppressed when it should be.

Fig. 7. Some example weighted classification decisions with child faces blurred - green text indicates the correct class is chosen, and red text indicates an incorrect class is chosen. The region colour (blue or purple) indicates the status of the occupant detection, classification for childlock and classification for airbag suppression systems (explained in individual image captions). Parts (a), (b) and (c) show examples where the class, and thus system status, is chosen correctly. Parts (d) and (e) show examples where a wrong class is chosen, but the correct decision for enabling or suppressing a system is made. Part (c) shows an example where a wrong class is chosen which results in an incorrect system status. (Color figure online)

class problem in Table 1. In the rear offside seat, the proposed method performs better when not including any transferred information, although the JDA baseline scores highest. The most likely explanation is that multiple unrelated, and dissimilar in feature space, classes could have an equal pull during the transfer

Table 1. Unweighted classification percentage accuracy over all eight classes, and four superclasses, for unweighted classifiers.

Method	8 classes					4 classes				
	FP	RN	RM	RO	Avg.	FP	RN	RM	RO	Avg.
Proposed (non tans)	85.1	77.3	72.0	71.7	76.5	87.2	89.7	85.2	86.3	87.1
Proposed (JDA)	85.8	81.4	**77.3**	**71.8**	**79.1**	89.8	90.9	90.7	**88.5**	90.0
Proposed	**86.0**	**81.9**	**77.3**	70.2	78.9	**90.1**	**91.2**	**91.1**	88.3	**90.2**
Zhang [23]	74.7	64.3	68.0	56.5	65.9	80.8	76.8	75.6	57.8	72.7
Classic approach	80.5	72.5	71.8	71.0	74.0	88.1	80.1	82.0	80.6	82.7

process. These results suggest that the proposed method is more well suited to weighted classification problems (for which it was designed) than unweighted.

6 Conclusion

A framework was introduced to perform the tasks of occupant detection and occupant classification for automatic child locks and airbag suppression. The method works for all passenger seats using a single overhead NIR camera. To make full use of the training data from all seats, a cost-based transfer learning technique was introduced to adjust source features in a reduced space, whilst taking into account the cost of misclassifying these features as certain classes. An evaluation was performed on a large dataset consisting of 40 adult and 60 child volunteers in real vehicles, and the transferred features contributed to an improved classification accuracy.

There are a number of possible avenues for future work. One is to investigate if the proposed method can also incorporate information from standard object datasets [44] to recognize valuable objects left behind in the cabin, and issue warnings if necessary. Here, costs can be attributed to certain items - a laptop would have a higher cost of being left behind than a book, for instance. Other possibilities are to investigate if occupants' postures are suitable for airbag deployment, or if occupants are wearing seatbelts correctly.

References

1. Bottazzi, V.S., Borges, P.V.K., Jo, J.: A vision-based lane detection system combining appearance segmentation and tracking of salient points. In: IEEE Intelligent Vehicles Symposium, pp. 443–448 (2013)
2. Hanwell, D., Mirmehdi, M.: Detection of lane departure on high-speed roads. In: International Conference on Pattern Recognition Applications and Methods (2012)
3. Bonnin, S., Weisswange, T.H., Kummert, F., Schmuedderich, J.: Pedestrian crossing prediction using multiple context-based models. In: International Conference on Intelligent Transportation Systems (2014)

4. Monwar, M.M., Vijaya Kumar, B.V.K.: Vision-based potential collision detection for reversing vehicle. In: IEEE Intelligent Vehicles Symposium, pp. 88–93 (2013)
5. Sivaraman, S., Trivedi, M.M.: Looking at vehicles on the road: a survey of vision-based vehicle detection, tracking, and behavior analysis. IEEE Trans. Intell. Transp. Syst. **14**, 1773–1795 (2013)
6. Vicente, F., Huang, Z., Xiong, X., Torre, F., Zhang, W., Levi, D.: Driver gaze tracking and eyes off the road detection system. IEEE Trans. Intell. Transp. Syst. **16**, 1–14 (2015)
7. Garcia, I., Bronte, S., Bergasa, L.M., Almazan, J., Yebes, J.: Vision-based drowsiness detector for real driving conditions. In: IEEE Intelligent Vehicles Symposium (2012)
8. Huang, S.S.: Discriminatively trained patch-based model for occupant classification. IET Intell. Transp. Syst. **6**, 132–138 (2012)
9. Huang, S.S., Jian, E., Hsiao, P.: Occupant classification invariant to seat movement for smart airbag. In: IEEE International Conference on Vehicular Electronics and Safety (2011)
10. Gao, Z., Duan, L.: Vision detection of vehicle occupant classification with Legendre moments and support vector machine. In: IEEE International Congress on Image and Signal Processing (2010)
11. Goktuk, S.B., Rafii, A.: An occupant classification system eigen shapes or knowledge-based features. In: Computer Vision and Pattern Recognition (2005)
12. Glass, R.J., Segui-Gomez, M., Graham, J.D.: Child passenger safety: decisions about seating location, airbag exposure, and restraint use. Risk Anal. **20**, 521–527 (2000)
13. Technologies, Challenges, and Research and Development Expenditures for Advanced Air Bags. Report to the Chairman and Ranking Minority Member, Committee on Commerce, Science, and Transportation, U.S. Senate (2001)
14. Mehney, M.A., McCarthy, M.C., Fullerton, M.G., Malecke, F.J.: Vehicle occupant weight sensor apparatus (2000)
15. Seip, R.: Linear ultrasound transducer array for an automotive occupancy sensor system (2002)
16. George, B., Zangl, H., Bretterklieber, T., Brasseur, G.: A combined inductive capacitive proximity sensor for seat occupancy detection. IEEE Trans. Instrum. Meas. **59**, 1463–1470 (2010)
17. Wallace, M.W.: Vehicle occupant classification system and method (2003)
18. Cheng, S.Y., Trivedi, M.M.: Vision-based infotainment user determination by hand recognition for driver assistance. IEEE Trans. Intell. Transp. Syst. **11**, 759–764 (2010)
19. Kong, H., Sun, Q., Bauson, W., Kiselewich, S., Ainslie, P., Hammoud, R.: Disparity based image segmentation for occupant classification. In: Computer Vision and Pattern Recognition Workshop (2004)
20. Cheng, S.Y., Trivedi, M.M.: Human posture estimation using voxel data for "smart" airbag systems: issues and framework. In: IEEE Intelligent Vehicles Symposium (2004)
21. Alefs, B., Clabian, M., Painter, M.: Occupant classification by boosting and PMD-technology. In: IEEE Intelligent Vehicles Symposium (2008)
22. Farmer, M.E., Jain, A.K.: Occupant classification system for automotive airbag suppression. In: Computer Vision and Pattern Recognition (2003)
23. Zhang, Y., Kiselewich, S.J., Bauson, W.A.: A monocular vision-based occupant classification approach for smart airbag deployment. In: IEEE Intelligent Vehicles Symposium (2005)

24. Devarakota, P.R.: Occupant classification using range images. IEEE Trans. Veh. Technol. **56**, 1983–1993 (2007)
25. Wang, S.: A new transfer learning boosting approach based on distribution measure with an application on facial expression recognition. In: International Joint Conference on Neural Networks (2014)
26. Shao, H., Tong, B., Suzuki, E.: Extended MDL principle for feature-based inductive transfer learning. Knowl. Inf. Syst. **35**, 365–389 (2012)
27. Farajidavar, N.: Adaptive transductive transfer machines. In: British Machine Vision Conference (2014)
28. Campos, T., Khan, A., Yan, F., Farajidavar, N., Windridge, D., Kittler, J., Christmas, W.: A framework for automatic sports video annotation with anomaly detection and transfer learning. In: Machine Learning and Cognitive Science (2013)
29. Rohrbach, M., Ebert, S., Schiele, B.: Transfer learning in a transductive setting. In: Neural Information Processing Systems (2013)
30. Pan, Z., Li, Y., Zhang, M., Sun, C., Guo, K., Tang, X., Zhou, S.Z.: IEEE Virtual Reality Conference (2010)
31. Garcke, J., Vanck, T.: Importance weighted inductive transfer learning for regression. In: Calders, T., Esposito, F., Hüllermeier, E., Meo, R. (eds.) ECML PKDD 2014, Part I. LNCS (LNAI), vol. 8724, pp. 466–481. Springer, Heidelberg (2014). doi:10.1007/978-3-662-44848-9_30
32. Xu, J., Ramos, S., Vazquez, D., Lopez, A.M.: Cost-sensitive structured SVM for multi-category domain adaptation. In: International Conference on Pattern Recognition (2014)
33. Farmer, M.E., Jain, A.K.: Smart automotive airbags: occupant classification and tracking. IEEE Trans. Veh. Technol. **56**, 60–80 (2007)
34. Scaramuzza, D., Martinelli, A., Siegwart, R.: A toolbox for easily calibrating omnidirectional cameras. In: IEEE International Conference on Intelligent Robots and Systems (2006)
35. Brox, T., Bruhn, A., Papenberg, N., Weickert, J.: High accuracy optical flow estimation based on a theory for warping. In: Pajdla, T., Matas, J. (eds.) ECCV 2004. LNCS, vol. 3024, pp. 25–36. Springer, Heidelberg (2004). doi:10.1007/978-3-540-24673-2_3
36. Gretton, A.: A kernel two-sample test. J. Mach. Learn. Res. **13**, 723–773 (2012)
37. Kim, B., Pineau, J.: Maximum mean discrepancy imitation learning. In: Robotics: Science and Systems (2013)
38. Pan, S.J., Kwok, J.T., Yang, Q.: Transfer learning via dimensionality reduction. In: AAAI Conference on Artificial Intelligence (2008)
39. Long, M., Wang, J., Ding, G., Sun, J., Yu, P.S.: Transfer feature learning with joint distribution adaptation. In: International Conference on Computer Vision (2013)
40. Lagarias, J.C., Reeds, J., Wright, M.H., Wright, P.E.: Convergence properties of the nelder-mead simplex method in low dimensions. SIAM J. Optim. **9**, 112–147 (1998)
41. Chang, C., Lin, C.: LIBSVM: a library for support vector machines. ACM Trans. Intell. Syst. Technol. **2**, 1–27 (2011)
42. Lin, H.: National Taiwan University, Technical report (2010)
43. Huang, S., Hsiao, P.Y.: Occupant classification for smart airbag using Bayesian filtering. In: International Conference on Green Circuits and Systems (2010)
44. Deng, J.D.J., Dong, W., Socher, R., Li, L., Li, K., Fei-Fei, L.: ImageNet: a large-scale hierarchical image database. In: British Machine Vision Conference (2009)

The World Is Changing:
Finding Changes on the Street

Kuan-Ting Chen, Fu-En Wang, Juan-Ting Lin, Fu-Hsiang Chan,
and Min Sun$^{(\boxtimes)}$

Department of Electrical Engineering, National Tsing Hua University,
Hsinchu, Taiwan
winterdaphne104@gmail.com, tdk356ubuntu@gmail.com, brade31919@gmail.com,
corgi1205@gmail.com, sunmin@ee.nthu.edu.tw

Abstract. We propose to find changes in the constantly changing world,
given visual observations at street-level. In particular, we identify "long-
term" changes between Google Street View images and dashcam videos
captured at different months or even years. This is a challenging task,
since (1) dashcam frames are not localized in world coordinate, and (2)
there are many changes introduced by moving objects. We propose a
robust sequence alignment method to align dashcam sequence to Street
View images. Our method outperforms a strong baseline method [1] by
12% mean Average Precision (AP). We also propose a novel change detection method designed to detect long-term changes. Our change detection method (13.54%) outperforms a baseline method without handling
car interior and moving objects (11.70%) by 1.84% (relatively 13.6%)
in mean AP. In a controlled experiment, given manually aligned high
quality Street View images, our change detection method achieves a significantly better mean AP (45.57%).

1 Introduction

Geotagged image collections, such as Google Street View, allow users to experience the street scene of a remote location. This service allows users to find stores,
offices, etc. before physically visiting a place. Google Street View is gleaned by
a fleet of vehicles equipped with expensive cameras and sensors. This approach
can cover many cities at a specific moment in time (e.g., panorama captured on
Nov., 2012 in Fig. 1). However, the fleet is not large enough to keep up with the
changes in the real world. For instance, there is a new building and two billboard
ads in the image captured by a dashcam on Feb., 2014 in Fig. 1.

Another popular system which also captures the street scene is a dashboard
camera (later referred to as dashcam). A dashcam is a consumer device which
can be easily installed by users themselves. Due to its low cost and ease of use,
it has become popular in many countries like Taiwan, Russia, etc. As a result,
the number of vehicles with dashcam is significantly larger than the size of the
Google Street View fleet. Theoretically, it is possible that most changes in a street

© Springer International Publishing AG 2017
C.-S. Chen et al. (Eds.): ACCV 2016 Workshops, Part I, LNCS 10116, pp. 420–435, 2017.
DOI: 10.1007/978-3-319-54407-6_28

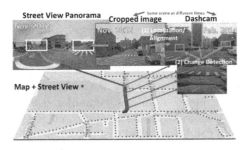

Fig. 1. Illustration of our system. We compare Street View images (taken on Nov., 2012) with dashcam videos (taken on Feb., 2014) to identify changes: a new building and two billboard ads (green boxes). Our system first (1) localizes the dashcam video on the map, then (2) identifies changed rectangle regions. (Color figure online)

scene will be recorded by a dashcam. This implies that fusing the information from dashcam and Google Street View could be mutually beneficial.

We propose to identify "long-term" changes between Google Street View images and dashcam videos captured at different months or even years (Fig. 1). This is a challenging task, since (1) dashcam frames are not precisely localized in world coordinate, and (2) there are many changes introduced by moving objects. To achieve this, we first propose a robust sequence alignment method to localize dashcam frames on the coordinate of Street View images. Our method needs to handle significant structural changes in the scene accumulated through months or years, whereas most localization methods (such as [2]) assume that the changes are mainly due to lighting condition and slight viewpoint changes. We address the challenge by utilizing reliable matches from frames with less changes to geometrically align the whole video sequence to the world coordinate. Given pairs of Street View images and frames, a novel change detection method is proposed to predict a number of rectangle changed regions, which likely correspond to a new building, a billboard ads, etc. (Fig. 1). The change detection method is explicitly designed to ignore (1) "short-term" changes introduced by moving objects (e.g., cars) and different weather conditions (Fig. 6(a)), and (2) changes corresponding to the interior of a vehicle (Fig. 6(a)-Right).

Our system aims to handle dashcam videos in the wild. Therefore, we harvest dashcam videos captured at a diverse set of location (across a radius of 155 km) and time (across about three years) for evaluating our system. Our sequence alignment method is shown to outperform a baseline method [1] by 12% in mean AP. Moreover, our change detection method, which considers the properties of dashcam videos, is also 1.83% (relatively 13.6%) better in mean AP than a generic baseline method based on dense SIFT matching [3]. By inspecting failure examples, we found that failures are typically due to severe viewpoint difference between Street View images and dashcam frames, or less ideal dashcam frame quality. Hence, we further conduct a controlled experiment using manually aligned high quality Street View images. In the control experiment, our change detection method achieves a significantly better mean AP (45.57%).

2   Related Work

In Computer Vision, visual localization of images has been widely studied. We summarize the related work in three groups.

Landmark localization. [4,5] are early work showing the ability to match query images to a set of reference images. Schindler et al. [6] improve the performance to handle city-scale localization. Hays and Efros [7] further demonstrate that query images can be matched to a collection of 6 million GPS-tagged images dataset (within 200 km) at a global scale. Zamir and Shah [8] propose to use Street View images as reference images with GPS-tags, and match SIFT keypoints in a query image efficiently to SIFT keypoints in reference images by using a tree structure. A voting scheme is also introduced to jointly localize a set of nearby query images (within 300 m). As a result of voting, it is able to outperform [6]. Vaca-Castano et al. [1] propose to estimate the trajectory of a moving camera in the longitude and latitude coordinate using Bayesian filtering to incorporate the map topology information. Similar to [1,8], we also use Street View images as reference images. However, unlike [8], we assume a video sequence is captured across an arbitrary distance (not restricted to within 300 m). Unlike [1], we use not only the topology of the map, but also the relative 3D position of the dashcam frames to improve the localization accuracy. Cao and Snavely [9] propose to match a query image to reference images with a graph-based structure to reliably retrieve a sub-group of images corresponding to a representative landmark. This method can be used to improve single image matching accuracy at the first stage of our method. Bettadapura et al. [10] also use Street View images as reference images to match the point-of-view images captured by a cellphone camera. Moreover, the method utilizes accelerometers, gyroscopes and compasses on the cellphone to improve the matching accuracy. The final estimated point-of-view is used as an approximation of the users' attention in applications such as egocentric video tours at museums.

Localization from point clouds. Accurate image-based camera 6 Degrees of Freedom (DoF) pose estimation can also be achieved by utilizing 3D point clouds of a scene [11–14]. However, it is challenging to obtain 3D point clouds representation given a sparse dataset like Google Street view. Moreover, for change detection, 6 DoF pose estimation is not required. Our method essentially estimates a 2D rigid transformation to align dashcam with Street View, since dashcam videos typically have common pitch and roll angles (i.e., parallel to the ground plane), a restricted yaw angle (i.e., tangent to the vehicle trajectory), and a fixed height.

Vehicle Localization. For egocentric vehicle localization, many methods combining image-based sensors with other sensors or map information have been proposed. Given the vehicle speed at all time and the rough initial vehicle location, Badino et al. [2] use Bayesian filter and a per-frame-based visual feature to align a current frame to pre-recorded frames in the database while considering the candidate location of previous frames. Taneja et al. [15] propose a similar but lightweight method requiring images sparsely sampled in space (in average

(a) Field-of-View (L:wide; R:narrow)   (b) Resolution (L:high; R:low)

(c) Weather Condition (L:sunny; R:rainy)   (d) Lighting Condition (L:day; R:night)

Fig. 2. Examples of dashcam videos in the wild. We focus on harvesting videos with good quality (highlighted by green bounding boxes). (Color figure online)

every 7 m). Lategahn et al. [16] combine frame-based visual matching and Inertial Measurement Unit (IMU) for localization. However, it also assumes the GPS information of the first frame is also given. Both [17,18] utilize relative position information from visual odometry with map information from OpenStreetMaps to globally localize a vehicle. However, these methods require the vehicle trajectory to be complex enough to be uniquely identified on the map (i.e., many turns, etc.). Dashcam videos on YouTube do not come with additional speed or initial location information. Moreover, a video is typically less than 5 min with simple trajectories. In contrast, our method does not require an initial location or any extract sensor, and it can even localize vehicles with simple trajectories using both map information and Street View images.

Our main application for dashcam localization is to detect changes between Street View images and dashcam frames. We summarize the related work of change detection below.

Change Detection. Many works have been proposed to compare images of a scene captured at different times [19]. However, their results are typically pixelwise change map, which is often noisy. When additional information of the scene is given, different change detection methods have been proposed. Pollard and Mundy [20] propose a change detection method for 3D scenes observed from an aerial vehicle. Taneja et al. [21] propose geometric change detection by comparing 3D reconstructions built from videos. Recently, Matzen and Snavely [22] propose to detect rectangle regions corresponding to changed billboard and Graffiti from a large number of internet images. Their results are impressive, since they utilize the 3D structure of the scene and assume images are captured densely in time. In our case, the task is much more challenging, since neither the 3D reconstruction of the Street View scene is available nor the dashcam videos is captured densely in time at the same location. In fact, we typically match images captured at different years.

3   Our Methods

We first describe how we harvest many dashcam videos in the wild. Then, we mention how to obtain relative camera position in dashcam video, and how to query Street View images as reference images for localizing cameras in world coordinate (i.e., longitude and latitude). Next, we introduce our robust sequence alignment method. Finally, we describe our novel change detection approach which identifies changed rectangle regions by considering the properties of dashcam videos.

3.1   Data Harvesting

In countries like Taiwan, Russian, etc., dashcam are commonly installed on cars. Through crawling YouTube, we can locate many dashcam videos with rough longitude and latitude positions annotated by the users who uploaded the videos. These videos are extremely diverse. However, not all of them are equally suitable for applying computer vision algorithm due to the following reasons:

Hardware Spec

- *Field of View.* The field of view of the cameras could range from 70° narrow angle to 200° wide angle (Fig. 2a).
- *Resolution.* The resolution of the videos could range from low resolution such as 640×480 pixels to high resolution such as 1280×720 pixels (Fig. 2b).

Environmental Condition

- *Weather Condition.* Dashcam videos recorded in raining and foggy weather have very low quality (Fig. 2c).
- *Lighting Condition.* There are also many videos recorded in low-light condition at night or in tunnels (Fig. 2d).

In order to harvest dashcam videos suitable for camera localization and change detection, we develop several semi-automatic methods to efficiently identify different conditions. Low resolution and small horizontal field-of-view videos can be approximately identified by checking the image dimension and aspect ratio (i.e., a large horizontal field-of-view video typically has width much larger than height), respectively. Low-light condition videos can be removed according to the color histogram with modest precision and high recall. As a result, we efficiently select many videos with good lighting condition and at least 1280×720 pixels (later referred to as high quality videos). Nevertheless, our selected videos are not guaranteed to have high signal to noise ratio or low image distortion. Hence, we still need to address these challenges while developing our method. The videos will be publicly available once the paper is published.

3.2   Relative Camera Position from Dashcam

A dashcam video can be interpreted as sequential images captured by a moving camera. Hence, we could apply classical sequential Structure-from-Motion (SfM) techniques to obtain relative camera positions with respect to the first frame. However, our problem is not a classical Visual Odometry (VO) problem, since we do not have the camera intrinsic information of the harvested dashcam video. Moreover, most of the dashcam videos are captured by a wide-angle or even a fish-eye camera, which introduces an unknown camera distortion in raw video frames. Therefore, it is critical to estimate the camera intrinsic including parameters to mitigate the distortion. We applied an efficient sequential Structure-from-Motion (SfM) method [23,24] to jointly estimate the camera intrinsic and extrinsic, and sparse 3D point cloud representation of the scene. The method estimates the first-order radial distortion parameter which significantly improves

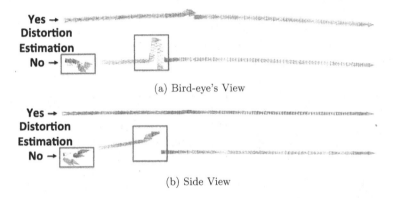

Fig. 3. Comparison between 3D trajectory of the camera with and without distortion estimation from bird-eye's and side view in panal (a) and (b), respectively. Note that without distortion estimation, the estimated 3D trajectory is disconnected and erroneously scattered (large errors indicated by red bounding boxes). (Color figure online)

Fig. 4. Illustration of the queried Street View panoramic image with a 180° vertical field-of-view and a 360° horizontal field-of-view. The road information is illustrated by overlaid yellow paths. We crop two images (later referred to as reference images) along the road direction indicated by the red rectangles. (Color figure online)

the quality of the camera's 3D trajectory reconstruction compared to the reconstruction from camera assuming zero distortion (Fig. 3). For consistency, we refer the coordinate system of the reconstruction as the "model coordinate" in the following sections.

3.3 Reference Images from Street View

Similar to [8], we use Street View images as reference images for localizing the path of a dashcam video. For each dashcam video, we query Street View images within a rectangle region centered at the user provided rough longitude and latitude position[1].

Cropping Street View Images. Note that raw Street View images are panoramic images with a 180° vertical field-of-view and a 360° horizontal field-of-view. In order to efficiently match Street View images to dashcam videos, we propose to crop a panoramic image so that the cropped image is likely to be similar to a frame in the dashcam video. Unlike [8] which cannot assume a common camera orientation, we can assume that the orientation of a dashcam camera is always parallel to the road direction. Hence, we simply crop two images from Street View panorama where their orientations are aligned with the road direction (i.e., one forward and one backward). For the field-of-view of the crop, we simply crop a images with 36% vertical field of view and 72% horizontal field of view (i.e., one fifth of the field of view of a panoramic image). We treat these cropped images as "reference images" which will be matched to the frames in the dashcam video (Fig. 4). The matching results will let us retrieve a subset of reference images along the path of the dashcam video.

3.4 Dashcam Global Localization

Given the reference images and extracted frames in a dashcam video, we first use a baseline image retrieval approach with geometric verification (similar to [6]) to establish matches between them. Then, we propose a new sequence alignment method to align the reconstructed dashcam path to the locations of the reference images in longitude and latitude coordinate.

Image Matching. In the following, we describe our image retrieval pipeline to establish baseline image matches.

Holistic Features Representation. For both the reference images and frames, we detect sparse SIFT keypoints and represent each keypoint using SIFT descriptor [25]. The number of keypoints in each images can be very different depending on the texture in the scene and the image quality. Typically, there are more keypoints in the reference images than in the video frames. In order to represent all images with a fixed feature dimension, we use fisher vector encoding [26] to

---

[1] It is very common that the rough GPS location of a dashcam video is described in the video description for the purpose of reporting accident.

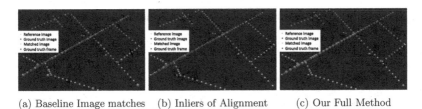

(a) Baseline Image matches   (b) Inliers of Alignment   (c) Our Full Method

Fig. 5. Illustration of the baseline image matching (Panel (a)), and our sequence alignment (Panel (b,c)) approach. Note that our matched images are well aligned to the ground truth trajectory (Panel (c)), but the baseline matched images are spreading on the map with many outliers (Panel (a)). (Color figure online)

generate our holistic feature representation for all images. It has been shown in [26] that the fisher vector representation can be used to efficiently and accurately retrieve visually similar pairs of reference images and frames.

Geometric Verification. The similarity between a pair of reference image and frame can be more reliably confirmed by applying geometric verification method consisting of (1) raw SIFT keypoint matches with ratio test (i.e., the ratio between the distance of the top match and the distance of the second top match), and (2) RANSAC matching with Epipolar geometric verification. However, geometric verification is more computational expensive than calculating similarity using holistic representation. Therefore, we limit the number of geometric verification in our pipeline as described below.

Candidate Reference Images Along the Path. In our application, we assume the reference images densely cover a square region on the map containing the path of the dashcam. Therefore, only a small set of reference images is along the path. We aim to find a set of "candidate reference images" along the path. For each frame, we first use its holistic feature to retrieve the top K similar reference images by considering cosine similarity. Then, we apply geometric verification only on the retrieved images to re-order them according to the number of inlier matches which is referred to as the "confidence score". The reference image with the largest confidence score is considered as the candidate reference image. At this point, we obtain many pairs of frames and their corresponding candidate reference images (green dots in Fig. 5a).

The image matching approach has two drawbacks:

- Candidate reference images might miss some truth reference images along the path (i.e., low recall).
- The relative location of candidate reference images can be inconsistent to the relative location of dashcam "sequence" (i.e., low precision).

Hence, we propose to utilize sequence information to increase (i) the precision by removing inconsistent pairs, and (ii) the recall by introducing more candidate reference images.

Sequence Alignment. Our sequence alignment method utilizes the following information:

- The relative locations of frames in the model 3D coordinate (as described in Sect. 3.2).
- The relative locations of reference images in the world 2D coordinate (i.e., longitude and latitude).

We aim at estimating the transformation between these two coordinates, given the noisy pairs of candidate reference image and frame. We reduce the model 3D coordinate to a 2D coordinate using Principle Component Analysis (PCA), since dashcam trajectory can be well approximated in a 2D coordinate system. Hence, we estimate a 2D rigid transformation (i.e., rotation, translation, and scale) between reduced model 2D coordinate and world 2D coordinate.

We propose a modified RANSAC to estimate the rigid transformation as follow.

- Guided sampling. We sample a pair with probability proportional to $\sqrt[n]{score}$, where *score* is the confidence score of a pair of a reference image and frame, and n is the order of the root. In this way, pairs with higher confidence will be sampled more often.
- Transformation Selection. Instead of using the number of inliers to select the Transformation, we use the sum of confidence scores of the inliers. In this way, transformation that preserves pairs with high confidence as inliers will be selected with a higher chance.

The estimated inliers (green dots in Fig. 5b) typically exclude incorrect pairs which are not consistent with the best alignment result. Hence, the precision typically increase while recall is still low.

Increasing Recall. Given the rigid transformation between two coordinate systems, we can transform all frames into the world coordinate. For any unmatched reference image which contains at least one aligned frame within Q meters, we propose it as a new candidate reference image. In this way, we can recall much more candidate reference images. For each new candidate reference image, we retrieve the top K closest aligned frames and apply "geometric verification" to find the best matched frame. At the end, we retrieve much more pairs, while maintaining the precision (green dots in Fig. 5c).

Quality Prediction. In a few rare cases, our method slightly decrease the performance. We propose the following criteria to automatically decide whether to use our results or not. We use our results only if (i) the ratio between number of inlier over the total number of pairs is above a threshold λ; or (ii) the ratio between the average of the confidence scores of our method over the average of the confidence scores of the baseline method is higher than a threshold γ. When our method is not used, we have the same results as the baseline method (see dots along the diagonal line in Fig. 7).

3.5   Change Detection

Many dashcam videos are captured at different months or years with respect to the reference images. This implies that the street scene might have significant changes. Change detection gives us a potential way to automatically update Street View images by processing many dashcam videos. The global localization method described in Sect. 3.4 has simplified the task of detecting changes between a set of images (Street View) and a video (dashcam) into detecting changes in pairs of matched images (i.e., cropped Street View images and dashcam frames). However, there are more challenges as follows,

Fig. 6. Pipeline of change detection. Panel (a) shows a pair of Street View image (Left) and dashcam frame (Right). Panel (b) shows the Initial reconstruction error. Panel (c) shows the error after removing car interior. Panel (d) shows the error after removing moving objects, sky, and road segments. Panel (e) shows the detected rectangle changed region (white-box) and ground truth changed region (green-box). (Color figure online)

- Street View images and dashcam videos are typically captured from notably different viewpoints (Fig. 6(a)).
- Dashcam videos typically capture the interior of the car (Fig. 6(a)-Right).
- Changes often are introduced by many moving objects (e.g., cars) and different weather conditions (Fig. 6(a)).

Our method aims to overcome these challenges and detect changed rectangle regions by applying the following pipeline.

Robust Pixel-wise Matches. To overcome the differences in viewpoints, we apply an efficient dense SIFT matching method [3] to find the best match of every dashcam pixel in the Street View image. By using the matched Street View pixels to reconstruct a dashcam frame, we can compute the pixel-wise reconstruction error (Fig. 6(b)). A high reconstruction error implies a potential change between the dashcam frame and the Street View image.

Identify Car Interior. Since the car interior typically differs from the Street View scene, we apply the same method to find the reverse match (i.e., best match of every Street View pixel in the dashcam frame.). Then, we consider every unmatched dashcam pixel as car interior and assign zero reconstruction error for these dashcam pixels (Fig. 6(c)).

Moving Objects, Sky, and Road. We further use state-of-the-art semantic segmentation methods [27,28] to estimate foreground (e.g., car, motorbike, people, etc.), sky and road segments, respectively. We assign zero reconstruction error for these segments in the dashcam frame (Fig. 6(d)). Note that we give special treatment for sky and road, since the appearance of sky and road changes significantly in different weather conditions.

Changed Rectangle Detection. Since structural changes in street scene typically corresponds to a new billboard, a new building, etc. We use candidate object hypotheses [29] as change candidates. In order to rank each candidate, we first convert reconstruction error into binary change map using the $T\%$ percentile of the whole reconstruction error as threshold (i.e., -1 for pixels with error $<$ threshold(T) and $+1$ for pixels with error $\geq$ threshold(T)). Then, we rank the candidates according to the accumulated values of the binary change map within a candidate object hypothesis (white-box in Fig. 6(e)).

3.6   Implementation Detail

Localization. We set K to 15, since the performance gain for $K > 15$ is not significant. We set Q to 10 m which is twice the 5 m error metric (Sect. 4) in order to recall most of the missing candidate reference images. We set $n = 3, \lambda = 0.2, \gamma = 0.4$ and we show in Sect. 4 that the performance is not sensitive to these three parameters.

Change Detection. For our method, we set $T = 90$ and obtain reasonable results as in Fig. 6(e).

4   Experiment Results

We evaluate our proposed dashcam localization and change detection methods on a subset of our harvest data in the wild with manually labeled ground truth locations and changes.

4.1   Dashcam Localization

In order to quantitatively evaluate our method's performance, we select 45 high quality videos captured as early as on Jun, 2012 and as late as on May 2015. For each video, we label the best matched frame corresponding to each reference image on the path of the dashcam.

Evaluation Metric. Given a few manually labeled pairs of reference images and matched frames, we align all frames to the longitude and latitude coordinate using alignment method described in Sect. 3.4. In this way, each frame is assigned with a longitude and latitude coordinate, which is treated as its ground truth longitude and latitude. We also obtain the ground truth reference images along the path of dashcam. Note that longitude and latitude coordinate can be converted to metric units such as meters. Now, given a predicted pair of reference image and frame, we can calculate the distance in meters between the predicted reference image and the ground truth location of the predicted frame. We further evaluate the precision and recall of the predicted pairs as follows. We consider a predicted pair is correct if the following two conditions are both true: (1) the predicted reference image is in the set of ground truth reference images; (2) the predicted reference location is within 5 m from the ground truth frame location. The precision is the percentage of correctly predicted pairs among all the predicted pairs. The recall is the percentage of ground true reference images which are retrieved by the correct prediction. Note that we only allow each reference image to exist once in all predicted pairs so that there won't be multiple predicted reference images corresponding to the same frame. Moreover, since each predicted pair is associated to a confidence score, we can remove pairs with scores below a threshold

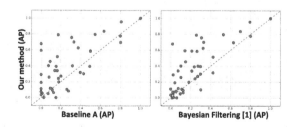

Fig. 7. Average Precision (AP) comparison between our method and baseline methods: a baseline image matching (Baseline A) and a baseline Bayesian filtering [1].

to control the number of predicted pairs. In this way, we can calculate a precision v.s. recall curve for each video sequence. The mean Average Precision (mAP) over all videos is used for comparison.

Sequence Alignment v.s. Baseline Methods. Figure 7 shows two scatter plots of AP comparison between our sequence alignment method (y-axis) and two baseline methods (x-axis): baseline image matching (Baseline A) and baseline Bayesian filtering, respectively. Our method achieves better AP for most of the videos (i.e., dots scatter in upper-left triangle). Note that the baseline Bayesian filtering is our implementation of [1]. Our proposed alignment method (37% mAP) significantly outperforms the baseline A (23.4% mAP) and Bayesian filtering (24.7% mAP). Qualitative results of our method in Fig. 8 shows that we can reliably align dashcam locations onto longitude and latitude coordinate. The 2nd row example in Fig. 8 even shows that our method can handle Street View images captured in bad weather condition.

Fig. 8. Our qualitative results. For each row, we first show alignment result, where the cyan, blue, red, and yellow dots represent all reference images, matched reference images, ground truth aligned frames, and predicted dashcam frames, respectively. Then, we show pairs of reference images (left) and predicted frames (right). Note that our method obtains impressive results even when Street View images are captured in bad weather condition (2nd row). (Color figure online)

Sensitivity Analysis. There are three main parameters involved in our system: the nth root, and thresholds λ and γ. For each parameter, we test 10 different values while fixing all other parameters. In Fig. 9, we show that our performance is not sensitive to the value changes when $\lambda < 0.4$ and $\gamma < 0.5$.

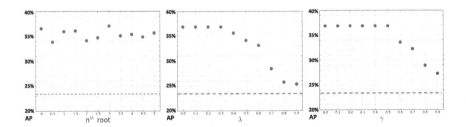

Fig. 9. Sensitivity Analysis: we evaluate 10 different values for parameters n^{th} root, λ, and γ as shown in the left, middle, and right panels, respectively. The blue dash-line indicates the "baseline A" performance (23.4% mAP). When $\lambda < 0.4$ and $\gamma < 0.5$, the performance of our method is stable and consistently better than the baseline. (Color figure online)

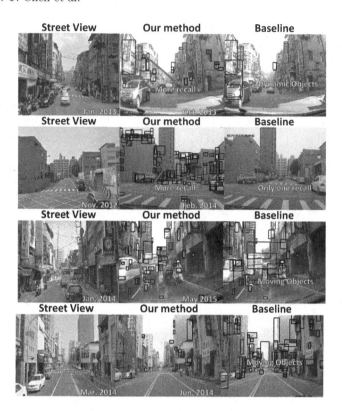

Fig. 10. Change Detection Results on dashcam videos. For each example (three images in a row), we show the Street View image, the detected changes of our method, and the detected changes of the baseline, from left to right, respectively. The month and year of the scene is overlaid on each image. The ground truth changes are marked in blue. True positive predicted changes are marked in red. False positive predicted changes are marked in black. We show the predicted changes from high confidence to low confidence up the one which gives the maximum recall. Hence, the more recalled true positive, the likely more false positive shown. (Color figure online)

4.2   Change Detection

Among the 45 videos, we manually confirm that there are 40 predicted pairs of Street View images and frames containing "long-term" changes. Then, we manually annotate ground truth changed rectangle regions in 40 predicted frames. We use the ground truth to evaluate the change detection performance. We consider a predicted rectangle region as a truth change if it overlaps[2] with a ground truth region more than 30%[3]. Our full method (13.54%) outperforms a baseline method without handling car interior and moving objects (11.70%) by 1.84% (relatively 13.6%) in mean AP. Typical examples

---

[2] We calculate the intersection over union area.

[3] 30% is used, since ground truth changes are typical irregular and possibly consists of more than one objects. It is very challenging to precisely detect the ground truth change rectangle.

Fig. 11. Change Detection Results on Time-machine images. For each example (two images in a row), we show the Street View image, the detected changes of our method, from left to right, respectively. The month and year of the scene is overlaid on each image. The predicted change and ground truth changes are visualized the same way as in Fig. 10. (Color figure online)

are shown in Fig. 10. By inspecting failure examples, we found that failures are typically due to severe viewpoint difference between Street View images and dashcam frames, or less ideal dashcam frame quality. Hence, we further conduct a controlled experiment using manually aligned high quality Street View images.

Controlled experiment. In order to directly evaluate the performance of change detection (not effected by the quality of alignment), we collected 192 aligned pairs of StreetView images from StreetView Time-machine. Images at the same geo-location in StreetView Time-machine are typically captured at different years. Hence, they are ideal to benchmark "long-term" change detection performance. The ground truth change rectangles are manually labeled for evaluation. Our proposed method (45.57%) outperforms the baseline method (43.67%) in mean AP. Typical examples are shown in Fig. 11. We believe that the overall performance is increased due to the ideal image alignment and much better image quality from Street View than dashcam frames.

5   Conclusion

We demonstrate that consumer-level dashcam videos in the wild can be used to identify changes in high quality Google Street View images. We propose a robust sequence alignment method which overcomes the challenges caused by significant structural changes in the scene, and outperforms the baseline method [1] by 12% mean AP. We also propose a novel change detection method especially designed to detect long-term changes and handle properties in dashcam videos. Our method (13.54%) outperforms a baseline method without handling car interior and moving objects (11.70%) by 1.84% (relatively 13.6%) in mean AP. Moreover, given manually aligned high quality Street View images, our method achieves a significant higher mean AP 45.75%. This suggest that good alignment and image quality are important to reliably detect "long-term" changes. In the future, we will focus on improving alignment quality and extending the change detection method to jointly consider all frames in a video sequence.

Acknowledgement. We thank Industrial Technology Research Institute (ITRI) project grants and MOST 103-2218-E-007-025 and MOST 104-3115-E-007-005 in Taiwan for their support.

References

1. Vaca-Castano, G., Zamir, A., Shah, M.: City scale geo-spatial trajectory estimation of a moving camera. In: CVPR (2012)
2. Badino, H., Huber, D., Kanade, T.: Real-time topometric localization. In: ICRA (2012)
3. Kim, J., Liu, C., Sha, F., Grauman, K.: Deformable spatial pyramid matching for fast dense correspondences. In: CVPR (2013)
4. Robertson, D., Cipolla, R.: An image-based system for urban navigation. In: BMVC (2004)
5. Zhang, W., Kosecka, J.: Image based localization in urban environments. In: 3DPVT (2006)
6. Schindler, G., Brown, M., Szeliski, R.: City-scale location recognition. In: CVPR (2007)
7. Hays, J., Efros, A.A.: Im2gps: estimating geographic information from a single image. In: CVPR (2008)
8. Zamir, A.R., Shah, M.: Accurate image localization based on google maps street view. In: Daniilidis, K., Maragos, P., Paragios, N. (eds.) ECCV 2010. LNCS, vol. 6314, pp. 255–268. Springer, Heidelberg (2010). doi:10.1007/978-3-642-15561-1_19
9. Cao, S., Snavely, N.: Graph-based discriminative learning for location recognition. In: CVPR (2013)
10. Bettadapura, V., Essa, I., Pantofaru, C.: Egocentric field-of-view localization using first-person point-of-view devices. In: WACV (2015)
11. Irschara, A., Zach, C., Frahm, J., Bischof, H.: From structure-from-motion point clouds to fast location recognition. In: CVPR (2009)
12. Li, Y., Snavely, N., Huttenlocher, D.P.: Location recognition using prioritized feature matching. In: Daniilidis, K., Maragos, P., Paragios, N. (eds.) ECCV 2010. LNCS, vol. 6312, pp. 791–804. Springer, Heidelberg (2010). doi:10.1007/978-3-642-15552-9_57
13. Sattler, T., Leibe, B., Kobbelt, L.: Fast image-based localization using direct 2d-to-3d matching. In: ICCV (2011)
14. Li, Y., Snavely, N., Huttenlocher, D., Fua, P.: Worldwide pose estimation using 3D point clouds. In: Fitzgibbon, A., Lazebnik, S., Perona, P., Sato, Y., Schmid, C. (eds.) ECCV 2012. LNCS, vol. 7572, pp. 15–29. Springer, Heidelberg (2012). doi:10.1007/978-3-642-33718-5_2

15. Taneja, A., Ballan, L., Pollefeys, M.: Never get lost again: vision based navigation using streetview images. In: Cremers, D., Reid, I., Saito, H., Yang, M.-H. (eds.) ACCV 2014. LNCS, vol. 9007, pp. 99–114. Springer, Heidelberg (2015). doi:10.1007/978-3-319-16814-2_7

16. Lategahn, H., Schreiber, M., Ziegler, J., Stiller, C.: Urban localization with camera and inertial measurement unit. In: Intelligent Vehicles Symposium (IV) (2013)

17. Floros, G., van der Zander, B., Leibe, B.: OpenStreetSLAM: global vehicle localization using openstreetmaps. In: ICRA (2013)

18. Brubaker, M., Geiger, A., Urtasun, R.: Lost! leveraging the crowd for probabilistic visual self-localization. In: CVPR (2013)

19. Radke, R., Andra, S., Al-Kofahi, O., Roysam, B.: Image change detection algorithms: a systematic survey. TIP **14**, 294–307 (2005)

20. Pollard, T., Mundy, J.: Change detection in a 3-d world. In: CVPR (2007)

21. Taneja, A., Ballan, L., Pollefeys, M.: City-scale change detection in cadastral 3d models using images. In: CVPR (2013)

22. Matzen, K., Snavely, N.: Scene chronology. In: Fleet, D., Pajdla, T., Schiele, B., Tuytelaars, T. (eds.) ECCV 2014. LNCS, vol. 8695, pp. 615–630. Springer, Heidelberg (2014). doi:10.1007/978-3-319-10584-0_40

23. Wu, C.: Towards linear-time incremental structure from motion. In: 3DV (2013)

24. Wu, C.: Visualsfm: a visual structure from motion system. http://ccwu.me/vsfm/

25. Lowe, D.G.: Distinctive image features from scale-invariant keypoints. Int. J. Comput. Vis. **60**, 91–110 (2004)

26. Jegou, H., Perronnin, F., Douze, M., Snchez, J., Perez, P., Schmid, C.: Aggregating local image descriptors into compact codes. TPAMI **34**, 1704–1716 (2011)

27. Zheng, S., Jayasumana, S., Romera-Paredes, B., Vineet, V., Su, Z., Du, D., Huang, C., Torr, P.: Conditional random fields as recurrent neural networks. In: ICCV (2015)

28. Long, J., Shelhamer, E., Darrell, T.: Fully convolutional networks for semantic segmentation. In: CVPR (2015)

29. Zitnick, C.L., Dollár, P.: Edge boxes: locating object proposals from edges. In: Fleet, D., Pajdla, T., Schiele, B., Tuytelaars, T. (eds.) ECCV 2014. LNCS, vol. 8693, pp. 391–405. Springer, Heidelberg (2014). doi:10.1007/978-3-319-10602-1_26

Pedestrian and Vehicle Detection and Tracking with Object-Driven Vanishing Line Estimation

Yi-Ming Chan[1]([⊠]), Li-Chen Fu[1], Pei-Yung Hsiao[2], and Shin-Shinh Huang[3]

[1] Computer Science and Information Engineering, National Taiwan University,
Taipei City, Taiwan
iming@sgi.csie.ntu.edu.tw
[2] Electrical Engineering, National University of Kaohsiung, Kaohsiung, Taiwan
[3] Computer and Communication Engineering,
National Kaohsiung First University of Science and Technology,
Kaohsiung, Taiwan

Abstract. To robustly detect people and vehicle on the road in a video sequence is a challenging problem. Most researches focus on detecting or tracking of specific targets only. On the contrary, instead of detecting vehicle or pedestrian individually, an integration framework combining the geometric information is proposed. The camera's pitch angle is estimated with a novel vanishing line estimator. Not only detecting the vanishing point using line intersection approach, but also the object information from tracker are considered. Specifically, the detected vehicle or pedestrian will cast votes for the hypothesized horizon line. The vanishing line can be estimated even when the scenes are cluttered or crowded, and thus the geometric information can be estimated under challenging circumstance. In turn, such information of scene can help the system refine our detection results through Bayes' network. Finally, to verify the performance of the system, comprehensive experiments have been conducted with the KITTI dataset. It is quite promising that the state-of-the-art detector, in our case, Regionlet detector, can be improved.

1 Introduction

Driver assistance systems are getting popular due to the advance of the sensing and recognition technologies. One of the most highly active research areas-computer vision-is getting ready for the commercial products. A world leading company in automotive business, Volvo, even has deployed vision based technologies, such as lane departure warning, road sign recognition, and pedestrian detection, into their high-end vehicles. Although the company has also implemented pedestrian and cyclist detection system using multiple sensors including camera and radar, it remains extremely challenging for building a pedestrian or vehicle detection system with camera only.

The traditional object detectors focus on the discriminability of one class of objects from their background might fail when other type of objects appear. For our interested pedestrian detection system, to improve its robustness, we

© Springer International Publishing AG 2017
C.-S. Chen et al. (Eds.): ACCV 2016 Workshops, Part I, LNCS 10116, pp. 436–451, 2017.
DOI: 10.1007/978-3-319-54407-6_29

Fig. 1. Vanishing line and the object detection in crowded scene. The vanishing line can be detected with the help of the object context. Traditional vanishing point detector tends to fail in this case due to the lack of major line intersection point. The red line shows the result of the estimated horizon using object "voting". (Color figure online)

might thus incorporate contextual information, including lane boundaries and camera's pitch angle. In this regard, Hoiem *et al.* [1] have introduced an unified framework to model the geometry information among object detectors, where the contextual relationships among objects were modeled within 3D world instead of 2D image plane. Nieto *et al.* [2] have modeled the camera configuration and image perspective transform for the lane and vehicle tracking. Geiger *et al.* [3] have modeled the road scene as the contextual information to improve the tracking results. These works all rely on the accuracy of stochastic "guess" of the camera pose; if there is a correct guess of the camera's pitch angle among all random guesses, the detector and tracker will surely perform better.

Unlike traditional vanishing point estimation algorithms which focus on how to infer major lines on the plane and estimate the intersection points, Richardson *et al.* [4] have proposed a novel approach of inferring the scene geometry from moving object. In a surveillance video, the moving objects (vehicles) will move along the directions of the lane markings on the ground plane. Thus, the object trajectories will intersect at the major vanishing point. Nevertheless, it is not the case in the on-road driving scenarios, which will be explained below. These approaches suffer from the noise of the line feature (edge pixels) and cannot work under the situation of absence of dominant lines. Figure 1 shows an example of the failure scene. Bad conditions in the clutter scene, such as that with strong random edge pixels on the ground from tree shadows, or one lane boundary occluded by heavy traffic, may lead to failure of vanishing point detection. Thus the horizon line and camera pitch angle estimation will also fail.

To cope with it, a new method to generate vanishing point hypothesis using the results of object detection is proposed. The vanishing point is estimated using the size and position information of the detected pedestrians and vehicles in the sequence of images. With the prior knowledge of the real world size of the objects, the vanishing point can be estimated reasonably precise. In Fig. 1, the vanishing line can be estimated using the information of the vehicles.

The main contributions of this paper are two folds. First, the vanishing points are estimated using not only raw image feature but also object information. On the other hand, such integrated information also helps to improve accuracy of estimating camera's pitch angle. Both together makes the overall system more

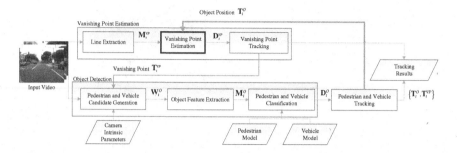

Fig. 2. The system overview. The track of vanishing point can provide the estimation of camera pitch angle, which is a variable external parameter in on-road driving scenarios. The candidate windows of the objects can be derived using the camera internal and external parameters and flat ground assumption. It turns out that the track of object can also help to improve the vanishing point estimation, since objects share the same ground plane on which the major vanishing point lies.

robust to the environmental variations. Second, a Bayes' network system aims at detecting and tracking pedestrians and vehicles while incorporating the estimated vanishing point information is proposed. Generally speaking, by employing appropriate trackers to encode the temporal information, the system tends to be more robust to the clutter background.

The system overview is shown in Fig. 2. The lower part shows the part of pedestrian and vehicle detection. The track of vanishing point $\mathbf{T}_t^{VP}$ until time t facilitates the generation of candidate widows $\mathbf{W}_t^O$ where the extracted features $\mathbf{M}_t^O$ will be taken into account for distinguishing pedestrians from vehicles. The pedestrian and vehicle tracking module will estimate the tracking results of objects $\mathbf{T}_t^O$, and in turn such tracked object position $\mathbf{T}_t^O$ will be used for the vanishing point detection $\mathbf{D}_t^{VP}$. On the other hand, the upper part shows the procedures of vanishing point extraction. Long lines in the images are extracted as the features of the vanishing point. Unlike the traditional work, the vanishing point is estimated using both line feature and object positions. Thus, the detected vanishing point will be then tracked in the tracking module.

The rest of the paper is organized as follows. The related work is discussed in the Sect. 2. Section 3 shows the scene modeling for the on-road applications. Vanishing point estimation using object information is introduced in Sect. 4. Object detection is described in Sect. 5. Sections 6 and 7 are the experiments and the conclusion, respectively.

2   Related Work

To detect pedestrians on the road, discriminative model is getting popular since the system performance can be improved using the model but at no price of slowing down the system. One of the most successful approaches is the histograms of oriented gradients (HOG) [5] with Support Vector Machine (SVM).

Nevertheless, when the on-road scenes [6] are tested, problems with unsatisfactory performance in face of clutter background or various gestures still remain unsolved.

Ess *et al.* [7] have considered Tracking-by-Detection tracker with camera pose estimation at the same time. Wojek *et al.* [8] have modeled this 3D scene context across multiple frames with Hidden Markov Model. Sudowe and Leibe [9] have improved the efficiency of the sliding window detector. In their case, the ground plane is assumed to be fixed and can be estimated offline. The geometric information is taken as the parameter to constraint the sliding window area to speed up the detection. On the contrary, our method tries to derive the relationship between object and the ground plane using vanishing point estimation.

Most researches [3,10] detect the main vanishing point using the intersection point of the major lines. There are many research teams have developed vanishing point estimation systems [4,10–15]. These techniques can be categorized into two folds, RANSAC-based or voting based. The RANSAC-based methods [10, 13–15] uses statistics techniques to sample possible major parallel lines, then estimate the vanishing point by the intersection points. The time complexity of RANSAC-based vanishing point algorithms thus is low and can run at real-time. In the situations of noisy road or absent of major lines, however, these algorithms may generate a wrong horizon. Voting-based methods [12,16,17] considering all possible lines in the voting space. The voting space can be Hough space or parallel coordinate space [17]. To exhaustively vote the candidates, the computational complexity of these types of vanishing point methods is usually high.

Visual tracking is important in a robust object detection system. In the field of visual tracking, the target is usually assumed to be detected before tracking. The strong assumption on successful initialization with detected object is now realized by the advanced object detectors. This is the reason why pedestrian tracking nowadays becomes more and more popular after the HOG feature based detectors is successfully applied. To track pedestrians using generative model [18,19], which generally is a reference model using histograms, is built a priori for the target. Relative to the object detection task, such reference model is usually too weak to differentiate the target from various backgrounds robustly, which is normally faced during the tracking process. On the contrary, tracking-by-detection approaches [7,20–23] are more robust to various kinds of environment by using sophisticated object detectors, where the tracker takes the advantages of various object detectors to solve the data association problem.

3   Scene Modeling and Extracting Using Bayes' Network

3.1   Object Tracking with Scene Modeling

Inspired by Geiger *et al.* [3], the contextual information including object positions, the ground plane information, and the temporal information are modeled in our work. In the scenarios of on-road driving, the contextual information can be derived from the existence of objects and the road plane geometry. The geometry context from the major vanishing point thus can help the system improve

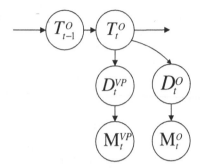

Fig. 3. The Bayes' network of the scene model. $\mathbf{T}_t^O$ stands for the object tracking results at time instance t. $\mathbf{M}_t^{VP}$, $\mathbf{M}_t^O$ are the measurements of the vanishing point, objects, respectively.

the detection and tracking results. Based on the assumptions that the plane is flat and the intrinsic parameters of camera are known, the road plane geometry can be derived using the major vanishing point from the intersection point of the lane markings or road boundaries. Since pedestrian or vehicles will appear on the ground plane in normal driving situations, the obtained road plane geometry can be utilized to determine the physical size and position of the detected objects. The objects with unreasonable size can be found and removed.

Instead of modeling geometric context and object context separately, an integrated system is proposed. This contextual information is modeled by tracks of the major vanishing point, pedestrians and vehicles on road. In our model, there are two different sources of contextual information, one from object and the other from the vanishing point. To model the tracking result of the object, the track of vehicle $\mathbf{T}_t^{O^v}$ or pedestrian $\mathbf{T}_t^{O^p}$ depend on the previous tracks $\mathbf{T}_{t-1}^{O^v}$ or $\mathbf{T}_{t-1}^{O^p}$ and the detection of vehicle $\mathbf{D}_t^{O^v}$ or pedestrian $\mathbf{D}_t^{O^p}$, and the detection of vanishing point $\mathbf{D}_t^{VP}$. The relationship between vanishing point and object location is strong and important. To take the advantage of the knowledge of vanishing point and object locations, the track of vanishing point $\mathbf{T}_t^{VP}$ is modeled using the previous track $\mathbf{T}_{t-1}^{VP}$ and the detected vehicle $\mathbf{D}_t^{O^v}$ and pedestrian $\mathbf{D}_t^{O^p}$, and the current vanishing point $\mathbf{D}_t^{VP}$. With the model of the strong contextual information from objects (vehicles or pedestrians), the robustness of the vanishing point estimation can be improved even when the background is cluttered. By considering both the object information and the vanishing point information, the graphical model of the system can be shown as Fig. 3.

The integration system with tracking can be modeled using Tracking-by-Detection techniques as follows. Given a set of detection results $\mathbf{D}_{1:t}^O$ and $\mathbf{D}_{1:t}^{VP}$, let their associated tracks (tracked targets, *i.e.*, pedestrian or vehicles) at time $t-1$ be denoted as $\mathbf{T}_{t-1}^O$, and we then construct a set of tracks $\mathbf{T}_t^O$ associated with the sequence of measurements $\mathbf{M}_{1:t}$ to maximize the posterior probability. By assuming the independence among detection results at time t and $t-1$

(that is, the detection results can be viewed as "logical" measurements), the posterior probability can be further derived as:

$$p(\mathbf{T}_t^O|\mathbf{D}_{1:t}^O, \mathbf{D}_{1:t}^{VP}, \mathbf{M}_{1:t}) = \alpha \cdot p(\mathbf{D}_t^O, \mathbf{D}_t^{VP}, \mathbf{M}_t|\mathbf{T}_t^O) \cdot p(\mathbf{T}_t^O|\mathbf{D}_{1:t-1}^O, \mathbf{D}_{1:t-1}^{VP}, \mathbf{M}_{1:t-1}) \ (1)$$

where α is a normalization constant. The rightmost term of (1) can be further derived as

$$p(\mathbf{T}_t^O|\mathbf{D}_{1:t-1}^O, \mathbf{D}_{1:t-1}^{VP}, \mathbf{M}_{1:t-1}) =$$

$$\int \underbrace{p(\mathbf{T}_t^O|\mathbf{T}_{t-1}^O)}_{\text{transition \quad model}} \cdot \underbrace{p(\mathbf{T}_{t-1}^O|\mathbf{D}_{1:t-1}^O, \mathbf{D}_{1:t-1}^{VP}, \mathbf{M}_{1:t-1})}_{\text{previous \quad posterior}} \cdot d\mathbf{T}_{t-1}^O \,(2)$$

where $p(\mathbf{T}_t^O|\mathbf{T}_{t-1}^O)$ is the transition probability of tracks. In our case, the transition probability of tracks $p(\mathbf{T}_t^O|\mathbf{T}_{t-1}^O)$ is estimated using constant velocity model. The measurement $\mathbf{M}_t = \{\mathbf{M}_t^O, \mathbf{M}_t^{VP}\}$ is a set of feature vectors for objects and vanishing point. The features can be HOG, Canny edges for example. The goal of the tracking is equivalent to searching the candidate with the maximum prior and likelihood probability in (1).

By using Naïve Bayes' technique, i.e., assuming the conditional independence among the detection results of objects, the vanishing point, and the measurements given the track of target objects, the likelihood term can be further derived as shown below:

$$p(\mathbf{D}_t^O, \mathbf{D}_t^{VP}, \mathbf{M}_t|\mathbf{T}_t^O) = p(\mathbf{D}_t^O|\mathbf{T}_t^O)p(\mathbf{D}_t^{VP}|\mathbf{T}_t^O)p(\mathbf{M}_t^O|\mathbf{T}_t^O)p(\mathbf{M}_t^{VP}|\mathbf{T}_t^O) \quad (3)$$

Although the conditional independence may not hold in some applications, this assumption has been made in many applications [24,25]. Such simplification does yield acceptable results even when the independence hypothesis fails, because the dependences cancel out in some cases. Kittler et al. [24] argue that in many application, this assumption provides a plausible approximation.

By (3) and Fig. 3, the likelihood term can be therefore derived as follows:

$$p(\mathbf{D}_t^O, \mathbf{D}_t^{VP}, \mathbf{M}_t|\mathbf{T}_t^O) \propto p(\mathbf{D}_t^O|\mathbf{T}_t^O)p(\mathbf{D}_t^{VP}|\mathbf{T}_t^O)p(\mathbf{M}_t^O|\mathbf{D}_t^O)p(\mathbf{M}_t^{VP}|\mathbf{D}_t^{VP}) \quad (4)$$

where $p(\mathbf{D}_t^O|\mathbf{T}_t^O)$ and $p(\mathbf{D}_t^{VP}|\mathbf{T}_t^O)$ are the confidence of detectors given the current track, whereas $p(\mathbf{M}_t^O|\mathbf{D}_t^O)$ and $p(\mathbf{M}_t^{VP}|\mathbf{D}_t^{VP})$ are the likelihood. To reduce the time complexity (4) is evaluated for each individual target, i.e., a pedestrian, vehicle or vanishing point. To find good hypotheses of tracks, the particle filter [26] is used, and the weight of each particle is assigned according to (4). The details of each term will be discussed in the following sections.

Note that $p(\mathbf{D}_t^O|\mathbf{T}_t^O)$ and $p(\mathbf{D}_t^O|\mathbf{T}_t^{VP})$ are modeled as Gaussian functions with their means equal to the respective associated detection results. Thus, if the detection state vector coincides with the tracking particle, its weight will be high.

$$p(\mathbf{D}_t^O|\mathbf{T}_t^O) \propto N(\left\| \mathbf{D}_t^O - \mathbf{T}_t^O \right\|, \sigma^O) \quad (5)$$

For each particle, the shorter the distance to the associated detection, the higher weight the particle will possess. The standard deviation of the object is denoted as σ^o. In this case, only the track and detection results from the same class are compared. The $p(\mathbf{D}_t^{VP}|\mathbf{T}_t^O)$ is modeled using the Gaussian function for the physical size constrain of target. The details will be discussed in Sect. 4.

On the other hand, $p(\mathbf{M}_t^O|\mathbf{D}_t^O)$ and $p(\mathbf{M}_t^{VP}|\mathbf{D}_t^{VP})$ model the likelihood. In particular, $p(\mathbf{M}_t^O|\mathbf{D}_t^O)$ is modeled using a sigmoid function with detector score S from object detector:

$$p(\mathbf{M}_t^O|\mathbf{D}_t^O) \propto \frac{1}{1+e^{-S}} \tag{6}$$

Please note that, given a tracking target, only the associated detection result is considered in our case. $p(\mathbf{M}_t^{VP}|\mathbf{D}_t^{VP})$ models the likelihood of the vanishing point. The details of the model will be discussed in the following section.

4   Vanishing Point Estimation with Object Information

4.1   Vanishing Point Detection

For on-road driving scenarios, we estimate the vanishing point using RANSAC-based detection as well as tracking technique. The measurement vector of the vanishing points $\mathbf{M}_t^{VP}$ is the set of lines found by progressive probabilistic Hough transform [27]. Then, the vanishing points $\mathbf{D}_t^{VP}$ are detected using the approach from Nieto and Salgado [15]. The likelihood $p(\mathbf{M}_t^{VP}|\mathbf{D}_t^{VP})$ is modeled using the ratio of edge pixels of the line and the line length. The more edge pixels on the line, the higher the likelihood. The details of the equation can be shown as follows:

$$p(\mathbf{M}_t^{VP}|\mathbf{D}_t^{VP}) \propto \frac{EdgePixels}{LineLength} \tag{7}$$

where the *EdgePixels* is the number of edge pixels over the detected line pair, and the *LineLength* is the total length along with the detected line pair from the vanishing point to the bottom of images.

4.2   The Probability of Vanishing Point Given Tracked Objects

Different depths of an object result in different sizes of the corresponding 2-D object images on the 2-D image plane. With the information of the camera parameters and the flat ground plane assumption, the possible 2D windows can be confined into certain physical region on the ground. The relationship of vehicle and the camera coordinates are shown in Fig. 4.

Using pinhole camera model, the foot of the pedestrian will be projected to the line at the principal point whose V-coordinate is equal to v_0. Note that the vanishing line of the ground plane will be the parallel line pass through v_{VP}.

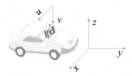

Fig. 4. The relationship of the vehicle coordinate and camera coordinate. The vehicle coordinate (3D) is aligned with the ground plane, whose origin is directly below the principal point of the camera.

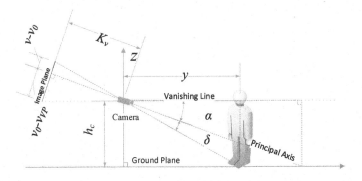

Fig. 5. The general case of side view of a pedestrian standing on the ground plane.

Let the focal length of the camera in pixel be denoted as K_v. Thus, given the camera height h_c, the distance y can be derived as follows:

$$\tan \alpha = \frac{v_0 - v_{VP}}{K_v} \tag{8}$$

$$y = \frac{h_c}{\tan(\alpha + \delta)}, \tan \delta = \frac{v - v_0}{K_v} \tag{9}$$

Figure 5 illustrates the variable definition in a schematic form for a general case. Using (8) and (9), the distance of the object can be estimated as follows:

$$y = \frac{K_v h_c - (v - v_0)h_c \tan \alpha}{(v - v_0) + K_v \tan \alpha} \tag{10}$$

where is the tilt angle of the camera, h_c is the height of the camera, K_u, K_v are scaling factors for U-axis and V-axis of the camera, (x, y) is the position of the vehicle coordinate, (u, v) is the position in the image coordinate, (u_0, v_0) is the principal point of the camera in the image coordinate.

Similarly, the lateral offset of an object on the ground plane can be estimated as follows:

$$x = \frac{y(u - u_0)}{K_u} \tag{11}$$

According to (11), the lateral and longitudinal distances of a pedestrian can be evaluated with the given camera parameters. Accordingly, the physical width of a pedestrian, whose image is enclosed in a bounding box with its bottom aligned with the image segment from (u_l, v) to (u_r, v), is derived as follows:

$$P_{width} = x_r - x_l = \frac{y(u_r - u_0)}{K_u} - \frac{y(u_l - u_0)}{K_u} = \frac{y(u_r - u_l)}{K_u} \tag{12}$$

$$(u_r - u_l) = \frac{K_u \cdot P_{width}}{y} \tag{13}$$

With (8), (10) and (12) and the physical world width, the vanishing point will be

$$\begin{cases} y = \frac{K_v \cdot h_c - (v - v_0) \cdot h_c \cdot \tan \alpha}{(v - v_0) + K_v \cdot \tan \alpha} \\ \tan \alpha = \frac{(v_0 - v_{VP})}{K_v} \end{cases} \tag{14}$$

or more elaborately

$$v_{VP} = v_0 - \frac{K_v^2 h_c - (v - v_0) \cdot K_v \cdot K_u \cdot P_{width} / (u_r - u_l)}{(v - v_0) h_c + K_v \cdot K_u \cdot P_{width} / (u_r - u_l)} \tag{15}$$

Please note that each tracked object will have a range of physical width P_{width} corresponding to the object class, *i.e.*, either pedestrian or vehicle. With the prior knowledge of the range of the physical size of tracked objects (*e.g.*, 1.8 m 2.5 m width for vehicle), the possible horizon line can be derived using (15). The v coordinate estimated using the mean of the object width is denoted as μ^{VP^v}. Then, the probability of the vanishing point given tracked objects, $p(\mathbf{D}_t^{VP} | \mathbf{T}_t^O)$ is computed using a zero mean Gaussian function whose variance equals to σ^{O^v} (depend on the object class) with the difference of the v coordinate of the vanishing point $\mathbf{D}_t^{VP^v}$ and the mean of vanishing line μ^{VP^v} as input:

$$p(\mathbf{D}_t^{VP} | \mathbf{T}_t^O) \propto N(D_t^{VP^v} - \mu^{VP^v}, \sigma^{O^v}) \tag{16}$$

4.3   Vanishing Point Tracking with Object Information

In order to take advantage of the object information in the process of vanishing point estimation, the tracks of objects are considered. Instead of using the trajectories of moving objects, the size and location are taken as features of vanishing point tracking.

Similar to the derivation of track of objects, the posterior of the vanishing point is

$$p(\mathbf{T}_t^{VP} | \mathbf{D}_{1:t}^O, \mathbf{D}_{1:t}^{VP}, \mathbf{M}_{1:t}) = \\ \alpha \cdot p(\mathbf{D}_t^O, \mathbf{D}_t^{VP}, \mathbf{M}_t | \mathbf{T}_t^{VP}) \cdot p(\mathbf{T}_t^{VP} | \mathbf{D}_{1:t-1}^O, \mathbf{D}_{1:t-1}^{VP}, \mathbf{M}_{1:t-1}) \tag{17}$$

where $p(\mathbf{T}_t^{VP} | \mathbf{D}_{1:t-1}^O, \mathbf{D}_{1:t-1}^{VP}, \mathbf{M}_{1:t-1}) =$
$$\int p(\mathbf{T}_t^{VP} | \mathbf{T}_{t-1}^{VP}) \cdot p(\mathbf{T}_{t-1}^{VP} | \mathbf{D}_{1:t-1}^O, \mathbf{D}_{1:t-1}^{VP}, \mathbf{M}_{1:t-1}) \cdot d\mathbf{T}_{t-1}^{VP} \tag{18}$$

which can be estimated using constant velocity model of the vanishing point from previous estimation.

Similar to the (4), by assuming the conditional assumptions and the dependency of Fig. 3, the likelihood term $p(\mathbf{D}_t^O, \mathbf{D}_t^{VP}, \mathbf{M}_t | \mathbf{T}_t^{VP})$ can be further derived as follow:

$$p(\mathbf{D}_t^O, \mathbf{D}_t^{VP}, \mathbf{M}_t | \mathbf{T}_t^{VP}) \propto p(\mathbf{D}_t^O | \mathbf{T}_t^{VP}) p(\mathbf{D}_t^{VP} | \mathbf{T}_t^{VP}) p(\mathbf{M}_t^O | \mathbf{D}_t^O) p(\mathbf{M}_t^{VP} | \mathbf{D}_t^{VP}) \quad (19)$$

As shown in (19), $p(\mathbf{D}_t^O | \mathbf{T}_t^{VP})$ and $p(\mathbf{D}_t^{VP} | \mathbf{T}_t^{VP})$ are object and vanishing point probability, respectively, given the tracked vanishing point. In particular, $p(\mathbf{D}_t^{VP} | \mathbf{T}_t^{VP})$ is computed by taking a 1D Gaussian function center at the v-coordinate of the tracked vanishing point. The variance of the Gaussian function is estimated using the training set.

$$p(\mathbf{D}_t^{VP} | \mathbf{T}_t^{VP}) \propto N(\mathbf{D}_t^{VP^v} - \mathbf{T}_t^{VP^v}, \sigma^v) \quad (20)$$

The probability of $p(\mathbf{D}_t^O | \mathbf{T}_t^{VP})$ is computed using the similarity between estimated real world width from (12), $P_{width}(\mathbf{D}_t^O, \mathbf{T}_t^{VP})$, and the modeled width of the target. For example, the width of vehicle is modeled within the range of 1.8 m 2.5 m. The probability of $p(\mathbf{D}_t^O | \mathbf{T}_t^{VP})$ is estimated by a zero mean Gaussian function with standard deviation of the width σ_{width}. The mean width μ_{width} is, e.g., $(1.8+2.5)/2=2.15$ in the case of cars. The probability of $p(\mathbf{D}_t^O | \mathbf{T}_t^{VP})$ can be derived as follow:

$$p(\mathbf{D}_t^O | \mathbf{T}_t^{VP}) \propto N(P_{width}(\mathbf{D}_t^O, \mathbf{T}_t^{VP}) - \mu_{width}, \sigma_{width}) \quad (21)$$

The larger the difference of the estimated width between the models, the lower the probability $p(\mathbf{D}_t^O | \mathbf{T}_t^{VP})$ is.

5   Pedestrian and Vehicle Detection

The proposed detection module consists of three stages, namely, candidate generation, pedestrian classification, and the non-maximum suppression. The details are discussed as follows.

5.1   Object Candidate Generation

In R-CNN detector, the candidate windows are generated from the selective search algorithm. To improve the efficiency while reducing false positives, the physical height of an average pedestrian is restricted within the range from 1.5 m to 2.4 m. Given an upright pedestrian, the aspect ratio is roughly 1:3 in a detecting window [5], i.e., the width of a pedestrian is 1/3 of its height. Thus, it is not hard to show that the range of width satisfies $0.5\,m < P_{width} < 0.80\,m$. Similar to the pedestrian candidates, car candidates are constrained with the height range from 1.4 m to 3.0 m. With the knowledge of the range of object size and given the size of scanning widow in pixels, the bottom row can be estimated by using (12) to estimate the distance first then use (9) to calculate the coordinate. After that, the candidate sliding windows are generated within the range of possible rows.

5.2 Pedestrian Classification

One of the most promising object detector is the R-CNN detector [28]. With the help of the advance of GPU performance and the improvement of learning technique, a deep convolutional neural network (CNN) thus can learn features from huge amount of training data automatically. The selected CNN feature works so well that defeat other features in PASCAL VOC challenges as well as other computer vision competitions. Please note that R-CNN is designed for generic objects, and thus can be applied to pedestrian detection as well as vehicle detection. Please also note that, the object detectors are not limited to R-CNN in our framework. In the KITTI benchmarks, the Regionlets detector [29] is chosen as our baseline object candidate detector. The Regionlets detectors combine multiple variable size region histograms to form boost classifiers. The deformation and aspect ratio are explicitly solved in the framework using region normalization. The detection results are promising and provided in the KITTI benchmarks.

5.3 Non-maximum Suppression

For a true pedestrian in an image, there are many positive windows after applying a sliding window approach. Normally, two assumptions for multiple detected windows of pedestrian are made as follows; a strong detector yields positive scores near a human, and a non-human object is not consistently positive. Thus, the detection windows near the local maximum window are suppressed supposing there are multiple detecting windows generated around a pedestrian.

6 Experiments

6.1 System Parameters

To test the performance of detecting the on-road vehicles and pedestrians, the KITTI vision benchmark suite [30] is used. The benchmark suite contains different datasets, including those for object detection, object tracing, road segmentation and other vision related sets. The tracking benchmark is chosen in the experiments. There are 21 training sequences and 29 testing sequences. The "car" and "pedestrian" are included in the evaluations. There are about 8,000 and 11,000 images in the training and testing set, respectively. Figure 6 shows the exemplar images. There are shadows on the road which may produce noisy edge pixels on the ground. Also, the bright sun light causes the overexposure condition on the road surface.

6.2 Horizon Line Estimation Analysis

To verify the performance of vanishing point estimation, we use KITTI tracking dataset. In our case the horizon line, the v-coordinate of the vanishing point is most important for object detection and tracking, thus we focus on the analysis

Fig. 6. The KITTI benchmark tracking training dataset examples. From the (a) to (d) shows sequence 1, 12, 14 and 19 in training set.

of the position of the horizon line. We use Nieto's method to generate 3 vanishing points for each frame. The canny edges thresholds are (60, 180). A 5 by 5 blur filter is used before the canny edge detector. Our method generates one vanishing point using object information for each frame. We define "Success" as if there is any vanishing point located within 20 pixels to the horizon line. Please note that this threshold is only valid for short range applications (less than 40 m).

In KITTI Tracking test dataset (Table 1), only 53.68% of frames can use Nieto's approach to find vanishing points. While we can achieve 95.40% success rate overall, there are some case we cannot solve. In the sequence 17, there is no clear lane markings for the vanishing point estimation. In sequence 11, many frames do not contain any object or lane markings, thus result in bad performance compare to the other sequence.

6.3 Object Tracking Performance Analysis

In KITTI tracking benchmark, only training sequence labels are provided. Thus, the following analysis is conducted in training sequence. The type of objects are 'Car', 'Van', 'Truck', 'Pedestrian', 'Person_sitting', 'Cyclist', 'Tram', 'Misc' or 'DontCare'. Here, one should note that the "Van" is not considered as a "Car" in the evaluation.

Table 2 shows the performance concerning recall and precision of the pedestrian tracking in the KITTI benchmark. The performance of our implemented CEM [23] using the detection results from Regionlet are chosen as the baseline, where the original CEM uses the L-SVM DPM detection results. Although the precision of the detection results is high, the miss rate of the pedestrian detection is high in comparison with that of Regionlet detector. Thus, we also choose the Regionlet detection results as our detection input. To compare with the state-of-the-art tracking result on the KITTI tracking benchmark, the MDP from Xiang [31] is also compared. With the help of the geometry information, the number of

Table 1. Vanishing point estimation performance

Training set	Nieto		Our		Testing set	Nieto		Our	
	Success	Success rate (%)	Success	Success rate		Success	Success rate (%)	Success	Success rate
0	83	53.90	152	98.70	0	145	31.18	448	96.34
1	196	43.85	401	89.71	1	6	4.08	138	93.88
2	164	70.39	217	93.13	2	201	82.72	242	99.59
3	45	31.25	143	99.31	3	114	44.36	230	89.49
4	214	68.15	313	99.68	4	160	38.00	412	97.86
5	241	81.14	297	100.00	5	637	78.74	797	98.52
6	228	84.44	254	94.07	6	102	89.47	114	100.00
7	304	38.00	634	79.25	7	146	67.91	215	100.00
8	351	90.00	385	98.72	8	100	60.61	161	97.58
9	510	63.51	715	89.04	9	225	64.47	317	90.83
10	263	89.46	294	100.00	10	842	71.60	1129	96.00
11	286	76.68	373	100.00	11	182	23.51	599	77.39
12	7	8.97	64	82.05	12	254	36.60	690	99.42
13	188	55.29	292	85.88	13	100	65.79	150	98.68
14	41	38.68	100	94.34	14	519	61.06	830	97.65
15	223	59.31	345	91.76	15	230	32.81	647	92.30
16	130	62.20	203	97.13	16	204	40.00	476	93.33
17	116	80.00	116	80.00	17	50	16.39	265	86.89
18	257	75.81	321	94.69	18	40	22.22	171	95.00
19	421	39.75	932	88.01	19	280	69.31	403	99.75
20	657	78.49	830	99.16	20	132	76.30	173	100.00
					21	111	54.68	202	99.51
					22	306	70.18	436	100.00
					23	305	70.93	429	99.77
					24	214	67.72	315	99.68
					25	127	72.16	175	99.43
					26	110	64.71	162	95.29
					27	33	38.82	84	98.82
					28	81	46.29	175	100.00
Total	4925	61.50	7381	**92.17**	Total	5956	53.68	10585	**95.40**

false detections can be reduced, as a result, the tracking precision improve 7.5% while detection rate decrease 3.8%.

Table 3 shows the CLEAR MOT [32] metrics of the pedestrian tracking results. MOTA stands for multiple object tracking accuracy, whereas MOTP is the abbreviation of the multiple object tracking precision. MODA and MODP

Table 2. Pedestrian tracking recall/precision

	Recall	Precision	F1 Score
CEM	0.3654	0.6708	0.4731
MDP	0.5880	0.7247	0.6493
Ours	0.7235	0.8603	0.7861

is the abbreviation of multiple object detection accuracy and multiple object detection precision, respectively. The accuracy of our system is improved not only in detection but also in detection with the help of the context and geometric information.

Table 3. Pedestrian tracking performance

	MOTA	MOTP	MODA	MODP
CEM	0.1818	0.6848	0.1860	0.1759
MDP	0.3591	0.7037	0.3629	0.2228
Ours	0.5979	0.7725	0.6041	0.2203

Table 4 shows the Car tracking performance. In our system, the precision can be improved while the recall reduced slightly, thus can obtain a better F1 score. Although our tracking accuracy is not high in these cases, the number of tracked targets are more than that in the original CEM. The MT and PT in the table are the abbreviations of the mostly tracked and partly tracked target rate. The higher the number is, the more the targets are tracked in the video sequence.

Table 4. Vehicle tracking performance

	Recall	Precision	F1 Score	MOTA	MOTP	MODA	MODP	MT	PT
CEM	0.5549	0.8547	0.6729	0.4431	0.7711	0.4467	0.6071	0.195	0.491
MDP	0.7985	0.9076	0.8496	0.6935	0.8210	0.6974	0.7119	0.514	0.355
Ours	0.8316	0.9042	0.8664	0.7079	0.8120	0.7165	0.7096	0.575	0.305

7   Conclusion

With the help of the obstacle objects and context integration, the robustness and efficiency of the system can be improved. One of the state-of-the-art detector, the Regionlet object detector, can be improved using our framework. The scene geometric information is estimated using the flat ground plane assumption and the camera parameters. A novel vanishing line estimator using object detection results is proposed. With the knowledge of physical size of objects, the vanishing line distribution can be estimated even when major parallel lines on the ground are missing. The success of the horizon estimation can improve the object tracking system in clutter scenes. To extend the system to no-flat ground plane will be the future work.

References

1. Hoiem, D., Efros, A., Hebert, M.: Putting objects in perspective. Int. J. Comput. Vision **80**, 3–15 (2008)
2. Nieto, M., Laborda, J.A., Salgado, L.: Road environment modeling using robust perspective analysis and recursive bayesian segmentation. Mach. Vis. Appli. **22**, 927–945 (2011)
3. Geiger, A., Lauer, M., Wojek, C., Stiller, C., Urtasun, R.: 3D traffic scene understanding from movable platforms. IEEE Trans. Pattern Anal. Mach. Intell. **36**, 1012–1025 (2014)
4. Richardson, E., Peleg, S., Werman, M.: Scene geometry from moving objects. In: IEEE International Conference on Advanced Video and Signal Based Surveillance, pp. 13–18 (2014)
5. Dalal, N., Triggs, B.: Histograms of oriented gradients for human detection. In: Computer Vision and Patter Recognition, vol. 1, pp. 886–893 (2005)
6. Dollar, P., Wojek, C., Schiele, B., Perona, P.: Pedestrian detection: an evaluation of the state of the art. IEEE Trans. Pattern Anal. Mach. Intell. **34**, 743–761 (2012)
7. Ess, A., Leibe, B., Schindler, K., Van Gool, L.: A mobile vision system for robust multi-person tracking. In: IEEE Conference on Computer Vision and Pattern Recognition, pp. 1–8 (2008)
8. Wojek, C., Roth, S., Schindler, K., Schiele, B.: Monocular 3D scene modeling and inference: understanding multi-object traffic scenes. In: Daniilidis, K., Maragos, P., Paragios, N. (eds.) ECCV 2010. LNCS, vol. 6314, pp. 467–481. Springer, Heidelberg (2010). doi:10.1007/978-3-642-15561-1_34
9. Sudowe, P., Leibe, B.: Efficient use of geometric constraints for sliding-window object detection in video. In: Crowley, J.L., Draper, B.A., Thonnat, M. (eds.) ICVS 2011. LNCS, vol. 6962, pp. 11–20. Springer, Heidelberg (2011). doi:10.1007/978-3-642-23968-7_2
10. Moghadam, P., Starzyk, J.A., Wijesoma, W.S.: Fast vanishing-point detection in unstructured environments. IEEE Trans. Image Process. **21**, 425–430 (2012)
11. Geiger, A., Wojek, C., Urtasun, R.: Joint 3D estimation of objects and scene layout. In: Advances in Neural Information Processing Systems, pp. 1467–1475 (2011)
12. Lezama, J., Grompone von Gioi, R., Randall, G., Morel, J.M.: Finding vanishing points via point alignments in image primal and dual domains. In: IEEE Conference on Computer Vision and Pattern Recognition, pp. 509–515 (2014)
13. Yang, W., Luo, X., Fang, B., Zhang, D., Tang, Y.Y.: Fast and accurate vanishing point detection in complex scenes. In: IEEE Conference on Intelligent Transportation Systems, pp. 93–98 (2014)
14. Wildenauer, H., Hanbury, A.: Robust camera self-calibration from monocular images of manhattan worlds. In: IEEE Conference on Computer Vision and Pattern Recognition, pp. 2831–2838 (2012)
15. Nieto, M., Salgado, L.: Real-time robust estimation of vanishing points through nonlinear optimization. In: SPIE Photonics Europe, pp. 772402–772402-14. International Society for Optics and Photonics (2010)
16. Quan, L., Mohr, R.: Determining perspective structures using hierarchical hough transform. Pattern Recogn. Lett. **9**, 279–286 (1989)
17. Dubska, M., Herout, A., Havel, J.: Pclines - line detection using parallel coordinates. In: IEEE Conference on Computer Vision and Pattern Recognition, pp. 1489–1494 (2011)

18. Vermaak, J., Doucet, A., Perez, P.: Maintaining multimodality through mixture tracking. In: IEEE International Conference on Computer Vision, vol. 2, pp. 1110–1116 (2003)
19. Lanz, O.: Approximate bayesian multibody tracking. IEEE Trans. Pattern Anal. Mach. Intell. **28**, 1436–1449 (2006)
20. Kuo, C.H., Huang, C., Nevatia, R.: Multi-target tracking by on-line learned discriminative appearance models. In: IEEE Conference on Computer Vision and Pattern Recognition, pp. 685–692 (2010)
21. Breitenstein, M.D., Reichlin, F., Leibe, B., Koller-Meier, E., Van Gool, L.: Online multiperson tracking-by-detection from a single, uncalibrated camera. IEEE Trans. Pattern Anal. Mach. Intell. **33**, 1820–1833 (2011)
22. Leibe, B., Schindler, K., Cornelis, N., Van Gool, L.: Coupled object detection and tracking from static cameras and moving vehicles. IEEE Trans. Pattern Anal. Mach. Intell. **30**, 1683–1698 (2008)
23. Milan, A., Roth, S., Schindler, K.: Continuous energy minimization for multitarget tracking. IEEE Trans. Pattern Anal. Mach. Intell. **36**, 58–72 (2014)
24. Kittler, J., Hatef, M., Duin, R.P.W., Matas, J.: On combining classifiers. IEEE Trans. Pattern Anal. Mach. Intell. **20**, 226–239 (1998)
25. Bota, S., Nedesvchi, S.: Multi-feature walking pedestrians detection for driving assistance systems. IET Intel. Trans. Syst. **2**, 92–104 (2008)
26. Isard, M., Blake, A.: Condensation-conditional density propagation for visual tracking. Int. J. Comput. Vision **29**, 5–28 (1998)
27. Matas, J., Galambos, C., Kittler, J.: Robust detection of lines using the progressive probabilistic hough transform. Comput. Vis. Image Underst. **78**, 119–137 (2000)
28. Girshick, R., Donahue, J., Darrell, T., Malik, J.: Rich feature hierarchies for accurate object detection and semantic segmentation. In: IEEE Conference on Computer Vision and Pattern Recognition, pp. 580–587 (2014)
29. Wang, X., Yang, M., Zhu, S., Lin, Y.: Regionlets for generic object detection. In: IEEE International Conference on Computer Vision, pp. 17–24 (2013)
30. Geiger, A., Lenz, P., Urtasun, R.: Are we ready for autonomous driving? the kitti vision benchmark suite. In: IEEE Conference on Computer Vision and Pattern Recognition, pp. 3354–3361 (2012)
31. Xiang, Y., Alahi, A., Savarese, S.: Learning to track: Online multi-object tracking by decision making. In: Proceedings of the IEEE International Conference on Computer Vision, pp. 4705–4713 (2015)
32. Bernardin, K., Stiefelhagen, R.: Evaluating multiple object tracking performance: the clear mot metrics. J. Image Video Process. **2008**, 1–10 (2008)

Recognition of Texting-While-Walking by Joint Features Based on Arm and Head Poses

Fumito Shinmura[1,2]([envelope]), Yasutomo Kawanishi[3], Daisuke Deguchi[4], Ichiro Ide[3], Hiroshi Murase[3], and Hironobu Fujiyoshi[5]

[1] Institute of Innovation for Future Society (MIRAI), Nagoya University, Nagoya, Japan
`shinmuraf@murase.m.is.nagoya-u.ac.jp`
[2] JST/COI, Nagoya, Japan
[3] Graduate School of Information Science, Nagoya University, Nagoya, Japan
[4] Information Strategy Office, Nagoya University, Nagoya, Japan
[5] Department of Robotics Science and Technology, Chubu University, Kasugai, Japan

Abstract. Pedestrians "texting-while-walking" increase the risk of traffic accidents, since they are often not paying attention to their surrounding environments and fails to notice approaching vehicles. Thus, the recognition of texting-while-walking from an in-vehicle camera should be helpful for safety driving assistance. In this paper, we propose a method to recognize a pedestrian texting-while-walking from in-vehicle camera images. The proposed approach focuses on the characteristic relationship between the arm and the head poses observed during a texting-while-walking behavior. In this paper, Pose-Dependent Joint HOG feature is proposed as a novel feature, which uses parts locations as prior knowledge and describes the cooccurrence of the arm and the head poses. To show the effectiveness of the proposed method, we constructed a dataset and evaluated it.

1 Introduction

Recently, pedestrian recognition methods using an in-vehicle camera have been studied widely to reduce traffic accidents with pedestrians. In the case of Advanced Emergency Braking System, it first detects obstacles in front of the vehicle and then actuates the brake to avoid collisions with pedestrians. However, this system provides limited support for avoiding collisions since it does not actuate the brake until a pedestrian runs out in the road. In order to realize safer driving without traffic accidents, predicting the probable risks of traffic accidents will be important. In the case of risk prediction against pedestrians, it relies on the prediction of their behavior. For prediction of pedestrians' behavior, not only the detection of their positions but also the recognition of their attributes (e.g. body orientation, etc.) becomes important. Although some methods for pedestrians' orientation estimation have been studied to predict their path directions [4], various attributes should be considered for risk prediction.

© Springer International Publishing AG 2017
C.-S. Chen et al. (Eds.): ACCV 2016 Workshops, Part I, LNCS 10116, pp. 452–462, 2017.
DOI: 10.1007/978-3-319-54407-6_30

Pedestrian's carelessness is also an important issue. Pedestrians who are not paying attention to their surrounding environments often fail to notice approaching vehicles. Such pedestrians can be considered that they have a higher risk of running out into the road. Recently, dangerousness of pedestrian's "texting-while-walking" has been reported [5,9]. Nintendo's Pokémon GO[1] player is a typical example of a pedestrian texting-while-walking. This is a behavior that a pedestrian's focus is on a hand-held electronic device such as a smartphone, while walking. Such a pedestrian tends to immerse him/herself into the operation of the device and is most likely not paying attention to his/her surrounding environment. Therefore, the recognition of a pedestrian's texting-while-walking behavior should contribute to the risk prediction of the pedestrian to be involved in an accident. In this paper, we propose a method to recognize whether a pedestrian is texting-while-walking or not from pedestrian images.

A pedestrian, who is texting-while-walking, holds a smartphone in his/her hand and looks down at the screen. Figure 1 shows examples of pedestrians texting-while-walking. Most of them take almost the same characteristic pose; bending the elbow and looking down.

Fig. 1. Examples of pedestrians texting-while-walking.

However, there is a problem that poses similar to that of texting-while-walking may be observed while walking normally. For example, a pose merely bending the elbow, which occurs when a pedestrian waves his/her arms, appears similar to that of texting-while-walking. In order to prevent false recognitions on such cases, the proposed method focuses on the poses of both bending the elbow and looking down simultaneously.

In addition, pedestrians tend to walk in the same pose while texting-while-walking, in which case, the features of poses are almost constant for several seconds. On the other hand, when a pedestrian takes a similar pose momentarily, the features should vary widely during the same period. Thus, the proposed method sequentially observes the poses of pedestrian images during several seconds.

In summary, for accurate recognition of texting-while-walking, the proposed method takes the following approach:

---

[1] http://pokemongo.nianticlabs.com/.

- Focusing on the cooccurrence of poses of the arm and the head, we propose Pose-Dependent Joint HOG features, which is a variant of the Joint HOG features [7]. The proposed method selects features based on prior knowledge of parts locations.
- The proposed method prevents false recognition by observing the continuousness of the same pose.

2 Related Works

Various methods have been proposed for pedestrian detection and pedestrian attributes recognition. Many works that combine image feature extraction and supervised learning have been reported. Dalal et al. proposed a method that combined Histogram of Oriented Gradients (HOG) features and the Support Vector Machine (SVM) classifier [1], and Dollar et al. proposed a method that combined the Aggregate Channel Features (ACF) and the Boosted Trees [2] for pedestrian detection. For pedestrian attributes recognition, Shimizu and Poggio proposed a method for pedestrian's orientation estimation that combined the Haar-wavelet features and the SVM classifier [8]. Gandhi and Trivedi proposed a method that combined the HOG features and the SVM classifier [4].

Several methods for pedestrian detection and attributes recognition based on body parts features also have been proposed and proved to perform well. These methods extract image features not from the whole body but from body parts. Felzenszwalb et al. proposed a method using Deformable Part Models (DPM) [3] for pedestrian detection. Their models consist of a set of parts filters, and a pedestrian is detected based on the shape of parts and their positional relations. Tao and Klette proposed a method using Random Forest as a classifier and used body parts selectively for training [10] for pedestrian's orientation estimation. Their method extracted image features from part areas selected randomly, and estimated their orientation using a Decision Tree for each part.

In order to recognize texting-while-walking, the proposed method captures the features of the arm and the head poses. The methods introduced above based on image feature extraction can capture features of the arm and the head, but they may misrecognize a pedestrian with a similar pose to that of texting-while-walking.

To overcome this problem, the arm and the head poses need to be focused simultaneously. Hence, observing the cooccurrence between the arm and the head should be effective for the recognition of texting-while-walking. Mitsui and Fujiyoshi proposed joint features based on HOG (Joint HOG) [7] in order to represent the cooccurrence of appearances. Their joint features are obtained by combining the HOG features for several different local areas by means of AdaBoost, which allows to capture shape symmetry and edge continuity. In their method, more effective local areas for recognition are selected automatically by AdaBoost, although areas of the arm and the head may not always be selected. In the case of the recognition of texting-while-walking, selecting local areas from the arm and the head areas is more effective than selecting them from other

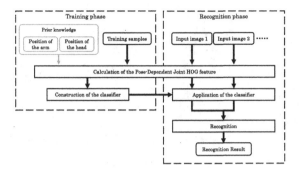

Fig. 2. Process flow of the proposed method.

areas. Therefore, the proposed method is improved to preferentially select local areas from the arm and the head areas.

3   Recognition of Texting-While-Walking by Joint Features

This paper proposes a method to recognize texting-while-walking from pedestrian images. Pedestrians are assumed to be detected beforehand by an arbitrary detection method. We have observed that a pedestrian's pose when he/she uses a smartphone tends to be as follows:

- Bending the elbow to hold a smartphone.
- Looking down at the screen of a smartphone.

Thus, the proposed method extracts image features that represent both the arm and the head poses, and constructs a classifier to recognize texting-while-walking using the extracted features. The process flow of the proposed method is shown in Fig. 2.

The pose of pedestrians simply bending their elbow or looking down are similar to the pose of texting-while-walking as shown in Fig. 3, even if they are not texting-while-walking. To distinguish between these poses, the proposed method needs to capture both the arm and the head poses simultaneously, and thus focuses on the cooccurrence of the arm and the head poses; It uses a joint feature of the arm and the head poses.

This method uses the Joint HOG features [7] to describe the cooccurrence of the arm and the head poses. It uses AdaBoost in two different contexts; (1) to calculate the joint features, and (2) to construct the classifier to recognize texting-while-walking. Since the conventional Joint HOG features select feature pairs by means of the second AdaBoost, the feature pairs are not always selected from the arm and the head areas. Thus, we propose the Pose-Dependent Joint HOG feature, which is a variant of the Joint HOG feature, that is given a prior knowledge of parts locations to select feature pairs. This joint feature combines

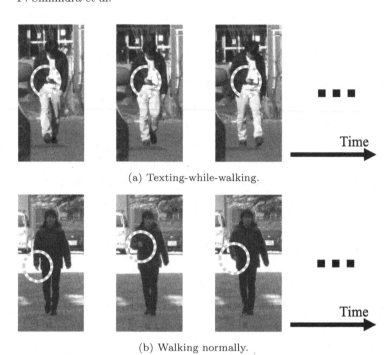

(a) Texting-while-walking.

(b) Walking normally.

Fig. 3. Examples of a pedestrian texting-while-walking and that walking normally. (Color figure online)

the HOG feature extracted from each of the arm and the head areas, which can describe the cooccurrence of their shapes.

Additionally, the proposed method sequentially observes the poses of pedestrians to reduce false recognition. Pedestrians texting-while-walking keep the same pose, although those walking normally varies their poses by waving their arms. Figure 3 shows examples of a pedestrian texting-while-walking and that walking normally during several frames. The red circles in the figure indicate the arm positions. A pedestrian texting-while-walking keeps his/her arm still, although that walking normally moves his/her arm. Thus, we consider that pedestrians recognized as in a texting-while-walking state for several seconds are actually texting-while-walking. On the contrary, pedestrians momentarily recognized as in a texting-while-walking state are not actually texting-while-walking. Thus, the proposed method assesses by majority voting of the recognition results in a two seconds period.

3.1   Extraction of the Pose-Dependent Joint HOG Features

In order to represent the pose of the arm and the head, this method uses the HOG feature proposed by Dalal et al. [1] as a low-level feature. First, an input pedestrian image is resized to 100×200 pixels, and then divided into local areas called cells with a size of 10×10 pixels. A histogram of gradient orientations containing nine orientation bins is created for each cell. The HOG features are calculated by normalizing these histograms in a block with a size of 3×3 cells. This is repeated by sliding the block one cell after another. One HOG feature is represented by a normalized histogram in a cell.

Next, two HOG features for different cells are selected to calculate the joint features. Since this method uses features of the arm and the head, the arm and the head positions are given as prior knowledge as shown in Fig. 4. One of the HOG features is selected from a cell in the arm area, and the other is selected from a cell in the head area.

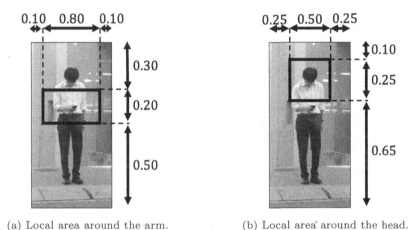

(a) Local area around the arm.   (b) Local area around the head.

Fig. 4. Pre-determined local areas used as prior knowledge.

From the two selected HOG features, the proposed method describes the cooccurrence of the two HOG features. The value of a binary symbol s is determined with the following equation:

$$s(\boldsymbol{V}) = \begin{cases} 1 & \text{if } pv_o > p\theta \\ 0 & \text{otherwise} \end{cases}, \tag{1}$$

where $\boldsymbol{V} = [v_1, v_2, \cdots, v_9]$ is the HOG feature, θ is the threshold value, $v_o \in \boldsymbol{V}$ is the value of a histogram of orientation gradient, o is the orientation of gradient, p is a parity indicating the direction of the inequality sign and takes the values $p \in \{+1, -1\}$. The value of a certain histogram of orientation gradient v_o is selected from $\boldsymbol{V}$ and used for the calculation of $s(\boldsymbol{V})$. o, p and θ are determined when the AdaBoost learns as described below. The binary symbol s represents whether the target pedestrian is texting-while-walking or not. Two binary symbols that are calculated from cells around the arm and the head are obtained. The cooccurrence feature is generated by combining these two symbols.

It takes four values based on the combination of the values of the symbols from the two areas.

The features effective for discrimination are selected from the cooccurrence features by the Real AdaBoost algorithm. When a set of N labeled training samples $(x_1, y_1), \ldots, (x_N, y_N)$, where $y_i \in \{+1, -1\}$ is the class label associated with a training sample x_i, is given, the strong classifier in the AdaBoost is constructed with the following equation:

$$H(x) = \sum_{t=i}^{T} h_t(x), \tag{2}$$

where $h_t(x)$ is the weak classifier in the AdaBoost. When a cooccurrence feature $J(x) = j$ is observed, $h_t(x)$ is expressed as follows:

$$h_t(x) = \frac{1}{2} \ln \frac{P_t(y = +1|j) + \varepsilon}{P_t(y = -1|j) + \varepsilon}, \tag{3}$$

where $t = 1, \ldots, T$ is the number of training rounds, $\varepsilon = 10^{-7}$ is a very small value to prevent division by zero. $P_t(y = +1|j)$ and $P_t(y = -1|j)$ are the respective conditional probability distributions calculated with the following equations:

$$P_t(y = +1|j) = \sum_{i}^{N} I(y = +1)D_t(i), \tag{4}$$

$$P_t(y = -1|j) = \sum_{i}^{N} I(y = -1)D_t(i), \tag{5}$$

where $I \in \{1, 0\}$ is an indicator that takes $I = 1$ when the condition is satisfied. $D_t(i)$ is a weight of the training sample x_i calculated with the following equation

$$D_{t+1}(i) = D_t(i) \exp(-y_i h_t(x_i)). \tag{6}$$

The weights are initialized by $D_1(i) = 1/N$.

The process described above is applied to all combinations of cells. When two different cells are expressed as c_m and c_n, the strong classifier is expressed as follows:

$$H^{c_m, c_n}(x) = \sum_{t=i}^{T} h_t^{c_m \in C_{\mathrm{arm}}, c_n \in C_{\mathrm{head}}}(x), \tag{7}$$

where C_{arm} and C_{head} are the cells in the arm and the head areas, respectively, and $H^{c_m, c_n}(x)$ is the joint feature. The joint features are generated for all combinations of cells. Here, since there were 108 cells for the arm and 81 cells for the head, 8,748 Pose-Dependent Joint HOG features were generated in total.

3.2   Construction of a Classifier to Recognize Texting-While-Walking

A classifier is constructed by Real AdaBoost using the calculated Pose-Dependent Joint HOG features. The features effective for discrimination are selected by training of the AdaBoost. This means that the classifier learns the positions and combinations of cells effective for discrimination.

Therefore, the classifier for recognition of texting-while-walking from an input pedestrian image is constructed.

3.3   Recognition of Texting-While-Walking

This method recognizes whether a pedestrian is texting-while-walking or not from an input pedestrian image sequence by using the constructed classifier.

First, the Pose-Dependent Joint HOG features are calculated from a pedestrian image detected from each frame. Next, the constructed classifier is applied, and the recognition result is obtained for each frame. Finally, whether the pedestrian is texting-while-walking or not is determined by majority voting of the recognition results of all the frames in a two seconds window.

4   Experiments

An experiment on recognition of texting-while-walking was conducted in order to evaluate the effectiveness of the proposed method.

4.1   Dataset

For the experiment, we prepared images captured in outdoor environments by a commercially available camera (Point Grey Grasshopper®3) in daytime. The bounding boxes of the pedestrians were manually annotated beforehand. The resolution of the images were $1,920 \times 1,440$ pixels, and the sizes of the cropped pedestrian images ranged from 275×550 to 409×818 pixels. Samples from the dataset are shown in Fig. 5.

The prepared dataset consists of 3,960 pedestrian images casted by eleven individuals. It contains four kinds of poses as follows:

– Texting-while-walking, namely, using a smartphone.
– Looking ahead.
– Bending the elbow without using a smartphone.
– Looking down without using a smartphone.

All pedestrians were captured from the frontal side.

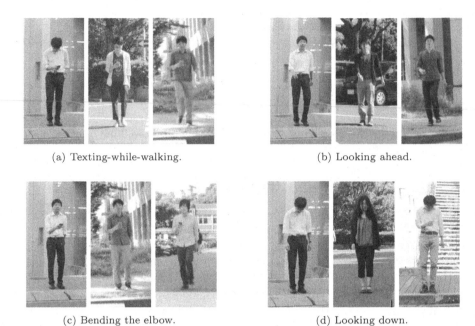

(a) Texting-while-walking.   (b) Looking ahead.

(c) Bending the elbow.   (d) Looking down.

Fig. 5. Samples from the dataset used in the experiment.

4.2 Experiments and Results

The prepared dataset was used for an experiment on recognition of texting-while-walking. In the experiment, data for ten subjects (3,600 images) were used for training and those for one other subject (360 images) were used for testing. The evaluation was repeated eleven times with a different subject for testing, and then the accuracy for each evaluation was averaged.

The experimental result is shown in Fig. 6 in a ROC curve. We compared the proposed method with two methods. Comparison method 1 was a method using the conventional Joint HOG features [7]. Comparison method 2 was a method using AlexNet that is a variety of the Convolutional Neural Network (CNN) [6], where we modified the output of its final layer to binary output for texting-while-walking recognition. As shown in Fig. 6, the proposed method using the cooccurrence features and majority voting achieved the best performance.

4.3 Discussion

Comparison method 1 also combined two HOG features and should have selected the effective pairs for recognition. It could also describe the cooccurrence features, but its accuracy was lower than that of the proposed method. Thus, we considered that the feature pairs truly effective for the recognition were not selected. In order to select truly effective feature pairs for recognition by this method, a training data with large variety is required. Likewise, comparison

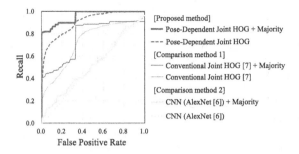

Fig. 6. Experimental results.

method 2 also requires a large variety of training data to train the CNN. In this experiment, the size of training data was insufficient to sufficiently train a CNN. If there were huge amount of training data, the results of CNN would be better. However, collecting many pedestrian images of texting-while-walking is even more difficult than collecting many pedestrian images for pedestrian detection, so it is better if we can cope with a small number of training data.

In the proposed method, by giving prior knowledge of parts location, we succeeded to select effective feature pairs for the recognition of texting-while-walking without a large training dataset.

5 Conclusions

This paper proposed a method to recognize a pedestrian texting-while-walking by focusing on the arm and the head poses. The characteristic feature for recognition of texting-while-walking is simultaneously bending the arm and looking down. The proposed method improved the Joint HOG feature to capture the cooccurrence of the two poses.

Our future work will include improvement to deal with pedestrian's orientation. Since the appearance changes according to his/her orientation, the recognition performance will vary according to the pedestrian's orientation. We also plan to conduct an experiment with a larger dataset including various pedestrian orientations.

Acknowledgement. This research is partially supported by the Center of Innovation Program (Nagoya-COI) from Japan Science and Technology Agency and Grant-in-Aid for Scientific Research from The Ministry of Education, Culture, Sports, Science and Technology.

References

1. Dalal, N., Triggs, B.: Histograms of oriented gradients for human detection. In: Proceedings of 2005 IEEE Computer Society Conference on Computer Vision and Pattern Recognition, pp. 886–893 (2005)

2. Dollar, P., Appel, R., Belongie, S., Perona, P.: Fast feature pyramids for object detection. IEEE Trans. Pattern Anal. Mach. Intell. **36**, 1532–1545 (2014)
3. Felzenszwalb, P., Girshick, R., McAllester, D., Ramanan, D.: Object detection with discriminatively trained part based models. IEEE Trans. Pattern Anal. Mach. Intell. **32**, 1627–1645 (2010)
4. Gandhi, T., Trivedi, M.: Image based estimation of pedestrian orientation for improving path prediction. In: Proceedings of 2008 IEEE Intelligent Vehicles Symposium, pp. 506–511 (2008)
5. Haga, S., Sano, A., Sekine, Y., Sato, H., Yamaguchi, S., Masuda, K.: Effects of using a smart phone on pedestrians' attention and walking. Procedia Manufact. **3**, 2574–2580 (2015)
6. Krizhevsky, A., Sutskever, I., Hinton, G.: Imagenet classification with deep convolutional neural networks. In: Proceedings of 26th Annual Conference on Neural Information Processing Systems, pp. 1097–1105 (2012)
7. Mitsui, T., Fujiyoshi, H.: Object detection by joint features based on two-stage boosting. In: Proceedings of 12th IEEE International Conference on Computer Vision Workshops, pp. 1169–1176 (2009)
8. Shimizu, H., Poggio, T.: Direction estimation of pedestrian from multiple still images. In: Proceedings of 2004 IEEE Intelligent Vehicles Symposium, pp. 596–600 (2004)
9. Stavrinos, D., Byington, K., Schwebel, D.: Distracted walking: cell phones increase injury risk for college pedestrians. J. Saf. Res. **42**, 101–107 (2011)
10. Tao, J., Klette, R.: Part-based RDF for direction classification of pedestrians, and a benchmark. In: Workshop on Intelligent Vehicles with Vision Technology in the 12th Asian Conference on Computer Vision, w11-p2 (2014)

A Motion Robust Remote-PPG Approach to Driver's Health State Monitoring

Bing-Fei Wu[1], Yun-Wei Chu[1(✉)], Po-Wei Huang[2],
Meng-Liang Chung[1], and Tzu-Min Lin[3]

[1] Institute of Electrical Control Engineering,
National Chiao Tung University, Hsinchu, Taiwan
yunwei@cssp.cn.nctu.edu.tw
[2] Department of Electrical and Computer Engineering,
National Chiao Tung University, Hsinchu, Taiwan
[3] Division of Rheumatology, Immunology and Allergy,
Department of Internal Medicine,
Taipei Medical University Hospital, Taipei, Taiwan

Abstract. With the surging significance of personal health care, driver's physiological state is no longer negligible nowadays. Among all the indicators of health state in human, heart rate (HR) is one of the most cardinal indicators. The commonly used HR measurement is contact-type, might result in driver's distraction and discomfort in the vehicle applications. To cope with this problem, remote photoplethysmography (rPPG) is utilized to monitor HR at a distance via a web camera. Nevertheless, the rPPG is not without its flaw. The main concern of the rPPG technique is the potential not-robustness result from the arbitrary motion. Consequently, the contribution of this paper is to conquer the motion noise when the car is driving and the driver's health state is well monitored to enhance the public safety. The proposed algorithm is investigated in not only the indoor environment but as well the outdoor driving, which contains much more unpredictable motion. With k-nearest neighbor (kNN) classifier on chrominance-based features, the mean square error can be reduced from 30.6 to 2.79 bpm, approaching the medical instrument level. The proposed method can be applied to human improving driving safety for Advanced Driver Assistance Systems.

1 Introduction

Advanced Driver Assistance Systems (ADAS) are utilized to assist drivers in their driving process and have gathered increasing attention in the public recently. With external and internal imaging process algorithms, applications of computer vision in ADAS provide a more convenient and effective way to improve the intelligence of vehicles.

The imaging processing to detect the driving behavior and/or healthy condition of a driver has generated relatively little discussion in the considerable amount of research work for vision-based ADAS around the vehicle. Health care of drivers, as a typical issue of driver's behavior, can be of enormous value to public traffic safety. Consequently, this

© Springer International Publishing AG 2017
C.-S. Chen et al. (Eds.): ACCV 2016 Workshops, Part I, LNCS 10116, pp. 463–476, 2017.
DOI: 10.1007/978-3-319-54407-6_31

work is to address the monitoring the driver's health state by a camera to enhance the public safety.

Along with the surging significance of personal health care, driver's physiological state is no longer negligible nowadays. Among all the indicators of health state in human, heart rate (HR) is one of the most significant indicators. In contrast with conventional techniques such as Electrocardiography (ECG), Photoplethysmography (PPG) is one of the technique that is capable of not only avoiding the skin irritation but as well providing a non-invasive and user-friendly recording. PPG, first described by Alrick Hertzman in 1937 [1], provides an electro-optic technique that detects the transmitted light on human's skin. The measurement sensed the periodic variation of light, which caused by the absorption difference when blood flows through the vessel under the skin (Zijlstra [2], Cheang [3], Allen [4]). Owing to PPG's simple, low-cost and non-invasive properties, contact sensors are commonly used to detect the variation of blood volume (e.g. the finger clipping device). Nevertheless, PPG is not without its flaw. As the skin contact is still required in practice, it may not suit for all scenario in real life (e.g. burn or contusion patient). Furthermore, applying PPG in vehicles to detect drivers' HR might cause drivers' distraction and discomfort.

Recently, several work have reported that even with a distance, blood volume variations during the cardiac cycle can also be detected from human's face with a web camera. This non-contact, image-based method for monitoring driver's HR is an exceptional approach that provides driver better flexibility and comfort experience during driving. In this paper, we apply the computer vision method on vehicles in order to estimate the driver's health and elevate the driving safety without interfering driver.

Lately, a non-contact and remote heart rate measurement has published by Poh et al. [5] who analyzed cardiac pulse from the three channel color video based on Independent Component Analysis (ICA), which is known as a Blind Source Separation (BSS) technique, leading to remote PPG (rPPG). The basic thought of BSS method has been applied to recover a clean signal (or called source) from a set of observations that are the linear combinations of the implicit source. This concept was expanded by Lewandowska et al. [6], who utilized Principal Component Analysis (PCA), a BSS method as well, claimed to reduce computational complexity in compared with ICA method. From that time, ample relative BSS based work has been reported to evaluate the retrieved HR signal more robustly from the video (Benjamin D Holton [7], Daniel Wedekind [8]). Mannapperuma [9] illustrated a comparison of different ICA methods (Fast ICA, JADE ICA, RADICAL ICA) and recorded the limitations of ICA-based HR detection in rPPG. Nonetheless, the accuracy of rPPG technique was highly affected by the motion of the subject up to now. To conquer the arbitrary motion challenge, Wenjin Wang [10] implemented face detection and tracking technique, Farneback dense optical flow algorithm [11] and PCA method to overcome the influence of global motion (due to face shifting and the rotating) and local motion (due to blinking and talking). Hamed Monkaresi [12] replicated Poh's ICA method and implemented the machine learning technique aimed to improve the robustness of subject motion in three different scenarios (resting, naturalistic HCI, in door cycling). Though these BSS-based methods retrieved independent and clean sources from the compound observations, there was no immediate way to determine which separated signal is the HR signal. All previously published BSS-based literatures selected the most periodic signal from a collection of the independent signal as HR signal, resulting in distortion when arbitrary motion occurred.

de Haan and Jeanne [13] presented a chrominance-based rPPG (CrPPG) that focused on improving motion robustness. They eliminated the component selection issue, which mattered in the BSS-based method, by constructing a linear combination of the normalized color signals orthogonal to the assumed distortions, regardless the color of the illumination. After that time, abound relative works (de Hann [14], van Gastel M [15]) have continued improving motion robustness based on CrPPG in indoor cycling and stepping device scenario. Ren-You Huang [16] extended the concept of signal recovery from CrPPG and integrated with the ICA method, trying to separate a cleaner HR signal in indoor treadmill scenario from observation of face's x-position, y-position and CrPPG. Yung-Chien Hsu [17] also based on CrPPG method and utilized support vector regression (SVR), a machine learning technique, for predicting HR more accurately in indoor naturalistic HCI scenario.

Most of the previous work aimed to enhance the motion robustness of rPPG or CrPPG in the indoor and controllable scenario; nonetheless, the restriction of the scenario environment will limit the application of this promising technique. Consequently, the aim of this paper is to provide an enhanced motion-robust rPPG technique in different scenarios, especially the outdoor driving vehicles scenario for monitoring driver's heart rate in real time. In this paper, we reduce the noise result from artificial motion by conducting Empirical Mode Decomposition (EMD) based on CrPPG method and predict the HR by k-Nearest Neighbors (kNN), a machine learning technique for classification problem. The results confirm that the proposed algorithm can reduce the error between the predicted HR and actual HR, which is detected by Scosche Rhythm+ [18] during the experiment as ground truth. The accuracy of the proposed algorithm reduce the error from 30.6 to 2.79 bpm in the outdoor and uncontrollable sunlight environment during driving. The proposed application of rPPG on vehicles can be applied in intelligent vehicles system and monitor the driver's heart rate continuously, ensuring driver's health and safety without drivers' awareness.

The remainder of this paper is organized as follows. Sections 2 and 3 introduce our HR method and the experimental scenario respectively. The experimental results and comparisons are illustrated in Sect. 4. Ultimately, conclusions are summarized in Sect. 5.

2 Our Approach

In this section, the details of our algorithms are depicted. In addition, the estimation problems and corresponding solutions are introduced as well. The diagram of the proposed algorithm is shown in Fig. 1 First, a video sequence containing the participants' face is regarded as the input images. Then, a time domain raw signal, CrPPG, was obtained from the linear combination of the input's RGB channel. For reducing noise of raw signal, bandpass filter and Empirical Mode Decomposition (EMD) were utilized. Ultimately, k-nearest neighbor (kNN) classification, the machine learning based method, estimates HR from the frequency domain features. Each algorithm is

Fig. 1. Flowchart of proposed machine learning method for HR extraction from video recording.

discussed in detail in the following sections. All the experimental videos were recorded by a web camera (Logitech C920R).

2.1   Face Detection and HR Signal Recovery

Heart rate detection based on rPPG requires monitoring the skin region, especially the face. Dlib [19], the open toolkit commonly used in both industry and academia for a wide range of domains, was used for frontal face detection. At first, the whole frame is detected to figure out the entire face region as our region of interest (ROI). To lessen the computational load, the face region of the latter frame is based on the 20% expansion of the previous face ROI (fROI). The following HR estimation regard proportion of the fROI, which denoted as pROI, as input images. Then we compute the mean value of all pixels in pROI on R, G, B channels, which denoted as $\mu(R)_i$, $\mu(G)_i$, $\mu(B)_i$, where $i = 1, 2, 3, \ldots$ representing the different frames. A realization of de Haan and Jeanne's CrPPG method linearly combines the three channels as follows:

$$X_i = 3\mu(R)_i - 2\mu(G)_i \text{ and} \tag{1}$$

$$Y_i = 1.5\mu(R)_i + \mu(G)_i - 1.5\mu(B)_i \tag{2}$$

HR signal, denoted as S, computed by Eqs. (1) and (2) and illustrated as follows:

$$S = X_f - \alpha Y_f \tag{3}$$

with

$$\alpha = \frac{\sigma(X_f)}{\sigma(Y_f)} \tag{4}$$

where $\sigma(X_f)$ and $\sigma(Y_f)$ are the standard deviations of X_f and Y_f, respectively, and X_f and Y_f are the length-64 collections of X_i and Y_i. Nevertheless, the algorithm suffers from several limitations of face detection. First, the angle of the face rotation might exceed the limitation of the face detection algorithm. To cope with this kind of missing face problem, we use skin detection based on the previous fROI to ensure the HR signal's continuity. Secondly, as a result of arbitrary motion, the pROI is not certainly the exact same portion of face. Consequently, the noise reduction algorithms are implemented.

2.2   Filtering and Noise Reduction

The original CrPPG signal is frequently polluted by unpredictable channel noise, especially in arbitrary motion scenario. Consequently, the noise reduction process is a must-have. Two steps to eliminate the noise are summarized as follow. First, since human HR is in the range between 0.7 Hz and 3Hz, an FIR Band Pass Filter (BPF) along with Hamming Window is utilized to extract the HR signal. Secondly, as the noise of which frequency band is the same as HR cannot be eliminated by the BPF, Empirical Mode Decomposition (EMD) is utilized to split the polluted signal into HR and noise.

For the design of BPF, the cutoff frequencies are 0.7 Hz and 3Hz with 128 order. The time domain and frequency domain results are illustrated in Fig. 2(a). If there were little noise, the peak frequency would represent the HR. Nonetheless, although the BPF singles out the signal of which frequency energy lay on the range from 0.7 to 3 Hz, there could still be some short duration but strong noise in the same frequency band, as shown in Fig. 2(b). This sort of noise appears frequently in the motion scenario and would confuse the choice of peaks frequency, leading to low estimation accuracy.

So as to reduce the unpredictable noise caused by arbitrary motion, EMD serves as a nice filter to extract the main component of the signal. EMD is used to decompose the signal to several unique intrinsic mode functions (IMFs) and one residue function. The iteration is shown in Eq. (5).

$$\tilde{g}(t) = \sum_{i=1}^{n} IMF_i(t) + R_n(t) \tag{5}$$

Where $IMF_i(t)$ is the ith IMF at time step t, $R_n(t)$ is the residue function at time step t, and n is the number of EMD iterative. The EMD ensures that every IMF is

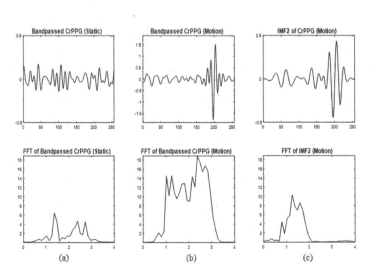

Fig. 2. The comparison of band passed signal and IMF in static and motion scenarios.

symmetrical with zero mean and the same amounts of peaks and zero-crossing points. With the EMD process, the main component of signal in the Fig. 2(b) is extracted as illustrated in Fig. 2(c), making the peak frequency more lucid. As a result, compared with the original band passed signal, the extracted IMF signal serves as more reliable input variables of the following kNN process.

2.3   k-nearest Neighbor Classification and Prediction

Most of the previous work estimate their accuracy by comparing the max peak among all frequency band (MPA, Maximum Peak among All) as an indicator. Nevertheless, the channel noise in motion would distort the frequency response, resulting in several pseudo peak frequencies. The kNN classifier [20] which figures out the closest distance between the testing data and the categories of the training data is implemented in our method. In contrast with MPA, kNN based frequencies selection is more resistant to the distortion of signal. Due to kNN's high accuracy and strong ability to classify unknown and non-Gaussian distribution data, it is of high suitability to our study. Also, its simple and instant property enhances the feasibility of abound real-time applications such as ADAS.

The top five peaks of the FFT spectrum are regarded as the features for the kNN classifier (k = 1). Since kNN is a supervised learning model, the real HR as well as frequency features are required for training. With the assistance of EMD and bandpass filter, the better features are extracted in the frequency band and enhance our classifier. For each volunteer, 40000–45000 training data and 12000–13000 testing data are recorded. With the predicted HR, a Kalman filter is utilized to achieve the smooth and optimal estimation of real HR and avoid the unreasonable huge variation between short time duration. Ultimately, the accuracy of our method is investigated by mean square error (MSE), maximum error (ME) and root mean squared error (RMSE).

3   Experiment Setup

Our experiment contains three scenarios. The first and the second scenarios were evaluated in an indoor environment. First, the participants were allowed to sit in front of the computer casually. In the second study, participants were asked to sway their bodies that caused strong artificial motion. The third study was measured in the driving, at which encountered unpredictable artificial motion.

3.1   First scenario—Sitting Casually

Seven volunteers (five males and two females, age between 22 and 25) participated in our first and second studies. All participants are seated in front of the same computer running Windows 10 in an indoor environment. Video recording was carried out using a web camera (Logitech C920R) mounted on the screen in front of the participants at 80 cm (see Fig. 3). The only illumination source was the ambient ceiling light. All videos were recorded at 30fps in 24-bit RGB color at 1280 × 720 resolution and saved

Fig. 3. The experimental environment of the first and the second studies. The camera was mounted on the computer in front of the participants. Scosche Rhythm+ is banded on participants' wrist to get the heart rate data simultaneously when being video recorded.

in BitMaP (BMP) format. During the experiment, video sequences and the real heart rate were recorded simultaneously. Participants wore Scosche Rhythm+, which inseparably banded on their wrist and transmitted with Bluetooth in 30 Hz, to get the heart rate data when being video recorded. The participants were allowed to either talking on the cell-phone or eating in front of the computer during the measurement. The experiment duration was about 10 min for each participant.

3.2   Second scenario—Strong Artificial Motion

This study was conducted using the same instruments used in the first study. All the participants were asked to swing their body so as to conduct a strong motion scenario. This study was conducted using the same instruments used in the first study. The movement even caused the face's location to be variable. The location of the face was recorded as X-axis (horizontal position) and Y- axis (vertical position). We measured the range and the standard deviation of the face's position as the indicator of the motion. This experiment duration was approximately 10 min for each participant.

3.3   Third scenario—Car Driving

In this study, two males from above studies participated in the car driving with different road section. Same camera was mounted on instrument panel below the front of the driver at the distance of 80 cm. The frames captured by the camera were the driver's face with a slight elevation. Participants as well wore Scosche Rhythm+ to get the heart rate data when being recorded (see Fig. 4). The driving involved flat and slightly uphill asphalt road section in Hsinchu. Driver's face swung while turning the vehicle or sometimes driving through a bumpy road. The illumination source was sunlight that shines in at the windshield. Two participants drove 4.4 and 5.2 km respectively. We divided the driving process to serval intervals and classified in two states, one is the

Fig. 4. The experimental environment of the third study. The camera was mounted on instrument panel below the front of the driver. Scosche Rhythm+ is also banded on participants' wrist to get the heart rate data simultaneously when being video recorded.

Table 1. Frames of each interval and the traffic states for the first participant are recorded as below.

	# of frames	State
Interval I	5290	Driving
Interval II	4580	Stops at the traffic light
Interval III	4220	Driving
Interval IV	3810	Stops at the traffic light
Interval V	2500	Driving

Table 2. Frames of each interval and the traffic states for the second participant are recorded as below.

	# of frames	State
Interval VI	2700	Driving
Interval VII	800	Stops at the traffic light
Interval VIII	7800	Driving
Interval IX	1700	Stops at the traffic light
Interval X	2100	Driving
Interval XI	800	Stops at the traffic light
Interval XII	9159	Driving

interval while driving and the other is the interval while waiting at the traffic light. Two participants drove in different road section and divided to separate intervals recorded in Tables 1 and 2, respectively.

4 Experimental Results

For each participants, the kNN model consisted of 200000–225000 training data. The testing data used the user-dependent training model to estimate the HR. According to the experimental setup, there are three scenario leaving for discussion. The mean square error (MSE), maximum error (ME) and the root mean square error (RMSE) between the predicted HR and the ground truth are regarded as performance criteria. In the meantime, the comparison with MPA, which is commonly used in other rPPG works to compare the efficiency, is illustrated.

4.1 Result for Sitting Casually

For the first scenario, participants were asked to act normally like eating or talking on the cellphone in front of the camera, which is mounted on top of the computer. The video sequence of each participant consists of 8800–12300 frames as testing data. The standard deviation of the x and y coordinates, representing face's movement of all participants were 11.1 and 2.85 pixels respectively. Though the face movement of each participant was quite small, the MSE of the MPA method still can up to 13.05. The noise reduction method can eliminate the wrong feature for training the kNN model. MSE of the model can be reduced to 2–4 bpm, compared to 3–13 bpm of MPA (see Table 3).

Table 3. This table shows the motion level by recorded the face movement of each participant in the first scenario. Therefore, the comparison of our proposed method and MPA was measured in MSE, ME and RMSE.

ID	#of frames	Range of x	Std of x	Range of y	Std of y	Method	MSE	ME	RMSE
1	9201	94	6.4	96	1.7	MPA	5.00	55.21	0.02
						kNN	3.23	14.5	0.008
2	12310	203	16.3	123	2.4	MPA	13.05	42.01	0.02
						kNN	4.32	12.41	0.048
3	8801	196	13.2	158	3.4	MPA	10.30	49.62	0.047
						kNN	3.30	11.23	0.041
4	9101	179	10.2	193	4.2	MPA	3.61	15.62	0.037
						kNN	2.00	6.89	0.024
5	9880	211	10.6	154	3.8	MPA	9.81	50.34	0.03
						kNN	3.41	11.3	0.007
6	10300	262	11.3	169	2.9	MPA	11.61	44.50	0.04
						kNN	2.29	12.68	0.013
7	9760	164	9.8	106	1.6	MPA	10.16	38.51	0.069
						kNN	2.91	10.67	0.037

4.2 Result for Strong Artificial Motion

For the second scenario, participants were asked to swing their body so as to conduct a strong motion scenario with the same instruments in first scenario. The goal of this experimental setup was to evaluate the robustness of our HR method with strong artificial motion. The video sequence of each participant regards 9100–10500 frames as testing data. The standard deviation of the x and y coordinates representing the face's movement of all participants were 46.1 and 14.37 pixels respectively, which were much stronger than that in the first scenario. The strong motion of the face would lead FFT spectrum of the time sequences into chaos. MPA regards the strongest but the wrong peak in the distorted spectrum as HR and the MSE could arise up to 26 bpm. With the user-dependent model, kNN method predicts the HR that's more suitable for corresponding user. The MSE of our proposed method can reduced to 2–8 bpm in comparison with 15–26 bpm of the MPA method (see Table 4).

Table 4. This table shows the stronger motion level compared with the first scenario. Therefore, the comparison of our proposed method and the MPA method was measured in MSE, ME and RMSE.

ID	#of frames	Range of x	Std of x	Range of y	Std of y	Method	MSE	ME	RMSE
1	10565	229	21.3	73	6.7	MPA	20.78	49.21	0.247
						kNN	8.95	23.81	0.102
2	10400	361	49.6	196	15.3	MPA	15.19	54.01	0.179
						kNN	2.75	13.10	0.03
3	9135	413	51.2	217	13.4	MPA	26.24	65.81	0.298
						kNN	2.59	11.56	0.036
4	9700	395	56.1	226	16.2	MPA	15.23	59.21	0.214
						kNN	7.79	27.12	0.116
5	10100	334	53.9	234	19.7	MPA	24.31	67.21	0.243
						kNN	2.44	10.43	0.041
6	9665	449	54.6	236	17.0	MPA	15.62	51.30	0.164
						kNN	6.62	19.6	0.093
7	9890	294	36.4	116	12.3	MPA	24.3	49.80	0.261
						kNN	2.23	10.62	0.039

4.3 Result for Car Driving

In the third scenario, two participants drove the same vehicle in different road section. One drove 4.4 km and the other drove 5.2 km, and the recorded video contains 20501 and 25326 frames respectively. Therefore, the driving was divided into serval intervals and two states, one state is the intervals recorded in driving scenario and the other is the interval recorded while waiting at the traffic light. The MSE, ME and RMSE for each interval are regarded as the performance criteria. The same devices of the previous scenarios are used, but the position of the camera was below the front of the participant that mounted on instrument panel. The goal of this experimental setup was to evaluate the feasibility of the proposed algorithm in an outdoor driving scenario.

The standard deviation of the x-direction movement of the first and the second participant were 20.54 and 13.7 pixels, and 6.59 and 10.6 pixels in y-direction respectively. We used the same training model as the previous scenarios, and the recorded data were tested by participants' own model for predicting HR. With the proposed algorithm, the result shows the advance of robustness in uncontrollable motion during car driving. The MSE can reduce to 2–6 and 3–7 bpm for the first and the second participant respectively. The corresponding intervals and the MPA method's MSE are recorded in Tables 5 and 6. Figures 5 and 6 depict the ground truth which

Table 5. We divided five intervals for the first participant's car driving. The error measurements for each interval were reported in this table.

Interval	#of frames	Range of x	Std of x	Range of y	Std of y	Method	MSE	ME	RMSE
I	5290	143	22.6	38	6.47	MPA	30.6	68.0	0.46
						kNN	2.79	7.27	0.04
II	4580	161	16.7	36	4.81	MPA	22.9	51.2	0.39
						kNN	4.1	11.5	0.07
III	4220	130	20.4	38	5.97	MPA	28.1	64.2	0.50
						kNN	3.57	12.6	0.06
IV	3810	145	21.2	86	9.32	MPA	20.5	60.2	0.41
						kNN	5.62	14.0	0.10
V	2500	155	21.8	46	6.41	MPA	24.1	55.4	0.54
						kNN	6.61	13.4	0.14

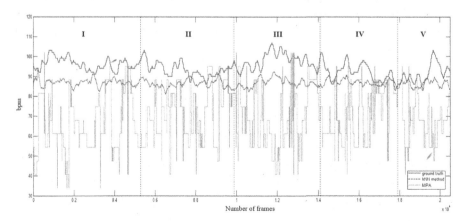

Fig. 5. The red bold line represents the first driver's real HR and the blue dotted line is the HR predicted by our method. The compared method, MPA, is illustrated in green solid line and suffered abruptly drop and climb during the process (Color figure online).

Table 6. We divided seven intervals for the first participant's driving. The error measurements for each interval were reported in this table.

Interval	#of frames	Range of x	Std of x	Range of y	Std of y	Method	MSE	ME	RMSE
VI	2700	100	12.9	54	9.19	MPA	11.3	40.4	0.27
						kNN	6.48	17.3	0.13
VII	800	82	10.6	51	9.67	MPA	12.6	46.1	0.31
						kNN	5.39	13.5	0.11
VII	7800	68	9.68	55	10.03	MPA	11.2	50.2	0.16
						kNN	3.58	13.2	0.05
IX	1700	128	21.2	66	11.3	MPA	10.0	29.4	0.32
						kNN	4.15	8.70	0.11
X	2100	74	11.0	64	10.9	MPA	14.3	33.4	0.36
						kNN	3.87	11.2	0.10
XI	800	34	8.21	53	12.4	MPA	17.9	38.4	0.70
						kNN	7.53	16.6	0.32
XII	9195	128	22.9	64	10.9	MPA	14.0	44.4	0.17
						kNN	4.35	17.0	0.05

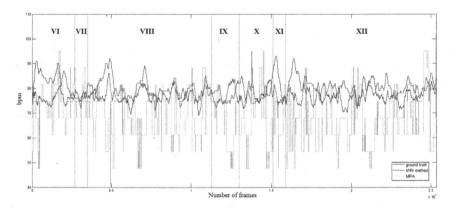

Fig. 6. The red bold line represents the first driver's real HR and the blue dotted line is the HR predicted by our method. Our method can predict the trend of HR with enhanced accuracy in compared with MPA method, which is illustrate in green solid line (Color figure online).

denoted as the red bold line, the HR predicted by our method and MPA method which denoted as the blue dotted line and green solid line respectively for each participants. According to the figures, the HR predicted by the MPA method suffered some suddenly drop and climb during the car driving. While the HR evaluated by our method was not exactly equal to the ground truth, it closely follows the trend of real HR with higher accuracy. The final result and monitoring user interface was illustrated in Fig. 7.

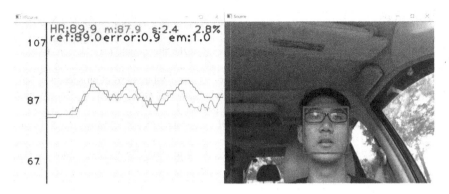

Fig. 7. This figure shows the user interface of the proposed HR monitoring system.

5   Conclusions and Future Work

In this paper, rPPG and our approach are investigated in three different scenarios: an indoor room with the participants sitting static and with strong motion respectively, and the outdoor car driving. To transfer the HR monitoring system from the indoor scenario to an outdoor vehicle requires the stronger anti-interference ability. Consequently, the goal of this paper is to improve motion robustness and increased the feasibility of rPPG method in vehicle application, in which the motion is stronger and unpredictable. To cope with the uncontrollable noise, the user-dependent kNN classifier can customize the HR for each user and reduce abruptly drop and climb phenomena as seen in the MPA method. The new approach of the machine learning based rPPG yields the more accurate and reliable result, reducing the RMSE from 0.208 to 0.036.

Although the accuracy of the proposed approach is increased, we still need to evaluate HR more precisely in other environment conditions. That is, the light changes as well affect the time domain signal during the car driving. So our next investigation will also determine the illuminance source variation in the outdoor environment. In addition, kNN is user-dependent classifier; therefore, the training process for new user is still required. The total solution of the classification model for every user should be considered in our future progress.

In conclusion, with the proposed internal imaging process, the algorithm is beneficial for ADAS. In contrast with contact-type HR sensors, the proposed rPPG can estimate the HR without the concern of distraction and discomfort. The application of computer vision in monitoring driver's heart rate can ensure driver's health and enhance the safety in public.

Acknowledgement. This work was supported by Ministry of Science and Technology under Grand no. MOST105-2622-E-009-013-CC2.

References

1. Hertzman, A.B.: Photoelectric plethysmography of the fingers and toes in man. Exp. Biol. Med. **37**, 290–292 (1937)
2. Zijlstra, W., Buursma, A.: Spectrophotometry of hemoglobin: absorption spectra of bovine oxyhemoglobin, deoxyhemoglobin, carboxyhemoglobin, and methemoglobin. Comp. Biochem. Physiol. B **118**, 743–749 (1997)
3. Cheang, P.Y.S., Smith, P.R.: An overview of non-contact photoplethysmography. Electron. Syst. Control Div. Res, 57–59 (2003)
4. Allen, J: Photoplethysmography and its application in clinical physiological measurement. Physiol. Meas., **28**, R1 (2007)
5. Poh, M.-Z., McDuff, D.J., Picard, R.W.: Non-contact, automated cardiac pulse measurements using video imaging and blind source separation. Opt. Express **18**, 10762–10774 (2010)
6. Lewandowska, M., Ruminski, J., Kocejko, T., Nowak, J.: Measuring pulse rate with a webcam—a non-contact method for evaluating cardiac activity. In: (IEEE) Proceedings of Federated Conference Computer Science and Information Systems, pp. 405–410 (2011)
7. Holton, B.D., Mannapperuma, K., Lesniewski, P.J., Thomas, J.C.: Signal recovery in imaging photoplethysmography. Physiol. Meas. **34**, 1499–1511 (2013)
8. Wedekind, D., Trumpp, A., Andreotti, F., Gaetjen, F., Rasche, S., Matschke, K., et al.: Assessment of source separation techniques to extract vital parameters from videos. In: 23rd European Signal Processing Conference, pp. 434–438 (2015)
9. Mannapperuma, K., Holton, B.D., Lesniewski, P.J., Thomas, J.C.: Performance limits of ICA-based heart rate identification techniques in imaging photoplethysmography. Physiol. Meas. **36**(1), 67–83 (2015)
10. Wang, W.: Exploiting spatial redundancy of image sensor for motion robust rPPG. IEEE Trans. Biomed. Eng. **62**(2), 415–425 (2015)
11. Farnebäck, G.: Two-frame motion estimation based on polynomial expansion. In: Bigun, J., Gustavsson, T. (eds.) SCIA 2003. LNCS, vol. 2749, pp. 363–370. Springer, Heidelberg (2003). doi:10.1007/3-540-45103-X_50
12. Monkaresi, H., Calvo, R.A., Yan, H.: A machine learning approach to improve contactless heart rate monitoring using a webcam. IEEE J. Biomed. Health Inf. **13**, 1153–1160 (2014)
13. de Haan, G., Jeanne, V.: Robust pulse rate from chrominance-based rPPG IEEE Trans. Biomed. Eng. **60**, 2878–2886 (2013)
14. de Haan, G., van Leest, A.: Improved motion robustness of remote-PPG by using the blood volume pulse signature. Physiological measurement **35**(9), 1913–1926 (2014)
15. van Gastel, M.: Motion robust remote-PPG in infrared. IEEE Trans. Biomed. Eng. **62**(5), 1425–1433 (2015)
16. Huang, R.Y., Dung, L.R.: A motion-robust contactless photoplethysmography using chrominance and adaptive filtering, 1–4 (2015)
17. Hsu, Y.: Learning-based heart rate detection from remote photoplethysmography features. In: Proceedings of IEEE International Conference Acoustics Speech Signal Processing, pp. 4433–4437 (2014)
18. SCOSCHE, Rhythm+ (2015). http://www.scosche.com/rhythm-plus-heart-rate-monitor-armband
19. Dlib library, imaging processing toolbox (2016), http://dlib.net/imaging.html
20. Aha, D., Kibler, D., Albert, M.: Instance-based learning algorithms. Mach. Learn. **6**(1), 37–66 (1991)

Instance-Level Segmentation of Vehicles by Deep Contours

Jan van den Brand[1][(✉)], Matthias Ochs[1], and Rudolf Mester[1,2]

[1] VSI Lab, Goethe University, Frankfurt am Main, Germany
vandenbrand@vsi.cs.uni-frankfurt.de
[2] Computer Vision Laboratory, ISY, Linköping University, Linköping, Sweden

Abstract. The recognition of individual object instances in single monocular images is still an incompletely solved task. In this work, we propose a new approach for detecting and separating vehicles in the context of autonomous driving. Our method uses the *fully convolutional network (FCN)* for semantic labeling and for estimating the boundary of each vehicle. Even though a contour is in general a one pixel wide structure which cannot be directly learned by a CNN, our network addresses this by providing areas around the contours. Based on these areas, we separate the individual vehicle instances. In our experiments, we show on two challenging datasets (Cityscapes and KITTI) that we achieve state-of-the-art performance, despite the usage of a subsampling rate of two. Our approach even outperforms all recent works w.r.t. several rating scores.

1 Introduction

One of the important and still unresolved tasks in many computer vision problems is precise object recognition in monocular images. There are several different applications where this information is urgently needed. For example, if a robot wants to grasp a specific object out of group of those objects, like a single cup from a cupboard, the vision algorithm must first detect the objects and subsequently segment the objects into their individual instances. But not only in robotics the task of detection and segmentation of objects is important. In the context of autonomous driving, a full scene understanding of the environment is also crucial, because objects must be recognized for collision avoidance and path planning. Furthermore, it is not sufficient to just classify each pixel into different kinds of object categories (semantic scene labeling) like vehicles, ground, buildings or persons. This is since every individual instance of an object, like a car, can act in a completely different way. Therefore, the detection and segmentation of objects into their single instances is an important task on the path towards full understanding of traffic scenes.

With the rise of deep learning techniques [1] in recent years, the performance of pixel-wise semantic image labeling made a huge leap forward. Deep learning in combination with convolutional neural networks nearly doubled the detection rate compared to previous works, which were mainly based on different types of

© Springer International Publishing AG 2017
C.-S. Chen et al. (Eds.): ACCV 2016 Workshops, Part I, LNCS 10116, pp. 477–492, 2017.
DOI: 10.1007/978-3-319-54407-6_32

Fig. 1. Two example results of our car instance separation approach.

hand crafted features, e.g. in the PASCAL-VOC challenge [2]. Recently, there are new approaches which perform even better. Most of these state-of-the-art methods are employ the fully convolutional network (FCN) [3] and some kind of post-processing step like the usage of Markov random fields (MRF) or conditional random fields (CRF). Other recent approaches combine the strengths of the CNNs and such CRFs. An approximation of the CRF as a Recurrent Neural Network (RNN) is plugged in the CNN. Hence, an end-to-end learning including an approximated CRF is possible.

But all of those approaches have in common that they provide only a pixel-wise semantic labeling, which is not enough for many applications in robotics. Often a semantic segmentation of objects into their individual instances is needed as well. Early approaches separated these detected objects with a simple rectangle boundary box. However, it is much more valuable if the instance-level segmentation yielded a pixel-wise segmentation mask for each instance. Recent works on instance-level segmentation show that a CNN in combination with additional computer vision methods is able to separate instances at pixel-level.

In this work, we focus on automotive vision. Our approach separates only vehicles (cars, buses and trucks) into their individual entities, but it can be

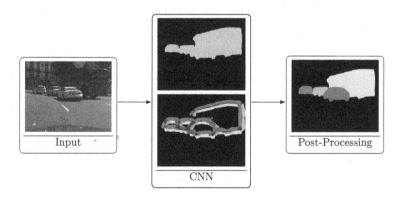

Fig. 2. Our CNN detects vehicles and labels areas left, right, above and below of the contours of the vehicle. With this information we are able to reconstruct the vehicle instances using a floodfill-based algorithm.

also generalized to arbitrary objects if the necessary training data is provided (Fig. 1). The main contribution of this work is contour based separation of the vehicles into their singles instances at pixel-level. The input of our CNN is a single monocular image. Therefore, we adapt the last layers of the FCN in such a way that it provides five channels, where the first channel gives us the semantic class label. The other four channels classify each pixel based on the alignment of the contour (Fig. 2). Thereby, a contour can be aligned as a left, right, bottom or a top contour. Hence, a single vehicle should be characterized by one of each contours, which gives a closed contour around the vehicle. This assumption may not hold in some few cases, because there can be also small missing gaps within the contour. Thus, we introduce a very simple method to close those gaps. Finally, a floodfill-like algorithm yields the individual instances of the vehicles. To train our CNN, we use the fine polygonal chains laid around the vehicles in the Cityscapes training dataset [4].

We have evaluated our approach on two challenging datasets: the Cityscapes [4] and the KITTI [5] datasets. Our method achieves state-of-the-art performance without any depth information and computationally intensive usage of MRFs or CRFs. Additionally, we use a subsampling rate of two during testing in contrast to other approaches.

2   Related Work

Before the deep learning approaches became popular and increased the performance enormously in recent years, many previous works were built on different types of hand-crafted features, like deformable part models, followed by a support vector machine for classification [6,7]. Other common approaches for detecting the individual object instances are based on minimizing a global energy function and CRF models [8] or on shape mask and template matching methods [9].

In contrast to these approaches, our method belongs to the deep learning techniques and can be subdivided into two main tasks. First, a semantic scene labeling of the image is needed. Based on this intermediate classification, the single pixel-wise instances of the object are separated. Our method uses the contour of the object for instance-level segmentation; how to do this is the main contribution of this paper. We review following three basic steps:

2.1   Semantic Scene Labeling

Many approaches [10–12] which use CNNs for semantic scene labeling are built on the fully convolutional network (FCN) from Shelhamer et al. [3] and are mostly followed by some rather complex random-field optimization. Our CNN is an adaptation of the FCN as well. We borrow the first few layers and modify the structure of the last layers for detecting the vehicles themselves (semantic scene labeling) and the their boundaries.

The FCN is also the basis for many other recent works. For example one of the methods [13] top ranked on the Cityscapes dataset combines the FCN output with a multi-resolution approach with Laplacian pyramids for refinement of the semantic labels.

But there are also several other networks which do not take the FCN as a basis. For instance, Lin et al. [14] designed their own 'FeatMap-Net', which has learned semantic correlations between patches. Additionally contextual information is modelled by a CRF. Those CRFs can be also fully integrated as an approximation into a CNNs as a Recurrent Neural Network [15]. The advantage of these networks is that an end-to-end training is possible and the CRF does not need to be integrated as a post-processing step.

2.2   Instance-Level Segmentation

The task of instance-level segmentation can be subdivided into two different categories. The first class generates some kind of a proposal. Based on this proposal, the instances of an object are separated. For example, Gupta et al. [16] detect region proposals in RGB-D data by a CNN, and those regions are subsequently classified into single instances by a random decision forest and finally smoothed with superpixels. This approach was improved by [17,18] to monocular color images. Moreover, the detected regions serve as input for a second CNN, which computes features for the final classification into the instances with a support vector machine. A rather similar approach is pursued by the authors of [19]. Additionally, their CNNs provide semantic labels and boundary box proposals for each instance, which are then refined by bootstrapping. Ren and Zemel [20] also take a boundary box as a proposal. In combination with a recurrent attentive process implemented as a RNN, they get the instances for each object.

The work of Liang et al. [21] can clearly be included in the second category of proposal-free approaches. Besides the semantic labeling, their CNN also provides the pixel-wise position of a boundary box around an object instance. Furthermore, they assign an instance label to each pixel. Finally a standard clustering algorithm returns the individual instances. In the approaches of Zhang et al. [10,11], the CNN also predicts instance labels for every pixels based on a stride of local image patches. This rough estimate of instances is refined by a Markov random field. Uhrig et al. [12] propose a extended FCN, which additionally to the semantic labels returns the discrete depth information and the direction to the center of each object instance. Based on this information the single instances are inferred by template matching.

Our approach cannot be uniquely assigned to one of these two categories. It can be considered as a hybrid method since we do not have clear proposals for each instance, like a box or an image region, neither do we get instances directly from our network. Our CNN provides a set of different directions of contours, from which we conclude the instances.

2.3   Contour Detection

Before different types of learning methods became popular in the field of contour detection, contours were mainly identified by the analysis of local image information, like brightness, color or texture. In [22], all these visual cues are combined into a model to detect object boundaries. Mairal et al. [23] proposed a discriminative sparse image model, which has got the ability to obtain class-specific edges. Based on these edges, the boundary of an object is estimated by a classifier. Dollar et al. [24] introduced a probabilistic boosting tree for edge and contour detection. This supervised learning algorithm implicitly combines textual image information. Texture information and image gradients are also the basis of the inverse object detector of [25]. They fuse the information of a bottom-up contour detector and an object detectors to get the semantic object instances.

Another contour detector was introduced by Arbelaez et al. [26]. Their work also uses multiple local image cues for detecting the contours. Furthermore the output of this or any other contour detector is plugged into a hierarchical region tree for segmentation of the objects. This idea was generalized and extended to RGB-D images in [27].

Kokkinos [28], Rupprecht et al. [29] and Shen et al. [30] showed that object boundaries can also be detected with deep learning techniques. Our FCN also identifies the contour of vehicles, although this boundary is typically only one pixel wide.

3   Approach

In this work, we propose a method for detecting and segmenting vehicles into their individual instances based on the contours of the vehicles. These contours are estimated by an adaptation of the well known fully convolutional network (FCN) [3]. The instance-level segmentation is performed by first using the strong classification capabilities of the FCN to detect the contours. In a following step, we separate the vehicle into their instances by using a floodfill-like algorithm which detects the regions that are separated by contours.

3.1   Detecting Contours with CNNs

Contours of an object are typically a one pixel-wide structure around the border of an object. This means, that those contours are usually not trainable with the typical approaches of the FCN. The FCN usually labels each pixel with a n-dimensional vector, where n is the number of categories and the i-th entry of the vector is the probability score of the pixel belonging to category number i.

Likewise, the ground truth is a pixel-wise labeling with vectors from $\{0,1\}^n$ with an entry being one, if the pixel belongs to the corresponding category. During the training phase, the distance of the label vector of each pixel to the ground truth vector is minimized. This typical approach, however, cannot be used to directly train the labeling of contours for multiple reasons:

– The typical error measure punishes slight shifts of the contour. If some parts of the contour which are generated by the CNN are slightly shifted, the result

can still be considered as good — especially compared to the case where a part of the contour is completely missing. In this case, the floodfill algorithm can still detect individual instances if the areas are separated by contours. If some parts of the contours are missing, however, the two areas become connected and the floodfill algorithm is not able to separate the two instances anymore.
- Moreover, the typical error measures used for classification tend to prefer missing contours over slightly shifted ones. At the same time other pixels which do belong to the real contour are not classified as such. So the error measures will treat a shifted contour as false positives **and** false negatives.
- If some parts of the contour are indeed missing, then the pixels which are not labeled as contour, are only considered to be false negatives, see Fig. 3. As a result, these classification error measures are not suited to measure the quality of the contour classifications.
- The contours are a one pixel-wide structure. However, the FCN uses pooling operations which reduce the resolution of the generated feature maps. As a result, the network can classify only areas and not individual pixels.
- Unfortunately these pooling operations cannot be omitted, because the kernels used by the network would need to be larger which would lead to overfitting. For example, the output of the *pool 4* layer of the FCN is downscaled by a factor of $2^4 = 16$ compared to the input image. The FCN convolves the output of *pool 4* with kernels of size 3×3. If the pooling operations are omitted, then these kernels would need to be of size 48×48 in order to get the information of a region of the same size. So each of these kernels would have $\frac{48^2}{3^2} = 256$ times as many parameters which would lead to serious overfitting.

Due to these reasons, we do not train directly the network on contours. Instead we train it to detect the area directly adjacent to a contour. Each pixel is labeled with a 5 dimensional vector. The first four entries are used to classify the area left, right, above and below a contour. With this information, we can reconstruct the contours by detecting the edges of these areas, e.g. the right edge of the area

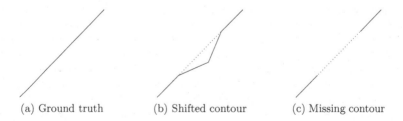

(a) Ground truth   (b) Shifted contour   (c) Missing contour

Fig. 3. The left image shows the ground truth contour which is also displayed as a dotted line in the middle and right image. The center image shows a slightly shifted contour line. An error measure for classification will consider the shifted part as false positives and the dotted part as false negatives. On the right image, the dotted part will also be considered as a false negative, but there are no parts which were wrongfully classified as contour, so a classification measure will prefer the right image over the centered one.

(a) Detection of vehicles (red) (b) Left (blue) and right (red) areas next to the contour (c) Above (blue) and below (red) areas next to the contour (d) The full detected boundaries of the vehicles

Fig. 4. Illustration of five dimensional output vector of our network. CNN detects vehicles and the areas left/right/above and below contours of the vehicles. To obtain the complete vehicle boundary, we detect the edges of these areas. (Color figure online)

labeled as left of a contour. The fifth entry of the labeling vector is used to encode whether a pixel belongs to the corresponding vehicle category (Fig. 4).

3.2 Architecture of the CNN

We use the FCN as a basis for our network and replace the last three convolutional layers with our own. With this approach the network achieves a pixel stride of 16, while using less memory and convolutions than the comparable *FCN-16s*. We will first describe the architecture of most top layers of the *FCN-16s*. Later, we describe why our approach uses less convolutions and memory.

In the *FCN-32s* the *pool 5* layer is followed by three convolutional layers *fc1*, *fc2* and *score*. The layer *fc1* uses 7×7 large kernels for its convolutions, while the kernels of the other two layers are of size 1×1. The output of the *score* layer is upscaled to the resolution of the input image, followed by a sigmoid or softMax layer to obtain classification values in $[0, 1]$.

The network has a pixel stride of $32 = 2^5$, as the feature maps were pooled five times. In order to decrease the pixel stride for a finer classification, the *FCN-16s* uses the same structure of the *fc1*, *fc2* and *score* convolutional layers again, but with the *pool 4* layer as input. This way two independent score and feature maps are generated, whose outputs are simply added pixel-wise.

So in order to decrease the pixel stride by a factor of two, three more convolutions were needed. We propose a different approach which does not need any more convolutions than the *FCN-32s*.

Our top layers, see Fig. 5, receive *pool 4* and *pool 5* layers of the FCN as input. Like the *FCN-32s*, our *fc1* layer convolves the output of the *pool 5* layer with kernels of the size 7×7. This results in an output that contains a lots of abstract information regarding the surrounding pixels.

As the next *fc2* layer of the *FCN-32s* is using only 1×1 kernels, we can safely upscale the result of the previous *fc1* layer by a factor of two, without having to increase the following sizes of the kernels. The upscaled feature maps generated

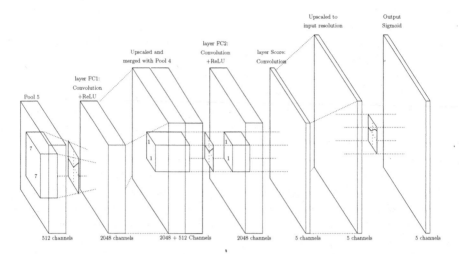

Fig. 5. Our network is based on the *FCN-32s*, where we replaced the top layers with the displayed architecture above. The *pool 5* layer of the FCN is followed by a convolutional layer with 2048 kernels and a ReLU activation. The result is upscaled by a factor of two and merged with the *pool 4* layer. The merged feature maps are then convolved by another 2048 kernels followed with a ReLU activation. The resulting feature maps are convolved again by five kernels to obtain a score value for each classification category. The score map is upscaled to the input resolution of the image and mapped to the interval $[0, 1]$ by a sigmoid activation function.

by the *fc1* layer, thus have the same resolution as the feature maps generated by the *pool 4* layer. So we can simply merge them. In this way, the next two convolutions by the layers *fc2* and *score* can combine the abstract contextual informations of *fc1* with the higher resolved information from *pool 4*. As a result the network has a pixel stride of 16 and can reliably label structures which are at least 16 pixels wide, similar to the *FCN-16s*.

3.3   Reconstructing Instances with Contours

For contours without missing sections, the floodfill algorithm can be used to generate a region of adjacent pixels which are bounded by the contours. If some parts of the contours between two instances are missing, however, the floodfill algorithm is not capable to separate the two instances. As the boundaries generated by the CNN might still have some errors, we need to adapt the floodfill algorithm to separate two instances, even if there is not a complete contour separating them. The general idea is to slightly blur the image representing the contours, as this will close small missing parts in the contour images.

In practice, we simply merge the contour image with the vehicle classification image and use a 8×8 pooling operations with pixel-stride of 8 on the resulting image. If there is at least one pixel classified as contour or not classified as vehicle

(a) A 8 × 8 pooled image of pixels classified as vehicle and not classified as contour.

(b) Instance image which is computed by using the flood-fill algorithm.

(c) Upsampled instance image with a less pooled vehicle/instance image.

(d) Result of expanding the instance image.

(e) Fully finalized instance-level segmentation.

Fig. 6. Multiple steps for reconstructing the instances based on the detected contours and vehicle classification.

in the 8 × 8 block, the whole block gets classified as contour or as not belonging to the vehicle class (see Fig. 6a).

Afterwards, we apply the floodfill algorithm on the new downsampled contour/vehicle image to obtain the instances. As a starting point for the floodfill algorithm, we choose a pixel that is classified as vehicle, not classified as a contour and not yet assigned to an instance. This is repeated until no valid starting point exists anymore which means every pixel that was classified as a vehicle and is not a part of a contour, belongs to an instance (see Fig. 6b).

The result is an image of the vehicle instances with low resolution. To obtain a pixel-wise labeling of the instances with the same resolution as the input image, the instance image is repeatedly upsampled and expanded. The image representing, the instances, is upsampled by a factor of two with a nearest neighbor filter. We also downsample the contour/vehicle image with a pooling operation to the same resolution as the now upsampled instance image (see Fig. 6c).

Next, the instances are expanded. For each instance, we select all directly adjacent pixels which do not belong to an instance, yet. From these selected pixels, we add each pixel to their adjacent instance, if the pixel is classified as vehicle and not as contour. If a pixel is adjacent to multiple instances, it is added to only one of the instances. This is repeated until none of the instances can be expanded anymore (see Fig. 6d).

This upsampling and expansion of the instances are repeated three times until the instance image has the same resolution as the input image. In the last step we expand all instances again, though this time the expansion may add pixels classified as contour to an instance. In this way, we obtain an instance segmentation in which all pixels classified as vehicle are assigned to an instance (see Fig. 6e).

Note that the repeated upscaling by a factor of 2 and then expanding the instances is important. This should not be omitted by directly upscaling the instance image to the input resolution. By using a higher resolved contour image, the chances of having a small hole in a contour are increased. These small holes in the contours result in two areas no longer being separated. In such a case the expansion of the instances can result in an instance 'leaking' into the previously separated area of another instance. By repeatedly upscaling and expanding the instances, this 'leaking' effect is reduced.

4   Experiments

The experiments are performed with the Caffe framework [31]. For training the network, we use the Cityscapes training set, which was downscaled by a factor of 2 and consists of 4.486 pixel-wise annotated images. The learning rate was constantly 10^{-9}. The training is performed until the loss function on the Cityscape evaluation set converged to a minimum.

For evaluation, we use of the *Cityscapes dataset* [4] the evaluation (500 images) and the test set (1.525 images) as well as the annotated test set (144 images) from [10], which is based on the *KITTI object detection* benchmark [5].

4.1   Metrics

In the work of [10,12], several different metrics are proposed to evaluate the results on the different datasets. We give a brief overview over these metrics:

- *IoU* (intersection over union):
  The *IoU* is a measure for the overlap. For a ground truth instance and prediction instance the *IoU* is the ratio of true positive pixels to the sum of true positive, false positive and false negative pixels.
- *MWCov and MUCov* (mean weighted/unweighted coverage):
  For each ground truth instance, we select the predicted instance with the best IoU. These best IoU values are then averaged across the image. For the weighted coverage value, the IoUs are weighted by the size of instances.
- *AvgPr and AvgRe* (average precision/recall):
  For each predicted instance, we count the pixels belonging to the *vehicle* class. Then, we divide this number by the number of pixels belonging to the predicted instance to obtain the precision value of the instance. Averaging these values results in the *AvgPr* value. For the recall value, we calculate for each ground truth instance the ratio of detected pixels to the number of total pixels belonging to the ground truth instance. Averaging these values results in the *AvgRe* value.
- *AvgFP and AvgFN* (average false positives/negatives):
  If no pixel of a predicted instance belongs to the *vehicle* class, the instance

is considered as false positive. Likewise, if a ground truth instance is not overlapped by any predicted instances, it is considered as a false negative. Averaging the numbers across all images yields the *AvgFP* and *AvgFN* score.

- *InsPr, InsRe and InsF1* (instance precision/recall/F1 score):
  By selecting the ground truth prediction pairs with an IoU of more than 50%, each ground truth/prediction instance is paired with at most one prediction/ground truth instance. The ratio of these detected pairs to the number of total instance prediction yields the *InsPr*, while the ratio to the number of total ground truth instances yields the *InsRe*. *InsF1* is the F1 score of the instance precision and instance recall.

- *AP, AP50%, AP100m* and *AP50m* (average precision across recall values):
  The *AP* is another average precision value, however this time it is the average in relation to the recall, meaning it is the area below the precision-recall-curve. This average precision is calculated multiple times with different *IoU* thresholds for the prediction/ground truth pairs and then averaged across the thresholds. The thresholds range from 0.5 to 0.95 in steps of 0.05.
  The *AP50%* value is the average precision value for the 50% *IoU* threshold and *AP100m* and *AP50m* are the *AP* score restricted to instances within 100 m or 50 m distance.

4.2 Evaluation

Cityscapes: Our results are better than the combination of MCG and R-CNN approach [4] and slightly worse than the results of [12]. The performance difference to [12] becomes a lot smaller for close vehicles. It is important to note that the approach of [12] uses a network with pixel-stride of 8 and the evaluation was performed on fully resolved images. Our approach uses a network with pixel-stride of 16 instead and the evaluation images were downsampled by a factor of two. With these technical limitations, it is not possible for our network to properly segment small instances, i.e. vehicles in the distance. This can be seen in the last image of Fig. 7 and also in the quantitative results in Tables 1 and 2, where restricting the evaluation to close vehicles greatly improves the performance. For close cars, for instance, our results in Table 1 are comparable with those of [12].

Table 1. Quantitative results using the datasets and metrics described in [4]. The *vehicles* class is the union of the instances of *car,truck,bus,caravan* and *trailer*.

Method	Dataset	Labels	AP ↑	AP50% ↑	AP100m ↑	AP50m ↑
[12]	CS test	car	22.5	37.8	36.4	40.7
[4]	CS test	car	10.5	26.0	17.5	21.2
Ours	CS test	car	18.2	29.2	31.0	39.0
Ours	CS val	car	19.3	30.6	31.0	39.7
Ours	CS val	vehicles*	19.6	31.4	31.5	41.1

| Input image | Ground truth | Our Prediction |

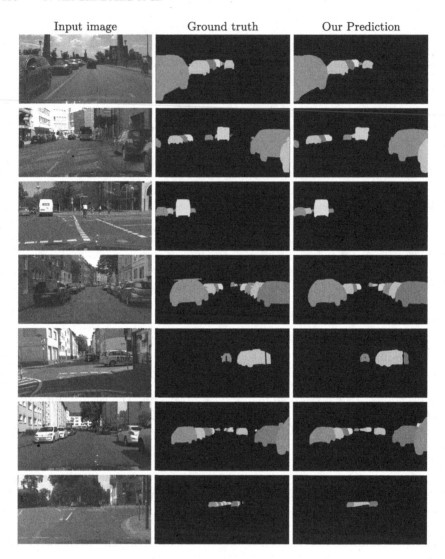

Fig. 7. Example results on the Cityscapes validation set. The last three examples represent typical failcases: Oversegmentation when an instance consists of disconnected areas, oversemgentation of small structures (i.e. side-view mirrors), not being able to separate small instances.

KITTI: Compared to other approaches, we see in Table 2 that the recall of our approach is very good, while the precision is worse compared to [10] and [12]. Our instance precision is only slightly outperformed by [10] while our recall is so much better that the overall instance detection score *InsF1* clearly outperforms [10], too. Besides of the two recall values *AvgRe* and *InsRe* we are also able to outperform [12] in regards to the *AvgFN*. When comparing the coverage values, we outperform [10] in both the weighted and unweighted case, while having comparable results to [20] and a very similar *MWCov* value compared to [12].

Besides of the same types of failure cases that can also be seen on the Cityscapes examples, there is a new failure case for KITTI data: We were not able

Table 2. Quantitative results using the KITTI test set from [10] and the Cityscapes validation set [4]. All results are in percent and the best values are highlighted. The last two rows only uses instances within a distance of 100 m or 50 m respectively. We used the median depth of an instance for the distance threshold.

Method	Dataset	MWCov ↑	MUCov ↑	AvgPr ↑	AvgRe ↑	AvgFP ↓	AvgFN ↓	InsPr ↑	InsRe ↑	InsF1 ↑
[12]	KITTI test	79.7	**75.8**	**85.6**	82.0	0.201	0.159	**86.3**	74.1	**79.7**
[11]	KITTI test	67.0	49.8	82.0	61.3	0.479	0.840	48.9	43.8	46.2
[10]	KITTI test	69.7	51.8	83.9	57.5	0.375	1.139	65.3	50.0	56.6
[20]	KITTI test	**82.2**	67.7	n/a	n/a	**0.188**	0.382	n/a	n/a	n/a
Ours	KITTI test	80.8	70.1	69.2	**91.6**	0.986	**0.097**	63.2	**78.0**	69.8
Ours	CS val	73.0	43.4	86.1	71.3	0.13	1.136	68.6	45.0	54.3
Ours	CS val 100 m	76.2	59.0	87.4	82.8	0.090	0.228	71.7	63.0	67.1
Ours	CS val 50 m	81.1	69.5	89.8	88.9	0.074	0.088	79.1	75.5	77.3

Input image	Ground truth	Our prediction

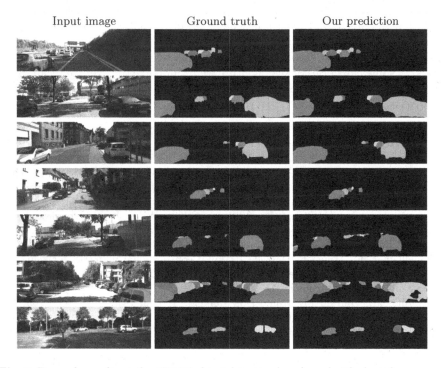

Fig. 8. Example results at the *KITTI object detection* benchmark. The last three examples represent typical fail cases: untrained perspective distortions result in various errors such as false positives, not being able to separate instances and missing patches of vehicles. Also disconnected areas of a single instance are oversegmented.

to train on KITTI data, as a result the perspective distortions by the different camera settings are unknown to our network. Especially the strong distortions close to the left and right border of the image result in false positive instances (5th image in Fig. 8) and other errors as seen in the 6th image of Fig. 8.

5   Summary and Conclusion

In this work, we proposed a new hybrid approach for instance-level segmentation of vehicles which can also be generalized to arbitrary objects. Our method does not rely on proposals, like a boundary box or a specific given region around each instance. It identifies each individual vehicle based on its contour. Although this is generally one pixel-wide boundary around an object and thus it cannot be trained in network, it is ensured by the output of our fully convolutional network. Thus, the first channel represents the semantic class label and the other four specify the contour area around the object. Based on this boundary, single object instances can be obtained by simple post-processing algorithms.

Our approach yields comparable results to state-of-the-art methods on two challenging datasets: Cityscapes [4] and KITTI [5]. On the Cityscapes benchmark for instance-level segmentation, our method achieves the second best result. It is only slightly worse than the current leader [12] in context of car labels, however, we use a subsampling of two in contrast to them. Although we do not train our network at the KITTI dataset, we obtain similar and comparable results to other works. In some categories of the quantitative results, like the average recall, false negative and instance recall score, we even outperform all other current available approaches.

References

1. Krizhevsky, A., Sutskever, I., Hinton, G.E.: Imagenet classification with deep convolutional neural networks. In: Advances in Neural Information Processing Systems (NIPS), pp. 1097–1105 (2012)
2. Everingham, M., Van Gool, L., Williams, C.K.I., Winn, J., Zisserman, A.: The pascal visual object classes (VOC) challenge. Int. J. Comput. Vis. (IJCV) **88**, 303–338 (2010)
3. Shelhamer, E., Long, J., Darrell, T.: Fully convolutional networks for semantic segmentation. Trans. Pattern Anal. Mach. Intell. (PAMI) (2016). http://ieeexplore. ieee.org/abstract/document/7478072/
4. Cordts, M., Omran, M., Ramos, S., Rehfeld, T., Enzweiler, M., Benenson, R., Franke, U., Roth, S., Schiele, B.: The cityscapes dataset for semantic urban scene understanding. In: Conference on Computer Vision and Pattern Recognition (CVPR) (2016)
5. Geiger, A., Lenz, P., Urtasun, R.: Are we ready for autonomous driving? the KITTI vision benchmark suite. In: Conference on Computer Vision and Pattern Recognition (CVPR), pp. 3354–3361 (2012)
6. Tighe, J., Niethammer, M., Lazebnik, S.: Scene parsing with object instances and occlusion ordering. In: Conference on Computer Vision and Pattern Recognition (CVPR), pp. 3748–3755 (2014)

7. Felzenszwalb, P.F., Girshick, R.B., McAllester, D., Ramanan, D.: Object detection with discriminatively trained part-based models. Trans. Pattern Anal. Mach. Intell. (PAMI) **32**, 1627–1645 (2010)
8. Ladický, Ľ., Sturgess, P., Alahari, K., Russell, C., Torr, P.H.S.: What, where and how many? combining object detectors and CRFs. In: Daniilidis, K., Maragos, P., Paragios, N. (eds.) ECCV 2010. LNCS, vol. 6314, pp. 424–437. Springer, Heidelberg (2010). doi:10.1007/978-3-642-15561-1_31
9. He, X., Gould, S.: An exemplar-based CRF for multi-instance object segmentation. In: Conference on Computer Vision and Pattern Recognition (CVPR), pp. 296–303 (2014)
10. Zhang, Z., Fidler, S., Urtasun, R.: Instance-level segmentation for autonomous driving with deep densely connected MRFs. In: Conference on Computer Vision and Pattern Recognition (CVPR) (2016)
11. Zhang, Z., Schwing, A.G., Fidler, S., Urtasun, R.: Monocular object instance segmentation and depth ordering with CNNs. In: International Conference on Computer Vision (ICCV), pp. 2614–2622 (2015)
12. Uhrig, J., Cordts, M., Franke, U., Brox, T.: Pixel-level encoding and depth layering for instance-level semantic labeling. In: Rosenhahn, B., Andres, B. (eds.) GCPR 2016. LNCS, vol. 9796, pp. 14–25. Springer, Cham (2016). doi:10.1007/978-3-319-45886-1_2
13. Ghiasi, G., Fowlkes, C.C.: Laplacian pyramid reconstruction and refinement for semantic segmentation. In: Leibe, B., Matas, J., Sebe, N., Welling, M. (eds.) ECCV 2016. LNCS, vol. 9907, pp. 519–534. Springer, Cham (2016). doi:10.1007/978-3-319-46487-9_32
14. Lin, G., Shen, C., van den Hengel, A., Reid, I.: Efficient piecewise training of deep structured models for semantic segmentation. In: Conference on Computer Vision and Pattern Recognition (CVPR) (2016)
15. Zheng, S., Jayasumana, S., Romera-Paredes, B., Vineet, V., Su, Z., Du, D., Huang, C., Torr, P.: Conditional random fields as recurrent neural networks. In: International Conference on Computer Vision (ICCV), pp. 1529–1537 (2015)
16. Gupta, S., Girshick, R., Arbeláez, P., Malik, J.: Learning rich features from RGB-D images for object detection and segmentation. In: Fleet, D., Pajdla, T., Schiele, B., Tuytelaars, T. (eds.) ECCV 2014. LNCS, vol. 8695, pp. 345–360. Springer, Cham (2014). doi:10.1007/978-3-319-10584-0_23
17. Hariharan, B., Arbeláez, P., Girshick, R., Malik, J.: Simultaneous detection and segmentation. In: Fleet, D., Pajdla, T., Schiele, B., Tuytelaars, T. (eds.) ECCV 2014. LNCS, vol. 8695, pp. 297–312. Springer, Cham (2014). doi:10.1007/978-3-319-10584-0_20
18. Girshick, R., Donahue, J., Darrell, T., Malik, J.: Region-based convolutional networks for accurate object detection and segmentation. Trans. Pattern Anal. Mach. Intell. (PAMI) **38**, 142–158 (2016)
19. Wu, Z., Shen, C., van den Hengel, A.: Bridging Category-level and Instance-level Semantic Image Segmentation. arXiv:1605.06885 [cs.CV] (2016)
20. Ren, M., Zemel, R.S.: End-to-End Instance Segmentation and Counting with Recurrent Attention. arXiv:1605.09410 [cs.LG] (2016)
21. Liang, X., Wei, Y., Shen, X., Yang, J., Lin, L., Yan, S.: Proposal-free network for instance-level object segmentation. arXiv:1509.02636 [cs.CV] (2015)
22. Martin, D.R., Fowlkes, C.C., Malik, J.: Learning to detect natural image boundaries using local brightness, color, and texture cues. Trans. Pattern Anal. Mach. Intell. (PAMI) **26**, 530–549 (2004)

23. Mairal, J., Leordeanu, M., Bach, F., Hebert, M., Ponce, J.: Discriminative sparse image models for class-specific edge detection and image interpretation. In: Forsyth, D., Torr, P., Zisserman, A. (eds.) ECCV 2008. LNCS, vol. 5304, pp. 43–56. Springer, Heidelberg (2008). doi:10.1007/978-3-540-88690-7_4

24. Dollar, P., Tu, Z., Belongie, S.: Supervised learning of edges and object boundaries. In: Conference on Computer Vision and Pattern Recognition (CVPR), pp. 1964–1971 (2006)

25. Hariharan, B., Arbeláez, P., Bourdev, L., Maji, S., Malik, J.: Semantic contours from inverse detectors. In: International Conference on Computer Vision (ICCV), pp. 991–998 (2011)

26. Arbelaez, P., Maire, M., Fowlkes, C., Malik, J.: Contour detection and hierarchical image segmentation. Trans. Pattern Anal. Mach. Intell. (PAMI) **33**, 898–916 (2011)

27. Gupta, S., Arbeláez, P., Girshick, R., Malik, J.: Indoor scene understanding with RGB-D images: bottom-up segmentation, object detection and semantic segmentation. Int. J. Comput. Vis. (IJCV) **112**, 133–149 (2015)

28. Kokkinos, I.: Pushing the boundaries of boundary detection using deep learning. In: International Conference on Learning Representations (ICLR) (2016)

29. Rupprecht, C., Huaroc, E., Baust, M., Navab, N.: Deep Active Contours. arXiv:1607.05074 [cs.CV] (2016)

30. Shen, W., Wang, X., Wang, Y., Bai, X., Zhang, Z.: Deepcontour: a deep convolutional feature learned by positive-sharing loss for contour detection. In: Conference on Computer Vision and Pattern Recognition (CVPR), pp. 3982–3991 (2015)

31. Jia, Y., Shelhamer, E., Donahue, J., Karayev, S., Long, J., Girshick, R., Guadarrama, S., Darrell, T.: Caffe: Convolutional Architecture for Fast Feature Embedding. arXiv preprint arxiv:1408.5093 (2014)

STFCN: Spatio-Temporal Fully Convolutional Neural Network for Semantic Segmentation of Street Scenes

Mohsen Fayyaz[1(✉)], Mohammad Hajizadeh Saffar[1], Mohammad Sabokrou[1], Mahmood Fathy[2], Fay Huang[3], and Reinhard Klette[4]

[1] Malek-Ashtar University of Technology, Tehran, Iran
mohsen.fayyaz89@gmail.com
[2] Iran University of Science and Technology, Tehran, Iran
[3] National Ilan University, Yilan, Taiwan
[4] Auckland University of Technology, Auckland, New Zealand

Abstract. This paper presents a novel method to involve both spatial and temporal features for semantic segmentation of street scenes. Current work on *convolutional neural networks* (CNNs) has shown that CNNs provide advanced spatial features supporting a very good performance of solutions for the semantic segmentation task. We investigate how involving temporal features also has a good effect on segmenting video data. We propose a module based on a *long short-term memory* (LSTM) architecture of a recurrent neural network for interpreting the temporal characteristics of video frames over time. Our system takes as input frames of a video and produces a correspondingly-sized output; for segmenting the video our method combines the use of three components: First, the regional spatial features of frames are extracted using a CNN; then, using LSTM the temporal features are added; finally, by deconvolving the spatio-temporal features we produce pixel-wise predictions. Our key insight is to build *spatio-temporal convolutional networks* (spatio-temporal CNNs) that have an end-to-end architecture for semantic video segmentation. We adapted fully some known convolutional network architectures (such as FCN-AlexNet and FCN-VGG16), and dilated convolution into our spatio-temporal CNNs. Our spatio-temporal CNNs achieve state-of-the-art semantic segmentation, as demonstrated for the Camvid and NYUDv2 datasets.

1 Introduction

Semantic segmentation of video data is a fundamental task for scene understanding. For many computer vision applications, semantic segmentation is considered as being (just) a pre-processing task. Consequently, the performance of semantic segmentation has a direct effect on subsequent computer vision solutions which depend on it. Self-driving cars is one of the areas in technology that has received much attention recently. These cars can detect surroundings using advanced driver assistance systems (ADAS) that consist of many different systems such as

© Springer International Publishing AG 2017
C.-S. Chen et al. (Eds.): ACCV 2016 Workshops, Part I, LNCS 10116, pp. 493–509, 2017.
DOI: 10.1007/978-3-319-54407-6_33

radar, GPS, computer vision, and in-car networking to bring safety to driving and roads. One of the main processes for the computer vision part of these systems can be identified as being semantic segmentation of all objects in surroundings to transmit accurate and complete information to the ADAS system such that the system can make the best decision to avoid accidents.

Segmentation is typically approached as a classification problem. First, using a set of labeled video frames, the characteristics of all segments (classes) are learned. These characteristics are used for labeling the pixels of test frames [1,48]. Recently, deep learning methods, especially CNNs, ensured state-of-the-art performance in different areas of computer vision, such as in image classification [23], object detection [13], or activity recognition [38].

We consider the application of advanced features, extracted by using CNNs, for semantic video segmentation. Semantic segmentation methods use both given image data at selected locations as well as a semantic context. A set of pixels is usually predicted as defining one class (or even one segment) if connected, and also referring to one particular semantic interpretation.

Previous methods for video segmentation have efficiently exploited CNNs, but they did not use temporal features; of course, temporal features can be useful for interpreting a video semantically. For example, the authors of [1,48] represented and interpreted video frames using a deep learning method, but the main disadvantage of their methods is that they consider those frames as being independent from each other. Neglecting the time dimension in video data basically means that the given raw data are down-sampled without using fully given information. Using temporal features can help the system to distinguishing, for example, between two objects of different classes having the same spatial features but showing differences in the time feature dimension.

Consequently, we propose a method which uses a similar paradigm for extracting spatial features (as in the cited papers), but which differs by also using temporal features (i.e., features of a continues sequences of frames). We propose to identify components which can be embedded "on top" of spatially extracted features maps in individual frames. Such a component can be seen as being equipped with a set of memory cells which save the assigned regions in previous frames. This allows us that relations between regions, available in previous frames, can be used to define temporal features. We process the current video frame by using the spatio-temporal output features of our processing modules.

Similar to other segmentation methods, we use then some fully convolutional layers to perform regional semantic classification. In our method, these fully convolutional layers perform spatio-temporal classifications. Finally, we use a deconvolution procedure for mapping (i.e., scaling) the obtained predictions into the original carrier (i.e., the image grid) of the given frames for having a pixelwise prediction. See Fig. 1.

CNN-based methods usually combine two components, where one is for describing and inferring a class of different regions of a video frame as a feature map, and another one for performing an up-sampling of the labeled feature

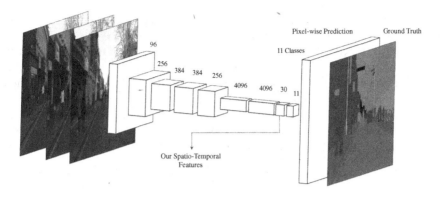

Fig. 1. A spatio-temporal fully convolutional Alexnet architecture, later also to be discussed in Sect. 4.1.

maps to the size of the given video frames. An advantage of our method is that we can adjust and embed our proposed module into the end of the first component (before inferring the labels) of current CNN-based methods as an end-to-end network. We show that the proposed changes in the network lead to an improvement in the performance of state-of-the-art methods, such as, FCN-8 [29] and dilated convolution [48].

The main contributions of this paper are as follows:

– The proposed method can be easily adapted for enhancing already published state-of-the art methods for improving their performance.
– We propose an end-to-end network for semantic video segmentation in respect to both spatial and temporal features.
– We propose a module for transforming traditional, fully convolutional networks into spatio-temporal CNNs.
– We outperformed state-of-the art methods on two standard benchmarks.

The rest of this paper is organized as follows. Top-ranked related work on semantic video segmentation is reviewed in Sect. 2. Section 3 introduces the proposed method. The performance of our method is shown in Sect. 4. Section 5 concludes the paper.

2 Related Work

There is a wide range of approaches that have been published so far for video segmentation. Some of them have advantages over others. These approaches can be categorized based on the kind of data that they operate on, the method that is used to classify the segments, and the kind of segmentation that they can produce.

Some approaches focus on binary classes such as foreground and background segmentation [2,4]. This field includes also some work that has a focus on anomaly detection [34,35] since authors use a single-class classification scheme and constructed an outlier detection method for all other categories. Some other approaches concentrate on multi-class segmentation [6,26,27,43].

Recently created video datasets provide typically image data in RGB format. Correspondingly, there is no recent research on gray-scale semantic video segmentation; the use of RGB data is common standard, see [11,21,26,27,46]. There are also some segmentation approaches that use RGB-D datasets [16,17,30].

Feature selection is a challenging step in every machine learning approach. The system's accuracy is very much related to the set of features that are chosen for learning and model creation. Different methods have been proposed for the segmentation-related feature extraction phase.

2.1 Feature Extraction

We recall briefly some common local or global feature extraction methods in the semantic segmentation field. These feature extraction methods are commonly used after having super-voxels extracted from video frames [27].

Pixel color features are features used in almost every semantic segmentation system [11,21,26,27,30]. Those includes three channel values for RGB or HSV images, and also values obtained by histogram equalization methods. The *histogram of oriented gradients* (HOG) defines a set of features combining at sets of pixels approximated gradient values for partial derivatives in x or y direction [21,26]. Some approaches also used other histogram definitions such as the hue color histogram or a texton histogram [46].

Further appearance-based features are defined as across-boundary appearance features, texture features, or spatio-temporal appearance features; see [11,21,26,27]. Some approaches that use RGB-D datasets, also include 3-dimensional (3D) positions or 3D optical flow features [17,30]. Recently, some approaches are published that use CNNs for feature extraction; using pre-trained models for feature representation is common in [1,16,47].

After collecting a set of features for learning, a model must be chosen for training a classifier for segmentation. Several methods have been provided already for this purpose, and we recall a few.

2.2 Segmentation Methods

Some researches wanted to propose a (very) general image segmentation approach. For this reason, they concentrated on using unsupervised segmentation. This field includes clustering algorithms such as k-means and mean-shift [28], or graph-based algorithms [11,17,21,45].

A *random decision forest* (RDF) can be used for defining another segmentation method that is a kind of a classifier composed of multiple classifiers which are trained and enhanced by using randomness extensively [15,33]. The *support vector machine* (SVM) [41] or a *Markov random field* (MRF) [36,44] are further methods used for segmentation but not as popular as the *conditional random field* (CRF) that is in widespread use in recent work [5,26,32].

Neural networks are a very popular method for image segmentation, especially with the recent success of using convolutional neural network in the semantic segmentation field. Like for many other vision tasks, neural networks have become very useful [1,13,16,19,29,47].

Fully convolutional networks (FCNs) are one of the topics that interest researchers recently. An FCN is based on the idea of extending a *convolutional network* (ConvNet) for arbitrary-sized inputs [29]. On the way of its development, it has been used for 1-dimensional (1D) and 2-dimensional (2D) inputs [31,42], and for solving various tasks such as image restoration, sliding window detection, depth estimation, boundary prediction, or semantic segmentation. In recent years, many approaches use ConvNets as feature extractor [1,16,47]. Some approaches turn ConvNets into FCNs by discarding the final classifier layer, and convert all fully connected layers into convolutions. By this change, authors use a front-end module for solving their vision tasks [1,13,16,19,29,47].

Recently, a new convolutional network module has been introduced by Yu and Fisher [48] that is especially designed for dense prediction. It uses dilated convolutions for multi-scale contextual information aggregation, and achieves some enhancements in semantic segmentation compared to previous methods. Kundu and Abhijit [24] optimized the mapping of pixels into a Euclidean feature space; they achieve even better results for semantic segmentation than [48] by using a graphical CRF model.

Many approaches that have been introduced in this field have not yet used temporal features, especially in the field of deep CNNs [11,17,21,26,27,46]. These approaches cannot be identified as being end-to-end methods, which points to an essential disadvantage when applying these approaches. Some approaches use deep CNNs [16,24] by introducing an end-to-end architecture for also using spatio-temporal features for semantic labeling. However, none of them can change the size of time windows dynamically.

Long short-term memory (LSTM) is a memory cell module that was introduced by [12,18]. It has many advantages such as the ability to support very large time windows, the ability to change time windows dynamically, the ability to handle noise, distributed representations, continuous values, and so forth. We propose for the first time an approach that uses a deep CNN network with LSTM modules as an end-to-end trainable architecture for semantic video segmentation and labeling.

3   The Proposed Method

3.1   Overall Scheme

We have four key steps in our method as shown in Fig. 2. We feed the frame I_t (i.e., the tth frame of a video), into a FCN network. This network down-samples the input images and describes a frame I_t, defined on an image grid Ω of size $W \times H$, as a features set $S_t^{1..m}$ in m different maps. The input is I_t and the output of the latest layer (i.e., of lowest resolution) of the FCN is $S_t^{1..m}$ of size $W' \times H'$, where $W' \ll W$ and $H' \ll H$. As a result, frame I_t is represented as

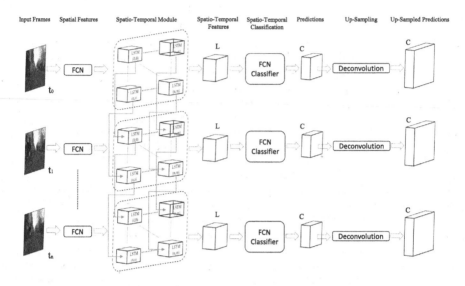

Fig. 2. Overall scheme for our proposed end-to-end network architecture. The LSTMs are used for inferring the relations between spatial features which are extracted from the frames of the video. L is the number of LSTM units hidden nodes, and C is the number of classes.

a feature set $\{S_t^{1..m}\}$. Every point (i,j), with $1 \leq i \leq W'$ and $1 \leq j \leq H'$, in $S_t^{1..m}$ is a descriptor of size m for a region (receptive field) in I_t.

We put our *spatio-temporal module* on top of the final convolutional layer. So, feature set $\{S_t^{1..m}\}$ will be represented as a spatio-temporal feature set of $\{ST_t^{1..m}\}^{(i,j)}$ by our *spatio-temporal module*. By applying an FCN classifier layer on top of these features, we predict the semantic classes of these regions in the video. Finally, we up-sample these predictions to the size of the I_t frame. In following subsections, the methodologies that have been used in this approach, will be described.

3.2   Fully Convolutional Network

Convolutional neural networks (CNNs) are applied for a large set of vision tasks. Some researchers improve CNNs by changing its basic architecture and introducing new architectures. Recently, fully convolutional networks (FCNs) have been introduced by discarding the final classifier layer, and by converting all fully connected layers into convolutional layers. We follow this principle.

3.3   LSTM

A long short-term memory (LSTM) network is a special kind of *recurrent neural networks* (RNNs) that have been introduced by [18] to solve the vanishing gradient problem and for remembering information over long periods. For an example

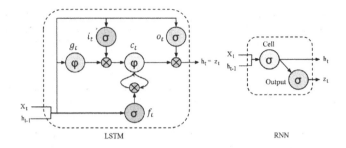

Fig. 3. An example of a basic LSTM cell (*left*) and a basic RNN cell (*right*). Figure follows a drawing in [8].

of a basic RNN and an LSTM cell, see Fig. 3. LSTMs are not confined to fixed-length inputs or outputs, and this advantage makes them powerful for solving sequential problems.

Each LSTM module consists of a memory cell and a number of input and output gates that control the information flow in a sequence and prevent it from loosing important information in a time series. Assuming S_t as the input of an LSTM module at time t, the cell activation is as formulated in the following equations:

$$i_t = \sigma(W_{xi}x_t + W_{hi}h_{t-1} + b_i) \tag{1}$$
$$f_t = \sigma(W_{xf}x_t + W_{hf}h_{t-1} + b_f) \tag{2}$$
$$o_t = \sigma(W_{xo}x_t + W_{ho}h_{t-1} + b_o) \tag{3}$$
$$g_t = \phi(W_{xc}x_t + W_{hc}h_{t-1} + b_c) \tag{4}$$
$$c_t = f_t \odot c_{t-1} + i_t \odot g_t \tag{5}$$
$$h_t = o_t \odot \phi(c_t) \tag{6}$$

where σ and ϕ are symbols for a sigmoid and the tanh function, respectively. Symbol $h_t \in R^N$ denotes a hidden state with N units, and $c_t \in R^N$ is the memory cell. By $i_t \in R^N$, $f_t \in R^N$, $o_t \in R^N$, and $g_t \in R^N$ we denote the input gate, forget gate, output gate, and input modulation gate at time t, respectively. Symbol $\odot$ stands for element-wise multiplication.

3.4 Spatio-Temporal Module

In regards to every $W' \times H'$ region of I_t, which is described by an FCN as an Ω grid, an LSTM is embedded (see Sect. 3.1). Thus we have altogether $W' \times H'$ LSTMs. Element $\{S_t^{1..m}\}^{(i,j)}$ defines a spatial characteristics of a region in the I^t frame. These characteristics are given to LSTM$^{(i,j)}$ for processing; it infers a relation with spatial features of equivalent regions in frames previous to frame I_t. With this "trick", both spatial and temporal features of a frame are considered. (Note that LSTM$^{(i,j)}(\{S_t^{1..m}\}^{(i,j)}) = \{ST_t^{1..m}\}^{(i,j)}$) where S and ST are spatial and spatio-temporal features, respectively).

We embed one LSTM for each region. Equation (7) shows a representation of frame I_t with respect to our suggested spatial and temporal features:

$$\Omega'_t(i,j) = (LSTM^{(i,j)}(\Omega_t(i,j)) \tag{7}$$

where the size of Ω' is equal to that of Ω, and value m specifies a map which assigns spatio-temporal features to every point for describing an equivalent region (i.e., a segment) in I_t. Now, similar to other methods [1, 29, 48], the labels for points in Ω' are predicted and up-sampled to the frame at the original size. The overall update function can be briefly specified as follows:

$$\Omega'_t(i,j) = \sigma(W_{xo}x_t + W_{ho}h_{t-1} + b_o) \odot \phi(f_t \odot c_{t-1} + i_t \odot g_t) \tag{8}$$

Altogether, we introduced an operator layer to several LSTMs for properly representing the temporal features.

This proposed network executes and processes the input frames as an end-to-end network. Figure 2 shows the overall scheme of our method.

3.5   Deconvolution

Interpolation is a common method for mapping outputs into dense pixels. There are several interpolation (or upsampling) algorithms such as bilinear, non-linear, cubic, and so forth. Up-sampling by a factor k can be considered as being a convolution with a fractional input stride of $1/k$. As a result, a convolution operator with input stride of $1/k$ can be applied backward (called deconvolution) with a stride of k [29].

4   Experimental Results

For implementing our *spatio-temporal fully convolutional network* (STFCN) we use the standard Caffe distribution [20] and a modified Caffe library with an LSTM implementation.[1] We merged this LSTM implementations into the Caffe standard distribution and released our modified Caffe distribution to support new FCN layers that have been described in [29]. Our code has been tested on NVIDIA TITAN, and NVIDIA TITAN-X GPUs.[2]

To show the performance of our modified version of FCNs we use their implemented models for two cases, with and without our spatio-temporal module. We tested our STFCN networks on Camvid[3] and NYUDv2[4] datasets. Our evaluation methodology is as in other state-of-the-art semantic segmentation tests, such as in [1, 29].

---

[1] Available at https://github.com/junhyukoh/caffe-lstm.

[2] Our modified Caffe distribution and STFCN models are publicly available at https://github.com/MohsenFayyaz89/STFCN.

[3] Available at mi.eng.cam.ac.uk/research/projects/VideoRec/CamVid/.

[4] Available at https://cs.nyu.edu/~silberman/datasets/nyu_depth_v2.html.

In the following, first we describe the way how we embed our spatio-temporal module into FCNs and dilation convolution networks. Then we describe the metrics used in the evaluation process. After that we report our experiments on CamVid and NYUDv2. Finally, we discuss the performance of our method.

4.1 Embedding the Spatio-Temporal Module in FCN Networks

FCN-8 and FCN-32 [29] are fully convolutional versions of VGG-16 with some modifications to combine features of shallow layers with more precise spatial information with features of deeper layers which have more precise semantic information.

As mentioned in Sect. 3, it is of benefit to embed the spatio-temporal module on top of the deepest layers. Thus we embed our spatio-temporal module on top of the $fc7$ layer of FCN-8 and FCN-32. The $fc7$ is the deepest fully convolutional layer which has large corresponding receptive fields in the input image. This layer extracts features which represent more semantic information in comparison to shallower layers.

An example of this modification of an FCN-Alexnet is shown in Fig. 1. After embedding our spatio-temporal module in FCN-8 and FCN-32 networks, we call them STFCN-8 and STFCN-32. Our spatio-temporal module consists of LSTMs with 30 hidden nodes and 3 time-steps for the CamVid dataset. We fine-tuned our STFCN networks from pre-trained weights on PASCAL VOC [9] provided by [29]. We used a momentum amount of 0.9, and a learning rate of 10e-5.

4.2 Embedding Our Module in Dilated Convolution Networks

A dilated convolution network is an FCN network which benefits from some modifications such as reducing down-sampling layers and using a context module which uses dilated convolutions. This module brings multi-scale ability to the network [48].

The dilated8 network [48] consists of two modules, front-end and context. The front-end module is based on a VGG-16 network with some modifications. The context layer is connected on top of this module. The $fc7$ layer of the front-end layer provides the main spatial features with 4,096 maps. This network has an input of size $900 \times 1,100$. Because of removing some of its down-sampling layers, the $fc7$ layer has an output of size 66×91 which defines a high dimension for spatio-temporal computations. For overcoming this complexity problem, we down-sampled the output of this layer by a convolution layer to the size of 21×30, and fed it to our spatio-temporal module. Then, the spatio-temporal features are fed to a convolutional layer to decrease their maps to the size of the *final* layer of the front-end module.

After resizing the maps, features are fed to a deconvolution layer to up-sample them to the size of the *final* layer output (66×91). Finally, we fuse them with the front-end *final* layer by an element-wise sum operation over all features.

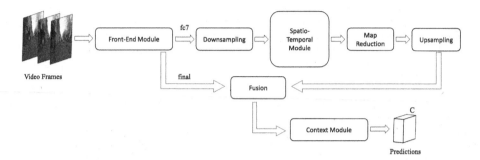

Fig. 4. Our STDilation8 model architecture

The fused features are fed to the context module. Let STDilated8 be the modified version of dilated8; see Fig. 4. The spatio-temporal module of STDilated8 consists of 30 hidden nodes of LSTMs with a time-step of 3. For training this network, we fixed the front-end module and fine-tuned the spatio-temporal and context modules with dilation8 pre-trained weights on CamVid. We used a momentum amount of 0.9, and a learning rate of 10e-5.

For better performance of the spatio-temporal module, we down-sampled the output of the $fc7$ layer of the dilation8 front-end module and fed it to the spatio-temporal module. Then we reduced the feature maps by a fully convolutional layer for a better description of the spatio-temporal features and make them the same size as the final layer of the front-end module. Finally we up-sample and fuse the spatio-temporal features with the final layer output and feed them into the context module.

4.3   Quality Measures for Evaluation

There are already various measures available for evaluating the accuracy of semantic segmentation. We describe most commonly used measures for accuracy evaluation which we have used to evaluate our method.

Mean Intersection over Union. Mean IU is a segmentation performance measure that quantifies the overlap of two objects by calculating the ratio of the area of intersection to the area of unions [22,46]. This is a popular measure since it penalizes both over-segmentation and under-segmentation separately [36]. It is defined as follows:

$$\frac{1}{n_{cl}} \cdot \sum_i \frac{n_{ii}}{t_i + \sum_j n_{ji} - n_{ii}} \tag{9}$$

where n_{ii} is the number of pixels of class i that is predicted correctly as belonging to class i, t_i is the total number of pixels in class i, and n_{cl} is the number of classes.

4.4 CamVid

The *Cambridge-driving labelled video database* (CamVid) [3] is a collection of videos with object-class semantic labels, complete with meta-data. The database provides ground truth labels that associate each pixel with one of 32 semantic classes. Like in [39], we partitioned the dataset into 367 training images, 100 validation, and 233 test images. Eleven semantic classes are used in the selected images.

For FCN-8, FCN-32, STFCN-8, and STFCN-32, the images are downsampled to 400 × 400. For dilation8 and STDilation8, the images are down-sampled to 640 × 480. As mentioned before, we used time-step 3 for our spatio-temporal module which means that we feed a sequence of 3 frames to our spatio-temporal networks.

The reason for choosing number 3 is that the annotated frames of CamVid have a distance of 30 frames to each other. In fact when we use 3 frames as a sequence, the first and last frame of the sequence have a distance of 90 frames. Using more annotated frames is computationally possible because of the given LSTM abilities, but it is semantically wrong because of the high amount of changes in the frames.

Our results of FCNs and STFCNs tests on CamVid are shown in Table 1. It appears that adding our spatio-temporal module into FCN networks shows an improvement of their performance by close to one percent. Results for dilation8 and STDilation8 tests on CamVid are shown in Table 2. The effect of the spatio-temporal module is here an improvement by 0.8%. Improvements are in both cases not "dramatic" but consistent. Note that reports about improvements in the semantic segmentation area are typically in the sub-one-percent range [24,29,48].

Table 1. Evaluating FCNs and STFCNs for video semantic segmentation on Camvid (i.e., without or with our spatio-temporal module)

	FCN-32s	STFCN-32s	FCN-8s	STFCN-8s
Mean IU	46.1%	46.9%	49.7%	50.6%

Table 2. Evaluating dilated convolution networks, without or with our module on Camvid

	Dilation8	STDilation8 (90 Frames)
Mean IU	65.3%	65.9%

Dilation8 achieves the best results in comparison to other work, and this is due to the power of multi-scale semantic segmentation. STDilation8 achieves even slightly better results because of benefits from temporal features. Detailed results on the CamVid test set are reported in Table 3. Our model outperforms prior state-of-the-art work.

Table 3 shows that some approaches are competitive to related work such as Liu and He [26] with a performance superiority by 0.8 percent compared to SegNet [1]. In contrast, other approaches with a new base architecture achieved a better performance. Since our approach is based on FCN [29] or Dilation8 [48] methodologies, with our introduced spatio-temporal module, performance enhancement is close to one percent on FCN network, and close to 0.8 percent on

Table 3. Our STDilation8 improves Dilation8 and outperforms prior work on Camvid

	Building	Tree	Sky	Car	Sign	Road	Pedestrian	Fence	Pole	Sidewalk	Bicyclist	Mean IU
ALE [25]	73.4	70.2	91.1	64.2	24.4	91.1	29.1	31.0	13.6	72.4	28.6	53.6
SuperParsing [40]	70.4	54.8	83.5	43.3	25.4	83.4	11.6	18.3	5.2	57.4	8.90	42.0
Liu and He [26]	66.8	66.6	90.1	62.9	21.4	85.8	28.0	17.8	8.3	63.5	8.50	47.2
SegNet [1]	68.7	52.0	87.0	58.5	13.4	86.2	25.3	17.9	16.0	60.5	24.8	46.4
STFCN-8	73.5	56.4	90.7	63.3	17.9	90.1	31.4	21.7	18.2	64.9	29.3	50.6
DeepLab-LFOV [7]	81.5	74.6	89.0	82.2	42.3	92.2	48.4	27.2	14.3	75.4	50.1	61.6
Dilation8 [48]	82.6	76.2	89.9	84.0	46.9	92.2	56.3	35.8	23.4	75.3	55.5	65.3
Dilation + FSO [24]	84.0	77.2	91.3	85.6	49.9	92.5	59.1	37.6	16.9	76.0	57.2	66.1
STDilation8	83.4	76.5	90.4	84.6	50.4	92.4	56.7	36.3	22.9	75.7	56.1	65.9

Fig. 5. Outputs on CamVid. *Top to bottom rows:* Test samples, ground truth, Dilation8 [48], and STDilation8.

Dilation8 architecture; both can be considered as a being a noticable enhancement. Dilation + FSO [24] has been published recently based on Dilation8 architecture and became state-of-the-art video semantic segmentation method. Our approach differs from FSO in several ways:

- Our approach does not need any pre-processing or feature optimization for result enhancement. In contrast, FSO has used optical flow as a feature set to be used by a CRF model. This is a computational operation which can be considered as a weakness for a semantic segmentation method. Computation efficiency and speed is very crucial in some tasks, such as, online video processing in advanced driver assistance systems. Some researches are ongoing to resolve optical flow computational cost by using convolutional networks [10].
- We used time-step 3 for our spatio-temporal module to use the CamVid dataset annotations as-is without any preprocessing. This simplicity in design and configuration, is one of the strengths of our work.
- Our approach proposes an end-to-end network for semantic video segmentation which consists of spatial and temporal features altogether.
- In our approach we proposed a neural network based module for transforming traditional, fully convolutional networks into spatio-temporal CNNs. It can also be used for other related video processing tasks.

Also, we embedded our spatio-temporal module into FCN-Alexnet and evaluated its performance with and without our spatio-temporal module. Our spatio-temporal module improved its performance on CamVid dataset. Because the basic FCN-Alexnet has a low performance for semantic segmentation as described in [29], so we decided not to include details into this paper.

4.5 NYUDv2

The NYU-Depth V2 data set is comprised of video sequences from a variety of indoor scenes recorded with an RGB-Depth camera [37]. It features 1,449 densely labelled pairs of aligned RGB and depth images, including 464 new scenes taken at three cities, and 407,024 new unlabeled frames (Fig. 5).

We selected this dataset to evaluate the effect of multi-modal learning on our spatio-temporal module. Also, we tested our method on two totally different datasets (outdoor vs. indoor) to evaluate its flexibility. One problem of this dataset is that its annotated frames vary in length of sequences per subject or location. Thus, for this dataset, we do not use a constant time step for the spatio-temporal module. We fed sequences of different lengths based on their location. This problem showed its effect on results by decreasing the amount of improvements compared to none-temporal models.

Gupta et al. [15] coalesced NYU-Depth V2 into 40 classes. Similar to [29] we report results on a standard split into 795 training images and 654 test images. We selected our models based on [29] to be able to evaluate an embedding of our spatio-temporal module into their models. We use FCN-32s RGB and FCN-32s

Table 4. Evaluating STFCNs, FCNs, and dilated convolution networks for semantic video segmentation on NYUDv2

Network	Pixel accuracy	Mean accuracy	Mean IU
Gupta *et al.* [14]	60.3%	–	28.6%
FCN-32s RGB [29]	60.0%	42.2%	29.2%
FCN-32s RGBD [29]	61.5%	42.4%	30.5%
STFCN-32s RGB	60.9%	42.3%	29.5%
STFCN-32s RGBD	62.1%	42.6%	30.9%

RGBD models to embed our spatio-temporal module in the way as explained before. Tests on NYUDv2 data are reported in Table 4.

Results show in this case the enhancement effect of the spatio-temporal module on FCN-32s RGB and FCN-32s RGBD compared to the related networks FCN-32s RGB and FCN-32s RGBD, respectively.

4.6 Discussion

We showed the power of our spatio-temporal module by embedding it into other known spatial, fully convolutional networks. In fact we introduced a spatio-temporal, fully convolutional network for extracting spatio-temporal features from video data and evaluated it based on two semantic segmentation case studies.

Our module benefits from the LSTM characteristics and is able to handle long-short term sequences. In our tests we were only able to use a limited number of video frames as being one sequence because of the limited number of available annotated frames. The method should also be tested on datasets with more extensive sets of annotated frames to check the effect of sequence length on the performance of the system. It is possibly also of value to check the effect of involving unannotated frames into input sequences by using prior or posterior annotated frames in the system.

5 Conclusions

In this paper we proposed a new architecture for spatio-temporal feature extraction from video. We designed and used this architecture for semantic video segmentation. First, a pre-trained CNN model was turned into an FCN model by changing classification layers into fully convolutional layers. In this phase, spatial features from input frames can be used for classification. But, in semantic video segmentation, relationships between frames can provide very useful information and enhance the accuracy of the segmentation program. Therefore, LSTM modules have been used to take advantage of temporal features. This architecture has been proposed as an end-to-end trainable model and can also be used for other vision tasks. Also, it does not need to be a pre-processing or post-processing module only, as we have seen in some other approaches. We illustrated the performance of our architecture by embedding our spatio-temporal module into some state-of-the-art fully convolutional networks, such as FCN-VGG, and dilation convolution. Other types of LSTM modules have been proposed recently and have shown promising results for some vision tasks. Applying these newly proposed modules may enhance further the architecture of our spatio-temporal module, e.g. for scene understanding, anomaly detection in video, video captioning, object tracking, activity recognition, and so forth.

References

1. Badrinarayanan, V., Kendall, A., Cipolla, R.: SegNet: a deep convolutional encoder-decoder architecture for image segmentation. arXiv preprint arXiv:1511.00561 (2015)
2. Bittel, S., Kaiser, V., Teichmann, M., Thoma, M.: Pixel-wise segmentation of street with neural networks. arXiv preprint arXiv:1511.00513 (2015)
3. Brostow, G.J., Shotton, J., Fauqueur, J., Cipolla, R.: Segmentation and recognition using structure from motion point clouds. In: Forsyth, D., Torr, P., Zisserman, A. (eds.) ECCV 2008. LNCS, vol. 5302, pp. 44–57. Springer, Heidelberg (2008). doi:10.1007/978-3-540-88682-2_5
4. Carreira, J., Sminchisescu, C.: Constrained parametric min-cuts for automatic object segmentation. In: CVPR, pp. 3241–3248 (2010)
5. Chang, F.J., Lin, Y.Y., Hsu, K.J.: Multiple structured-instance learning for semantic segmentation with uncertain training data. In: CVPR, pp. 360–367 (2014)
6. Chen, A.Y., Corso, J.J.: Propagating multi-class pixel labels throughout video frames. In: Image Processing Workshop (WNYIPW), pp. 14–17 (2010)
7. Chen, L.C., Yang, Y., Wang, J., Xu, W., Yuille, A.L.: Attention to scale: Scale-aware semantic image segmentation. arXiv preprint arXiv:1511.03339 (2015)
8. Donahue, J., Anne Hendricks, L., Guadarrama, S., Rohrbach, M., Venugopalan, S., Saenko, K., Darrell, T.: Long-term recurrent convolutional networks for visual recognition and description. In: CVPR, pp. 2625–2634 (2015)
9. Everingham, M., Van Gool, L., Williams, C.K.I., Winn, J., Zisserman, A.: The PASCAL visual object classes challenge 2011 (2011). www.pascal-network.org/challenges/VOC/voc2011/workshop/index.html
10. Fischer, P., Dosovitskiy, A., Ilg, E., Häusser, P., Hazirbas, C., Golkov, V., van der Smagt, P., Cremers, D., Brox, T.: Flownet: Learning optical flow with convolutional networks. arXiv preprint arXiv:1504.06852 (2015)
11. Galasso, F., Keuper, M., Brox, T., Schiele, B.: Spectral graph reduction for efficient image and streaming video segmentation. In: CVPR, pp. 49–56 (2014)
12. Gers, F.A., Schmidhuber, J., Cummins, F.: Learning to forget: continual prediction with LSTM. In: Neural computation, pp. 2451–2471 (2000)
13. Girshick, R., Donahue, J., Darrell, T., Malik, J.: Rich feature hierarchies for accurate object detection and semantic segmentation. In: CVPR, pp. 580–587 (2014)
14. Gupta, S., Arbelaez, P., Malik, J.: Perceptual organization and recognition of indoor scenes from RGB-D images. In: CVPR, pp. 564–571 (2013)
15. Gupta, S., Girshick, R., Arbeláez, P., Malik, J.: Learning rich features from RGB-D images for object detection and segmentation. In: Fleet, D., Pajdla, T., Schiele, B., Tuytelaars, T. (eds.) ECCV 2014. LNCS, vol. 8695, pp. 345–360. Springer, Cham (2014). doi:10.1007/978-3-319-10584-0_23

16. He, Y., Chiu, W.C., Keuper, M., Fritz, M.: RGBD semantic segmentation using spatio-temporal data-driven pooling. arXiv preprint arXiv:1604.02388 (2016)
17. Hickson, S., Birchfield, S., Essa, I., Christensen, H.: Efficient hierarchical graph-based segmentation of RGBD videos. In: CVPR, pp. 344–351 (2014)
18. Hochreiter, S., Schmidhuber, J.: Long short-term memory. Neural Comput. **12**, 1735–1780 (1997)
19. Hong, S., Noh, H., Han, B.: Decoupled deep neural network for semi-supervised semantic segmentation. In: NIPS, pp. 1495–1503 (2015)
20. Jia, Y., Shelhamer, E., Donahue, J., Karayev, S., Long, J., Girshick, R., Guadarrama, S., Darrell, T.: Caffe: convolutional architecture for fast feature embedding. In: Proceedings of the ACM International Conference Multimedia, pp. 675–678 (2014)
21. Khoreva, A., Galasso, F., Hein, M., Schiele, B.: Classifier based graph construction for video segmentation. In: CVPR, pp. 951–960 (2015)
22. Klette, R., Rosenfeld, A.: Digital Geometry. Morgan Kaufmann, San Francisco (2004)
23. Krizhevsky, A., Sutskever, I., Hinton, G.E.: ImageNet classification with deep convolutional neural networks. In: Advances Neural Information Processing Systems, pp. 1097–1105 (2012)
24. Kundu, A., Vineet, V., Koltun, V.: Feature space optimization for semantic video segmentation. In: CVPR (2016)
25. Russell, C., Kohli, P., Torr, P.H.: Associative hierarchical CRFs for object class image segmentation. In: ICCV, pp. 739–746 (2009)
26. Liu, B., He, X.: Multiclass semantic video segmentation with object-level active inference. In: CVPR, pp. 4286–4294 (2015)
27. Liu, X., Tao, D., Song, M., Ruan, Y., Chen, C., Bu, J.: Weakly supervised multiclass video segmentation. In: CVPR, pp. 57–64 (2014)
28. Liu, Y., Liu, J., Li, Z., Tang, J., Lu, H.: Weakly-supervised dual clustering for image semantic segmentation. In: CVPR, pp. 2075–2082 (2013)
29. Long, J., Shelhamer, E., Darrell, T.: Fully convolutional networks for semantic segmentation. In: CVPR, pp. 3431–3440 (2015)
30. Martinovic, A., Knopp, J., Riemenschneider, H., Van Gool, L.: 3D all the way: semantic segmentation of urban scenes from start to end in 3D. In: CVPR, pp. 4456–4465 (2015)
31. Matan, O., Burges, C.J., LeCun, Y., Denker, J.S.: Multi-digit recognition using a space displacement neural network. In: NIPS, pp. 488–495 (1991)
32. Mottaghi, R., Fidler, S., Yao, J., Urtasun, R., Parikh, D.: Analyzing semantic segmentation using hybrid human-machine CRFs. In: CVPR, pp. 3143–3150 (2013)
33. Richmond, D.L., Kainmueller, D., Yang, M.Y., Myers, E.W., Rother, C.: Relating cascaded random forests to deep convolutional neural networks for semantic segmentation. arXiv preprint arXiv:1507.07583 (2015)
34. Sabokrou, M., Fathy, M., Hoseini, M., Klette, R.: Real-time anomaly detection and localization in crowded scenes. In: CVPR, Workshops, pp. 56–62 (2015)
35. Sabokrou, M., Fathy, M., Hoseini, M.: Video anomaly detection and localisation based on the sparsity and reconstruction error of auto-encoder. Electron. Lett. **52**, 1122–1124 (2016)
36. Sharma, A., Tuzel, O., Jacobs, D.W.: Deep hierarchical parsing for semantic segmentation. In: CVPR, pp. 530–538 (2015)
37. Silberman, N., Hoiem, D., Kohli, P., Fergus, R.: Indoor segmentation and support inference from RGBD images. In: Fitzgibbon, A., Lazebnik, S., Perona, P., Sato, Y.,

Schmid, C. (eds.) ECCV 2012. LNCS, vol. 7576, pp. 746–760. Springer, Heidelberg (2012). doi:10.1007/978-3-642-33715-4_54

38. Simonyan, K., Zisserman, A.: Two-stream convolutional networks for action recognition in videos. In: Advances Neural Information Processing Systems, pp. 68–576 (2014)

39. Sturgess, P., Alahari, K., Ladicky, L., Torr, P.H.: Combining appearance and structure from motion features for road scene understanding. In: BMVC (2012)

40. Tighe, J., Lazebnik, S.: SuperParsing: scalable nonparametric image parsing with superpixels. In: Daniilidis, K., Maragos, P., Paragios, N. (eds.) ECCV 2010. LNCS, vol. 6315, pp. 352–365. Springer, Heidelberg (2010). doi:10.1007/978-3-642-15555-0_26

41. Volpi, M., Ferrari, V.: Semantic segmentation of urban scenes by learning local class interactions. In: CVPR, pp. 1–9 (2015)

42. Wolf, R., Platt, J.C.: Postal address block location using a convolutional locator network. In: NIPS, pp. 745–745 (1994)

43. Yang, Y., Hallman, S., Ramanan, D., Fowlkes, C.C.: Layered object models for image segmentation. IEEE Trans. PAMI **34**, 1731–1743 (2012)

44. Zheng, C., Wang, L.: Semantic segmentation of remote sensing imagery using object-based Markov random field model with regional penalties. IEEE J. Sel. Top. Appl. Earth Observations Remote Sens. **8**, 1924–1935 (2015)

45. Zhang, L., Song, M., Liu, Z., Liu, X., Bu, J., Chen, C.: Probabilistic graphlet cut: exploiting spatial structure cue for weakly supervised image segmentation. In: CVPR, pp. 1908–1915 (2013)

46. Zhang, Y., Chen, X., Li, J., Wang, C., Xia, C.: Semantic object segmentation via detection in weakly labeled video. In: CVPR, pp. 3641–3649 (2015)

47. Zhu, Y., Urtasun, R., Salakhutdinov, R., Fidler, S.: Segdeepm: exploiting segmentation and context in deep neural networks for object detection. In: CVPR, pp. 4703–4711 (2015)

48. Yu, F., Koltun, V.: Multi-scale context aggregation by dilated convolutions. In: ICLR (2016)

Spontaneous Facial Behavior Analysis

LBP-TOP: A Tensor Unfolding Revisit

Xiaopeng Hong, Yingyue Xu, and Guoying Zhao$^{(\boxtimes)}$

Center for Machine Vision and Signal Analysis, University of Oulu, Oulu, Finland
gyzhao@ee.oulu.fi

Abstract. Local Binary Pattern histograms from Three Orthogonal Planes (LBP-TOP) has shown its promising performance on facial expression recognition as well as human activity analysis, as it extracts features from spatial-temporal information. Originally, as the calculation of LBP-TOP has to traverse all the pixels in the three dimensional space to compute the LBP operation along XY, YT and XT planes respectively, the frequent use of loops in implementation shapely increases the computational costs. In this work, we aim to fasten the computational efficiency of LBP-TOP on spatial-temporal information and introduce the concept of tensor unfolding to accelerate the implementation process from three-dimensional space to two-dimensional space. The spatial-temporal information is interpreted as a 3-order tensor, and we use tensor unfolding method to compute three concatenated big matrices in two-dimensional space. LBP operation is then performed on the three unfolded matrices. As the demand for loops in implementation is largely down, the computational cost is substantially reduced. We compared the computational time of the original LBP-TOP implementation to that of our fast LBP-TOP implementation on both synthetic and real data, the results show that the fast LBP-TOP implementation is much more time-saving than the original one. The implementation code of the proposed fast LBP-TOP is now publicly available (The implementation code of the proposed fast LBP-TOP can be downloaded at http://www.ee.oulu.fi/ research/imag/cmvs/files/code/Fast_LBPTOP_Code.zip).

1 Introduction

Feature extraction, for a long time, plays an important role in image processing and pattern recognition. Specifically, feature extraction constructs dimensionality-reduced values from the large amount of original data to describe the statistical characters, and facilitates recognition tasks such as facial expression recognition, object detection, texture classification, *etc.* There are a variety of methods for feature description, including geometric feature-based methods, shape-based methods and appearance-based methods [1], among which appearance-based methods are the most widely used ones. Appearance-based methods extract the image features in spatial domain and learn the feature extraction scheme based on the relationship between the components or points within the space [2].

X. Hong and Y. Xu—These two authors contributed equally.

© Springer International Publishing AG 2017
C.-S. Chen et al. (Eds.): ACCV 2016 Workshops, Part I, LNCS 10116, pp. 513–527, 2017.
DOI: 10.1007/978-3-319-54407-6_34

Local Binary Pattern (LBP) [1] is one of the most popular and efficient appearance-based feature descriptors [3,4]. It has proven to be highly discriminative and its invariance to monotonic gray-level discrepancies makes it a robust feature descriptor in two-dimensional space. In general, an LBP operator measures each pixel of a given image by thresholding its neighborhood with the value of the center pixel and forms the results into a binary pattern. Then, the occurrence histogram based on the resulted binary patterns can be computed over an image or a region of the image, which is proven to be a powerful feature descriptor. Now, due to its discriminative power and computational simplicity, LBP is broadly utilized in image pattern recognition. Moreover, it receives tremendous success in facial expression recognition as it is insensitive to illumination variations and well describes subtle appearance details of the local features on human faces [5].

As LBP is proven to be of high performance as well as low computational cost, it is frequently applied in pattern recognition on static images. Following the steps of LBP, Zhao et al. [6] started to explore appearance-based feature descriptors on dynamic or temporal information, which combines appearance and motion. Hence, Local Binary Pattern histograms from Three Orthogonal Planes (LBP-TOP) was proposed. LBP-TOP is an extension of LBP from two-dimensional space to three-dimensional space including spatial and time domain. More specifically, LBP-TOP regards the pixel in a three dimensional space with spatial and temporal properties and computes the LBP of each pixel on three orthogonal planes, and finally formulates three occurrence histograms corresponding to the three orthogonal planes. LBP-TOP not only inherits merits from LBP which is insensitive to illumination variations, translations or rotations, but also extends its applications to high dimensional video feature analysis. Moreover, LBP-TOP has shown its promising performance on facial expression recognition as well as human activity analysis, as it is capable to analyse appearance changes from a sequence of images [7–11].

However, as LBP-TOP operates on three dimensional spatial-temporal information, the computational cost also sharply increases compared to LBP on two dimensional static images. Suppose there is an image with frame width W and height H, the total number of LBP operations over all the pixels on the image is $(W \times H)$ times. Now we have another sequence of dynamic images with the same image size of width W and height H and the number of frames T, LBP-TOP needs to be applied to this three dimensional matrix. Then the total number of LBP operations over all the pixels climbs to $(3 \times W \times H \times T)$ times. Moreover, as LBP-TOP traverses all the pixels to compute LBP operations on XY, YT and XT planes respectively, it results in a frequent usage of nested loops in implementation. The use of complicated nested loops in implementation already heavily affects the computational time of LBP-TOP, let along the steep rise in computational costs, especially when we need to analyse a long sequence of images.

In this paper, we aim to fasten the computational efficiency of LBP-TOP for feature extraction on spatial-temporal information and introduce the concept of tensor unfolding to accelerate the implementation process from three-dimensional space to two-dimensional space. Therefore, we propose a fast LBP-TOP implementation method which unfolds the 3-order tensor of spatial-temporal information to 2-order tensors and performs LBP operations over the 2-order tensors. The proposed fast LBP-TOP implementation method benefits from optimized codes with reduced nested loops and largely saves the computational costs compared to the original implementation.

The contribution of this paper are two folds:

1. We propose a fast LBP-TOP implementation method that takes the advantage of tensor unfolding method to simplify the LBP-TOP implementation from a three-dimensional space to a two-dimensional space. The tensor unfolding method largely reduces the demand of loops in codes, such that the computational cost is substantially reduced.
2. We perform experimental comparisons of the proposed fast LBP-TOP implementation method to the original one. The results show significant improvements in terms of computational cost.

2 Related Work

2.1 Local Binary Patterns

The local binary patterns (LBP) has proven to be a simple yet very efficient operator for feature description. The LBP operator is derived from a general definition of texture in a local neighborhood, and was firstly proposed by Ojala et al. in 1996 [12]. It can be regarded as a unifying methodology other than traditionally divergent statistical and structural models of texture analysis. Now it is broadly utilized in fields such as face expression recognition and analysis [13,14], biomedical image processing [15] and texture analysis [16].

Given a texture or an image, suppose that we computes the LBP operation over all the pixels. For each pixel, a certain range of its neighborhood is pre-defined for the LBP operator. In this work, P counts for the number of neighboring pixels around the central one, while R refers to the radius of a circle with the P equally spaced neighbors surrounding the central pixel. Figure 1 presents the distributions of the neighborhood of a given pixel with different settings of P and R.

Now, we perform a basic LBP operation over a pixel with its neighborhood setting as $P = 8$ and $R = 1$ as is shown in the example in Fig. 2. Then, we assume that the intensity value of the center pixel is g_c, while the intensity values of its circular neighborhood are $g_p(p = 0, \ldots, P - 1)$. Here we threshold the neighbourhood by the intensity value of its center, namely

$$B = b(s(g_0 - g_c), s(g_1 - g_c), \ldots, s(g_{P-1} - g_c)), \tag{1}$$

where

$$s(n) = \begin{cases} 1, & n \geq 0 \\ 0, & n < 0 \end{cases} \tag{2}$$

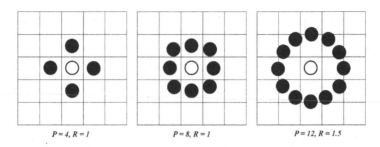

Fig. 1. Examples of central pixel with different settings of circular neighborhoods. Each grid represents a pixel and the pixel values are bilinearly interpolated if the sampling point is not right in the center of a pixel. The white points represent the central pixel and the black points refer to its neighboring pixels being selected.

7	6	3
8	7	2
7	9	8

Example

1	0	0
1		0
1	1	1

Thresholded

1	2	4
128		8
64	32	16

Weights

Binary: 11110001
Decimal: 241

Fig. 2. LBP operation with neighborhood settings as $P = 8$ and $R = 1$. The example block is shown on the left with intensity values and the neighborhood points are thresholded by the central pixel. Then the results are formulated into a bit-wise binary pattern, and the corresponding weights are used to compute the decimal value of the local binary pattern.

After thresholding the neighborhood using Eqs. 1 and 2, we formulate the results into a bit-wise binary pattern as shown in Fig. 2. Thus, this $(P = 8, R = 1)$ circular neighborhood results in a final binary pattern as 11110001. Then we calculate the decimal value of the local binary pattern as follow:

$$LBP_{P,R} = \sum_{p=0}^{P-1} s(g_p - g_c)2^p \tag{3}$$

After computing all the LBP decimal values over all the pixels on the image, we finally construct the occurrence histogram with all the results. The computed LBP histogram adopts uniform patterns so that the histogram has a separate bin for every uniform pattern and all nonuniform patterns are assigned to a single bin.

2.2   Local Binary Patterns from Three Orthogonal Planes

Following the steps of LBP on static image analysis, Zhao *et al.* [6] proposed the Local Binary Patterns from Three Orthogonal Planes (LBP-TOP) to analyse videos with motions.

The spatial-temporal information can be regarded as a set of volumes in the (X, Y, T) space, where X and Y represent the spatial coordinates, while T denotes the frame index (time) in temporal domain. Hence, the neighborhood of each pixel no longer fall in a two dimensional space, where LBP operation can be used to extract features into histograms. Instead, we need to compute feature descriptor in the three dimensional space (X, Y, T). Thus, LBP-TOP is proposed to describe the spatial-temporal information in the three dimensional space.

Similar to LBP, LBP-TOP also computes the local binary patterns of a center pixel by thresholding its neighborhood. However, as the spatial-temporal information falls in a three dimensional space, LBP-TOP decomposes the three dimensional volume into three orthogonal planes: XY, XT and YT as is shown in Fig. 3. The XY plane represents the appearance feature in spatial domain, while the XT plane describes a visual impression of one row changing with time and YT captures the features of motion for one column in temporal space. Then, LBP values are extracted for all pixels from the XY, XT and YT planes, denoted as XY-LBP, XT-LBP and YT-LBP. In such a representation, a sequence of images are encoded by the appearance (XY-LBP) and two spatial temporal (XT-LBP and YT-LBP) co-occurrence statistics. Finally, the LBP-TOP is computed by concatenating the histograms from all the three orthogonal planes including XY, XT and YT into a single histogram.

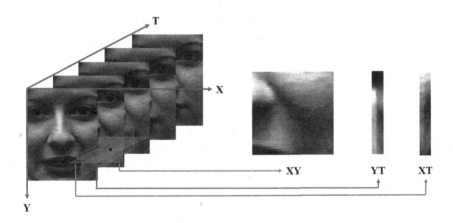

Fig. 3. Visualization of XY, XT and YT orthogonal planes on the spatial-temporal information with volume $183 \times 229 \times 11$ based on LBP-TOP algorithm. The image in XY plane is a 70×70 clip, the image in XT plane is a 11×70 clip when $y = 140$, and the image in YT plane is a 11×70 clip when $x = 140$ (the starting point is on the left-top of the image).

Note that when computing LBP-TOP over the spatial-temporal information, we also define a neighborhood in the three dimensional space for LBP coding. R_X, R_Y and R_T denote the radius of the sampling points surrounding the central

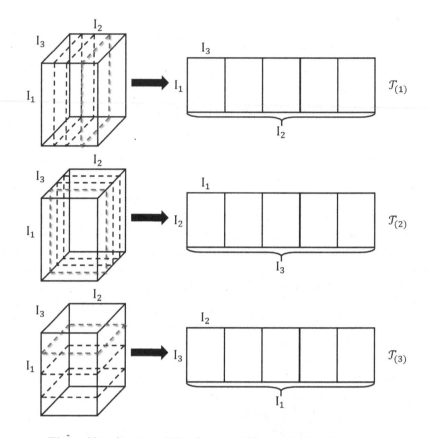

Fig. 4. Visualization of the three unfoldings of a 3-order tensor.

pixel along X, Y and T direction respectively. And P_{XY}, P_{XT} and P_{YT} count for the number of neighborhood points sampled on XY plane, XT plane and YT plane respectively. In the experiments in Sect. 3, we will use these denotations to differentiate various parameter settings of the defined neighborhood for LBP-TOP implementation.

2.3   Revisit LBP-TOP from a Tensor Point of View

A tensor refers to a multi-dimensional matrix, or array of numbers. The order of a tensor is defined by the number of dimensionality of the matrix required to describe the data [17,18]. That is, an one-dimensional array is an 1-order tensor, a two-dimensional matrix is a 2-order tensor and so forth.

Given a sequence of images with spatial-temporal information, they are usually regarded as a three-dimensional matrix, of which two coordinates label the spatial information and the third represents the time span. We can also consider the three-dimensional matrix as a 3-order tensor, of which the components include spatial information as well as temporal information.

This facilitates spatial-temporal information analysis from a tensor point of view. Many previous works achieve video analysis with tensor theory, for instance, Wang *et al.* [19,20] proposed a tensor independent color space to analyze micro-expression recognition on spatial-temporal information over different color channels, while Kim *et al.* [21] proposed tensor canonical correlation analysis for action classification.

Given an N-order tensor $\mathcal{T} \in \mathbb{R}^{I_1 \times I_2 \times \cdots \times I_N}$, the element inside $\mathcal{T}$ is represented as $\tau_{i_1 i_2 \cdots i_N}$, where $i_1, i_2, \cdots, i_N$ denote the coordinate positions. In practice, the spatial-temporal information are usually a video with a sequence of frames. Thus, we denote the video as a 3-order tensor $\mathcal{T}$, which is a $I_1 \times I_2 \times I_3$ matrix. I_1 is the height of the frame, I_2 refers to the width of the frame, and I_3 denotes the number of frames in temporal domain. Then, LBP-TOP traverses all the elements $\tau_{i_1 i_2 i_3}$ inside the tensor $\mathcal{T}$ to compute LBP on the three orthogonal planes $I_1 I_2$, $I_2 I_3$ and $I_1 I_3$ respectively.

2.4 Fast LBP-TOP Implementation Based on Tensor Unfolding

Given an N-order tensor $T \in \mathbb{R}^{I_1 \times I_2 \times \cdots \times I_N}$, the tensor unfolding [22–24] $\mathcal{T}_n \in \mathbb{R}^{I_n \times (I_1 I_2 \cdots I_{n-1} I_{n+1} I_{n+2} \cdots I_N)}$ contains the element $\tau_{i_1 i_2 \cdots i_n i_{n+1} \cdots i_N}$ at the position with row number i_n and column number that is equal to $[(i_{n+1} - 1)I_{n+2} \cdots I_P I_1 \cdots I_{n-1}] + [(i_{n+2} - 1)I_{n+3} \cdots I_P I_1 \cdots I_{n-1}] + \cdots + [(i_2 - 1)I_3 I_4 \cdots I_{n-1}] + \cdots + i_{n-1}$. Figure 4 visualizes the unfolded results of the 3-order tensor. Tensor unfolding is one of the simplest way to reduce the number of dimensions of a matrix.

Then we apply tensor unfolding on the 3-order matrix, and receive three 2-order unfolded tensors as visualized in Fig. 4.

Apparently, the unfolded 2-order tensor $\mathcal{T}_{(1)}$ concatenates all the YT planes into a large two-dimensional matrix. Similarly, the tensor $\mathcal{T}_{(2)}$ connects all the XY planes and the tensor $\mathcal{T}_{(3)}$ concatenates all the XT planes.

Originally, LBP-TOP has to traverse all the pixels in the 3-order tensor to compute the local binary patterns along XY, YT and XT planes respectively. The huge demand of nested looping in implementation sharply increases the computation costs. In this work, as we unfold the 3-order tensor to three two-dimensional concatenated matrices, we can effectively accelerate the computation process by optimizing the codes with vectorization. Then, we reformulate the unfolded tensors back to 3-order tensor and computes occurrence histograms for each orthogonal planes. The pseudo code of the original LBP-TOP implementation is illustrated in Algorithm 1, while that of the proposed fast LBP-TOP implementation is shown in Algorithm 2. Apparently, the proposed fast LBP-TOP implementation largely reduces the usage of nested loops and optimizes the codes through vectorization that uses matrix and vector operations.

3   Experiments

Usually, LBP-TOP is performed over the spatial-temporal information for feature extraction and the extracted features are utilized for further pattern

---

Algorithm 1. Original LBP-TOP implementation.

Data: Video data V, where $[H, W, T] = size(V)$. R_X, R_Y and R_T are radius of neighborhood along X, Y and T direction respectively, and P_{XY}, P_{XT} and P_{YT} are numbers of neighborhood points on XY, XT and YT plane respectively.

Result: LBP occurrence histograms on XY, XT and YT plane, namely $HIST_{XY}$, $HIST_{XT}$, and $HIST_{YT}$ respectively.

for $i = R_T$ **to** $T - R_T - 1$ **do**
    for $j = -R_T$ **to** R_T **do**
       | $Frame_{H \times W \times (j+R_T+1)} = V_{H \times W \times t}$, where $t = i + j + 1$;
    end
    for $yc = 1$ **to** H **do**
       for $xc = 1$ **to** W **do**
          $CenterVar = Frame(yc, xc, R_T)$;
          // Compute LBP on XY plane for center pixel with
             neighborhood (P_{XY}, R_X, R_Y).
          $LBP_{CenterVar} = $ LBP_PIXEL$(CenterVar, P_{XY}, R_X, R_Y)$;
          Update $HIST_{XY}$ with the value of $LBP_{CenterVar}$;
          // Compute LBP on XT plane for center pixel with
             neighborhood (P_{XT}, R_X, R_T).
          $LBP_{CenterVar} = $ LBP_PIXEL$(CenterVar, P_{XY}, R_X, R_T)$;
          Update $HIST_{XT}$ with the value of $LBP_{CenterVar}$;
          // Compute LBP on YT plane for center pixel with
             neighborhood (P_{YT}, R_Y, R_T).
          $LBP_{CenterVar} = $ LBP_PIXEL$(CenterVar, P_{XY}, R_Y, R_T)$;
          Update $HIST_{YT}$ with the value of $LBP_{CenterVar}$;
       end
    end
end

---

recognition or classification. In most of the cases, the extracted LBP-TOP features facilitate data training process for pattern recognition. Thus, in the experiment, we imitate the feature extraction process for data training to evaluate the performance of the original LBP-TOP implementation and the proposed fast LBP-TOP implementation. More specifically, we extract LBP-TOP features on spatial-temporal information with the same extraction strategy but different implementations (original LBP-TOP and fast LBP-TOP), to evaluate the performance improvements of the proposed fast LBP-TOP implementation. We implement the fast LBP-TOP on Matlab R2014b. The original implementation of LBP-TOP is provided by the corresponding authors in Matlab version[1].

We compared the computational time of the original LBP-TOP implementation and our fast LBP-TOP implementation on both synthetic and real data. Firstly, we compared the computational time of the original LBP-TOP

---

[1] The original implementation code of LBP-TOP method can be downloaded at http://www.cse.oulu.fi/CMV/Downloads/LBPMatlab.

Algorithm 2. Fast LBP-TOP implementation.

Data: Video data V, where $[H, W, T] = size(V)$. R_X, R_Y and R_T are radius of neighborhood along X, Y and T direction respectively, and P_{XY}, P_{XT} and P_{YT} are numbers of neighborhood points on XY, XT and YT plane respectively.

Result: LBP occurrence histograms on XY, XT and YT plane, namely $HIST_{XY}$, $HIST_{XT}$, and $HIST_{YT}$ respectively.

```
// Simple zero padding is used but not limited in this pseudo code.

// 1. Operations on XY plane.
```
$Plane_{XY}$ =zeros $(H, 2 \times R_T + W \times R)$;
```
// Unfold the 3-order tensor to 2-order tensor.
```
$Plane_{XY}(:, R_T + 1 : end - R_T) = $ Unfold $(V, [H, W \times T])$;
```
/* Unfolding can be done through permute and reshape in Matlab for
   example.                                                          */
// Compute LBP on XY plane with neighborhood (P_XY, R_X, R_Y).
```
LBP_{XY} =LBP_PLANE$(Plane_{XY}, P_{XY}, R_X, R_Y)$;
```
/* LBP on all the pixels on a 2D plane is implemented through
   vectorization instead of loops.                                  */
// Reformulate the 2-order tensor to 3-order tensor.
```
LBP_{XY} =Reformulate$(LBP_{XY}, [H - 2 \times R_Y, W, T])$;
$HIST_{XY}$ =Histogram(LBP_{XY});
```
// 2. Operations on YT plane.
```
$Plane_{YT}$ =zeros $(T, 2 \times R_X + W \times H)$;
$Plane_{YT}(:, R_X + 1 : end - R_X) = $ Unfold $(V, [T, W \times H])$;
```
// Compute LBP on YT plane with neighborhood (P_YT, R_Y, R_T).
```
LBP_{YT} =LBP_PLANE$(Plane_{YT}, P_{YT}, R_Y, R_T)$;
LBP_{YT} =Reformulate$(LBP_{YT}, [T - 2 \times R_T, H, W])$;
$HIST_{YT}$ =Histogram(LBP_{YT});
```
// 3. Operations on XT plane.
```
$Plane_{XT}$ =zeros $(W, 2 \times R_H + W \times T)$;
$Plane_{XT}(:, R_H + 1 : end - R_H) = $ Unfold $(V, [W, T \times H])$;
```
// Compute LBP on XT plane with neighborhood (P_XT, R_X, R_T).
```
LBP_{XT} =LBP_PLANE$(Plane_{XT}, P_{XT}, R_X, R_T)$;
LBP_{XT} =Reformulate$(LBP_{XT}, [T - 2 \times R_T, W, H])$;
$HIST_{XT}$ =Histogram(LBP_{XT});

---

implementation to our fast LBP-TOP implementation on synthetic data. Firstly, we evaluate the computational time by averaging 50 randomly simulated video samples with varying settings of the neighborhood, as is shown in Table 1. Apparently, the proposed fast LBP-TOP implementation effectively improves the computational costs, especially when the sizes of the videos are large or the neighborhood settings are complicated. Further, we randomly formulate video clips of different frame size and time length to evaluate the computational cost of the original LBP-TOP implementation and the proposed fast LBP-TOP implementation. The results are illustrated in Fig. 5. From Fig. 5(a), it

Table 1. Average computational time of the original LBP-TOP implementation and the fast LBP-TOP (FLBP-TOP) on 50 randomly simulated video samples. The numbers of sampled points of the neighborhood P_{XY}, P_{YT} and P_{XT} are set equal (denoted as P), and the radius R_X, R_Y and R_T are set equal as well (denoted as R). 'Inc.' refers to the increasing rate of computational time from LBP-TOP to FLBP-TOP.

W	H	T	P	R	LBP-TOP(s)	FLBP-TOP(s)	Inc.
64	64	30	8	1	1.58	0.06	26.85
64	64	30	8	2	1.38	0.05	26.59
64	64	30	8	3	1.19	0.05	22.56
64	64	30	16	2	7.66	0.10	78.72
64	64	30	16	3	7.43	0.10	72.69
64	64	60	8	1	3.17	0.18	17.59
64	64	60	8	2	2.88	0.17	16.91
64	64	60	8	3	2.60	0.17	15.71
64	64	60	16	2	9.65	0.38	25.36
64	64	60	16	3	9.28	0.39	23.57
64	64	180	8	1	9.59	0.56	17.21
64	64	180	8	2	8.92	0.56	15.94
64	64	180	8	3	8.23	0.54	15.31
64	64	180	16	2	17.42	1.21	14.42
64	64	180	16	3	16.61	1.21	13.75
128	128	30	8	1	6.25	0.36	17.41
128	128	30	8	2	5.69	0.35	16.15
128	128	30	8	3	5.05	0.32	15.92
128	128	30	16	2	13.21	0.77	17.19
128	128	30	16	3	12.45	0.69	18.14
128	128	60	8	1	12.88	0.74	17.43
128	128	60	8	2	12.09	0.71	17.07
128	128	60	8	3	11.31	0.72	15.67
128	128	60	16	2	21.66	1.54	14.07
128	128	60	16	3	20.77	1.53	13.56
128	128	180	8	1	39.71	2.26	17.60
128	128	180	8	2	37.64	2.18	17.29
128	128	180	8	3	36.15	2.17	16.68
128	128	180	16	2	54.82	4.66	11.76
128	128	180	16	3	52.93	4.63	11.43
256	256	500	8	1	455.80	29.71	15.34
256	256	500	8	2	448.61	27.09	16.56
256	256	500	8	3	438.66	25.74	17.04
256	256	500	16	2	586.64	55.33	10.60
256	256	500	16	3	578.37	54.78	10.56

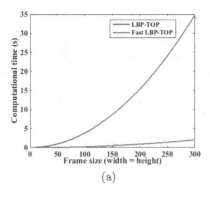 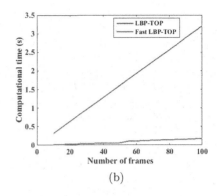

(a)   (b)

Fig. 5. Comparisons of computational time of the original LBP-TOP implementation and the fast LBP-TOP implementation. (a) shows the computational time of LBP-TOP and Fast LBP-TOP when frame size increases, of which the x-axis represents the equal lengths of width and height, and the neigborhood settings are $P_{XY} = P_{XT} = P_{YT} = 8$, $R_X = R_Y = R_T = 1$, and the video length $T = 30$. (b) illustrates the computational time of LBP-TOP and Fast LBP-TOP when video length increases, where the neighborhood settings are $P_{XY} = P_{XT} = P_{YT} = 8$, $R_X = R_Y = R_T = 1$ and the frame size is 30×30.

is easy to perceive that as the frame size increases, the computational time of the original LBP-TOP implementation sharply increases while that of the fast LBP-TOP implementation has no remarkable change. Similarly, in Fig. 5(b), when the length of the video clip increases, our fast LBP-TOP implementation reveals significant advances on computational time than the original LBP-TOP implementation. Hence, the fast LBP-TOP implementation is proven to be much more time-saving than the original LBP-TOP implementation in all cases especially for spatial-temporal information of high volume.

Then, we evaluate the computational time on real data based on three databases: CASME II [25], SMIC [26] and Cohn-Kanade [27]. All of the three databases are video clips with spatial-temporal information. CASME II is a micro-expression database of micro facial movements with high temporal and spatial resolution. SMIC is a spontaneous micro-expression database for analyzing people's deceitful behaviors. The Cohn-Kanade AU-Coded Facial Expression database is for research in automatic normal facial expression analysis and synthesis and for perceptual studies.

On each database, we randomly select 5 patches from each facial expression video clip of each subject, and Table 2 shows the computational time of the original LBP-TOP implementation and the proposed fast LBP-TOP implementation. In the table, we evaluate the performance of the original LBP-TOP implementation versus the proposed fast LBP-TOP implementation on three databases with different settings of frame width and height. As we perform LBP-TOP on all the video clips over each database, the temporal lengths are always the same such that we eliminate the setting of parameter T in the table. The parameters of

Table 2. Total computational time of the original LBP-TOP implementation and the fast LBP-TOP (FLBP-TOP) implementation on three databases: CASME II [25], SMIC [26] and Cohn-Kanade [27]. As LBP-TOP is performed by selecting 5 patches on every video clip of each database, the temporal lengths are the same for each comparison group such that video lengths T are not listed. The numbers of sampled points of the neighborhood P_{XY}, P_{YT} and P_{XT} are set equal (denoted as P), and the radius R_X, R_Y and R_T are set equal as well (denoted as R). The column 'Clips' shows the total number of video clips of each database. 'Inc.' refers to the increasing rate of computational time from LBP-TOP to FLBP-TOP.

Database	W	H	Clips	P	R	LBP-TOP(s)	FLBP-TOP(s)	Inc.
CASME II	30	30	1285	4	1	834.14	20.58	20.58
	30	30	1285	4	2	704.42	18.74	18.74
	30	30	1285	8	1	1001.94	38.37	38.37
	30	30	1285	8	2	850.17	35.88	35.88
	50	50	1285	4	1	2357.59	59.02	59.02
	50	50	1285	4	2	2130.38	55.33	55.33
	50	50	1285	8	1	2805.47	138.59	138.59
	50	50	1285	8	2	2495.50	128.60	128.60
Cohn-Kanade	30	30	2415	4	1	462.69	12.71	12.71
	30	30	2415	4	2	367.43	11.07	11.07
	30	30	2415	8	1	569.51	25.79	25.79
	30	30	2415	8	2	465.13	22.53	22.53
	50	50	2415	4	1	1174.95	28.14	28.14
	50	50	2415	4	2	970.90	25.08	25.08
	50	50	2415	8	1	1402.63	54.44	54.44
	50	50	2415	8	2	1154.99	48.33	48.33
SMIC	30	30	1640	4	1	527.54	13.71	13.71
	30	30	1640	4	2	434.42	11.82	11.82
	30	30	1640	8	1	638.04	25.77	25.77
	30	30	1640	8	2	533.85	23.56	23.56
	50	50	1640	4	1	1429.49	32.33	32.33
	50	50	1640	4	2	1244.76	30.48	30.48
	50	50	1640	8	1	1705.84	62.66	62.66
	50	50	1640	8	2	1471.51	58.51	58.51

neighborhood including the number of sampling points P_{XY}, P_{XT} and P_{YT} and radius R_X, R_Y and R_T are set according to Table 2. From Table 2, it is obvious that the fast LBP-TOP implementation dramatically improves the computational time by 31.19 times on average to the original LBP-TOP implementation.

4   Discussions

The proposed fast LBP-TOP implementation takes the advantage of tensor unfolding to reformulate the tree-dimensional matrix to two-dimensional matrices. In future, this tensor unfolding method can be applied to any other descriptors that extract features from spatial-temporal information, such as LGBP-TOP [28] and SIFT-TOP [29]. Moreover, we implement the fast LBP-TOP codes on Matlab as an example. Further, the fast LBP-TOP can be implemented on other platforms that optimize codes through vectorization instead of nested loops, such as Python, Octave, R, *etc.*

5   Conclusions

In this work, we propose a fast LBP-TOP implementation method to fasten the computation efficiency of LBP-TOP for feature extraction on spatial-temporal information. We introduce the concept of tensor unfolding to accelerate the implementation process from three-dimensional space to two-dimensional space. The proposed fast LBP-TOP implementation method benefits from the optimization of codes with less nested loops and largely reduces the computational cost compared to the original implementation. We compare the computational time of the original LBP-TOP implementation and our fast LBP-TOP implementation on both synthetic and real data. The results show that our fast LBP-TOP implementation is quite time-saving than the original one. In future, as parallel computing can be performed on two dimensional matrices, the computational time of the fast LBP-TOP implementation can be further saved.

Acknowledgement. This work is sponsored by the Academy of Finland, Infotech Oulu, the post-doc fellow position of Infotech Oulu, and Tekes Fidipro Program. Moreover, Xiaopeng Hong is partly supported by the Natural Science Foundation of China under the contract No. 61572205. Also, we gratefully acknowledge the support of NVIDIA Corporation with the donation of the Tesla K40 GPU used for this research.

References

1. Wolf, L.: Face recognition, geometric vs. appearance-based. In: Encyclopedia of Biometrics, pp. 495–500 (2015)
2. Gross, R., Matthews, I., Baker, S.: Appearance-based face recognition and light-fields. IEEE Trans. Pattern Anal. Mach. Intell. **26**, 449–465 (2004)
3. Ojala, T., Pietikainen, M., Maenpaa, T.: Multiresolution gray-scale and rotation invariant texture classification with local binary patterns. IEEE Trans. Pattern Anal. Mach. Intell. **24**, 971–987 (2002)
4. Ojala, T., Pietikäinen, M., Mäenpää, T.: Gray scale and rotation invariant texture classification with local binary patterns. In: Vernon, D. (ed.) ECCV 2000. LNCS, vol. 1842, pp. 404–420. Springer, Heidelberg (2000). doi:10.1007/3-540-45054-8_27

5. Huang, X., Wang, S.J., Zhao, G., Piteikainen, M.: Facial micro-expression recognition using spatiotemporal local binary pattern with integral projection. In: Proceedings of the IEEE International Conference on Computer Vision Workshops, pp. 1–9 (2015)
6. Zhao, G., Pietikainen, M.: Dynamic texture recognition using local binary patterns with an application to facial expressions. IEEE Trans. Pattern Anal. Mach. Intell. **29**, 915–928 (2007)
7. de Freitas Pereira, T., Anjos, A., De Martino, J.M., Marcel, S.: *LBP − TOP* based countermeasure against face spoofing attacks. In: Park, J.-I., Kim, J. (eds.) ACCV 2012. LNCS, vol. 7728, pp. 121–132. Springer, Heidelberg (2013). doi:10.1007/978-3-642-37410-4_11
8. Kellokumpu, V., Zhao, G., Pietikäinen, M.: Human activity recognition using a dynamic texture based method. In: BMVC, vol. 1, p. 2 (2008)
9. Wang, Y., Yu, H., Stevens, B., Liu, H.: Dynamic facial expression recognition using local patch and lbp-top. In: 2015 8th International Conference on Human System Interaction (HSI), pp. 362–367. IEEE (2015)
10. Chen, Y., Guo, X., Klein, D.: Orthogonal combination of local binary patterns for dynamic texture recognition. In: Ninth International Symposium on Multispectral Image Processing and Pattern Recognition (MIPPR2015), 98130R. International Society for Optics and Photonics(2015)
11. Qi, X., Li, C.G., Zhao, G., Hong, X., Pietikäinen, M.: Dynamic texture and scene classification by transferring deep image features. Neurocomputing **171**, 1230–1241 (2016)
12. Ojala, T., Pietikäinen, M., Harwood, D.: A comparative study of texture measures with classification based on featured distributions. Pattern Recogn. **29**, 51–59 (1996)
13. Ahonen, T., Hadid, A., Pietikainen, M.: Face description with local binary patterns: application to face recognition. IEEE Trans. Pattern Anal. Mach. Intell. **28**, 2037–2041 (2006)
14. Ahonen, T., Hadid, A., Pietikäinen, M.: Face recognition with local binary patterns. In: Pajdla, T., Matas, J. (eds.) ECCV 2004. LNCS, vol. 3021, pp. 469–481. Springer, Heidelberg (2004). doi:10.1007/978-3-540-24670-1_36
15. Murala, S., Wu, Q.J.: Local mesh patterns versus local binary patterns: biomedical image indexing and retrieval. IEEE J. Biomed. Health Inform. **18**, 929–938 (2014)
16. Mäenpää, T.: The local binary pattern approach to texture analysis: extensions and applications. Oulun yliopisto (2003)
17. Kolecki, J.C.: An introduction to tensors for students of physics and engineering (2002)
18. Kolda, T.G., Bader, B.W.: Tensor decompositions and applications. SIAM Rev. **51**, 455–500 (2009)
19. Wang, S., Yan, W.J., Li, X., Zhao, G., Fu, X.: Micro-expression recognition using dynamic textures on tensor independent color space. In: ICPR, pp. 4678–4683. Citeseer (2014)
20. Wang, S.J., Yan, W.J., Li, X., Zhao, G., Zhou, C.G., Fu, X., Yang, M., Tao, J.: Micro-expression recognition using color spaces. IEEE Trans. Image Process. **24**, 6034–6047 (2015)
21. Kim, T.K., Wong, S.F., Cipolla, R.: Tensor canonical correlation analysis for action classification. In: 2007 IEEE Conference on Computer Vision and Pattern Recognition, pp. 1–8. IEEE (2007)
22. Meyer, C.D.: Matrix Analysis and Applied Linear Algebra, vol. 2. SIAM, Philadelphia (2000)

23. Kuang, L., Hao, F., Yang, L.T., Lin, M., Luo, C., Min, G.: A tensor-based approach for big data representation and dimensionality reduction. IEEE Trans. Emerg. Topics Comput. **2**, 280–291 (2014)

24. Manolopoulos, Y., Symeonidis, P., Nanopoulos, A.: Tag recommendations based on tensor dimensionality reduction (2008)

25. Yan, W.J., Li, X., Wang, S.J., Zhao, G., Liu, Y.J., Chen, Y.H., Fu, X.: Casme ii: an improved spontaneous micro-expression database and the baseline evaluation. PloS One **9**, e86041 (2014)

26. Li, X., Pfister, T., Huang, X., Zhao, G., Pietikäinen, M.: A spontaneous micro-expression database: Inducement, collection and baseline. In: 2013 10th IEEE International Conference and Workshops on Automatic Face and Gesture Recognition (FG), pp. 1–6. IEEE (2013)

27. Tian, Y.I., Kanade, T., Cohn, J.F.: Recognizing action units for facial expression analysis. IEEE Trans. Pattern Anal. Mach. Intell. **23**, 97–115 (2001)

28. Almaev, T.R., Valstar, M.F.: Local gabor binary patterns from three orthogonal planes for automatic facial expression recognition. In: 2013 Humaine Association Conference on Affective Computing and Intelligent Interaction (ACII), pp. 356–361. IEEE (2013)

29. Sun, B., Li, L., Zhou, G., He, J.: Facial expression recognition in the wild based on multimodal texture features. J. Electron. Imaging **25**, 061407 (2016)

3D Convolutional Neural Networks for Facial Expression Classification

Wenyun Sun[1], Haitao Zhao[2], and Zhong Jin[1(✉)]

[1] School of Computer Science and Engineering,
Nanjing University of Science and Technology, Nanjing, China
zhongjin@njust.edu.cn
[2] School of Information Science and Engineering,
East China University of Science and Technology, Shanghai, China

Abstract. In this paper, the general rules of designing 3D Convolutional Neural Networks are discussed. Four specific networks are designed for facial expression classification problem. Decisions of the four networks are fused together. The single networks and the ensemble network are evaluated on the extended Cohn-Kanade dataset, achieve accuracies of 92.31% and 96.15%. The performance outperform the state-of-the-art. A reusable open source project called 4DCNN is released. Based on this project, implementing 3D Convolutional Neural Networks for specific problems will be convenient.

1 Introduction

Research on facial expression was started by psychologists. Facial Action Coding System (FACS) [1] and Emotional Facial Action Coding System (EMFACS) [2] were proposed by Ekman and Friensen. In their studies, facial expressions are defined as serval action units (AUs) associated with 6 basic emotions. In the community of computer vision, quite a few quantitatively studies have been devoted to analyse expressions in images or videos [3]. One of the most important method is Active Appearance Models (AAMs) [4] in which a statistical model was defined using 68 fiducial points. Fiducial points are easy to be understood and manipulated. Action units and emotional labels can be inferred from these fiducial points by rules [5]. Besides the fiducial point method, using global or local appearance of the face is another way to recognize the expressions [6–11].

In the research of facial expression recognition, performance can be improved by making the most use of the structures in 3D space. Recently, 3D Gabor filters are used to extract features from 3D scanning data, keeping invariant to head pose, clutter and lighting condition [12]. Local Binary Pattern histograms from Three Orthogonal Planes (LBP-TOP) are used to extract spatial-temporal features for recognizing facial expressions in movie clips [13]. The extra dimension plays an important role in these studies.

In this paper, the general rules of building 3D Convolutional Neural Networks are discussed. Four networks are proposed for facial expression classification problem. After they are trained separately, decisions of the four networks are

© Springer International Publishing AG 2017
C.-S. Chen et al. (Eds.): ACCV 2016 Workshops, Part I, LNCS 10116, pp. 528–543, 2017.
DOI: 10.1007/978-3-319-54407-6_35

fused together. Experiment result shows that these networks work well. The single networks and the ensemble network are evaluated on the extended Cohn-Kanade dataset, achieve accuracies of 92.31% and 96.15%. The performance outperform the state-of-the-art [6–10].

The remainder of this paper is organized as follows. In Sect. 2, related work on deep learning and expression recognition is surveyed. Section 3 gives the definition of the 3D convolutional layers. In Sect. 4, the general rules for designing the 3D networks are discussed. In Sect. 5, four networks are proposed to solve the expression classification problem. Results are analysed and compared with previous work. Finally, conclusions are presented in Sect. 6.

2   Related Work

2.1   Convolutional Neural Networks

In a standard Convolutional Neural Network (CNN) [14], the element at position (x, y) in the c-th feature map of the l-th convolutional layer, denoted as $v_{l,c}^{x,y}$, is given by

$$v_{l,c}^{x,y} = tanh(\sum_{C=1}^{C_{l-1}} \sum_{p=1}^{P_l} \sum_{q=1}^{Q_l} W_{l,c,C}^{p,q} v_{(l-1),C}^{(x+p-1),(y+p-1)} + b_{l,c}), \tag{1}$$

where $tanh$ is the hyperbolic tangent activation function, $b_{l,c}$ is the bias for the c-th feature map. C_{l-1} is the number of feature maps in the $(l-1)$-th layer. C indexes over the set of feature maps in the $(l-1)$-th layer connected to the current feature map. $W_{l,c,C}^{p,q}$ is the value at the position (p, q) of the kernel connected to the C-th feature map. P_l and Q_l are the height and width of the kernel, respectively.

2.2   3D Convolutional Neural Networks

3D Convolutional Neural Networks were proposed by Ji et al. [15] for solving the human action recognition problem. Considering the 3rd dimension as the temporal dimension of a video data, motion information can be captured by 3D convolution. Formally, the element at position (x, y, z) in the c-th feature map of the l-th convolutional layer, denoted as $v_{l,c}^{x,y,z}$, is reformulated by

$$v_{l,c}^{x,y,z} = tanh(\sum_{C=1}^{C_{l-1}} \sum_{p=1}^{P_l} \sum_{q=1}^{Q_l} \sum_{r=1}^{R_l} W_{l,c,C}^{p,q,r} v_{(l-1),C}^{(x+p-1),(y+q-1),(z+r-1)} + b_{l,c}), \tag{2}$$

where R_l is the size of the 3D kernel along the 3rd dimension. $W_{l,c,C}^{p,q,r}$ is the value at the position (p, q, r) of the kernel connected to the C-th feature map in the previous layer.

In this paper, full-connected layer and 1D/2D/3D convolutional layers are reformulated in a general form which is different from Eq. (2). The close relationship among them is discovered.

2.3   CNNs for Expression Recognition

Facial expression is an interesting research object in the community of computer vision. There are unsolved problems like classification, detection, manipulation and transfer. In this paper, our attention is focused on solving the problem of classifying faces into 6 basic emotional categories using Convolutional Neural Networks.

As we know, CNNs were proposed in the 1990s. A piece of interesting work about expression recognition based on CNNs has already been done in the early days by Matsugu et al. [5]. After Krizhevsky et al. won the Imagenet Large Scale Visual Recognition Challenge 2012 (ILSVRC-2012) [16] using a novel network called Alex-Net [17]. Deep CNNs attract more attentions than before. Sun et al. designed an expression classifier [11] based on the modern CNN proposed by Krizhevsky et al. Byeon et al. designed an expression classifier [18] based on 3DCNN proposed by Ji et al. [15]. This paper aims at proposing general definitions and designing rules for 3DCNNs other than doing specific experiments on facial tasks.

3   3D Convolutional Layers

The 3D convolutional layers are extended from the full-connected layers. Let's start from the full-connected layer. Denote the l-th full-connected layer's activation as $\mathbf{v}_l \in \mathbb{R}^n$ and the previous layer's activation as $\mathbf{v}_{l-1} \in \mathbb{R}^m$. A full-connected layer can be defined as follows:

$$\mathbf{v}_l = tanh(g(\mathbf{v}_{l-1})), \tag{3}$$

where

$$g(\mathbf{x}) = \mathbf{W}\mathbf{x} + \mathbf{b}. \tag{4}$$

The function g plays as the role of full-connecting operator. It converts a vector of m dimension to a vector of n dimension by linear transformation $\mathbf{W} \in \mathbb{R}^{n \times m}$. After that, the converted vector is added by bias $\mathbf{b} \in \mathbb{R}^n$, then activated by an element-wise nonlinear activation function $tanh$ (or $sigmoid$, $max(\cdot, 0)$, etc.).

The naive definition of full-connected layer can be generalized to convolutional layer. Firstly, denote the elements of $g(\mathbf{x})$, $\mathbf{W}$, $\mathbf{x}$ and $\mathbf{b}$ as g_i, $W_{i,j}$, x_j and b_i, ($i \in \{1, 2, \cdots, n\}, j \in \{1, 2, \cdots, m\}$) respectively. The vectorized full-connecting operator g can be rewritten as

$$g_i = (\sum_{j=1}^{m} W_{i,j} \cdot x_j) + b_i, i \in \{1, 2, \cdots, n\}. \tag{5}$$

Secondly, Eq. (5) is modified by replacing the scalar multiplication operator "$\cdot$" by vector/matrix/tensor convolution operator "$\bullet$".

The scalar elements g_i, $W_{i,j}$, x_j and b_i are also replaced by vector/matrix/tensor. Then we get

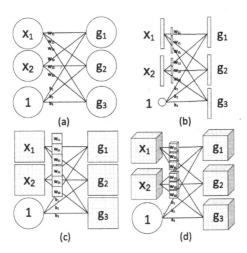

Fig. 1. Examples of convolutional layer without element-wise activating. (a) full-connected layer, (b) 1D convolutional layer, (c) 2D convolutional layer, (d) 3D convolutional layer.

$$\mathbf{g}_i = (\sum_{j=1}^{m} \mathbf{W}_{i,j} \bullet \mathbf{x}_j) + \mathbf{b}_i, i \in \{1, 2, \cdots, n\}, \tag{6}$$

where $\mathbf{g}_i \in \mathbb{R}^{dim(\mathbf{g}_i)_1 \times dim(\mathbf{g}_i)_2 \times \cdots \times dim(\mathbf{g}_i)_D}$, $\mathbf{W}_{i,j} \in \mathbb{R}^{dim(\mathbf{W}_{i,j})_1 \times dim(\mathbf{W}_{i,j})_2 \times \cdots \times dim(\mathbf{W}_{i,j})_D}$, $\mathbf{x}_j \in \mathbb{R}^{dim(\mathbf{x}_j)_1 \times dim(\mathbf{x}_j)_2 \times \cdots \times dim(\mathbf{x}_j)_D}$, $\mathbf{b}_i \in \mathbb{R}^{dim(\mathbf{b}_i)_1 \times dim(\mathbf{b}_i)_2 \times \cdots \times dim(\mathbf{b}_i)_D}$, $i \in \{1, 2, \cdots, n\}$, $j \in \{1, 2, \cdots, m\}$. $\mathbf{b}_i$ is a constrained vector/matrix/tensor in which all the elements are equal.

Full-connected layer and 1D/2D/3D convolutional layers can be defined by this general definition in the following specific cases:

- When $D = 1$ and $dim(\mathbf{g}_i)_1 = dim(\mathbf{x}_j)_1 = dim(\mathbf{W}_{i,j})_1 = 1$, Eq. (6) is specialized to Eq. (4)/Eq. (5). It describes the full-connected layer (see Fig. 1(a)).
- When $D = 1$ and $dim(\mathbf{g}_i)_1, dim(\mathbf{x}_j)_1, dim(\mathbf{W}_{i,j})_1 \geqslant 2$, its describe the 1D convolutional layer in Time Delay Neural Networks (TDNNs) [19] (see Fig. 1(b)).
- When $D = 2$, it describes the commonly used 2D convolutional layer (see Fig. 1(c)). $\mathbf{x}$ contains m channels of $[\mathbf{x}_1, \mathbf{x}_2, \cdots, \mathbf{x}_m]^{\mathrm{T}}$. Each channel is a spatial image of $dim(\mathbf{x}_j)_1 \times dim(\mathbf{x}_j)_2$. $\mathbf{g}$ contains n channels of $[\mathbf{g}_1, \mathbf{g}_2, \cdots, \mathbf{g}_n]^{\mathrm{T}}$. Each channel is a spatial image of $dim(\mathbf{g}_i)_1 \times dim(\mathbf{g}_i)_2$. $\mathbf{W}$ contains $n \times m$ convolutional kernels, whose sizes are $dim(\mathbf{W}_{i,j})_1 \times dim(\mathbf{W}_{i,j})_2$. From this point of view, the convolutional layer is a full-connected layer with convolutional connections.
- When $D \geqslant 3$, it describes the D-dimensional convolutional layer which processes D-th order tensor data (see Fig. 1(d)).

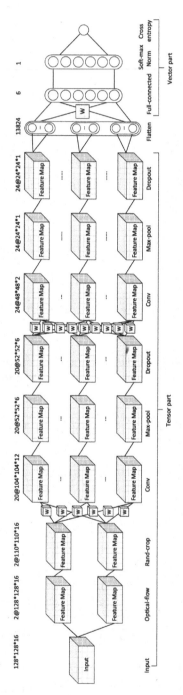

Fig. 2. An example of 3DCNN.

4   General 3DCNNs

A 3DCNN consists of layers including convolutional, max-pooling, full-connected, dropout, Gabor, optical-flow, flatten, soft-max norm, cross entropy loss, mean squared error loss, etc. [14,17]. As an example illustrated in Fig. 2, a flatten layer is placed in the middle of the network. It reshapes and concatenates multiple tensors into a single vector. The flatten layer divides the whole network into 2 parts: the convolutional part on the left and the full-connected part on the right.

The input and output data of the layers on the left are sets of 3rd-order tensors, which are commonly called feature maps or channels. An optical-flow layer or a Gabor layer is often placed at the beginning of the network. They calculate low-level dense image features. Cropping layers randomly crop a sub-tensor of specified size for data augmentation. Stacked convolutional and max-pooling layer pairs learn mid-level and high-level feature maps from labeled data.

Layers on the right have the same functions as the ones in full-connected neural networks. Full-connected layers are good at logical reasoning based on the high-level feature maps produced by convolutional layers or max-pooling layers. According to Eq. 6, the full-connected layers can be considered as specific cases of the general convolutional layers, in which spatial size of the activation is $1 \times 1 \times \cdots \times 1$ and spatial size of the convolution kernel is $1 \times 1 \times \cdots \times 1$. For better understanding, the full-connected layer illustrated in Fig. 2 follow the regular form in [14]. It is intrinsically equivalent to the proposed definition. Soft-max normalization layers, cross-entropy loss layers are often attached to the end of a 3DCNN for solving classification problem. Sometimes, mean squared error layers are used for solving regression problem. Dropout layers can be inserted after full-connected layers, convolutional layers or max-pooling layers for improving the quality of the middle representations.

The training procedure is minimizing the loss with respect to the parameters. By making sure all these layers are derivable, all gradients of the loss with respect to the parameters can be calculated using the chain rule. Then the gradient descent method can be applied to solve this problem. The Stochastic Gradient Descent (SGD) with mini-batches method is good at making full use of the parallel computing ability of the modern computer. It is widely used for training neural networks on large datasets.

5   Experiments

5.1   Network Designments

As listed in Tables 1, 2, 3 and 4, four different networks are proposed to solve the expression classification problem, namely 3DCNN-A, 3DCNN-B, 3DCNN-C and 3DCNN-D. They are wished to make complementary predictions. The major difference between the four networks is their low-level feature extractors. 3D Gabor layers are used in 3DCNN-A and 3DCNN-B. Optical-flow layers are used in 3DCNN-C and 3DCNN-D. It means that their predictions are made according to different views of the data, namely the texture and the motion. 3DCNN-A and 3DCNN-B are intended to classify facial appearances, and 3DCNN-C and 3DCNN-D are intended to classify facial motions. Moreover, there are serval minor differences including the counts of convolutional/full-connected layers, the receptive field sizes of the convolutional kernels, the output channel counts of the convolutional kernels, and the activation functions.

3D Gabor layers and optical-flow layers are used to extract the low-level features in spatial-temporal space. 80 Gabor filters are created by varying standard deviation, wave length, yaw angle and pitch angle of the Gabor function. The standard deviation is selected from {4, 8, 16, 32}. The wave length is the same as standard deviation. The yaw and pitch angles are selected manually. Horn-Schunck method [20] is used to calculate vertical and horizontal optical-flow feature maps between each frames.

Although four networks can work well independently, the performance can be further improved by decision-level fusion. After all the networks were trained on different view of the data separately, decisions supported by each network (activations of normalized soft-max layer, the probabilities of 6 categories) are fused together to make the final prediction.

The symmetry of the face should also be considered. Decisions according to frontal view and mirrored view are also fused together. The pipeline is outlined in Fig. 3.

Table 1. 3DCNN-A.

Layer name	Output dimension	Parameter number
Input	128*128*5*1	-
3D Gabor filter	118*118*1*80	-
2D convolutional	112*112*1*8	7*7*1*80*8 + 8*1
2D max-pool	56*56*1*8	-
2D convolutional	52*52*1*16	5*5*1*8*16 + 16*1
2D max-pool	26*26*1*16	-
2D convolutional	22*22*1*16	5*5*1*16*16 + 16*1
2D max-pool	11*11*1*16	-
Flatten	1*1*1*1936	-
Full-connected	1*1*1*25	25*1936 + 25*1
Scalar dropout	1*1*1*25	-
Full-connected	1*1*1*6	6*25 + 6*1
Soft-max norm	1*1*1*6	-
Cross entropy loss	1	-
Total	-	89,581

Table 2. 3DCNN-B.

Layer name	Output dimension	Parameter number
Input	128*128*5*1	-
3D Gabor filter	118*118*1*80	-
Randomly cropping	116*116*1*80	-
2D max-pool	58*58*1*80	-
2D convolutional	52*52*1*16	7*7*1*80*16 + 16*1
2D max-pool	26*26*1*16	-
2D convolutional	20*20*1*16	7*7*1*16*16 + 16*1
2D max-pool	10*10*1*16	-
2D convolutional	6*6*1*32	5*5*1*16*32 + 32*1
2D max-pool	3*3*1*32	-
2D convolutional	1*1*1*64	3*3*1*32*64 + 64*1
Scalar dropout	1*1*1*64	-
Full-connected	1*1*1*6	6*64 + 6*1
Soft-max norm	1*1*1*6	-
Cross entropy loss	1	-
Total	-	107,014

Table 3. 3DCNN-C.

Layer name	Output dimension	Parameter number
Input	128*128*16*1	-
Optical-flow	128*128*16*2	-
Randomly cropping	108*108*16*2	-
3D convolutional	104*104*12*20	5*5*5*2*20 + 20*1
3D max-pool	52*52*6*20	-
3D dropout	52*52*6*20	-
3D convolutional	48*48*2*24	5*5*5*20*24 + 24*1
3D max-pool	24*24*1*24	-
3D dropout	24*24*1*24	-
Flatten	1*1*1*13824	-
Full-connected	1*1*1*6	6*13824 + 6*1
Soft-max norm	1*1*1*6	-
Cross entropy loss	1	-
Total	-	147,994

Table 4. 3DCNN-D.

Layer name	Output dimension	Parameter number
Input	128*128*16*1	-
Optical-flow	128*128*16*2	-
Randomly cropping	110*110*16*2	-
3D convolutional	104*104*12*20	$7*7*5*2*20 + 20*1$
3D max-pool	52*52*6*20	-
3D dropout	52*52*6*20	-
3D convolutional	48*48*2*24	$5*5*5*20*24 + 24*1$
3D max-pool	24*24*1*24	-
3D dropout	24*24*1*24	-
Flatten	1*1*1*13824	-
Full-connected	1*1*1*6	$6*13824 + 6*1$
Soft-max norm	1*1*1*6	-
Cross entropy loss	1	-
Total	-	152,794

5.2   Extended Cohn-Kanade Dataset

The extended Cohn-Kanade dataset (CK+) [21] is employed to evaluate the proposed deep networks. It is one of the most commonly used datasets in the studies of facial expression recognition. There are 97 subjects in it. 1–9 video sequences are recorded for each subject. Each sequence contains a motion vary from a normal expression to a stable posed expression. The length of the sequences are different. Totally, the number of sequences is 487. We use the corresponding manual annotation data [22] which provide coordinates of 59 facial fiducial points and 6 basic emotional labels.

As shown in Table 5, all the sequences are divided into training set, validation set and test set by 78.6%, 10.7% and 10.7%. From another point of view, the sequences are roughly divided into 6 balanced categories. The training set, validation set and test set containing sequence belonging to subjects 58–138, subjects 42–57 and subjects 10–37 respectively. There is no overlap between the training and testing subjects. The validation set is used to design networks, select meta-parameters and observe the training progress.

5.3   Preprocesses

All faces were aligned and cropped by locating the outer fiducial points. Backgrounds were removed. The length of each sequence was rescaled to 16 by using 3D cubic spline interpolation method. We named these sequences as CK+II. Frames 1 to 11 of each sequence were removed subsequently. Only the last 5 frames containing stable expressions were kept. These data were named as CK+I. The dimensions of samples in CK+I/CK+II were 128*128*5/128*128*16.

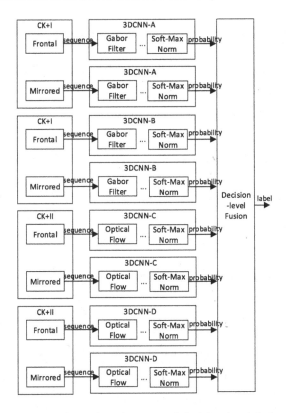

Fig. 3. Pipeline of predicting and decision-level fusion.

5.4 Training and Test

The training meta-parameters were different for four networks. 3DCNN-A and 3DCNN-B were trained by using SGD algorithm with a batch size of 128 examples, learning rate of 0.01, momentum of 0.9, and weight decay of 0.0005. 3DCNN-C and 3DCNN-D were trained by using SGD algorithm with a batch size of 32 examples, learning rate of 0.005. Momentum and weight decay were not used in training 3DCNN-C and 3DCNN-D. All the networks was trained on a single workstation with Intel Xeon E5-2620 CPU. Each of them roughly cost 2–4 days to converge.

After all networks were converged, four decision-level fusion strategies including Nearest Neighbor (1-NN), Support Vector Machine (SVM), probability maximum and probability averaging were applied. All performances of the single networks and the ensemble network were evaluated.

Table 5. Subsets division of CK+.

Category	Size		
	Training[1]	Validation[2]	Test[3]
Anger	59	9	9
Fear	65	7	8
Disgust	30	7	9
Sadness	63	10	8
Happiness	84	10	9
Surprise	82	9	9
Total	383	52	52

5.5   Results and Analysis

The results including the accuracies of single networks and ensemble network are summarized in Table 6. The accuracies of 3DCNN-A, 3DCNN-B, 3DCNN-C and 3DCNN-D are 90.39%, 90.39%, 92.31%, 92.31% respectively.

Table 6. Classification and fusion results on CK+.

Data	Low-level feature	Network	Data view	Accuracy	Accuracy by fusing frontal & mirrored views	Accuracy by fusing 4 networks
CK+I	3D Gabor	3DCNN-A	Frontal	92.31%	90.39%	**96.15%**
			Mirrored	90.39%		
		3DCNN-B	Frontal	92.31%	90.39%	
			Mirrored	90.39%		
CK+II	Optical-flow	3DCNN-C	Frontal	90.39%	92.31%	
			Mirrored	92.31%		
		3DCNN-D	Frontal	92.31%	92.31%	
			Mirrored	90.39%		

As listed in Table 7, four decision-level fusion strategies have been tried. After fusing four networks together, the best accuracy of 96.15% was achieved by probability averaging method. The improvement was strongly associated with the combination of different views and low-level features which contains useful complementary information. Fusing the frontal and mirrored view is a convenient trick used in many facial tasks, but the effect is not so obvious here ($\approx \pm 1\%$). By comparison, after fusing the decisions supported by four networks, the performance was improved greatly ($\approx 4\%$). The proposed networks used two kinds of low-level features, the optical-flow feature of the whole sequence and the 3D Gabor feature of the stable frames. They are totally different. They contain much

Table 7. Results of different decision-level fusion strategies on CK+.

Fusion strategies	Accuracy
1-Nearest Neighbor	80.77%
SVM with RBF Kernel	94.23%
Probability maximum	92.31%
Probability averaging	**96.15%**

Table 8. Confusion matrix of ensemble network on CK+.

	Anger	Fear	Disgust	Sadness	Happiness	Surprise	Total
Anger	8	0	0	1	0	0	9
Fear	0	8	0	0	0	0	8
Disgust	1	0	9	0	0	0	10
Sadness	0	0	0	7	0	0	7
Happiness	0	0	0	0	9	0	9
Surprise	0	0	0	0	0	9	9
Total	9	8	9	8	9	9	52

more complementary information than the frontal face and its mirrored view. Success is due to the ensemble of them.

The confusion matrix (see Table 8) shows that the networks almost perfectly separate each classes from others, except for the pairs of anger-disgust and anger-sadness which are naturally hard to distinguish.

As it is concluded in [17], data augmentation (mirroring, randomly cropping) and dropout were important to prevent over-fitting especially for tasks with small sample size limitation. In this experiment, new findings about the over-fitting problem are discovered. As shown in Table 9, the gap between performances on training set and test set was very small in the case of 3DCNN-C and 3DCNN-D (even without using weight decay). We guess that optical-flow feature may play an important role in reducing the impact of over-fitting. Although there is no sufficient proof except this result, the advantages of using motion information are undoubted. Optical-flow is a widely used method for processing video data. By integrating optical-flow calculator into a deep convolutional neural network, the motion information can be explicitly extracted. Since the optical-flow calculator is designed to work in temporal space, the motion signals are more strong and stable than the classic edge or texture detector like Gabor filters which usually work in spatial space. It maybe significant for some pure motion analysing tasks such as human action recognition. We suggest that the optical-flow features could be widely used in practice when the distinguishable motion is available. And it will be better if complementary texture features are provided too.

Table 9. The gaps between performances evaluated on training set and test set.

Low-level feature	Network	Accuracy on training set	Accuracy on test set	Gap of accuracies	Data augment	Dropout	Weight decay
3D Gabor	3DCNN-A	98.96%	90.39%	8.57%	Used	Used	Used
	3DCNN-B	100.0%	90.39%	9.61%	Used	Used	Used
Optical-flow	3DCNN-C	93.21%	92.31%	0.90%	Used	Used	Not used
	3DCNN-D	91.65%	92.31%	−0.66%	Used	Used	Not used

5.6   Comparing with Previous Work

In order to compare our results with some previous studies using Areas Under ROC Curve (AUC) as their performance measurement, 6 stand-alone expression detectors were built by making a slight modification to the classifiers. The detector makes a CNN feed forward pass first, fetches the probabilities (output of the normalized soft-max layer), compares the probability of interest with a threshold, predicts its binary result. By varying the thresholds, Receiver Operating Characteristic (ROC) curves (see Fig. 4) and Areas Under ROC Curve (AUC) (see Table 10) were obtained.

As listed in Table 10, the proposed method outperforms the compared methods. One of the compared methods uses the still frame [10], and the others use the video data [6–9]. Compared to these methods, the proposed method has two important advantages.

- Firstly, both static and dynamic information is explicitly extracted and integrated to make the final prediction. The compared methods use only one of them.

Table 10. Performance comparison with previous work on CK+.

Method	Accuracy of classifier	AUC of detector						
		Anger	Fear	Disgust	Sadness	Happiness	Surprise	Average
Yang et al. [6]	-	0.973	0.916	0.941	0.978	0.991	0.998	0.966
Long et al. [7]	-	0.933	0.964	0.988	0.991	0.993	0.999	0.978
Jeni et al. [8]	-	0.990	0.980	1.000	0.990	1.000	0.990	0.992
Lorincz et al. [9]	-	0.991	0.987	0.994	0.995	0.999	0.996	0.994
Zheng et al. [10]	91.90%	-	-	-	-	-	-	-
3DCNN-A	90.39%	0.9535	0.9716	1.0000	0.9886	0.9948	1.0000	0.9848
3DCNN-B	90.39%	0.9897	0.9943	0.9922	0.9915	1.0000	1.0000	**0.9946**
3DCNN-C	**92.31%**	0.9793	1.0000	0.9767	0.9858	1.0000	1.0000	0.9903
3DCNN-D	**92.31%**	0.9793	1.0000	0.9871	0.9886	1.0000	1.0000	0.9925
Ensemble net	**96.15%**	0.9922	1.0000	0.9922	0.9972	1.0000	1.0000	**0.9969**

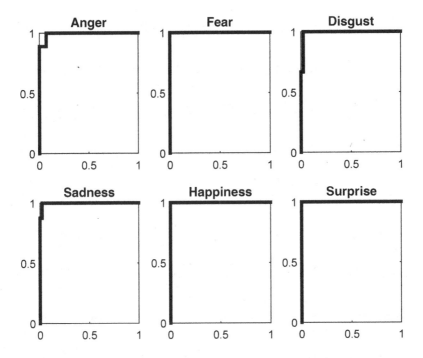

Fig. 4. ROC curves of ensemble network on CK+.

- Secondly, the mid-level representations and high-level representations of CNNs learnt by supervised learning play important roles in distinguishing the subtle differences of expressions between categories. The compared methods use the features without learning (like Haar, Gabor) or 1D linear features (like ICA).

6 Conclusion

The result shows that 3DCNNs are capable of classifying facial expressions into 6 basic emotional categories. The key to achieving good performance is making use of complementary information in a video sequence by explicitly extracting, processing and fusing multi-view data. To achieve this goal, 3D convolutional layers are defined. Four networks are designed for processing spatial-temporal data.

Inspiring by ideas from some previous work like caffe [23] and theano [24,25], we create two open source projects called 2DCNN [26] and 4DCNN [27]. The projects implement the layers and algorithms of Convolutional Neural Networks used in this paper including convolutional layer, full-connected layer, max-pooling layer, dropout layer, soft-max regression layer, Gabor layer, optical-flow layer, SGD solver with mini-batches, etc. The 4DCNN project is derived from the 2DCNN project for processing 4th order tensor data. The training/testing codes, preprocessed data and trained networks are provided in the open source package.

Our open source implementation already have the ability of processing 4D data like BU-4DFE [28] and BP4D-Spontaneous [29]. But after some preliminary studies, it is found that processing 4D data is more challenging. The applications based on 4DCNNs needs to be further studied.

Another unsolved problem is how to control the optimal capacity of 3DCNNs by varying their depth and breadth. Optimal capacity may be determined by many factors such as tasks, low-level features, number of the training samples, resolution of the image, memory limitation, etc. The proposed networks are designed by experience. Until now, the guiding method of designing 3DCNNs for arbitrary tasks is not so clear. More theoretical research and practical experience are needed.

References

1. Ekman, P., Friensen, E.: Facial Action Coding System (FACS): Manual. Consulting Psychologists Press, Palo Alto (1978)
2. Friensen, W., Ekman, P.: Emfacs-7: emotional facial action coding system. Technical report, University of California at San Francisico (1983)
3. Zeng, Z., Pantic, M., Roisman, G., Huang, T.S., et al.: A survey of affect recognition methods: audio, visual, and spontaneous expressions. IEEE Trans. Pattern Anal. Mach. Intell. **31**, 39–58 (2009)
4. Cootes, T.F., Edwards, G.J., Taylor, C.J.: Active appearance models. IEEE Trans. Pattern Anal. Mach. Intell. **23**, 681–685 (2001)
5. Matsugu, M., Mori, K., Mitari, Y., Kaneda, Y.: Subject independent facial expression recognition with robust face detection using a convolutional neural network. Neural Netw. **16**, 555–559 (2003)
6. Yang, P., Liu, Q., Metaxas, D.N.: Boosting encoded dynamic features for facial expression recognition. Pattern Recogn. Lett. **30**, 132–139 (2009)
7. Long, F., Wu, T., Movellan, J.R., Bartlett, M.S., Littlewort, G.: Learning spatiotemporal features by using independent component analysis with application to facial expression recognition. Neurocomputing **93**, 126–132 (2012)
8. Jeni, L., Girard, J.M., Cohn, J.F., De La Torre, F., et al.: Continuous AU intensity estimation using localized, sparse facial feature space. In: 2013 10th IEEE International Conference and Workshops on Automatic Face and Gesture Recognition (FG), pp. 1–7. IEEE (2013)
9. Lorincz, A., Jeni, L., Szabo, Z., Cohn, J.F., Kanade, T., et al.: Emotional expression classification using time-series kernels. In: 2013 IEEE Conference on Computer Vision and Pattern Recognition Workshops (CVPRW), pp. 889–895. IEEE (2013)
10. Zheng, H.: Facial expression analysis. Technical report, School of Computer Science and Engineering, Southeast University, Nanjing, China (2014)
11. Sun, W., Jin, Z.: Facial expression classification based on convolutional neural networks. In: Advances in Face Image Analysis: Theory and Applications. Bentham Science Publishers, Sharjah (2015, in press)
12. Yun, T., Guan, L.: Human emotional state recognition using real 3D visual features from gabor library. Pattern Recogn. **46**, 529–538 (2013)

13. Dhall, A., et al.: Collecting large, richly annotated facial-expression databases from movies (2012)
14. LeCun, Y., Bottou, L., Bengio, Y., Haffner, P.: Gradient-based learning applied to document recognition. Proc. IEEE **86**, 2278–2324 (1998)
15. Ji, S., Xu, W., Yang, M., Yu, K.: 3D convolutional neural networks for human action recognition. IEEE Trans. Pattern Anal. Mach. Intell. **35**, 221–231 (2013)
16. Russakovsky, O., Deng, J., Su, H., Krause, J., Satheesh, S., Ma, S., Huang, Z., Karpathy, A., Khosla, A., Bernstein, M., et al.: Imagenet large scale visual recognition challenge. Int. J. Comput. Vis. **115**, 1–42 (2014)
17. Krizhevsky, A., Sutskever, I., Hinton, G.E.: Imagenet classification with deep convolutional neural networks. In: Advances in Neural Information Processing Systems, pp. 1097–1105 (2012)
18. Byeon, Y.H., Kwak, K.C.: Facial expression recognition using 3D convolutional neural network. Int. J. Adv. Comput. Sci. Appl. **5**, 107–112 (2014)
19. Waibel, A., Hanazawa, T., Hinton, G., Shikano, K., Lang, K.J.: Phoneme recognition using time-delay neural networks. IEEE Trans. Acoust. Speech Signal Process. **37**, 328–339 (1989)
20. Horn, B.K., Schunck, B.G.: Determining optical flow. In: 1981 Technical Symposium East, pp. 319–331. International Society for Optics and Photonics (1981)
21. Lucey, P., Cohn, J.F., Kanade, T., Saragih, J., Ambadar, Z., Matthews, I.: The extended Cohn-Kanade dataset (ck+): a complete dataset for action unit and emotion-specified expression. In: 2010 IEEE Computer Society Conference on Computer Vision and Pattern Recognition Workshops (CVPRW), pp. 94–101. IEEE (2010)
22. Regianini, L.: Manual annotations of facial fiducial points on the cohn-kanade database (2015). http://lipori.dsi.unimi.it/download.html
23. Jia, Y., Shelhamer, E., Donahue, J., Karayev, S., Long, J., Girshick, R., Guadarrama, S., Darrell, T.: Caffe: convolutional architecture for fast feature embedding. arXiv preprint arXiv:1408.5093 (2014)
24. Bastien, F., Lamblin, P., Pascanu, R., Bergstra, J., Goodfellow, I.J., Bergeron, A., Bouchard, N., Bengio, Y.: Theano: new features and speed improvements. In: Deep Learning and Unsupervised Feature Learning NIPS 2012 Workshop (2012)
25. Bergstra, J., Breuleux, O., Bastien, F., Lamblin, P., Pascanu, R., Desjardins, G., Turian, J., Warde-Farley, D., Bengio, Y.: Theano: a CPU and GPU math expression compiler. In: Proceedings of the Python for Scientific Computing Conference (SciPy). Oral Presentation (2010)
26. Sun, W., Jin, Z.: The 2DCNN project (2015). https://github.com/anders0821/2DCNN
27. Sun, W., Jin, Z.: The 4DCNN project (2015). https://github.com/anders0821/4DCNN
28. Yin, L., Chen, X., Sun, Y., Worm, T., Reale, M.: A high-resolution 3D dynamic facial expression database. In: 8th IEEE International Conference On Automatic Face & Gesture Recognition, FG 2008, pp. 1–6. IEEE (2008)
29. Zhang, X., Yin, L., Cohn, J.F., Canavan, S., Reale, M., Horowitz, A., Liu, P., Girard, J.M.: BP4D-spontaneous: a high-resolution spontaneous 3D dynamic facial expression database. Image Vis. Comput. **32**, 692–706 (2014)

Suppression of Alpha Oscillation During Micro-expression Recognition

Ming Zhang[1,2], Yu-Hsin Chen[1,3], and Xiaolan Fu[1,4(✉)]

[1] State Key Laboratory of Brain and Cognitive Science, Institute of Psychology, Chinese Academy of Sciences, Beijing, China
fuxl@psych.ac.cn
[2] Department of Psychology, Dalian Medical University, Dalian, China
[3] Institute of Psychology and Behavior Sciences, Wenzhou University, Wenzhou, China
[4] University of Chinese Academy of Sciences, Beijing, China

Abstract. Behavioral evidence shows that the recognition of micro-expression is impaired by emotional context, especially negative context. However, the neural oscillatory features of such emotional context effect on recognition of micro-expression remain unclear. The present study used time-frequency analysis to explore the event-related spectral perturbation (ERSP) characteristics reflected in the processing of micro-expression followed emotional contexts. In the occipital-central region, we found that emotional context modulated micro-expression spectral power in alpha band and a strong suppression of alpha oscillation was observed in negative context. This study provided spectral dynamic evidence in support of the effect of emotional context during micro-expression recognition.

Keywords: Emotional context · Micro-expression · Alpha oscillation · ERSP · Time-frequency analysis

1 Introduction

Micro-expression is a quick facial expression that typically lasts for 1/25 s to 1/5 s and always occurs in the flow of facial expressions [1,2]. The facial expressions appearing before and after the micro-expression, i.e. emotional context, influenced the recognition of micro-expressions [3]. It has found that emotional context impaired the recognition of micro-expression, especially the negative context regardless of the duration of micro-expression [3]. Previous study only provided behavioral evidence for the effect of emotional context on the recognition of micro-expression. However, it remains unknown exactly how the effect of emotional context reflects in micro-expression processing.

Previous event-related potentials (ERPs) studies on facial expression recognition have provided supporting evidence for the effect of emotional context [4,5]. It is found that the emotional valence of context can modulate ERP outcomes [6]. Positive and negative pictures compared with neutral control pictures

© Springer International Publishing AG 2017
C.-S. Chen et al. (Eds.): ACCV 2016 Workshops, Part I, LNCS 10116, pp. 544–551, 2017.
DOI: 10.1007/978-3-319-54407-6_36

produce overall larger positive-going ERPs [7,8], for instance, the amplitudes of P1 and P2 are greater after negative pictures than that after neutral stimuli in the centro-frontal and the parieto-occipital region [9,10]. Moreover, this emotional context modulation mainly reflects in the early process of facial expression recognition [11–13].

Oscillatory activity is considered to provided more complex information than ERP can provide, revealing connections between different brain areas to a specific task [14]. One of the indices typically assessed in previous studies are event-related spectral perturbation (ERSP), means change in spectral power from baseline [15]. It is found that increased alpha synchronization activities are associated with differences induced by images of varying emotional valence [16]. Alpha band (8–12 Hz) oscillation is commonly evaluated for changes induced by different emotions [17–19]. The differentiation between negative and positive facial expressions was observed in the alpha frequencies, that spectral power in the alpha band responses upon angry face was significantly higher than it upon happy face at posterior locations [20]. There are lateralization effects overall changes in alpha power, that is, a relative right frontal activation associated with negative emotions and a relatively greater the left frontal activation associated with positive emotions [21]. In addition, previous EEG studies have reported that alpha rhythms in posterior region were also related with working memory [22], which was assumed to be capable of attention focus [23].

In order to investigate the alpha oscillatory properties reflecting the effect of emotional context in micro-expression processing, we employed time-frequency analysis and adopted synthesized micro-expressions task [3] which modified from Micro-Expression Training Tool (METT) [24] in present study. Based on previous study, subject pays attention to emotional stimuli [25], we expect that the judgment of micro-suppression will be biased and the frequency band will be suppressed.

2   Materials and Methods

2.1   Participants

Ten male and 11 female undergraduate students (age: 22.86 ± 1.65 years) participated in the formal EEG study as paid volunteers for RMB 80 Yuan. All participants were right-handed and had normal or corrected-to-normal vision. No participant had a neurologic or psychiatric history. All the participants provided written, informed consent prior to the experiment. This study was approved by the IRB of the Institute of Psychology, Chinese Academy of Sciences.

2.2   Stimuli

The stimuli used in this study were the same as those in previous study [3]. A total of 120 images (10 female models, 10 male models) were selected from the NimStim database [26] as the experiment materials, with the criterion of mean accuracy rate more than 85% in the former stimuli estimate experiment. The images of facial expressions (anger, happiness, and neutral) with the mouth closed were serving as emotional contexts, and the remaining images of facial expressions (anger, happiness, and neutral) with the mouth opened were serving as target micro-expressions. Images (visual angle $11.8° \times 15.1°$) were displayed at the center of the screen with a uniform silver-gray background on a 17-inch cathode-ray tube (CRT) monitor (frequency 100 Hz, resolution 1024×768).

2.3 Procedure

Stimuli presentations and manual response measurements were controlled using E-Prime 2.0 (Psychological Software Tools, Inc., Pittsburgh, PA, USA). On each trial (see Fig. 1), a black fixation cross was first presented for 500 ms, followed by emotional context (either anger, neutral, or happiness) for 1000 ms. After that, one of three target micro-expressions (anger, neutral, or happiness) was presented for 60 ms. Then, the same emotional context was again presented for 1000 ms. Finally, the labels of the three target expressions (anger, happiness, and neutral) were presented on the screen. Participants were asked to respond by using the mouse to click one of the emotion labels in the lower portion of the screen. The locations of three labels on each trial were randomly presented. There were three types of emotional contexts for each of three target expressions, for a total of 9 conditions. Participants performed one practice block and nine test blocks of 60 trials intermixed randomly. Participants were asked to rest for approximately 1-min break after one block and were instructed to respond as quickly and accurately as possible.

Fig. 1. Experimental setup of each trial

2.4 EEG Recording

EEG data was acquired from Scan 4.5 Acquisition Software (Compumedics Neuroscan, Inc., Texas, USA) and a 32-channel NuAmps Quickcap, 40-channel NuAmps DC amplifier, which was placed according to the specifications of the International 10–20 system. Vertical and horizontal electrooculogram (EOG) were recorded at supraorbital and suborbital to the left eye, and the outer canthi of both eyes to monitor eye blinks and eye movements. Both EEG and EOG were sampled at 1000 Hz, and electrode impedance was maintained below 5 kΩ. All of the scalp electrodes were referenced to the left mastoid online.

2.5   Data Analysis

Time-frequency analysis was performed on MATLAB (The Mathworks, Inc., Natick, MA, USA) using an open source toolbox of EEGLAB [27]. The EEG data were band-pass filtered between 1 and 30 Hz and were re-referencing offline to A1 and A2 (average mastoid reference). Each epoch began 1500 ms before context expression and ended 2000 ms after context expression onset. Trials that associated with an error response were excluded. Using newtimef function in matlab, spectral analyses were computed on the entire epochs, then baseline corrected to a 500 ms pre-stimulus (context expression) interval. Forty frequency points were calculated ranged from 4 to 30 Hz. Significant event-related context effect was assessed by parametric statistical with Bonferroni correction. We focused on the change in spectral power relative to target micro-expression with different emotional context of 200 ms in the range of 8–12 Hz (alpha-band) according to previous studies [28]. Signals were averaged separately across trials which induced alpha response for each target to create the following nine category-specific conditions: contexts (negative, positive, and neutral) × target micro-expressions (anger, happiness, and neutral). Repeated-measures ANOVAs with context and target variables were performed on the largest ERSP effect, which focused on respective frequency modulation of emotional context on target micro-expression with time windows 1000 ms to 1200 ms.

3   Results

3.1   Behavioral Performance

The two-way repeated ANOVA with emotional context and target micro-expression as the within-subject variables was conducted (Fig. 2). It revealed

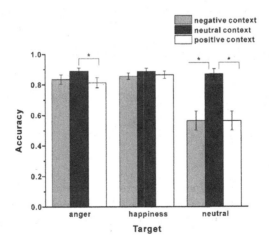

Fig. 2. Mean accuracy rates for target micro-expressions followed different emotional contexts.

a significant effect of emotional context, $F(2,40) = 26.53$, $p < 0.001$, $\eta_p^2 = 0.57$. It also revealed a significant effect of target micro-expression, $F(2,40) = 19.53$, $p < 0.001$, $\eta_p^2 = 0.49$. The interaction of context and target reached significance, $F(4,80) = 7.91$, $p < 0.001$, $\eta_p^2 = 0.28$. Additional analysis revealed that the accuracy rate for neutral was significantly higher with neutral context than it with negative or positive context, $t(19) = 4.26$, $p < 0.001$, $d = 0.97$; $t(19) = 4.31$, $p < 0.001$, $d = 0.99$; the accuracy rate for anger was significantly higher with neutral context than it with positive context, $t(19) = 3.26$, $p < 0.05$, $d = 0.75$, but there were no significant differences for happiness followed different emotional contexts (all $ps > 0.05$).

3.2 ERSP Results

Repeated-measures ANOVAs with context and target variables were performed on the largest ERSP. The largest ERSP effect on the alpha band was observed at Oz site and the alpha band effect had an occipital central distribution. For the baseline- corrected power at alpha frequencies (8–12 Hz) from 1000 ms to 1200 ms, the main effects of context and target were not significant, $F(2,19) = 2.00$, $p = 0.16$; $F(2,19) = 1.68$, $p = 0.21$, the interaction of context and target reached significance, $F(4,17) = 3.02$, $p < 0.05$, $\eta_p^2 = 0.42$. Further analysis showed a significant decrease of alpha power was observed for target anger with negative context compared with neutral context (see Fig. 3), $t(19) = -3.14$, $p < 0.05$, $d = 0.72$. No other effects were significant (all $ps > 0.05$).

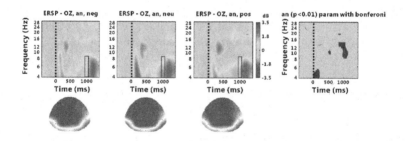

Fig. 3. Time-frequency representation at Oz site of alpha band for anger followed different emotional contexts.

4 Discussion

The present study compared oscillation properties of micro-expressions followed different emotional contexts, which provided spectral dynamic evidence in support of the effect of emotional context on micro-expression recognition. Participants' behavioral performance for neutral micro-expression was significantly poorer followed emotional context than that followed neutral context, which is consistent with previous findings [3]. A significant suppression of alpha spectral power was observed in negative context compared with neutral context in the occipital-central region, indicating the modulation effect of emotional context on spectral power response to micro-expression.

Emotional context, as a kind of emotional stimuli, may modulate attention allocation [29, 30] and capture more attention in facial expression recognition [31]. More attention resources were directed to emotional context [32] and were not transferred immediately to target micro-expression, thus poor performance for target neutral and anger micro-expressions recognition followed negative or positive context was observed. Yet, due to stronger emotional arousal [33, 34], we did not find such significant effect for target happiness micro-expression.

Time-frequency analysis found that the effect of negative context compared with the effect of positive or neutral context reduced occipital-central alpha spectral power. Alpha suppression over occipital regions has been involved in an increase of visual attention [35]. Negative context, as a kind of highly biologically salient stimuli [36], lead to increased attention load [37] and might trigger a long period of effect on subsequent micro-expression. Thus, the decreased spectral power response to micro-expression anger was observed when micro-expression followed negative context.

Our results indicated that alpha band was influenced by emotional context in the processing of micro-expression recognition. Moreover, alpha band is associated with working memory [38], which was involved in attention-demanding process [23]. Working memory was needed in our experimental task, further provided that emotional context can modulate attention allocation [39].

5 Conclusion

Emotional context modulated micro-expression spectral power in the alpha band and a strong suppression of alpha spectral power was observed in negative context in the occipital-central region. Our results not only provided spectral dynamic evidence for the effect of emotional context, but also provided insight into the possible existence of neural mechanism for the effect of emotional context on micro-expression recognition.

Acknowledgments. We thank Dr. Kai Wang for the early suggestions on data analysis of the study. This research was supported by grants from the National Natural Science Foundation of China (61375009).

References

1. Ekman, P., Friesen, W.V.: Nonverbal leakage and clues to deception. Psychiatry **32**, 88–106 (1969)
2. Ekman, P.: Lie catching and micro expressions. In: The Philosophy of Deception, pp. 118–133 (2009)
3. Zhang, M., Fu, Q.F., Chen, Y.C., Fu, X.L.: Emotional context influences micro-expression recognition. PloS One **9**, e95018 (2014)

4. Werheid, K., et al.: Priming emotional facial expressions as evidenced by event-related brain potentials. Int. J. Psychophysiol. **55**, 209–219 (2005)
5. Hietanen, J.K., Astikainen, P.: N170 response to facial expressions is modulated by the affective congruency between the emotional expression and preceding affective picture. Biol. Psychol. **92**, 114–124 (2013)
6. Bernat, E., Bunce, S., Shevrin, H.: Event-related brain potentials differentiate positive and negative mood adjectives during both supraliminal and subliminal visual processing. Int. J. Psychophysiol. **42**, 11–34 (2001)
7. Cuthbert, B.N., et al.: Brain potentials in affective picture processing: covariation with autonomic arousal and affective report. Biol. Psychol. **52**, 95–111 (2000)
8. Dolcos, F., Cabeza, R.: Event-related potentials of emotional memory: encoding pleasant, unpleasant, and neutral pictures. Cogn. Affect. Behav. Neurosci. **2**, 252–263 (2002)
9. Carretié, L., et al.: Automatic attention to emotional stimuli: neural correlates. Hum. Brain Mapp. **22**, 290–299 (2004)
10. Franchini, M.: Emotional face processing: an ERP study on visual selective attention of emotional faces presented subliminally and supraliminally. University of Geneva (2011)
11. Righart, R., de Gelder, B.: Context influences early perceptual analysis of faces–an electrophysiological study. Cereb. Cortex **16**, 1249–1257 (2006)
12. Righart, R., de Gelder, B.: Rapid influence of emotional scenes on encoding of facial expressions: an ERP study. Soc. Cogn. Affect. Neurosci. **3**, 270–278 (2008)
13. Morel, S., et al.: Very early modulation of brain responses to neutral faces by a single prior association with an emotional context: evidence from MEG. Neuroimage **61**, 1461–1470 (2012)
14. Ma, J., Liu, C., Chen, X.: Emotional conflict processing induce boosted theta oscillation. Neurosci. Lett. **595**, 69–73 (2015)
15. Wang, K., et al.: Temporal and spectral profiles of stimulus-stimulus and stimulus-response conflict processing. Neuroimage **89**, 280–288 (2014)
16. Aftanas, L., et al.: Event-related synchronization and desynchronization during affective processing: emergence of valence-related time-dependent hemispheric asymmetries in theta and upper alpha band. Int. J. Neurosci. **110**, 197–219 (2001)
17. Davidson, R.J.: Anterior cerebral asymmetry and the nature of emotion. Brain Cogn. **20**, 125–151 (1992)
18. Davidson, R.J.: Cerebral asymmetry and emotion: conceptual and methodological conundrums. Cogn. Emot. **7**, 115–138 (1993)
19. Gotlib, I.H.: EEG alpha asymmetry, depression, and cognitive functioning. Cogn. Emot. **12**, 449–478 (1998)
20. Güntekin, B., Basar, E.: Emotional face expressions are differentiated with brain oscillations. Int. J. Psychophysiol. **64**, 91–100 (2007)
21. Jaušovec, N., Jaušovec, K., Gerlič, I.: Differences in event-related and induced EEG patterns in the theta and alpha frequency bands related to human emotional intelligence. Neurosci. Lett. **311**, 93–96 (2001)
22. Jensen, O., et al.: Oscillattons in the alpha band (9–12 Hz) increase with memory load during retention in a short-term memory task. Cereb. Cortex **12**, 877–882 (2002)
23. Baddeley, A.: Working memory: theories, models, and controversies. Annu. Rev. Psychol. **63**, 1–29 (2012)
24. Ekman, P.: METT. Micro expression training tool. CD-ROM. Oakland (2003)
25. Ratcliff, N.J., et al.: The scorn of status: a bias toward perceiving anger on high-status faces. Social Cogn. **30**, 631–642 (2012)

26. Tottenham, N., et al.: The NimStim set of facial expressions: judgments from untrained research participants. Psychiat. Res. **168**, 242–249 (2009)
27. Delorme, A., Makeig, S.: EEGLAB: an open source toolbox for analysis of single-trial EEG dynamics including independent component analysis. J. Neurosci. Meth. **134**, 9–21 (2004)
28. Klimesch, W.: EEG-alpha rhythms and memory processes. Int. J. Psychophysiol. **26**, 319–340 (1997)
29. Ito, T.A.: Reflections on social neuroscience. Social Cogn. **28**, 686–694 (2010)
30. Vuilleumier, P., et al.: Effects of attention and emotion on face processing in the human brain: an event-related fMRI study. Neuron **30**, 829–841 (2001)
31. Alpers, G.W.: Eye-catching: right hemisphere attentional bias for emotional pictures. Laterality **13**, 158–178 (2008)
32. Wilson, J.P., Hugenberg, K.: Shared signal effects occur more strongly for salient outgroups than ingroups. Social Cogn. **31**, 636–648 (2013)
33. Adolphs, R., Russell, J.A., Tranel, D.: A role for the human amygdala in recognizing emotional arousal from unpleasant stimuli. Psychol. Sci. **10**, 167–171 (1999)
34. Hugenberg, K., Sczesny, S.: On wonderful women and seeing smiles: social categorization moderates the happy face response latency advantage. Social Cogn. **24**, 516–539 (2006)
35. Jessen, S., Kotz, S.A.: The temporal dynamics of processing emotions from vocal, facial, and bodily expressions. Neuroimage **58**, 665–674 (2011)
36. Hess, U., Adams, R.B., Kleck, R.E.: The categorical perception of emotions and traits. Social Cogn. **27**, 320–326 (2009)
37. Phelps, E.A., LeDoux, J.E.: Contributions of the amygdala to emotion processing: from animal models to human behavior. Neuron **48**, 175–187 (2005)
38. Klimesch, W., Schack, B., Sauseng, P.: The functional significance of theta and upper alpha oscillations. Exp. Psychol. **52**, 99–108 (2005)
39. Meaux, E., et al.: Event-related potential and eye tracking evidence of the developmental dynamics of face processing. Eur. J. Neurosci. **39**, 1349–1362 (2014)

Distinguishing Posed and Spontaneous Smiles by Facial Dynamics

Bappaditya Mandal[1(✉)], David Lee[2], and Nizar Ouarti[3,4]

[1] Visual Computing Department,
Institute for Infocomm Research, Singapore, Singapore
bmandal@i2r.a-star.edu.sg
[2] Electrical and Computer Engineering,
National University of Singapore, Singapore, Singapore
a0097074@u.nus.edu
[3] Université Pierre et Marie Curie, Paris, France
nizar.ouarti@ipal.cnrs.fr
[4] Sorbonne Universités, Paris, France

Abstract. Smile is one of the key elements in identifying emotions and present state of mind of an individual. In this work, we propose a cluster of approaches to classify posed and spontaneous smiles using deep convolutional neural network (CNN) face features, local phase quantization (LPQ), dense optical flow and histogram of gradient (HOG). Eulerian Video Magnification (EVM) is used for micro-expression smile amplification along with three normalization procedures for distinguishing posed and spontaneous smiles. Although the deep CNN face model is trained with large number of face images, HOG features outperforms this model for overall face smile classification task. Using EVM to amplify micro-expressions did not have a significant impact on classification accuracy, while the normalizing facial features improved classification accuracy. Unlike many manual or semi-automatic methodologies, our approach aims to automatically classify all smiles into either 'spontaneous' or 'posed' categories, by using support vector machines (SVM). Experimental results on large UvA-NEMO smile database show promising results as compared to other relevant methods.

1 Introduction

In the past, the research on human affect was focused on basic emotions which include happiness, sadness, fear, disgust, anger, and surprise [1]. Past research conducted in this area largely focused on posed facial expressions. Primarily, because of difficulty in obtaining and capturing spontaneous facial expression that lead to unavailability of such databases. However, these discrete emotions (posed facial expressions) fail to describe the wide range of emotions that occur in natural social interactions. Recent studies have shown that spontaneous facial expressions reveal more information about the emotions of the person, and spontaneous expressions differ greatly from its posed counterpart [2]. For example, a spontaneous smile is generally interpreted to show enjoyment or happiness,

© Springer International Publishing AG 2017
C.-S. Chen et al. (Eds.): ACCV 2016 Workshops, Part I, LNCS 10116, pp. 552–566, 2017.
DOI: 10.1007/978-3-319-54407-6_37

but it can also arise out of frustration [3]. As a result, more emphasis has been placed on spontaneous facial expressions in the recent years. Smile is one of the key elements in identifying emotions and state of mind from facial expressions, as it is frequently exhibited to convey emotions like amusement, politeness and embarrassment [4], and is also used to mask other emotional expressions [5]. Since smile is the easiest expression to pose [5], it is important for the machines (and also humans) to distinguish when it is posed and when it is a genuine smile of enjoyment.

This work contributes to the field of affective computing by solving the problem of distinguishing posed and spontaneous smiles. Development in this field can be applied commercially to enhance our daily lives. For example, automatic human affect recognition can be applied to wearable devices like Google Glass to help children with autism who have difficulty reading expressions to understand emotions in a social environment [6–9]. It can also be installed in vehicles to detect fatigue levels of the driver and prevent fatigue-causing accidents from occurring. Affective computing can also be applied in the academic fields like psychology, psychiatry, behavioral science and neuroscience, to reduce the time-consuming task of labeling human affects and can improve human lives.

1.1   Related Work

In early years of research, it was thought that the morphological features of the face were good indicators of a spontaneous smile. The Facial Action Coding System (FACS) [10] defines Action Units (AU), which are the contraction or relaxation of one or more muscles. It is commonly used to code facial expressions, and can be used to identify emotions. In FACS, a smile corresponds to AU12, which is the contraction of the zygomatic major muscle that raises the lip corners. A genuine smile of joy is thought to include AU6, also known as the Duchenne Marker, which is the contraction of the orbicularis oculi (pars lateralis) muscle that raises the cheek, narrows the eye aperture, and forms wrinkles on the external sides of the eyes. However, recent research casts doubt on the reliability of Duchenne marker in identifying true feelings of enjoyment [11]. Another possible marker of spontaneous smiles is the symmetry of the smile. Initial studies claim that smile symmetry is a factor in identifying spontaneous smiles, where spontaneous smiles are more symmetrical than posed smiles [12]. However, later studies report no significant differences of symmetry [13,14].

Recently, more attention has been paid to dynamical properties of smiles such as the duration, amplitude, speed, and acceleration instead of static features like smile symmetry or the AUs. To analyze these properties, the smile is generally broken up into three different phases - onset, apex, and offset. Spontaneous smiles tend to have a smaller amplitude, a slower onset [15], and a shorter total duration [16]. The eye region is analyzed as well - the eyebrow raise in posed smiles have a higher maximum speed, larger amplitude and shorter duration than spontaneous ones [14]. Most techniques extract dynamic properties of smiles that are known to be important factors in classifying smiles [1,17]. Apart from these properties of smiles, facial dynamics can reveal other useful information for the classification of smiles, such as the subject's age.

Currently, the method with the best performance is the one proposed by Dibeklioglu *et al.* [13]. In their method, 11 facial feature points are tracked using a Piecewise Bezier Volume Deformation (PBVD) tracker, which was proposed by Tao and Huang [18]. The duration, amplitude, speed and acceleration of various features in the eyes, cheeks and mouth regions are calculated across the 3 different phases of the smile, and a mid-level fusion is used to concatenate the features. Using these features, the classification accuracy on the UvA-NEMO Smile Database was 87.0%. An optical flow along with various face component based features is proposed in [19], where it is shown that even optical based features can perform similar to face component based features. However, their tracking is initialized by manually annotated facial landmarks.

This work aims to discover other additional features from facial dynamics that can be useful for classification. To do so, the entire face region should be processed and used for smile classification instead of extracting the known features of a smile described above. A cluster of approaches are developed to extract features from a video and use a classifier to determine whether the smile in the video is posed or spontaneous. The system first pre-processes the video by tracking the face and extracting the face region from the video after normalization. 3 different normalization techniques are tested to determine its efficacy. Micro-expressions in the face region are amplified to test their impact on smile classification. Then, several image processing techniques are used to extract features from the face region. The features are then post-processed to reduce the dimensionality of the features and to normalize the number of features per video for the classifier to work. Finally, the processed features are given as an input to a support vector machines (SVM) to classify between posed and spontaneous smiles.

In the rest of this paper, in Sect. 2, a cluster of methodologies is proposed to test the effectiveness of three normalization techniques, and the effectiveness of HOG, LPQ, dense optical flow and pre-trained convolutional neural network features for smile classification. A relatively new technique for amplifying micro-expressions is also tested to analyze the impact of micro-expressions in classifying smiles. In Sect. 3, experimental results and analysis are presented. Conclusions are drawn in Sect. 4.

2   Proposed Methodologies

Figure 1 shows the flow of our proposed approaches. The incoming videos/images are pre-processed to extract the face and eye regions from the image frames. Faces are then tracked using these detected regions so as to understand the global movement of the faces. 3 separate normalization techniques are used to normalize the face in the extracted image sequence, and each technique is tested to see how effective each one is. Eulerian video magnification (EVM) is used in the pre-processing step to amplify micro-expressions in the face region to test the

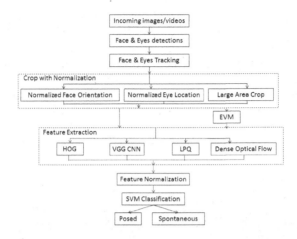

Fig. 1. Block diagram of the proposed system.

impact of micro-expressions in classification. After the face region is extracted, the videos are processed to extract features for classification. Different feature extraction techniques are used to test the effectiveness of each method. The features are then post-processed to reduce the dimensionality of the features and to normalize the number of features per video for the classifier to work. Finally, the processed features are given as an input to the SVM to classify between posed and spontaneous smiles. Each of these blocks are discussed below.

2.1 Pre-processing Techniques

The incoming image frames are pre-processed using three different methodologies which are described below:

Face and Eye Detections: The face and eyes are first detected to be used for tracking in the next step. Initially, a cascade object detector from Matlab's computer vision toolbox was implemented to detect the face and both eyes in the first frame of the video. However despite fine-tuning, it was inaccurate as there were many cases of false positives. Such errors will propagate throughout the entire system and cause anomalous results in the classification. Therefore this implementation was discarded and replaced by a more accurate method by Mandal *et al.* [20,21]. This method uses a fusion of OpenCV face and eye detectors [22] and Integration of Sketch and Graph patterns (ISG) eye detectors developed for human-robot-interaction by Yu *et al.* in [23]. Through the integration of both eye detectors [24], we are able to achieve high success rate of 96.3% eye localization in the smile face images for both frontal and semi-frontal faces at various scales with large global motions. The output of this step is 3 bounding boxes containing the face, the left eye and the right eye, separately.

Face and Eye Tracking: A Kanade-Lucas-Tomasi (KLT) tracker [25] was implemented to track the face and eyes in the video. The tracker searches for good feature points to track within the face and eyes bounding boxes and tracks these points across the video. These points are marked as crosses (+) in the faces shown in Fig. 2. An affine transformation is estimated based on the movement of the feature points from one frame to the next, and the bounding box is then warped according to the transformation to keep track of the face region. Figure 2 shows the original bounding box in the first frame of the video, and the warped bounding box in a later frame. The output of this step is the coordinates of the bounding box vertices at every frame.

Fig. 2. KLT tracker in action.

Cropping Strategies: The KLT tracker is used to track the face region in the video. Then, three different methods of normalization are tested.

1. *Normalize eye location:* This method normalizes the location of the eyes in the video, as described by Mandal *et al.* in [20,26]. The location of the eyes are tracked, and the video is rotated and resized such that both eyes are always on the same horizontal axis, and there is a fixed pixel distance between both eyes (234 pixels apart for a face image of 400×500, similar to [27–29]). Figure 3 shows a sample image of three subjects from UvA-NEMO smile database.

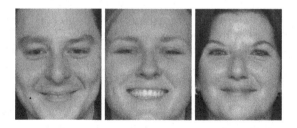

Fig. 3. Cropped face regions after eye location normalization.

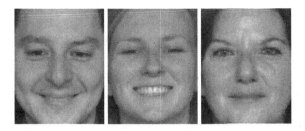

Fig. 4. Cropped face regions after rotation normalization.

2. *Normalize face orientation:* This method normalizes the orientation of the face. The orientation of the face is obtained from the KLT tracking data, and the video is rotated to compensate for the face rotation such that the face appears upright. Figure 4 shows the corresponding samples.
3. *No normalization:* This is a control experiment to observe the effects of normalization. Figure 5 shows the corresponding samples.

After normalization from each of these techniques, the face region is cropped out of the video for feature extraction. The first two methods produce a 400×500 pixel region, while the third method produces a 720×900 pixel region.

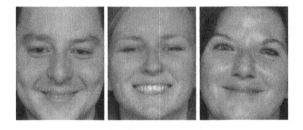

Fig. 5. Cropped face regions with no normalization.

2.2   Micro-expression Amplification

Eulerian Video Magnification (EVM) is a Eulerian method to amplify small motions and color variations in videos. It was first developed by Wu *et al.* [30], and the motion amplification was later improved with a phase-based pipeline by Wadhwa *et al.* [31]. It is used to amplify micro-expressions in the extracted face region. This method of motion amplification has shown good results when used in a system to detect spontaneous micro-expressions [32]. The algorithm first decomposes an input video into different spatial frequency bands and orientations using complex steerable pyramids, and the amplitude of local wavelets are separated from their phase. The phases at each location, orientation and scale are then independently filtered and the phase of a specified frequency range is amplified (or attenuated) and de-noised. Finally, the video is reconstructed, which now has some motions amplified, depending on the frequencies selected.

2.3   Feature Extraction

In this subsection details of the feature extraction and their corresponding normalization techniques are discussed. Features are extracted from the cropped face region. Four different image-processing techniques are used to extract features.

1. *Local Phase Quantization:* LPQ was originally used for texture description [33], which outperformed the Local Binary Pattern operator in texture classification and has been extended to face recognition [34]. LPQ features are insensitive to motion blurs, out of focus blurs, and atmospheric turbulence blurs. The algorithm computes a 2D Discrete Fourier Transform (DFT) over a neighborhood at each pixel using 2D convolutions. Then, the covariance matrix of the DFT is obtained, de-correlated and quantized to obtain the local phase information. LPQ descriptors are less sensitive to image blurs which may arise from the interpolation of the normalization techniques, and can be useful for smile recognition.

2. *Histogram-of-Oriented-Gradients:* HOG is typically used for objection detection, and has been used for human detection in videos [35]. The idea behind HOG is that an object's shape can be well-characterized by the distribution of local intensity gradients (edge directions) without precise knowledge of their positions. The algorithm divides the image into spatial regions called cells, and a histogram of gradient directions is computed for each pixel in the cell. Cells are grouped to form larger spatial regions called blocks, and the local histogram 'energy' of a block is used to normalize the contrast in all the cells in that block. These normalized blocks are the HOG descriptors. The 'UoCTTI' variant of HOG [36] is used as it has a 31 features per cell compared to the original 36 features per cell, which is a significant reduction in dimensionality. A 4×4 pixel cell with 8 orientations is used as early experiments show that it has the best results for the given problem.

3. *Dense Optical Flow:* Optical flow is the pattern of apparent motion of objects in a visual scene caused by a relative movement between an observer and the scene. Optical flow can be computed for a sparse feature set to track the movement of the feature points, or for a dense feature set whereby every pixel in an image sequence is tracked. Dense optical flow is used to calculate the movement of each pixel throughout the video. This work uses a dense optical flow field determined by a differential method proposed by Farneback [37]. Optical flow works on the assumptions that the pixel intensities of a scene do not change between consecutive frames, and that neighboring pixels have similar motion. Farneback's algorithm approximates an image signal to be a polynomial expansion and solves for the displacement of that polynomial assuming a translation occurred. The solution is made to be more robust by introducing assumptions to constrain the system. The main assumptions made are that the displacement field is slowly varying, and that the displacement field can be parameterized according to a motion model.

4. *Pre-trained Convolutional Neural Network (CNN) Features:* Another emerging class of features are the pre-trained deep CNN features. Studies have shown that deep convolutional neural network (CNN) model trained for certain application can be applied to similar classification problems perform well [38,39]. We use the pre-trained deep CNN model for face recognition, which is trained with 2.6 million images comprising of over 2600 people for face smile classification. Most existing models related to face or object recognition is applied to images, not videos. To accommodate this method for video-processing, each frame of the video is processed and features are extracted from a deep fully-connected layer. These features are post-processed to combine them and used for classification with a SVM. This work uses the VGG face recognition model [39], and 4096 features with L2 normalized per frame are extracted from the 35th layer of the CNN network.

The techniques mentioned above extract features frame-by-frame. Since the video samples have a varying number of frames, there is also a varying number of features per sample. Our learning algorithms require a fixed number of features per sample, so the number of features per sample is normalized. To do so, the features of each video are concatenated across time and transformed to the cosine domain via Discrete Cosine Transform, such that the number of cosine coefficients is equal to the number of frames in the video. Then, the number of coefficients are normalized by either cropping or padding them such that there is an equal number of cosine coefficients per sample video; this is equivalent to normalizing the number of frames per video. The number of frames in the videos range between 80–500 frames, and the average number of frames is 150. A suitable number of frames for normalization would be 150, however computational constraints limit the number of features that each sample can have, thus the number of frames per video is normalized to 100 for the experiments.

2.4   Classification

A SVM is a machine learning algorithm that can be used for classification, regression or other tasks by constructing a hyperplane or a set of hyperplanes. The modern implementation of SVMs was developed by Cortes and Vapnik in [40]. Recent study shows that when applied to distinguishing posed and spontaneous smiles, the linear SVM classifier outperforms other classifiers like the Polynomial Kernel SVM, RBF Kernel SVM, Linear Discriminant, Logistic Regression, k-Nearest Neighbour and Naïve Bayes [41]. Therefore, linear SVM is used for classification in this work.

3   Experiments

The UvA-NEMO Smile Database is used to analyze the dynamics of posed and spontaneous smiles [13]. This is the largest database of videos of spontaneous and posed smiles to date [19], with 643 posed smiles and 597 spontaneous

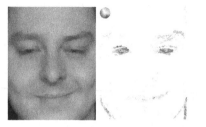 

(a) Optical flow of a subject blinking while smiling.   (b) Region extraction.

Fig. 6. Left two: optical flow of a blink, Right: extracted regions

smiles from 400 subjects. The videos were recorded with a Panasonic HDC-HS700 3MOS camcorder that was placed on a monitor approximately 1.5 m away from the recorded subjects. The videos have a resolution of 1920 × 1080 pixels and a frame rate of 50 frames per second. For posed smiles, each subject was asked to pose an enjoyment smile after being shown a demonstration video. For spontaneous smiles, each subject was shown a set of short, funny videos for approximately 5 min. Two trained annotators segmented the recorded video of their smiles to obtain the smile video.

Similar to [13, 19], the accuracy of the system is measured using 10-fold cross validation, where the data set is divided into 10 folds specified by the smile database. 9 folds are used for training and the remaining fold is used for testing. This is done 10 times, such that each fold is used for testing once in order to represent the entire data set, and the average of the results is used as the measure of accuracy. The classification accuracy is recorded into a confusion matrix. To simplify the presentation of results, only the accuracy of true positives and negatives are displayed. The large number of features from HOG coupled with the larger face region size from not normalizing the face results in a large dimensionality that could not be computed using the SVM classifier. Therefore, the results for that case is omitted.

Figure 6a shows the optical flow output, where the hue of the pixel denotes the flow direction, while the saturation of the pixel denotes the flow magnitude. The drawback of dense optical flow is that a large number of features is produced. For a face region of 400 × 500 pixels, each pixel will have 2 features for optical flow in the x- and y-directions. This results in 400,000 features per frame, and approximately 30 million features per video. The dimensionality for classification is too high, requiring a very long time and a large memory to process the training data. Therefore, the dimensionality must be reduced in order for this method to be feasible. To do so, the optical flow data from the eyes, nose and mouth regions are extracted and used for classification, while the other regions are discarded since they are not likely to have a large impact on classification. Figure 6b shows the bounding boxes of the regions that are extracted.

The eye regions are obtained from the tracking data. The nose and mouths regions are estimated based on the location of the eyes. The result is a dimen-

Table 1. True positive (posed smiles) and true negative (spontaneous smiles) classification accuracy (%) of our system with varying feature extraction methodologies and normalization.

	Norm. eye location		Norm. face orientation		No normalization	
	EVM	No EVM	EVM	No EVM	EVM	No EVM
HOG	81.7, 74.3	82.8, 72.6	82.4, 71.6	81.7, 71.7	-	-
LPQ	83.9, 65.1	80.1, 70.8	84.7, 65.1	79.9, 70.8	84.3, 65.3	82.6, 66.0
VGG	82.9, 70.6	82.9, 70.3	82.7, 69.7	82.2, 71.6	82.5, 69.6	82.4, 69.3

Table 2. Overall classification accuracy (%) of our system with varying feature extraction methodologies and normalization.

	Norm. eye location		Norm. face orientation		No normalization	
	EVM	No EVM	EVM	No EVM	EVM	No EVM
HOG	78.14	77.89	77.20	76.89	-	-
LPQ	74.85	75.62	75.26	75.52	75.15	74.61
VGG	76.98	76.83	76.44	77.10	76.29	76.09

sionality of approximately 52,800 features per 21 frame, and approximately 8 million features per video. However, the accuracy obtained using 10-fold cross validation (see Sects. 3 and 4) for dense optical flow is bad, with 59.9% accuracy for classifying posed smiles correctly, and 70.3% accuracy for classifying spontaneous smiles. Due to the long time that it takes to extract dense optical flow features and the large dimensionality of the features, it was deemed that this method is not worth further tests and is omitted from the discussions hereafter.

Table 1 shows the classification accuracy of the system by varying the normalization methods, feature extraction techniques, and the use of EVM for microexpression amplification. For example, the top-left entry of the table shows that there is a 81.7% accuracy for classifying posed smiles, and 74.3% accuracy for classifying spontaneous smiles, when the eye location is normalized, EVM is used to amplify micro-expressions, and HOG features are used. The table abbreviates the pre-trained VGG convolutional neural network model features to 'VGG'. From Table 1, it can be seen that the classification of posed smiles is always more accurate than spontaneous smiles, regardless of normalization or feature extraction method. This may be a result of having more prominent visual appearances of the posed smiles as compared to the spontaneous ones. Perhaps, this allows the SVM to have better support vectors to define posed smiles, whereas lesser visual appearances of spontaneous smiles could mean that the support vectors are less well defined and the hyperplane produced is not a good true separator of boundaries.

Table 2 shows that overall accuracy of our proposed approaches. Although VGG face model is trained with millions of face images, features from such deep CNN network may not be good for face smile classification. It is evident that HOG features with normalization using the eye locations along with the magnified micro-expression using EVM performs best in most of the cases as compared to LPQ and VGG features.

3.1   Comparison with Other Methods

Correct classification rates (%) using various methods on UvA-NEMO are shown in Table 3. Most of the existing methodologies involve semi-automatic processes for face feature extraction, whereas our method is fully automatic and does not require any manual intervention. It is evident from the table that our proposed approach is quite competitive as compared to the other state-of-the-arts methodologies.

Table 3. Correct classification rates (%) on UvA-NEMO database.

Method	Process	Correct Classification Rate (%)
Pfister *et al.* [42]	Semi-automatic	73.1
Dibeklioglu *et al.* [43]	Semi-automatic	71.1
Cohn & Schmidt [14]	Semi-automatic	77.3
Eyelid Features [13]	Semi-automatic	85.7
Eye + Lips + dense optical flow [19]	Semi-automatic	80.4
Our proposed method	Fully automatic	78.1

3.2   Discussions

Table 1 shows that the classification of posed smiles is always more accurate than spontaneous smiles, regardless of normalization or feature extraction method. This is probably because posed smiles may have more prominent visual appearances as compared to the spontaneous ones. Figure 7 represents the results in Table 2 in a histogram by grouping the feature extraction methods together. It can be seen that HOG outperforms the other two feature extraction methods, LPQ and VGG, for most of the normalization techniques, except for one case where using HOG has a 0.21% lesser accuracy than using the pre-trained NN model features when normalizing the face orientation, and not using EVM. Similarly, using the pre-trained VGG CNN model outperforms LPQ as a feature extraction method as the classification accuracy using VGG is higher than LPQ for all cases.

From the experiments, it can be said that among the 3 feature extraction methods, HOG features are the best, and LPQ features are the worst. Features from pre-trained VGG model using large number of face images for face recognition has not been able to generalize well for smile classification. It is interesting to note that HOG is capable of capturing the fine grained facial features that helps in distinguishing posed and spontaneous smiles.

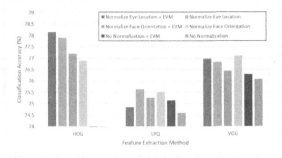

Fig. 7. Comparison of accuracy with feature extraction methods.

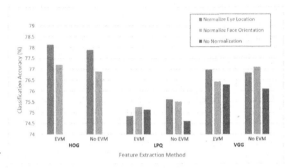

Fig. 8. Comparison of classification accuracy with normalization methods.

Figure 8 shows the comparison of classification accuracy between normalization methods, which are grouped according to the parameters. Generally, having no face normalization results in a lower classification accuracy as compared to having either normalization method. The only exception is the case of using LPQ features with EVM, where the classification accuracy is 0.3% higher without normalization than normalizing the eye location. Normalizing the eye location performs the best in 4 out of the 6 cases, thus it seems like it is the better normalization. The most significant difference between normalization techniques is seen when using HOG descriptors, where there is 1% difference in classification accuracy between normalizing eye location vs. normalizing face orientation.

4   Conclusions

In this work, a cluster of methodologies is proposed to distinguish posed and spontaneous smiles. It involves four feature extraction methods, three normalization methods and the use of EVM for micro-expression amplification, but was unable to improve the state-of-the-art performance. The best classification accuracy obtained was 78.14%. Using EVM to amplify micro-expressions did not have a significant impact on classification accuracy, while the normalizing facial features improved classification accuracy. The advantage of our proposed approaches as compared to other methods is that they are fully automatic. The effectiveness of the feature extraction methods in smile classification is ranked from most to least effective as follows: HOG, a pre-trained VGG CNN model features, LPQ and dense optical flow. This work succeeded in identifying techniques which are helpful and detrimental to smile classification. Experimental results on large UvA-NEMO smile database show promising results as compared to other relevant methods.

References

1. Zeng, Z., Pantic, M., Roisman, G.I., Huang, T.S.: A survey of affect recognition methods: audio, visual, and spontaneous expressions. PAMI **31**, 39–58 (2009)
2. Ekman, P., Rosenberg, E.: What the Face Reveals: Basic and Applied Studies of Spontaneous Expression Using the Facial Action Coding System, 2nd edn. Oxford University Press, New York (2005)
3. Hoque, M., McDuff, D., Picard, R.: Exploring temporal patterns in classifying frustrated and delighted smiles. IEEE Trans. Affect. Comput. **3**, 323–334 (2012)
4. Ambadar, Z., Cohn, J., Reed, L.: All smiles are not created equal: morphology and timing of smiles perceived as amused, polite, and embarrassed/nervous. J. Nonverbal Behav. **33**, 17–34 (2009)
5. Ekman, P.: Telling Lies: Cues To Deceit in the Marketplace, Politics, and Marriage. WW. Norton & Company, New York (1992)
6. Hadwin, J., Baron-Cohen, S., Howlin, P., Hill, K.: Can we teach children with autism to understand emotion, belief, or pretense? Dev. Psychopathol. **8**, 345–365 (1996)
7. Xu, Q., Ching, S., Mandal, B., Li, L., Lim, J.H., Mukawa, M., Tan, C.: Socio glass: social interaction assistance with face recognition on google glass. J. Sci. Phone Apps Mob. Devices **2**, 1–4 (2016)
8. Mandal, B., Lim, R.Y., Dai, P., Sayed, M.R., Li, L., Lim, J.H.: Trends in machine and human face recognition. In: Kawulok, M., Celebi, M.E., Smolka, B. (eds.) Advances in Face Detection and Facial Image Analysis, pp. 145–187. Springer, Cham (2016). doi:10.1007/978-3-319-25958-1_7
9. Mandal, B., Wang, Z., Li, L., Kassim, A.A.: Performance evaluation of local descriptors and distance measures on benchmarks and first-person-view videos for face identification. Neurocomputing **184**, 107–116 (2016)
10. Ekman, P., Friesen, W.: The Facial Action Coding System: A Technique for the Measurement of Facial Movement. Consulting Psychologists Press Inc., San Francisco (1978)
11. Krumhuber, E.G., Manstead, A.S.: Can duchenne smiles be feigned? new evidence on felt and false smiles. Emotion **9**, 807–820 (2009)
12. Ekman, P., Hager, J., Friesen, W.: The symmetry of emotional and deliberate facial actions. Psychophysiology **18**, 101–106 (1981)
13. Dibeklioğlu, H., Salah, A.A., Gevers, T.: Are you really smiling at me? spontaneous versus posed enjoyment smiles. In: Fitzgibbon, A., Lazebnik, S., Perona, P., Sato, Y., Schmid, C. (eds.) ECCV 2012. LNCS, vol. 7574, pp. 525–538. Springer, Heidelberg (2012). doi:10.1007/978-3-642-33712-3_38

14. Schmidt, K., Bhattacharya, S., Denlinger, R.: Comparison of deliberate and spontaneous facial movement in smiles and eyebrow raises. J. Nonverbal Behav. **33**, 35–45 (2009)
15. Cohn, J., Schmidt, K.: The timing of facial motion in posed and spontaneous smiles. Int. J. Wavelets, Multiresolut. Inf. Process. **2**, 1–12 (2004)
16. Schmidt, K., Ambadar, Z., Cohn, J., Reed, I.: Movement differences between deliberate and spontaneous facial expressions: zygomaticus major action in smiling. J. Nonverbal Behav. **30**, 37–52 (2006)
17. Valstar, M.F., Pantic, M., Ambadar, Z., Cohn, J.F.: Spontaneous vs. posed facial behavior: automatic analysis of brow actions. In: Proceedings of ACM International Conference on Multimodal Interaction, pp. 162–170 (2006)
18. Tao, H., Huang, T.: Explanation-based facial motion tracking using a piecewise bézier volume deformation model. In: Proceedings of IEEE Conference on Computer Vision and Pattern Recognition, vol. 1, pp. 611–617 (1999)
19. Mandal, B., Ouarti, N.: Spontaneous vs. posed smiles - can we tell the difference? In: International Conference on Computer Vision and Image Processing (CVIP), vol. 460, pp. 261–271. Roorkee, India (2016)
20. Mandal, B., Chia, S.-C., Li, L., Chandrasekhar, V., Tan, C., Lim, J.-H.: A wearable face recognition system on google glass for assisting social interactions. In: Jawahar, C.V., Shan, S. (eds.) ACCV 2014. LNCS, vol. 9010, pp. 419–433. Springer, Cham (2015). doi:10.1007/978-3-319-16634-6_31
21. Mandal, B., Li, L., Chandrasekhar, V., Lim, J.H.: Whole space subclass discriminant analysis for face recognition. In: IEEE International Conference on Image Processing (ICIP), Quebec, Canada, pp. 329–333 (2015)
22. OpenCV: Open source computer vision (2014). http://opencv.org/
23. Yu, X., Han, W., Li, L., Shi, J.Y., Wang, G.: An eye detection and localization system for natural human and robot interaction without face detection. In: Groß, R., Alboul, L., Melhuish, C., Witkowski, M., Prescott, T.J., Penders, J. (eds.) TAROS 2011. LNCS, vol. 6856, pp. 54–65. Springer, Heidelberg (2011). doi:10.1007/978-3-642-23232-9_6
24. Mandal, B., Zhikai, W., Li, L., Kassim, A.A.: Evaluation of descriptors and distance measures on benchmarks and first-person-view videos for face identification. In: Jawahar, C.V., Shan, S. (eds.) ACCV 2014. LNCS, vol. 9008, pp. 585–599. Springer, Cham (2015). doi:10.1007/978-3-319-16628-5_42
25. Tomasi, C., Kanade, T.: Detection and tracking of point features. Carnegie Mellon University Technical Report CMU-CS-91-132 (1991)
26. Mandal, B., Eng, H.L.: 3-parameter based eigenfeature regularization for human activity recognition. In: 35th IEEE International Conference on Acoustics Speech and Signal Processing (ICASSP), pp. 954–957 (2010)
27. FERET: Feret normalization (2005). http://www.cs.colostate.edu/evalfacerec/data/normalization.html
28. Mandal, B., Jiang, X.D., Kot, A.: Verification of human faces using predicted eigenvalues. In: 19th International Conference on Pattern Recognition (ICPR), Tempa, Florida, USA, pp. 1–4 (2008)
29. Jiang, X.D., Mandal, B., Kot, A.: Face recognition based on discriminant evaluation in the whole space. In: IEEE 32nd International Conference on Acoustics, Speech and Signal Processing (ICASSP 2007), Honolulu, Hawaii, USA, pp. 245–248 (2007)
30. Wu, H.Y., Rubinstein, M., Shih, E., Guttag, J., Durand, F., Freeman, W.T.: Eulerian video magnification for revealing subtle changes in the world. ACM Trans. Graph. **31**, 65 (2012)

31. Wadhwa, N., Rubinstein, M., Durand, F., Freeman, W.T.: Phase-based video motion processing. ACM Trans. Graph. **32** (2013)

32. Li, X., Hong, X., Moilanen, A., Huang, X., Pfister, T., Zhao, G., Pietikäinen, M.: Reading hidden emotions: spontaneous micro-expression spotting and recognition. CoRR abs/1511.00423 (2015)

33. Ojansivu, V., Heikkilä, J.: Blur insensitive texture classification using local phase quantization. Image Signal Process. **5099**, 236–243 (2008)

34. Ahonen, T., Rahtu, E., Ojansivu, V., Heikkilä, J.: Recognition of blurred faces using local phase quantization. In: 19th International Conference on Pattern Recognition, pp. 1–4 (2008)

35. Dalal, N., Triggs, B.: Histograms of oriented gradients for human detection. In: IEEE Computer Society Conference on Computer Vision and Pattern Recognition, vol. 1, pp. 886–893 (2005)

36. Felzenszwalb, P.F., Girshick, R.B., McAllester, D., Ramanan, D.: Object detection with discriminatively trained part based models. IEEE Trans. Pattern Anal. Mach. Intell. **32**, 1627–1645 (2009)

37. Farnebäck, G.: Two-frame motion estimation based on polynomial expansion. In: Bigun, J., Gustavsson, T. (eds.) SCIA 2003. LNCS, vol. 2749, pp. 363–370. Springer, Heidelberg (2003). doi:10.1007/3-540-45103-X_50

38. Hu, G., Yang, Y., Yi, D., Kittler, J., Christmas, W.J., Li, S.Z., Hospedales, T.M.: When face recognition meets with deep learning: an evaluation of convolutional neural networks for face recognition. CoRR abs/1504.02351 (2015)

39. Parkhi, O.M., Vedaldi, A., Zisserman, A.: Deep face recognition. In: Proceedings of the British Machine Vision Conference (BMVC), vol. 41, no. 1–41, p. 12 (2015)

40. Cortes, C., Vapnik, V.: Support-vector networks. Mach. Learn. **20**, 273–297 (1995)

41. Dibeklioglu, H., Salah, A., Gevers, T.: Recognition of genuine smiles. IEEE Trans. Multimedia **17**, 279–294 (2015)

42. Pfister, T., Li, X., Zhao, G., Pietikainen, M.: Differentiating spontaneous from posed facial expressions within a generic facial expression recognition framework. In: ICCV Workshop, pp. 868–875 (2011)

43. Dibeklioglu, H., Valenti, R., Salah, A., Gevers, T.: Eyes do not lie: spontaneous versus posed smiles. In: ACM Multimedia, pp. 703–706 (2010)

Affective Gait Recognition and Baseline Evaluation from Real World Samples

Vili Kellokumpu[(⊠)], Markus Särkiniemi, and Guoying Zhao

Center for Machine Vision and Signal Analysis, University of Oulu, Oulu, Finland
{kello,misarkin,gyzhao}@ee.oulu.fi

Abstract. Over the years a lot of research efforts have been put into recognizing human emotions from facial expressions. However, in many scenarios access to suitable face data is difficult, and therefore there is a need for methodology that can be used when people are observed from a distance. A potential modality for this is human gait. Early attempts to recognize human emotion from gait have been limited to acted data. Furthermore, in these approaches the data has been captured in controlled settings. This paper presents the first experiments for automated affective gait recognition using non acted real world samples. A database of 96 subjects affected by positive or negative feedback is collected and two baseline methods are used to recognize the affective state of a person. The baseline results are promising and encourage further study in this domain.

1 Introduction

Human emotion recognition is one of the most active areas of computer vision research. It is driven by its importance in human-machine interfaces. It also has application potential in other domains such as social sciences, health care and security. Most of the research efforts so far have focused on facial expression recognition. This limits the applicability of emotion recognition techniques because suitable frontal face data of sufficient quality cannot be captured when subjects are observed from a distance.

It is evident that humans can observe a great deal of information from human movement. For example, recognizing friends based on the way they move even from a distance is very natural. In computer vision research this phenomena is used in gait recognition: human movement is used as biometric signature to identify people based on the way they walk. In gait recognition the aim is to discriminate between the walking styles of different people. However, human walking movement conveys even more subtle information, such as cues about feelings or mood of a person. It has been shown that human observers are able to identify emotions from gait [6]. Another study suggests that humans are able to recognize emotions using walking speed and posture [8].

Earlier attempts at automated emotion recognition from human body movement have been limited to analyzing acted data. In this work, our hypothesis is that we can automatically recognize the affective state of a person by observing

© Springer International Publishing AG 2017
C.-S. Chen et al. (Eds.): ACCV 2016 Workshops, Part I, LNCS 10116, pp. 567–575, 2017.
DOI: 10.1007/978-3-319-54407-6_38

Fig. 1. When subjects are observed from a distance and facial expressions are not available, analyzing gait offers additional cues for estimating the affective state of a person.

their gait using real non acted data. This is in contrast to previous works that perform emotion analysis from subjects artificially behaving and posing their feelings. So far the availability of non acted data has been a limiting factor and in this work tackles this limitation by collecting data from tv-broadcast material (Fig. 1).

The rest of the paper is organized as follows. Related work on gait is reviewed in Sect. 2. The data collection for our experiments is described in Sect. 3. Section 4 describes the chosen baseline methods and Sect. 5 presents the obtained experimental results. Finally, Sect. 6 concludes the paper.

2   Related Work

A lot of research on recognizing human emotion from gait has been done over the years. The work is typically restricted by data that is based on acted emotions. Montepare et al. showed that identifying emotions from gait is possible for human observers [6]. In the experiments ten subjects observed five walkers that were instructed to express four different emotions: sadness, anger, happiness and anger. The results suggest that gait characteristics such as the amount of arm swing, stride length, heavyfootedness, and walking speed differentiated the emotions expressed by the walkers.

Crane and Gross [3] used autobiographical memory tasks to record video and motion capture data of people feeling anger, sadness, content, joy or neutral. In their data collection participants completed a self-report of felt emotions. The data was then viewed and classified by human observers. It was argued that emotions can be successfully elicited in the laboratory setting, emotions can be recognized in the body movements of others, and that body movements are affected by felt emotions.

Roether et al. [8] investigated the influence of posture and dynamic cues for the expression and recognition of emotions in gait. To identify emotion-specific features they motion-captured the neutral and emotionally expressive (anger, happiness, sadness, fear) gaits of 25 subjects. They found that in terms of kinematics, size and speed of movements were among the most important features driving the perception of emotional gaits.

Instead of gait, Atkinson et al. [1] used a database of people expressing emotions or performing simple actions with whole-body movement. The actors in the database were free to interpret and express the emotions as they saw fit, with only minimal guidance as to the sorts of situations in which people might experience those emotions. Human observers were then shown either a full-light or patch-light stimuli of the movement. It was then discovered that kinematics help distinguish emotions expressed by body gestures.

It is evident that earlier work suffers from a low number of subjects. Also, the data is not typically recorded in an authentic manner, i.e. the subjects are asked to act as if feeling an instructed emotion.

Apart from emotion recognition, a lot of work has been devoted for human gait analysis in computer vision. Many methods have been proposed for gait recognition aiming at biometric recognition [11]. The methods and databases used for gait recognition traditionally rely on using silhouettes as input.

One of the simplest yet effective features for gait recognition is the average silhouette proposed by Liu and Sarkar [5]. In this approach silhouettes are aligned and averaged into a single image template without considering the dynamics between frames. On the contrary, Kellokumpu et al. [4] treated the human gait pattern as a dynamic texture. The method calculates Local Binary Patterns from Three Orthogonal Planes (LBP-TOP) and thus utilizes both appearance and dynamics of human movements for gait recognition. The average silhouette and dynamic texture based methods are selected as baseline methods in this work and are briefly introduced in Sect. 4.

3   Data from Real Life Samples

As discussed above, earlier work suffers from a low number of subjects. Furthermore, the subjects are asked to act as if feeling an instructed emotion when data is captured. To overcome these limitations, we collect affective gait data from 96 subjects after they have received positive or negative feedback. Table 1 compares the databases used for affective gait analysis. The data used in our experiments is made publicly available

Table 1. Comparison of emotional databases in literature. Symbols: X = No, Y = Yes.

	Subjects	Emotions	Sensors/Markers	Acted
Atkinson [1]	32	5	X	Y
Pollick et al. [7]	2	19	Y	Y
Montepare et al. [6]	5	4	X	Y
Wallbott [10]	16	14	X	Y
Crane & Gross [3]	42	5	Y	Y
Roether et al. [8]	25	5	Y	Y
Our work	**96**	**2**	**X**	**X**

The bases for the data collection are various TV-shows where candidates audition or present their talent in order to compete for attractive prizes. After an audition, the judges of the show determine if a candidate will advance in the competition or if he/she is eliminated. The starting point of the data capture is the moment when the result of a performance is published and the candidate leaves the stage affected by the positive or negative verdict.

The data is collected from different shows and different seasons. While the clips are selected so that the subject is captured from the side view, the capturing conditions vary between shows and seasons. Furthermore, the collected samples include variation also within the videos as they are captured with cameras mounted on moving studio cranes.

Altogether videos of 96 subjects are collected. The data contains an equal number of samples (48) of both positive and negative class. The cuts in the broadcast data are typically rather fast and capturing long gait sequences is challenging. Therefore, the number of half cycles, i.e. steps, in the collected samples varies from one to four.

Since most of the methods and databases used for biometric gait recognition rely on silhouettes, we also adopt this approach. The TV broadcast data that is used for the data collection is captured with moving cameras and it may also contain dynamic lighting variations. This makes automatic extraction of silhouettes difficult. Instead, we use user assisted graph cut algorithm [2] to do the segmentation. The resulting silhouettes are realistic and comparable to what a background subtraction algorithm should produce from a video captured with a static camera.

Figure 2 illustrates the user input for an example frame and gives a few examples of the silhouette quality. It is evident that the data contains a lot of natural variation in the subjects body shape, size, clothing etc.

One limitation of this approach to data collection is that the ground truth for the affective state can only be categorized either as positive or negative instead of a more detailed categorization. As discussed above, in the acted scenarios a more detailed ground truth label may be set. Also, the data collection does not allow to estimate the intensity of the felt reaction.

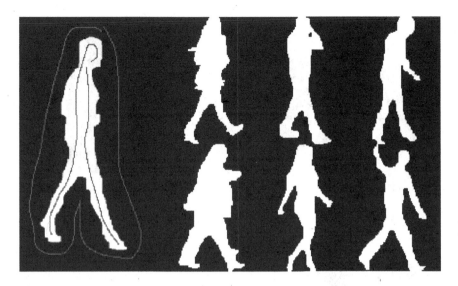

Fig. 2. Illustration of a the user annotation for segmentation and example silhouettes from the database

4   Methods

Since gait recognition methods are capable of capturing small differences in human movement to recognize people from the way they walk, we adopt a similar approach in this work. We use existing features originally proposed for gait recognition and apply them with a Support Vector Machine to learn a classifier for affective gait recognition.

In the preprocessing stage, the silhouette sequence is first normalized and stacked into a 3D volume. In feature extraction either average silhouette or LBP-TOP features are computed. Finally the sequence is classified based on the features into a positive or negative class.

4.1   Preprocessing

We treat gait as a spatiotemporal volume. In order to build the volume, silhouettes are scaled and aligned. First, the silhouette size is normalized by scaling it into a fixed height. The height is set to 128 pixels similar to [5,9]. Then, vertical alignment is performed by aligning the silhouettes based on their centroids. Finally the silhouettes are stacked into a 3D volume.

4.2   Features

Based on the spatiotemporal volume, two feature representations are calculated: average silhouette [5] and LBP-TOP [4]. The reason for choosing these features is twofold. Firstly, the average silhouette is extensively used as a baseline method

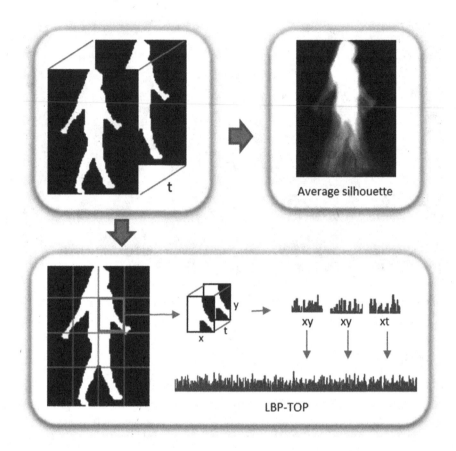

Fig. 3. Illustration of preprocessing and feature calculation for average silhouettes and LBP-TOP

in gait recognition. Secondly, average silhouette only captures appearance information whereas LBP-TOP representation captures motion information in addition to appearance information, which is important according to the previous studies [6,8].

Average silhouette is computed simply as an average along time axis for each pixel. The dynamic texture features are extracted by calculating the LBP texture features on three orthogonal planes of the volume: xy, xt and yt. Figure 3 illustrates the feature calculation for both feature types from the spatiotemporal volume.

We follow the implementation details as presented in [4,5]. The parameters we used for LBP-TOP on all planes are as follows: radius is set to one and the number of sampling points is eight. Also, uniform patterns are used. It should be noted that larger values for radiuses were used in [4], but since our data volume size is much smaller, small radius is more practical. Furthermore we divide the volume into 4×4 blocks to preserve more spatial structure.

4.3 Classification

Since the aim of gait recognition is to differentiate people from each other, it means that matching sequences directly based on the gait features will result in matching sequences to people moving similarly and not to people with similar feelings. For affective gait recognition this means that such simple matching should fail to work. Therefore we need to learn the differences in the feature representation that indicate differences in the affective state.

For classification we consider two methods. Firstly nearest neighbor (NN) classification is considered as a sanity check. Euclidean distance is used as a distance measure for the average silhouettes. For LBP-TOP histogram intersection is used as a similarity measure. Secondly, in order learn the important cues that enable emotion recognition, we use Support Vector Machines to learn the difference between positive and negative classes.

5   Experiments

We used the leave one out cross validation to test the average silhouette and LBP-TOP methods for affective gait recognition on the collected data. Thus one sample is used for testing while the others are used for training. This is repeated for all samples. The results are summarized in Table 2.

Figure 4 shows the SVM classification results with various parameters (powers of 2). It can be seen that fairly good results can be achieved with a range of parameter combinations.

Nearest neighbor based classification typically used in gait recognition leads to matching people to similar looking people. A biometric cue should be invariant to emotion, and this is a likely reason why the nearest neighbor classification basically fails to generalize for emotion recognition and achieves only slightly above random performance. In contrast to this, SVM seems to be able to do better in learning the differences caused by the affective state of the subject from the training data.

Also an interesting observation from the results is that it seems that the dynamic texture based description performs better than the average silhouette. It seems that the dynamics of gait are a good cue for analyzing the affective state of a person. This is in line with the observations of Montepare [6] and Roether [8].

Table 2. Classification results for affective gait recognition with average silhouettes and LBP-TOP.

Method (feature + classifier)	Performance
Average silhouette + NN	59%
LBP-TOP + NN	64%
Average silhouette + rbf SVM	72%
LBP-TOP + rbf SVM	78%

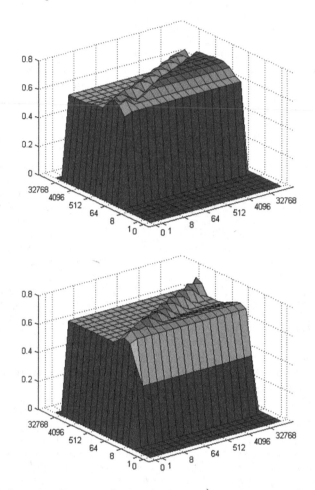

Fig. 4. (top) Average silhouette, (bottom) LBP-TOP. Performance plots with various parameter combinations for rbf SVM

6   Conclusions

This work describes the first video based experiments for recognizing affect from human gait patterns based on real world data. Furthermore we provide the community with the first affective video gait dataset captured in real world conditions.

Although the experimental results suggest that it is indeed possible to recognize affect from human gait, the problem is far from solved. One important aspect to consider as future work is the evaluation of the level of the observed affect and not just the classification to positive and negative. Also, for practical applications, it would be useful to study the problem using the image data directly instead of silhouettes.

Acknowledgement. This work was sponsored by the Academy of Finland, Infotech Oulu and Nokia Visiting Professor grant.

References

1. Atkinson, A.P., Tunstall, M.L., Dittrich, W.H.: Evidence for distinct contributions of form and motion information to the recognition of emotions from body gestures. Cognition **104**, 59–72 (2007)
2. Boykov, Y., Kolmogorov, V.: An experimental comparison of Min-Cut/Max-Flow Algorithms for energy minimization in vision. IEEE Trans. Pattern Anal. Mach. Intell. **26**(9), 1124–1137 (2004)
3. Crane, E., Gross, M.: Motion capture and emotion: affect detection in whole body movement. Affect. Comput. Intell. Interact. **4738**, 95–101 (2007)
4. Kellokumpu, V., Zhao, G., Li, S., Pietikäinen, M.: Dynamic texture based gait recognition. In: International Conference on Biometrics, ICB 2009 (2009)
5. Liu, Z., Sarkar, S.: Simplest representation yet for gait recognition: averaged silhouette In: Proceedings of the 17th International Conference on Pattern Recognition, vol. 4, pp. 211–214 (2004)
6. Montepare, J.M., Goldstein, S.B., Clausen, A.: The identification of emotions from gait information. J. Nonverbal Behav. **11**(1), 33–42 (1987)
7. Pollick, F.E., Paterson, H.M., Bruderlin, A., Sanford, A.J.: Perceiving affect from arm movement. Cognition **82**, B51–B61 (2001)
8. Roether, C.L., Omlor, L., Christensen, A., Giese, M.A.: Critical features for the perception of emotion from gait. J. Vis. **9**, 1–32 (2009)
9. Sarkar, S., Phillips, P.J., Liu, Z., Robledo, I., Grother, P., Bowyer, K.W.: The human ID gait challenge problem: data sets, performance, and analysis. IEEE Trans. Pattern Anal. Mach. Intell. **27**(2), 162–177 (2005)
10. Wallbott, H.G.: Bodily expression of emotion. Eur. J. Soc. Psychol. **28**, 879–896 (1998)
11. Wang, J., She, M., Nahavandi, S., Kouzani, A.: A review of vision-based gait recognition methods for human identification. In: International Conference on Digital Image Computing: Techniques and Applications (DICTA), pp. 320–327 (2010)

Weighted Non-locally Self-similarity Sparse Representation for Face Deblurring

Lei Tian[1], Chunxiao Fan[1], Yue Ming[1(✉)], and Xiaopeng Hong[2]

[1] Beijing Key Laboratory of Work Safety Intelligent Monitoring,
School of Electronic Engineering,
Beijing University of Posts and Telecommunications,
Beijing, People's Republic of China
myname35875235@126.com
[2] Department of Computer Science and Engineering,
University of Oulu, Oulu, Finland

Abstract. The human face is one of the most interesting subjects in various computer vision tasks. In recent years, significant progress has been made for generic image deblurring problem, but existing popular sparse representation based deblurring methods are not able to achieve excellent results on blurry face images. The failure of these methods mainly stems from the lack of local/non-local self-similarity prior knowledge. There are many similar non-local patches in the neighborhood of a given patch in a face image, therefore, this property should be effectively exploited to obtain a good estimation of the sparse coding coefficients. In this paper, we introduce the current weighted non-locally self-similarity (WNLSS) method [1], which is originally proposed to remove the noise for natural images, into the face deblurring model. There are two terms in the WNLSS sparse representation model, data fidelity term and regularization term. Based on the theoretical analysis, we show the properties of data fidelity term and regularization term also can fit well for face deblurring problem. The results also demonstrate that WNLSS method can achieve excellent performance in terms of both synthetic and real blurred face dataset.

1 Introduction

Restoring a clear image from one blurred image has many important applications, such as video surveillance, remote sensing and so on. The process under a spatially-invariant model can be generally formulated as

$$B = k * I + \varepsilon, \tag{1}$$

where I is the latent sharp image, B is the blurred (observed) image, k and ε denote the blurred kernel and the noise term, respectively. In the past decades, extensive studies have been conducted on the natural image deblurring problem [2–8]. Since the ill-posed nature of image deblurring, additional prior knowledge

© Springer International Publishing AG 2017
C.-S. Chen et al. (Eds.): ACCV 2016 Workshops, Part I, LNCS 10116, pp. 576–589, 2017.
DOI: 10.1007/978-3-319-54407-6_39

is need to constrain the solutions of Eq. (1), such as heavy-tailed gradient distributions [3,9], regularization-based techniques [6,10] and sparsity constraints [11,12]. Though the above prior knowledge and models work well for natural images, they can not perform well when faced with specific object category, such as face image. Compared with natural image, the face image has special non-local self-similarity (NSS) and less texture information, the work [13] exploits the NSS patches to reconstruct the visual light face images from the near infrared images. Since face is one of the most interesting objects in various computer vision application, we focus on face image deblurring in this paper.

The success of state-of-the-art image deblurring methods [2,14,15] mainly depends on extraction of salient edges for kernel estimation. However, the edge prediction step usually does not work well when faced with those blurred images with less texture, such as face image, which is demonstrated in the work [8]. As an alternative, the sparsity representation based deblurring methods have achieved excellent results for natural image deblurring. The sparse representation of a clear image I and an observed image B can be denoted as following:

$$\min_{a_I} \left\{ \|I - Da_I\| + \lambda \|a_I\|_p \right\} \tag{2}$$

and

$$\min_{a_B} \left\{ \|B - k * Da_B\| + \lambda \|a_B\|_p \right\} \tag{3}$$

where $\|\bullet\|_p$ is p-norm term (p is set to be 0 or 1) and it is expected that the coding coefficients a_I and a_B are sparse as much as possible. D denotes a learning-based over-completed dictionary and the parameter λ controls the contribution of regularization item. It is impossible for us to obtain the I and its coding coefficients a_I, so we hope the coding coefficients of observed image a_B which is learned from our deblurring model could be close enough to a_I. Therefore, the reconstructed version of original image $\hat{I} = Da_B$ is considered as the final deblurring result. Due to l_0-minimization problem is a NP-hard problem, the l_1-minimization problem is usually used in sparse representation model.

1.1 Motivation

For a blurred face image, it is very challenging to solve the sparse codes a_B of observed image B just by using sparsity constraint prior, just like Eq. (3). The universal prior knowledge easily leads to the solution of Eq. (3) entrap into a local minimum and we need more specific prior knowledge to design the objective function of facial deblurring model. In this paper, we apply the existing WNLSS sparse representation model [1] for the face image deblurring task. It is well known that there are lots of non-local similar patches in a face image, as shown in Fig. 1. The self-similarity based coding coefficients a_z can be computed, so that the more accurate estimation of sparse coding coefficients a_B can be obtained by exploiting these non-local redundancies. Moreover, many existing sparse representation based deblurring methods [16,17] take observed image B as a long vector and obtain a global coding coefficients for the whole image by solving the Eq. (3). However, the solution of this kind of deblurring model has been proved is unstable and time-consuming [18]. In order to make the computational process stable and fast, we should provide smaller input signal for deblurring model. Due to the face image has the NSS property, a given local patch can be reconstructed well by referencing its similar patches. We intend to extract the dense local patches from the observed image and fed them into our sparse representation model. So the input size of sparse representation model is relatively small and we can obtain the stable and effective coding coefficients for each local patch. To further reduce the patch-wise residual, the weighted encoding technique [1] is introduced into the data fidelity term of WNLSS method. The large pixel residual will lead to the reconstructed local patch produce the ghost or ringing visual artifacts. The pixel i which has large pixel residual is expected to assign small weights, so that this pixel's effect on the encoding of observed image over over-completed dictionary can be reduced.

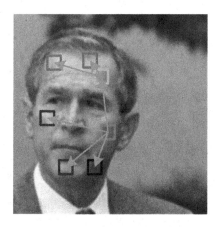

Fig. 1. The illustration of non-local self-similarity in a face image. This property brings us an alternative approach to estimate the sparse coding coefficients of reconstruction signal. For solving the real coding coefficients of a given blurred patch, we just need to search a set of similar patches and reconstruct this patch by linearly combining these similar patches. Then, we estimate the real coding coefficients of a given blurred patch by using the coding coefficients of the reconstructed patch.

The rest of the paper is organized as follows. we describe the deblurring method in the Sect. 2 and investigate the performance of our method on both synthetic and real blurry face datasets in Sect. 3 and conclude this paper in Sect. 4.

2 Weighted Non-locally Self-similarity (WNLSS) Sparse Representation

For an image $I \in \mathbb{R}^{M \times N}$ and I_i denotes a local patch extracted at pixel i with size $m \times n$. The i-th lcoal patch I_i of I can be sparsely encoded as $I_i = D a_{I,i}$.

In order to deblur I from B, B need to be sparsely coded with the learned dictionary D by solving the Eq. (3). Table 1 summary some notations which are frequently used in our paper.

Table 1. The notation table

Notations	Descriptions
I	The latent sharp (clear) image.
B	The observed (blurred) image.
D	The dictionary of sparse representation model.
a_I	The sparse coding coefficients of clear image I.
a_B	The sparse coding coefficients of observed image B.
a_z	The estimated sparse coding coefficients based on facial NSS redundancies.
k	The blur kernel (known in this paper).
e	The residual vector between original signal and reconstructed signal.
W	The weight matrix of weighted coding based data fidelity term

2.1 The Dictionary Learning

In this section, we first discuss the dictionary D learning problem. Many existing deblurring methods need some example images to learn dictionary [6,8] and this kind of example images are required to choose elaborately, so that the learned dictionary can contain indispensable texture units. Therefore, the usage of this kind of deblurring methods is restricted. Compared with this kind of methods, we intend to learn dictionary D from the input image itself. If we learn an universal dictionaries from the whole original image by using K-SVD [19], the such over-completed dictionary may leads to expensive computational cost and unstable solutions, this observation has been proved by the work [12]. How to obtain the stable and properly sparse solution is a key problem in the dictionary learning process. Inspired by the work [20], all patches of the input image are clustered into K centers by the traditional K-Means algorithm. Other complicated clustering algorithms are also investigated in our experiment, such as Gaussian Mixture Model, mean-shift and spectral clustering, but there is no significant improvement of deblurring performance. So we select the K-Means algorithm due to its simplicity. The problem of dictionary selection can be denoted as follow:

$$\min_{D_k, C_k} \left\{ \|X_k - D_k C_k\|_2^2 + \lambda \|C_k\|_1 \right\}, \tag{4}$$

where X_k denotes the kth clustered set and C_k is the coding coefficient matrix over cluster-based dictionary D_k. Because of the elements in each cluster X_k is similar, it is not necessary to form an over-completed dictionary in order to describe the variation of X_k. So the compact representation is better than over-completed representation due to the computational cost of dictionary learning. For each X_k, we compute the principal components by PCA algorithm. To compactly represent the discriminative pattern of each cluster set X_k, we just select L eigenvectors $[p_1, \cdots, p_l, \cdots, p_L]$ corresponding to the largest eigenvalues to construct the dictionary D_k.

For each patch need to be coded, only one sub-dictionary is adaptively selected to code it and the coding coefficients of other all sub-dictionaries is set to be 0. Therefore, the sparse constraint in the existing sparsity-based deblurring is also satisfied.

2.2   Formulation

In order to make the Eq. (3) can work well on face image deblurring problem, the weighted coding technique [1] and facial non-local self-similarity prior knowledge is introduced into the data fidelity term $\|(B - k * Da)\|_2^2$ and regularization term $R(a)$ of the WNLSS method, respectively. We will particularly describe these two terms in the following parts.

Weighted Coding Based Data Fidelity Term. It is well known that the residual between the deblurred image and observed image is directly related to the performance of reconstruction method. It is naturally expected that the effect of those pixels with large residual can be decreased. In order to minimize the data fitting residual $\|B - k * Da\|_2^2$ for all pixels, the weighted coding technique is employed to weight the residual. Assume the overall residual vector is defined as following:

$$e = [e_1, e_2, \cdots, e_N] = B - k * Da, \tag{5}$$

where $e_i = B_i - k * Da_i$ and $e_1, e_2, \cdots, e_N$ are $i.i.d$ samples. According to modern robust estimation technique, each residual should be assigned a proper weight, like $\bar{e}_i = w_i^{1/2} e_i$. Therefore, Eq. (3) can be re-written as following:

$$\min_{W, a_B} \left\{ \left\| W^{1/2} (B - k * Da_B) \right\|_2^2 + \lambda \|a_B\|_p \right\}, \tag{6}$$

where W is a diagonal weight matrix with $W_{ii} = w_i$. Obviously, the weight w_i and the magnitude of residual e_i^2 should have inverse property, so the relationship between these two items can be simply formulated as following:

$$w_i = \exp\left(-c_0 e_i^2\right) \tag{7}$$

where c_0 is the penalization parameters of the residual e_i.

Non-local Self-similarity Based Regularization Term. As above mentioned, we expect a_B is close to a_I as much as possible, but a_I is unknown for face deblurring task. Fortunately, there is a great number of non-local redundancies in face images, regardless of clear images or blurred images. So it can be expected that providing a good estimation a_z for a_I by exploiting the self-similarity redundancies prior knowledge.

For a given patch I_i, a set of similar non-local patches I_i^P are collected in a large window around I_i. We select P most similar patches I_i^P to I_i and assign them different weight to predict the $\hat{I}_i$ as $\hat{I}_i = \sum_{p=1}^{P} d_i^p I_i^p$, where P denotes the number of similar patches, and the weight d_i^p can be computed as a traditional regularized least-square problem:

$$\min_{d_i} \|I_i - Id_i\|_2^2 + \eta \|d_i\|_2^2, \tag{8}$$

where $I = \left[I_i^1, I_i^2, \cdots, I_i^P\right]$ and $d_i = \left[d_i^1, d_i^2, \cdots, d_i^P\right]^T$. The regularization term $\|d_i\|_2^2$ is applied to enhance the stability of least-square solution and η is the regularization parameter. Therefore, each patch I_i can be reconstructed by a set of non-local similar patches I_i^p. In other words, for an observed images, each patch I_i and its NSS based reconstruction $\hat{I}_i$ can be denoted as $I_i = D_i a_{B,i}$ and $\hat{I}_i = D_i a_{z,i}$. Accordingly, the $a_{z,i}$ for the patch $\hat{I}_i$ can be obtained as:

$$a_{z,i} = D_i^T \hat{I}_i. \tag{9}$$

In order to make the reconstruction $\hat{I}_i$ is close enough to I_i, obviously, the coding coefficients of non-local reconstruction $a_{z,i}$ should be similar as that of observation $a_{B,i}$. Therefore, the NSS information is introduced into the objective function and the Eq. (6) can be formulated as following:

$$\min_{W,a_B} \left\{ \left\|W^{1/2}\left(B - k * Da_B\right)\right\|_2^2 + \lambda \|a_B - a_z\|_1 \right\}. \tag{10}$$

It is the final objective function in our method, and the optimization process is detailed in the next section.

2.3   Optimization

Assumed the blurring kernel k and dictionary D are given, there are two variables W and a_B in the proposed formulation, we can effectively solve them by updating one variable and fixing another.

Updating a_B and fixing W. When W is fixed, the objective function of our method becomes the ℓ_1-norm sparse coding problem. This problem can be effectively solved via the iterative re-weighted scheme [21]. According to the deduction of work [21], the $(k + 1)$th sparse coding $a_B^{(k+1)}$ can be denoted as:

$$a_B^{(k+1)} = \left(D^T W D + V^{(k+1)}\right)^{-1}\left(D^T W B - D^T W Da_z\right) + a_z, \tag{11}$$

where V is a diagonal matrix and it can be initialized as an identity matrix. And the diagonal element $V_{ii}^{(k+1)}$ of auxiliary matrix $V^{(k+1)}$ can be updated as:

$$V_{ii}^{(k)} = \lambda / \left(\left(a_{B,i}^{(k)} - a_{z,i}^{(k)} \right)^2 + \varepsilon^2 \right)^{1/2}, \tag{12}$$

where ε is used to avoid the singular problem of solving process. The $a_B^{(k)}$ and $a_z^{(k)}$ are the i-th element in the k-th iteration.

Updating W and fixing a_B. Due to we have updated the observed coding coefficients a_B and NSS based coding coefficients a_z in the above section, therefore, the residual $e^{(k)}$ can be computed as $e^{(k)} = B - \hat{I}^{(k)}$. Further, the weight elements $w_{ii}^{(k)}$ in the k-iterations can be updated by the Eq. (7).

Algorithm 1 summarizes the whole procedure of our deblurring method.

Algorithm 1. Weighted Non-Locally Self-Similarity (WNLSS) sparse representation method for face deblurring

---
Input: Blurred image B, Dictionary D, the iteration stop criterion σ.
Output: Deblurring image I.
1: **Step 1 (Initialization):**
2:    Initialize the residual e by $\hat{e}^{(0)} = B - \hat{I}^{(0)}$;
3:    Initialize the weights W by Eq (7);
4:    Initialize the NSS sparse coding coefficients a_z as the all-zeros vector;
5: **Step 2 (Optimization):**
6: **Loop:**
7:    For j do
8:       **Step 2.1**: Fix W and update a_B by using Eq (11) and (12);
9:       **Step 2.2**: Update the NSS coding coefficients a_z by Eq (9);
10:      **Step 2.3**: Compute the residual e by $e^{(j)} = B - \hat{I}^{(j)}$;
11:      **Step 2.4**: Fix a and update W by using Eq (7);
12:   **Until** $\left\| Da_B^{(j+1)} - Da_B^{(j)} \right\|_2 / \left\| Da_B^{(j)} \right\|_2 \leq \sigma$
13: **Loop end**
14: **Return**: The deblurred face image $I = \Phi a_B^{(J)}$.

---

3   Experiments

In this section, we first investigate the performance of the WNLSS model on both synthetic and real blurred face images in Sects. 3.1 and 3.2, respectively. To evaluate the quality of deblurring face image, we compute the Peak Signal-to-Noise Ratio (PSNR) and perceptual quality metric FSIM [22] in this section. The basic parameters of WNLSS are set to be as follows: The size of local patch $[m, n]$ is $[7, 7]$, the dictionary size K is 200, the number of eigenvectors L is 70, the regularization parameter λ is 0.1 and the iteration number J is 10.

3.1   Evaluate on Synthetic Blur Images

For synthetic blurred image, we first use two widely applied blur kernels (i.e., uniform kernel and Gaussian kernel) to convolute the clear face images. Then the synthetic blurred images with $r \times r$ uniform blur kernel and 2D Gaussian blur kernel with standard deviation s can be obtained. In order to demonstrate the robustness of the WNLSS method when faced with noise, we add some additive white Gaussian noise (AWGN) into the blurred images. As the section of related work mentioned, we mainly compare the three kinds of deblurring method, i.e., (1) the universal sparse representation based deblurring methods [14,15,23,24]; (2) face image based deblurring methods [8] and (3) natural image based deblurring methods [2–4,25].

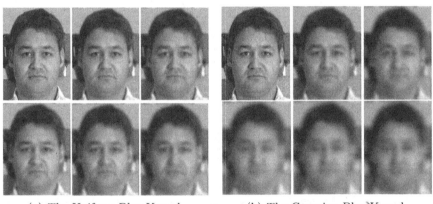

(a) The Uniform Blur Kernel   (b) The Gaussian Blur Kernel

Fig. 2. Examples of synthetic blurred face images with different blur parameters. (a) The blurred face image with uniform kernel. From left to right and top to bottom is that: original face image, the size of uniform kernel is 5×5, 7×7, 9×9, 11×11, 13×13. (b) The blurred face image with Gaussian kernel. From left to right and top to bottom is that: original face image, blurred face image with standard deviation 3, 5, 7, 9, 11.

Results on the Uniform Kernel. We vary the size of uniform kernel $r \times r$ from 5 to 13 with step size 2. The blurred samples with uniform kernel are listed in Fig. 2(a). The value of PSNR and FSIM on these blurred images of different deblurring methods are reported in Table 2. The deblurring results of 9×9 uniform kernel are also listed in Fig. 3.

We can observe that the WNLSS method and the work [23] can achieve large PSNR and FSIM improvement compared with other deblurring methods from Table 2 and the uniform blur is basically removed in the face image from Fig. 3. The WNLSS method achieves better performance than other all methods on all uniform blur cases. Though the work [8] is specially designed for face deblurring problem, it requires a large enough exemplars dataset for the match of face exemplar structure and its performance directly depends on the initialization of predicted salient edges. When the exemplar dataset is not large enough, the poor predicted edges may lead to the failure of kernel estimation from exemplar structure. The other works also introduce the prior knowledge of sparsity and edge selection into their models, but they do not achieve excellent performances as our method. On the one hand, they do not consider the special structural priors of face image and ignore the non-local self-similarity information. On the other hand, compared with other sparse based deblurring methods, the weighted coding technique is introduced to further optimize the overall residual of a face

Table 2. The values of PSNR (FSIM) by different methods with different size of uniform kernel

Kernel Size	5	7	9	11	13
Xu and Jia [15]	37.61 (0.9100)	35.30 (0.8438)	34.68 (0.8244)	33.61 (0.7524)	33.05 (0.7517)
Krishnan [14]	37.11 (0.8947)	35.96 (0.8540)	34.66 (0.8044)	33.85 (0.7605)	32.97 (0.7384)
Portilla [24]	36.53 (0.8737)	34.70 (0.8004)	33.62 (0.7373)	32.82 (0.6838)	32.17 (0.6365)
Kheradmand [23]	36.96 (0.8152)	36.67 (0.8096)	36.17 (0.8187)	36.13 (0.8267)	36.13 (0.8325)
Pan [8]	39.55 (**0.9167**)	37.50 (**0.8845**)	35.56 (0.8350)	34.51 (0.8017)	34.34 (0.7997)
Shan [3]	36.88 (0.8642)	35.60 (0.8291)	34.74 (0.8012)	33.34 (0.7202)	32.63 (0.6651)
Cho [2]	36.81 (0.8504)	34.63 (0.8032)	33.83 (0.7410)	33.56 (0.7017)	33.04 (0.6653)
Xu [4]	37.99 (0.9137)	35.86 (0.8651)	35.48 (0.8479)	34.77 (0.8336)	34.06 (0.8051)
Beck [25]	37.16 (0.8210)	36.84 (0.8285)	36.12 (0.8234)	35.50 (0.8141)	34.90 (0.8058)
Ours	**39.86** (0.8966)	**38.78** (0.8836)	**37.86 (0.8703)**	**37.29 (0.8617)**	**36.61 (0.8500)**

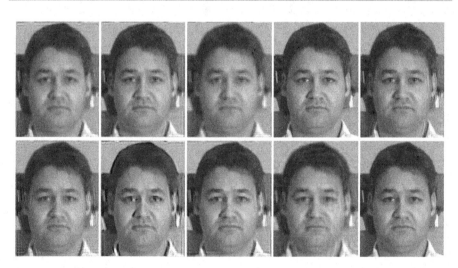

Fig. 3. The comparison of deblurring performance on the face image with 9×9 uniform kernel. From left to right and top to bottom is that: [2–4, 8, 14, 15, 23–25] and our WNLSS.

image. Therefore, the WNLSS achieve better deblurring performance on face image.

Results on the Gaussian Kernel. In this subsection, we set the size of Gaussian kernel to be 25×25 and the standard deviation from 3 to 11 with step size 2. The blurred samples with Gaussian kernel are listed in Fig. 2(b). The value of PSNR and FSIM on these blurred images of different deblurring methods are reported in Table 3. The deblurring results of Gaussian kernel with standard deviation 3 are also listed in Fig. 4.

Different from the observation of uniform kernel, our method work better than other methods on small Gaussian blur and the work [23] can achieve better performance on large blur cases. For a noisy and blurred image, the work [23] first need to remove the contribution of the noise and then perform the deblurring operation. And the performance of denoising is directly related to the initialization of blurring matrix. However, our method need not to explicitly remove the AWGN of the face image and the objective function of WNLSS is naturally robust to the AWGN. What is more, the computational cost in terms of time complexity of the work [23] can also not be ignored. Compared with the single loop in the WNLSS model, the work [23] has the outer loop and inner loop. Therefore, it takes more computational time for large blur, regardless of encountering the uniform kernel or Gaussian kernel. For the Gaussian blur kernel with standard deviation 5, the work [23] takes 26.94 s and our method only takes 13.08 s in a PC with 2.0 GHz CPU, the WNLSS method just use half computational time to achieves similar performance with the work [23]. Compared with other deblurring methods except the work [23], the WNLSS method also achieves better PSNR and FSIM performance due to the introduction of weighted coding and non-local self-similarity priors.

Table 3. The values of PSNR (FSIM) on the Gaussian kernel with different standard deviation of face image

Standard Deviation	3	5	7	9	11
Xu and Jia [15]	34.69 (0.8381)	32.21 (0.7256)	31.79 (0.6965)	31.45 (0.6708)	31.20 (0.6513)
Krishnan [14]	35.44 (0.8452)	33.63 (0.7692)	32.46 (0.7179)	31.65 (0.6787)	31.28 (0.6511)
Portilla [24]	33.57 (0.7523)	31.95 (0.6787)	31.27 (0.6430)	30.97 (0.6262)	30.82 (0.6163)
Kheradmand [23]	37.16 (0.8804)	**35.10** (0.8230)	**34.04** (0.7908)	**34.25** (0.8030)	**34.36** (0.8081)
Pan [8]	36.48 (0.8711)	34.29 (0.7995)	32.80 (0.7401)	32.34 (0.7298)	32.06 (0.7215)
Shan [3]	34.97 (0.8271)	33.44 (0.7643)	32.09 (0.6991)	31.47 (0.6516)	31.30 (0.6216)
Cho [2]	34.97 (0.7805)	32.35 (0.6611)	30.10 (0.6426)	31.08 (0.6145)	31.07 (0.6207)
Xu [4]	34.06 (0.8126)	33.32 (0.7557)	32.47 (0.7265)	32.12 (0.7181)	31.91 (0.7151)
Beck [25]	36.28 (0.8314)	34.51 (0.7884)	33.18 (0.7392)	32.91 (0.7309)	32.76 (0.7219)
Ours	**37.29** (0.8674)	34.75 (0.7983)	33.44 (0.7561)	33.59 (0.7693)	33.51 (0.7725)

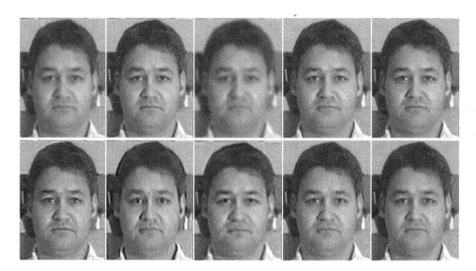

Fig. 4. The comparison of deblurring performance on the face image with Gaussian kernel of standard deviation 5. From left to right and top to bottom is that: [2–4, 8, 14, 15, 23–25] and our WNLSS.

3.2 Evaluate on Real Blur Images

For real blurred image, we select two real blurred image from the Point-and-Shoot Cameras (PaSC) [26]. The face images from PaSC dataset are shot by the point-and-shoot camera, therefore, there are not only motion blur but also poor focus. Since the focus of our work is to restore the blurred image by the known kernel, we borrow the motion blur kernel estimation from the work [27] to estimate the blur kernel. What is more, it impossible for us to obtain the original clear face image from the PaSC dataset, therefore, we can not compute the PSNR and FSIM values like the above section. **We just evaluate the deblurring performance by visual observation.** The deblurring performance of the work [8], the work [23] and our work is significantly better than other methods. Therefore, for a more difficult deblurring task, we just compare these methods on real blur face images.

We show the original blurred face image and deblurring results by the work [8], the work [23] and our work in Figs. 5 and 6. We can observe that the face image are deblurred by the WNLSS method are much clear than other methods. Since the WNLSS method can accurately restructure a local patch by referencing other similar patches and further reduce the coding residual by using the weighted coding methodology, therefore, it can restructure more detail information on real blurred and noisy face images. The estimated kernels from [27] have some bias with the real unknown blur kernels, but it still achieves excellent performances in this case, it further demonstrates the WNLSS method is robust to the kernel estimation error. Though the work [8] is specially designed for face deblurring problem, it requires a large enough exemplars dataset for the match of face exemplar structure and its performance directly depends on the initialization of predicted salient edges. For the blurred Fig. 5, we manually locate the initial contours of blurred face images and fed the contours information into their model, and for the blurred Fig. 6, we search its best matched contours in the author provided exemplar dataset. The work [8] produces some visual ringing and ghost artifacts in the Fig. 6, and the deblurring result are much worse than other two methods when there is a mismatch between the predicted salient edges and real edges (e.g. Fig. 6). For another work [23], since it is designed by using normalized graph Laplacian, it works well when faced with the structural blur kernel (i.e., Gaussian kernel). However, it can not restore more facial detail information by using the pre-defined structural prior knowledge when faced with real blur image. Moreover, this work [23] comprises of outer and inner loop and takes a large amount of time to deblur the complicated blurred image.

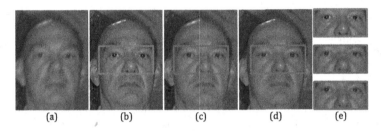

(a)   (b)   (c)   (d)   (e)

Fig. 5. The deblurring performance comparison on real blurred face images. From left to right is that: original real image, deblurred image by [8], deblurred image by [23], deblurred image by our WNLSS and close-up view for detail presentation.

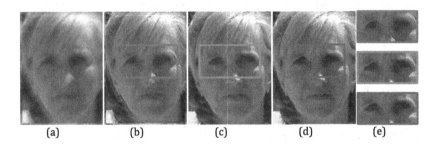

(a)   (b)   (c)   (d)   (e)

Fig. 6. The deblurring performance comparison on real blurred face images. From left to right is that: original real image, deblurred image by [8], deblurred image by [23], deblurred image by our WNLSS and close-up view for detail presentation.

4 Conclusion

In this paper, we introduce the existing weighted non-local self-similarity sparse model [1] into face deblurring task and verify that the model is suitable for face image deblurring. In theory, on the one hand, the weighted coding technology can effectively reduce the coding residual of sparse representation model, so that the pixels with large residual have smaller effect on the whole deblurring process. On the other hand, the estimation of real unknown coding coefficients can be learned by exploiting the NSS property of face image. Extensive results on synthetic and real blurred face image show the WNLSS achieves competitive performance compared with other state-of-the-art methods and demonstrates the effectiveness and robustness of the WNLSS method.

Acknowledgement. The work presented in this paper was supported by the National Natural Science Foundation of China (Grants No. NSFC-61402046), Fund for Beijing University of Posts and Telecommunications (No.2013XZ10, 2013XD-04), Fund for the Doctoral Program of Higher Education of China (Grants No.20120005110002).

References

1. Jiang, J., Zhang, L., Yang, J.: Mixed noise removal by weighted encoding with sparse nonlocal regularization. IEEE Trans. Image Process. **23**, 2651–2662 (2014)
2. Cho, S., Lee, S.: Fast motion deblurring. In: ACM SIGGRAPH Asia 2009 Papers, SIGGRAPH Asia 2009, pp. 145:1–145:8, ACM, New York (2009)
3. Shan, Q., Jia, J., Agarwala, A.: High-quality motion deblurring from a single image. ACM Trans. Graph. **27**, 73:1–73:10 (2008)
4. Xu, L., Jia, J.: Two-phase kernel estimation for robust motion deblurring. In: Daniilidis, K., Maragos, P., Paragios, N. (eds.) ECCV 2010. LNCS, vol. 6311, pp. 157–170. Springer, Heidelberg (2010). doi:10.1007/978-3-642-15549-9_12
5. Mairal, J., Bach, F., Ponce, J., Sapiro, G., Zisserman, A.: Non-local sparse models for image restoration. In: 2009 IEEE 12th International Conference on Computer Vision, pp. 2272–2279. IEEE (2009)
6. Dong, W., Zhang, L., Shi, G., Wu, X.: Image deblurring and super-resolution by adaptive sparse domain selection and adaptive regularization. IEEE Trans. Image Process. **20**, 1838–1857 (2011)
7. Sun, L., Cho, S., Wang, J., Hays, J.: Edge-based blur kernel estimation using patch priors. In: 2013 IEEE International Conference on Computational Photography (ICCP), pp. 1–8. IEEE (2013)
8. Pan, J., Hu, Z., Su, Z., Yang, M.-H.: Deblurring face images with exemplars. In: Fleet, D., Pajdla, T., Schiele, B., Tuytelaars, T. (eds.) ECCV 2014. LNCS, vol. 8695, pp. 47–62. Springer, Cham (2014). doi:10.1007/978-3-319-10584-0_4
9. Levin, A., Weiss, Y., Durand, F., Freeman, W.T.: Efficient marginal likelihood optimization in blind deconvolution. In: 2011 IEEE Conference on Computer Vision and Pattern Recognition (CVPR), pp. 2657–2664. IEEE (2011)
10. Oliveira, J.P., Bioucas-Dias, J.M., Figueiredo, M.A.: Adaptive total variation image deblurring: a majorization-minimization approach. Sig. Process. **89**, 1683–1693 (2009)
11. Cai, J.F., Ji, H., Liu, C., Shen, Z.: Framelet-based blind motion deblurring from a single image. IEEE Trans. Image Process. **21**, 562–572 (2012)
12. Dong, W., Shi, G., Hu, X., Ma, Y.: Nonlocal sparse and low-rank regularization for optical flow estimation. IEEE Trans. Image Process. **23**, 4527–4538 (2014)
13. Chen, J., Yi, D., Yang, J., Zhao, G., Li, S.Z., Pietikainen, M.: Learning mappings for face synthesis from near infrared to visual light images. In: IEEE Conference on Computer Vision and Pattern Recognition, CVPR 2009, pp. 156–163. IEEE (2009)

14. Krishnan, D., Tay, T., Fergus, R.: Blind deconvolution using a normalized sparsity measure. In: 2011 IEEE Conference on Computer Vision and Pattern Recognition (CVPR), pp. 233–240. IEEE (2011)
15. Xu, L., Zheng, S., Jia, J.: Unnatural L0 sparse representation for natural image deblurring. In: Proceedings of the IEEE Conference on Computer Vision and Pattern Recognition, pp. 1107–1114 (2013)
16. Mairal, J., Sapiro, G., Elad, M.: Learning multiscale sparse representations for image and video restoration. Multiscale Model. Simul. **7**, 214–241 (2008)
17. Rubinstein, R., Bruckstein, A.M., Elad, M.: Dictionaries for sparse representation modeling. Proc. IEEE **98**, 1045–1057 (2010)
18. Elad, M., Yavneh, I.: A plurality of sparse representations is better than the sparsest one alone. IEEE Trans. Inf. Theor. **55**, 4701–4714 (2009)
19. Aharon, M., Elad, M., Bruckstein, A.: K-SVD: an algorithm for designing overcomplete dictionaries for sparse representation. IEEE Trans. Sig. Process. **54**, 4311–4322 (2006)
20. Dong, W., Zhang, L., Shi, G., Li, X.: Nonlocally centralized sparse representation for image restoration. IEEE Trans. Image Process. **22**, 1620–1630 (2013)
21. Daubechies, I., DeVore, R., Fornasier, M., Güntürk, C.S.: Iteratively reweighted least squares minimization for sparse recovery. Commun. Pure Appl. Math. **63**, 1–38 (2010)
22. Zhang, L., Zhang, L., Mou, X., Zhang, D.: FSIM: a feature similarity index for image quality assessment. IEEE Trans. Image Process. **20**, 2378–2386 (2011)
23. Kheradmand, A., Milanfar, P.: A general framework for regularized, similarity-based image restoration. IEEE Trans. Image Process. **23**, 5136–5151 (2014)
24. Portilla, J.: Image restoration through L0 analysis-based sparse optimization in tight frames. In: 2009 16th IEEE International Conference on Image Processing (ICIP), pp. 3909–3912. IEEE (2009)
25. Beck, A., Teboulle, M.: Fast gradient-based algorithms for constrained total variation image denoising and deblurring problems. IEEE Trans. Image Process. **18**, 2419–2434 (2009)
26. Beveridge, J.R., Phillips, J., Bolme, D.S., Draper, B., Givens, G.H., Lui, Y.M., Teli, M.N., Zhang, H., Scruggs, W.T., Bowyer, K.W., et al.: The challenge of face recognition from digital point-and-shoot cameras. In: 2013 IEEE Sixth International Conference on Biometrics: Theory, Applications and Systems (BTAS), pp. 1–8. IEEE (2013)
27. Danielyan, A., Katkovnik, V., Egiazarian, K.: BM3D frames and variational image deblurring. IEEE Trans. Image Process. **21**, 1715–1728 (2012)

Fiducial Points Detection of a Face Using RBF-SVM and Adaboost Classification

Shreyank N. Gowda[✉]

Indian Institute of Technology-Madras, Chennai, India
kini5gowda@gmail.com

Abstract. Fiducial points are points that are used as points of reference or measure. Determining of fiducial points can be a fundamental step to recognize a face. A few important fiducial points are the eyes, lip edges, nose, chin etc. Using the fiducial points we can either obtain an outline of the entire face or develop a relationship between the fiducial points themselves to act as a medium to recognize a face. Lot of research has been ongoing in this regard. In this paper the Fiducial points and their existing relationships are studied using a Support Vector Machine with a Radial basis Function kernel. New images when tested showed a high accuracy of correct results in terms of the actual positions of the fiducial points in the image. Further classification of the fiducial points is done using an Adaboost classification to improve the accuracy.

1 Introduction

Studying face recognition is a research that has been ongoing for many years. There are a number of applications for face recognition that range from commercial to security. However, a system has to be trained to recognise a face. Even then there are a multiple number of factors that inhibit the opportunity to recognise a face [1]. These include: lighting conditions of the image, image quality etc.

There are many different methods available for face recognition as mentioned by Zhao et al. [2]. Two main approaches in these are the holistic approach and feature extraction approach.

In case of holistic approaches, face recognition is done by making use of a single feature vector that represents the whole face image. Examples of holistic approaches are eigenfaces as proposed by Turker et al. [3], linear discriminant analysis as proposed by Martinez et al. [4], discriminative common vector as proposed by Cevikalp et al. [5], bayesian intrapersonal classifier as proposed by Moghaddam et al. [6], and classifiers trained by Neural networks as proposed by Rowley et al. [7].

These methods have been extremely accurate for classification of frontal face views, without any occlusions. However, the global features are highly sensitive to translations, scale, rotations and appearance variation.

These restrictions imposed by the global methods have resulted in a shift towards feature-based approaches which have, in turn, proved to perform better as shown by Heisele et al. [8].

© Springer International Publishing AG 2017
C.-S. Chen et al. (Eds.): ACCV 2016 Workshops, Part I, LNCS 10116, pp. 590–598, 2017.
DOI: 10.1007/978-3-319-54407-6_40

The feature-based approaches require localizing of some features such as eyes, nose, mouth etc. in order to apply a local-based classification. Within this category, we can further distinguish between the various approaches that require the localization of rough facial features only, and the ones which need precise fiducial point estimation for classification.

Examples of the first class are: the template-based approaches as proposed by Brunelli et al. [9], the Local Feature Analysis proposed by Penev et al. [10], while the Elastic Bunch Graph Matching (EBGM) proposed by Wiskott et al. [11] is an example of the second class.

The remainder of this paper is organized as follows. In the next section a background on research about fiducial point detection is shown. Then the proposed algorithm is followed in the next section. In that section details about the algorithm and the reasons for that are explained. Then, the experimental results and analysis section follows. The last section is the conclusion section explaining the results and findings of the experiment.

2   Background

The amount of research work being done on detecting fiducial points on the face is constantly increasing as shown by Waldir et al. [12].

In recent times, the approaches could be divided into two main types: local and global. In local methods, individual fiducial points are detected and processed and no additional information is used.

The global methods are characterized by detecting more fiducial points in comparison using deformable models, less susceptible to pose and illumination variations than local methods. The design of a classifier is the most important stage of a fiducial point detection algorithm. In this stage, several machine learning algorithms can be used as shown by Jahanbin et al. [13].

Particularly, some studies use a classification method called Support Vector Machine. The mathematical formulation of SVM is obtained by optimization problem with restrictions.

There are many recent papers about this theme. For example, Silva et al. propose in [12] a face recognition subsystem framework that makes use of a fiducial points detection. The detection of the fiducial points is a combination of two techniques.

The first is using Gabor filters coefficients for local detection and then following is the use of a human face anthropometric measurements. The system proposed by Araujo et al. [14] explores the same problem. The authors use classifiers based on Inner Detector Product correlation filters. These filters are designed by making use of principal components.

In [15] Eduardo et al. proposed the use of a SVM mathematical formulation called C-SVC (Support Vector Classification) [16], for fiducial point detection.

3 Proposed Algorithm

The SVM, as proposed by Vapnik et al. in [17], is used to solve the classification and regression problems.

The SVM algorithm, provides an optimal separating hyperplane with a maximum margin. The SVM can be formulated to even consider mislabelled samples. In this case, this technique is called soft margin [16].

The use of the soft margin provides a generalization method of the SVM to deal with the problems of separating classes.

The mathematical formulation is written as follows. First, consider the hyperplane:

$$y_i(w^t x_i) + b \geq 1 - \epsilon_i \tag{1}$$

where 'i' varies from 1 to n, i is non negative slack variable, y_i represents the class labels, n is the total number of class elements and the superscript t represents the transpose of the vector.

A radial basis function kernel or RBF kernel is used as the kernel to support the SVM in terms of its classification.

Equation (2) represents an RBF kernel on two samples x and x'.

$$K(x, x') = exp(-(\|x - x'\|^2 / 2\sigma^2)) \tag{2}$$

$$d = \|x - x'\|^2 \tag{3}$$

where 'd' is the Euclidean distance between 2 feature vectors.

The detection of Fiducian points follows after two steps: training and testing.

First the SVM is trained using a set of 931 images taken from the BioID [20] database.

Next a new image is taken as a testing image and then the testing is done. In any case the first step is the pre-processing step.

The pre-processing step is composed of two stages.

First the face is detected using the Viola Jones algorithm as proposed by Viola et al. [18]. Next the face is made to scale to a 320×240 image to obtain a clearer image consisting only of the facial features.

In the training stage first, the image is pre-processed. To determine the position of Fiducian points we find areas of high probability of its occurrence by using a Gaussian Mixture Model inspired by the mathematical derivations in [12].

This Gaussian Mixture model (GMM) consists of 4 models developed around the Gaussian prior model (GPM) by changing the parameters of the prior model to accommodate a larger set of points into the classification with higher accuracy.

Table 1 shows the comparison of accuracy of classification with the number of different models that were used and tested.

It could be seen that using 4 different models to make the Mixture Model gave best results in terms of accuracy of classification.

When a single Gaussian Priori model was used an accuracy of around 74% was obtained. This accuracy increased till the number of models were 4 and then started decreasing.

Table 1. Different number of prior model results comparison

Total number of models used	Accuracy of classification
Single Gaussian prior model	74.45%
GMM with 2 GPM	78.26%
GMM with 3 GPM	80.19%
GMM with 4 GPM	**82.26%**
GMM with 5 GPM	81.14%
GMM with 6 GPM	79.86%

A candidate to any fiducial point with a label q in the image, having coordinates equal to y, is considered to be inside the ellipse that is defined by the Mahalanobis distance to the average of all the fiducial points in the training set, given by (4):

$$Max_{label(w)=q} \left\{ 1.05|w - \mu|^2 \right\} \geq (y - \mu)^t \sum_x{}^{-1} (y - \mu) \tag{4}$$

X is the random vector whose realisation is equal to the ficudial points, is the vector mean and X is the covariance matrix of X.

In the training phase a set of images are taken from database and the Viola Jones algorithm is applied to detect the face of the person in the image.

Next as shown above a Gaussian Mixture Model is developed to determine areas with high probability of containing fiducial points.

Next we use a Wiener filter to remove some noise and focus on each facial feature to obtain the fiducial points of that feature [19].

Each block is represented as Az and is usually of the size 13 × 13. Alternative filters could also be used such as Linear or median filters for the same purpose, however, a Wiener filter was chosen for convenience.

We add an adaboost algorithm to further increase the strength of the classification. In fact as we can later see in the experimental analysis section the adaboost increases the classification accuracy by 12.

In the case of this algorithm as can be seen in the results section, the eyes gives us 3 fiducial points, the nose gives us 3 fiducial points and the mouth gives us 4 fiducial points (2 on edges of lips and the top and bottom of the lip also count as fiducial points).

Workflow of the algorithm can be seen in Fig. 1.

Further in Fig. 2 we can see a sample picture being taken and the resulting output of each phase can also be seen.

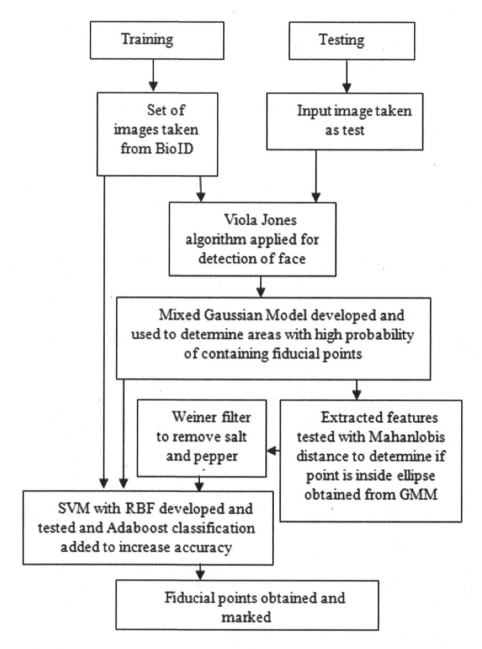

Fig. 1. Workflow of algorithm.

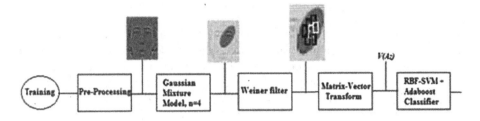

Fig. 2. Training block diagram.

4   Experimental Results and Analysis

In the training phase, first the pre-processing is done which includes the Viola Jones algorithm combined with the illumination correction. The images were taken from the BioID database. Next the fiducial points were added to these images manually and the SVM was trained using this data.

Next 4 Gaussian prior models were developed and hence a Gaussian mixture model was formed by using the given data to form a model that would, in case of testing purposes give an elliptical region which would be having a high probability of occurrence of a fiducial point.

The elliptical region was focused and any salt and pepper noise obtained was removed using a Weiner filter. The fiducial point in case of the eyes was the edges of the eye and the centre of the eyeball. Similarly the training was done to determine 13 fiducial points in total.

The entire code was written in Python OpenCV and the training was done using 827 images and 104 images were later going to be used for testing purposes.

For the testing phase, a pixel was considered true positive if the distance between automatic and manual annotations is less than 10.

Table 2 compares the results of all the algorithms in terms of false positive and true positives. The higher the value of true positives the better the algorithm and similarly lower the value of the false positive, better the algorithm.

Fiducial point 0 refers to left eye's left corner, 1 refers to centre of left eye, 2 refers to left eye's right corner.

Similarly 3, 4 and 5 are for the right eye. 6 refers to the left of the nose, 7 to the centre of the nose and 8 to the right of the nose.

9 and 10 refer to the left and right side of the mouth, while 11 and 12 refer to the top and bottom of the mouth respectively.

In total the algorithm gives us 13 Fiducial points, before a maximum of 11 was being obtained.

It can also be seen that the algorithm provides much better accuracy than any other previous algorithm on average.

The average being spoken about refers to the average classification success for each and every fiducial point.

Table 2. Comparison of different algorithms

Fiducial point number	FP1	TP1	FP2	TP2	FP3	TP3	FP4	TP4	FP5	TP5
0	1.09	92.24	0.05	96.18	0.22	21.12	0.76	99.65	0.11	98.24
1	0.32	99.64	0.10	95.26	0.30	51.15	0.40	92.43	0.07	99.43
2	1.58	96.56	0.24	95.33	0.29	53.71	0.78	96.67	0.21	96.54
3	0.46	93.27	0.08	95.57	0.31	65.14	0.92	94.64	0.12	89.26
4	2.46	97.71	0.11	95.19	0.40	61.25	0.51	95.56	0.08	92.34
5	2.27	98.26	0.05	97.83	0.24	70.31	0.33	92.46	0.15	94.26
6	1.40	99.11	0.14	97.58	0.24	60.25	1.97	96.67	0.19	95.65
7	0.06	47.78	0.18	93.32	0.40	55.24	0.59	55.89	0.13	98.81
8	0.40	95.63	0.17	94.63	0.27	51.37	0.75	96.52	0.12	95.53
9	2.91	96.67	0.18	92.75	0.11	43.51	2.09	91.68	0.17	97.36
10	1.40	98.17	0.18	87.62	0.14	22.65	1.34	95.34	0.14	98.81
11	N/A	N/A	N/A	N/A	N/A	N/A	N/A	N/A	0.17	94.35
12	N/A	N/A	N/A	N/A	N/A	N/A	N/A	N/A	0.11	91.51

FP1 and TP1 represent False Positives and True Positives of SVM-L respectively. Similarly 2, 3, 4, 5 represent DF-PCA, DF, C-SVC and the proposed algorithm respectively.

The first advantage the algorithm provides is that no other algorithm works for detecting 13 fiducial points.

Also if we consider the average of true positives of the proposed algorithm and compare it with the average of the DF-PCA true positives we see that the average true positives of the proposed algorithm (95.54) is greater than that of the DF-PCA algorithm (94.66), hence the classification on the whole seems better in case of the proposed algorithm.

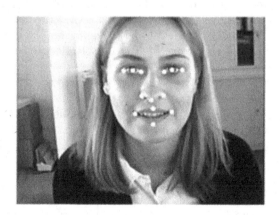

Fig. 3. Sample output.

In fact among all the algorithms the proposed algorithm performs much better and also has the additional facility of detecting 2 extra fiducial points. The worst performing algorithm was the DF algorithm which was one of the first algorithms made in order to detect fiducial points.

Figure 3 shows us a sample output when given a test image. Even though the points are not explicitly marked, they are marked in the code as explained above from 0–12.

5 Conclusion and Future Works

The proposed algorithm after testing could be seen on average to perform better than any other recent algorithms in terms of the false positives and true positives classification. Also the proposed algorithm gives us two additional fiducial points in comparison to the 11 given by other algorithms.

However, despite the benefits, the algorithm is comparatively slower to most other algorithms. This is due to the dual classification happening along with the formation of a Gaussian mixture model.

In that regard the algorithm could be improved and that is the goal of the future works. Also since the accuracy is not perfect there is always scope for improvement in determining the accuracy as well.

References

1. Wagner, A., Wright, J., Ganesh, A., Zhou, Z., Mobahi, H., Ma, Y.: Toward a practical face recognition system: robust alignment and illumination by sparse representation. IEEE Trans. Pattern Anal. Mach. Intell. **34**(2), 372–386 (2012)
2. Zhao, W., Chellappa, R., Phillips, P.J., Rosenfeld, A.: Face recognition: a literature survey. ACM Comput. Surv. (CSUR) **35**(4), 399–458 (2003)
3. Turk, M.A., Pentland, A.P.: Face recognition using eigenfaces. In: Proceedings of IEEE Computer Society Conference on Computer Vision and Pattern Recognition, pp. 586–591 (1991)
4. Martínez, A.M., Kak, A.C.: PCA versus LDA. IEEE Trans. Pattern Anal. Mach. Intell. **23**(2), 228–233 (2001)
5. Cevikalp, H., Neamtu, M., Wilkes, M., Barkana, A.: Discriminative common vectors for face recognition. IEEE Trans. Pattern Anal. Mach. Intell. **23**(2), 4–13 (2005)
6. Moghaddam, B., Nastar, C., Pentland, A.: A Bayesian similarity measure for direct image matching. In: Proceedings of the 13th IEEE International Conference on Pattern Recognition, pp. 350–358 (1996)
7. Rowley, H.A., Baluja, S., Kanade, T.: Neural network-based face detection. IEEE Trans. Pattern Anal. Mach. Intell. **20**(1), 23–38 (1998)
8. Heisele, B., Ho, P., Poggio, T.: Face recognition with support vector machines: global versus component-based approach. In: Proceedings of the 13th IEEE International Conference on Computer Vision, pp. 688–694 (2001)
9. Brunelli, R., Poggio, T.: Face recognition: features versus templates. IEEE Trans. Pattern Anal. Mach. Intell. **15**(10), 1042–1052 (1993)

10. Penev, P.S., Atick, J.J.: Local feature analysis: a general statistical theory for object representation. Netw. Comput. Neural Syst. **7**(3), 477–500 (1996)
11. Wiskott, L., Fellous, J.M., Kuiger, N., Von Der Malsburg, C.: Face recognition by elastic bunch graph matching. IEEE Trans. Pattern Anal. Mach. Intell. **19**(7), 775–779 (1997)
12. Waldir, S.S., Araújo, G.M., da Silva, E.A., Goldenstein, S.K.: Facial fiducial points detection using discriminative filtering on principal components. In: Proceedings of the 13th IEEE International Conference on Image Processing, pp. 2681–2684 (2010)
13. Jahanbin, S., Choi, H., Bovik, A.C.: Passive multimodal 2-D + 3-D face recognition using Gabor features and landmark distances. IEEE Trans. Inform. Forensics Secur. **6**(4), 1287–1304 (2011)
14. Araujo, G.M., Júnior, W.S., Silva, E.A., Goldenstein, S.K.: Facial landmarks detection based on correlation filters. In: Proceedings of the 13th IEEE International Telecommunication Symposium (2010)
15. Silva, L.E.S., Júnior, P.D.T., Santos, K.V., Junior, W.S.S.: Fiducial points detection using SVM linear classifiers. In: CS and IT-CSCP, pp. 23–31 (2014)
16. Chang, C.C., Lin, C.J.: LIBSVM: a library for support vector machine (2001). http://www.csie.ntu.edu.tw/~cjlin/libsvm
17. Vapnik, V.: The Nature of Statistical Learning Theory. Springer, New York (2000)
18. Viola, P., Jones, M.: Robust real-time object detection. J. Comput. Vis. **57**(2), 137–154 (2001)
19. Chen, J., Benesty, J., Huang, Y., Doclo, S.: New insights into the noise reduction Wiener filter. IEEE Trans. Audio Speech Lang. Process. **14**(4), 1218–1234 (2006)
20. Jesorsky, O., Kirchberg, K.J., Frischholz, R.W.: Robust face detection using the Hausdorff distance. In: Proceedings of International Conference on Audio-and Video-Based Biometric Person Authentication, pp. 90–95 (2001)

Real-Time Head Pose Estimation on Mobile Devices

Zhengxin Cheng$^{(\boxtimes)}$ and Fangyu Bai

College of Information Science and Technology,
Beijing Normal University, Beijing, People's Republic of China
roboman@foxmail.com

Abstract. There are a lot of actual application scenarios of head pose estimation, such as live detection, human interaction, gesture-based equipment, VR devices. This paper proposes a method based on linear regression to get the pose angle. The method utilizes supervised gradient descent method to get the facial feature points, and estimates the head pose by least square regression. In regression modeling, we only use seven feature point positions, which decrease the model size and the computing load. A data normalization technique is employed to eliminate the effect of the camera imaging parameters on pose estimation. Moreover, the training data set is derived from a group of 3D face models according to the camera imaging model, which can provide precise pose parameters. Experiments validate the proposed method, and show it can be run in real time. The algorithm can be easily transplanted to the mobile terminals. We have released our source code at https://github.com/RoboPai/sdm.

1 Introduction

Head pose estimation is the process of determining the face orientation relative to a global coordinate system from image data. It is a classical computer vision problem, and has been widely used in many face related applications such as face recognition [1], human computer interaction [2], gaze tracking [3] etc. Like face recognition, pose estimation faces some discouraging factors such as illumination, occlusion, expression, etc. Especially with the development of the mobile Internet and the popularization of smart phones, high quality images can be obtained easily, but the limited computing power of smart phones cannot meet the real time requirements in real applications.

Many approaches have been proposed for head pose estimation using three kinds of data sources, including 2D images [4], 3D range data [5] and the combination of 2D image and 3D range data [6]. The advantages of the methods using 3D range data lie in its insensitiveness to variation of illumination, expression, and other factors. However, specific sensing devices, such as laser radar, depth cameras, are always needed to obtain range data, so the additional equipment cost is large and its application scenarios are limited greatly. Methods based on 2D images can be roughly divided into facial appearance based [4,7] et al. and feature points based methods. The former methods map a rectangular

© Springer International Publishing AG 2017
C.-S. Chen et al. (Eds.): ACCV 2016 Workshops, Part I, LNCS 10116, pp. 599–609, 2017.
DOI: 10.1007/978-3-319-54407-6_41

region of image pixels to a pose space. They firstly extract the information of the images, and then use pattern recognition or machine learning methods to estimate the face pose. For example, Sherrah et al. [7] conducted a Convolution with a complex Gabor wavelet to extract the directed features of images, and then measured the similarity between the image and a set of templates with corresponding pose labels after the convolution. Nonlinear Regression (NR) is the commonly used technique for building the mappings from images to the head pose. Fanelli et al. [8] proposed a real-time head pose estimation technique using random regression forests. Y. Li et al. [9] utilize dimension-reduction method to express face images in PCA space, and then use SVR to establish the regression models. Alioua [10] proposed a orientation-based head descriptor by fusing four most relevant orientation-based head descriptors, and used the support vector machine (SVM) to classify the head pose variations. Besides NR methods, some literatures [11,12] use Manifold Embedding Methods such as Isomap and LLE to project face images to a low-dimensional manifold constrained by pose variations. Majority of appearance based methods rely on robust head detection, and are easily influenced by such factors like.

Illumination, occlusion and identity. Feature points based methods firstly locate the feature points of the face images, and then use the point information to estimate the head pose. These methods can be divided into two classes: learning based [13,14] and geometry based methods [15,16]. The former uses the facial features computed at these points, while the latter directly uses the point locations. For example, Ma et al. [14] extracted sparse features after localizing 20 key facial points, and then utilized Relevance Vector Machine RVM to build the mapping from the sparse representation to the pose parameters. They presented two different feature sets in these points. One is the concatenation of the normalized positions of these points, and the other is composed of the wavelet features extracted at some sampling points around each facial point. Geometric methods usually compute the head pose by the perspective projection camera imaging model with some facial priors such as symmetry, coplanar, parallel. Wang et al. [15] assume that three line segments between the outer eye corners, the inner eye corners, and the mouth corners are parallel, and derive an analytic solution when the ratio of the length of the eye-line segment to the length of the mouth-line segment is known. Similarly, Duan et al. [16] also present an analytic solution with the known camera intrinsic parameters and this parallel assumption. Feature points based methods, especially geometric methods are very sensitive to the localization precision of the feature points.

Majority of these methods are not fit for applications on mobile devices, since they usually need a heavy computing load. In this paper, we propose a real time head pose estimation method using facial feature points. We first locate the feature points by supervised descent method SDM, and then estimate the head pose by least square regression. The contributions are three folds: Firstly, we use only seven feature point positions, i.e. the four corners of left and right eyes, nasal tip point and the two corners of the mouth, and this decreases the model size and the computing load; Secondly, we use a data normalization technique which can eliminate the effect of the camera imaging parameters on pose estimation; Thirdly, the training data set is derived from a group of 3D face models according to the camera imaging model, which can provide precise pose parameters. The paper is organized as follows: Sect. 2. describes the process of the head pose estimation. Experimental results are shown in Sect. 3. Finally we give some conclusions.

2 Head Pose Estimation

In this section, we describe the process of the head pose estimation. We first locate the facial feature points by SDM [17], secondly, we use a data normalization technique which can eliminate the effect of the camera imaging parameters on pose estimation, and then estimate the head pose by least square regression.

2.1 Facial Feature Points Localization

Usually, face detection is the first priority of the facial feature points localization. Adaboost [18] algorithm is an adaptive algorithm where weak classifiers are combined to generate a strong classifier, and it is often used in face detection. Here we also use it for this aim. After face detection, SDM algorithm, proposed by xiong et al. [19], is utilized to search for the key facial feature points. SDM is a supervised learning method for solving Non-linear Least Squars (NLS) problems, and can overcomes the deficiency of second order optimization schemes, such as nondifferentiability and expensive computation of the Jacobians and Hessians. It is extremely fast and accurate. Since the localization of facial feature points is vital for head pose estimation, here we briefly introduce the algorithm, and for detailed process please refer to [19]. Given a training image set $\{(I_i, S_i)\}_i^n$, where I_i denotes the i_{th} image, and S_i is a set of manually annotated feature points of image i, SDM learns generic descent direction in a supervised manner as follows:

$$r^t = \arg\min_r \sum_{i=1}^n \|y_i^t - r(I_i, \hat{S}_i^{t-1})\|^2 \tag{1}$$

where $y_i^t = S_i - \hat{S}_i^t$ is the position variance between the regressed and target feature point, namely the gradient of optimization objective, r^t is the regressor in each iteration. In testing phase, these regressors are used to obtain feature points in cascade way.

As shown in Fig. 1, 68 feature points are tracked successfully in real time by SDM. In order to reduce processing time on the mobile terminals, e.g. iPhone, we decrease the number of feature points to 32, and among them, we select 7 points (see Sect. 3.1) to achieve real-time head pose regression. Its enough to describe head pose with these seven points, and at the same time, these seven points are more accurate than others in the process of feature point location.

(a)   (b)

Fig. 1. Feature points located by SDM

2.2   Data Normalization

Since each image has different camera imaging parameters, the position and size of the 2D projection of each head are different. This is not good for statistical analysis of pose information. Data normalization can provide normalized data for statistical analysis. Since Hartley [17] proposed data normalization in fundamental matrix estimation, it is used in many computer vision problems. data normalization can improve the stability when solving a linear system. The normalization suggested by Hartley is a translation and isotropic scaling of each image so that the centroid of the input data points is brought to the origin and the RMS distance of the points from the origin is equal to $sqrt2$. The data normalization we use is as follows,

$$u' = \frac{u - \bar{u}}{u_{max} - u_{min}}, v' = \frac{v - \bar{v}}{u_{max} - u_{min}} \tag{2}$$

where u, v denote the 2D coordinate of a feature point in the image, $\bar{u}$, $\bar{v}$ denote the average values of two directions. Assume $\alpha = u_{max} - u_{min}$, thus data normalization can be expressed below,

$$\begin{bmatrix} u' \\ v' \\ 1 \end{bmatrix} = \begin{bmatrix} \alpha & 0 & -\alpha \cdot \bar{u} \\ 0 & \alpha & -\alpha \cdot \bar{v} \\ 0 & 0 & 1 \end{bmatrix} \cdot \begin{bmatrix} u \\ v \\ 1 \end{bmatrix} \tag{3}$$

It is important to note that the goal of data normalization is to ensure the scaling uniformity so that one dimension is not dominated. This two-step preprocessing is crucial for the accuracy of final head pose regression. Proposition 1: The data normalization by Eq. (3) does not change the head pose. Proof. The imaging process of general cameras including mobile phones can be modeled by the pinhole camera model, which describes the mathematical relationship between the coordinates of a 3D point and its projection onto the image plane of by pinhole imaging model. We consider the pinhole imaging model of camera without distortion as follows:

$$
\lambda \begin{bmatrix} u \\ v \\ 1 \end{bmatrix} = \begin{bmatrix} f_x & 0 & c_x \\ 0 & f_y & c_y \\ 0 & 0 & 1 \end{bmatrix} [R|t] \begin{bmatrix} X \\ Y \\ Z \\ 1 \end{bmatrix} \Rightarrow \lambda m = K [R|t] M \tag{4}
$$

In Eq. (4), K is the camera's intrinsic parameters, (X, Y, Z) is the coordinate of a space point in the world coordinate system, (u, v) is coordinate of its projection on the image plane, and R, t represent rotation and translation matrices from the world coordinate system into camera coordinate system, respectively. By combining Eqs. (3) and (4), after normalization process, the imaging model becomes

$$
\lambda \begin{bmatrix} u' \\ v' \\ 1 \end{bmatrix} = \begin{bmatrix} \alpha \cdot f_x & 0 & \alpha(c_x - \bar{u}) \\ 0 & \alpha \cdot f_y & \alpha(c_y - \bar{v}) \\ 0 & 0 & 1 \end{bmatrix} [R|t] \begin{bmatrix} X \\ Y \\ Z \\ 1 \end{bmatrix} = K' [R|t] M \tag{5}
$$

From Eq. (5), the data normalization by Eq. (3) is equivalent to changing the camera's internal parameters K, and the actual pose of 3D objects in the world coordinate system is not affected. So the proposition holds.

2.3 Pose Estimation with Linear Regression

In this section, we will generate a regressor to predict head pose in a image. Given 7 face landmarks, we provide two solutions: One is to estimate the rotation matrix, the other is to estimate the euler angles. That is, given a set of 7face landmarks, $S(x, y)$, it is mapping to euler angles θ or rotation matrix R. Then we predict rotation matrix R or euler angles θ for a new image with the linear regressor. Assume we have m images with known pose.

Solution 1: Rotation Matrix Regression. One solution to get head pose is regressing the corresponding rotation matrix from the world coordinate system into camera coordinate system. To minimize the mean square error, we define the cost function as,

$$
J(W) = \frac{1}{2m} \sum_{i=1}^{m} (h_W(S_i(x, y)) - R_i)^2 \tag{6}
$$

where

$$
h_W(S_i(x, y)) = W S_i(x, y) + \xi \tag{7}
$$

where $S_i(x, y) = (x_1, y_1, \ldots, x_7, y_7)$ denotes 7 facial feature points described in Sect. 3.1. And $R_i = (r_1, r_2, r_3)$ denotes a column vector that the rotation matrix rearranged. In the sense of maximum likelihood estimation, to achieve minimum $J(\boldsymbol{W})$, the optimal solution is,

$$\hat{\boldsymbol{W}} = (\boldsymbol{S}^T \boldsymbol{S})^{-1} \boldsymbol{S}^T \boldsymbol{R} \tag{8}$$

In order to guarantee the computation stability, in practice, $\boldsymbol{W}$ is computed according to following regularization formulation,

$$\hat{\boldsymbol{W}} = (\boldsymbol{S}^T \boldsymbol{S} + \lambda \boldsymbol{E})^{-1} \boldsymbol{S}^T \boldsymbol{R} \tag{9}$$

where $\boldsymbol{E}$ denotes the identity matrix, and λ is a regularization factor. When input a group of feature points $S(x, y)$ of an image, the pose regression model is as follows,

$$\boldsymbol{R}(S(x, y)) = \hat{\boldsymbol{W}} S(x, y). \tag{10}$$

Finally, the pose vector $\boldsymbol{R}(S(x, y))$ can be rearranged into the rotation matrix we need. When we get the rotation matrix by MLS regression, the three euler angles of estimated head pose can be calculated by matrix decomposition of the rotation matrix.

Solution 2: Euler Angle Regression. Unlike the solution in the previous section, we can directly regress the three euler angles by MLS regression with the same principle, that is, switch the matrix R to the corresponding euler angle, the optimal solution of parameters is

$$\hat{\boldsymbol{W}} = (\boldsymbol{S}^T \boldsymbol{S} + \lambda \boldsymbol{E})^{-1} \boldsymbol{S}^T \boldsymbol{\theta} \tag{11}$$

After the optimal fitting parameters are obtained, real-time head pose estimation on mobile terminals could be achieved successfully.

3   Experiment

3.1   Generation of Train Data

In fact, in practical training phase, it is hard for us to get an actual head pose of one color RGB image. In this paper some three-dimensional face models are selected to replace actual human head, and 7 space points, that is the four corners of left and right eyes, nasal tip point and the two corners of the mouth, are picked to generate random rotation and transformation, as shown in Fig. 2. We collected 100 3D face models using Artec Eva, and the 3D point cloud registration method described in [20] is adopted to establish the point correspondence across these face models. We divide these face models into 10 groups, and each includes 10 face models. By taking the average of each group, we get 10 average face models. For each model, we calibrate its seven key points, and generate rotation and translation randomly for 300 times. Then the pinhole camera imaging model is used to project the space points to the image plane, thus a series of planar point sets are obtained. That is, 3000 sets of 2D face landmarks with known pose parameters are generated,

$$\{S_i(x, y) \to \boldsymbol{\theta}_i \Leftrightarrow \boldsymbol{R}_i, i = 1, \ldots, 3000\} \tag{12}$$

We use these data as training data to establish the regressor.

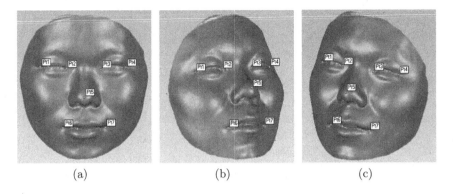

(a)   (b)   (c)

Fig. 2. Schematic diagram of random rotation

3.2 Results and Analysis

The method proposed in this paper is tested on the Biwi database, which includes some RGBD images of 20 persons (14 men and 6 women) who have different head rotation directions. There are 24 video sequence data in the database because a few people recorded twice. All of the head position and rotating angle in the images are already calibrated. We compare our method with random forests method [8] using this data set, as shown in Table 1.

Table 1. Comparison of head estimation results on the Biwi Kinect dataset

Method	Pitch	Yaw	Roll	Mean
Random forests [20]	8.50	9.20	8.00	8.60
OUR-METHOD1	6.38	**6.97**	**4.13**	5.83
OUR-METHOD2	**5.15**	7.29	4.66	**5.70**

As can be seen from the table, the proposed method has improved the precision significantly, meanwhile, we can find that the second method to regress euler angles has a slight advantage in accuracy. We select the 13th sample in the database and compare the real angle values of 320 frames with the second method shown in the following Figs. 3, 4 and 5. The horizontal axis presents video frame sequence index, and vertical axis is the head pose angles with the unit^o.

It can be seen from the experiments that our method on the Biwi data set performs well, and compared with other existing methods, the estimation precision is improved. At the same time it achieves real-time processing in the mobile terminals. From the results, the regression accuracy will decrease when angles become large, and one of the reasons is that the performance of SDM algorithm is not good enough for larger angles. At the same time, our method

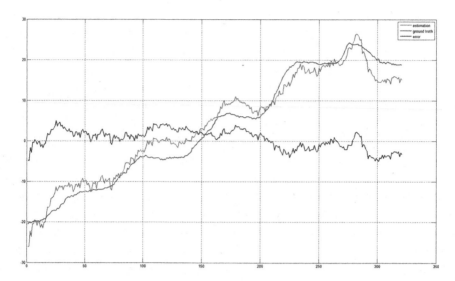

Fig. 3. Error curve of Pitch (red curve is estimation, blue curve is ground truth, black curve is error) (Color figure online)

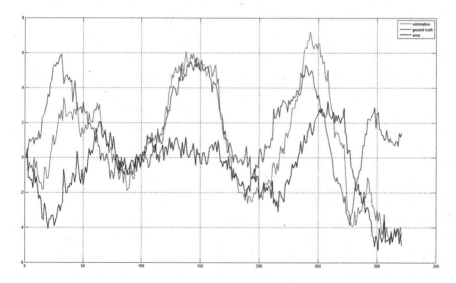

Fig. 4. Error curve of Roll (red curve is estimation, blue curve is ground truth, black curve is error) (Color figure online)

in this paper couldn't get smooth angle information, and it can fluctuate within a certain range. For testing the time of the method, we apply the head pose estimation method to iPhone 6s, Fig. 6 shows a sample. The prediction of head pose from a frame costs 38.8 ms. It is real time in mobile devices and can be used in same motion sensing game.

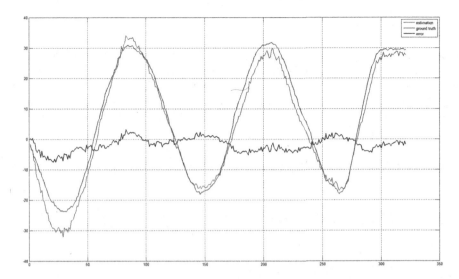

Fig. 5. Error curve of Yaw (red curve is estimation, blue curve is ground truth, black curve is error) (Color figure online)

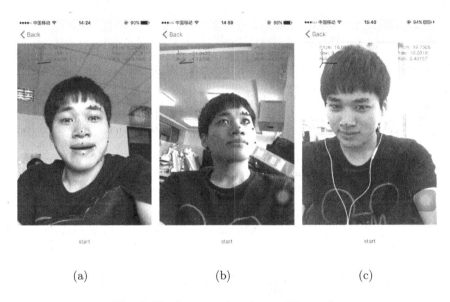

(a)   (b)   (c)

Fig. 6. Head pose estimation on iPhone 6s

Actually, for a video the average running time of 38.8 ms only include the SDM for facial feature points location and linear regression for head pose estimation. It is not necessary for every frame to detect where is face, and SDM algorithm is able to get current correct landmark with previous frames landmark result. Source code of our method is available at https://github.com/RoboPai/sdm.

4 Conclusion

Head pose estimation has been widely used in many face related applications. In this paper, we propose a real time head pose estimation method using facial feature points. We first locate the feature points by supervised descent method SDM, and then estimate the head pose by least square regression. The data normalization technique we used can eliminate the effect of the camera imaging parameters on pose estimation. The training data set is derived from a general 3D face model according to the camera imaging model. This can provide precise pose parameter. Experimental results show that this algorithm can be running in a real time way, and be transplanted to the mobile terminals.

Acknowledgements. This work was partially supported by the National Natural Science Foundation of China under Grant No. 61572078, and Beijing Municipal Natural Science Foundation under Grant No. 4152028.

References

1. Zhu, X., Lei, Z., Yan, J., Yi, D., Li, S.Z.: High-fidelity pose and expression normalization for face recognition in the wild. In: Proceedings of the IEEE Conference on Computer Vision and Pattern Recognition, pp. 787–796 (2015)
2. Kim, J.B., Kim, H.I., Park, R.H.: Head pose estimation using a coplanar face model for human computer interaction. In: 2014 IEEE International Conference on Consumer Electronics (ICCE), pp. 560–561. IEEE (2014)
3. Valenti, R., Sebe, N., Gevers, T.: Combining head pose and eye location information for gaze estimation. IEEE Trans. Image Process. **21**, 802–815 (2012)
4. Kim, H., Lee, S.H., Sohn, M.K., Kim, D.J.: Illumination invariant head pose estimation using random forests classifier and binary pattern run length matrix. Hum. Centric Comput. Inf. Sci. **4**, 1–12 (2014)
5. de Farias Macedo, M.C., Apolinário, A.L., dos Santos Souza, A.C.: A robust real-time face tracking using head pose estimation for a markerless ar system. In: 2013 XV Symposium on Virtual and Augmented Reality (SVR), pp. 224–227. IEEE (2013)
6. Kaymak, S., Patras, I.: Exploiting depth and intensity information for head pose estimation with random forests and tensor models. In: Park, J.-I., Kim, J. (eds.) ACCV 2012. LNCS, vol. 7729, pp. 160–170. Springer, Heidelberg (2013). doi:10.1007/978-3-642-37484-5_14
7. Sherrah, J., Gong, S., Ong, E.J.: Face distributions in similarity space under varying head pose. Image Vis. Comput. **19**, 807–819 (2001)
8. Fanelli, G., Weise, T., Gall, J., Gool, L.: Real time head pose estimation from consumer depth cameras. In: Mester, R., Felsberg, M. (eds.) DAGM 2011. LNCS, vol. 6835, pp. 101–110. Springer, Heidelberg (2011). doi:10.1007/978-3-642-23123-0_11
9. Li, Y., Gong, S., Sherrah, J., Liddell, H.: Support vector machine based multi-view face detection and recognition. Image Vis. Comput. **22**, 413–427 (2004)
10. Alioua, N., Amine, A., Rogozan, A., Bensrhair, A., Rziza, M.: Driver head pose estimation using efficient descriptor fusion. EURASIP J. Image Video Process. **2016**, 1–14 (2016)

11. Raytchev, B., Yoda, I., Sakaue, K.: Head pose estimation by nonlinear manifold learning. In: Proceedings of the 17th International Conference on Pattern Recognition, ICPR 2004, vol. 4, pp. 462–466. IEEE (2004)
12. Hu, N., Huang, W., Ranganath, S.: Head pose estimation by non-linear embedding and mapping. In: IEEE International Conference on Image Processing 2005, vol. 2, pp. II–342. IEEE (2005)
13. Wu, J., Trivedi, M.M.: A two-stage head pose estimation framework and evaluation. Pattern Recogn. **41**, 1138–1158 (2008)
14. Ma, Y., Konishi, Y., Kinoshita, K., Lao, S., Kawade, M.: Sparse bayesian regression for head pose estimation. In: 18th International Conference on Pattern Recognition (ICPR 2006), vol. 3, pp. 507–510. IEEE (2006)
15. Wang, J.G., Sung, E.: Em enhancement of 3d head pose estimated by point at infinity. Image Vis. Comput. **25**, 1864–1874 (2007)
16. Duan, F., Wu, F., Hu, Z.: Pose determination and plane measurement using a trapezium. Pattern Recogn. Lett. **29**, 223–231 (2008)
17. Hartley, R.I.: In defense of the eight-point algorithm. IEEE Trans. Pattern Anal. Mach. Intell. **19**, 580–593 (1997)
18. Freund, Y., Schapire, R.E.: A desicion-theoretic generalization of on-line learning and an application to boosting. In: Vitányi, P. (ed.) EuroCOLT 1995. LNCS, vol. 904, pp. 23–37. Springer, Heidelberg (1995). doi:10.1007/3-540-59119-2_166
19. Xiong, X., Torre, F.D.L.: Supervised descent method and its applications to face alignment, vol. 9, pp. 532–539 (2013)
20. Hu, Y., Duan, F., Yin, B., Zhou, M., Sun, Y., Wu, Z., Geng, G.: A hierarchical dense deformable model for 3d face reconstruction from skull. Multimedia Tools Appl. **64**, 345–364 (2013)

3D Modelling and Applications

DNA-SLAM: Dense Noise Aware SLAM for ToF RGB-D Cameras

Oliver Wasenmüller[1]([✉]), Mohammad Dawud Ansari[2], and Didier Stricker[1,2]

[1] German Research Center for Artificial Intelligence (DFKI),
Kaiserslautern, Germany
{oliver.wasenmueller,didier.stricker}@dfki.de
[2] University of Kaiserslautern, Kaiserslautern, Germany
m_ansari15@cs.uni-kl.de

Abstract. SLAM with RGB-D cameras is a very active field in Computer Vision as well as Robotics. Dense methods using all depth and intensity information showed best results in the past. However, usually they were developed and evaluated with RGB-D cameras using *Pattern Projection* like the Kinect v1 or Xtion Pro. Recently, *Time-of-Flight* (ToF) cameras like the Kinect v2 or Google Tango were released promising higher quality. While the overall accuracy increases for these ToF cameras, noisy pixels are introduced close to discontinuities, in the image corners and on dark/glossy surfaces. These inaccuracies need to be specially addressed for dense SLAM. Thus, we present a new Dense Noise Aware SLAM (DNA-SLAM), which considers explicitly the noise characteristics of ToF RGB-D cameras with a sophisticated weighting scheme. In a rigorous evaluation on public benchmarks we show the superior accuracy of our algorithm compared to the state-of-the-art.

1   Introduction

Simultaneous Localization And Mapping (SLAM) is the process of determining the pose (position and orientation) of a camera and creating a map of the environment by analyzing the associated camera images. This is a very active research field in both *Computer Vision* and *Robotics*. When using monocular cameras, sparse features [1,2] or dense pixels [3,4] are tracked in the images to estimate the camera pose. Since the depth of all pixels is unknown, this is a challenging task, which comprehends several inaccuracies or requires extensive computations. With the release of low-cost RGB-D sensors (e.g. Microsoft Kinect) the possibility for dense depth measurements of the environment was given. There are two common approaches of depth measurement: *Pattern Projection* and *Time-of-Flight* (ToF). Cameras with *Pattern Projection*, such as Microsoft Kinect v1, Asus Xtion Pro or Occipital Structure, project

---

Electronic supplementary material The online version of this chapter (doi:10.1007/978-3-319-54407-6_42) contains supplementary material, which is available to authorized users.

C.-S. Chen et al. (Eds.): ACCV 2016 Workshops, Part I, LNCS 10116, pp. 613–629, 2017.
DOI: 10.1007/978-3-319-54407-6_42

a known pattern into the scene and estimate the depth from its distortion. Recently, ToF cameras, such as Microsoft Kinect v2 [5] or Google Tango [6], became very popular, since they claim a higher accuracy in general. ToF cameras measure depth by estimating the time emitted light takes from the camera to the object and back. RGB-D cameras spurred a bulk of research on SLAM. While first approaches extended only sparse monocular algorithms [7,8], best performing algorithms in the literature use dense depth and intensity measurements [9,10]. These dense algorithms try to estimate the rigid camera motion between two images by minimizing the photometric and geometric error. This holds under the assumption of a photometric and geometric consistency. Previous methods were mainly developed and evaluated with RGB-D cameras using *Pattern Projection*, where these assumptions hold although the overall accuracy is lower. However, in some experiments with ToF cameras we detected that the geometric consistency is often violated due to the sensor noise, leading to inaccurate trajectories. Thus, we propose a *Dense Noise Aware Simultaneous Localization And Mapping* (DNA-SLAM), which considers explicitly the noise characteristics of ToF RGB-D cameras with a sophisticated weighting scheme. More precisely, our main contributions in this paper are:

- A robust reliability estimation for each single pixel based on the noise characteristics of ToF cameras.
- The integration of the individual reliabilities into a sophisticated weighting scheme for dense motion estimation.
- A systematic and rigorous evaluation of our algorithm with well-known publicly available benchmarks with real image data.

2   Related Work

Many state-of-the-art methods establish correspondences between sparse features to estimate the cameras motion [8,11,12]. Recently, several dense algorithms [9,10,13,14] emerged for SLAM systems showing better accuracy. Steinbrücker et al. [15] and Audras et al. [16] proposed to minimize the photometric error between consecutive RGB-D frames. This concept was extended by Kerl et al. [9] for their DVO algorithm by weighting photometric errors according to the t-distribution. Klose et al. [17] presented a efficient second order minimization scheme for motion estimation. DVO was extended by a minimization of the geometric error by Kerl et al. [18], since only intensity was used so far. Mailland et al. [19] used for their motion estimation super-resolved keyframes, and Gutierrez et al. [10] parametrized the geometric error by the inverse depth in addition. Several extensions and optimizations of these algorithms exist in the literature, for example for rolling shutter cameras [20], planar scenes [13] and many more. In contrast to all related work, we explicitly consider the noise characteristics of *Time-of-Flight* cameras in a sophisticated weighting. This leads to a superior accuracy as we show in our evaluation in Sect. 5.

3   Basic Idea

The basic idea of DNA-SLAM is to estimate the reliability of each single pixel in a ToF RGB-D image in order to weight its influence on the dense camera motion estimation. In state-of-the-art papers [10,14,18], the camera motion is estimated based on the assumptions of photometric and geometric consistency. This means that neither the scene color nor the scene geometry changes over time in captured images. To verify these assumptions, we analyze in Fig. 1 RGB-D images of a given scene with two different types of cameras, namely a *Time-of-Flight* camera (Microsoft Kinect v2) and a *Pattern Projection* camera (Microsoft Kinect v1). We place both cameras on a stable tripod with very similar viewpoints and capture two consecutive RGB-D images (I_1, D_1) and (I_2, D_2). Ideally, the two consecutive images should not contain any differences in intensity and depth. Thus, the differences between the intensity images and the depth images respectively should be zero. As visible in Fig. 1 the differences in intensity are very small and consequently negligible for both cameras types. The depth images of

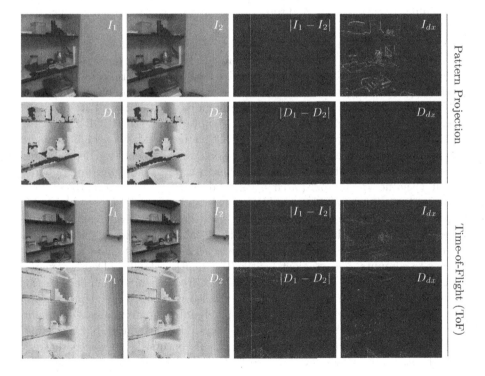

Fig. 1. The basic assumption of dense motion estimation is the photometric and geometric consistency. To verify this assumption we place a *Pattern Projection* camera (top) and a *Time-of-Flight* (ToF) camera (bottom) on a stable tripod and capture two consecutive RGB-D frames (I_1, D_1) and (I_2, D_2). We discover that ToF cameras violate the geometric consistency assumption $|D_1 - D_2| = 0$. However, the depth derivative D_{dx} coincides with these violations. See the supplementaries for large-scale images.

the *Pattern Projection* camera disclose only for some rare pixels bigger differences in depth. However, the ToF camera discloses violations of the geometric consistency assumption for numerous pixels as shown in the lowest row of Fig. 1. These violations for ToF cameras will lead to inaccuracies in the dense motion estimation, unless they are not especially treated. Thus, we propose a new sophisticated weighting scheme in this paper that incorporates the reliability of each single ToF pixel.

In the above mentioned experiment we can detect the violations quite easily by looking at the differences in depth, since the camera motion is zero. However, for unknown camera motions this is not possible. Therefore, we must define a measurement for the reliability of a pixel, which can be estimated on a single image without knowing the camera motion. In our experiments in Fig. 1 we discovered that for ToF cameras the derivatives of the depth image and the violations of the geometric consistency assumption coincide for most pixels. This means, ToF pixels with a high depth derivative are likely to violate the consistency assumption, whereas pixels with a low depth derivative can be considered as reliable. We repeated this experiment for numerous different scenes and all show similar results. We selected the example of Fig. 1 so that it is representative for other scenes.

ToF cameras have several noise sources [21]: dark and glossy scene colors, large scene distances, pixels close to the image boundaries, *flying pixels* close to depth discontinuities, etc. All these noises lead to violations of the geometric consistency assumption. Thus, in the ToF images of Fig. 1 the black shelfs and the four image corners violate the assumption for example. They all are precisely detected by the local derivatives. This is explicable for ToF RGB-D images, because of their noise characteristic [21]. ToF cameras exhibit a high frequent per-pixel noise. This means, in case of imprecise measurements the depth values of neighboring pixels strongly differ. In contrast, *Pattern Projection* cameras exhibit their noise as a per-patch noise or distortions over the surface. This means, neighboring pixels have similar values and errors. The per-pixel noise of ToF cameras can be easily detected by local derivatives, whereas per-patch noise can not. The particular thing is that the local derivatives are able to detect the noise positions independent of its source. This qualifies them to serve as an easy-to-compute measurement for the ToF depth reliability. Depth cameras with *Pattern Projection* enclose much less high-frequent noise, but lack in the overall accuracy [21]. In the supplementary video we show a live depth stream that illustrates the difference between per-pixel and per-patch noise. One might also claim that high derivatives are usual on depth discontinuities, but this is not possible due to the sensor technology. On depth discontinuities the sensor delivers either *flying pixels* (which are detected by the derivatives) or invalid values. In the following section we present a way to transform depth derivatives into a weighting scheme and to integrate them into the camera motion estimation.

4   DNA-SLAM

In this section, we present our DNA-SLAM algorithm. The goal is to estimate the camera motion ξ between consecutive RGB-D images composed of intensity images I_1 and I_2 and depth images D_1 and D_2. In DNA-SLAM we estimate the camera motion ξ according to the basic concept of state-of-the-art papers [10,15,16,18] by minimizing residuals r_i for all n valid pixels by

$$\xi = \arg\min_{\xi} \sum_{i}^{n} w(r_i)(r_i(\xi))^2, \tag{1}$$

where w_i are individual weights for each pixel. In the following sections we describe step-by-step how Eq. 1 is built and solved, while explicitly considering the noise characteristics of ToF cameras. Figure 2 gives an overview of the whole procedure. After describing some preliminaries in Sect. 4.1, we define in Sect. 4.2 the residuals r_i, which contain the photometric and geometric error of a given camera motion ξ. In Sect. 4.3, the noise aware weighting is described, which weights each pixel individually according to its reliability. As motivated in Sect. 3 we use (amongst others) the depth derivatives to compute robust weights. In Sect. 4.4 we detail how the camera motion ξ is computed by solving Eq. 1 with an iterative re-weighted least square algorithm. To enhance the accuracy of the motion estimation, we use a coarse-to-fine strategy.

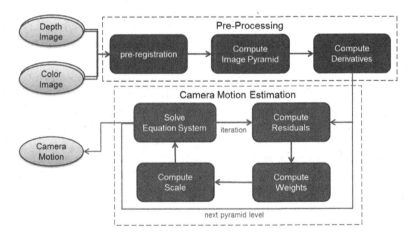

Fig. 2. Overview of DNA-SLAM. As an input the algorithm requires two image tuples (color and depth). After a pre-processing, the iterative motion estimation is performed as described in Sect. 4. For a short summary see also the supplementary video.

4.1   Preliminaries

In this section, we define the basics of the RGB-D camera images I and D, its derivatives $I_{dx,dy}$ and $D_{dx,dy}$ as well as the camera motion ξ. A point p in the 3D space is defined by homogeneous coordinates as $p = (X, Y, Z, 1)^{\top}$. The 2D pixel coordinates $x = (x, y)^{\top}$ of the 3D point p are defined by the standard pinhole camera model and the projection function π as

$$x = \pi(p) = \left(\frac{X f_x}{Z} + c_x, \frac{Y f_y}{Z} + c_y \right)^{\top},$$

(2)

where f_x, f_y are the focal lengths and c_x, c_y is the camera center. Together with the depth $Z = D(x)$ a 3D point p can be reconstructed using the inverse projection function π^{-1} defined by

$$p = \pi^{-1}(x, D(x)) = \left(\frac{x - c_x}{f_x} D(x), \frac{y - c_y}{f_y} D(x), D(x), 1 \right)^{\top}.$$

(3)

An essential assumption for Eq. 3 is, that the intensity image I and the depth image D coincide. This means that corresponding pixels in the two images are located on the same image coordinates. For some devices (e.g. Kinect v2) this assumption does not hold for the raw images and an additional pre-processing step is necessary. We detail our pre-processing in the Supplementary Material.

The derivatives of an image $A \in \{I, D\}$ are defined via the local neighborhoods. This means, for each pixel coordinate $x = (x, y)^{\top}$ the derivative can be computed for x and y direction by writing forward-backward differences as

$$A_{dx}(x, y) = \frac{A(x - 1, y) - A(x + 1, y)}{2} \quad \text{and}$$

$$A_{dy}(x, y) = \frac{A(x, y - 1) - A(x, y + 1)}{2}.$$

(4)

When applying Eq. 4 on the depth images, one must be careful because of invalid depth measurements. For some pixels the depth sensor is (independent of *Pattern Projection* or ToF) not able to measure the depth leading to invalid values often represented as zero. In that case the step size must be locally increased or decreased in order to achieve correct derivative values.

A 3D point p is transformed to the coordinate frame of the second camera according to the camera motion ξ, which we define as a rigid body motion $\xi \in \mathrm{SE}(3)$. A rigid body motion comprises a rotation $R \in \mathrm{SO}(3)$ and a translation $t \in \mathbb{R}^3$. The rotation R can be expressed by a 3×3 matrix and the translation t by a 3×1 matrix leading to a 4×4 transformation matrix

$$T_{4 \times 4} = \begin{pmatrix} R_{3 \times 3} & t_{3 \times 1} \\ 0 & 1 \end{pmatrix}.$$

(5)

Hence, a point p' in the frame of the second camera is given as

$$p' = \xi(p) = T p.$$

(6)

Since T has twelve parameters, whereas a rigid body motion has only six Degrees Of Freedom (DOF) (three DOF of rotation and three DOF for translation), we use – like many others [9,10,17,18] – a minimal parametrization of ξ using twist coordinates given by the Lie algebra. For more information about the Lie algebra we refer to [22]. Summarizing Eqs. 2, 3 and 6 a warping function τ is given by

$$\tau(\xi, \mathrm{x}) = \pi\left(\xi\left(\pi^{-1}\left(\mathrm{x}, D(\mathrm{x})\right)\right)\right), \tag{7}$$

which maps a pixel x from the first image to the second image using the camera motion ξ.

4.2 Residual Definition

Given the warping function τ we can define – like many state-of-the-art papers [9,13] – for each pixel x the residuals r_I and r_D for the intensity image I and the depth image D respectively as

$$r_I(\mathrm{x}, \xi) = I_2(\tau(\xi, \mathrm{x})) - I_1(\mathrm{x}) \quad \text{and}$$
$$r_D(\mathrm{x}, \xi) = D_2(\tau(\xi, \mathrm{x})) - [\xi\left(\pi^{-1}\left(\mathrm{x}, D_1(\mathrm{x})\right)\right)]_Z, \tag{8}$$

where $[.]_Z$ is the z component of a 3D point, which is equivalent to the depth. Note, the depth residual r_D resembles the geometric error in variants of ICP [23]. In DNA-SLAM we want to explicitly use the derivatives of the depth image to derive a indicator for the reliability of pixels. In general we would have to consider four values, namely $D_{dx,1}$, $D_{dy,1}$ for the first image and $D_{dx,2}$, $D_{dy,2}$ for the second image. To reduce the complexity we also define the influence of the derivatives as residuals r_{Dx} and r_{Dy} by

$$r_{Dx}(\mathrm{x}, \xi) = D_{dx,2}(\tau(\xi, \mathrm{x})) - D_{dx,1}(\mathrm{x}) \quad \text{and}$$
$$r_{Dy}(\mathrm{x}, \xi) = D_{dy,2}(\tau(\xi, \mathrm{x})) - D_{dy,1}(\mathrm{x}), \tag{9}$$

where D_{dx} and D_{dy} are the derivatives as defined in Eq. 4. This has the advantage of considering only two values representing the derivatives for the weight computation. If an unreliable point (indicated by a high derivative) occurs at one of the pixel coordinates x or $\tau(\xi, \mathrm{x})$, the residual will be high too. The case that both pixel coordinates are unreliable is extremely rare as asserted during our tests. Thus, the residuals r_{Dx} and r_{Dy} of Eq. 9 are an authentic indicator for the reliability of the given pixels.

At the end, the residuals are used for both weight computation and motion computation. Since we want to use the derivatives only for the weight computation, we define two residuals vector R and Ω containing the residuals r_I, r_D, r_{Dx} and r_{Dy} of Eqs. 8 and 9 by

$$R(\mathrm{x}, \xi) = \begin{bmatrix} r_I(\mathrm{x}, \xi) \\ r_D(\mathrm{x}, \xi) \end{bmatrix} \quad \text{and} \quad \Omega(\mathrm{x}, \xi) = \begin{bmatrix} r_I(\mathrm{x}, \xi) \\ r_D(\mathrm{x}, \xi) \\ r_{Dx}(\mathrm{x}, \xi) \\ r_{Dy}(\mathrm{x}, \xi) \end{bmatrix}, \tag{10}$$

where $R \in \mathbb{R}^2$ is a two-dimensional vector and $\Omega \in \mathbb{R}^4$ is a four-dimensional vector.

4.3   Noise Aware Weighting and Scale Definition

In this section, we want to determine a weight w for each pixel coordinate $x = (x, y)^\top$, which states how reliable the pixel at x is. As stated in Sect. 4.2 the weight should be determined based on the residual vector Ω. A straightforward approach would be to combine the single residuals in Ω linearly. State-of-the-art approaches use therefore heuristic [24] or manually chosen weights [14]. However, we want to limit the influence of pixels, where exactly one or combinations of the following holds: high derivatives, high intensity error, or high depth error. Kerl et al. [9] showed already that intensity and depth errors follow the t-distribution with $\nu = 5$ DOF. Inspired by [18] we model the residual vector Ω as a multivariate t-distribution. In Sect. 5.1 we show that also the derivate residual follows the t-distribution with $\nu = 5$ DOF. The weights can then be computed by

$$w(x, \xi) = \frac{\nu + 1}{\nu + \Omega(x, \xi)^\top \Sigma_\Omega(\xi)^{-1} \Omega(x, \xi)}, \tag{11}$$

where Σ_Ω is the covariance matrix of the residual vector Ω. It is defined for all n valid pixels in an image as

$$\Sigma_\Omega(\xi) = \frac{1}{n} \sum_i^n \Omega(i, \xi) \Omega(i, \xi)^\top \frac{\nu + 1}{\nu + \Omega(i, \xi) \Sigma'_\Omega(\xi)^{-1} \Omega(i, \xi)^\top}, \tag{12}$$

where Σ'_Ω is the covariance matrix of the previous iteration. We will detail the iterative process in Sect. 4.4. Note, while state-of-the-art algorithms [18] work with a 2×2 covariance matrix, we use a 4×4 covariance matrix containing also the derivatives. The covariance matrix Σ_Ω has the task to automatically scale the weights, so that heuristic [24] or manually chosen weights [14] are not necessary.

4.4   Iterative Re-weighted Least Square

In this section, we demonstrate how the camera motion ξ can be estimated with the help of an iterative re-weighted least square. Equation 1 already shows the basic principle: the camera motion ξ is computed by minimizing residuals. For DNA-SLAM we follow state-of-the-art algorithms [13, 18, 19] and minimize the photometric and geometric error, which is summarized in the residuals $R(x, \xi)$. Minimizing the residuals $\Omega(x, \xi)$ is not useful here, because the depth derivatives are not precise enough; even with our sophisticated weighting of Sect. 4.3. The basic formula for the motion estimation in Eq. 1 changes for the multivariate case that we use in DNA-SLAM to

$$\xi = \arg \min_\xi \sum_i^n w(i, \xi) R(i, \xi)^\top \Sigma_R(\xi)^{-1} R(i, \xi). \tag{13}$$

Note, the 2×2 covariance matrix $\Sigma_R(\xi)$ is used here, since also the residuals $R(x, \xi)$ are used. However, the weights $w(x, \xi)$ are computed with the residuals $\Omega(x, \xi)$ and covariance matrix $\Sigma_\Omega(\xi)$ as stated in Eq. 11.

The residuals $R(\mathrm{x}, \xi)$ in Eq. 10 are non-linear in the camera motion ξ. Therefore, we follow [9,15,18] and linearize Eq. 13 around ξ using the first order Taylor expansion. This reformulation is possible, because the motion between the frames is very small as the frame rate of ToF cameras is usually rather high (e.g. Kinect v2: 30 Hz). In addition, we utilize a coarse-to-fine strategy to ensure small camera motion as detailed in Sect. 4.4. The resulting normal equations of the non-linear least square problem are [18]:

$$\sum_i^n w(i,\xi) J(i,\xi)^\top \Sigma_R(\xi)^{-1} J(i,\xi) \Delta\xi = -\sum_i^n w(i,\xi) J(i,\xi)^\top \Sigma_R(\xi)^{-1} R(i,\xi),$$

(14)

where $J(i,\xi)$ is the 2×6 matrix containing the derivatives of $R(i,\xi)$ with respect to ξ. The linear equation system in (14) can be efficiently solved for the motion increments $\Delta\xi$ using Cholesky decomposition [9]. In each iteration the equation system is re-weighted by re-estimating the weights $w(i,\xi)$ and the scale matrix $\Sigma_\Omega(\xi)$ as shown in Fig. 2. The trajectory of the camera is determined by concatenating the motion estimations between single frames.

For the Taylor expansion in Eq. 14 we must ensure small camera motions. Thus, we utilize a coarse-to-fine strategy by building an image pyramid. Therefore, we subsample each image tuple $(I_i,\ D_i)$ by halving the image resolution for each level. For subsampling the depth images it is important to consider the invalid depth values. The camera motion ξ is estimated in the coarsest level first and then propagated to finer levels as a prior. In our experiments we were using four pyramid levels. The finest level is not used, since the results are almost unchanged, but runtime increases significantly.

5   Evaluation

In order to obtain meaningful evaluation results we make use of two publicly available benchmarks [25,26]. First, we use the CoRBS benchmark [26], which was recently published and is the only dataset providing real ToF image data of the Kinect v2 together with ground truth trajectories for the camera and ground truth reconstructions of the scene. The ground truth trajectory of CoRBS was acquired with an external motion capture system with sub-millimeter precision. This is an ideal basis for the evaluation of our new algorithm. In addition, we also use the TUM RGB-D benchmark [25], which uses the Kinect v1 and also provides a ground truth camera trajectory. We use this dataset in order to show the applicability of our new algorithm to Kinect v1. Furthermore, many previous algorithms were benchmarked on this dataset.

For the evaluation of the estimated camera trajectory we follow [25,26] and use the Relative Pose Error (RPE). It measures the drift of the estimated trajectory compared to a given ground truth over a fixed time interval Δ in m/s. For more details on that measurement please refer to the Supplementary Material.

5.1   Noise Aware Weights

One of the key contributions in this paper is to estimate the reliability of each pixel individually and integrate it into a sophisticated weighing scheme. As motivated in Sect. 3 we use the depth derivatives to judge the reliability of pixels. Together with the photometric and depth residuals they are transformed to weights as defined in Eq. 11. In Fig. 3(a), (b) and (c) the inputs for the weighting function of our DNA-SLAM are depicted for an exemplary frame of the CoRBS [26] dataset. While the intensity r_I and depth r_D residuals are high close to edges, the new r_{Dx} is high in unreliable regions. These are e.g. points on the reflective floor, in the image corners and flying pixels. In Fig. 3(d) the weights without the new reliability estimation r_{Dx} are shown as used in DVO [18]. Here, only points close to edges are down-weighted, whereas the remaining point have almost equally high weights. In Fig. 3(e) the weights with the new reliability estimation r_{Dx} are shown as applied in DNA-SLAM. Here, the points close to edges are down-weighted too, but also regions like e.g. the floor and the image corners are down-weighted. In total, the different weights are much more distributed and selective compared to Fig. 3(d). From a first impression one might claim the weight values w_R are only lower by a constant value or factor

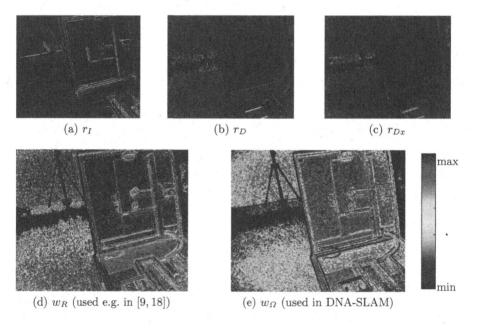

(a) r_I (b) r_D (c) r_{Dx}

(d) w_R (used e.g. in [9, 18])   (e) w_Ω (used in DNA-SLAM)

Fig. 3. Our DNA-SLAM uses a noise aware weighting scheme. Weights are computed based on (a) intensity residuals, (b) depth residuals and (c) depth derivatives. While (d) the weights in state-of-the-art algorithms are quite clustered, (e) DNA-SLAM estimates the weights for each pixel very selective, representing the individual reliability. Points on the reflective floor, in the image corners and *flying pixels* are clearly down-weighted. (Color figure online)

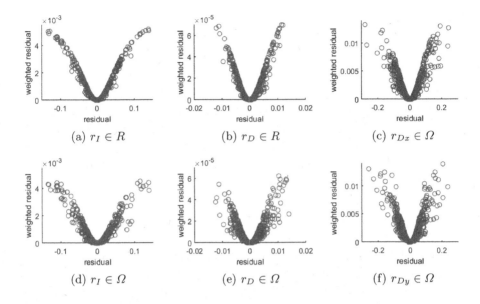

Fig. 4. In DNA-SLAM we weight residuals r_i by Eq. 1 to $w(r_i)r_i^2$. (a) and (b) show the weighted residuals of R, in contrast (d) and (e) use our new residuals Ω including depth derivatives (cp. Eq. 10). (c) and (f) show that also the derivatives follow the t-distribution.

compared to w_Ω. However, this is a non-linear transformation. For example, the cab of the electrical cabinet is transformed from dark red to red, whereas the glossy floor is transformed much stronger from red to green/blue. This shows the effectiveness of the new proposed weighting. Another argument is that the weights are scaled by the (co-)variance (see Eq. 12). A scaling of the weights by a certain value would be neutralized by the variance. The effects of this new weighting on the trajectory accuracy is shown in Sect. 5.2. We also tested binary weights for down-weighting unreliable pixels, but the results were poor.

The definition of weights (cp. Eq. 11) assumes a t-distribution of the used derivative residuals r_{Dx} and r_{Dy}. In Fig. 4 we verified this assumption. (a) and (b) show the weighted residuals, if only R (cp. Eq. 10) is utilized for weight computation as applied in DVO [18]. The plots verify that $r_I, r_D \in R$ follow clearly the t-distribution. If we use our new Ω (cp. Eq. 10) for weight computation, the plots change as illustrated in Fig. 4(d) and (e). The residuals $r_I, r_D \in \Omega$ still follow the t-distribution, but some values are down-weighted. These are values which were detected as unreliable. In addition, Fig. 4(c) and (f) show that our new residuals $r_{Dx}, r_{Dy} \in \Omega$ follow the t-distribution.

5.2 SLAM Results

In this section, we evaluate our DNA-SLAM with public datasets and compare it against state-of-the-art algorithms. For the implementation of DNA-SLAM we

Table 1. RMSE of the translational and rotational drift (RPE) in m/s and deg/s respectively for different sequences of the CoRBS dataset [26] using the Kinect v2 ToF RGB-D camera. Best result are bold. DNA-SLAM performs superior in most sequences or is at least on par.

Algorithm	D1		D2		D4	
	E_{trans}	E_{rot}	E_{trans}	E_{rot}	E_{trans}	E_{rot}
ICP [23]	0.0443	2.0819	0.0317	1.9360	0.0420	2.3651
DVO [18]	0.0596	2.5411	0.0410	2.1356	0.0335	1.5915
DNA-SLAM (*ours*)	**0.0266**	**0.9702**	**0.0209**	**1.1008**	**0.0223**	**0.9988**
Algorithm	E1		E4		E5	
	E_{trans}	E_{rot}	E_{trans}	E_{rot}	E_{trans}	E_{rot}
ICP [23]	0.0770	4.4267	0.0569	2.9581	0.0453	4.1031
DVO [18]	**0.0335**	1.5915	0.0335	**1.5915**	0.0309	1.7145
DNA-SLAM (*ours*)	0.0349	**1.4264**	**0.0264**	1.7547	**0.0143**	**0.7128**
Algorithm	H1		H2		H3	
	E_{trans}	E_{rot}	E_{trans}	E_{rot}	E_{trans}	E_{rot}
ICP [23]	0.0559	2.3807	0.0412	2.3018	0.0676	2.9232
DVO [18]	0.0616	2.3214	0.0251	1.5564	0.0827	3.2376
DNA-SLAM (*ours*)	**0.0196**	**0.7248**	**0.0124**	**0.6778**	**0.0300**	**1.4728**

build on top of the open source implementation of DVO [18]. First, we make use of the CoRBS dataset [26], which uses the Kinect v2 ToF RGB-D camera. We run the DNA-SLAM algorithm with several sequences and measure the relative translational as well as the rotational error of the estimated trajectories. In order to evaluate the effect of the new weighting, we compare against the state-of-the-art algorithm DVO [18] as well as ICP [23,27] which is used e.g. in KinectFusion [28] and others [14,29]. We apply DVO and ICP without keyframes, since we want to measure the frame-to-frame drift of the algorithms. The quantitative evaluation in Table 1 verifies the superior accuracy of our DNA-SLAM. The trajectories estimated with ICP exhibit a lower accuracy than the new DNA-SLAM. Compared to DVO the relative translational as well as the rotational error was substantially reduced with DNA-SLAM in most sequences. For the sequences E1 and E4 we achieve an accuracy on par. Figure 6 gives a visual impression of the trajectories. In all sequences DNA-SLAM is closer to the ground truth (GT) than DVO. Figure 5 shows the mappings of two exemplary sequences. The maps of DNA-SLAM are much more accurate, which is a logical consequence of the reduced drift compared to DVO. Summarized, we can conclude that our new weighting scheme substantially reduces the drift in dense motion estimation with ToF cameras.

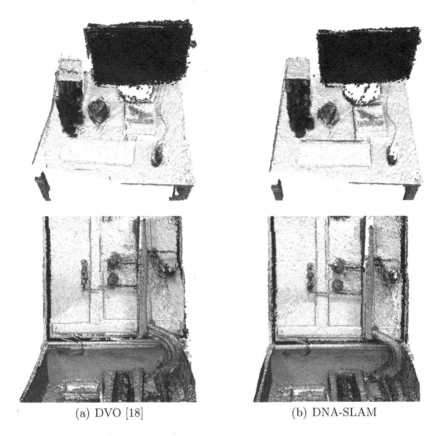

(a) DVO [18]   (b) DNA-SLAM

Fig. 5. Visual comparison of the mapped geometry with DVO [18] and DNA-SLAM using the CoRBS [26] sequences D1 and E5. DNA-SLAM delivers a higher accuracy.

For the sake of completeness we tested DNA-SLAM also with the TUM dataset [25], which utilizes the Kinect v1 using *Pattern Projection* for depth estimation. Since this dataset is already available for some years, several state-of-the-art algorithms [8,10,14,17–19] used it for evaluation. In Table 2 we state the relative translational errors. Here, we are slightly better than DVO, which is the most similar algorithm. Some state-of-the-art algorithms [10,19] deliver more accurate results, but they are also using keyframes adulterating comparison. Summarized, the evaluation shows that our algorithm is also applicable with Kinect v1 delivering state-of-the-art results. However, when using *Time-of-Flight* cameras like the Kinect v2 our algorithm shows clearly superior accuracy.

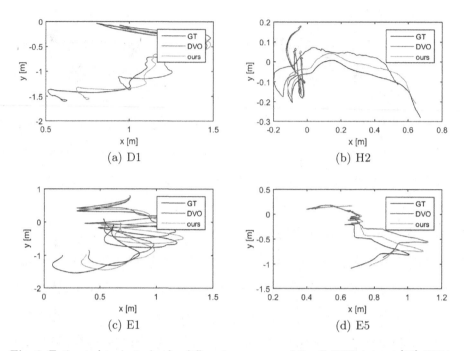

Fig. 6. Estimated trajectories for different sequences of the CoRBS dataset [26]. DNA-SLAM is closer to the ground truth (GT) trajectories than the state-of-the-art algorithm DVO [18]. See Table 1 for a quantitative evaluation.

Table 2. RMSE of the translational drift (RPE) in m/s for selected sequences of the TUM dataset [25] using the Kinect v1 with *Pattern Projection*. DNA-SLAM achieves state-of-the-art results even with this camera while being designed for ToF cameras; only algorithms with keyframes (marked with *) perform better.

Algorithm	Author	fr1/desk	fr1/desk2	fr1/room	fr2/desk
FOVIS	Huang et al. [8]	0.0604*	-	0.0642*	0.0136*
	Whelan et al. [14]	0.0393	-	0.0622	0.0208
	Klose et al. [17]	0.0302	0.0526	0.0397	0.0147
VP	Meilland et al. [19]	0.0259*	-	0.0351*	0.0147*
	Gutierrez et al. [10]	0.0260*	0.0387*	0.0491*	0.0121*
DVO	Kerl et al. [18]	0.036	0.049	0.058	-
DNA-SLAM	*ours*	0.0333	0.0482	0.0498	0.0195

6   Conclusion and Future Work

In this paper, we proposed our new DNA-SLAM algorithm, which addresses especially ToF RGB-D cameras. We discovered that the noise in the depth image leads to inaccuracies in the camera motion estimation, unless not especially treated. In our experiments we verified that the local depth derivatives are a good indicator for the depth pixel reliability. Thus, DNA-SLAM estimates the reliability of each pixel individually and transforms that into a weighting scheme. In the evaluation with the public CoRBS benchmark [26] we showed the substantially reduced drift of DNA-SLAM compared to state-of-the-art algorithms. Thus, we demonstrated that a sophisticated weighting scheme can compensate the errors introduced by ToF depth cameras leading to superior localization and mapping results. For a short summary see also the supplementary video.

In future work, we will integrate a keyframe selection and extend the tracking from frame-to-frame to frame-to-keyframe. This is a straightforward procedure and was already applied in several works [10, 18, 19]. We expect a further improvment in accuracy, which is required for several applications [30]. In addition, we will speed-up the implementation, since our current CPU version requires around 100 ms for a single motion estimation. A major speed-up can be achieved by porting the algorithm either to SSE [18] or to a GPU [31] as already shown in the literature.

References

1. Davison, A.J.: Real-time simultaneous localisation and mapping with a single camera. In: International Conference on Computer Vision (ICCV), pp. 1403–1410. IEEE (2003)
2. Klein, G., Murray, D.: Parallel tracking and mapping for small AR workspaces. In: International Symposium on Mixed and Augmented Reality (ISMAR), pp. 225–234. IEEE (2007)
3. Newcombe, R.A., Lovegrove, S.J., Davison, A.J.: DTAM: Dense tracking and mapping in real-time. In: International Conference on Computer Vision (ICCV), pp. 2320–2327. IEEE (2011)
4. Engel, J., Schöps, T., Cremers, D.: LSD-SLAM: large-scale direct monocular SLAM. In: Fleet, D., Pajdla, T., Schiele, B., Tuytelaars, T. (eds.) ECCV 2014. LNCS, vol. 8690, pp. 834–849. Springer, Cham (2014). doi:10.1007/978-3-319-10605-2_54
5. Microsoft: (Kinect v2). www.microsoft.com/en-us/kinectforwindows/
6. Google: (Tango). www.google.com/atap/project-tango/
7. Engelhard, N., Endres, F., Hess, J., Sturm, J., Burgard, W.: Real-time 3D visual slam with a hand-held RGB-D camera. In: RGB-D Workshop on 3D Perception in Robotics at the European Robotics Forum, vol. 180 (2011)
8. Huang, A.S., Bachrach, A., Henry, P., Krainin, M., Maturana, D., Fox, D., Roy, N.: Visual odometry and mapping for autonomous flight using an RGB-D camera. In: International Symposium on Robotics Research (ISRR), vol. 2 (2011)
9. Kerl, C., Sturm, J., Cremers, D.: Robust odometry estimation for RGB-D cameras. In: IEEE International Conference on Robotics and Automation (ICRA), pp. 3748–3754. IEEE (2013)
10. Gutierrez-Gomez, D., Mayol-Cuevas, W., Guerrero, J.: Dense RGB-D visual odometry using inverse depth. Robot. Auton. Syst. **75**, 571–583 (2016)

11. Brunetto, N., Fioraio, N., Stefano, L.: Interactive RGB-D SLAM on mobile devices. In: Jawahar, C.V., Shan, S. (eds.) ACCV 2014. LNCS, vol. 9010, pp. 339–351. Springer, Cham (2015). doi:10.1007/978-3-319-16634-6_25

12. Belter, D., Nowicki, M., Skrzypczyński, P.: On the performance of pose-based RGB-D visual navigation systems. In: Cremers, D., Reid, I., Saito, H., Yang, M.-H. (eds.) ACCV 2014. LNCS, vol. 9004, pp. 407–423. Springer, Cham (2015). doi:10.1007/978-3-319-16808-1_28

13. Ma, L., Kerl, C., Stueckler, J., Cremers, D.: CPA-SLAM: consistent plane-model alignment for direct RGB-D slam. In: International Conference on Robotics and Automation (ICRA) (2016)

14. Whelan, T., Johannsson, H., Kaess, M., Leonard, J.J., McDonald, J.: Robust real-time visual odometry for dense RGB-D mapping. In: International Conference on Robotics and Automation (ICRA), pp. 5724–5731. IEEE (2013)

15. Steinbruecker, F., Sturm, J., Cremers, D.: Real-time visual odometry from dense RGB-D images. In: International Conference on Computer Vision Workshop (ICCV Workshop) (2011)

16. Audras, C., Comport, A., Meilland, M., Rives, P.: Real-time dense appearance-based slam for RGB-D sensors. In: Australasian Conference on Robotics and Automation (ACRA) (2011)

17. Klose, S., Heise, P., Knoll, A.: Efficient compositional approaches for real-time robust direct visual odometry from RGB-D data. In: International Conference on Intelligent Robots and Systems (IROS), pp. 1100–1106. IEEE (2013)

18. Kerl, C., Sturm, J., Cremers, D.: Dense visual slam for RGB-D cameras. In: International Conference on Intelligent Robot Systems (IROS) (2013)

19. Meilland, M., Comport, A.I.: On unifying key-frame and voxel-based dense visual slam at large scales. In: International Conference on Intelligent Robots and Systems (IROS), pp. 3677–3683. IEEE (2013)

20. Kerl, C., Stueckler, J., Cremers, D.: (Dense continuous-time tracking and mapping with rolling shutter RGB-D cameras)

21. Wasenmüller, O., Stricker, D.: Comparison of kinect v1 and v2 depth images in terms of accuracy and precision. In: Chen, C.-S., Lu, J., Ma, K.-K. (eds.) ACCV 2016 Workshops, Part II. LNCS, vol. 10116, pp. 34–45. Springer, Cham (2017)

22. Ma, Y., Soatto, S., Kosecka, J., Sastry, S.S.: An invitation to 3-d vision: from images to geometric models, vol. 26. Springer Science & Business Media, New York (2012)

23. Besl, P.J., McKay, N.D.: Method for registration of 3-d shapes. In: Robotics-DL Tentative, pp. 586–606. International Society for Optics and Photonics (1992)

24. Tykkälä, T., Audras, C., Comport, A.I.: Direct iterative closest point for real-time visual odometry. In: International Conference on Computer Vision Workshops (ICCV Workshops), pp. 2050–2056. IEEE (2011)

25. Sturm, J., Engelhard, N., Endres, F., Burgard, W., Cremers, D.: A benchmark for the evaluation of RGB-D slam systems. In: IEEE/RSJ International Conference on Intelligent Robots and Systems (IROS), pp. 573–580. IEEE (2012)

26. Wasenmüller, O., Meyer, M., Stricker, D.: CoRBS: comprehensive RGB-D benchmark for slam using kinect v2. In: IEEE Winter Conference on Applications of Computer Vision (WACV). IEEE (2016)

27. Rusinkiewicz, S., Levoy, M.: Efficient variants of the ICP algorithm. In: IEEE International Conference on 3D Digital Imaging and Modeling, pp. 145–152. IEEE (2001)

28. Newcombe, R.A., Izadi, S., Hilliges, O., Molyneaux, D., Kim, D., Davison, A.J., Kohi, P., Shotton, J., Hodges, S., Fitzgibbon, A.: Kinectfusion: real-time dense

surface mapping and tracking. In: IEEE International Symposium on Mixed and Augmented Reality (ISMAR) (2011)

29. Lin, Y.C., Chen, C.Y., Huang, S.W., Huang, P.S., Chen, C.F.: Registration and merging of large scale range data using an improved ICP algorithm approach. In: International Conference Image and Vision Computing (IVCNZ) (2011)

30. Wasenmüller, O., Meyer, M., Stricker, D.: Augmented reality 3D discrepancy check in industrial applications. In: IEEE International Symposium on Mixed and Augmented Reality (ISMAR), pp. 125–134. IEEE (2016)

31. Lee, D., Kim, H., Myung, H.: Gpu-based real-time RGB-D 3D slam. In: International Conference on Ubiquitous Robots and Ambient Intelligence (URAI), pp. 46–48. IEEE (2012)

3D Shape Retrieval via Irrelevance Filtering and Similarity Ranking (IF/SR)

Xiaqing Pan[(✉)], Yueru Chen, and C.-C. Jay Kuo

Ming-Hsieh Department of Electrical Engineering,
University of Southern California, Los Angeles, CA 90089-2564, USA
xiaqingp@usc.edu

Abstract. A novel solution for the content-based 3D shape retrieval problem using an unsupervised clustering approach, which does not need any label information of 3D shapes, is presented in this work. The proposed shape retrieval system consists of two modules in cascade: the irrelevance filtering (IF) module and the similarity ranking (SR) module. The IF module attempts to cluster gallery shapes that are similar to each other by examining global and local features simultaneously. However, shapes that are close in the local feature space can be distant in the global feature space, and vice versa. To resolve this issue, we propose a joint cost function that strikes a balance between two distances. Irrelevant samples that are close in the local feature space but distant in the global feature space can be removed in this stage. The remaining gallery samples are ranked in the SR module using the local feature. The superior performance of the proposed IF/SR method is demonstrated by extensive experiments conducted on the popular SHREC12 dataset.

1 Introduction

Content-based 3D shape retrieval [1] has received a lot of attention in recent years due to an rapidly increasing number of 3D models over the Internet (e.g., Google sketchup and Yobi3D). Applications of 3D shape retrieval technologies include: 3D model repository management, mechanical components retrieval, medical organ model analysis, etc. Given a 3D shape model as the query, a content-based 3D shape retrieval system analyzes the query shape and retrieves ranked 3D shapes from the gallery set according to a similarity measure. Its performance is evaluated by consistency between ranked shapes and human interpretation. A robust and efficient 3D shape retrieval system is needed for users to access and exploit large 3D datasets effectively.

Recently, convolutional neural-network (CNN) based solutions achieved impressive performance by training a network using either multiple views of 3D shapes [2–6] or the 3D volumetric data [7–9]. However, their training procedure demands a large amount of labeled data, which is labor-intensive. In this work, we address the 3D shape retrieval problem using an unsupervised learning approach. It has broader applications since no labeled data are needed.

© Springer International Publishing AG 2017
C.-S. Chen et al. (Eds.): ACCV 2016 Workshops, Part I, LNCS 10116, pp. 630–646, 2017.
DOI: 10.1007/978-3-319-54407-6_43

The main challenge in 3D shape retrieval lies in a wide range of shape variations. A generic 3D shape dataset such as SHREC12 [10] includes both rigid and non-rigid shapes. Shape variations can be categorized into inter-class similarities and intra-class variations. For the latter, we have articulation, surface deformation, noise, etc.

Global and/or local features can be used to measure the similarity between two 3D shapes. The rotation invariant spherical harmonics (RISH) [11] and the D2 shape distribution [12] are two representative global features. They capture object surface properties using the frequency decomposition and the vertex distance histogram, respectively. The retrieval performance using global features only may degrade due to the loss of fine shape details. To overcome this limitation, research efforts in recent years have focused on developing more discriminative local features. They can be categorized into surface-based and view-based features. Surface-based local features [13–17] describe a local surface region to achieve pose oblivion, scale and orientation invariance. Although surface-based retrieval methods are effective in handling non-rigid shape retrieval [18,19], they are not robust against shape artifacts that do occur in generic shape datasets. Retrieval methods using view-based local features are favored for this reason.

View-based methods project a 3D shape into multiple views. Generally speaking, an adequate number of view samples can represent a 3D shape well. The light field descriptor (LFD) method [20] and the multi-view depth line approach (MDLA) [21] represent each view by Zernike moments plus polar Fourier descriptors and depth lines, respectively. The similarity between two shapes is measured by enumerating multiple rotation setups. The salient local visual feature (SLVF) method [22] extracts the SIFT points [23] from each range view. After constructing a codebook from the training pool, one feature of a 3D shape can be represented by the histogram of SIFT points from all views using the Bag of Words (BoW) approach. The DG1SIFT method [24] extends the SLVF method by extracting three types of SIFT points from each view. They are dense, global and one SIFTs. The depth buffered super vector coding (DBSVC) method [25] uses a dense power SURF feature and the super vector coding algorithm to improve the feature discriminability.

Although local features achieve a better performance than global features when being tested against several generic 3D shape datasets [10,26–28], their discriminative power is restricted in the global scale. In particular, they may retrieve globally irrelevant 3D shapes in high ranks. To illustrate this point, we show five retrieval results by applying the DG1SIFT method to the SHREC12 3D shape dataset in Fig. 1(a)–(e). With the five query shapes in the leftmost column, the top 10 retrieved results are presented in the first row of each subfigure Obviously, errors in these retrieval results are counter to human intuition. Being motivated by the observation, we propose a more robust 3D shape retrieval system which is called the irrelevance filtering and similarity ranking (IF/SR) method. Its retrieved results are shown in the second row of each subfigure. All mistakes generated by DG1SIFT are corrected by our method. Clearly, the proposed IF/SR system has a more robust performance as shown in these examples.

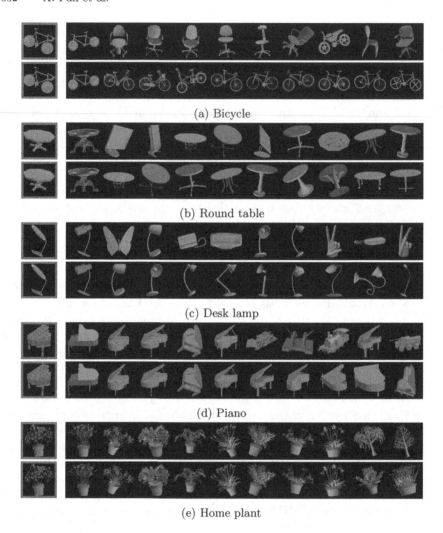

(a) Bicycle

(b) Round table

(c) Desk lamp

(d) Piano

(e) Home plant

Fig. 1. Comparison of retrieved shapes using the DG1SIFT method (the first row) and the proposed IF/SR method (the second row) against five query shapes (from top to bottom): (a) bicycle, (b) round table, (c) desk lamp, (d) piano, and (e) home plant.

There are two main contributions of this work. First, we develop more powerful and robust global features to compensate for the weaknesses of local features. Feature concatenation are often adopted by traditional methods to combine local and global features. However, proper feature weighting and dimension reduction remain to be a problem. For the second contribution, we propose a robust shape retrieval system that consists of two modules in cascade: the irrelevance filtering (IF) module and the similarity ranking (SR) module. The IF module attempts to cluster gallery shapes that are similar to each other by examining global and local features simultaneously. However, shapes that are close in the local feature space can be distant in the global feature space, and vice versa. To resolve this issue, we propose a joint cost function that strikes a balance between two distances. In particular, irrelevant samples that are close in the local feature space but distant in the global feature space can be removed in this stage. The remaining gallery samples are ranked in the SR module using the local feature.

The rest of this paper is organized as follows. The proposed IF/SR method is explained in Sect. 2. Experimental results are shown in Sect. 3. Finally, concluding remarks are given in Sect. 4.

2   Proposed IF/SR Method

2.1   System Overview

The flow chart of the proposed IF/SR method is shown in Fig. 2. The IF module is trained in an off-line process with the following three steps.

1. **Initial label prediction.** All gallery samples are assigned an initial label in their local feature space using an unsupervised clustering method.
2. **Local-to-global feature association.** Samples close to each cluster centroid are selected as the training data. A random forest classifier is trained based on their global features. All gallery samples are re-predicted by the random forest classifier to build an association from the local feature space to the global feature space.
3. **Label refinement.** We assign every gallery sample a set of relevant cluster indices based on a joint cost function. The joint cost function consists of two assignment scores. One score reflects the relevant cluster distribution of the query sample itself while the other is the mean of the relevant cluster distributions of its local neighbors. The ultimate relevant cluster indices are obtained by thresholding the cost function.

In the on-line query process, we extract both global and local features from a query shape and proceed with the following two steps.

1. **Relevance prediction.** we adopt the same scheme in the label refinement step to assign relevant cluster indices to a given query.
2. **Similarity ranking.** The similarity between the query and all relevant gallery samples is measured in the local feature space. In this step, a post-processing technique can also be adopted to enhance retrieval accuracy.

An exemplary query, desk lamp, is given in Fig. 2 to illustrate the on-line retrieval process. In the dashed box "retrieval without Stage I", the traditional local feature (DG1SIFT) retrieves erroneous shapes such as butterfly, desk phone and keyboard in the top five ranks. They are apparently irrelevant to the query shape and successfully removed in the relevance prediction step in the IF stage (Stage I). The retrieved top five samples in the SR stage (Stage II) are all desk lamp shapes. We will explain the processing of Stages I and II in detail below.

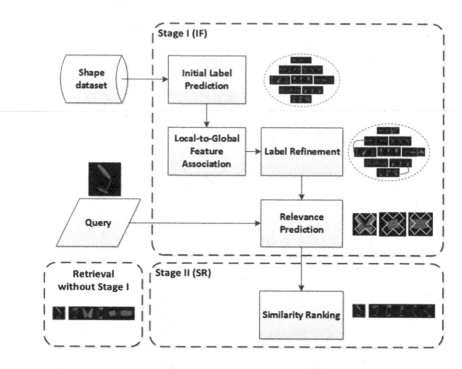

Fig. 2. The flow chart of the proposed IF/SR system.

2.2 Stage I: Irrelevance Filtering

3D Shape Preprocessing. We have two preprocessing steps: (1) model representation conversion and (2) 3D shape normalization. For model representation conversion, since we extract features from both mesh models and volumetric models, we adopt the parity count method [29] to convert a mesh model into a volumetric model. Each volumetric model has resolution $256 \times 256 \times 256$. 3D shape normalization aims to align shapes of the same class consistently to achieve translational, scaling and rotational invariance.

Translational invariance is achieved by aligning the center of mass with the origin. For scale invariance, we re-scale a shape to fit a unit sphere. For rotational invariance, we adopt the reflective symmetry axial descriptor [30] to calculate the nearly complete symmetry function for each 3D shape. The PCA on the symmetry function extracts three principal axes to form three principal planes. To determine the order of three principal planes, we project the shape into each plane and the projection views with the first and second largest areas are aligned with the XOY plane and the ZOX plane, respectively. Finally, the YOZ plane is determined automatically. Figure 3 shows some normalization results using the above-mentioned method.

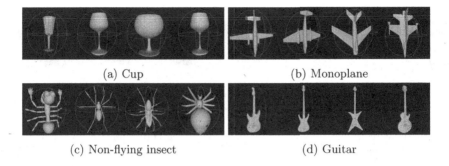

(a) Cup   (b) Monoplane

(c) Non-flying insect   (d) Guitar

Fig. 3. Shape normalization results of four 3D shape classes.

Global Features. To capture the global properties of a 3D shape, we describe it using three feature types: (1) surface features (f_s), (2) wavelet features (f_w) and (3) geometrical features (f_g).

The 3D surface features, denoted by f_s, are a generalization of the 2D polar Fourier descriptor [31]. N rays are emitted from the origin of a normalized 3D shape. Each ray has the orientation $\mathbf{r} = (cos\phi cos\theta, cos\phi sin\theta, sin\phi)$ with two directional parameters (θ, ϕ), where θ and ϕ are uniformly sampled from intervals $[0, \pi)$ and $[0, 2\pi)$, respectively, with step size $\frac{\pi}{6}$. For each ray, the Euclidean distance from the origin to its intersected point on a face forms a function $g(\theta, \phi)$. If a ray intersects with multiple faces, we consider the farthest one only. In this way, we convert the original surface function $f(x, y, z)$ into a 2D distance function parameterized by $g(\theta, \phi)$. Then, we calculate the Fourier coefficients of the 2D distance function. The magnitude information forms a 72-D feature vector denoted by f_s. The Fourier descriptors of four shapes belonging to two classes are visualized in Fig. 4, where each subfigure contains an original shape in the left and its surface feature in the right. We see intra-class consistency and inter-class discrimination from this figure.

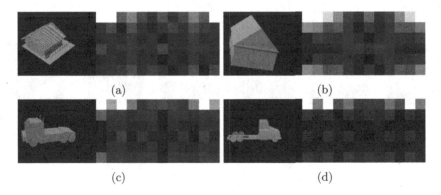

(a)   (b)

(c)   (d)

Fig. 4. Visualization of surface features of four shapes, where (a) and (b) provide two house shapes while (c) and (d) provide two truck shapes.

Fig. 5. Illustration of the seven-band Haar filters.

For the wavelet features denoted by f_w, we adopt the generalized 3D Haar-like filters [32]. Seven bands of 3D Haar-like filters as shown in Fig. 5 are applied to a normalized and voxelized model. The first three filters capture the left-right, top-bottom, front-back symmetry properties. The last four filters analyze diagonal sub-regions. The responses from these seven filters form a 7D wavelet feature vector.

Furthermore, we incorporate four geometrical features: (1) the aspect ratio, (2) xyz-invariance, (3) $\alpha\beta\gamma$-invariance and (4) rectilinearity [33]. The aspect ratio is a 3D feature based on three side lengths - l_x, l_y, l_z of the bounding box of a normalized shape. It is expressed as

$$AR = [\frac{l_x}{l_x + l_y + l_z}, \frac{l_y}{l_x + l_y + l_z}, \frac{l_z}{l_x + l_y + l_z}]. \tag{1}$$

The xyz-variance and $\alpha\beta\gamma$-variance are adopted to examine the variance of cut-planes of a normalized volumetric model. To measure the xyz-variance, we extract all cut-planes orthogonal to the X-axis, the Y-axis and the Z-axis, respectively. The variances of three groups of cut-planes form a 3D feature. Similarly, the $\alpha\beta\gamma$-variance measures the variance of groups of rotated cut-planes centered at the X-axis, the Y-axis and the Z-axis, repectively. The robust rectilinearity measure from [33] is used to obtain the rectilinearity feature. It calculates the ratio between the total surface area and the sum of projected triangle areas on the XOY, the ZOX and the YOZ planes. Finally, the geometrical feature, denoted by f_g, is a 10-D feature vector. The geometric features of six examples are shown in Fig. 6 in boxes of black (xyz-invariance), red ($\alpha\beta\gamma$-variance) and blue (rectilinearity), respectively.

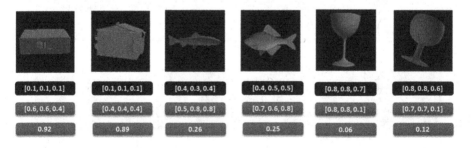

Fig. 6. The xyz-invariance (black box), $\alpha\beta\gamma$-variance (red box) and rectilinearity (blue box) values of six examples from three classes: apartment house, fish and cup. (Color figure online)

Initial Label Prediction. In traditional 3D shape retrieval formulation, all shapes in the dataset are not labeled. Under this extreme case, we select the spectral clustering algorithm [34] to reveal the underlying relationship between gallery samples. The local feature is strong at grouping locally similar shapes but it is sensitive to local variances as discussed in Sect. 1. In contrast, the global feature is powerful at differentiating global dissimilar shapes but weak at finding locally similar shapes. Thus, the combination of the two in this early stage tends to cause confusion and lower the performance. For this reason, we use the local feature only to perform clustering.

For the SHREC12 dataset, shapes in several clusters using the DG1SIFT feature are shown in Fig. 7. Some clusters look reasonable while others do not. Actually, any unsupervised clustering method will encounter two challenges. First, uncertainty occurs near cluster boundaries so that samples near boundaries have a higher probability of being wrongly clustered. Second, the total number of shape classes is unknown. When the cluster number is larger than the class number in the database, the clustering algorithm creates sub-classes or even mixed classes. We address the first challenge in the local-to-global feature association step and the second challenge in the label refinement step.

Fig. 7. Several clusterd SHREC12 shapes using the spectral clustering method using the DG1SIFT feature.

Local-to-Global Feature Association. We extract N_k samples closest to the centroid of the kth cluster and assign them a cluster label. Clearly, samples sharing the same cluster label are close to each other in the feature space. There is a trade-off in choosing a proper value of N_k. A smaller N_k guarantees higher clustering accuracy but fewer gallery samples will be assigned cluster labels. Empirically, we set the value of N_k to one half of the size of the kth cluster. Then, we convert the gallery samples from the local feature space to a global feature space. We will correct clustering errors in the global feature space at a later stage. Furthermore, samples that come from the same class but are separated in the local feature space can be merged by their global features. To build the association, labeled samples are used to train a random forest classifier [35] with global features. Finally, all gallery shapes are treated as test samples.

The random forest classifier is used to predict the probability of each cluster type by voting. In this way, samples clustered in the local feature space can be linked to multiple clusters probabilistically due to the similarity in the global feature space.

Label Refinement. The output of the IF module includes: (1) a set of indexed clusters, and (2) soft classification (or multi-labeling) of all gallery samples. For item #1, we use unsupervised spectral clustering to generate clusters as described above. If the class number is known (or can be estimated), it is desired that the cluster number is larger than the class number. Each of these clusters is indexed by a cluster ID. For item #2, we adopt soft classification so that each sample can be associated with multiple clusters. This is done for two reasons. If two sub-classes belong to the same ground truth class, we need a mechanism to re-group them together. Clearly, a hard classification process does not allow this to happen. Second, a hard classification error cannot be easily compensated while a soft classification error is not as fatal and it is likely to be fixed in the SR module (stage II).

We consider two relevant cluster assignment schemes below.

(1) *Direct Assignment*
We apply the random forest classifier to both training and testing samples based on their global features. Then, the probability for the ith shape sample (denoted by y_i) belonging to the kth cluster (denoted by c_k) can be estimated by the following normalized voting result:

$$P_{rf}(y_i \in c_k) = \frac{v_k}{\sum_j v_j}, \tag{2}$$

where v_k is the number of votes claiming that y_i belongs to c_k. Equation (2) associates y_i to its relevant clusters directly.

(2) *Indirect Assignment*
Intuitively, a good cluster relevance assignment scheme should take both global and local features into account. For query sample, y_i, we find its K nearest neighbors (denoted by x_j) using a certain distance function in a local feature space (e.g. the same feature space used in DG1SIFT). Then, the probability of y_i belonging to c_k can be estimated by the weighted sum of the probability in Eq. (2) in form of

$$P_{knn}(y_i \in c_k) = \frac{\sum_{x_j \in knn(y_i)} P_{rf}(x_j \in c_k)}{\sum_{c_m} \sum_{x_j \in knn(y_i)} P_{rf}(x_j \in c_m)}. \tag{3}$$

Equation (3) associates y_i to its relevant clusters indirectly. That is, the assignment is obtained by averaging the relevant clusters assignment of its K nearest neighbors. Empirically, we choose K to be 1.5 times the average cluster size in the experiments.

We show an example that assigns a query desk lamp shape to its relevant clusters in Fig. 8(a), whose x-axis and y-axis are the negative log functions of Eqs.

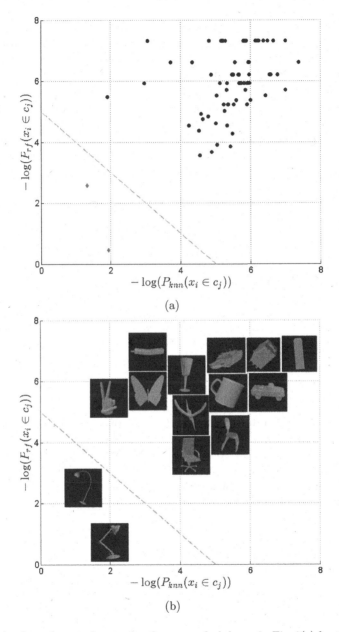

(a)

(b)

Fig. 8. Selecting relevant clusters for the query desk lamp in Fig. 1(c) by thresholding a cost function shown in Eq. 4.

(2) and (3), respectively. Every dot in Fig. 8(a) represents a cluster after shape clustering. To visualize shapes represented by a dot, we plot a representative sample of each cluster in Fig. 8(b).

We see that the distance between the hand cluster and the desk lamp cluster is small in the x-axis but large in the y-axis. This is because that samples of the desk lamp and hand clusters are interleaved in the local feature space as shown in the retrieval results of DG1SIFT in Fig. 1(c). However, the desk lamp and the hand clusters have little intersection in the global feature space. In contrast, the wheel chair and desk lamp clusters have large intersection in the global feature space. Yet, their distance is far in the local feature space. It is apparent that Eqs. (2) and (3) provide complementary relevance assignment strategies for query sample y_i. It is best to integrate the two into one assignment scheme. For example, we can draw a line to separate relevant and irrelevant clusters with respect to the query apple shape in this plot.

Mathematically, we can define the following cost function

$$\mathbb{J}(y_i, c_k) = -\log(P_{knn}(y_i \in c_k)P_{rf}(y_i \in c_k))$$
$$= -[\log(P_{knn}(y_i \in c_k)) + \log(P_{rf}(y_i \in c_k))]. \qquad (4)$$

We compute $\mathbb{J}(y_i, c_k)$ for all clusters c_k. If

$$\mathbb{J}(y_i, c_k) < \epsilon, \qquad (5)$$

where ϵ is a pre-selected threshold. We say that cluster c_k is a relevant cluster for query y_i. Otherwise, it is irrelevant.

2.3 Stage II: Similarity Ranking

In the SR module, we rank the similarity between a given query and gallery samples in the retrieved relevant clusters using a local-features-based matching scheme (e.g., DG1SIFT). Additionally, we adopt the Local Constrained Diffusion Process (LCDP) [36] in the post-processing step. The diffusion process is slightly modified with the availability of relevant clusters in the IF/SR system since the diffusion process can be conducted on a more reasonable manifold due to the processing in Stage I.

3   Experimental Results

We demonstrate the retrieval performance of the proposed IF/SR method by conducting experiments on the generic 3D shape dataset of SHREC12 [10]. It contains 1200 3D shapes in 60 independent classes. Samples are uniformly distributed so that each class has 20 shape samples. The retrieval performance is measured by five standard metrics. They are: Nearest-Neighbor (NN), First-Tier score (FT), Second-Tier score (ST), E-measurement (E), Discounted Cumulative Gain (DCG).

We compare the proposed IF/SR method with five state-of-the-art methods:

- LSD-sum [37]. It uses a local surface-based feature that considers local geodesic distance distribution and Bag-of-Words.
- ZFDR [38]. It adopts a hybrid feature that integrates the Zernike moment, the Fourier descriptor, the ray-based features.
- 3DSP_L2_1000_chi2 [10]. It employs a local surface-based feature that computes the 3D SURF descriptor under the spatial pyramid matching scheme.
- DVD+DB+GMR [10]. It adopts a hybrid feature that contains a dense voxel spectrum descriptor and a depth-buffer shape descriptor.
- DG1SIFT [24]. It uses a view-based feature that extracts three types of SIFT features (Dense SIFT, Grid SIFT and One SIFT) per view.

The IF/SR method adopts DG1SIFT as the local feature for shape clustering. We show the first-tier (FT) scores of the IF/SR method using a different cluster number M for shape clustering in Table 1. Generally speaking, the performance degrades when M is small due to the loss of discriminability in larger cluster sizes. The retrieval performance improves as the cluster number increases up to 64. After that, the performance saturates and could even drop slightly. That means that we lose the advantage of clustering when the cluster size is too small. For the remaining experimental results, we choose $M = 64$.

Table 1. Comparison of the First-Tier (FT) scores with different cluster numbers for the IF/SR method in the SHREC12 dataset. The best score is shown in bold.

M	16	32	48	64	80	96	112
FT	0.666	0.672	0.709	**0.720**	0.717	0.717	0.715

Table 2. Comparison of the NN, FT, ST, E and DCG scores of five state-of-the-art methods, the proposed IF/SR method, and the IF/SR method with LCDP postprocessing for the SHREC12 dataset. The best score for each measurement is shown in bold.

Method	NN	FT	ST	E	DCG
LSD-sum	0.517	0.232	0.327	0.224	0.565
ZFDR	0.818	0.491	0.621	0.442	0.776
3DSP_L2_1000_chi2	0.662	0.367	0.496	0.346	0.678
DVD+DB+GMR	0.828	0.613	0.739	0.527	0.833
DG1SIFT	0.879	0.661	0.799	0.576	0.871
IF/SR	**0.896**	0.720	0.837	0.608	0.891
IF/SR+LCDP	0.893	**0.734**	**0.858**	**0.620**	**0.899**

We compare the performance of seven 3D shape retrieval methods with five measures in Table 2. Clearly, the proposed IF/SR method (with or without LCDP postprocessing) outperforms the other five benchmarking methods. The IF/SR method with postprocessing improves the result of DG1SIFT by around 7% in the First-Tier score. Since DG1SIFT adopts the manifold ranking process

Table 3. Comparison of top 20, 25, 30, 35, 40 retrieval accuracy for the SHREC12 dataset, where the best results are shown in bold.

N	20	25	30	35	40
LSD-sum	0.232	0.260	0.286	0.310	0.327
ZFDR	0.491	0.539	0.575	0.603	0.621
3DSP_L2_1000_chi2	0.367	0.411	0.446	0.476	0.496
DVD+DB+GMR	0.613	0.656	0.691	0.719	0.739
DG1SIFT	0.661	0.718	0.756	0.783	0.799
IF/SR	0.720	0.775	0.802	0.824	0.837
IF/SR+LCDP	**0.734**	**0.786**	**0.817**	**0.841**	**0.858**

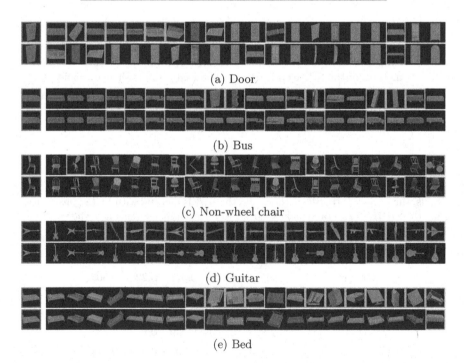

(a) Door

(b) Bus

(c) Non-wheel chair

(d) Guitar

(e) Bed

Fig. 9. Comparison of retrieved top 20 rank-ordered shapes. For each query case given in the leftmost column, retrieved results of DG1SIFT and the proposed IF/SR method are shown in the first and second rows of all subfigures, repectively.

in its similarity measurement, the gap between the IF/SR method before and after LCDP is relatively small.

Since each SHREC12 shape class contains 20 shape samples, the measure of correctly retrieved samples from the top 20 (FT) and 40 (ST) ranks cannot reflect the true power of the proposed IF/SR method. To push the retrieval performance further, we compare the accuracy of retrieved results from the top 20, 25, 30, 35 and 40 ranks of the IF/SR method and five benchmarking methods in Table 3, whose first and last columns correspond the FT and ST scores reported in Table 2. The superiority of the IF/SR method stands out clearly in this table.

According to the top 20 retrieval performance, the IF/SR method still makes mistakes for some queries. We conduct error analysis and show the results of DG1SIFT and the IF/SR method in Figs. 9(a)–(e). For each query case given in the leftmost column, retrieved results of DG1SIFT and the IF/SR method are shown in the first and second rows of all subfigures, respectively. Each erroneous result is enclosed by a thick frame. The errors of DG1SIFT are obvious. They are far away from human experience. The IF/SR method makes mistakes between door/keyboard, bus/truck, non-wheel chair/wheel chair, guitar/violin and bed/rectangle table (see the second row of all subfigures). These mistakes are more excusable since they are closer to each other based on human judgment.

Finally, we show the precision-and-recall curves of the IF/SR method and several methods in Fig. 10. We see from the figure that the IF/SR method outperforms all other methods by a significant margin.

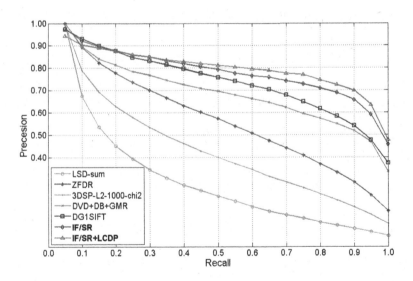

Fig. 10. Comparison of precision and recall curves of the proposed IF/SR method and several benchmarking methods for the SHREC12 dataset.

4 Conclusion

The IR/SF method was proposed to solve the unsupervised 3D shape retrieval problem. In the IF stage, irrelevant shape clusters are removed for each query shape. In the SR stage, the system can focus on the matching and ranking in a much smaller subset of shapes. It superior retrieval performance was evaluated on the popular SHREC12 dataset.

References

1. Tangelder, J.W., Veltkamp, R.C.: A survey of content based 3d shape retrieval methods. Multimedia Tools Appl. **39**, 441–471 (2008)
2. Su, H., Maji, S., Kalogerakis, E., Learned-Miller, E.: Multi-view convolutional neural networks for 3D shape recognition. In: Proceedings of the IEEE International Conference on Computer Vision, pp. 945–953 (2015)
3. Xie, J., Fang, Y., Zhu, F., Wong, E.: Deepshape: Deep learned shape descriptor for 3D shape matching and retrieval. In: Proceedings of the IEEE Conference on Computer Vision and Pattern Recognition, pp. 1275–1283 (2015)
4. Shi, B., Bai, S., Zhou, Z., Bai, X.: Deeppano: deep panoramic representation For 3-D shape recognition. IEEE Sig. Process. Lett. **22**, 2339–2343 (2015)
5. Savva, M., Yu, F., Su, H., Aono, M., Chen, B., Cohen-Or, D., Deng, W., Su, H., Bai, S., Bai, X., et al.: Shrec'16 track large-scale 3D shape retrieval from shapenet core55
6. Bai, S., Bai, X., Zhou, Z., Zhang, Z., Jan Latecki, L.: Gift: a real-time and scalable 3D shape search engine. In: The IEEE Conference on Computer Vision and Pattern Recognition (CVPR) (2016)
7. Wu, Z., Song, S., Khosla, A., Yu, F., Zhang, L., Tang, X., Xiao, J.: 3D shapenets: a deep representation for volumetric shapes. In: Proceedings of the IEEE Conference on Computer Vision and Pattern Recognition, pp. 1912–1920 (2015)
8. Maturana, D., Scherer, S.: Voxnet: a 3D convolutional neural network for real-time object recognition. In: 2015 IEEE/RSJ International Conference on Intelligent Robots and Systems (IROS), pp. 922–928. IEEE (2015)
9. Qi, C.R., Su, H., Niessner, M., Dai, A., Yan, M., Guibas, L.J.: Volumetric and multi-view CNNs for object classification on 3D data. arXiv preprint arXiv:1604.03265 (2016)
10. Li, B., Godil, A., Aono, M., Bai, X., Furuya, T., Li, L., López-Sastre, R.J., Johan, H., Ohbuchi, R., Redondo-Cabrera, C., et al.: Shrec'12 track: generic 3D shape retrieval. In: 3DOR, pp. 119–126 (2012)
11. Kazhdan, M., Funkhouser, T., Rusinkiewicz, S.: Rotation invariant spherical harmonic representation of 3-D shape descriptors. In: Symposium on Geometry Processing, vol. 6, pp. 156–164 (2003)
12. Osada, R., Funkhouser, T., Chazelle, B., Dobkin, D.: Shape distributions. ACM Trans. Graph. (TOG) **21**, 807–832 (2002)
13. Smeets, D., Keustermans, J., Vandermeulen, D., Suetens, P.: meshsift: local surface features for 3D face recognition under expression variations and partial data. Comput. Vis. Image Underst. **117**, 158–169 (2013)
14. Bronstein, M.M., Kokkinos, I.: Scale-invariant heat kernel signatures for non-rigid shape recognition. In: 2010 IEEE Conference on Computer Vision and Pattern Recognition (CVPR), pp. 1704–1711. IEEE (2010)

15. Bronstein, A.M., Bronstein, M.M., Guibas, L.J., Ovsjanikov, M.: Shape google: geometric words and expressions for invariant shape retrieval. ACM Trans. Graph. (TOG) **30**, 1 (2011)
16. Gal, R., Shamir, A., Cohen-Or, D.: Pose-oblivious shape signature. IEEE Trans. Vis. Comput. Graph. **13**, 261–271 (2007)
17. Reuter, M., Wolter, F.E., Peinecke, N.: Laplace-Beltrami spectra as shape-DNA of surfaces and solids. Comput.-Aided Des. **38**, 342–366 (2006)
18. Lian, Z., Godil, A., Bustos, B., Daoudi, M., Hermans, J., Kawamura, S., Kurita, Y., Lavoua, G., Dp Suetens, P.: Shape retrieval on non-rigid 3D watertight meshes. In: Eurographics Workshop on 3D Object Retrieval (3DOR) (2011)
19. Lian, Z., Zhang, J., Choi, S., ElNaghy, H., El-Sana, J., Furuya, T., Giachetti, A., Guler, R., Isaia, L., Lai, L., et al.: Shrec'15 track: Non-rigid 3D shape retrieval. In: Proceedings of the Eurographics Workshop on 3D Object Retrieval (2015)
20. Chen, D.Y., Tian, X.P., Shen, Y.T., Ouhyoung, M.: On visual similarity based 3D model retrieval. Comput. Graph. Forum **22**, 223–232 (2003). Wiley Online Library
21. Chaouch, M., Verroust-Blondet, A.: A new descriptor for 2D depth image indexing and 3D model retrieval. In: IEEE International Conference on Image Processing, ICIP 2007, vol. 6, p. VI-373. IEEE (2007)
22. Ohbuchi, R., Osada, K., Furuya, T., Banno, T.: Salient local visual features for shape-based 3D model retrieval. In: IEEE International Conference on Shape Modeling and Applications, SMI 2008, pp. 93–102. IEEE (2008)
23. Lowe, D.G.: Object recognition from local scale-invariant features. In: The Proceedings of the Seventh IEEE International Conference on Computer Vision, vol. 2, pp. 1150–1157. IEEE (1999)
24. Ohbuchi, R., Furuya, T.: Distance metric learning and feature combination for shape-based 3D model retrieval. In: Proceedings of the ACM Workshop on 3D Object Retrieval, pp. 63–68. ACM (2010)
25. Li, B., Lu, Y., Li, C., Godil, A., Schreck, T., Aono, M., Chen, Q., Chowdhury, N.K., Fang, B., Furuya, T., et al.: Shrec'14 track: large scale comprehensive 3D shape retrieval. In: Eurographics Workshop on 3D Object Retrieval, vol. 2014, pp. 131–140 (2014)
26. Bronstein, A., Bronstein, M., Castellani, U., Dubrovina, A., Guibas, L., Horaud, R., Kimmel, R., Knossow, D., Von Lavante, E., Mateus, D., et al.: Shrec'10 track: correspondence finding. In: 3DOR 2010-Eurographics Workshop on 3D Object Retrieval, pp. 87–91. Eurographics Association (2010)
27. Dutagaci, H., Godil, A., Daras, P., Axenopoulos, A., Litos, G., Manolopoulou, S., Goto, K., Yanagimachi, T., Kurita, Y., Kawamura, S., et al.: Shrec'11 track: generic shape retrieval. In: Proceedings of the 4th Eurographics Conference on 3D Object Retrieval, pp. 65–69. Eurographics Association (2011)
28. Li, B., Lu, Y., Li, C., Godil, A., Schreck, T., Aono, M., Burtscher, M., Chen, Q., Chowdhury, N.K., Fang, B., et al.: A comparison of 3D shape retrieval methods based on a large-scale benchmark supporting multimodal queries. Comput. Vis. Image Underst. **131**, 1–27 (2015)
29. Nooruddin, F.S., Turk, G.: Simplification and repair of polygonal models using volumetric techniques. IEEE Trans. Vis. Comput. Graph. **9**, 191–205 (2003)
30. Kazhdan, M., Chazelle, B., Dobkin, D., Funkhouser, T., Rusinkiewicz, S.: A reflective symmetry descriptor for 3D models. Algorithmica **38**, 201–225 (2004)
31. Zhang, D., Lu, G.: An integrated approach to shape based image retrieval. In: Proceedings of 5th Asian Conference on Computer Vision (ACCV), Melbourne, Australia (2002)

32. Cui, X., Liu, Y., Shan, S., Chen, X., Gao, W.: 3D haar-like features for pedestrian detection. In: 2007 IEEE International Conference on Multimedia and Expo, pp. 1263–1266. IEEE (2007)

33. Lian, Z., Rosin, P.L., Sun, X.: Rectilinearity of 3D meshes. Int. J. Comput. Vis. **89**, 130–151 (2010)

34. Ng, A.Y., Jordan, M.I., Weiss, Y., et al.: On spectral clustering: analysis and an algorithm. Adv. Neural Inf. Process. Syst. **2**, 849–856 (2002)

35. Breiman, L.: Random forests. Mach. Learn. **45**, 5–32 (2001)

36. Yang, X., Koknar-Tezel, S., Latecki, L.J.: Locally constrained diffusion process on locally densified distance spaces with applications to shape retrieval. In: IEEE Conference on Computer Vision and Pattern Recognition, CVPR 2009, pp. 357–364. IEEE (2009)

37. Bai, X., Li, L., Zhang, S.: Software for 3D model retrieval using local shape distributions (2012)

38. Li, B., Johan, H.: 3D model retrieval using hybrid features and class information. Multimedia Tools Appl. **62**, 821–846 (2013)

Author Index